AF717530

List of Elements

Name	Symbol	Atomic Number	Atomic Mass
Actinium	Ac	89	[227]*
Aluminum	Al	13	26.9815386(8)
Americium	Am	95	[243]*
Antimony	Sb	51	121.760(1)
Argon	Ar	18	39.948(1)
Arsenic	As	33	74.92160(2)
Astatine	At	85	[210]*
Barium	Ba	56	137.327(7)
Berkelium	Bk	97	[247]*
Beryllium	Be	4	9.012182(3)
Bismuth	Bi	83	208.98040(1)
Bohrium	Bh	107	[267]*
Boron	B	5	10.811(7)
Bromine	Br	35	79.904(1)
Cadmium	Cd	48	112.411(8)
Calcium	Ca	20	40.078(4)
Californium	Cf	98	[251]*
Carbon	C	6	12.0107(8)
Cerium	Ce	58	140.116(1)
Cesium	Cs	55	132.9054519(2)
Chlorine	Cl	17	35.453(2)
Chromium	Cr	24	51.9961(6)
Cobalt	Co	27	58.933195(5)
Copper	Cu	29	63.546(3)
Curium	Cm	96	[247]*
Darmstadtium	Ds	110	[281]*
Dubnium	Db	105	[268]*
Dysprosium	Dy	66	162.500(1)
Einsteinium	Es	99	[252]*
Erbium	Er	68	167.259(3)
Europium	Eu	63	151.964(1)
Fermium	Fm	100	[257]*
Fluorine	F	9	18.9984032(5)
Francium	Fr	87	[223]*
Gadolinium	Gd	64	157.25(3)
Gallium	Ga	31	69.723(1)
Germanium	Ge	32	72.64(1)
Gold	Au	79	196.966569(4)
Hafnium	Hf	72	178.49(2)
Hassium	Hs	108	[269]*
Helium	He	2	4.002602(2)
Holmium	Ho	67	164.93032(2)
Hydrogen	H	1	1.00794(7)
Indium	In	49	114.818(3)
Iodine	I	53	126.90447(3)
Iridium	Ir	77	192.217(3)
Iron	Fe	26	55.845(2)
Krypton	Kr	36	83.798(2)
Lanthanum	La	57	138.90547(7)
Lawrencium	Lr	103	[262]*
Lead	Pb	82	207.2(1)
Lithium	Li	3	6.941(2)
Lutetium	Lu	71	174.967(1)
Magnesium	Mg	12	24.3050(6)
Manganese	Mn	25	54.938045(5)
Meitnerium	Mt	109	[276]*
Mendelevium	Md	101	[258]*
Mercury	Hg	80	200.59(2)
Molybdenum	Mo	42	95.94(2)
Neodymium	Nd	60	144.242(3)
Neon	Ne	10	20.1797(6)
Neptunium	Np	93	[237]*
Nickel	Ni	28	58.6934(2)
Niobium	Nb	41	92.90638(2)
Nitrogen	N	7	14.0067(2)
Nobelium	No	102	[259]*
Osmium	Os	76	190.23(3)
Oxygen	O	8	15.9994(3)
Palladium	Pd	46	106.42(1)
Phosphorus	P	15	30.973762(2)
Platinum	Pt	78	195.084(9)
Plutonium	Pu	94	[244]*
Polonium	Po	84	[209]*
Potassium	K	19	39.0983(1)
Praseodymium	Pr	59	140.90765(2)
Promethium	Pm	61	[145]*
Protactinium	Pa	91	231.03588(2)*
Radium	Ra	88	[226]*
Radon	Rn	86	[222]*
Rhenium	Re	75	186.207(1)
Rhodium	Rh	45	102.90550(2)
Roentgenium	Rg	111	[280]*
Rubidium	Rb	37	85.4678(3)
Ruthenium	Ru	44	101.07(2)
Rutherfordium	Rf	104	[267]*
Samarium	Sm	62	150.36(2)
Scandium	Sc	21	44.955912(6)
Seaborgium	Sg	106	[271]*
Selenium	Se	34	78.96(3)
Silicon	Si	14	28.0855(3)
Silver	Ag	47	107.8682(2)
Sodium	Na	11	22.98976928(2)
Strontium	Sr	38	87.62(1)
Sulfur	S	16	32.065(5)
Tantalum	Ta	73	180.94788(2)
Technetium	Tc	43	[98]*
Tellurium	Te	52	127.60(3)
Terbium	Tb	65	158.92535(2)
Thallium	Tl	81	204.3833(2)
Thorium	Th	90	232.03806(2)*
Thulium	Tm	69	168.93421(2)
Tin	Sn	50	118.710(7)
Titanium	Ti	22	47.867(1)
Tungsten	W	74	183.84(1)
Ununbium	Uub	112	[285]*
Ununhexium	Uuh	116	[293]*
Ununpentium	Uup	115	[288]*
Ununquadium	Uuq	114	[289]*
Ununtrium	Uut	113	[284]*
Uranium	U	92	238.02891(3)*
Vanadium	V	23	50.9415(1)
Xenon	Xe	54	131.293(6)
Ytterbium	Yb	70	173.04(3)
Yttrium	Y	39	88.90585(2)
Zinc	Zn	30	65.409(4)
Zirconium	Zr	40	91.224(2)

Source of data: Atomic weights of the elements 2001 (IUPAC Technical Report) as supplemented by the Table of Standard Atomic Weights 2005 (to be published in *Pure and Applied Chemistry*) on the IUPAC web site, and "Nuclear Data Sheets for A-266-294" (to be published in *Nuclear Data Sheets*) at http://www.nndc.bnl.gov/superheavy.pdf.

*Element has no stable isotope. A value enclosed in brackets, e.g. [209], indicates the mass number of the longest-lived isotope of the element. Three such elements (Th, Pa, and U), however, do have a characteristic terrestrial isotopic composition, and an atomic mass is given for them. An uncertainty in the last digit in the Atomic Mass column is shown by the number in parentheses; e.g., 1.00794(7) indicates ± 0.00007.

CHEMISTRY

Principles, Patterns, and Applications

VOLUME

2

Bruce Averill
Patricia Eldredge

With contributions by

C. Alton Hassell
Baylor University

Daniel J. Stasko
University of Southern Maine

PEARSON
Benjamin Cummings

San Francisco Boston New York
Cape Town Hong Kong London Madrid Mexico City
Montreal Munich Paris Singapore Sydney Tokyo Toronto

Editor-in-Chief Adam Black, Ph.D.
Publisher Jim Smith
Director of Development Kay Ueno
Art Developmental Editors Sonia DiVittorio, Hilair Chism, Blake Kim
Executive Managing Editor Erin Gregg
Managing Editor Corinne Benson
Marketing Manager Scott Dustan
Market Development Manager Josh Frost, Susan Winslow
Text Developmental Editors Moira Lerner, Ruth Steyn
Developmental Editor, Glossary and Appendices John Murdzek
Photo Editor Travis Amos
Photo Researcher Brian Donnelly
Art Editor Kelly Murphy
Project Editors Katie Conley, Lisa Leung, Kate Brayton
Assistant Editor Cinnamon Hearst
Editorial Assistant Kristin Rose Field
Managing Media Producer Claire Masson
Project Manager Crystal Clifton, Progressive Publishing Alternatives
Composition Progressive Information Technologies
Illustrators Imagineering Media Services, Inc.
Graph Designers New York Charts and Diagrams
Manufacturing Buyer Pam Augspurger
Text Designer Mark Ong
Cover Designer Mark Ong
Text and Cover Printer Quebecor World Dubuque
Cover Photo Credit Harry Taylor, © Dorling Kindersley, Courtesy of the Natural History Museum, London

Library of Congress Cataloging-in-Publication Data
Averill, Bruce.
Chemistry: principles, patterns, and applications / Bruce Averill, Patricia Eldredge; with contributions by C. Alton Hassell, Daniel J. Stasko.—1st ed.
v. cm.
Includes index.
ISBN 0-8053-3803-9
1. Chemistry—Textbooks. 2. Chemistry, Physical and theoretical—Textbooks. 3. Environmental chemistry—Textbooks. 4. Chemistry, Technical—Textbooks. I. Eldredge, Patricia. II. Title.
QD31.3.A94 2006
540—dc22

2005022980

ISBN 0-8053-8283-6 Volume 2 with MasteringGeneralChemistry (Student Edition)
ISBN 0-8053-8319-0 Volume 2 without MasteringGeneralChemistry (Student Edition)

PEARSON
Benjamin Cummings
www.aw-bc.com

1 2 3 4 5 6 7 8 9 10 — QWD—09 08 07 06

To Harvey, who opened the door

About the Authors

Bruce Averill

After growing up in New England, Bruce Averill received his B.S. with high honors in chemistry at Michigan State University in 1969, and his Ph.D. in inorganic chemistry at MIT in 1973. After three years as an NIH and NSF Postdoctoral Fellow at Brandeis University and the University of Wisconsin, he began his independent academic career at Michigan State University in 1976. He moved to the University of Virginia in 1982, and was promoted to Professor in 1988. In 1994, Dr. Averill moved to the University of Amsterdam in the Netherlands as Professor of Biochemistry. In 2001, he returned to the United States where he is a Distinguished University Professor at the University of Toledo. Dr. Averill's research focuses on the role of metal ions in biology.

While in Europe, Dr. Averill headed an EU-funded network of seven research groups from seven different European countries. In addition, he lead a 22-member team investigating biocatalysis within the E. C. Slater Institute of the University of Amsterdam.

From 2004 to 2005, Dr. Averill was a Jefferson Science Fellow at the U.S. State Department, where he acted as a senior science and technology adviser in the Bureau of Western Hemisphere Affairs and focussed on energy issues in Latin America. He has been asked to return to the State Department for 2006 and 2007 as a William C. Foster Fellow in the Bureau of Political-Military Affairs to work on R&D issues in the area of critical infrastructure protection and cybersecurity.

Dr. Averill's published papers are frequently cited by other researchers, and he has been invited to give more than 100 presentations at educational and research institutions and at national and international scientific meetings. Dr. Averill has been an Honorary Woodrow Wilson Fellow, an NSF Predoctoral Fellow, an NIH and NSF Postdoctoral Fellow, and an Alfred P. Sloan Foundation Fellow; he has also received an NSF Special Creativity Award.

Dr. Averill has published more than 140 articles on chemical, physical, and biological subjects in refereed journals, 15 chapters in books, and more than 80 abstracts from national and international meetings. In addition, he has co-edited a graduate text on catalysis and taught courses at all levels, including general chemistry, biochemistry, advanced inorganic, and physical methods.

Aside from his research program. Dr. Averill is an enthusiastic sailor and an avid reader. He also enjoys traveling with his family, and at some point in the future he would like to sail around the world in a classic wooden boat.

Patricia Eldredge

Having been raised in the U.S. diplomatic service, Patricia Eldredge has traveled and lived around the world. After receiving a B.A. in Spanish language and literature from Ohio State University, Dr. Eldredge developed an interest in chemistry while studying general chemistry at Kent State University. She obtained a B.S. in chemistry from the University of Central Florida. Following several years as an analytical research chemist in industry, she began her graduate studies at the University of Virginia and obtained her Ph.D. in inorganic chemistry from the University of North Carolina at Chapel Hill. In 1989, Dr. Eldredge was named the Science Policy Fellow for the American Chemical Society. While in Washington, D.C., she examined the impact of changes in federal funding priorities on academic research funding. She was awarded a Postdoctoral Research Fellowship with Oak Ridge Associated Universities, working with the U.S. Department of Energy on heterogeneous catalysis and coal liquefaction. Subsequently, she returned to the University of Virginia as a Research Scientist and a member of the General Faculty. In 1992, Dr. Eldredge moved to Europe for several years. While there, she studied advanced Maritime Engineering, Materials, and Oceanography at the University of Southampton in England, arising from her keen interest in naval architecture. Since her return to the United States in 2002, she has been a Visiting Assistant Professor and a Senior Research Scientist at the University of Toledo. Her current research interests include the use of protein scaffolds to synthesize biologically relevant clusters.

Dr. Eldredge has published more than a dozen articles dealing with synthetic inorganic chemistry and catalysis, including several seminal studies describing new synthetic approaches to metal–sulfur clusters. She has also been awarded a patent for her work on catalytic coal liquefaction. Her diverse teaching experience includes courses on chemistry for the life sciences, introductory chemistry, general, organic, and analytical chemistry.

When not writing scientific papers or textbooks, Dr. Eldredge enjoys traveling, reading political biographies, sailing high-performance vessels under rigorous conditions, and caring for her fourth child, her pet Havanese.

Preface to the Instructor

In this new millenium, as the world faces new and extreme challenges, the importance of acquiring a solid foundation in chemical principles has become increasingly central to understanding the challenges that lie ahead. Moreover, as the world becomes more integrated and interdependent, so too do the scientific disciplines. The divisions between fields such as chemistry, physics, biology, environmental sciences, geology, and materials science, among others, have become less clearly defined. This text addresses the closer relationships developing among various disciplines and shows the relevance of chemistry to contemporary issues in a friendly and approachable manner.

Because of the enthusiasm of the majority of first-year chemistry students for biologically and medically relevant topics, this text uses an integrated approach that includes explicit discussions of biological and environmental applications of chemistry. Topics relevant to materials science are also introduced in order to meet the more specific needs of engineering students. To integrate this material, simple organic structures, nomenclature, and reactions are introduced very early in the text, and both organic and inorganic examples are used wherever possible. This approach emphasizes the distinctions between ionic and covalent bonding, thus enhancing the students' chance of success in the organic chemistry course that traditionally follows general chemistry.

Our overall goal is to produce a text that introduces the students to the relevance and excitement of chemistry. Although much of first-year chemistry is taught as a service course, there is no reason that the intrinsic excitement and potential of chemistry cannot be the focal point of the text and the course. We emphasize the positive aspects of chemistry and its relationship to students' lives; this approach requires bringing in applications early and often. Unfortunately, we cannot assume that students in these courses today are highly motivated to study chemistry for its own sake. The explicit discussion of biological, environmental, and materials applications from a chemical perspective is intended to motivate the students and help them appreciate the relevance of chemistry to their lives. Material that has traditionally been relegated to boxes, and perhaps perceived as peripheral by the students, has been incorporated into the text to serve as a learning tool.

To begin the discussion of chemistry rapidly, the traditional first chapter introducing units, significant figures, conversion factors, dimensional analysis, and so on has been reorganized in this book. The material has been placed in the chapters where the relevant concepts are first introduced, thus providing three advantages: it eliminates the tedium of the traditional approach, which introduces mathematical operations at the outset, and thus avoids the perception that chemistry is a mathematics course; it avoids the early introduction of operations such as logarithms and exponents, which are typically not encountered again for several chapters and may easily be forgotten when they are needed; and it provides a review for those students who have already had relatively sophisticated high school chemistry and math courses, although the sections are designed primarily for students unfamiliar with the topics.

Our specific objectives include the following:

- To write the text at a level suitable for science majors, but using a less formal writing style that will appeal to modern students.
- To produce a *truly* integrated text that gives the student who takes only a single year of chemistry an overview of the most important subdisciplines of chemistry,

including organic, inorganic, biological, materials, environmental, and nuclear chemistry, thus emphasizing unifying concepts.

- To introduce fundamental concepts in the first two-thirds of the chapter and then applications relevant to the health sciences or engineers, thus providing a flexible text that can be tailored to the specific needs and interests of the audience.
- To ensure the accuracy of the material presented, which is enhanced by the authors' breadth of professional and research experience.
- To produce a spare, clean, uncluttered text that is not distracting to the student, one in which each piece of art serves as a pedagogical device.
- To introduce the distinction between ionic and covalent bonding and reactions early in the text, and to continue to build on this foundation in the subsequent discussions while emphasizing the relationship between structure and reactivity.
- To use established pedagogical devices to maximize students' ability to learn directly from the text. Copious worked examples in the text, problem-solving strategies, and similar unworked exercises with solutions are included. End-of-chapter problems are designed to ensure that students have grasped major concepts in addition to testing their ability to solve numerical problems. Problems emphasizing applications are drawn from many disciplines.
- To emphasize an intuitive and predictive approach to problem solving that relies on a thorough understanding of key concepts and recognition of important patterns rather than on memorization. Many patterns are indicated throughout the text by a "Note the Pattern" feature in the margin.

The text is organized by units that discuss introductory concepts, atomic and molecular structure, the states of matter, kinetics and equilibria, and descriptive inorganic chemistry. The text divides the traditional chapter on liquids and solids into two chapters in order to expand the coverage of important topics such as semiconductors and superconductors, polymers, and engineering materials. Part V, included in the complete edition of the text, is a systematic summary of the descriptive chemistry of the elements organized by position in the periodic table; it is designed to bring together the key concepts introduced in the preceding chapters: chemical bonding, molecular structure, kinetics, and equilibrium. A great deal of descriptive chemistry will have been introduced prior to this point, but only in ways that are germane to particular points of interest.

In summary, our hope is that this text represents a step in the evolution of the general chemistry textbook toward one that reflects the increasing overlap between chemistry and other disciplines. Most important, the text discusses exciting and relevant aspects of biological, environmental, and materials science that are usually relegated to the last few chapters, and it provides a format that allows the instructor to tailor the emphasis to the needs of the class. By the end of Part I (Chapter 5), the student will have been introduced to environmental topics such as acid rain, the ozone layer, and periodic extinctions, and to biological topics such as antibiotics and the caloric content of foods. Nonetheless, the new material is presented in a way that minimally perturbs the traditional sequence of topics in a first-year course, making the adaptation easier for instructors.

Supplements

A full set of print and media supplements is available for the student and the instructor, which are designed to enhance in-class presentations, to engage students in classroom discussion, and to assist students outside the classroom.

For the instructor:

Instructor Solutions Manual: Complete solutions to all of the end-of-chapter problems in the textbook.

Instructor Guide: Includes chapter overviews and outlines, suggestions for lecture demonstrations, teaching tips, and a guide to the print and media resources available for each chapter.

Test Bank (print and electronic formats): More than 1000 questions to use in creating tests, quizzes, and homework assignments.

Transparency Acetates: 300 full-color illustrations and tables to enhance classroom lectures.

Instructor Resource CD-ROM: All images and tables from the book in jpeg and PowerPoint format, a PowerPoint lecture outline for each chapter, Clicker Questions, interactive graphs, interactive 3-D molecular structures, and links to student resources online.

Clicker Questions (for use with Classroom Response Systems): Five questions per chapter written to inspire in-class discussion and test student understanding of material.

PowerPoint Lecture Outline: Hundreds of editable slides for in-class presentations.

MasteringGeneralChemistry™: The most advanced online homework and tutorial system available provides thousands of problems and tutorials with automatic grading, immediate wrong-answer specific feedback, and simpler questions upon request. Problems include randomized numerical and algebraic answers and dimensional analysis. Instructors can compare individual student and class results against data collected through pre-testing of students nationwide.

For the student:

Student Solutions Manual: Detailed, worked-out solutions to selected end-of-chapter problems in the textbook.

Chemistry Place for General Chemistry: Features chapter quizzes, interactive 3-D molecular structures, interactive graphs, and InterAct Math for General Chemistry.

Study Guide: Study and learning objectives, chapter overviews, problem-solving tips, and practice tests.

MasteringGeneralChemistry: The first adaptive-learning tutorial system that grades students' homework automatically and provides Socratic tutorials with feedback specific to errors, hints and simpler subproblems upon demand, and motivation with partial credit. Pre-tested on students nationally, the system is uniquely able to respond to students' needs, effectively tutoring and motivating their learning of the concepts and strengthening their problem-solving skills.

Acknowledgments

Although putting a text together is always a team effort, there are several individuals at Benjamin Cummings whom we would like to particularly acknowledge for their tireless efforts and commitment to this project. We would especially like to thank the following individuals: Jim Smith for believing so strongly in this project; Kay Ueno for her superhuman efforts in the face of excruciating deadlines (we still have to get out for a sail!); Linda Davis, who, despite all the twists and turns, made sure that it happened; Sonia DiVittorio, whose enormous dedication and attention to detail produced a stellar art program, while keeping track of where everything was at any given time; and Moira Nelson, a truly impressive master of the American English language, who taught us that Los Angeles and Washington, D.C., aren't so different after all. Special thanks, too, are due to Neil Weinstein, who helped enormously in scrubbing the page proofs of errors and inconsistencies in data, and to Mike Helmstadter, whose expertise with Excel was crucial in generating many of the plots used in the figures. We would also like to thank the rest of the team at Benjamin Cummings for creating such a cordial and supportive environment despite the pressures of production.

Finally, to Tonya McCarley, thanks.

Reviewers

Several drafts of the manuscript were informed by the meticulous and considered comments of the instructors listed here. Their review of our work was invaluable to us as we polished the book. We acknowledge and thank them for their generous efforts on our behalf.

Dawood Afzal
Truman State University

Thomas E. Albrecht-Schmitt
Auburn University

Jeffrey Appling
Clemson University

David Atwood
University of Kentucky

Jim D. Atwood
State University of New York at Buffalo

Debbie Beard
Mississippi State University

Kevin Bennett
Hood College

Richard Biagioni
Southern Missouri State University

Robert S. Boikess
Rutgers University, The State University of New Jersey

Steven Boone
Central Missouri State University

Allen Clabo
Frances Marion University

Carl David
University of Connecticut, Storrs

Sonja Davison
Tarrant County College, Northeast

Nordulf Debye
Towson University

Dru DeLaet
Southern Utah University

Patrick Desrochers
University of Central Arkansas

Jane DeWitt
San Francisco State University

Michael Doherty
East Stroudsburg University

W. Travis Dungan
Trinity Valley Community College

Delbert J. Eatough
Brigham Young University

Marly Eidsness
University of Georgia

Paul Farnsworth
Brigham Young University

Michael Freitas
Ohio State University

Roy Garvey
North Dakota State University

Jim Geiger
Michigan State University

Brian Gilbert
Linfield College

Alexander Golger
Boston University

Lara Gossage
Hutchinson Community College

John M. Halpin
New York University

Greg Hartland
University of Notre Dame

Dale Hawley
Kansas State University

Gregory Kent Haynes
Morgan State University

James W. Hershberger
Miami University

Carl Hoeger
University of California, San Diego

Don Hood
Saint Louis University

John B. Hopkins
Louisiana State University

Jayanthi Jacob
Indiana University–Purdue University

David Johnson
University of Dayton

Lori Jones
Guelph University

Jeff Keaffaber
University of Florida

Philip C. Keller
University of Arizona

Michael E. Ketterer
Northern Arizona University

Raj Khanna
University of Maryland

Paul Kiprof
University of Minnesota, Duluth

Patrick Kolniak
Louisiana State University

Jeremy Kua
University of San Diego

Jothi V. Kumar
North Carolina A&T State University

Donald Land
University of California, Davis

Alan Levine
University of Louisiana, Lafayette

Scott B. Lewis
James Madison University

Robley J. Light
Florida State University

Da-hong Lu
Fitchburg State College

Joel T. Mague
Tulane University

Pshemak Maslak
Penn State University

Hitoshi Masui
Kent State University

Elmo Mawk
Texas A&M University

C. Michael McCallum
University of the Pacific

Robert McIntyre
East Carolina University

Abdul K. Mohammed
North Carolina A&T State University

Kathy Nabona
Austin Community College, Northridge

Cheuk-Yiu Ng
University of California, Davis

Frazier W. Nyasulu
University of Washington

Gay L. Olivier-Lilley
Point Loma Nazarene University

Joseph Pavelites
Purdue University, North Central

James Pazun
Pfeiffer University

Earl Pearson
Middle Tennessee State University

Lee Pedersen
University of North Carolina at Chapel Hill

Cathrine E. Reck
Indiana University

James H. Reho
East Carolina University

Jeff Roberts
University of Minnesota

John Selegue
University of Kentucky

Robert Sharp
University of Michigan

Jerald Simon
Frostburg State University

J. T. (Dotie) Sipowska
University of Michigan

Sheila Smith
University of Michigan, Dearborn

Michael Sommer
University of Wyoming

James Stickler
Allegany College of Maryland

Michael Stone
Vanderbilt University

Shane Street
University of Alabama, Tuscaloosa

Donald Thompson
University of Missouri–Columbia

Douglas Tobias
University of California, Irvine

Bilin P. Tsai
University of Minnesota, Duluth

Tom Tullius
Boston University

John B. Vincent
The University of Alabama

Yan Waguespack
University of Maryland, Eastern Shore

Thomas Webb
Auburn University

Mark Whitener
Montclair State University

Marcy Whitney
University of Alabama

Kathryn Williams
University of Florida

Kurt Winkelmann
Florida Institute of Technology

Troy Wood
State University of New York, Buffalo

Catherine Woytowicz
George Washington University

David Young
Ohio University, Athens

Class Testers

In addition, we are grateful to these professors who class tested portions of the early manuscript and student class testers who provided insights into what students need to effectively learn chemistry. We wish to acknowledge these instructors and their students:

Jamie Adcock
University of Tennessee, Knoxville

Adegboye Adeyemo
Savannah State University

Lisa Arnold
South Georgia College

Dale Arrington
South Dakota School of Mines

Karen Atkinson
Bunker Hill Community College

John Barry
Houston Community College, Town and Country

Krishna Bhat
Philadelphia University

Christine Bilicki
Pasadena City College

Rose Boll
University of Tennessee, Knoxville

Bill Brescoe
Tulsa Community College

Steve Burns
St. Thomas Aquinas College

Timothy Champion
Johnson C. Smith University

Thomas Chasteen
Sam Houston State University

Walter Cleland
University of Mississippi

Zee Ding
Queensland University of Technology

Judy Dirbas
Grossmont College

Marly Eidsness
University of Georgia

Cristina Fermin-Ennis
Gordon College

Richard Frazee
Rowan University

Neal Gray
University of Texas, Tyler

William Griffin
Bunker Hill Community College

Greg Hale
University of Texas, Arlington

Jessica Harper
Antelope Valley College

Alton Hassell
Baylor University

Michael Hauser
St. Louis Community College

Scott Hendrix
University of Tampa

Daniel Huchital
Florida Atlantic University

Steven Hughes
Carl Albert State College

Lawrence Kennard
Walters State Community College

Jeffrey Kovac
University of Tennessee, Knoxville

James Lankford
Saint Andrew's Presbyterian College

Debra Leedy
Arizona State University

Larry Manno
Triton College

Carol Martinez
TVI Community College, Albuquerque

Lydia Martinez-Rivera
University of Texas, San Antonio

Graeme Matthews
Florida Community College, Jacksonville

Keith McCleary
Adrian College

Larry McRae
Berry College

Gary Mercer
Boise State University

Matt Merril
Florida State University

David Nachman
Mesa Community College

Bill Newman
Mohawk Valley Community College

Jason Overby
College of Charleston

Linda Pallack
Washington and Jefferson College

James Pazun
Pfeiffer University

John Penrose
Jefferson Community College

Rafaelle Perez
University of South Florida

Joanna Petridou-Fischer
Spokane Falls Community College

Dale Powers
Elmira College

David Prentice
Coastal Carolina University

Victoria Prevatt
Tulsa Community College

Lisa Price
Bennett College for Women

Laura Pytlewski
Triton Community College

Gerald Ramelow
McNeese State University

Mitch Rhea
Chattanooga State

Lyle Roelofs
Haverford College

Steven Rowley
Middlesex County College

J. B. Schlenoff
Florida State University

Raymond Scott
Mary Washington College

Lisa Seagraves
Haywood Community College

Shirish Shah
Towson University

George Smith
Herkimer County College

Zihan Song
Savannah State University

Paris Svoronos
Queensborough Community College

Richard Terry
Brigham Young University

John Vincent
University of Alabama

Kjirsten Wayman
Humboldt State University

John Weide
Mesa Community College

Neil Weinstein
Santa Fe Community College

Barry West
Trident Technical College

Drew Wolfe
Hillsborough Community College

Servet Yatin
Quincy College

Lynne Zeman
Kirkwood Community College

Xueli Zou
California State University, Chico

Forum Participants

Beneficial to our crafting of this text have been the numerous Chemistry Forum participants, who took time to read and remark thoughtfully on our manuscript as it was in progress and share their ideas about the future of chemical education. Participants at the Chemistry Forums held in cities across the country include the following people:

Michael Abraham
University of Oklahoma

William Adeniyi
North Carolina A&T State University

Ramesh Arasasingham
University of California, Irvine

Yiyan Bai
HCC, Central

Monica Baloga
Florida Institute of Technology

Mufeed Basti
North Carolina A&T State University

Rich Bauer
Arizona State University

Debbie Beard
Mississippi State University

Jo A. Beran
Texas A&M, Kingsville

Wolfgang Bertsch
University of Alabama

Christine Bilicki
Pasadena City College

Bob Blake
Texas Tech University

Rose Boll
University of Tennessee, Knoxville

Philip Brucat
University of Florida

Kenneth Busch
Baylor University

Marianna Busch
Baylor University

Donnie Byers
Johnson County Community College

Brandon Cruikshank
Northern Arizona University

Mapi Cuevas
Santa Fe Community College

Nordulf Debye
Towson University

Michael Denniston
Georgia Perimeter College, Clarkston

Patrick Desrochers
University of Central Arkansas

Deanna Dunlavy
New Mexico State University

Marly Eidsness
University of Georgia

Thomas Engel
University of Washington

Deborah Exton
University of Oregon

Steven Foster
Mississippi State University

Mark Freilich
University of Memphis

Elizabeth Gardner
University of Texas, El Paso

Roy Garvey
North Dakota State University

John Gelder
Oklahoma State University

Eric Goll
Brookdale Community College

John Goodwin
Coastal Carolina

Thomas Greenbowe
Iowa State University

Asif Habib
University of Wisconsin, Waukesha

Jerry Haky
Florida Atlantic University

Greg Hale
University of Texas, Arlington

C. Alton Hassell
Baylor University

Gregory Haynes
Morgan State University

Claudia Hein
Diablo Valley College

Louise Hellwig
Morgan State University

David Hobbs
Montana Tech at University of Montana

Jim Holler
University of Kentucky

John Hopkins
Louisiana State University

Susan Hornbuckle
Clayton College and State University

James Hovick
University of North Carolina, Charlotte

Thomas Huang
East Tennessee State University

Denley Jacobson
North Dakota State University

Phillip Keller
University of Arizona

Debbie Koeck
Texas State University

Jeffrey Kovack
University of Tennessee, Knoxville

Jothi Kumar
North Carolina A&T State University

Charles Kutal
University of Georgia

Robley Light
Florida State University

Larry Manno
Triton College

Pam Marks
Arizona State University

Carol Martinez
Albuquerque Technical Vocational College

Selah Massoud
Louisiana University, Lafayette

Graeme Matthews
Florida Community College, Jacksonville

Joe Elmo Mawk
Texas A&M University

Maryann McDermott Jones
University of Maryland

Robert A. McIntyre
East Carolina University

Abdul Mohammed
North Carolina A&T State University

John Nelson
University of Nevada, Reno

Frazier Nyasulu
University of Washington

Greg Oswald
North Dakota State University

Jason Overby
College of Charleston

Gholam Pahlavan
Houston Community College

Colleen Partigianoni
Ferris State University

Cindy Phelps
California State University, Chico

Shawn Phillips
Vanderbilt University

Louis Pignolet
University of Minnesota

Laura Pytlewski
Triton Community College

William Quintana
New Mexico State University

Catherine Reck
Indiana University

John Richardson
University of Louisville

Jill Robinson
Indiana University

Peter Roessle
Georgia Perimeter College, Decatur

Jimmy Rogers
University of Texas, Arlington

Svein Saebo
Mississippi State University

James Schlegel
Rutgers University, Newark

Pat Schroeder
Johnson County Community College

Jack Selegue
University of Kentucky

John Sheridan
Rutgers University

Don Siegel
Rutgers University

Brett Simpson
Coastal Carolina

Jerry Skelton
St. Johns River Community College

Sheila Smith
University of Michigan, Dearborn

Sherril A. Soman
Grand Valley State University

Mark Sulkes
Tulane University

Ann Sullivan
John Sergeant Reynolds Community College

Susan Swope
Plymouth University

Will Tappen
San Diego Mesa College

James Terner
Virginia Commonwealth University

Donald Thompson
Oklahoma State University

Michael Topp
University of Pennsylvania

John Turner
University of Tennessee

Julian Tyson
University of Massachusetts, Amherst

Chris Uzomba
Austin Community College

Robert Vergenz
University of North Florida

Ed Walters
University of New Mexico

Philip Watson
Oregon State University

Thomas Webb
Auburn University

Neil Weinstein
Santa Fe Community College

Steven Weitstock
Indiana University

M. Stanley Whittingham
State University of New York, Binghamton

Alex Williamson
North Carolina A&T State University

Kim Woodrum
University of Kentucky

Vaneica Young
University of Florida

Lin Zhu
University of Georgia

Problem Advisory Board Members

All of the end-of-chapter problems were examined for consistency, clarity, and relevance. We warmly thank the following instructors for their contributions. Their efforts helped us considerably.

Jamie Adcock
University of Tennessee

William Adeniyi
North Carolina A&T State University

Mufeed Basti
North Carolina A&T State University

Salah M. Blaih
Kent State University, Trumbull

Mary Joan Bojoan
Pennsylvania State University

Jerry Burns
Pellissippi State Technical Community College

Frank Carey
Wharton County Junior College

Karen Eichstadt
Ohio University

Marly Eidsness
University of Georgia

Gregory Kent Haynes
Morgan State University

Melissa Hines
Cornell University

Denley Jacobson
North Dakota State University

Neil Kestner
Louisiana State University

Charles Kirkpatrick
St. Louis University

John Kovacs
University of Tennessee

Mark Kubinec
University of California, Berkeley

Barbara Lewis
Clemson University

Lauren McMillis
Ohio University

Matthew J. Mio
University of Detroit Mercy

Daniel Moriarty
Siena College

Gregory Oswald
North Dakota State University

Dale Powers
Elmira College

Lydia J. Martinez Rivera
University of Texas at San Antonio

Theodore Sakano
Rockland Community College

Fred Safarowic
Passaic County Community College

Pat Schroeder
Johnson County Community College

Dotie Sipowska
University of Minnesota, Duluth

Sheila Smith
University of Michigan, Dearborn

Bilin Tsai
University of Minnesota, Duluth

Jon Turner
University of Tennessee

M. Stanley Whittingham
State University of New York at Binghamton

Kurt Winkleman
Florida Institute of Technology

We welcome feedback from colleagues and students who use this text. Please send your comments to the Chemistry Editor, Benjamin Cummings, 1301 Sansome Street, San Francisco, California 94111.

Bruce Averill
Patricia Eldredge

Brief Contents

Volume 1
Chapters 1–13

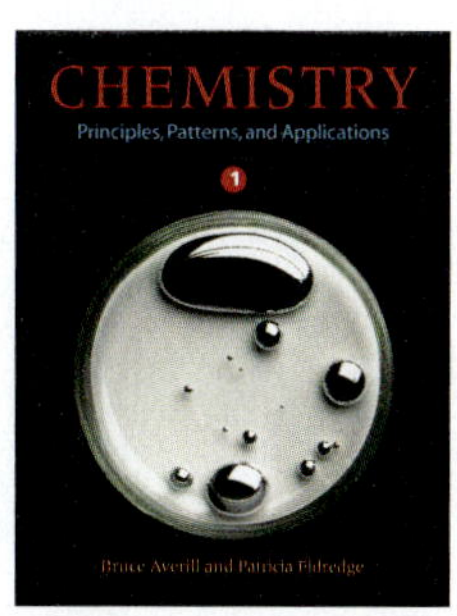

Volume 2
Chapters 14–20

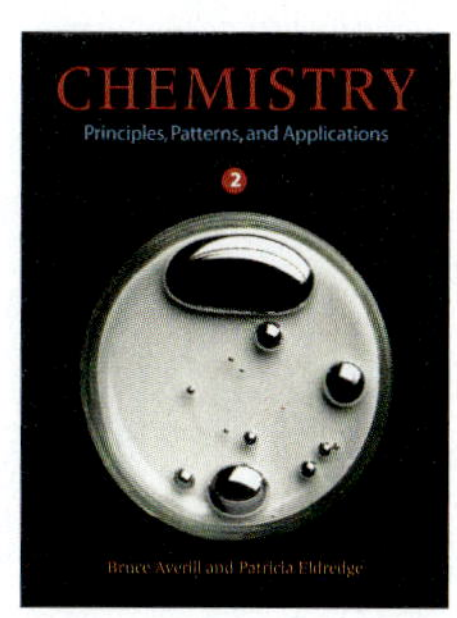

Complete Volume
Chapters 1–24

Detailed Contents

RELEVANT, INNOVATIVE, INFORMATIVE

Chemistry: Principles, Patterns, and Applications is rich in applications that make chemistry relevant to students' lives and provides an innovative approach to learning principles, recognizing patterns, and solving problems.

What the student needs to know

Chapter openers provide **Learning Objectives** and **Chapter Outlines**.

Selected chapters include **Essential Skill** sets, which help students build the requisite skills for understanding the equations and working the problems featured in the chapter.

1 Introduction to Chemistry

LEARNING OBJECTIVES

- To recognize the breadth, depth, and scope of chemistry
- To understand what is meant by the scientific method
- To be able to classify matter
- To understand the development of the atomic model
- To know the meaning of isotopes and atomic masses
- To become familiar with the periodic table and to be able to use it as a predictive tool

CHAPTER OUTLINE

1.1 Chemistry in the Modern World
1.2 The Scientific Method
1.3 A Description of Matter
1.4 A Brief History of Chemistry
1.5 The Atom
1.6 Isotopes and Atomic Masses
1.7 Introduction to the Periodic Table
1.8 Essential Elements

ESSENTIAL SKILLS 1

Units of Measurement
Scientific Notation
Significant Figures
Accuracy and Precision

As you begin your study of college chemistry, those of you who do not intend to become professional chemists may well wonder why you need to study chemistry. In fact, as you will soon discover, a basic understanding of chemistry is useful in a wide range of disciplines and career paths. You will also discover that an understanding of chemistry helps you make informed decisions about many issues that affect you, your community, and your world. A major goal of this text is to demonstrate the importance of chemistry in your daily life and in our collective understanding of both the physical world we occupy and the biological realm of which we are a part. The objective of this chapter is to introduce the breadth, the importance, and some of the challenges of modern chemistry, and to present some of the fundamental concepts and definitions you will need to understand how chemists think and work.

An atomic corral for electrons. A "corral" of 48 Fe atoms (yellow-orange) on a smooth Cu surface (cyan-purple) confine the electrons on the surface of the Cu, producing a pattern of "ripples" in the distribution of the electrons. Scientists assembled the 713-pm-diameter corral by individually positioning Fe atoms with the tip of a scanning tunneling microscope.

1

Contemporary applications embody modern chemistry ...

14.5 Half-Lives and Radioactive Decay Kinetics

Half-Lives

MGC Mastering Gen Chem

Another approach to describing reaction rates is based on the time required for the concentration of a reactant to decrease to one-half its initial value. This period of time is called the **half-life** of the reaction, written as $t_{1/2}$. Thus, the half-life of a reaction is the time required for the reactant concentration to decrease from $[A]_0$ to $[A]_0/2$. If two reactions have the same order, the faster reaction will have a shorter half-life and the slower reaction will have a longer half-life.

The half-life of a first-order reaction under a given set of reaction conditions is a constant. This is not true for zeroth- or second-order reactions. Most important, the half-life of a first-order reaction is *independent of the concentration of the reactants.* This becomes evident when we rearrange the integrated rate law for a first-order reaction (Equation 14.23) to produce the following equation:

$$\ln\frac{[A]_0}{[A]} = kt \qquad (14.31)$$

TABLE 14.7 Half-lives and applications of some radioactive isotopes

Radioactive Isotope	Half-Life	Typical Uses
Hydrogen-3 (tritium)	12.26 yr	Biochemical tracer
Carbon-11	20.39 min	PET scans (biomedical imaging)
Carbon-14	5730 yr	Dating of artifacts
Sodium-24	14.659 min	Cardiovascular system tracer
Phosphorus-32	14.3 days	Biochemical tracer
		Dating of rocks

EXAMPLE 14.10

The anticancer drug cisplatin hydrolyzes in water with a rate constant of 1.5×10^{-3} min^{-1} at pH 7.0 and 25°C. Calculate the half-life for the hydrolysis reaction under these conditions. If a freshly prepared solution of cisplatin has a concentration of 0.053 *M*, what will be the concentration of cisplatin after five half-lives? After 10 half-lives? What is the percent completion of the reaction after five half-lives? After 10 half-lives?

Given Rate constant, initial concentration, and number of half-lives

Asked for Half life, final concentrations, and percent completion

Contemporary applications in text discussions, illustrations, worked examples, and problem sets ...

Figure 8.1 A plot of potential energy vs. internuclear distance for the interaction between a gaseous Na^+ ion and a gaseous Cl^- ion. Note that the energy of the system reaches a minimum at a particular distance, r_0, where the attractive and repulsive interactions are balanced. Interactive Graph

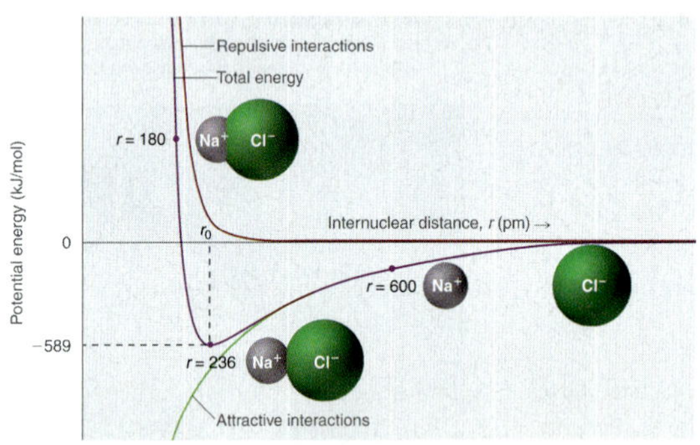

Key graphs are provided in an interactive format online at The Chemistry Place to enhance understanding.

The Ozone Layer and Chlorofluorocarbons

Each year since the mid-1970s, scientists have noted a disappearance of approximately 70% of the ozone (O_3) layer above Antarctica during the Antarctic spring, creating what is commonly known as the "ozone hole." In 2003, the Antarctic ozone hole measured 11.1 million square miles, slightly larger than the size of North America but smaller than the largest ever recorded, in the year 2000, when the hole covered 11.5 million square miles and for the first time extended over a populated area—the city of Punta Arenas, Chile (population 120,000).

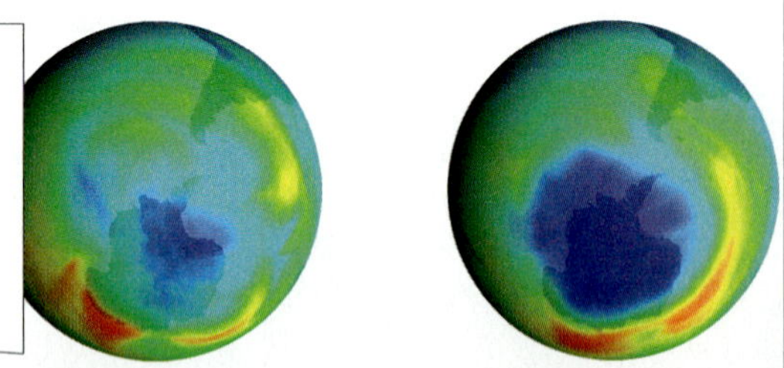

...showing students how and why chemistry is relevant

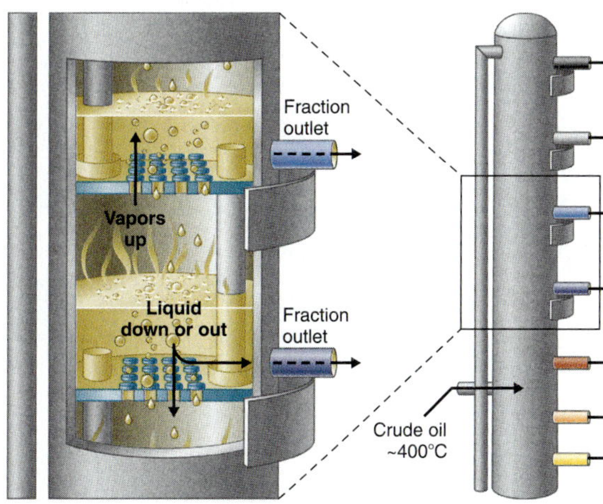

	Number of carbons	Boiling point range	Uses
Gases	1–4	0–30°C	Bottled and natural gas
Naphthas	5–10	30–180°C	Gasoline
Kerosenes	10–16	180–260°C	Kerosene for home heaters, jet fuel
Gas oils	16–60	260–350°C	Diesel fuel, feedstock for cracking
Lubricants	>60	350–575°C	Motor oil, feedstock for cracking
Fuel oil	>70	>490°C	Candles, fuel oil for ships and power stations
Asphalt	>80	>580°C	Roofing tar, road tar

(a) Petroleum distillation tower (b) Petroleum fractions

Figure 2.20 Distillation of petroleum.

Figure 4.20 Rust formation. The corrosion process involves an oxidation-reduction reaction in which metallic iron is converted to $Fe(OH)_3$, a reddish-brown solid.

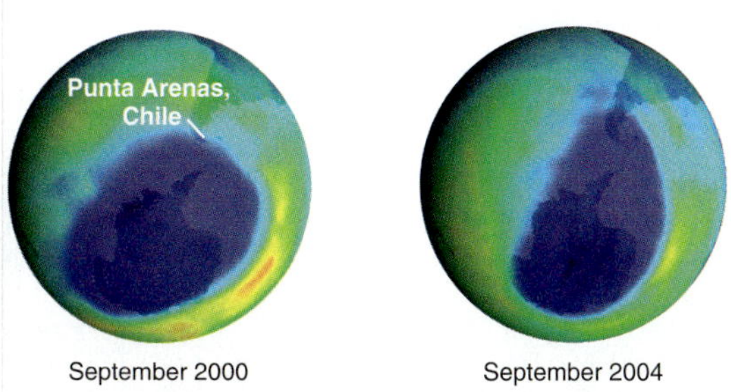

Figure 3.14 Satellite photos of Earth reveal the increased size of the Antarctic ozone hole with time. Dark blue colors correspond to the thinnest ozone, while light blue, green, yellow, orange, and red indicate progressively thicker ozone.

Build problem-solving and pattern recognition skills

Note the Pattern
Students learn to recognize important, recurring patterns encountered both in text discussions and in problems.

Worked examples avoid providing students with algorithmic solutions, and instead suggest strategies specific to each problem, thus promoting development of the analytical skills necessary for successful problem solving.

MasteringGeneralChemistry
Key concepts and problem-solving skills are reinforced through assignable interactive tutorials online at **www.masteringgeneralchemistry.com.**

Practice exercises reinforce problem-solving skills by enabling students to check their understanding of the strategy employed in the previous example.

Note the pattern
The lattice energy is usually the most important energy factor in determining the stability of an ionic compound.

MGC Introduction to Lewis Structures

If the formation of ionic lattices that contain multiply charged ions is so energetically favorable, why does CsF contain Cs^+ and F^- ions rather than Cs^{2+} and F^{2-} ions? If we assume that U for a $Cs^{2+}F^{2-}$ salt would be approximately the same as U for BaO, the formation of a lattice containing Cs^{2+} and F^{2-} ions would release 2291 kJ/mol (3048 kJ/mol − 756.9 kJ/mol) more energy than one containing Cs^+ and F^- ions. To form the Cs^{2+} ion from Cs^+, however, would require removing a 5p electron from a filled inner shell, which calls for a great deal of energy: I_2 = 2234.4 kJ/mol for Cs. Furthermore, forming an F^{2-} ion is expected to be even more energetically unfavorable than forming an O^{2-} ion. Not only is an electron being added to an already negatively charged ion, but because the F^- ion has a filled 2p subshell, the added electron would have to occupy an empty high-energy 3s orbital. Cesium fluoride, therefore, is not $Cs^{2+}F^{2-}$ because the energy cost of forming the doubly charged ions would be higher than the additional lattice energy that would be gained.

EXAMPLE 8.3

Use data from Figure 7.13, Tables 7.5, 8.2, and 8.3, and Appendix A to calculate the lattice energy of MgH_2.

Given Chemical compound, data from figures and tables

Asked for Lattice energy

Strategy

A Write a series of stepwise reactions for forming MgH_2 from the elements via the gaseous ions.

B Use Hess's law and data from the specified figures and tables to calculate the lattice energy.

Solution

A Hess's law allows us to use a thermochemical cycle (a Born–Haber cycle) to calculate the lattice energy for a given compound. We begin by writing reactions in which we form the component ions from the elements in a stepwise manner and then assemble the ionic solid:

(1)	$Mg(s) \longrightarrow Mg(g)$	$\Delta H_1 = \Delta H_{sub}(Mg)$
(2)	$Mg(g) \longrightarrow Mg^{2+}(g) + 2e^-$	$\Delta H_2 = I_1(Mg) + I_2(Mg)$
(3)	$H_2(g) \longrightarrow 2H(g)$	$\Delta H_3 = D(H_2)$
(4)	$2H(g) + 2e^- \longrightarrow 2H^-(g)$	$\Delta H_4 = 2EA(H)$
(5)	$Mg^{2+}(g) + 2H^-(g) \longrightarrow MgH_2(s)$	$\Delta H_5 = -U$
	$Mg(s) + H_2(g) \longrightarrow MgH_2(s)$	$\Delta H = \Delta H_f$

B Table 7.5 lists first and second ionization energies for the Period-3 elements [$I_1(Mg)$ = 737.7 kJ/mol, $I_2(Mg)$ = 1450.7 kJ/mol]. First electron affinities for all elements are given in Figure 7.13 [$EA(H)$ = −72.8 kJ/mol]. Table 8.2 lists selected enthalpies of sublimation [$\Delta H_{sub}(Mg)$ = 147.1 kJ/mol]. Table 8.3 lists selected bond dissociation energies [$D(H_2)$ = 436.0 kJ/mol]. Enthalpies of formation (ΔH_f = −75.3 kJ/mol for MgH_2) are listed in Appendix A. From Hess's law, ΔH_f is equal to the sum of the enthalpy changes for Reactions 1–5:

$$\begin{aligned} \Delta H_f &= \Delta H_1 + \Delta H_2 + \Delta H_3 + \Delta H_4 + \Delta H_5 \\ &= \Delta H_{sub}(Mg) + [I_1(Mg) + I_2(Mg)] + D(H_2) + 2EA(H) - U \\ -75.3\ \text{kJ/mol} &= 147.1\ \text{kJ/mol} + (737.7\ \text{kJ/mol} + 1450.7\ \text{kJ/mol}) + 436.0\ \text{kJ/mol} \\ &\quad + 2(-72.8\ \text{kJ/mol}) - U \\ U &= 2701.2\ \text{kJ/mol} \end{aligned}$$

EXERCISE 14.9

1,3-Butadiene (CH_2=CH–CH=CH_2) is a volatile and reactive organic molecule used in the production of rubber. Above room temperature it reacts slowly to form products. Concentrations of 1,3-butadiene (C_4H_6) as a function of time at 326°C are listed in the table along with ln[C_4H_6] and the reciprocal concentrations. Graph the data as concentration versus t. Then determine the order of the reaction in C_4H_6, the rate law, and the rate constant for the reaction.

Time (s)	$[C_4H_6]$ (M)	ln$[C_4H_6]$	1/$[C_4H_6]$ (M^{-1})
0	1.72×10^{-2}	−4.063	58.1
900	1.43×10^{-2}	−4.247	69.9
1800	1.23×10^{-2}	−4.398	81.3
3600	9.52×10^{-3}	−4.654	105
6000	7.30×10^{-3}	−4.920	137

Answer

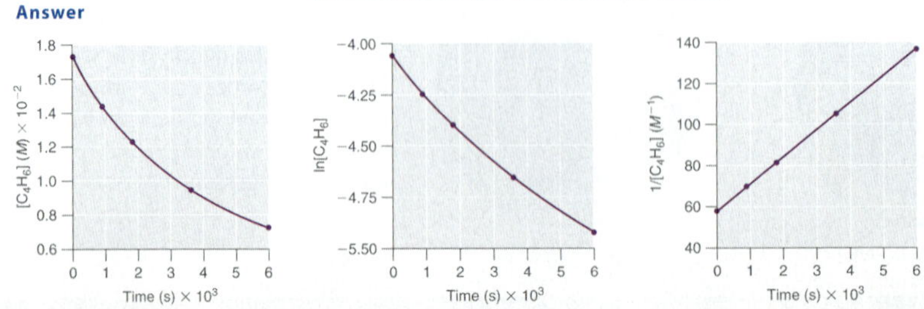

Second order in C_4H_6; rate = $k[C_4H_6]^2$; $k = 2.05 \times 10^3\ M^{-1} \cdot s^{-1}$

Reinforce learning with end-of-chapter materials

Review: Key Concepts, Key Terms, Key Equations

Practice: Conceptual, Numerical, and Applied Problems

SUMMARY AND KEY TERMS

14.0 Introduction (p. 1)

Chemical kinetics is the study of **reaction rates,** changes in the concentrations of reactants and products with time.

14.1 Factors That Affect Reaction Rates (p. 2)

Factors that influence the rates of chemical reactions include concentration of reactants, temperature, physical state of reactants and their dispersion, solvent, and presence of a catalyst.

14.2 Reaction Rates and Rate Laws (p. 4)

Reaction rates are reported either as the **average rate** over a period of time or as the **instantaneous rate** at a single time.

The **rate law** for a reaction is a mathematical relationship between the reaction rate and the concentrations of species in solution. Rate laws can be expressed either as a **differential rate law,** describing the change in reactant or product concentrations as a function of time, or as an **integrated rate law,** describing the actual concentrations of reactants or products as a function of time. The **rate constant,** k, of a rate law is a constant of proportionality between reaction rate and reactant

first-order reaction is a constant that is related to the rate constant for the reaction: $t_{1/2} = 0.693/k$.

Radioactive decay reactions are first-order reactions. The **rate of decay,** or **activity,** of a sample of a radioactive substance is the decrease in the number of radioactive nuclei per unit time.

14.6 Reaction Rates—A Microscopic View (p. 34)

A **reaction mechanism** is the microscopic path by which reactants are transformed into products. Each step is an **elementary reaction.** Species that are formed in one step and consumed in another are **intermediates.** Each elementary step can be described in terms of its **molecularity,** the number of molecules that collide in that step. The slowest step in a reaction mechanism is the **rate-determining step. Chain reactions** consist of three kinds of reaction: initiation, propagation, and termination. Intermediates in chain reactions are often called **radicals,** species that have an unpaired valence electron.

14.7 The Collision Model of Chemical Kinetics

KEY EQUATIONS

General definition of rate	$\text{rate} = \dfrac{\Delta[B]}{\Delta t} = -\dfrac{\Delta[A]}{\Delta t}$	(14.4)
General form of rate law	$\text{rate} = k[A]^m[B]^n$	(14.10)
Zeroth-order reaction	$\text{rate} = -\dfrac{\Delta[A]}{\Delta t} = k$	(14.16)
	$[A] = [A]_0 - kt$	(14.17)
First-order reaction	$\text{rate} = -\dfrac{\Delta[A]}{\Delta t} = k[A]$	(14.21)
	$[A] = [A]_0 e^{-kt}$	(14.22)
	$\ln[A] = \ln[A]_0 - kt$	(14.23)
Second-order reaction	$\text{rate} = -\dfrac{\Delta[A]}{\Delta t} = k[A]^2$	(14.25)
	$\dfrac{1}{[A]} = \dfrac{1}{[A]_0} + kt$	(14.26)
Half-life of first-order reaction	$t_{1/2} = \dfrac{0.693}{k}$	(14.32)
Radioactive decay	$A = kN$	(14.34)
Arrhenius equation	$k = Ae^{-E_a/RT}$	(14.43)

Approximately 100 problems covering conceptual, numerical, and applied material—building from simple to advanced—are found at the end of every chapter.

QUESTIONS AND PROBLEMS

Answers to odd-numbered problems are given in the Appendix. Complete solutions to even-numbered problems are provided in the Student Solutions Manual.

CONCEPTUAL

14.1 Factors That Affect Reaction Rates

1. What information can you obtain by studying the chemical kinetics of a reaction? Does a balanced chemical equation provide the same information? Why or why not?
2. If you were given the task of determining whether to proceed with a particular reaction in an industrial facility, why would studying the chemical kinetics of the reaction be important to you?
3. What is the ... of each of these factors with the rate of

reaction rates significantly relative to the same reaction run in the presence of a heterogeneous catalyst. What is the reason for anticipating that the relative rate will increase?

8. Water has a dielectric constant more than two times greater than that of methanol (dielectric constant = 80.1 for H_2O and 32.7 for CH_3OH). Which would be your solvent of choice for a substitution reaction between an ionic compound and a polar reagent, both of which are soluble in either methanol or water? Why?

14.2 Reaction Rates and Rate Laws

9. Explain why the rate of a reaction is generally fastest at early time intervals. For the reaction $A + B \rightarrow C$, what would a plot of concentration versus time look like?
10. Explain the differences between a differential rate law and an integrated rate law. ...

Paired problems reinforce key problem-solving skills.

Most can be assigned for online homework or tests with automatic grading at **www.masteringgeneralchemistry.com.**

NUMERICAL

This section includes "paired problems" (marked by brackets) that require similar problem-solving skills.

14.6 Reaction Rates—A Microscopic View

63. Cyclopropane, a mild anesthetic, rearranges to propylene via a collision that produces and destroys an energized species. Here are the important steps in this rearrangement.

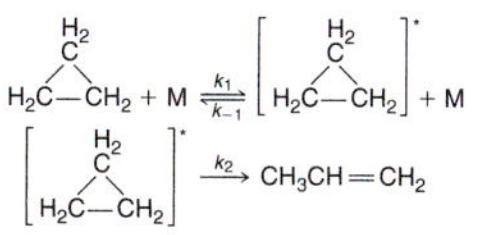

where M is any molecule, including cyclopropane. Only those cyclopropane molecules with sufficient energy (denoted with an asterisk) can rearrange to propylene. Which step determines the rate constant of the overall reaction?

64. Above approximately 500 K, the reaction between NO_2 and CO to produce CO_2 and NO follows the second-order rate law $\Delta[CO_2]/\Delta t = k[NO_2][CO]$. At lower temperatures, however, the rate law is $\Delta[CO_2]/\Delta t = k'[NO_2]^2$, for which it is known

Problems based on relevant applications from modern chemistry mirror the text and reinforce both the theoretical and the practical nature of chemistry.

APPLICATIONS

77. Atmospheric chemistry in the region below the clouds of Venus appears to be dominated by reactions of sulfur and carbon-containing compounds. Included in representative elementary reaction steps are the following:

$$SO_2 + CO \longrightarrow SO + CO_2$$
$$SO + CO \longrightarrow S + CO_2$$
$$SO + SO_2 \longrightarrow S + SO_3$$

For each elementary reaction, write an expression for the reaction rate in terms of changes in the concentrations of each reactant and product with time.

78. In acid, nitriles hydrolyze to produce a carboxylic acid and ammonium ion. For example, acetonitrile, a substance used to extract fatty acids from fish liver oils, is hydrolyzed to acetic acid via the reaction

$$CH_3C{\equiv}N + 2H_2O + H^+ \longrightarrow CH_3\overset{\displaystyle O}{\overset{\|}{C}}OH + NH_4^+$$

Express the rate of the reaction in terms of changes in the concentrations of each reactant and each product with time.

Teaching with art

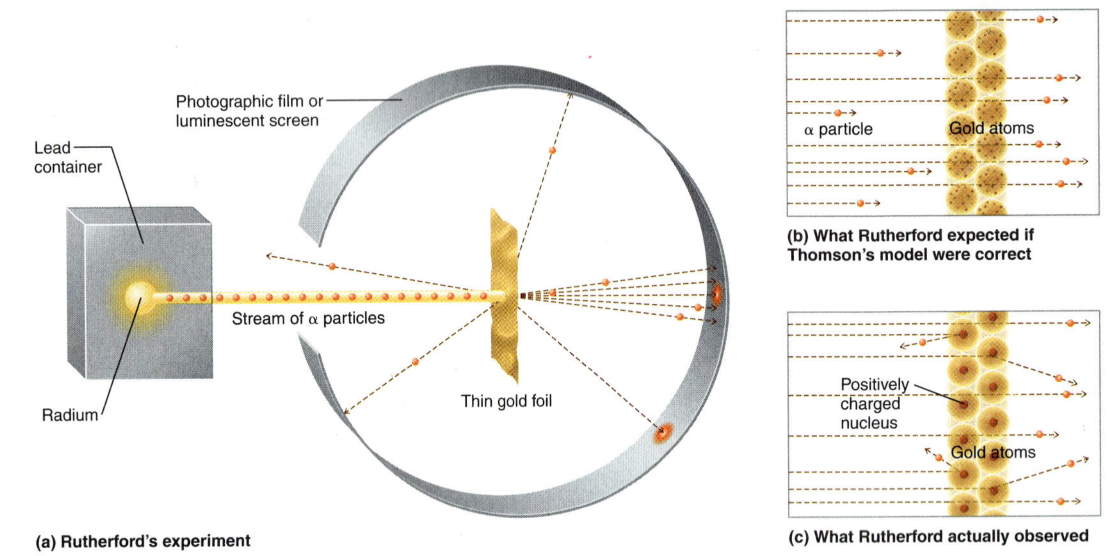

Figure 1.21 **A summary of Rutherford's experiments.**

An historical experiment illustrated and analyzed. Part (a) tells the experimental story, while parts (b) and (c) compare hypothesis versus observation at an atomic level, prompting students to think like an experimental scientist.

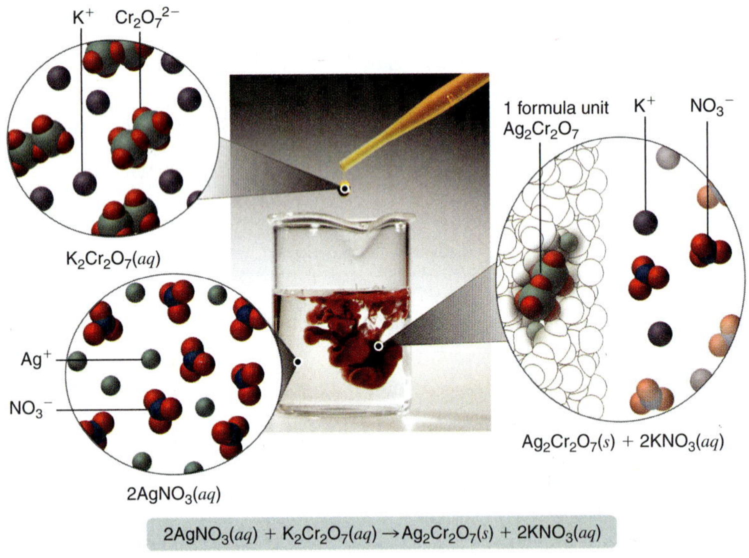

Figure 4.11 **Illustration of what happens at the molecular level when solutions of $AgNO_3$ and $K_2Cr_2O_7$ are mixed.** The Ag^+ and $Cr_2O_7^{2-}$ ions form a red precipitate of solid $Ag_2Cr_2O_7$, while the K^+ and NO_3^- ions remain in solution.

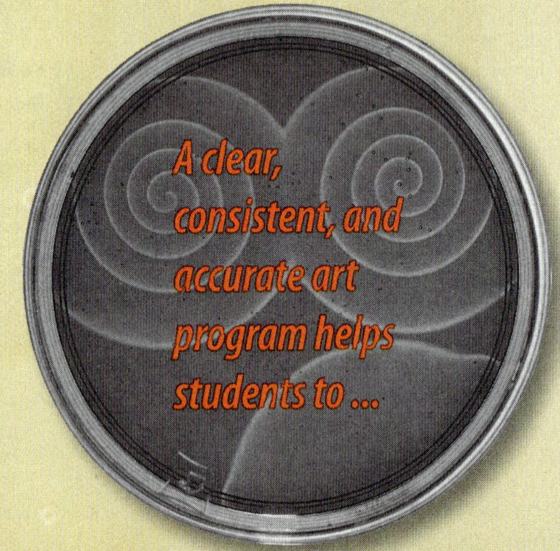

Molecular blowups help students visualize a precipitation reaction explored in the text. Highlighting of formula units and ions in the art makes the stoichiometrically related compositions of reactants and products very tangible.

Teaching with art

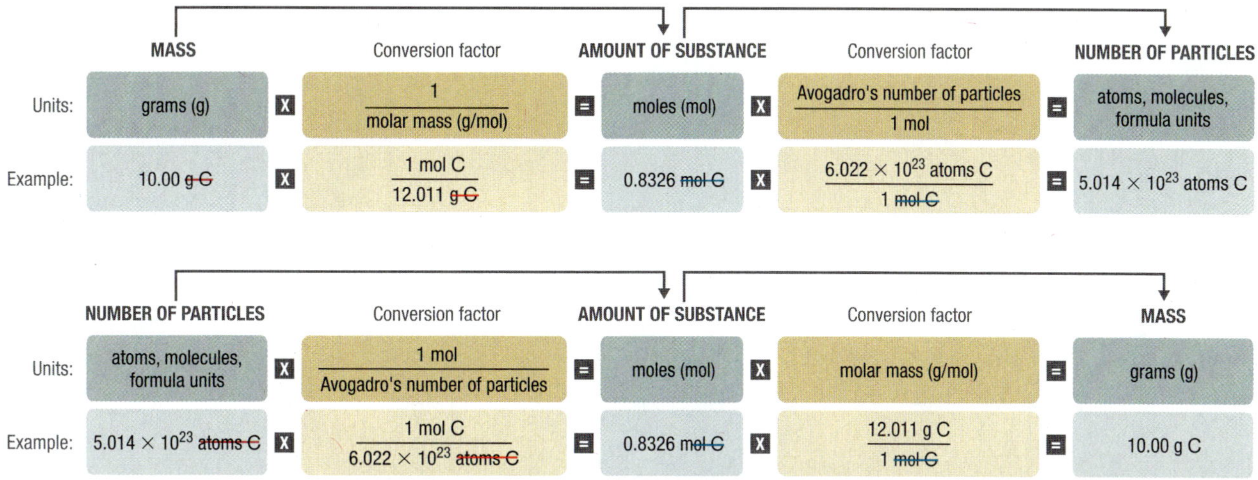

Figure 3.2 A flowchart for converting between mass, moles, and numbers of atoms, molecules, or formula units.

Mass/mole conversions are an essential skill. Flowcharts in Chapters 3 and 4 therefore guide students through every step. Paired examples allow students to confirm for themselves that the general procedure really works if properly implemented.

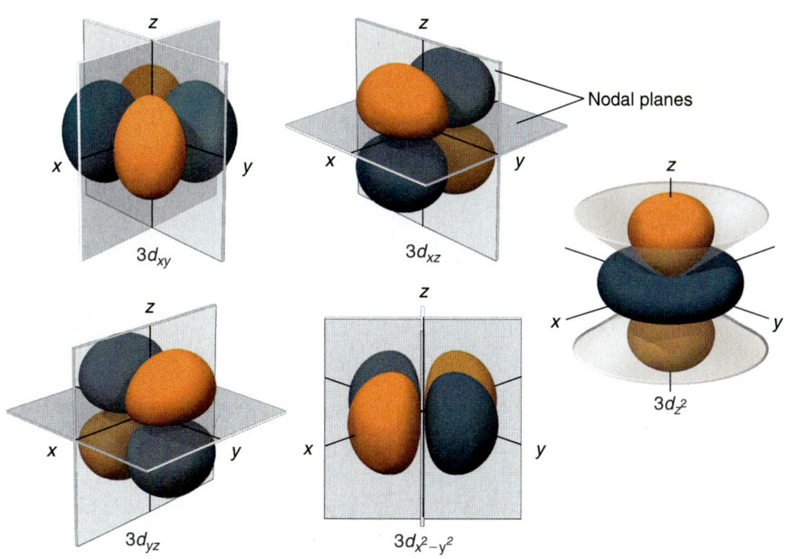

Figure 6.26 The five equivalent 3*d* orbitals of the hydrogen atom.

Orbitals in Chapter 6 are rendered based on Schrödinger's Equation to ensure that shapes and sizes of lobes are mathematically accurate—unlike the "elongated dumb-bells" still shown in some textbooks. Orange and blue are used to distinguish positive and negative phases of wave function, a distinction that will become important in later discussions of bonding and molecular orbital theory.

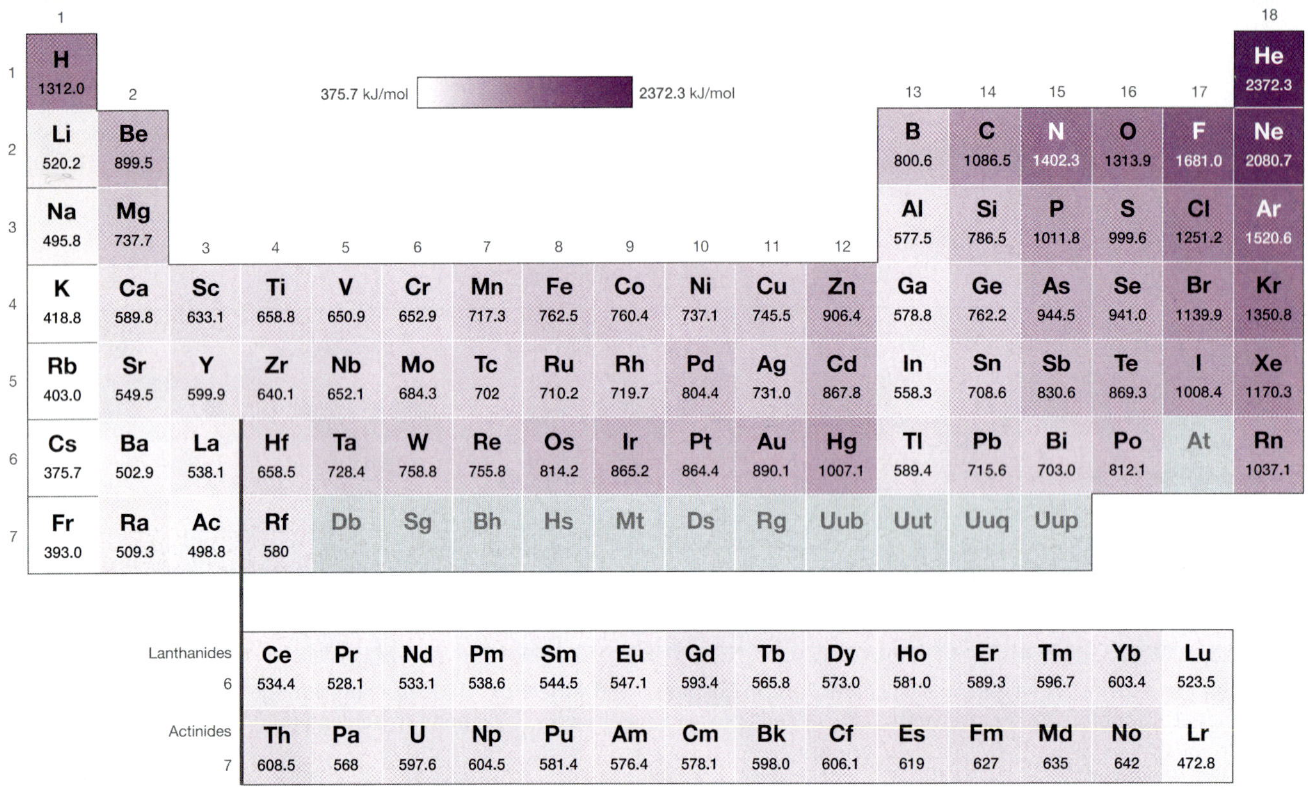

Figure 7.11 First ionization energies of the elements. The relative darkness of boxes illustrate the relative energies of the atoms.

All periodic trends are plotted from real data and consistently styled and color coded. Consistent presentations make it possible for students to easily compare the different trends covered in Chapter 7. Multiple styles of presentation accommodate differing visualization skills among students (some think most comfortably in 2D, others more comfortably in 3D) and are mutually reinforcing.

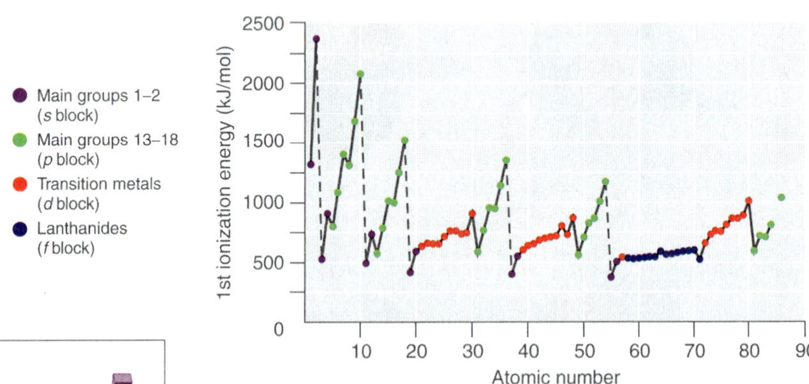

Figure 7.10 Plot of ionization energy versus atomic number for the first six rows of the periodic table. Note the decrease in first ionization energy within a group (most easily seen here for Groups 1 and 18).

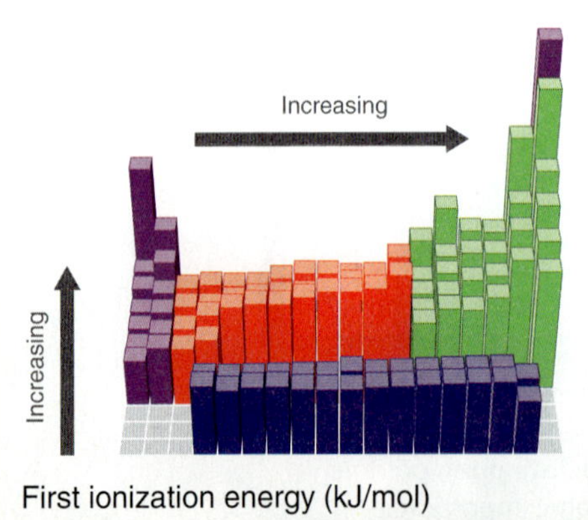

MasteringGeneralChemistry™ ... the most advanced and educationally effective online homework and tutorial system available

Students turn to **MasteringGeneral-Chemistry**™ as a personalized tutor available 24/7 to help them improve problem-solving skills and prepare for exams. Instructors assign homework problems and tutorials and can use data to pinpoint areas of difficulty for individual students or for the class as a whole.

Every tutorial and homework problem has been pre-tested on thousands of students to ensure that they are accurate and effective. Additional benefits to this student-data based approach:

- **Create rigorous and instructive homework assignments** that combine testing with tutoring. Multistep tutorials provide wrong-answer specific feedback, simpler problems upon request, and a variety of answer types.
- **Assign problems based on data drawn from pre-testing,** such as estimated time to complete each problem, level of difficulty, and answer type (for example, molecule drawing, dimensional analysis, orbital diagrams, algebraic solutions, and randomized numbers).
- **Adjust the focus and the pace of your course** based on data collected from homework problems. For example, data lets you see which problems stump students, as well as which steps taken to solve problems cause the most confusion.
- **Check the work of an individual student** in unprecedented detail, including time spent on each step of a problem, wrong answers submitted at every step, how much help was asked for, and how many practice problems a student worked.
- **Compare your results against the "national average"** problem by problem, step by step, class by class, and year by year.

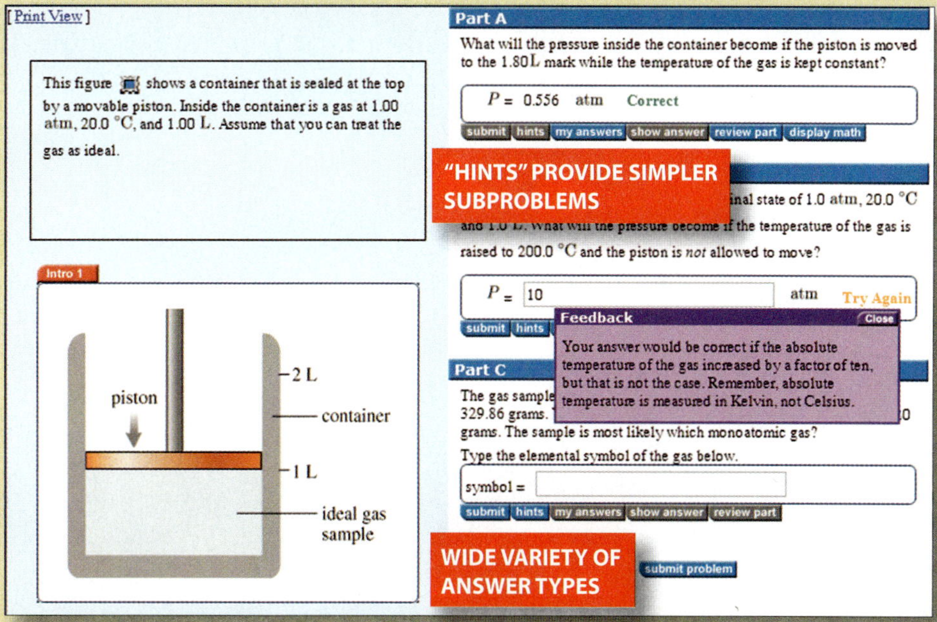

14 Chemical Kinetics

LEARNING OBJECTIVES

- To understand the factors that affect the rate of chemical reactions
- To be able to determine a reaction rate
- To understand the meaning of a rate law
- To be able to determine the concentration dependence of the rate of a chemical reaction
- To be able to predict the individual steps of a simple reaction
- To begin to understand why chemical reactions occur

CHAPTER OUTLINE

The gases, liquids, and solids you learned to describe quantitatively in Part III were systems whose chemical composition did not change with time. Now we will present a quantitative description of a far more common situation in which the chemical composition of a system is not constant with time. An example of such a system is the stratosphere, where chemicals rising from the ground level initiate reactions that lead to a decrease in the concentration of stratospheric ozone—the so-called ozone hole (see Chapter 3). Another example involves the production of polyethylene, in which the properties of the plastic are determined by the relative speeds of events that occur during the polymerization reaction (see Chapter 12). The techniques you are about to learn will enable you to describe the speed of many such changes and to predict how the composition of each system will change in response to changing conditions.

The Belousov-Zhabotinsky reaction, a chemical reaction that oscillates in time and space. When a very thin layer of an acidic solution containing $KBrO_3$, $(NH_4)_2Ce(NO_3)_6$, malonic acid ($HO_2CCH_2CO_2H$), and an indicator is poured into a shallow dish, local fluctuations in the concentration of the reactants and a complex series of reactions cause striking geometric patterns of concentric circles and spirals to propagate across the dish.

We begin Part IV with a discussion of **chemical kinetics**, the study of **reaction rates**, or the changes in the concentrations of reactants and products with time. As you learn about the factors that affect reaction rates, the methods chemists use for reporting and calculating those rates, and the clues that reaction rates provide about events at the molecular level, you will also discover the answers to questions such as these: How can normally stable substances such as flour and coal cause devastating explosions? How do archaeologists use isotopic composition to estimate the ages of ancient artifacts? How do the catalysts used in catalytic converters, some laundry detergents, and meat tenderizers work?

14.1 ◦ Factors That Affect Reaction Rates

Although a balanced chemical equation for a reaction describes the quantitative relationships between the amounts of reactants present and the amounts of products that can be formed, it gives us no information about whether or how fast a given reaction will occur. This information is obtained by studying the chemical kinetics of a reaction, which depend on various factors: reactant concentrations, temperature, physical states and surface areas of reactants, as well as solvent and catalyst properties if either are present. By studying the kinetics of a reaction, chemists gain insights into how to control reaction conditions to achieve a desired outcome.

Concentration Effects

Two substances cannot possibly react with each other unless their constituent particles (molecules, atoms, or ions) come into contact. If there is no contact, the rate of reaction will be zero. Conversely, the more reactant particles that collide per unit time, the more often a reaction between them can occur. Consequently, the rate of a reaction usually increases as the concentration of the reactants increases. One example of this effect is the reaction of sucrose (table sugar) with sulfuric acid, shown in Figure 14.1.

Figure 14.1 The effect of concentration on reaction rates. Mixing sucrose with *dilute* sulfuric acid in a beaker (a, *right*) produces a simple solution. Mixing the same amount of sucrose with *concentrated* sulfuric acid (a, *left*) results in a dramatic reaction (b) that eventually produces a column of black porous graphite (c) and an intense smell of burning sugar.

(a) Time = 0 seconds **(b) Time = 15 seconds** **(c) Time = 60 seconds**

Temperature Effects

You learned in Chapter 10 that increasing the temperature of a system increases the average kinetic energy of its constituent particles. As the average kinetic energy increases, the particles move faster, so they collide more frequently per unit time and possess greater energy when they collide. Both of these factors increase the rate of the reaction. Hence, the rate of virtually all reactions increases with increasing temperature. Conversely, the rate of virtually all reactions decreases with decreasing temperature. For example, refrigeration retards the rate of growth of bacteria in foods by decreasing the rates of the biochemical reactions that enable bacteria to reproduce. Figure 14.2 shows how temperature affects the light emitted by two chemiluminescent light sticks.

Figure 14.2 The effect of temperature on reaction rates. At high temperature, the reaction that produces light in a chemiluminescent light stick occurs more rapidly, producing more photons of light per unit time. Consequently, the light glows brighter in hot water *(left)* than in ice water *(right)*.

In systems where more than one reaction is possible, the same reactants can produce different products under different reaction conditions. For example, in the presence of dilute sulfuric acid and at temperatures around 100°C, ethanol is converted to diethyl ether:

$$2CH_3CH_2OH \xrightarrow{H_2SO_4} CH_3CH_2OCH_2CH_3 + H_2O \qquad (14.1)$$

At 180°C, however, a completely different reaction occurs, producing ethylene rather than diethyl ether as the major product:

$$CH_3CH_2OH \xrightarrow{H_2SO_4} C_2H_4 + H_2O \qquad (14.2)$$

Phase and Surface Area Effects

When two reactants are in the same fluid phase, their particles collide more frequently than when one or both of the reactants is solid (or when they are in different fluids that do not mix). If the reactants are uniformly dispersed in a single homogeneous solution, then the number of collisions per unit time depends on concentration and temperature, as we have just seen. If the reaction is heterogeneous, however, the reactants are in two different phases, and collisions between the reactants can occur only at interfaces between phases. The number of collisions between reactants per unit time is substantially reduced relative to the homogeneous case, and, hence, so is the reaction rate. The rate of a heterogeneous reaction depends on the surface area of the more condensed phase.

Automobile engines use surface area effects to increase reaction rates. Gasoline is injected into each cylinder, where it combusts upon ignition by a spark from the spark plug. The gasoline is injected in the form of microscopic droplets because in that form it has a much larger surface area and can burn much more rapidly than if it were fed into the cylinder as a stream. Similarly, a pile of finely divided flour burns slowly (or not at all), but spraying finely divided flour into a flame produces a vigorous reaction (Figure 14.3). Similar phenomena are partially responsible for the dust explosions that occasionally destroy grain elevators or coal mines.

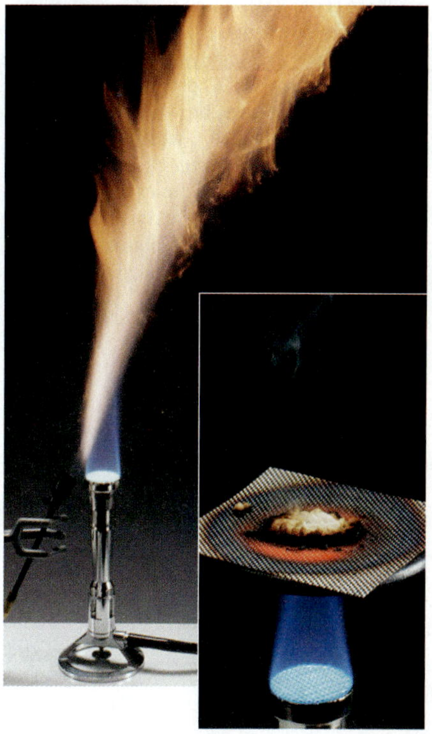

Figure 14.3 The effect of surface area on reaction rates. A pile of flour is only scorched by a flame *(right)*, but when the same flour is sprayed into the flame, it burns rapidly *(left)*.

Solvent Effects

The nature of the solvent can also affect the reaction rates of solute particles. For example, a sodium acetate solution reacts with methyl iodide in an exchange reaction to give methyl acetate and sodium iodide.

$$CH_3CO_2Na(soln) + CH_3I(l) \longrightarrow CH_3CO_2CH_3(soln) + NaI(soln) \qquad (14.3)$$

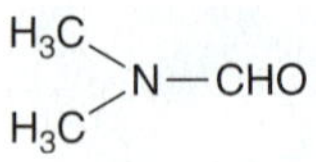

DMF (dimethylformamide)

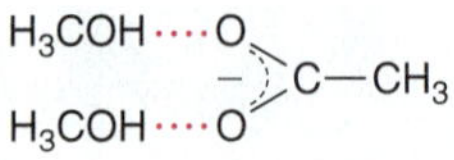

Methanol Acetate ion

This reaction occurs 10 million times more rapidly in the organic solvent DMF [$(CH_3)_2NCHO$] than it does in methanol (CH_3OH). Although both are organic solvents with similar dielectric constants (36.7 for DMF versus 32.6 for methanol), methanol is able to hydrogen bond with acetate ions, whereas DMF cannot. Hydrogen bonding reduces the reactivity of the oxygen atoms of the acetate ion.

Solvent viscosity is also important in determining reaction rates. In highly viscous solvents, dissolved particles diffuse much more slowly than in less viscous solvents and can collide less frequently per unit time. Thus, the rates of most reactions decrease rapidly with increasing solvent viscosity.

Figure 14.4 The effect of catalysts on reaction rates. A solution of hydrogen peroxide (H_2O_2) decomposes in water so slowly that the change is not noticeable *(left)*. Iodide ion acts as a catalyst for the decomposition of H_2O_2, producing oxygen gas. The solution turns brown because of the reaction of H_2O_2 with I^-, which generates small amounts of I_3^- *(center)*. The enzyme catalase is about 3 *billion* times more effective than iodide as a catalyst. Even in the presence of very small amounts of enzyme, the decomposition is vigorous *(right)*.

Catalyst Effects

You learned in Chapter 3 that a catalyst is a substance that participates in a chemical reaction and increases the rate of the reaction without undergoing a net chemical change itself. Consider, for example, the decomposition of hydrogen peroxide in the presence and absence of different catalysts (Figure 14.4). Because most catalysts are highly selective, they often determine the product of a reaction by accelerating only one of several possible reactions that could occur.

Most of the bulk chemicals produced in industry are formed through catalyzed reactions. Recent estimates indicate that about 30% of the gross national product of the United States and other industrialized nations relies either directly or indirectly on the use of catalysts.

14.2 Reaction Rates and Rate Laws

The factors discussed in Section 14.1 affect the rate of a chemical reaction, which in turn may determine whether a desired product is formed. In this section we will show you how to quantitatively determine the rate of a reaction.

Reaction Rates

MGC Reaction Rates

Reaction rates are usually expressed as the concentration of reactant consumed or the concentration of product formed per unit time. The units are thus moles per liter per unit time, written as *M*/s, *M*/min, or *M*/h. To measure reaction rates, chemists initiate the reaction, measure the concentration of the reactant or product at different times as the reaction progresses, perhaps plot the concentration as a function of time on a graph, and then calculate the *change* in the concentration per unit time.

The progress of a simple reaction ($A \longrightarrow B$) is shown in Figure 14.5, where the beakers are snapshots of the composition of the solution at 10-s intervals. The numbers of molecules of reactant (A) and product (B) are plotted as a function of time in the graph. Each point in the graph corresponds to one beaker in Figure 14.5. The reaction rate is the change in the concentration of either the reactant or the product over a period of time.

$$\text{rate} = \frac{\Delta[\text{B}]}{\Delta t} = -\frac{\Delta[\text{A}]}{\Delta t} \tag{14.4}$$

Just as in Chapters 4 and 5, square brackets indicate molar concentrations and the capital Greek delta (Δ) means "change in." The concentration of A decreases with time, while the concentration of B increases with time. Because chemists follow the

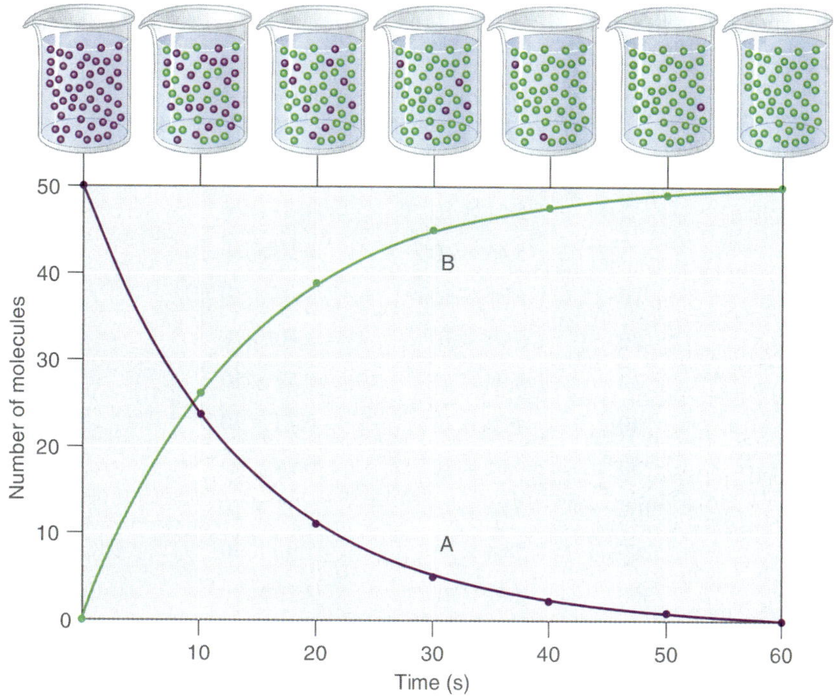

Figure 14.5 The progress of reaction A $\longrightarrow$ B over a 1-min period *(top)*. Note that the mixture initially contains only A molecules (purple). With increasing time the number of A molecules decreases and more B molecules (green) are formed.

A plot of the progress of the reaction *(bottom)*. This graph shows the change in the numbers of A and B molecules in the reaction depicted above as a function of time.

Interactive Graph

convention of expressing all reaction rates as positive numbers, however, a negative sign is inserted in front of $\Delta[\text{A}]/\Delta t$ to convert that expression to a positive number. Notice that the rate we would calculate for reaction A $\longrightarrow$ B using Equation 14.4 would be different for each interval (this is not true for every reaction, as you will see). A much greater change occurs in [A] and [B] during the first 10-s interval, for example, than during the last, which means that the rate of the reaction is fastest at first. This is consistent with the concentration effects described in the preceding section because the concentration of A is greatest at the beginning of the reaction.

Determining the Rate of Hydrolysis of Aspirin

We can use Equation 14.4 to determine the rate of hydrolysis of aspirin, probably the most commonly used drug in the world. (More than 25 million kilograms are produced annually worldwide.) Aspirin (acetylsalicylic acid) reacts with water (such as water in body fluids) to give salicylic acid and acetic acid.

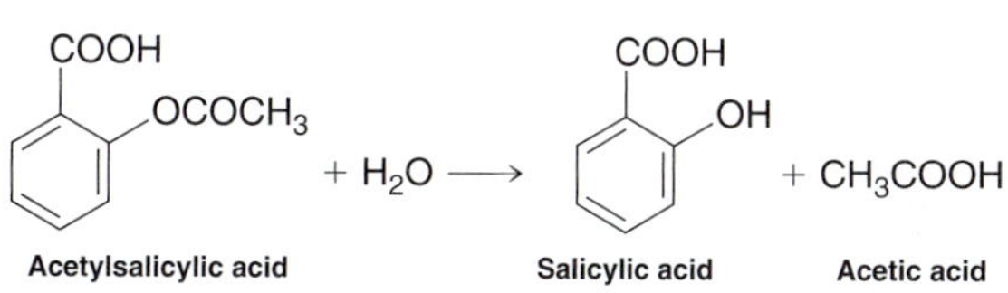

(14.5)

Because salicylic acid is the actual substance that relieves pain and reduces fever and inflammation, a great deal of research has focused on understanding this reaction and the factors that affect its rate. Data for the hydrolysis of a sample of aspirin are listed in Table 14.1 and are shown in the graph in Figure 14.6. These data were obtained by removing samples of the reaction mixture at the indicated times and analyzing them for the concentrations of the reactant (aspirin) and one of the products (salicylic acid).

We can calculate the **average reaction rate** for a given time interval from the concentrations of either the reactant or one of the products at the beginning of the interval (time $= t_0$) and at the end of the interval (t_1). Using salicylic acid, for example, we

TABLE 14.1 Data for the hydrolysis of aspirin in aqueous solution at pH 7.0 and 37°C[a]

Time (h)	[Aspirin], *M*	[Salicylic Acid], *M*
0	5.55×10^{-3}	0×10^{-3}
2.0	5.51×10^{-3}	0.040×10^{-3}
5.0	5.45×10^{-3}	0.10×10^{-3}
10	5.35×10^{-3}	0.20×10^{-3}
20	5.15×10^{-3}	0.40×10^{-3}
30	4.96×10^{-3}	0.59×10^{-3}
40	4.78×10^{-3}	0.77×10^{-3}
50	4.61×10^{-3}	0.94×10^{-3}
100	3.83×10^{-3}	1.72×10^{-3}
200	2.64×10^{-3}	2.91×10^{-3}
300	1.82×10^{-3}	3.73×10^{-3}

[a] Note that the reaction at pH 7.0 is very slow. It is *much* faster under acidic conditions, such as those found in the stomach.

find the rate of the reaction for the interval between $t = 0$ h and $t = 2.0$ h (recall that Δ is always calculated as "final minus initial"):

$$\text{rate}_{(t=0-2.0\text{ h})} = \frac{[\text{salicylic acid}]_2 - [\text{salicylic acid}]_0}{2.0\text{ h} - 0\text{ h}}$$

$$= \frac{0.040 \times 10^{-3}\,M - 0\,M}{2.0\text{ h}} = 2.0 \times 10^{-5}\,M\text{/h}$$

Figure 14.6 The hydrolysis of aspirin. This graph shows the concentrations of aspirin and salicylic acid as a function of time, based on the hydrolysis data in Table 14.1. The time dependence of the concentration of the other product, acetate, is not shown, but based on the stoichiometry of the reaction it is identical to the data for salicylic acid. Interactive Graph

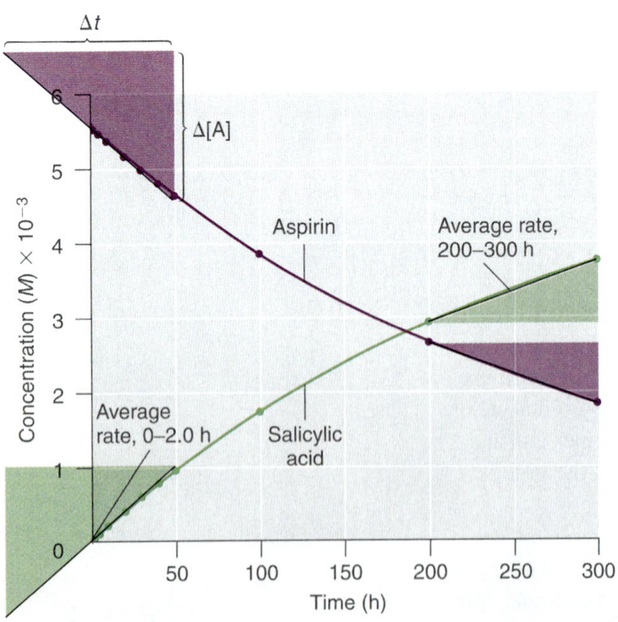

We can also calculate the rate of the reaction from the concentrations of aspirin at the beginning and the end of the same interval, remembering to insert a negative sign, because its concentration decreases:

$$\text{rate}_{(t=0-2.0\text{ h})} = -\frac{[\text{aspirin}]_2 - [\text{aspirin}]_0}{2.0\text{ h} - 0\text{ h}}$$

$$= -\frac{5.51 \times 10^{-3}\,M - 5.55 \times 10^{-3}\,M}{2.0\text{ h}}$$

$$= 2 \times 10^{-5}\,M\text{/h}$$

If we now calculate the rate during the last interval given in Table 14.1 (the interval between 200 h and 300 h after the start of the reaction), we find that the rate of the reaction is significantly slower than it was during the first interval ($t = 0\text{–}2.0$ h):

$$\text{rate}_{(t=200-300\text{ h})} = \frac{[\text{salicylic acid}]_{300} - [\text{salicylic acid}]_{200}}{300\text{ h} - 200\text{ h}}$$

$$= \frac{3.73 \times 10^{-3}\,M - 2.91 \times 10^{-3}\,M}{100\text{ h}}$$

$$= 8.2 \times 10^{-6}\,M\text{/h}$$

(You should verify from the data in Table 14.1 that you get the same rate using the concentrations of aspirin measured at 200 h and 300 h.)

Calculating the Rate of Fermentation of Sucrose

Notice in the preceding example that the stoichiometric coefficients in the balanced chemical equation are the same for all reactants and products; that is, the reactants and products all have the coefficient 1. Let us look at a reaction in which the coefficients are *not* all the same: the fermentation of sucrose to ethanol and carbon dioxide, which we encountered in Chapter 3.

$$\underset{\text{Sucrose}}{C_{12}H_{22}O_{11}(aq)} + H_2O(l) \longrightarrow 4C_2H_5OH(aq) + 4CO_2(g) \tag{14.6}$$

The coefficients show us that the reaction produces four molecules of ethanol and four molecules of carbon dioxide for every one molecule of sucrose that is consumed. As before, we can find the rate of the reaction by looking at the change in the concentration of any reactant or product. In this particular case, however, a chemist would probably use the concentration of either sucrose or ethanol because gases are usually measured as volumes and, as you learned in Chapter 10, the volume of CO_2 gas formed will depend on the total volume of the solution being studied and on the solubility of the gas in the solution, not just on the concentration of sucrose. The coefficients in the balanced equation tell us that the rate at which ethanol is formed is always four times faster than the rate at which sucrose is consumed:

$$\frac{\Delta[C_2H_5OH]}{\Delta t} = -\frac{4\Delta[\text{sucrose}]}{\Delta t} \tag{14.7}$$

Note again that the concentration of the reactant—in this case sucrose—*decreases* with increasing time, so the value of Δ[sucrose] is negative. Consequently, a minus sign is inserted in front of Δ[sucrose] in Equation 14.7 so that the rate of change of the sucrose concentration is expressed as a positive value. Conversely, the ethanol concentration *increases* with increasing time, so its rate of change is automatically expressed as a positive value.

Often the rate of the reaction is expressed in terms of the reactant or product that has the smallest coefficient in the balanced equation. The smallest coefficient in the sucrose fermentation reaction (Equation 14.6) corresponds to sucrose, so the rate is generally defined as

$$\text{rate} = -\frac{\Delta[\text{sucrose}]}{\Delta t} = \frac{1}{4}\left(\frac{\Delta[C_2H_5OH]}{\Delta t}\right) \tag{14.8}$$

EXAMPLE 14.1

Consider the thermal decomposition of gaseous N_2O_5 to NO_2 and O_2 via the equation

$$2N_2O_5(g) \xrightarrow{\Delta} 4NO_2(g) + O_2(g)$$

Write expressions for the reaction rate in terms of the rates of change in the concentrations of the reactant and each product with time.

Given Balanced chemical equation

Asked for Reaction rate expressions

Strategy

A Choose the species in the equation that has the smallest coefficient. Then write an expression for the rate of change of that species with time.

B For the remaining species in the equation, use molar ratios to obtain equivalent expressions for the rate of the reaction.

Solution

A Because O_2 has the smallest coefficient in the balanced equation for the reaction, we define the rate of the reaction as the rate of change in the concentration of O_2 and write that expression. B We know from the balanced equation that 2 mol of N_2O_5 must decompose for each 1 mol of O_2 produced and that 4 mol of NO_2 are produced for every 1 mol of O_2 produced. The molar ratios of O_2 to N_2O_5 and to NO_2 are thus 1:2 and 1:4, respectively. This means that we divide the rate of change of $[N_2O_5]$ and of $[NO_2]$ by its stoichiometric coefficient to obtain equivalent expressions for the rate of the reaction. For example, because NO_2 is produced at four times the rate of O_2, we must divide the rate of production of NO_2 by 4. The reaction rate expressions are thus

$$\text{rate} = \frac{\Delta[O_2]}{\Delta t} = \frac{\Delta[NO_2]}{4\Delta t} = -\frac{\Delta[N_2O_5]}{2\Delta t}$$

EXERCISE 14.1

The key step in the industrial production of sulfuric acid is the reaction of SO_2 with O_2 to produce SO_3.

$$2SO_2(g) + O_2(g) \longrightarrow 2SO_3(g)$$

Write expressions for the reaction rate in terms of the rate of change of the concentration of each species.

Answer $\text{rate} = -\frac{\Delta[O_2]}{\Delta t} = -\frac{\Delta[SO_2]}{2\Delta t} = \frac{\Delta[SO_3]}{2\Delta t}$

EXAMPLE 14.2

Using the reaction shown in Example 14.1, calculate the rate of the reaction from the following data taken at 56°C:

$2N_2O_5\,(g) \longrightarrow 4NO_2\,(g) + O_2(g)$

Time, s	$[N_2O_5], M$	$[NO_2], M$	$[O_2], M$
240	0.0388	0.0314	0.00792
600	0.0197	0.0699	0.0175

Given Balanced chemical equation and concentrations at specific times

Asked for Rate of the reaction

Strategy

A Using the equations in Example 14.1, subtract the initial concentration of a species from its final concentration, and substitute that value into the equation for that species.

B Substitute the value for the time interval into the equation. Check to make sure your units are consistent.

Solution

A We are asked to calculate the rate of the reaction in the interval between $t_1 = 240\ s$ and $t_2 = 600\ s$. From Example 14.1, we see that we can evaluate the rate of the reaction using any of three expressions:

$$\text{rate} = \frac{\Delta[O_2]}{\Delta t} = \frac{\Delta[NO_2]}{4\Delta t} = -\frac{\Delta[N_2O_5]}{2\Delta t}$$

Subtracting the initial concentration from the final concentration of N_2O_5 and inserting the corresponding time interval into the rate expression for N_2O_5 gives

$$\text{rate} = -\frac{\Delta[N_2O_5]}{2\Delta t} = -\frac{[N_2O_5]_{600} - [N_2O_5]_{240}}{2(600\text{ s} - 240\text{ s})}$$

B Substituting actual values into the expression gives

$$\text{rate} = -\frac{0.0197\ M - 0.0388\ M}{2(360\text{ s})} = 2.65 \times 10^{-5}\ M/\text{s}$$

Similarly, we can use NO_2 to calculate the rate:

$$\text{rate} = \frac{\Delta[NO_2]}{4\Delta t} = \frac{[NO_2]_{600} - [NO_2]_{240}}{4(600\text{ s} - 240\text{ s})} = \frac{0.0699\ M - 0.0314\ M}{4(360\text{ s})} = 2.67 \times 10^{-5}\ M/\text{s}$$

If we allow for experimental error, this is the same rate we obtained using the data for N_2O_5, as it should be because the rate of the reaction should be the same no matter which concentration is used. We can also use the data for O_2:

$$\text{rate} = \frac{\Delta[O_2]}{\Delta t} = \frac{[O_2]_{600} - [O_2]_{240}}{600\text{ s} - 240\text{ s}} = \frac{0.0175\ M - 0.00792\ M}{360\text{ s}} = 2.66 \times 10^{-5}\ M/\text{s}$$

Again, this is the same value we obtained from the N_2O_5 and NO_2 data. Thus, the rate of the reaction does not depend on which reactant or product is used to measure it.

EXERCISE 14.2

Using the data in the table, calculate the rate of the reaction of $SO_2(g)$ with $O_2(g)$ to give $SO_3(g)$.

$$2SO_2(g) + O_2(g) \longrightarrow 2SO_3(g)$$

Time, s	$[SO_2]$, M	$[O_2]$, M	$[SO_3]$, M
300	0.0270	0.0500	0.0072
720	0.0194	0.0462	0.0148

Answer $9.0 \times 10^{-6}\ M/\text{s}$

Instantaneous Rates of Reaction

So far we have determined average reaction rates over particular intervals of time. We can also determine the **instantaneous rate** of a reaction, which is the rate at any given point in time. As the period of time used to calculate an average rate of a reaction becomes shorter and shorter, the average rate approaches the instantaneous rate.*

Think of the distinction between the instantaneous and average rates of a reaction as being similar to the distinction between the actual speed of a car at any given time on a trip and the average speed of the car for the entire trip. Although you may travel for a long time at 65 mi/h on an interstate during a long trip, there may be times when you travel only 25 mi/h in construction zones, or 0 mi/h if you stop for meals or gas. Thus, your average speed on the trip may be only 50 mi/h, whereas your instantaneous speed on the interstate at a given moment may be 65 mi/h. Whether you are able to stop the car in time to avoid an accident depends on your instantaneous speed, not on your average speed. There are important differences between the speed of a car during a trip

* If you have studied calculus, you may recognize that the instantaneous rate of a reaction at a given time corresponds to the slope of a line tangent to the concentration-versus-time curve at that point—that is, the derivative of concentration with respect to time.

and the speed of a chemical reaction, however. The speed of a car may vary unpredictably over the length of the trip, and the initial part of the trip is often one of the slowest. In a chemical reaction, the initial interval normally has the fastest rate (though this is not always the case), and the rate generally changes smoothly over time.

In chemical kinetics, we generally focus on one particular instantaneous rate, $t = 0$, which is the initial rate of the reaction. Initial rates are determined by measuring the rate of the reaction at various times and then extrapolating a plot of rate versus time to $t = 0$.

Rate Laws

In the preceding section you learned that reaction rates generally decrease with time because reactant concentrations decrease as reactants are converted to products. You also learned that reaction rates generally increase when reactant concentrations are increased. We now examine the mathematical expressions called **rate laws**, which describe the relationships between reactant rates and reactant concentrations. Rate laws are laws as defined in Chapter 1; that is, they are mathematical descriptions of experimentally verifiable data.

Rate laws may be written from either of two different, but related, perspectives. A **differential rate law** expresses the rate of a reaction in terms of *changes* in the concentration of one or more reactants, $\Delta[R]$, over a specific time interval, Δt. In contrast, an **integrated rate law** describes the rate of a reaction in terms of the *initial* concentration, $[R]_0$, and the *measured* concentration of one or more reactants, [R], after a given amount of time, t; we will discuss integrated rate laws in the next section. The integrated rate law can be found by using calculus to integrate the differential rate law; the method of doing so is beyond the scope of this text. *Whether you use a differential or integrated rate law, always check to make sure that the rate law gives the proper units for the rate, usually* M/s.

Reaction Orders

For a reaction with the general equation

$$aA + bB \longrightarrow cC + dD \tag{14.9}$$

the experimentally determined rate law usually has the form

$$\text{rate} = k[A]^m[B]^n \tag{14.10}$$

The *proportionality constant, k*, is called the **rate constant**, and its value is characteristic of the reaction and reaction conditions. A given reaction has a particular value of the rate constant under a given set of conditions, such as temperature, pressure, and solvent; varying the temperature or the solvent usually changes the value of the rate constant. The numerical value of k, however, does *not* change as the reaction progresses under a given set of conditions.

Thus, the rate of a reaction depends on the rate constant for the given set of reaction conditions and on the concentration of A and B, raised to the powers m and n, respectively. The values of m and n are derived from experimental measurements of the changes in reactant concentrations over time and indicate the **reaction order**, the degree to which the rate of the reaction depends on the concentration of each reactant; m and n need not be integers. For example, Equation 14.10 tells us that Reaction 14.9 is mth order in reactant A and nth order in reactant B. It is important to remember that n and m are *not* related to the stoichiometric coefficients a and b in the balanced chemical equation but must be determined experimentally. The *overall reaction order* is the sum of all the exponents in the rate law, or $m + n$.

Although differential rate laws are generally used to describe what is occurring on a molecular level during a reaction, integrated rate laws are used for determining the reaction order and the value of the rate constant from experimental measurements.

(We present general forms for integrated rate laws in Section 14.3.) To illustrate how chemists interpret a differential rate law, we turn to the experimentally derived rate law for the hydrolysis of *t*-butyl bromide in 70% aqueous acetone. This reaction produces *t*-butanol according to the equation

$$(CH_3)_3CBr(soln) + H_2O(soln) \longrightarrow (CH_3)_3COH(soln) + HBr(soln) \qquad (14.11)$$

Combining the rate expression in Equations 14.4 and 14.10 gives us a general expression for the differential rate law:

$$rate = -\frac{\Delta[A]}{\Delta t} = k[A]^m[B]^n \qquad (14.12)$$

Inserting the identities of the reactants into Equation 14.12 gives the following expression for the differential rate law for the reaction:

$$\text{rate} = -\frac{\Delta[(CH_3)_3CBr]}{\Delta t} = k[(CH_3)_3CBr]^m[H_2O]^n \qquad (14.13)$$

Experiments done to determine the rate law for the hydrolysis of *t*-butyl bromide show that the rate of the reaction is directly proportional to the concentration of $(CH_3)_3CBr$ but is independent of the concentration of water. Thus, *m* and *n* in Equation 14.13 are 1 and 0, respectively, and

$$\text{rate} = k[(CH_3)_3CBr]^1[H_2O]^0 = k[(CH_3)_3CBr] \qquad (14.14)$$

Because the exponent for the reactant is 1, the reaction is *first order* in $(CH_3)_3CBr$. It is *zeroth order* in water because the exponent for $[H_2O]$ is 0 (recall that anything raised to the zeroth power equals 1). Thus, the overall reaction order is $1 + 0 = 1$. What the reaction orders tell us in practical terms is that doubling the concentration of $(CH_3)_3CBr$ doubles the rate of the hydrolysis reaction, halving the concentration of $(CH_3)_3CBr$ halves the rate, and so on. Conversely, increasing or decreasing the concentration of water has *no effect* on the rate of the reaction. (Again, remember when you work with rate laws that there is no simple correlation between the stoichiometry of the reaction and the rate law. The values of *k*, *m*, and *n* in the rate law *must* be determined experimentally.) Experimental data show that *k* has the value $5.15 \times 10^{-4}\ s^{-1}$ at 25°C. Note that the rate constant has units of reciprocal seconds (s^{-1}) because the rate of the reaction is defined in units of concentration per unit time (*M*/s). The units of a rate constant depend on the rate law for a particular reaction.

Under conditions identical to those for the *t*-butyl bromide reaction, the experimentally derived differential rate law for the hydrolysis of methyl bromide (CH_3Br) is

$$\text{rate} = -\frac{\Delta[CH_3Br]}{\Delta t} = k'[CH_3Br] \qquad (14.15)$$

This reaction also has an overall reaction order of 1, but the rate constant in Equation 14.15 is approximately 10^6 times smaller than that for *t*-butyl bromide. Thus, methyl bromide hydrolyzes about 1 million times more slowly than *t*-butyl bromide, and this information tells chemists how the reactions differ on a molecular level.

Frequently, changes in reaction conditions also produce changes in a rate law. In fact, chemists often change reaction conditions to obtain clues about what is occurring during a reaction. For example, when *t*-butyl bromide is hydrolyzed in aqueous acetone solution containing OH^- ions rather than in aqueous acetone alone, the differential rate law for the hydrolysis reaction does not change. For methyl bromide, in contrast, the differential rate law becomes rate $= k''[CH_3Br][OH^-]$, with an overall reaction order of 2. Thus, although the two reactions proceed similarly in neutral solution, they proceed very differently in the presence of a base, which again provides clues as to how the reactions differ on a molecular level.

EXAMPLE 14.3

We present three reactions and their experimentally determined differential rate laws. For each reaction: (a) give the *units* of the rate constant; (b) give the *order* of the reaction with respect to each reactant; (c) give the *overall order* of the reaction; and (d) predict what happens to the rate when the concentration of the first species in each chemical equation is doubled.

1. $2HI(g) \xrightarrow{Pt} H_2(g) + I_2(g)$ $\quad$ $\text{Rate} = -\frac{1}{2}\left(\frac{\Delta[HI]}{\Delta t}\right) = k[HI]^2$

2. $2N_2O(g) \xrightarrow{\Delta} 2N_2(g) + O_2(g)$ $\quad$ $\text{Rate} = -\frac{1}{2}\frac{\Delta[N_2O]}{\Delta t} = k$

3. $\text{Cyclopropane}(g) \longrightarrow \text{propane}(g)$ $\quad$ $\text{Rate} = -\frac{\Delta[\text{cyclopropane}]}{\Delta t} = k[\text{cyclopropane}]$

Given Balanced chemical equations and differential rate laws

Asked for Units of rate constant, reaction orders, and effect of doubling reactant concentration

Strategy

A Express the rate as mol/(L·s), or M/s. Then determine the units of each chemical species in the rate law. Divide the units for the rate of the reaction by the units for all species in the rate law to obtain the units for the rate constant.

B Identify the exponent of each species in the rate law to determine the order of the reaction with respect to that species. Sum all exponents to obtain the overall order of the reaction.

C Use the mathematical relationships as expressed in the rate law to determine the effect of doubling the concentration of a single species on the rate of the reaction.

Solution

1. **A** **(a)** Note that $[HI]^2$ will give units of M^2. For the reaction rate to have units of M/s, the rate constant must have units of $1/(M\cdot s)$:

$$kM^2 = \frac{M}{s} \quad k = \frac{M/s}{M^2} = \frac{1}{M\cdot s} = M^{-1}\cdot s^{-1}$$

B **(b)** The exponent in the rate law is 2, so the reaction is second order in HI. **(c)** Because HI is the only reactant and the only species that appears in the rate law, the reaction is also second order overall. **C** **(d)** If the concentration of HI is doubled, the rate will increase from $k[HI]_0^2$ to $k(2[HI])_0^2 = 4k[HI]_0^2$. The rate will therefore quadruple.

2. **A** **(a)** Because no concentration term appears in the rate law, the rate constant must have units of M/s in order for the rate to have units of M/s. **B** **(b, c)** The rate law tells us that the rate is constant and is independent of the N_2O concentration. That is, the reaction is zeroth order in N_2O and zeroth order overall. **C** **(d)** Because the rate is independent of the N_2O concentration, doubling the concentration will have no effect on the rate.

3. **A** **(a)** The rate law contains only one concentration term raised to the first power. Hence, the rate constant must have units of reciprocal seconds (s^{-1}) to have units of M/s for the rate: $M\cdot s^{-1} = M/s$. **B** **(b)** The only concentration in the rate law is that of cyclopropane, and its exponent is 1. This means that the reaction is first order in cyclopropane. **(c)** Cyclopropane is the only species that appears in the rate law, so the reaction is also first order overall. **C** **(d)** Doubling the initial cyclopropane concentration will increase the rate from $k[\text{cyclopropane}]_0$ to $2k[\text{cyclopropane}]_0$. This doubles the rate.

EXERCISE 14.3

Given the following two reactions and their experimentally determined differential rate laws: (a) determine the *units* of the rate constant if time is in seconds; (b) determine the *order* of the reaction with respect to each reactant; (c) give the *overall order* of the reaction; and (d) predict what will happen to the rate when the concentration of the first species in each equation is doubled.

1. $CH_3N{=}NCH_3(g) \longrightarrow C_2H_6(g) + N_2(g)$ $\quad$ $\text{Rate} = -\dfrac{\Delta[CH_3N{=}NCH_3]}{\Delta t} = k[CH_3N{=}NCH_3]$

2. $2NO_2(g) + F_2(g) \longrightarrow 2NO_2F(g)$ $\quad$ $\text{Rate} = -\dfrac{\Delta[F_2]}{\Delta t} = -\dfrac{1}{2}\left(\dfrac{\Delta[NO_2]}{\Delta t}\right) = k[NO_2][F_2]$

Answer

1. **(a)** s^{-1}; **(b)** first order in $CH_3N{=}NCH_3$; **(c)** first order overall; **(d)** doubling $[CH_3N{=}NCH_3]$ will double the rate
2. **(a)** $M^{-1} \cdot s^{-1}$; **(b)** first order in NO_2, first order in F_2; **(c)** second order overall; **(d)** doubling $[NO_2]$ will double the rate

14.3 • Methods of Determining Reaction Orders

MGC Reaction Order

In the examples in this text, the exponents in the rate law are almost always the positive integers 1 and 2 or even 0. Thus, the reactions are zeroth, first, or second order in each of the reactants. The common patterns used to identify the order of a reaction are described in this section, where we focus on characteristic types of differential and integrated rate laws and on how to determine reaction orders from experimental data.

Zeroth-Order Reactions

A **zeroth-order reaction** is one whose rate is independent of concentration; its differential rate law is rate = k. We refer to these reactions as *zeroth order* because we could also write their rate in a form such that the exponent of the reactant in the rate law is 0:

$$\text{rate} = -\frac{\Delta[A]}{\Delta t} = k[\text{reactant}]^0 = k(1) = k \qquad (14.16)$$

Because the rate is independent of reactant concentration, a graph of the concentration of any reactant as a function of time is a straight line with a slope of $-k$. The value of k is negative because the concentration of the reactant decreases with time. Conversely, a graph of the concentration of any product as a function of time is a straight line with a slope of k, a positive value.

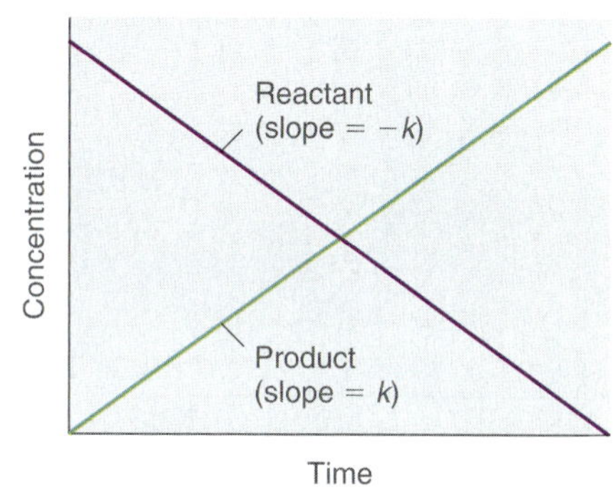

Shapes of zeroth-order curves. The change in concentration of reactant and product with time in a zeroth-order reaction.

The integrated rate law for a zeroth-order reaction also produces a straight line and has the general form

$$[A] = [A]_0 - kt \qquad (14.17)$$

where $[A]_0$ is the initial concentration of reactant A. (Notice that Equation 14.17 has the form of the algebraic equation for a straight line, $y = mx + b$, with $y = [A]$, $mx = -kt$, and $b = [A]_0$.) In a zeroth-order reaction, the rate constant must have the same units as the rate of the reaction, typically M/s.

Although it may seem counterintuitive for the rate of a reaction to be independent of the reactant concentration(s), such reactions are rather common. They occur most often when the reaction rate is determined by available surface area. An example is the decomposition of N_2O on a platinum surface to produce N_2 and O_2, which occurs at temperatures ranging from 200°C to 400°C:

$$2N_2O(g) \xrightarrow{\text{Pt}} 2N_2(g) + O_2(g) \qquad (14.18)$$

Without a platinum surface, the reaction requires temperatures higher than 700°C, but between 200°C and 400°C the only factor that determines how rapidly N_2O decomposes is the amount of Pt *surface* available (not the *amount* of Pt). As long as there is

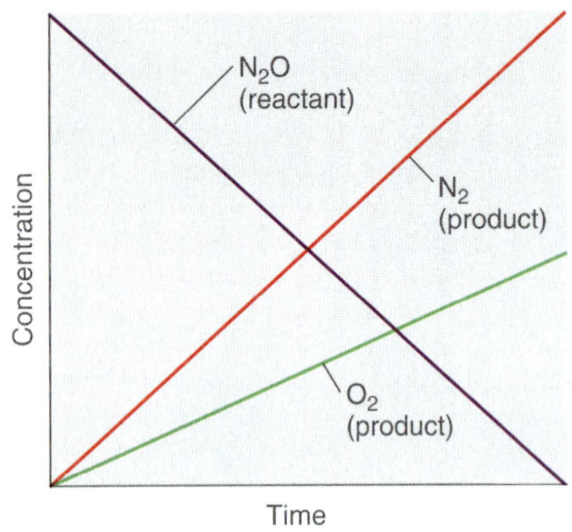

Figure 14.7 A zeroth-order reaction. This graph shows the concentrations of reactants and products vs. time for the zeroth-order catalyzed decomposition of N_2O to N_2 and O_2 on a platinum surface. Note that the change in the concentrations of all species with time is linear. Interactive Graph

Note the pattern

If a plot of reactant concentration versus time is linear, then the reaction is zeroth order in that reactant.

enough N_2O to react with the entire Pt surface, doubling or quadrupling the N_2O concentration will have no effect on the rate.* The rate of the reaction is

$$\text{rate} = -\frac{1}{2}\left(\frac{\Delta[N_2O]}{\Delta t}\right) = \frac{1}{2}\left(\frac{\Delta[N_2]}{\Delta t}\right) = \frac{\Delta[O_2]}{\Delta t} = k[N_2O]^0 = k \qquad (14.19)$$

Thus, the rate at which N_2O is consumed and the rates at which N_2 and O_2 are produced are independent of concentration. As shown in Figure 14.7, the change in the concentrations of all species with time is linear. Most important, the exponent (0) corresponding to the N_2O concentration in the experimentally derived rate law is *not* the same as the reactant's stoichiometric coefficient in the balanced equation (2). For this reaction, as for all others, *the rate law must be determined experimentally.*

A zeroth-order reaction that takes place in the human liver is the oxidation of ethanol (from alcoholic beverages) to acetaldehyde, catalyzed by the enzyme alcohol dehydrogenase. At high ethanol concentrations, this reaction is also a zeroth-order reaction. The overall reaction equation is

$$CH_3CH_2OH + NAD^+ \xrightarrow{\text{Alcohol dehydrogenase}} CH_3\overset{\overset{\displaystyle O}{\|}}{C}H + NADH + H^+ \qquad (14.20)$$

where NAD^+ and NADH are the oxidized and reduced forms of a species used by all organisms to transport electrons. When an alcoholic beverage is consumed, the ethanol is rapidly absorbed into the blood. Its concentration then decreases at a constant rate until it reaches zero (Figure 14.8a). An average 70-kg person typically takes about 2.5 h to oxidize the 15 mL of ethanol contained in a single 12-oz can of beer, a 5-oz glass of wine, or a shot of distilled spirits, such as whiskey or brandy. The actual rate, however, varies a great deal from person to person, depending on body size and the amount of alcohol dehydrogenase in the liver. The rate of the reaction does not increase if a greater quantity of alcohol is consumed over the same period of time because the rate is determined only by the amount of enzyme present in the liver.† When the ethanol has been completely oxidized and its concentration drops to essentially zero, the rate of oxidation also drops rapidly (Figure 14.8b).

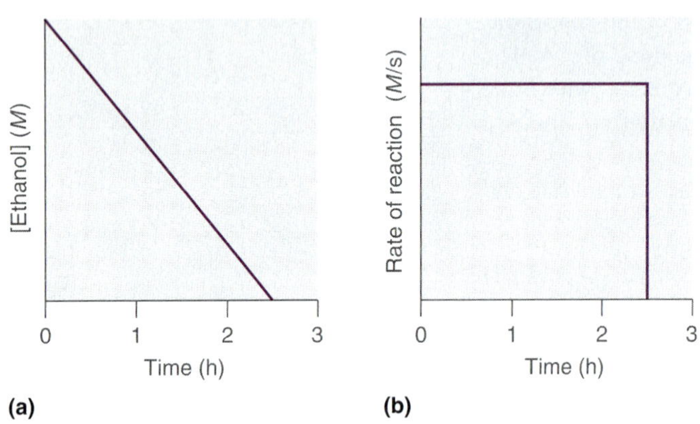

Figure 14.8 The catalyzed oxidation of ethanol. **(a)** The concentration of ethanol in human blood decreases linearly with time, which is typical of a zeroth-order reaction. **(b)** The rate at which ethanol is oxidized is constant until the ethanol concentration reaches essentially zero, at which point the rate of the reaction drops to zero. Interactive Graph

These examples illustrate two important points: (1) *In a zeroth-order reaction, the reaction rate does not depend on the reactant concentration;* and (2) *a linear change in concentration with time is a clear indication of a zeroth-order reaction.*

First-Order Reactions

In a **first-order reaction**, the reaction rate is directly proportional to the concentration of one of the reactants. First-order reactions often have the general form A $\longrightarrow$ products. The differential rate for a first-order reaction is

$$\text{rate} = -\frac{\Delta[A]}{\Delta t} = k[A] \qquad (14.21)$$

Thus, if the concentration of A is doubled, the rate of the reaction doubles; if the concentration of A is increased by a factor of 10, the rate increases by a factor of 10; and so forth. Because the units of the rate of the reaction are *M*/s, as always, the units of a first-order rate constant are inverse seconds, s^{-1}.

---

* At very low concentrations of N_2O, where there are not enough molecules present to occupy the entire available Pt surface, the rate of the reaction is dependent on the N_2O concentration.

† Contrary to popular belief, the caffeine in coffee is ineffective at catalyzing the oxidation of ethanol.

The integrated rate law for a first-order reaction can be written in two different ways: one using exponentials and one using logarithms. The exponential form is

$$[A] = [A]_0 e^{-kt} \qquad (14.22)$$

where $[A]_0$ is the initial concentration of reactant A at $t = 0$, k is the rate constant, and e is the base of the natural logarithms which has the value 2.718 (see Essential Skills 6), to three decimal places. Recall that an integrated rate law gives the relationship between reactant concentration and time. Equation 14.22 predicts that the concentration of A will decrease in a smooth exponential curve over time. By taking the natural logarithm of each side of Equation 14.22 and rearranging, we obtain an alternative logarithmic expression of the relationship between the concentration of A and t:

$$\ln[A] = \ln[A]_0 - kt \qquad (14.23)$$

Because Equation 14.23 has the form of the algebraic equation for a straight line, $y = mx + b$, with $y = \ln[A]$ and $b = \ln[A]_0$, a plot of ln[A] versus t for a first-order reaction should give a straight line with a slope of $-k$ and an intercept of $\ln[A]_0$. Either the differential rate law (Equation 14.21) or the integrated rate law (Equation 14.23) can be used to determine whether a particular reaction is first order.

First-order reactions are very common. In this chapter we have already encountered two examples of first-order reactions: the hydrolysis of aspirin (Equation 14.5) and the reaction of *t*-butyl bromide with water to give *t*-butanol (Equation 14.11). Another reaction that exhibits apparent first-order kinetics is the hydrolysis of the anticancer drug cisplatin.

Cisplatin, the first "inorganic" anticancer drug to be discovered, is unique in its ability to cause complete remission of the relatively rare but deadly cancers of the reproductive organs in young adults. The structures of cisplatin and its hydrolysis product are

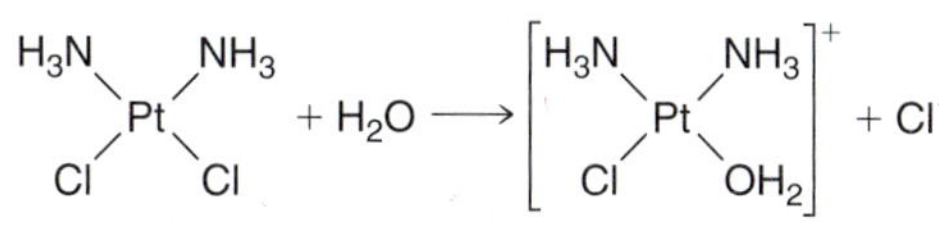

Cisplatin (14.24)

Both platinum compounds have four groups arranged in a square plane around a Pt(II) ion. The reaction shown in Equation 14.24 is important because cisplatin, the form in which the drug is administered, is *not* the form in which the drug is active. Instead, at least one chloride ion must be replaced by water to produce a species that reacts with DNA to prevent cell division and tumor growth. Consequently, the kinetics of the reaction in Equation 14.24 have been studied extensively to find ways of maximizing the concentration of the active species.

The rate law and reaction order of the hydrolysis of cisplatin are determined from experimental data such as those displayed in Table 14.2. The table lists initial rate data for four experiments in which the reaction was run at pH 7.0 and 25°C but with different initial concentrations of cisplatin. Because the rate increases with increasing cisplatin concentration, we know this cannot be a zeroth-order reaction. Comparing Experiments 1 and 2 in Table 14.2 shows that the rate of the reaction doubles

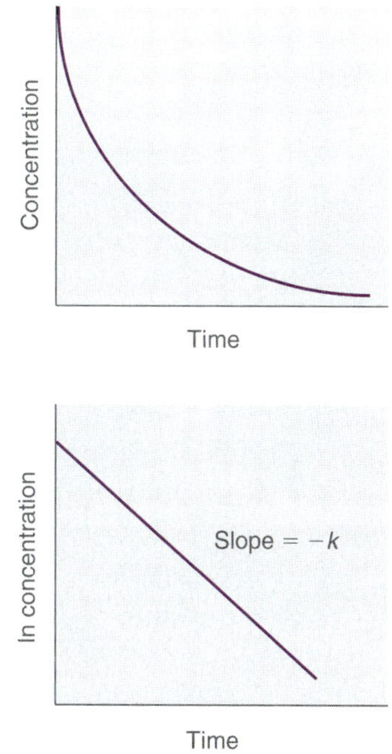

Shapes of first-order curves. The expected shapes of the curves for plots of reactant concentration vs. time *(top)* and the natural logarithm of reactant concentration vs. time *(bottom)* for a first-order reaction.

Note the pattern

If a plot of reactant concentration versus time is not linear but a plot of the natural logarithm of reactant concentration versus time is linear, then the reaction is first order.

TABLE 14.2 Rates of hydrolysis of cisplatin as a function of concentration at pH 7.0 and 25°C

Experiment	$[\text{Cisplatin}]_0$, *M*	Initial Rate, *M*/min
1	0.0060	9.0×10^{-6}
2	0.012	1.8×10^{-5}
3	0.024	3.6×10^{-5}
4	0.030	4.5×10^{-5}

$[(1.8 \times 10^{-5}\ M/\text{min}) \div (9.0 \times 10^{-6}\ M/\text{min}) = 2.0]$ when the concentration of cisplatin is doubled (from 0.0060 M to 0.012 M). Similarly, comparing Experiments 1 and 4 shows that the rate of the reaction increases by a factor of 5 $[(4.5 \times 10^{-5}\ M/\text{min}) \div (9.0 \times 10^{-6}\ M/\text{min}) = 5.0]$ when the concentration of cisplatin is increased by a factor of 5 (from 0.0060 M to 0.030 M). Because the rate of the reaction is directly proportional to the concentration of the reactant, the exponent of the cisplatin concentration in the rate law must be 1, and so the rate law is rate $= k[\text{cisplatin}]^1$. Thus, the reaction is first order. Knowing this, we can calculate the rate constant using the differential rate law for a first-order reaction and the data in any line of Table 14.2. For example, substituting the values for Experiment 3 into Equation 14.21 gives

$$3.6 \times 10^{-5}\ M/\text{min} = k(0.024\ M)$$
$$1.5 \times 10^{-3}\ \text{min}^{-1} = k$$

Knowing the rate constant for the hydrolysis of cisplatin and the rate constants for subsequent reactions that produce species that are highly toxic enables hospital pharmacists to provide patients with solutions that contain only the desired form of the drug.

EXAMPLE 14.4

At high temperatures, ethyl chloride produces HCl and ethylene by the reaction

$$CH_3CH_2Cl(g) \xrightarrow{\Delta} HCl(g) + C_2H_4(g)$$

Using the rate data for the reaction at 650°C presented in the table, calculate the reaction order with respect to the concentration of ethyl chloride, and determine the rate constant for the reaction.

Experiment	$[CH_3CH_2Cl]_0$, M	Initial Rate, M/s
1	0.010	1.6×10^{-8}
2	0.015	2.4×10^{-8}
3	0.030	4.8×10^{-8}
4	0.040	6.4×10^{-8}

Given Balanced chemical equation, initial concentrations of reactant, and initial rates of reaction

Asked for Reaction order and rate constant

Strategy

Ⓐ Compare the data from two experiments to determine the effect of changing the concentration of a species on the rate of the reaction.
Ⓑ Compare the observed effect with behavior characteristic of zeroth- and first-order reactions to determine the order of the reaction. Write the rate law for the reaction.
Ⓒ Use measured concentrations and rate data from any of the experiments to find the rate constant.

Solution

The order of the reaction with respect to ethyl chloride is determined by examining the effect of changes in the ethyl chloride concentration on the rate of the reaction. Ⓐ Comparing Experiments 2 and 3 shows that doubling the concentration doubles the rate, so the rate is proportional to $[CH_3CH_2Cl]$. Similarly, comparing Experiments 1 and 4 shows that quadrupling the concentration quadruples the rate, again indicating that the rate is directly proportional to $[CH_3CH_2Cl]$. Ⓑ This behavior is characteristic of a first-order reaction, for which the rate law is rate $= k[CH_3CH_2Cl]$. Ⓒ We can calculate the rate constant, k, using any line in the table. Selecting Experiment 1 gives

$$1.60 \times 10^{-8}\ M/\text{s} = k(0.010\ M)$$
$$1.6 \times 10^{-6}\ \text{s}^{-1} = k$$

EXERCISE 14.4

Sulfuryl chloride, SO_2Cl_2, decomposes to SO_2 and Cl_2 by the reaction

$$SO_2Cl_2(g) \longrightarrow SO_2(g) + Cl_2(g)$$

Data for the reaction at 320°C are listed in the table. Calculate the order of the reaction with regard to sulfuryl chloride, and determine the rate constant for the reaction.

Experiment	$[SO_2Cl_2]_0$, *M*	Initial Rate, *M*/s
1	0.0050	1.10×10^{-7}
2	0.0075	1.65×10^{-7}
3	0.0100	2.20×10^{-7}
4	0.0125	2.75×10^{-7}

Answer First order; $k = 2.2 \times 10^{-5}\ s^{-1}$

We can also use the integrated rate law to determine the reaction rate for the hydrolysis of cisplatin. To do this, we examine the change in the concentration of the reactant or the product as a function of time at a single initial cisplatin concentration. Figure 14.9a shows plots for a solution that originally contained 0.0100 *M* cisplatin and was maintained at pH 7 and 25°C. The concentration of cisplatin decreases smoothly with time, and the concentration of chloride ion increases in a similar way. When we plot the natural logarithm of the concentration of cisplatin versus time, we obtain the plot shown in Figure 14.9b. The straight line is consistent with the behavior of a system that obeys a first-order rate law. We can use any two points on the line to calculate the slope of the line, which gives us the rate constant for the reaction. Thus, taking the points from Figure 14.9a for t = 100 min ([cisplatin] = 0.0086 *M*) and t = 1000 min ([cisplatin] = 0.0022 *M*) gives

$$\text{slope} = \frac{\ln[\text{cisplatin}]_{1000} - \ln[\text{cisplatin}]_{100}}{1000\text{ min} - 100\text{ min}}$$

$$-k = \frac{\ln 0.0022 - \ln 0.0086}{1000\text{ min} - 100\text{ min}} = \frac{-6.12 - (-4.76)}{900\text{ min}} = -1.51 \times 10^{-3}\text{ min}^{-1}$$

$$k = 1.5 \times 10^{-3}\text{ min}^{-1}$$

Notice that the slope is negative because we are calculating the rate of disappearance of cisplatin. Also notice that the rate constant has units of min^{-1} because the times plotted on the horizontal axes in Figures 14.9a and 14.9b are in minutes rather than seconds.

The order of the reaction and the magnitude of the rate constant we obtain using the integrated rate law are exactly the same as those we calculated earlier using the differential rate law. This must be true if the experiments were carried out under the same conditions.

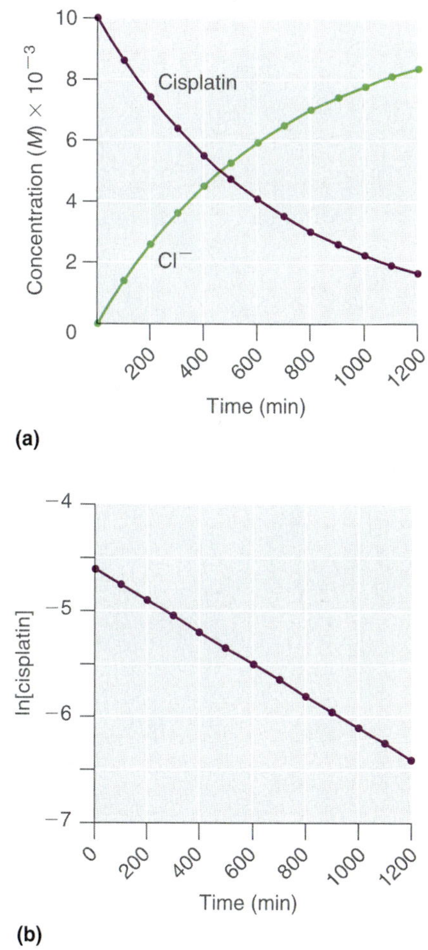

Figure 14.9 The hydrolysis of cisplatin, a first-order reaction. These plots show hydrolysis of cisplatin at pH 7.0 and 25°C as **(a)** the experimentally determined concentrations of cisplatin and chloride ions vs. time and **(b)** the natural logarithm of the cisplatin concentration vs. time. The straight line in (b) is expected for a first-order reaction. Interactive Graph

EXAMPLE 14.5

Refer back to Example 14.4. If a sample of ethyl chloride with an initial concentration of 0.0200 *M* is heated at 650°C, what is the concentration of ethyl chloride after 10 h? How many hours at 650°C must elapse in total for the concentration to decrease to 0.0050 *M*? (Recall that we calculated the rate constant for this reaction in Example 14.4.)

Given Initial concentration, rate constant, and time interval

Asked for Concentration at specified time and time required to obtain particular concentration

Strategy

Ⓐ Substitute values for the initial concentration, $[A]_0$, and the calculated rate constant for the reaction, *k*, into the integrated rate law for a first-order reaction. Calculate the concentration, [A], at the given time, *t*.

Ⓑ Given a concentration [A], solve the integrated rate law for time *t*.

Solution

The exponential form of the integrated rate law for a first-order reaction (Equation 14.22) is $[A] = [A]_0 e^{-kt}$. Ⓐ✓ Having been given the initial concentration of ethyl chloride, $[A]_0$, and having calculated the rate constant, *k*, in Example 14.4 ($k = 1.6 \times 10^{-6}\ s^{-1}$), we can use the rate law to calculate the concentration of the reactant at a given time *t*. Substituting the known values into the integrated rate law gives

$$
\begin{aligned}
[CH_3CH_2Cl]_{10\ h} &= [CH_3CH_2Cl]_0\, e^{-kt} \\
&= 0.0200\ M\{e^{-(1.6\times 10^{-6}\ s^{-1})[(10\ h)(60\ min/h)(60\ s/min)]}\} \\
&= 0.0189\ M
\end{aligned}
$$

We could also have used the logarithmic form of the integrated rate law (Equation 14.23), which gives

$$
\begin{aligned}
\ln[CH_3CH_2Cl]_{10\ h} &= \ln[CH_3CH_2Cl]_0 - kt \\
&= \ln 0.0200 - (1.6 \times 10^{-6}\ s^{-1})[(10\ h)(60\ min/h)(60\ s/min)] \\
&= -3.912 - 0.0576 = -3.970 \\
[CH_3CH_2Cl]_{10\ h} &= e^{-3.970}\ M \\
&= 0.0189\ M
\end{aligned}
$$

Ⓑ✓ To calculate the amount of time required to reach a given concentration, we must solve the integrated rate law for *t*. Equation 14.23 gives

$$
\begin{aligned}
\ln[CH_3CH_2Cl]_t &= \ln[CH_3CH_2Cl]_0 - kt \\
kt &= \ln[CH_3CH_2Cl]_0 - \ln[CH_3CH_2Cl]_t = \ln\frac{[CH_3CH_2Cl]_0}{[CH_3CH_2Cl]_t} \\
t &= \frac{1}{k}\left(\ln\frac{[CH_3CH_2Cl]_0}{[CH_3CH_2Cl]_t}\right) = \frac{1}{1.6\times 10^{-6}\ s^{-1}}\left(\ln\frac{0.0200\ M}{0.0050\ M}\right) \\
&= \frac{\ln 4.0}{1.6\times 10^{-6}\ s^{-1}} = 8.7\times 10^{5}\ s = 240\ h = 2.4\times 10^{2}\ h
\end{aligned}
$$

EXERCISE 14.5

In Exercise 14.4 you found that the decomposition of sulfuryl chloride (SO_2Cl_2) is first order, and you calculated the rate constant at 320°C. Use the form(s) of the integrated rate law to find the amount of SO_2Cl_2 that remains after 20 h if a sample with an original concentration of 0.123 *M* is heated at 320°C. How long would it take for 90% of the SO_2Cl_2 to decompose?

Answer 0.0252 *M*; 29 h

Second-Order Reactions

The simplest kind of **second-order reaction** is one whose rate is proportional to the square of the concentration of one reactant. These generally have the form $2A \longrightarrow$ products. A second kind of second-order reaction has a rate that is proportional to the *product* of the concentrations of two reactants. Such reactions generally have the form $A + B \longrightarrow$ products. An example of the former is a *dimerization reaction,* in which two smaller molecules, each called a *monomer*, combine to form a larger molecule (a *dimer*).

The differential rate law for the simplest second-order reaction in which 2A $\longrightarrow$ products is

$$\text{rate} = -\frac{\Delta[\text{A}]}{2\Delta t} = k[\text{A}]^2 \tag{14.25}$$

Consequently, doubling the concentration of A quadruples the rate of the reaction. In order for the units of the rate to be *M*/s, the units of a second-order rate constant must be $M^{-1} \cdot \text{s}^{-1}$. Because molarity is expressed as mol/L, the unit of the rate constant can also be written as L(mol · s).

For the reaction 2A $\longrightarrow$ products, the concentration of the reactant at a given time can be described by the following integrated rate law:

$$\frac{1}{[\text{A}]} = \frac{1}{[\text{A}]_0} + kt \tag{14.26}$$

Because Equation 14.26 has the form of an algebraic equation for a straight line, $y = mx + b$, with $y = 1/[\text{A}]$ and $b = 1/[\text{A}]_0$, a plot of $1/[\text{A}]$ versus t for a simple second-order reaction is a straight line with a slope of k and an intercept of $1/[\text{A}]_0$.

Simple second-order reactions are common. In addition to dimerization reactions, two other examples are the decomposition of NO_2 to NO and O_2 and the decomposition of HI to I_2 and H_2. Most examples involve simple inorganic molecules, but there are organic examples, too. We can follow the progress of the reaction described in the next paragraph by monitoring the decrease in the intensity of the red color of the reaction mixture.

Many cyclic organic compounds that contain two carbon–carbon double bonds undergo a dimerization reaction to give complex structures. One example is

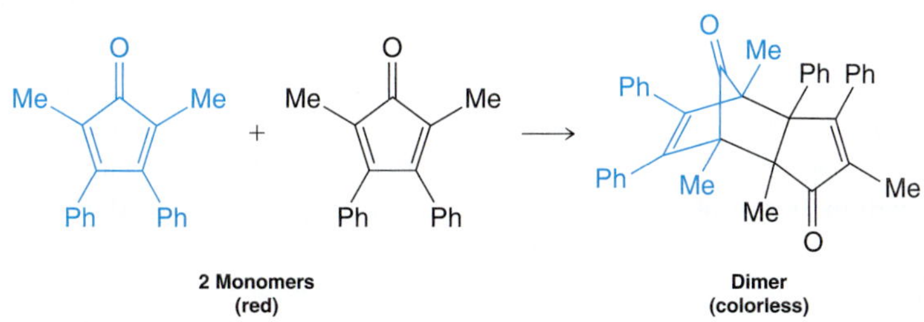

(14.27)

For simplicity, we will refer to this reactant and product as "monomer" and "dimer," respectively.* Because the monomers are the same, the general equation for this reaction is 2A $\longrightarrow$ product. This reaction represents an important class of organic reactions used in the pharmaceutical industry to prepare complex carbon skeletons for the synthesis of drugs. Like the first-order reactions studied in the preceding section, it can be analyzed using either the differential rate law (Equation 14.25) or the integrated rate law (Equation 14.26).

To determine the differential rate law for the reaction, we need data on how the rate of the reaction varies as a function of monomer concentrations, which are provided in Table 14.3. From the data, we see that the reaction rate is not independent of the monomer concentration, so this is not a zeroth-order reaction. We also see that the reaction rate is not proportional to the monomer concentration, so the reaction is not first order. Comparing the data in the second and fourth lines shows that the rate decreases by a factor of 2.8 when the monomer concentration decreases by a factor of 1.7:

$$\frac{5.0 \times 10^{-5}\ M/\text{min}}{1.8 \times 10^{-5}\ M/\text{min}} = 2.8 \quad \text{and} \quad \frac{3.4 \times 10^{-3}\ M}{2.0 \times 10^{-3}\ M} = 1.7$$

Because $(1.7)^2 = 2.9 \approx 2.8$, the rate of the reaction is approximately proportional to the *square* of the monomer concentration.

$$\text{rate} \propto [\text{monomer}]^2$$

* The systematic name of the monomer is 2,5-dimethyl-3,4-diphenylcyclopentadienone.

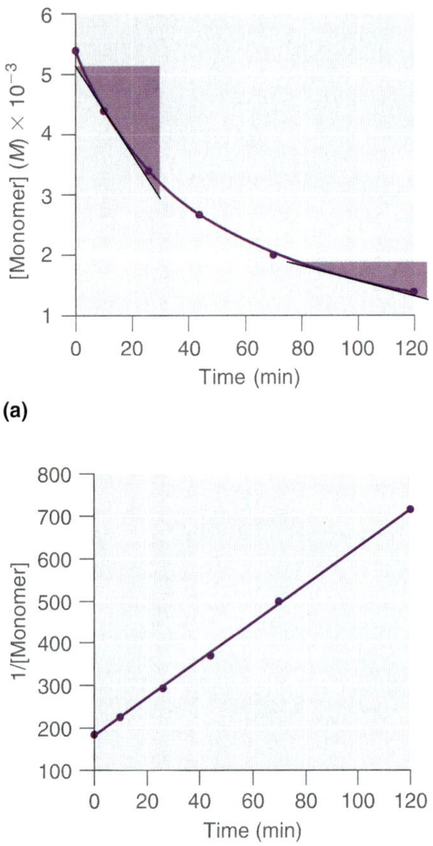

(a)

(b)

Figure 14.10 Dimerization of a monomeric compound, a second-order reaction. These plots correspond to dimerization of the monomer in Equation 14.27 as **(a)** the experimentally determined concentration of monomer vs. time and **(b)** 1/[monomer] vs. time. The straight line in (b) is expected for a simple second-order reaction. Interactive Graph

TABLE 14.3 Rates of reaction as a function of monomer concentration for an initial monomer concentration of 0.0054 *M*

Time, min	[Monomer], *M*	Instantaneous Rate, *M*/min
10	0.0044	8.0×10^{-5}
26	0.0034	5.0×10^{-5}
44	0.0027	3.1×10^{-5}
70	0.0020	1.8×10^{-5}
120	0.0014	8.0×10^{-6}

This means that the reaction is second order in the monomer. Using Equation 14.25 and the data from any line in Table 14.3, we can calculate the rate constant. Substituting values at time = 10 min, for example, gives

$$\text{rate} = k[\text{A}]^2$$
$$8.0 \times 10^{-5}\ M/\text{min} = k(4.4 \times 10^{-3}\ M)^2$$
$$4.1\ M^{-1} \cdot \text{min}^{-1} = k$$

We can also determine the reaction order using the integrated rate law. To do so, we use the decrease in the concentration of the monomer as a function of time for a single reaction, plotted in Figure 14.10a. The measurements show that the concentration of the monomer (initially 5.4×10^{-3} *M*) decreases with increasing time. This graph also shows that the *rate* of the reaction decreases smoothly with increasing time. According to the integrated rate law for a second-order reaction, a plot of 1/[monomer] versus *t* should be a straight line, as shown in Figure 14.10b. Any pair of points on the line can be used to calculate the slope, which is the second-order rate constant. In this example, $k = 4.1\ M^{-1} \cdot \text{min}^{-1}$, which is consistent with the result obtained using the differential rate equation. Although in this example the stoichiometric coefficient is the same as the reaction order, this is not always the case. *The reaction order must always be determined experimentally.*

Note that for two or more reactions of the *same order,* the reaction with the largest rate constant is the fastest. Because the units of the rate constants for zeroth-, first-, and second-order reactions are different, however, we cannot compare the magnitudes of rate constants for reactions that have different orders. The differential and integrated rate laws for zeroth-, first-, and second-order reactions and their corresponding graphs are shown in Table 14.6 in Section 14.4.

Note the pattern

If a plot of reactant concentration versus time is not linear but a plot of 1 over reaction concentration versus time is linear, then the reaction is second order.

EXAMPLE 14.6

At high temperatures, nitrogen dioxide decomposes to nitric oxide and oxygen.

$$2NO_2(g) \xrightarrow{\Delta} 2NO(g) + O_2(g)$$

Experimental data for the reaction at 300°C and four initial concentrations of NO_2 are listed in the table.

Experiment	$[NO_2]_0$, *M*	Initial Rate, *M*/s
1	0.015	1.22×10^{-4}
2	0.010	5.40×10^{-5}
3	0.0080	3.46×10^{-5}
4	0.0050	1.35×10^{-5}

Determine the order of the reaction and the rate constant.

Given Balanced chemical equation, initial concentrations, and initial rates

Asked for Reaction order and rate constant

Strategy

Ⓐ From the experiments, compare the changes in the initial rates of reaction with the corresponding changes in the initial concentrations. Determine whether the changes are characteristic of zeroth-, first-, or second-order reactions.

Ⓑ Determine the appropriate rate law. Using this rate law and data from any experiment, solve for the rate constant, k.

Solution

Ⓐ We can determine the order of the reaction with respect to nitrogen dioxide by comparing the changes in NO_2 concentrations with the corresponding reaction rates. Comparing Experiments 2 and 4 shows that doubling the concentration quadruples the rate $[(5.40 \times 10^{-5}) \div (1.35 \times 10^{-5}) = 4.0]$, which means that the rate is proportional to $[NO_2]^2$. Similarly, comparing Experiments 1 and 4 shows that tripling the concentration increases the rate by a factor of 9, again indicating that the rate is proportional to $[NO_2]^2$. This behavior is characteristic of a second-order reaction. Ⓑ We have rate $= k\,[NO_2]^2$. We can calculate the rate constant, k, using data from any experiment in the table. Selecting Experiment 2, for example, gives

$$\text{rate} = k[NO_2]^2$$
$$5.40 \times 10^{-5}\ M/\text{s} = k(0.010\ M)^2$$
$$0.54\ M^{-1} \cdot s^{-1} = k$$

EXERCISE 14.6

When the highly reactive species HO_2 forms in the atmosphere, one important reaction that then removes it from the atmosphere is

$$2HO_2(g) \longrightarrow H_2O_2(g) + O_2(g)$$

The kinetics of this reaction have been studied in the laboratory, and some initial rate data at 25°C are listed in the table.

Experiment	$[HO_2]_0$, M	Initial Rate, M/s
1	1.1×10^{-8}	1.7×10^{-7}
2	2.5×10^{-8}	8.8×10^{-7}
3	3.4×10^{-8}	1.6×10^{-6}
4	5.0×10^{-8}	3.5×10^{-6}

Determine the order of the reaction and the rate constant.

Answer Second order in HO_2; $k = 1.4 \times 10^9\ M^{-1} \cdot s^{-1}$

EXAMPLE 14.7

If a flask that initially contains 0.056 M NO_2 is heated at 300°C, what will be the concentration of NO_2 after 1.0 h? How long will it take for the concentration of NO_2 to decrease to 10% of the initial concentration? Use the integrated rate law for a second-order reaction (Equation 14.26) and the rate constant calculated in Example 14.6.

Given Balanced chemical equation, rate constant, time interval, and initial concentration

Asked for Final concentration and time required to reach specified concentration

Strategy

A Given k, t, and $[A]_0$, use the integrated rate law for a second-order reaction to calculate [A].
B Setting [A] equal to one-tenth $[A]_0$, use the same equation to solve for time t.

Solution

A We know k and $[NO_2]_0$, and we are asked to determine $[NO_2]$ at $t = 1$ h (3600 s). Substituting the appropriate values into Equation 14.26 gives

$$\frac{1}{[NO_2]_{3600}} = \frac{1}{[NO_2]_0} + kt = \frac{1}{0.056\ M} + [(0.54\ M^{-1} \cdot s^{-1})(3600\ s)]$$
$$= 2.0 \times 10^3\ M^{-1}$$

Thus, $[NO_2]_{3600} = 5.1 \times 10^{-4}\ M$. **B** In this case, we know k and $[NO_2]_0$, and we are asked to calculate at what time $[NO_2] = 0.1[NO_2]_0 = 0.1(0.056\ M) = 0.0056\ M$. To do this, we solve Equation 14.26 for t, using the concentrations given.

$$t = \frac{(1/[NO_2]) - (1/[NO_2]_0)}{k} = \frac{(1/0.0056\ M) - (1/0.056\ M)}{0.54\ M^{-1} \cdot s^{-1}} = 3 \times 10^2\ s = 5.0\ min$$

Notice that NO_2 decomposes very rapidly; under these conditions, the reaction is 90% complete in only 5 min.

EXERCISE 14.7

In Exercise 14.6 you calculated the rate constant for the decomposition of HO_2 as $k = 1.4 \times 10^9\ M^{-1} \cdot s^{-1}$. This high rate constant means that HO_2 decomposes rapidly under the reaction conditions given in the problem. In fact, the HO_2 molecule is so reactive that it is virtually impossible to obtain in high concentrations. Given a 0.0010 M sample of HO_2, calculate the concentration of HO_2 that remains after 1.0 h at 25°C. How long will it take for 90% of the HO_2 to decompose? Use the integrated rate law for a second-order reaction (Equation 14.26) and the rate constant calculated in Exercise 14.6.

Answer $2.0 \times 10^{-13}\ M$; 6.4×10^{-6} s

In addition to the simple second-order reaction and rate law we have just described, another very common second-order reaction has the general form A + B ⟶ products, in which the reaction is first order in A and first order in B. The differential rate law for this reaction is

$$\text{rate} = -\frac{\Delta[A]}{\Delta t} = -\frac{\Delta[B]}{\Delta t} = k[A][B] \tag{14.28}$$

Because the reaction is first order both in A and in B, it has an overall reaction order of 2. (The integrated rate law for this reaction is rather complex, so we will not describe it.) We can recognize second-order reactions of this sort because the rate is proportional to the concentrations of each of two reactants. We presented one example at the end of Section 14.2, the reaction of CH_3Br with OH^- to produce CH_3OH.

Determining the Rate Law of a Reaction

The number of fundamentally different mechanisms (sets of steps in a reaction) is actually rather small compared to the large number of chemical reactions that can occur. Thus, understanding reaction mechanisms can simplify what might seem to be a confusing variety of chemical reactions. The first step in discovering the mechanism of a reaction is to determine the reaction's rate law. This can be done by designing experiments that measure the concentration(s) of one or more of the reactants or

TABLE 14.4 Rate data for a hypothetical reaction of the form A + B $\longrightarrow$ products

Experiment	[A], *M*	[B], *M*	Initial Rate, *M*/min
1	0.50	0.50	8.5×10^{-3}
2	0.75	0.50	19×10^{-3}
3	1.00	0.50	34×10^{-3}
4	0.50	0.75	8.5×10^{-3}
5	0.50	1.00	8.5×10^{-3}

products as a function of time. For the reaction A + B $\longrightarrow$ products, for example, we need to determine the value of k and the exponents m and n in the equation

$$\text{rate} = k[\text{A}]^m[\text{B}]^n \tag{14.29}$$

To do this, we might keep the initial concentration of B constant while varying the initial concentration of A and calculating the initial rate of the reaction. This information would permit us to deduce the order of the reaction with respect to A. Similarly, we could determine the order of the reaction with respect to B by studying the initial reaction rate when the initial concentration of A is kept constant while the initial concentration of B is varied. In earlier examples we determined the order of a reaction with respect to a given reactant by comparing the different rates obtained when only the concentration of the reactant in question was changed. An alternative way of determining reaction orders is to set up a proportion using the rate laws for two different experiments.

Rate data for a hypothetical reaction of the type A + B $\longrightarrow$ products are given in Table 14.4. The general rate law for the reaction is given in Equation 14.29. We can obtain the value of m or n directly by using a proportion of the rate laws for two experiments in which the concentration of one of the reactants is the same, such as Experiments 1 and 3 in Table 14.4.

$$\frac{\text{rate}_1}{\text{rate}_3} = \frac{k[\text{A}_1]^m[\text{B}_1]^n}{k[\text{A}_3]^m[\text{B}_3]^n}$$

Inserting the appropriate values from Table 14.4 gives

$$\frac{8.5 \times 10^{-3}\ \cancel{M/\text{min}}}{34 \times 10^{-3}\ \cancel{M/\text{min}}} = \frac{\cancel{k}[0.50\ M]^m\cancel{[0.50\ M]^n}}{\cancel{k}[1.00\ M]^m\cancel{[0.50\ M]^n}}$$

Noting that 1.00 to any power is 1, we can cancel like terms to give $0.25 = [0.50]^m$, which can also be written as $1/4 = [1/2]^m$. Thus, we can conclude that $m = 2$ and that the reaction is second order in A. Note that by selecting two experiments in which the concentration of B is the same, we were able to solve for the value of m.

Conversely, by selecting two experiments in which the concentration of A is the same (for example, Experiments 5 and 1), we can solve for the value of n.

$$\frac{\text{rate}_1}{\text{rate}_5} = \frac{k[\text{A}_1]^m[\text{B}_1]^n}{k[\text{A}_5]^m[\text{B}_5]^n}$$

Substituting the appropriate values from Table 14.4 gives

$$\frac{8.5 \times 10^{-3}\ \cancel{M/\text{min}}}{8.5 \times 10^{-3}\ \cancel{M/\text{min}}} = \frac{\cancel{k[0.50\ M]^m}[0.50\ M]^n}{\cancel{k[0.50\ M]^m}[1.00\ M]^n}$$

Canceling leaves $1.0 = [0.50]^n$, which gives $n = 0$; that is, the reaction is zeroth order in B. The experimentally determined rate law is therefore

$$\text{rate} = k[\text{A}]^2[\text{B}]^0 = k[\text{A}]^2$$

We can now calculate the rate constant by inserting the data from any line of Table 14.4 into the experimentally determined rate law and solving for *k*. Using Experiment 2, we obtain

$$19 \times 10^{-3}\ M/\text{min} = k(0.75\ M)^2$$
$$3.4 \times 10^{-2}\ M^{-1} \cdot \text{min}^{-1} = k$$

You should verify that using data from any other line of Table 14.4 gives the same rate constant. This must be true as long as the experimental conditions, such as temperature and solvent, are the same.

EXAMPLE 14.8

Nitric oxide is produced in the body by several different enzymes and acts as a signal that controls blood pressure, long-term memory, and other critical functions. The major route for the removal of NO from biological fluids is via reaction with O_2 to give NO_2, which then reacts rapidly with water to give nitrous acid and nitric acid:

$$2NO + O_2 \longrightarrow 2NO_2 \xrightarrow{H_2O} HNO_2 + HNO_3$$

These reactions are important in maintaining steady levels of NO. The table lists kinetics data for the reaction of NO with O_2 at 25°C:

$$2NO(g) + O_2(g) \longrightarrow 2NO_2(g)$$

Determine the rate law for the reaction, and calculate the rate constant.

Experiment	$[NO]_0$, *M*	$[O_2]_0$, *M*	Initial Rate, *M*/s
1	0.0235	0.0125	7.98×10^{-3}
2	0.0235	0.0250	15.9×10^{-3}
3	0.0470	0.0125	32.0×10^{-3}
4	0.0470	0.0250	63.5×10^{-3}

Given Balanced chemical equation, initial concentrations, and initial rates

Asked for Rate law and rate constant

Strategy

A Compare the changes in initial concentrations with the corresponding changes in rates of reaction to determine the order of the reaction for each species. Write the rate law for the reaction.

B Using data from any experiment, substitute appropriate values into the rate law. Solve the rate equation for *k*, the rate constant.

Solution

A Comparing Experiments 1 and 2 shows that as $[O_2]$ is doubled at a constant value of $[NO_2]$, the rate approximately doubles. Thus, the rate of the reaction is proportional to $[O_2]^1$, so the reaction is first order in O_2. Comparing Experiments 1 and 3 shows that the rate of the reaction essentially quadruples when [NO] is doubled and $[O_2]$ is held constant. That is, the reaction rate is proportional to $[NO]^2$, which indicates that the reaction is second order in NO. Using these relationships, we can write the rate law for the reaction:

$$\text{rate} = k[NO]^2[O_2]$$

B The data in any line can be used to calculate the rate constant. Using Experiment 1, for example, gives

$$k = \frac{\text{rate}}{[NO]^2[O_2]} = \frac{7.98 \times 10^{-3}\ M/\text{s}}{(0.0235\ M)^2\,(0.0125\ M)} = 1.16 \times 10^3\ M^{-2} \cdot \text{s}^{-1}$$

The overall reaction order ($m + n$) is 3, so this is a third-order reaction. Notice that the units of the rate constant become more complex as the overall reaction order increases.

EXERCISE 14.8

The peroxydisulfate ion, $S_2O_8^{2-}$, is a potent oxidizing agent that reacts rapidly with iodide ion in water:

$$S_2O_8^{2-}(aq) + 3I^-(aq) \longrightarrow 2SO_4^{2-}(aq) + I_3^-(aq)$$

The table lists kinetics data for this reaction at 25°C. Determine the rate law, and calculate the rate constant.

Experiment	$[S_2O_8^{2-}]_0$, M	$[I^-]_0$, M	Initial Rate, M/s
1	0.27	0.38	2.05
2	0.40	0.38	3.06
3	0.40	0.22	1.76

Answer Rate $= k[S_2O_8^{2-}][I^-]$; $k = 20\ M^{-1} \cdot s^{-1}$

14.4 Using Graphs to Determine Rate Laws, Rate Constants, and Reaction Orders

In Section 14.3 you learned that the integrated rate law for each of the common types of reaction (zeroth, first, or second order in a single reactant) can be plotted as a straight line. The use of these plots offers an alternative to the methods described in that section for showing how reactant concentration changes with time and for determining the order of the reaction.

We will illustrate the use of these graphs by considering the thermal decomposition of NO_2 gas at elevated temperatures, which occurs according to the equation

$$2NO_2(g) \xrightarrow{\Delta} 2NO(g) + O_2(g) \qquad (14.30)$$

Experimental data for this reaction at 330°C are listed in Table 14.5; they are provided as $[NO_2]$, $\ln[NO_2]$, and $1/[NO_2]$ versus time to correspond to the integrated rate laws for zeroth-, first-, and second-order reactions, respectively. The actual concentrations of NO_2 are plotted versus time in Figure 14.11a. Because the plot of $[NO_2]$

TABLE 14.5 Concentration of NO_2 as a function of time at 330°C

Time, s	$[NO_2]$, M	$\ln[NO_2]$	$1/[NO_2]$, M^{-1}
0	1.00×10^{-2}	−4.605	100
60	6.83×10^{-3}	−4.986	146
120	5.18×10^{-3}	−5.263	193
180	4.18×10^{-3}	−5.477	239
240	3.50×10^{-3}	−5.655	286
300	3.01×10^{-3}	−5.806	332
360	2.64×10^{-3}	−5.937	379

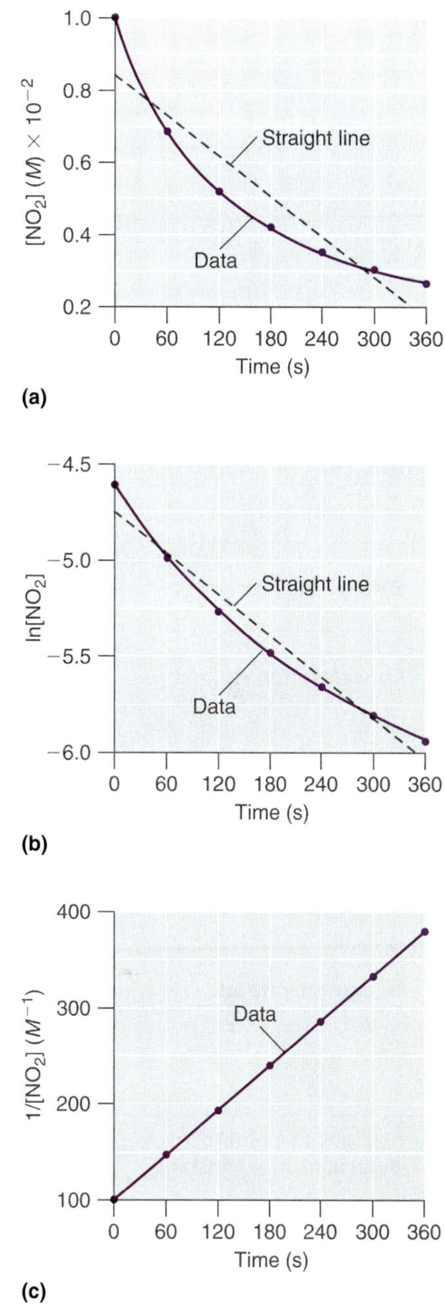

Figure 14.11 The decomposition of NO_2. These plots show the decomposition of a sample of NO_2 at 330°C as **(a)** the concentration of NO_2 vs. time, **(b)** the natural logarithm of $[NO_2]$ vs. time, and **(c)** $1/[NO_2]$ vs. time.

MGC Using Graphs to Determine Rate

versus t is not a straight line, we know the reaction is not zeroth order in NO_2. A plot of $\ln[NO_2]$ versus t (Figure 14.11b) shows us that the reaction is not first order in NO_2 either because a first-order reaction would give a straight line. Having eliminated zeroth-order and first-order behavior, we construct a plot of $1/[NO_2]$ versus t (Figure 14.11c). This plot is a straight line, indicating that the reaction is second order in NO_2.

Notice that we have just determined the order of a reaction using data from a single experiment by plotting the concentration of the reactant as a function of time. Because of the characteristic shapes of the lines shown in Table 14.6, the graphs can be used to determine the reaction order of an unknown reaction. In contrast, the method described in Section 14.3 required multiple experiments at different NO_2 concentrations as well as accurate initial rates of reaction, which can be difficult to obtain for rapid reactions.

TABLE 14.6 Properties of reactions that obey zeroth-, first-, and second-order rate laws

	Zeroth Order	First Order	Second Order
Differential rate law	$\text{Rate} = -\dfrac{\Delta[A]}{\Delta t} = k$	$\text{Rate} = -\dfrac{\Delta[A]}{\Delta t} = k[A]$	$\text{Rate} = -\dfrac{\Delta[A]}{\Delta t} = k[A]^2$
Concentration vs. time	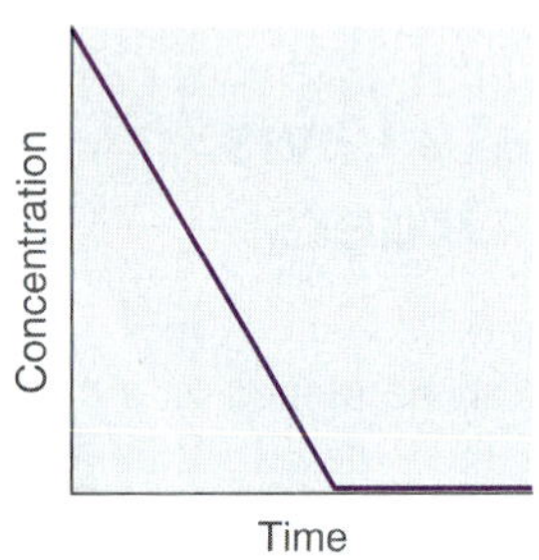	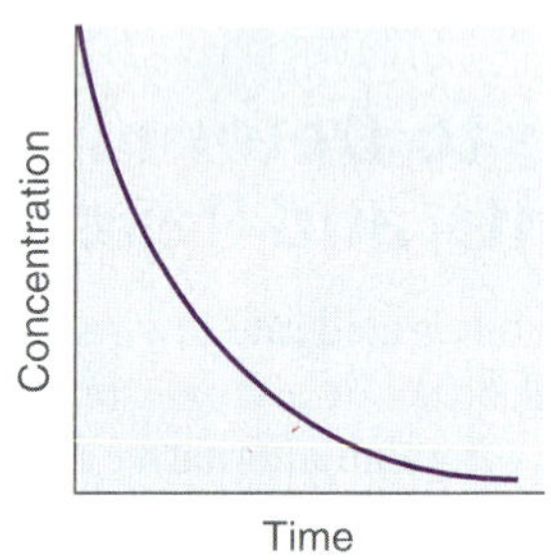	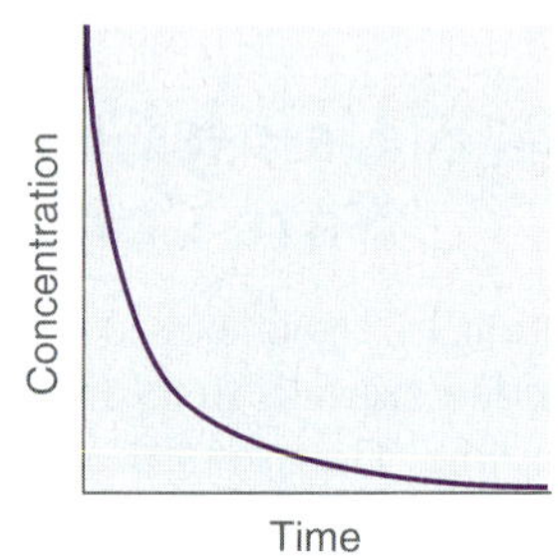
Integrated rate law	$[A] = [A]_0 - kt$	$[A] = [A]_0 e^{-kt}$ or $\ln[A] = \ln[A]_0 - kt$	$\dfrac{1}{[A]} = \dfrac{1}{[A]_0} + kt$
Straight-line plot to determine rate constant	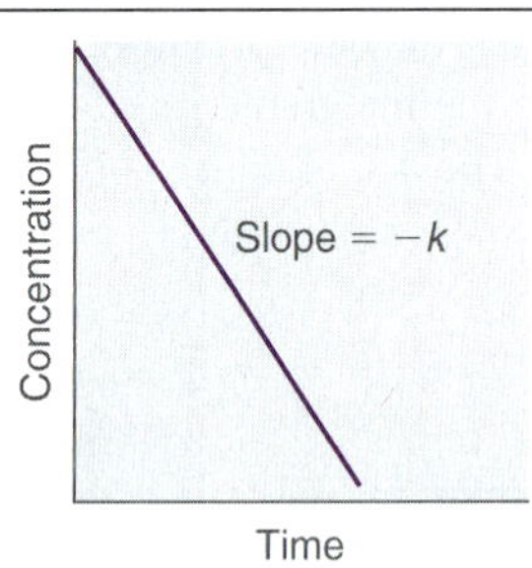	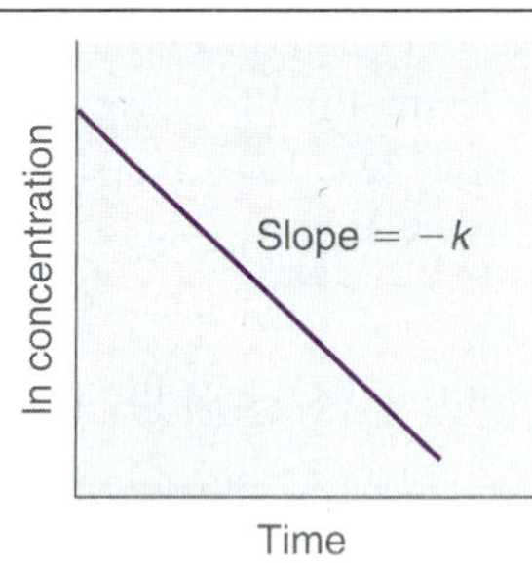	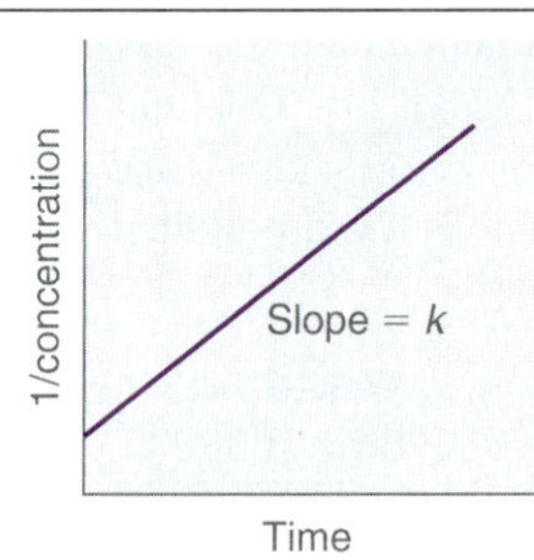
Relative rate vs. concentration	[A] (M) / Rate (M/s): 1 / 1 2 / 1 3 / 1	[A] (M) / Rate (M/s): 1 / 1 2 / 2 3 / 3	[A] (M) / Rate (M/s): 1 / 1 2 / 4 3 / 9
Half-life	$t_{1/2} = \dfrac{[A]_0}{2k}$	$t_{1/2} = \dfrac{0.693}{k}$	$t_{1/2} = \dfrac{1}{k[A]_0}$
Units of *k*, rate constant	M/s	1/s	$M^{-1} \cdot s^{-1}$

EXAMPLE 14.9

Dinitrogen pentoxide (N_2O_5) decomposes to NO_2 and O_2 at relatively low temperatures in the reaction

$$2N_2O_5(soln) \longrightarrow 4NO_2(soln) + O_2(g)$$

This reaction is carried out in a CCl_4 solution at 45°C. The concentrations of N_2O_5 as a function of time are listed in the table, together with the natural logarithms and reciprocal N_2O_5 concentrations. Plot a graph of the concentration versus *t*, ln concentration versus *t*, and 1/concentration versus *t*, and then determine the rate law and calculate the value of the rate constant.

Time, s	$[N_2O_5]$, *M*	$\ln[N_2O_5]$	$1/[N_2O_5]$, M^{-1}
0	0.0365	−3.310	27.4
600	0.0274	−3.597	36.5
1200	0.0206	−3.882	48.5
1800	0.0157	−4.154	63.7
2400	0.0117	−4.448	85.5
3000	0.00860	−4.756	116
3600	0.00640	−5.051	156

Note the pattern

For a zeroth-order reaction, a plot of reactant concentration vs time is a straight line. For a first-order reaction, a plot of the natural logarithm of reactant concentration is a straight line. For a second-order reaction, a plot of 1/reactant concentration vs time is a straight line.

Given Balanced chemical equation, reaction times, and concentrations

Asked for Graph of data, rate law, and rate constant

Strategy

A Use the data in the table to separately plot concentration, the natural logarithm of the concentration, and the reciprocal concentration (the vertical axis) versus time (the horiztonal axis). Compare the graphs with those shown in Table 14.6 to determine the order of the reaction.

B Write the rate law for the reaction. Using the appropriate data from the table and the linear graph corresponding to the rate law for the reaction, calculate the slope of the plotted line to obtain the rate constant for the reaction.

Solution

A Here are plots of $[N_2O_5]$ versus *t*, $\ln[N_2O_5]$ versus *t*, and $1/[N_2O_5]$ versus *t*:

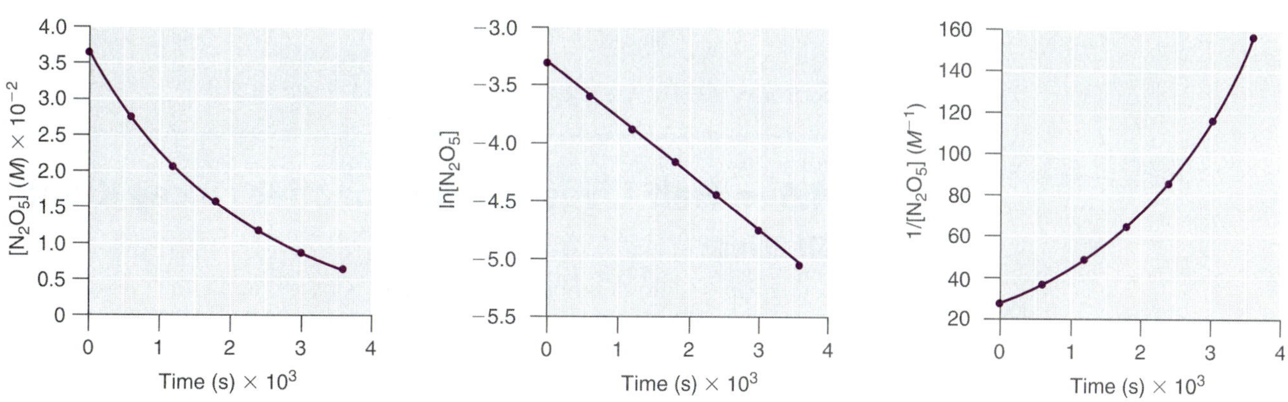

The plot of $\ln[N_2O_5]$ versus *t* gives a straight line, whereas the plots of $[N_2O_5]$ versus *t* and $1/[N_2O_5]$ versus *t* do not. This means that the decomposition of N_2O_5 is first order in $[N_2O_5]$.
B The rate law for the reaction is therefore

$$\text{rate} = k[N_2O_5]$$

Calculating the rate constant is straightforward because we know that the slope of the plot of ln[A] versus *t* for a first-order reaction is $-k$. We can calculate the slope using any two points that lie on the line in the plot of $\ln[N_2O_5]$ versus t. Using the points for $t = 0$ and 3000 s gives

$$\text{slope} = \frac{\ln[N_2O_5]_{3000} - \ln[N_2O_5]_0}{3000\text{ s} - 0\text{ s}} = \frac{(-4.756) - (-3.310)}{3000\text{ s}} = -4.820 \times 10^{-4}\text{ s}^{-1}$$

Thus, $k = 4.820 \times 10^{-4}\text{ s}^{-1}$.

EXERCISE 14.9

1,3-Butadiene ($CH_2{=}CH{-}CH{=}CH_2$) is a volatile and reactive organic molecule used in the production of rubber. Above room temperature it reacts slowly to form products. Concentrations of 1,3-butadiene (C_4H_6) as a function of time at 326°C are listed in the table along with $\ln[C_4H_6]$ and the reciprocal concentrations. Graph the data as concentration versus t, ln concentration versus t, and 1/concentration versus t. Then determine the order of the reaction in C_4H_6, the rate law, and the rate constant for the reaction.

Time, s	$[C_4H_6]$, M	$\ln[C_4H_6]$	$1/[C_4H_6]$, M^{-1}
0	1.72×10^{-2}	−4.063	58.1
900	1.43×10^{-2}	−4.247	69.9
1800	1.23×10^{-2}	−4.398	81.3
3600	9.52×10^{-3}	−4.654	105
6000	7.30×10^{-3}	−4.920	137

Answer

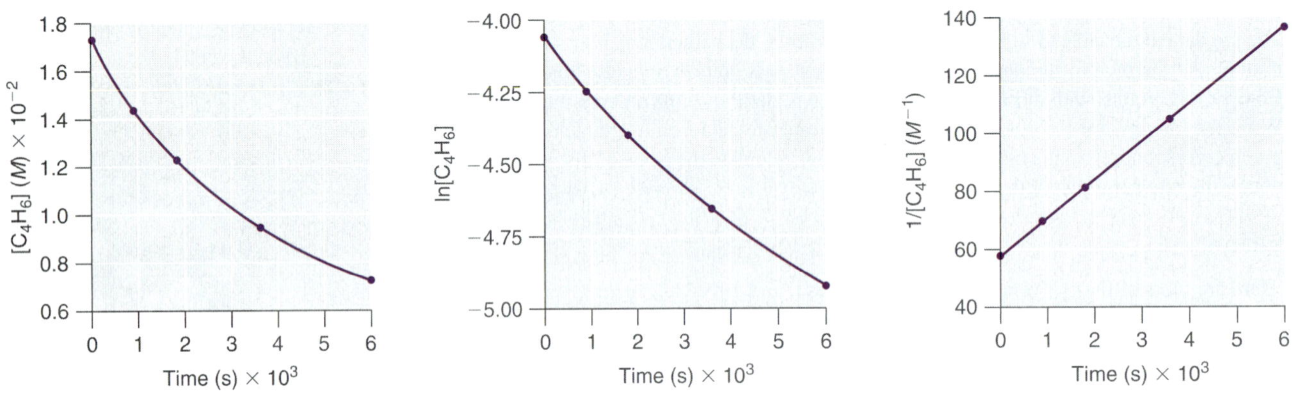

Second order in C_4H_6; rate $= k[C_4H_6]^2$; $k = 1.3 \times 10^{-2}\ M^{-1} \cdot \text{s}^{-1}$

14.5 ○ Half-Lives and Radioactive Decay Kinetics

Half-Lives

MGC Half-life

Another approach to describing reaction rates is based on the time required for the concentration of a reactant to decrease to one-half its initial value. This period of time is called the **half-life** of the reaction, written as $t_{1/2}$. Thus, the half-life of a reaction is the time required for the reactant concentration to decrease from $[A]_0$ to $[A]_0/2$. If two reactions have the same order, the faster reaction will have a shorter half-life and the slower reaction will have a longer half-life.

The half-life of a first-order reaction under a given set of reaction conditions is a constant. This is not true for zeroth- or second-order reactions. Most important, the half-life of a first-order reaction is *independent of the concentration of the reactants*. This becomes evident when we rearrange the integrated rate law for a first-order reaction (Equation 14.23) to produce the following equation:

$$\ln\frac{[A]_0}{[A]} = kt \tag{14.31}$$

Substituting $[A]_0/2$ for [A] and $t_{1/2}$ for t (to indicate a half-life) into Equation 14.31 gives

$$\ln\frac{[A]_0}{[A]_0/2} = \ln 2 = kt_{1/2}$$

The natural logarithm of 2 (to three decimal places) is 0.693. Substituting this value into the equation, we obtain the expression for the half-life of a first-order reaction:

$$t_{1/2} = \frac{0.693}{k} \qquad (14.32)$$

Thus, for a first-order reaction, each successive half-life is the same length of time, as shown in Figure 14.12, and is independent of [A].

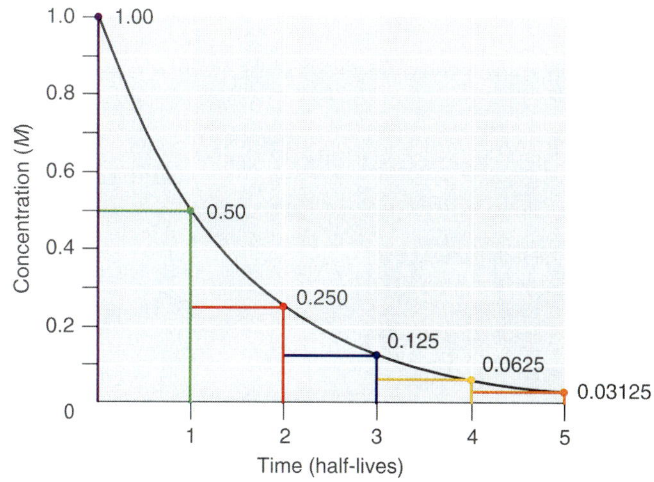

Figure 14.12 The half-life of a first-order reaction. This plot shows the concentration of the reactant in a first-order reaction as a function of time and identifies a series of half-lives, intervals in which the reactant concentration decreases by a factor of 2. In a first-order reaction, every half-life is the same length. Interactive Graph

If we know the rate constant for a first-order reaction, then we can use half-lives to predict how much time is needed for the reaction to reach a certain percent completion.

Number of Half-Lives	% Reactant Remaining	
1	$\frac{100\%}{2} = 50\%$	$\frac{1}{2}(100\%) = 50\%$
2	$\frac{50\%}{2} = 25\%$	$\frac{1}{2}\left(\frac{1}{2}\right)(100\%) = 25\%$
3	$\frac{25\%}{2} = 12.5\%$	$\frac{1}{2}\left(\frac{1}{2}\right)\left(\frac{1}{2}\right)(100\%) = 12.5\%$
n	$\frac{100\%}{2^n}$	$\left(\frac{1}{2}\right)^n(100\%) = \left(\frac{1}{2}\right)^n\%$

As you can see from the table, the amount of reactant left after n half-lives of a first-order reaction is $(1/2)^n$ times the initial concentration.

Note the pattern

For a first-order reaction, the concentration of the reactant decreases by a constant with each half-life and is independent of [A].

EXAMPLE 14.10

The anticancer drug cisplatin hydrolyzes in water with a rate constant of 1.5×10^{-3} min^{-1} at pH 7.0 and 25°C. Calculate the half-life for the hydrolysis reaction under these conditions. If a freshly prepared solution of cisplatin has a concentration of 0.053 *M*, what will be the concentration of cisplatin after five half-lives? After 10 half-lives? What is the percent completion of the reaction after five half-lives? After 10 half-lives?

Given Rate constant, initial concentration, and number of half-lives

Asked for Half life, final concentrations, and percent completion

Strategy

A Use Equation 14.32 to calculate the half-life of the reaction.

B Multiply the initial concentration by (0.50) to the power corresponding to the number of half-lives to obtain the remaining concentrations after those half-lives.

C Subtract the remaining concentration from the initial concentration. Then divide by the initial concentration, multiplying the fraction by 100 to obtain the percent completion.

Solution

A We can calculate the half-life of the reaction using Equation 14.32:

$$t_{1/2} = \frac{0.693}{k} = \frac{0.693}{1.5 \times 10^{-3}\ \text{min}^{-1}} = 4.6 \times 10^2\ \text{min}$$

Thus, it takes almost 8 h for half of the cisplatin to hydrolyze. After five half-lives (about 38 h), the remaining concentration of cisplatin will be

$$\frac{0.053\ M}{2^5} = \frac{0.053\ M}{32} = 0.0017\ M$$

After 10 half-lives (77 h), the remaining concentration of cisplatin will be

$$\frac{0.053\ M}{2^{10}} = \frac{0.053\ M}{1024} = 5.2 \times 10^{-5}\ M$$

The percent completion after five half-lives will be

$$\text{percent completion} = \frac{(0.053\ M - 0.0017\ M)\,(100)}{0.053\ M} = 97\%$$

The percent completion after 10 half-lives will be

$$\text{percent completion} = \frac{(0.053\ M - 5.2 \times 10^{-5}\ M)\,(100)}{0.053\ M} = 100\%$$

Thus, a first-order chemical reaction is 97% complete after five half-lives and to two significant figures is 100% complete after 10 half-lives.

EXERCISE 14.10

In Example 14.4 you found that ethyl chloride decomposes to ethylene and HCl in a first-order reaction that has a rate constant of $1.6 \times 10^{-6}\ s^{-1}$ at 650°C. What is the half-life for the reaction under these conditions? If a flask that originally contains 0.077 *M* ethyl chloride is heated at 650°C, what is the concentration of ethyl chloride after four half-lives?

Answer 4.3×10^5 s = 120 h = 5.0 days; $4.8 \times 10^{-3}\ M$

Radioactive Decay Rates

As you learned in Chapter 1, radioactivity, or radioactive decay, is the emission of a particle or a photon that results from the spontaneous decomposition of the unstable nucleus of an atom. The rate of radioactive decay is an intrinsic property of each radioactive isotope, independent of the chemical and physical form of the radioactive isotope. The rate is also independent of temperature. In this section we will describe radioactive decay rates and how half-lives can be used to monitor radioactive decay processes.

In any sample of a given radioactive substance, the number of atoms of the radioactive isotope must decrease with time as their nuclei decay to nuclei of a more stable isotope. Using N to represent the number of atoms of the radioactive isotope, we can define the **rate of decay** of the sample, which is also called its **activity**, A, as the decrease in the number of the radioisotope's nuclei per unit time:

$$A = -\frac{\Delta N}{\Delta t} \qquad (14.33)$$

Activity is usually measured in *disintegrations per second (dps) or disintegrations per minute (dpm).*

The activity of a sample is directly proportional to the number of atoms of the radioactive isotope in the sample:

$$A = kN \qquad (14.34)$$

Here, the symbol k is the *radioactive decay constant*, which has units of inverse time (for example, s^{-1}, yr^{-1}) and has a characteristic value for each radioactive isotope. If

TABLE 14.7 Half-lives and applications of some radioactive isotopes

Radioactive Isotope	Half-Life	Typical Uses
Hydrogen-3 (tritium)	12.32 yr	Biochemical tracer
Carbon-11	20.33 min	PET scans (biomedical imaging)
Carbon-14	5.70×10^3 yr	Dating of artifacts
Sodium-24	14.951 hr	Cardiovascular system tracer
Phosphorus-32	14.26 days	Biochemical tracer
Potassium-40	1.248×10^9 yr	Dating of rocks
Iron-59	44.495 days	Red blood cell lifetime tracer
Cobalt-60	5.2712 yr	Radiation therapy for cancer
Technetium-99*m*[a]	6.006 h	Biomedical imaging
Iodine-131	8.0207 days	Thyroid studies tracer
Radium-226	1.600×10^3 yr	Radiation therapy for cancer
Uranium-238	4.468×10^9 yr	Dating of rocks and Earth's crust
Americium-241	432.2 yr	Smoke detectors

[a]The *m* denotes metastable, where an excited state nucleus decays to the ground state of the same isotope.

we combine Equations 14.33 and 14.34, we obtain the relationship between the number of decays per unit time and the number of atoms of the isotope in a sample:

$$-\frac{\Delta N}{\Delta t} = kN \tag{14.35}$$

Equation 14.35 is the same as the equation for the rate of a first-order reaction (Equation 14.21), except that it uses numbers of atoms instead of concentrations. In fact, radioactive decay *is* a first-order process and can be described in terms of either the differential rate law (Equation 14.35) or the integrated rate law:

$$N = N_0\, e^{-kt}$$

$$\ln \frac{N}{N_0} = -kt \tag{14.36}$$

Because radioactive decay is a first-order process, the time required for half of the nuclei in any sample of a radioactive isotope to decay is a constant, called the *half-life of the isotope.* The half-life tells us how radioactive an isotope is (the number of decays per unit time); thus, it is the most commonly cited property of any radioisotope. For a given number of atoms, isotopes with shorter half-lives decay more rapidly, undergoing a greater number of radioactive decays per unit time than do isotopes with longer half-lives. The half-lives of several isotopes are listed in Table 14.7, along with some of their applications.

Radioisotope Dating Techniques

In our earlier discussion, we used the half-life of a first-order reaction to calculate how long the reaction had been occurring. Because nuclear decay reactions follow first-order kinetics and have a rate constant that is independent of temperature and chemical or physical environment, we can perform similar calculations using the half-lives of isotopes to estimate the ages of geological and archaeological artifacts. The techniques that have been developed for this application are known as *radioisotope dating* techniques.

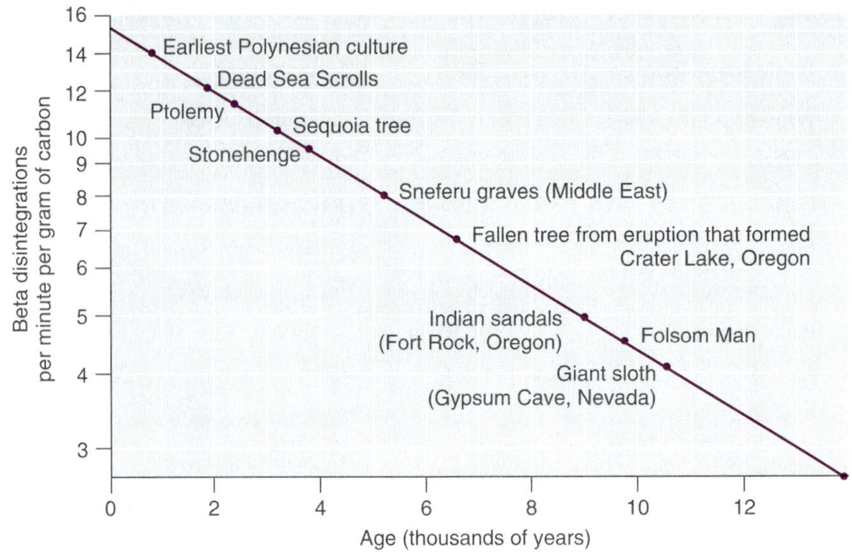

Figure 14.13 Radiocarbon dating. A plot of the specific activity of ^{14}C vs. age for a number of archaeological samples shows an inverse linear relationship between ^{14}C content (a log scale) and age (a linear scale).

The most common method for measuring the age of ancient objects is *carbon-14 dating.* The carbon-14 isotope, created continuously in the upper regions of Earth's atmosphere, reacts with atmospheric oxygen or ozone to form $^{14}CO_2$. As a result, the CO_2 that plants use as a carbon source for synthesizing organic compounds always includes a certain proportion of $^{14}CO_2$ molecules as well as nonradioactive $^{12}CO_2$ and $^{13}CO_2$. Any animal that eats a plant ingests a mixture of organic compounds that contains approximately the same proportions of carbon isotopes as the atmosphere. When the animal or plant dies, the carbon-14 nuclei in its tissues decay to nitrogen-14 nuclei by a radioactive process known as *beta decay*, which releases low-energy electrons (β particles) that can be detected and measured:

$$^{14}C \longrightarrow {}^{14}N + \beta^- \tag{14.37}$$

The half-life for this reaction is 5700 ± 30 yr.

The $^{14}C/^{12}C$ ratio in living organisms is 1.3×10^{-12}, with a decay rate of 15 dpm per gram of carbon (dpm/g carbon) (Figure 14.13). Comparing the disintegrations per minute per gram of carbon from an archaeological sample with those from a recently living sample enables scientists to estimate the age of the artifact, as illustrated in Example 14.11.*

EXAMPLE 14.11

In 1990 the remains of an apparently prehistoric man were found in a melting glacier in the Italian Alps. Analysis of the ^{14}C content of samples of wood from his tools gave a decay rate of 8.0 dpm/g carbon. How long ago did the man die?

Given Isotope and final activity

Asked for Elapsed time

Strategy

A Use Equation 14.34 to calculate N_0/N. Then substitute the value for the half-life of ^{14}C into Equation 14.32 to find the rate constant for the reaction.

---

* Using this method implicitly assumes that the $^{14}CO_2/^{12}CO_2$ ratio in the atmosphere is constant, which is not strictly correct. Other methods, such as tree-ring dating, have been used to calibrate the dates obtained by radiocarbon dating, and all radiocarbon dates reported are now corrected for minor changes in the $^{14}CO_2/^{12}CO_2$ ratio over time.

B Using the values obtained for N_0/N and the rate constant, solve Equation 14.36 to obtain the elapsed time, t.

Solution

We know the initial activity from the isotope's identity (15 dpm/g), the final activity (8.0 dpm/g), and the half-life, so we can use the integrated rate law for a first-order nuclear reaction (Equation 14.36) to calculate the elapsed time, t (the amount of time elapsed since the wood for the tools was cut and began to decay).

$$\ln \frac{N}{N_0} = -kt$$

$$t = \frac{\ln(N/N_0)}{k}$$

A✓ From Equation 14.34, we know that $A = kN$. We can therefore use the initial and final activities ($A_0 = 15$ dpm and $A = 8.0$ dpm) to calculate N_0/N:

$$\frac{A_0}{A} = \frac{kN_0}{kN} = \frac{N_0}{N} = \frac{15}{8.0}$$

Now we need only calculate the rate constant for the reaction from its half-life (5730 yr) using Equation 14.32:

$$t_{1/2} = \frac{0.693}{k}$$

This equation can be rearranged to give

$$k = \frac{0.693}{t_{1/2}} = \frac{0.693}{5730 \text{ yr}} = 1.22 \times 10^{-4} \text{ yr}^{-1}$$

B✓ Substituting into the equation for t gives

$$t = \frac{\ln(N_0/N)}{k} = \frac{\ln(15/8.0)}{1.22 \times 10^{-4} \text{ yr}^{-1}} = 5.2 \times 10^3 \text{ yr}$$

From our calculations, the man died 5200 years ago.

EXERCISE 14.11

It is believed that humans first arrived in the Western Hemisphere during the last Ice Age, presumably by traveling over an exposed land bridge between Siberia and Alaska. Archaeologists have estimated that this occurred about 11,000 yr ago, but some argue that recent discoveries in several sites in North and South America suggest a much earlier arrival. Analysis of a sample of charcoal from a fire in one such site gave a ^{14}C decay rate of 0.4 dpm/g of carbon. What is the approximate age of the sample?

Answer 30,000 yr

14.6 • Reaction Rates—A Microscopic View

One of the major reasons for studying chemical kinetics is to use measurements of the macroscopic properties of a system, such as the rate of change in the concentration of reactants or products with time, to discover the sequence of events that occur at the molecular level during a reaction. This molecular description is the *mechanism* of the reaction; it describes how individual atoms, ions, or molecules interact to form particular products. The stepwise changes are called the **reaction mechanism**.

MGC Mechanisms and Molecularity

In an internal combustion engine, for example, isooctane reacts with oxygen to give carbon dioxide and water.

$$2C_8H_{18}(l) + 25O_2(g) \longrightarrow 16CO_2(g) + 18H_2O(g) \qquad (14.38)$$

For this reaction to occur in a single step, 25 dioxygen molecules and 2 isooctane molecules would have to collide simultaneously and be converted to 34 molecules of product, which is very unlikely. It is more likely that a complex series of reactions takes place in a stepwise fashion. Each of the individual reactions, called **elementary reactions,** involves one, two, or (rarely) three atoms, molecules, or ions. The overall sequence of elementary reactions is the mechanism of the reaction. *The sum of the individual steps, or elementary reactions, in the mechanism must give the balanced chemical equation for the overall reaction.*

Molecularity and the Rate-Determining Step

To demonstrate how the analysis of elementary steps helps us determine the overall mechanism of a reaction, we will examine the much simpler reaction of carbon monoxide with nitrogen dioxide.

$$NO_2(g) + CO(g) \longrightarrow NO(g) + CO_2(g) \qquad (14.39)$$

From the balanced chemical equation, one might expect the reaction to occur via a collision of one molecule of NO_2 with a molecule of CO that results in the transfer of an oxygen atom from nitrogen to carbon. The experimentally determined rate law for the reaction, however, is

$$\text{rate} = k[NO_2]^2 \qquad (14.40)$$

The fact that the reaction is second order in $[NO_2]$ and independent of [CO] tells us that it does not occur by the simple collision of an NO_2 molecule with a CO molecule. If it did, its predicted rate law would be rate = $k[NO_2][CO]$.

The following two-step mechanism is consistent with the rate law if Step 1 is much slower than Step 2:

Step 1	$NO_2 + NO_2 \xrightarrow{\text{slow}} NO_3 + NO$	Elementary reaction
Step 2	$NO_3 + CO \longrightarrow NO_2 + CO_2$	Elementary reaction
Sum	$NO_2 + CO \longrightarrow NO + CO_2$	Overall reaction

According to this mechanism, the overall reaction occurs in two steps, or elementary reactions. Notice that summing Steps 1 and 2 and canceling on both sides of the equation gives the overall balanced chemical equation for the reaction. The NO_3 molecule is an **intermediate** in the reaction, a species that does not appear in the balanced equation for the overall reaction. It is formed as a product of the first step but then is consumed in the second step.

Using Molecularity to Describe a Rate Law

The **molecularity** of an elementary reaction is the number of molecules that collide during that step in the mechanism. If there is only a single reactant molecule in an elementary reaction, that step is designated as *unimolecular*; if there are two reactant molecules, it is *bimolecular*; and if there are three reactant molecules (a relatively rare situation), it is *termolecular*. Elementary reactions that involve the simultaneous collision of more than three molecules are highly improbable and have never been observed experimentally. (To understand why, try to make three or more marbles or pool balls collide with one another simultaneously!)

Writing the rate law for an elementary reaction is straightforward because we know how many molecules must collide simultaneously for the elementary reaction to occur; hence, the order of the *elementary reaction* is the same as its molecularity

TABLE 14.8 Common types of elementary reactions and their rate laws

Elementary Reaction	Molecularity	Rate Law	Reaction Order
$A \longrightarrow$ products	Unimolecular	Rate = $k[A]$	First
$2A \longrightarrow$ products	Bimolecular	Rate = $k[A]^2$	Second
$A + B \longrightarrow$ products	Bimolecular	Rate = $k[A][B]$	Second
$2A + B \longrightarrow$ products	Termolecular	Rate = $k[A]^2[B]$	Third
$A + B + C \longrightarrow$ products	Termolecular	Rate = $k[A][B][C]$	Third

(Table 14.8). In contrast, the rate law for the reaction cannot be determined from the balanced equation for the overall reaction. The general rate law for a unimolecular elementary reaction ($A \longrightarrow$ products) is rate = $k[A]$. For bimolecular reactions, the reaction rate depends on the number of collisions per unit time, which is proportional to the product of the concentrations of the reactants, as shown in Figure 14.14. Hence, for a bimolecular elementary reaction of the form $A + B \longrightarrow$ products, the general rate law is rate = $k[A][B]$.

Identifying the Rate-Determining Step

Note the important difference between writing rate laws for the elementary steps of a reaction and for the balanced chemical equation of the overall reaction. Because the balanced chemical equation does not necessarily reveal the individual elementary reactions by which the reaction occurs, we *cannot* obtain the rate law for a reaction from the overall balanced equation alone. In fact, the rate law for the overall reaction is the same as the rate law for the slowest step in the reaction mechanism, the **rate-determining step.*** The reason for this is that any process that occurs through a sequence of steps can take place no faster than the slowest step in the sequence. In an automotive assembly line, for example, a component cannot be used faster than it is produced. Similarly, blood pressure is regulated by the flow of blood through the smallest passages, the capillaries. Because movement through capillaries constitutes

Figure 14.14 The basis for writing rate laws of elementary reactions. This diagram illustrates how the number of possible collisions per unit time between two reactant species, A and B, depends on the number of A and B particles present. Note that the number of collisions between A and B particles increases as the *product* of the number of particles, not as the sum. This is why the rate law for an elementary reaction depends on the product of the concentrations of the species that collide in that step.

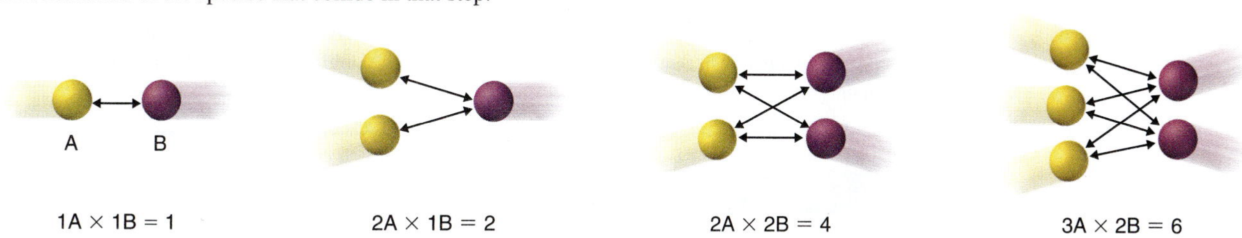

* This statement is true if one step is substantially slower than all the others, typically by a factor of 10 or more. If two or more slow steps have comparable rates, the experimentally determined rate laws can become complex. Our discussion is limited to reactions in which one step can be identified as being substantially slower than any other.

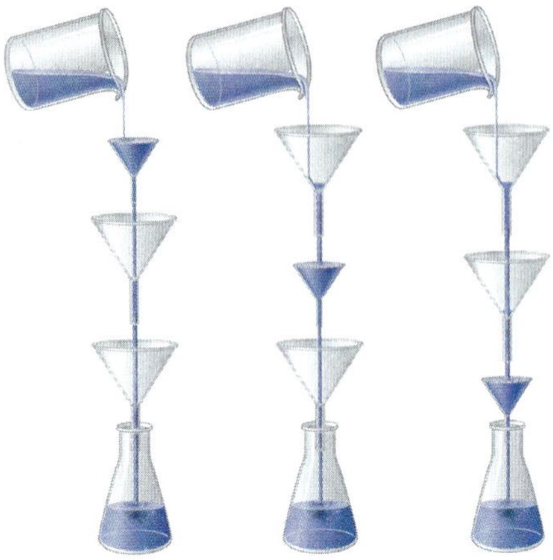

Rate-determining step. The phenomenon of a rate-determining step can be compared to a succession of funnels. The smallest-diameter funnel controls the rate at which the bottle is filled, whether it is the first or the last in the series. Pouring liquid into the first funnel faster than it can drain through the smallest results in an overflow.

the rate-determining step in blood flow, blood pressure can be regulated by medications that cause the capillaries to contract or dilate. A chemical reaction that occurs via a series of elementary reactions can take place no faster than the slowest step in the series of reactions.

Take a look at the rate laws for each elementary reaction in our example as well as for the overall reaction.

Step 1 $NO_2 + NO_2 \xrightarrow{k_1} NO_3 + NO$ Rate $= k_1[NO_2]^2$ (predicted)
Step 2 $NO_3 + CO \xrightarrow{k_2} NO_2 + CO_2$ Rate $= k_2[NO_3][CO]$ (predicted)
Sum $NO_2 + CO \xrightarrow{k} NO + CO_2$ Rate $= k[NO_2]^2$ (observed)

The experimentally determined rate law for the reaction of NO_2 with CO is the same as the predicted rate law for Step 1. This tells us that the first elementary reaction is the rate-determining step, so k for the overall reaction must equal k_1. That is, NO_3 is formed slowly in Step 1, but once it is formed it reacts very rapidly with CO in Step 2.

Sometimes chemists are able to propose two or more mechanisms that are consistent with the available data. If a proposed mechanism predicts the wrong experimental rate law, however, the mechanism must be incorrect.

EXAMPLE 14.12

In an alternative mechanism for the reaction of NO_2 with CO, N_2O_4 appears as an intermediate.

Step 1 $NO_2 + NO_2 \xrightarrow{k_1} N_2O_4$
Step 2 $N_2O_4 + CO \xrightarrow{k_2} NO + NO_2 + CO_2$
Sum $NO_2 + CO \longrightarrow NO + CO_2$

Write the rate law for each of the elementary steps. Is this mechanism consistent with the experimentally determined rate law (rate $= k[NO_2]^2$)?

Given Elementary steps of reaction

Asked for Rate law for each elementary step and overall rate law

Strategy

Ⓐ Determine the rate law for each elementary step in the reaction.
Ⓑ Determine which of these rate laws corresponds to the experimentally determined rate law for the reaction. This rate law is the one for the rate-determining step.

Solution

Ⓐ The rate law for Step 1 is rate $= k_1[NO_2]^2$; for Step 2 it is rate $= k_2[N_2O_4][CO]$. Ⓑ If Step 1 is slow (and therefore the rate-determining step), then the overall rate law for the reaction will be the same: rate $= k_1[NO_2]^2$. This is the same as the experimentally determined rate law. Hence, this mechanism, with N_2O_4 as an intermediate, and the one described previously, with NO_3 as an intermediate, are kinetically indistinguishable. In this case, further experiments are needed to distinguish between them. For example, the researcher could try to detect the proposed intermediates, NO_3 and N_2O_4, directly.

EXERCISE 14.12

Iodine monochloride, ICl, reacts with H_2 as follows:

$$2ICl(l) + H_2(g) \longrightarrow 2HCl(g) + I_2(s)$$

The experimentally determined rate law is rate $= k[ICl][H_2]$. Write a two-step mechanism for this reaction using only bimolecular elementary steps, and show that it is consistent with the experimental rate law. (*Hint:* HI is an intermediate.)

Answer

Step 1	$ICl + H_2 \longrightarrow HCl + HI$	Rate $= k_1[ICl][H_2]$ (slow)
Step 2	$HI + ICl \longrightarrow HCl + I_2$	Rate $= k_2[HI][ICl]$ (fast)
Sum	$2ICl + H_2 \longrightarrow 2HCl + I_2$	

This mechanism is consistent with the experimental rate law *if* the first step is the rate-determining step.

Chain Reactions

Many reaction mechanisms, like those discussed so far, consist of only two or three elementary reactions. Many others consist of long series of elementary reactions. The most common mechanisms are **chain reactions**, in which one or more elementary reactions that contain a highly reactive species repeat again and again during the reaction process. Chain reactions occur in fuel combustion, in explosions, in the formation of many polymers, and in the tissue changes associated with aging. They are also important in the chemistry of the atmosphere.

Chain reactions are described as having three stages. The first is *initiation*, a step that produces one or more reactive intermediates. Often these intermediates are **radicals**, species that have an unpaired valence electron. In the second stage, *propagation*, reactive intermediates are continuously consumed and regenerated while products are formed. Intermediates are also consumed but not regenerated in the final stage of a chain reaction, *termination*, usually by forming stable products.

Let us look at the reaction of methane with chlorine at elevated temperatures (400–450°C), a chain reaction used in industry to manufacture methyl chloride (CH_3Cl), dichloromethane (CH_2Cl_2), chloroform ($CHCl_3$), and carbon tetrachloride (CCl_4):

$$CH_4 + Cl_2 \longrightarrow CH_3Cl + HCl$$
$$CH_3Cl + Cl_2 \longrightarrow CH_2Cl_2 + HCl$$
$$CH_2Cl_2 + Cl_2 \longrightarrow CHCl_3 + HCl$$
$$CHCl_3 + Cl_2 \longrightarrow CCl_4 + HCl$$

Direct chlorination generally produces a mixture of all four carbon-containing products, which must then be separated by distillation. In our discussion we will examine only the chain reactions that lead to the preparation of CH_3Cl.

In the initiation stage of this reaction, the relatively weak Cl—Cl bond cleaves at temperatures of about 400°C to produce chlorine atoms ($Cl\cdot$):

$$Cl_2 \longrightarrow 2Cl\cdot$$

During propagation, a chlorine atom removes a hydrogen atom from a methane molecule to give HCl and $CH_3\cdot$, the methyl radical:

$$Cl\cdot + CH_4 \longrightarrow CH_3\cdot + HCl$$

The methyl radical then reacts with a chlorine molecule to form methyl chloride and another chlorine atom, $Cl\cdot$:

$$CH_3\cdot + Cl_2 \longrightarrow CH_3Cl + Cl\cdot$$

Note that the sum of the propagation reactions is the same as the overall balanced equation for the reaction:

$$\cancel{Cl} + CH_4 \longrightarrow \cancel{CH_3}\cdot + HCl$$
$$\cancel{CH_3} + Cl_2 \longrightarrow CH_3Cl + \cancel{Cl}$$
$$\overline{Cl_2 + CH_4 \longrightarrow CH_3Cl + HCl}$$

Without a chain-terminating reaction, the propagation reactions would continue until either the methane or the chlorine was consumed. Because radical species react rapidly with almost anything, however, including each other, they eventually form neutral compounds, thus terminating the chain reaction in any of three ways:

$$CH_3\cdot + Cl\cdot \longrightarrow CH_3Cl$$
$$CH_3\cdot + CH_3\cdot \longrightarrow H_3CCH_3$$
$$Cl\cdot + Cl\cdot \longrightarrow Cl_2$$

Here is the overall chain reaction, with the desired product (CH_3Cl) highlighted:

Initiation $$Cl_2 \longrightarrow 2Cl\cdot$$

Propagation $$Cl\cdot + CH_4 \longrightarrow CH_3\cdot + HCl$$
$$CH_3\cdot + Cl_2 \longrightarrow CH_3Cl + Cl\cdot$$

Termination $$CH_3\cdot + Cl\cdot \longrightarrow CH_3Cl$$
$$CH_3\cdot + CH_3\cdot \longrightarrow H_3CCH_3$$
$$Cl\cdot + Cl\cdot \longrightarrow Cl_2$$

The chain reactions responsible for explosions generally have an additional feature: one or more *chain branching steps*, in which one radical reacts to produce two or more radicals, each of which can then go on to start a new chain reaction. Repetition of the branching step has a cascade effect such that a single initiation step generates large numbers of chain reactions. The result is a very rapid reaction or an explosion. The reaction of H_2 and O_2, used to propel rockets, is an example of a chain branching reaction:

Initiation $$H_2 + O_2 \longrightarrow HO_2\cdot + H\cdot$$

Propagation $$HO_2\cdot + H_2 \longrightarrow H_2O + OH\cdot$$
$$OH\cdot + H_2 \longrightarrow H_2O + H\cdot$$

Branching $$H\cdot + O_2 \longrightarrow OH\cdot + \cdot O\cdot$$
$$\cdot O\cdot + H_2 \longrightarrow OH\cdot + H\cdot$$

Termination reactions occur when the extraordinarily reactive $H\cdot$ or $OH\cdot$ radicals react with a third species. The complexity of a chain reaction makes it unfeasible to write a rate law for the overall reaction.

14.7 • The Collision Model of Chemical Kinetics

MGC The Arrhenius Equation

In Section 14.6 you saw that it is possible to use kinetics studies of a chemical system, such as the effect of changes in reactant concentrations, to deduce events that occur on a microscopic scale, such as collisions between individual particles. Such studies have led to the *collision model of chemical kinetics*, which is a useful tool for understanding the behavior of reacting chemical species. According to the collision model, a chemical reaction can occur only when the reactant molecules, atoms, or ions collide with more than a certain amount of kinetic energy and in the proper orientation. The collision model explains why, for example, most collisions between molecules do *not* result in a chemical reaction. Nitrogen and oxygen molecules in a single liter of air at room temperature and 1 atm of pressure collide about 10^{30} times per second. If every collision produced two molecules of NO, the atmosphere would have been converted to NO and then NO_2 a long time ago. Instead, in most collisions the molecules simply bounce off one another without reacting, much as marbles bounce off each other when they collide. The collision model also explains why such chemical reactions occur more rapidly at higher temperatures. For example, the rates of many reactions that occur at room temperature approximately double with a temperature increase of only 10°C. In this section we will use the collision model to analyze this relationship between temperature and reaction rates.

Activation Energy

In Chapter 10 we discussed the kinetic molecular theory of gases, which showed that the average kinetic energy of the particles of a gas increases with increasing temperature. Because the speed of a particle is proportional to the square root of its kinetic energy, increasing the temperature will also increase the number of collisions between molecules per unit time. What the kinetic molecular theory of gases does not explain is why the rate of most reactions approximately doubles with a 10°C temperature increase. This result is surprisingly large considering that a 10°C increase in the temperature of a gas from 300 K to 310 K increases the kinetic energy of the particles by only about 4%, leading to an increase in molecular speed of only about 2% and a correspondingly small increase in the number of bimolecular collisions per unit time.

The collision model of chemical kinetics explains this behavior by introducing the concept of activation energy. We will define this concept using the reaction of NO with ozone, which plays an important role in the depletion of ozone in the ozone layer:

$$NO(g) + O_3(g) \longrightarrow NO_2(g) + O_2(g) \tag{14.41}$$

Increasing the temperature from 200 K to 350 K causes the rate constant for this particular reaction to increase by a factor of more than 10, whereas the increase in the frequency of bimolecular collisions over this temperature range is only 30%. Thus, something other than an increase in the collision rate must be affecting the reaction rate.

The rate of a reaction will vary with concentrations but not the rate constant. However, the rate constant does vary with temperature. Figure 14.15 shows a plot of the rate constant of the reaction of NO with O_3 at various temperatures. Notice that the relationship is not linear, but instead resembles the relationships seen in graphs of vapor pressure versus temperature (Chapter 11) and of conductivity versus temperature (Chapter 12). In all three cases, the shape of the plots results from a distribution of kinetic energy over a population of particles (electrons in the case of conductivity; molecules in the case of vapor pressure; and molecules, atoms, or ions in the case of reaction rates). Only a fraction of the particles have sufficient energy to overcome an energy barrier.

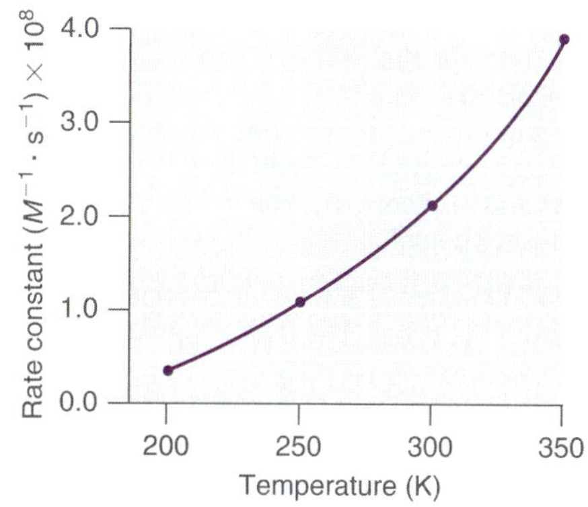

Figure 14.15 Rate constant vs. temperature for the reaction of NO with O_3. The nonlinear shape of the curve is caused by a distribution of kinetic energy over a population of molecules. Only a fraction of the particles have enough energy to overcome an energy barrier, but as the temperature is increased, the size of that fraction increases. Interactive Graph

In the case of vapor pressure, particles must overcome an energy barrier to escape from the liquid to the gas phase. This barrier corresponds to the energy of the intermolecular forces that hold the molecules together in the liquid. In conductivity, the barrier is the energy gap between the filled and empty bands. In chemical reactions, the energy barrier corresponds to the amount of energy the particles must have in order to react when they collide. This energy threshold, called the **activation energy** (**E_a**), was first postulated in 1888 by the Swedish chemist Svante Arrhenius (1859–1927, Nobel Prize 1903). It is the minimum amount of energy needed for a reaction to occur. Reacting molecules must have enough energy to overcome electrostatic repulsion, and a minimum amount of energy is required to break chemical bonds so that new ones may be formed. Molecules that collide with less than the threshold energy bounce off one another chemically unchanged, with only their direction of travel and their speed altered by the collision. Molecules that are able to overcome the energy barrier, and so are able to react, form an arrangement of atoms called the **activated complex** or the **transition state** of the reaction. The activated complex is *not* a reaction intermediate; it does not last long enough to be detected readily.

Note the pattern

Any phenomenon that depends on the distribution of thermal energy in a population of particles has a nonlinear temperature dependence.

Graphing the Energy Changes During a Reaction

We can graph the energy of a reaction by plotting the potential energy of the system as the reaction progresses. Figure 14.16 shows a plot for the NO—O_3 system in which the vertical axis is potential energy and the horizontal axis is the *reaction coordinate*, indicating the progress of the reaction with time. The activated complex is shown in

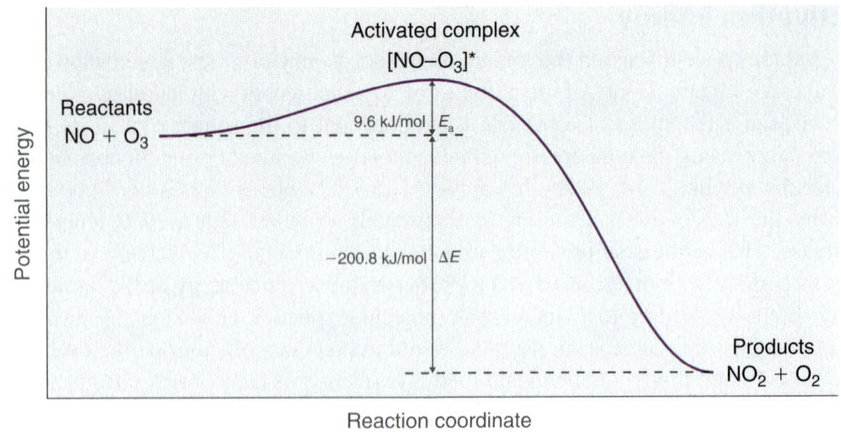

Figure 14.16 Energy of the activated complex for the NO—O_3 system. The diagram shows how the energy of this system varies as the reaction proceeds from reactants to products. Note the initial increase in energy required to form the activated complex.

brackets with an asterisk. The overall change in potential energy for the reaction (ΔE) is negative, which means that the reaction releases energy. (In this case ΔE is -200.8 kJ/mol.) In order to react, however, the molecules must overcome the energy barrier to reaction (E_a is 9.6 kJ/mol). That is, 9.6 kJ/mol must be put into the system as the activation energy. Below this threshold the particles do not have enough energy for the reaction to occur.

Figure 14.17a illustrates the general situation in which the products have a lower potential energy than the reactants. In contrast, Figure 14.17b illustrates the case in which the products have a higher potential energy than the reactants, so the overall reaction requires an input of energy; that is, it is energetically uphill and $\Delta E > 0$. Although the energy changes that result from a reaction can be positive, negative, or even zero, in all cases an energy barrier must be overcome before the reaction can occur. This means that the activation energy, E_a, is always positive.

Whereas ΔE is related to the tendency of a reaction to occur spontaneously (one of the central topics of Chapter 18), E_a gives us information about the rate of a reaction and how rapidly the rate changes with temperature. *For two similar reactions under comparable conditions, the reaction with the smallest E_a will occur more rapidly.*

Even when the energy of collisions between two reactant species is greater than E_a, however, most collisions do not produce a reaction. The probability of a reaction occurring depends not only on the collision energy but also on the spatial orientation of the molecules when they collide. For NO and O_3 to produce NO_2 and O_2, a terminal oxygen atom of O_3 must collide with the nitrogen atom of NO at an angle that allows O_3 to transfer an oxygen atom to NO to produce NO_2 (Figure 14.18). All other collisions produce no reaction. Because fewer than 1% of all possible orientations of NO and O_2 result in a reaction at kinetic energies greater than E_a, most collisions of

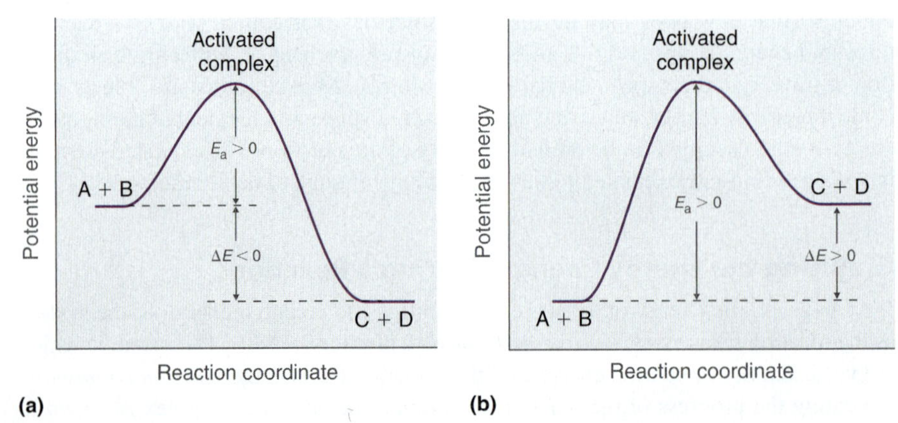

Figure 14.17 Differentiating between E_a and ΔE. The potential energy diagrams for a reaction with **(a)** $\Delta E < 0$ and **(b)** $\Delta E > 0$ illustrate the change in the potential energy of the system as reactants are converted to products. Note that E_a is always positive. For a reaction such as the one shown in **(b)**, E_a must be *greater* than ΔE. Interactive Graph

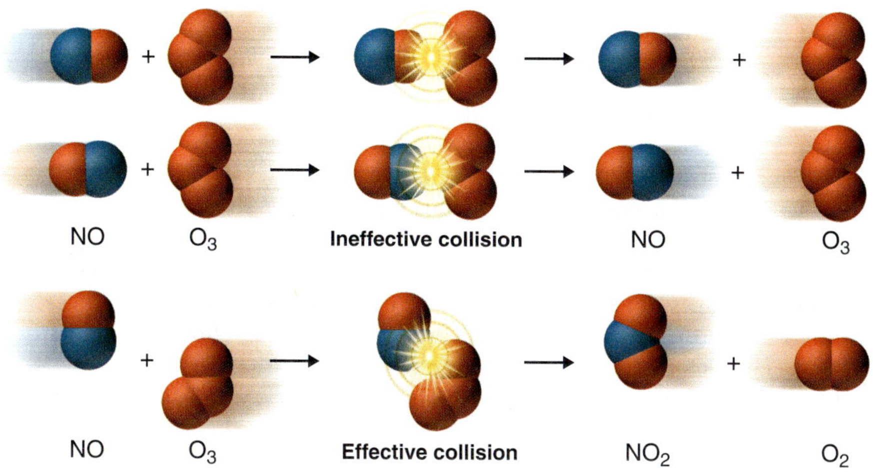

Figure 14.18 The effect of molecular orientation on reaction of NO and O_3. Most collisions of NO and O_3 molecules occur with an incorrect orientation for a reaction to occur. Only those collisions in which the N atom of NO collides with one of the terminal O atoms of O_3 are likely to produce NO_2 and O_2, even if the molecules collide with $E > E_a$.

NO and O_3 are unproductive. The fraction of orientations that result in a reaction is called the **steric factor**, ***p***, and, in general, its value can range from 0 (no orientations of molecules result in reaction) to 1 (all orientations result in reaction).

The Arrhenius Equation

Figure 14.19 shows both the kinetic energy distributions and a potential energy diagram for a reaction. The shaded areas show that at the lower temperature (300 K) only a small fraction of molecules collide with kinetic energy greater than E_a; however, at the higher temperature (500 K) a much larger fraction of molecules collide with kinetic energy greater than E_a. Consequently, the rate of the reaction is much slower at the lower temperature because only a relatively few molecules collide with enough energy to overcome the potential energy barrier.

For an A + B elementary reaction, all the factors that affect the reaction rate can be summarized in a single series of relationships:

$$\text{rate} = (\text{collision frequency})(\text{steric factor})(\text{fraction of collisions with } E > E_a)$$

where

$$\text{rate} = k[\text{A}][\text{B}] \tag{14.42}$$

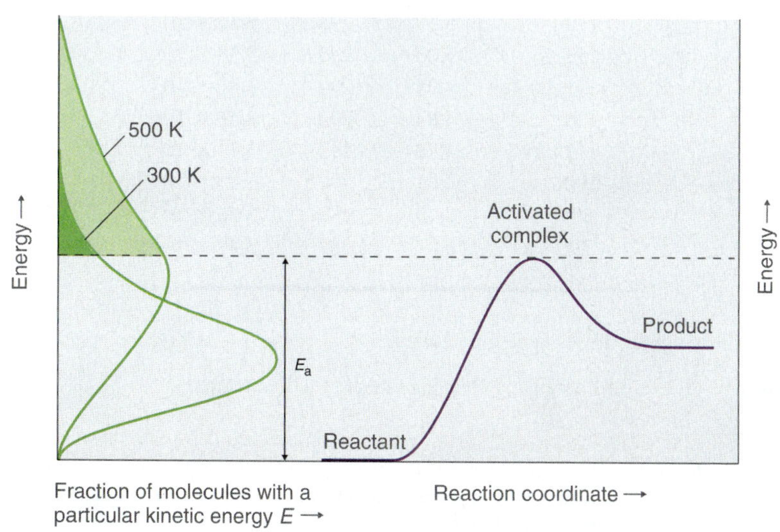

Figure 14.19 Surmounting the energy barrier to a reaction. This chart juxtaposes the energy distributions of lower-temperature (300 K) and higher-temperature (500 K) samples of a gas against the potential energy diagram for a reaction. Only those molecules in the shaded region of the energy distribution curve have $E > E_a$ and are therefore able to cross the energy barrier separating reactants and products. The fraction of molecules with $E > E_a$ is much greater at 500 K than at 300 K, and so the reaction will occur much more rapidly at 500 K. Interactive Graph

Arrhenius used these relationships to arrive at an equation that relates the magnitude of the rate constant for a reaction to the temperature, the activation energy, and the constant, A, called the **frequency factor**:

$$k = Ae^{-E_a/RT} \tag{14.43}$$

The frequency factor is used to convert concentrations to collisions per second.* Equation 14.43 is known as the **Arrhenius equation** and summarizes the collision model of chemical kinetics, where T is the absolute temperature (in K) and R is the ideal gas constant [8.314 J/(K · mol)]. The value of E_a indicates the sensitivity of the reaction to changes in temperature. The rate of a reaction with a large E_a increases rapidly with increasing temperature, whereas the rate of a reaction with a smaller E_a increases much more slowly with increasing temperature.

If we know the rate of a reaction at various temperatures, we can use the Arrhenius equation to calculate the activation energy. Taking the natural logarithm of both sides of Equation 14.43 gives

$$\ln k = \ln A + \left(-\frac{E_a}{RT}\right) = \ln A + \left[\left(-\frac{E_a}{R}\right)\left(\frac{1}{T}\right)\right] \tag{14.44}$$

Equation 14.44 is the equation of a straight line, $y = mx + b$, where $y = \ln k$ and $x = 1/T$. This means that a plot of $\ln k$ versus $1/T$ is a straight line with a slope of $-E_a/R$ and an intercept of $\ln A$. In fact, we need to measure the rate of a reaction at only two temperatures to estimate E_a.

Knowing the value of E_a at one temperature allows us to predict the rate of a reaction at other temperatures. This is important in cooking and food preservation, for example, as well as in controlling industrial reactions to prevent potential disasters. The procedure for determining E_a from reaction rates measured at several temperatures is illustrated in the next example.

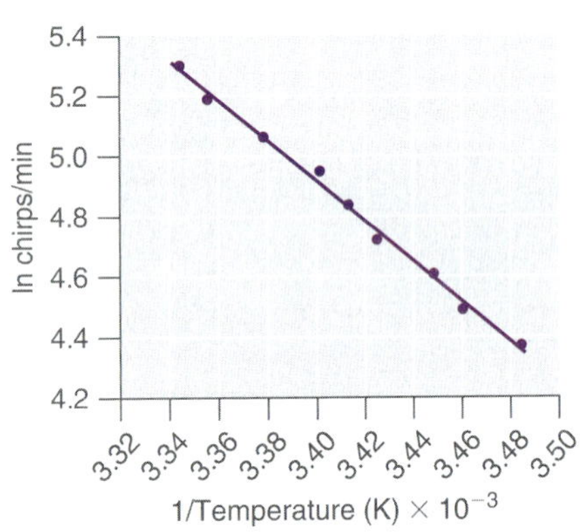

Figure 14.20 Graphical determination of E_a for tree cricket chirping. When the natural logarithm of the rate of tree cricket chirping is plotted vs. $1/T$, a straight line results. The slope of the line suggests that the chirping rate is controlled by a single reaction with an E_a of 55 kJ/mol.

EXAMPLE 14.13

Many people believe that the rate of a tree cricket's chirping is related to the temperature. To see whether this is true, biologists have carried out accurate measurements of the rate of tree cricket chirping (f) as a function of temperature. Use the data in the table, along with the graph of ln[chirping rate] versus $1/T$ in Figure 14.20, to calculate E_a for the biochemical reaction that controls cricket chirping. Then predict the chirping rate on a very hot night, when the temperature is 308 K.

Frequency, f, chirps/min	$\ln f$	T, K	$1/T$, K
200	5.30	299	3.34×10^{-3}
179	5.19	298	3.36×10^{-3}
158	5.06	296	3.38×10^{-3}
141	4.95	294	3.40×10^{-3}
126	4.84	293	3.41×10^{-3}
112	4.72	292	3.42×10^{-3}
100	4.61	290	3.45×10^{-3}
89	4.49	289	3.46×10^{-3}
79	4.37	287	3.48×10^{-3}

Given Chirping rate at various temperatures

Asked for Activation energy and chirping rate at specified temperature

---

* Because the frequency of collisions depends on the temperature, A is actually *not* constant. Instead, A increases slightly with temperature as the increased kinetic energy of molecules at higher temperatures causes them to move slightly faster and thus undergo more collisions per unit time.

Strategy

A From the plot of ln f versus $1/T$ in Figure 14.20, calculate the slope of the line ($-E_a/R$); then solve for the activation energy.

B Express Equation 14.44 in terms of k_1 and T_1, and then in terms of k_2 and T_2.

C Subtract the two equations, then rearrange the result to describe k_2/k_1 in terms of T_2 and T_1.

D Using measured data from the table, solve the equation to obtain the ratio k_2/k_1. Using the value listed in the table for k_1, solve for k_2.

Solution

A If cricket chirping is controlled by a reaction that obeys the Arrhenius equation, then a plot of ln f versus $1/T$ should give a straight line (Figure 14.20). Also, the slope of the plot of ln f versus $1/T$ should be equal to $-E_a/R$. We can use the two endpoints in Figure 14.20 to estimate the slope:

$$\text{slope} = \frac{\Delta \ln f}{\Delta(1/T)} = \frac{5.30 - 4.37}{3.34 \times 10^{-3}\ \text{K}^{-1} - 3.48 \times 10^{-3}\ \text{K}^{-1}}$$

$$= \frac{0.93}{-0.14 \times 10^{-3}\ \text{K}^{-1}} = -6.6 \times 10^{3}\ \text{K}$$

A computer best-fit line through all the points has a slope of -6.67×10^3, so our estimate is very close. We now use it to solve for the activation energy:

$$E_a = -(\text{slope})(R) = -(-6.6 \times 10^3\ \text{K})\left(\frac{8.314\ \text{J}}{\text{K} \cdot \text{mol}}\right)\left(\frac{1\ \text{KJ}}{1000\ \text{J}}\right) = \frac{55\ \text{kJ}}{\text{mol}}$$

B If the activation energy of a reaction and the rate constant at one temperature are known, then we can calculate the rate at any other temperature. We can use Equation 14.44 to express the known rate constant, k_1, at the first temperature, T_1, as

$$\ln k_1 = \ln A - \frac{E_a}{RT_1}$$

Similarly, we can express the unknown rate constant, k_2, at the second temperature, T_2, as

$$\ln k_2 = \ln A - \frac{E_a}{RT_2}$$

C These two equations contain four known quantities (E_a, T_1, T_2, and k_1) and two unknowns (A and k_2). We can eliminate the term A by subtracting the first equation from the second:

$$\ln k_2 - \ln k_1 = \left(\ln A - \frac{E_a}{RT_2}\right) - \left(\ln A - \frac{E_a}{RT_1}\right) = -\frac{E_a}{RT_2} + \frac{E_a}{RT_1}$$

Then

$$\ln \frac{k_2}{k_1} = \frac{E_a}{R}\left(\frac{1}{T_1} - \frac{1}{T_2}\right)$$

D To obtain the best prediction of chirping rate at 308 K (T_2), we try to choose for T_1 and k_1 the measured rate constant and corresponding temperature in the data table that is closest to the best-fit line in the graph. Choosing data for $T_1 = 296$ K, where $f = 158$, and using the value of E_a calculated above gives

$$\ln \frac{k_{T_2}}{k_{T_1}} = \frac{E_a}{R}\left(\frac{1}{T_1} - \frac{1}{T_2}\right) = \frac{55\ \text{kJ/mol}}{8.314\ \text{J/(K} \cdot \text{mol)}}\left(\frac{1000\ \text{J}}{1\ \text{kJ}}\right)\left(\frac{1}{296\ \text{K}} - \frac{1}{308\ \text{K}}\right) = 0.87$$

Thus, $k_{308}/k_{296} = 2.4$ and $k_{308} = (2.4)(158) = 380$, and the chirping rate on a night when the temperature is 308 K is predicted to be 380 chirps per minute.

EXERCISE 14.13

The equation for the decomposition of NO_2 to NO and O_2 is second order in NO_2:

$$2NO_2(g) \longrightarrow 2NO(g) + O_2(g)$$

Data for the rate of the reaction as a function of temperature are listed in the table. Calculate E_a for the reaction and the rate constant at 700 K.

T, K	*k*, $M^{-1} \cdot s^{-1}$
592	522
603	755
627	1700
652	4020
656	5030

Answer $E_a = 114$ kJ/mol; $k_{700} = 13{,}600\ M^{-1} \cdot s^{-1} = 1.9 \times 10^4\ M^{-1} \cdot s^{-1}$.

What value of E_a results in a doubling of the reaction rate with a 10°C increase in temperature from 20° to 30°C?

Answer About 51 kJ/mol.

14.8 • Catalysis

MGC Activation Energy and Catalysts

Chapter 3 described **catalysts** as substances that increase the rate of a chemical reaction without being consumed in the process. A catalyst, therefore, does not appear in the overall stoichiometry of the reaction it catalyzes, but it *must* appear in at least one of the elementary steps in the mechanism for the catalyzed reaction. The catalyzed pathway has a lower E_a, but the *net* change in energy that results from the reaction (the difference between the energy of the reactants and the energy of the products) is *not* affected by the presence of a catalyst (Figure 14.21). Nevertheless, because of its lower E_a, the rate of a catalyzed reaction is faster than the rate of the uncatalyzed reaction at the same temperature. Note that because a catalyst decreases the height of the energy barrier, its presence increases the rates of *both* the forward *and* the reverse reactions by the same amount. In this section we will examine the three major classes of catalysts: heterogeneous catalysts, homogeneous catalysts, and enzymes.

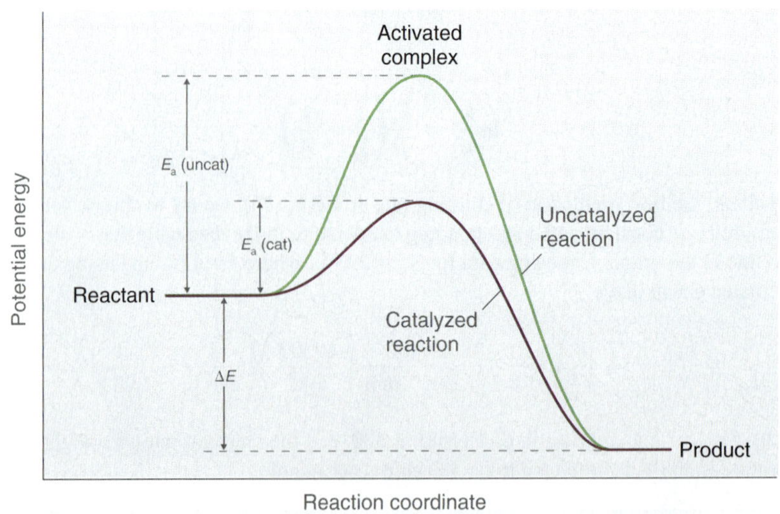

Figure 14.21 Lowering the activation energy of a reaction by a catalyst. This graph compares potential energy diagrams for a single step reaction in the presence and absence of a catalyst. Note that the only effect of the catalyst is to lower the activation energy of the reaction. The catalyst does not affect the energy of the reactants or products (and thus does not affect ΔE). Interactive Graph

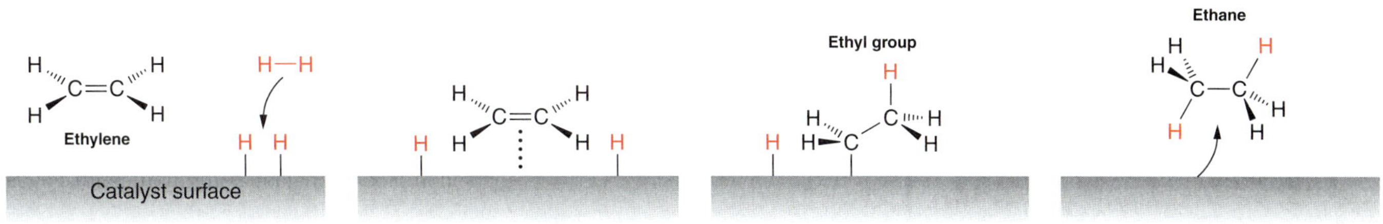

(a) Hydrogen (H_2) adsorbs to the catalyst surface (M) to form adsorbed H atoms.

(b) Ethylene (C_2H_4) adsorbs to the catalyst surface.

(c) Adsorbed ethylene reacts with an adsorbed H atom to give a bound ethyl group (C_2H_5).

(d) The ethyl group reacts with a second adsorbed H atom to give the product ethane (C_2H_6).

Figure 14.22 Hydrogenation of ethylene on a heterogeneous catalyst. When a molecule of hydrogen absorbs to the catalyst surface, the H—H bond breaks and new M—H bonds are formed. The individual H atoms are more reactive than gaseous H_2. When a molecule of ethylene interacts with the catalyst surface, it reacts with the H atoms in a stepwise process to eventually produce ethane, which is released.

Heterogeneous Catalysis

In **heterogeneous catalysis** the catalyst is in a different phase from the reactants. At least one of the reactants interacts with the solid surface in a physical process called **adsorption** in such a way that a chemical bond in the reactant becomes weak and then breaks. *Poisons* are substances that bind irreversibly to catalysts, preventing reactants from adsorbing and thus reducing or destroying the catalyst's efficiency.

An example of heterogeneous catalysis is the interaction of hydrogen gas with the surface of a metal such as Ni, Pd, or Pt. As shown in Figure 14.22a, the hydrogen–hydrogen bonds break and produce individual adsorbed hydrogen atoms on the surface of the metal. Because the adsorbed atoms can move around on the surface, two hydrogen atoms can collide and form a molecule of hydrogen gas that can then leave the surface in the reverse process, called *desorption.* Adsorbed H atoms on a metal surface are substantially more reactive than a hydrogen molecule. Because the relatively strong H—H bond (dissociation energy = 432 kJ/mol) has already been broken, the energy barrier for most reactions of H_2 is substantially lower on the catalyst surface.

Figure 14.22 shows a process called *hydrogenation*, in which hydrogen atoms are added to the double bond of an alkene, such as ethylene, to give a product that contains C—C single bonds, in this case ethane. Hydrogenation is used in the food industry to convert vegetable oils, which consist of long chains of alkenes, to more commercially valuable solid derivatives that contain alkyl chains. Hydrogenation of some of the double bonds in polyunsaturated vegetable oils, for example, produces margarine, a product with a melting point, texture, and other physical properties similar to those of butter.

Several important examples of industrial heterogeneous catalytic reactions are listed in Table 14.9. Although the mechanisms of these reactions are considerably more complex than the simple hydrogenation reaction described above, they all involve adsorption of the reactants onto a solid catalytic surface, chemical reaction of the adsorbed species (sometimes via a number of intermediate species), and finally desorption of the products from the surface.

Homogeneous Catalysis

In **homogeneous catalysis**, the catalyst is in the same phase as the reactant(s). The number of collisions between reactants and catalyst is at a maximum because the catalyst is uniformly dispersed throughout the reaction mixture. Many homogeneous catalysts in industry are transition metal compounds (Table 14.10), but recovering these expensive catalysts from solution has been a major challenge. As an added barrier to their widespread commercial use, many homogeneous catalysts can be used only at

TABLE 14.9 Some commercially important reactions that employ heterogeneous catalysts

Commercial Process	Catalyst	Initial Reaction	Final Product
Contact process	V_2O_5 or Pt	$2SO_2 + O_2 \longrightarrow 2SO_3$	H_2SO_4
Haber process	Fe, K_2O, Al_2O_3	$N_2 + 3H_2 \longrightarrow 2NH_3$	NH_3
Ostwald process	Pt and Rh	$4NH_3 + 5O_2 \longrightarrow 4NO + 6H_2O$	HNO_3
Water–gas shift reaction	Fe, Cr_2O_3, or Cu	$CO + H_2O \longrightarrow CO_2 + H_2$	H_2 for NH_3, CH_3OH, and other fuels
Steam reforming	Ni	$CH_4 + H_2O \longrightarrow CO + 3H_2$	H_2
Methanol synthesis	ZnO and Cr_2O_3	$CO + 2H_2 \longrightarrow CH_3OH$	CH_3OH
SOHIO process	Bismuth phosphomolybdate	$CH_2{=}CHCH_3 + NH_3 + \frac{3}{2}O_2 \longrightarrow CH_2{=}CHCN + 3H_2O$	$CH_2{=}CHCN$ **Acrylonitrile**
Catalytic hydrogenation	Ni, Pd, or Pt	$RCH{=}CHR' + H_2 \longrightarrow RCH_2{-}CH_2R'$	Partially hydrogenated oils for margarine, etc.

relatively low temperatures, and even then they tend to decompose slowly in solution. Despite these problems, a number of commercially viable processes have been developed in recent years. High-density polyethylene and polypropylene are produced by homogeneous catalysis.

Enzymes

Enzymes, catalysts that occur naturally in living organisms, are almost all protein molecules with typical molecular masses of 20,000–100,000 amu. Some are homogeneous catalysts that react in aqueous solution within a cellular compartment of an organism. Others are heterogeneous catalysts embedded within the membranes that separate cells and cellular compartments from their surroundings. The reactant in an enzyme-catalyzed reaction is called a **substrate**.

Because enzymes can increase reaction rates by enormous factors (up to 10^{17} times the uncatalyzed rate!) and tend to be very specific, typically producing only a single product in quantitative yield, they are the focus of active research. At the same time, enzymes are usually expensive to obtain, they often cease functioning at temperatures higher than 37°C, they have limited stability in solution, and they have such high specificity that they are confined to turning one particular set of reactants into one particular

TABLE 14.10 Some commercially important reactions that employ homogeneous catalysts

Commercial Process	Catalyst	Reactants	Final Product
Union Carbide	$[Rh(CO)_2I_2]^-$	$CO + CH_3OH$	CH_3CO_2H
Hydroperoxide process	Mo(VI) complexes	$CH_3CH{=}CH_2 + R{-}C{-}O{-}O{-}H$	$CH_3CH{-}CH_2$ (O bridging) $+ ROH$ **Propylene oxide**
Hydroformylation	Rh/PR_3 complexes	$RCH{=}CH_2 + CO + H_2$	RCH_2CH_2CHO
Adiponitrile process	Ni/PR_3 complexes	$2HCN + CH_2{=}CHCH{=}CH_2$	$NCCH_2CH_2CH_2CH_2CN$ used to synthesize nylon
Olefin polymerization	$(RC_5H_5)_2ZrCl_2$	$CH_2{=}CH_2$	$(CH_2CH_2)_n$ High-density polyethylene

product. This means that separate processes using different enzymes must be developed for chemically similar reactions, which is time-consuming and expensive. Thus far, enzymes have found only limited industrial applications, although they are used as ingredients in laundry detergents, contact lens cleaners, and meat tenderizers. The enzymes in these applications tend to be *proteases*, which are able to cleave the amide bonds that hold amino acids together in proteins. Meat tenderizers, for example, contain a protease called papain, which is isolated from papaya juice. It cleaves some of the long, fibrous protein molecules that make inexpensive cuts of beef tough, producing a tenderer piece of meat. Some insects, like the bombadier beetle, carry an enzyme capable of catalyzing the decomposition of hydrogen peroxide to water (Figure 14.23).

Figure 14.23 A catalytic defense mechanism. The scalding, foul-smelling spray emitted by this bombardier beetle is produced by the catalytic decomposition of H_2O_2.

Enzyme inhibitors cause a decrease in the rate of an enzyme-catalyzed reaction by binding to a specific portion of an enzyme and thus slowing or preventing a reaction from occurring. Irreversible inhibitors are therefore the equivalent of poisons in heterogeneous catalysis. One of the oldest and most widely used commercial enzyme inhibitors is aspirin, which selectively inhibits one of the enzymes involved in the synthesis of molecules that trigger inflammation. The design and synthesis of related molecules that are more effective, more selective, and less toxic than aspirin are important objectives of biomedical research.

SUMMARY AND KEY TERMS

14.0 Introduction (p. 617)

Chemical kinetics is the study of **reaction rates**, changes in the concentrations of reactants and products with time.

14.1 Factors That Affect Reaction Rates (p. 618)

Factors that influence the rates of chemical reactions include concentration of reactants, temperature, physical state of reactants and their dispersion, solvent, and presence of a catalyst.

14.2 Reaction Rates and Rate Laws (p. 620)

Reaction rates are reported either as the **average rate** over a period of time or as the **instantaneous rate** at a single time.

The **rate law** for a reaction is a mathematical relationship between the reaction rate and the concentrations of species in solution. Rate laws can be expressed either as a **differential rate law**, describing the change in reactant or product concentrations as a function of time, or as an **integrated rate law**, describing the actual concentrations of reactants or products as a function of time.

The **rate constant**, k, of a rate law is a constant of proportionality between reaction rate and reactant concentration. The power to which a concentration is raised in a rate law indicates the **reaction order**, the degree to which the rate of the reaction depends on the concentration of a particular reactant.

14.3 Methods of Determining Reaction Orders (p. 629)

The rate of a **zeroth-order reaction** is independent of the concentration of the reactants. The rate of a **first-order reaction** is directly proportional to the concentration of one of the reactants. The rate of a simple **second-order reaction** is proportional to the square of the concentration of one of the reactants.

14.4 Using Graphs to Determine Rate Laws, Rate Constants, and Reaction Orders (p. 641)

For a zeroth-order reaction, a plot of the concentration of any reactant versus time is a straight line with a slope of $-k$. For a first-order reaction, a plot of the natural logarithm of the concentration of a reactant versus time is a straight line with a slope of $-k$. For a second-order reaction, a plot of the inverse of the concentration of a reactant versus time is a straight line with a slope of k.

14.5 Half-Lives and Radioactive Decay Kinetics (p. 644)

The **half-life** of a reaction is the time required for the reactant concentration to decrease to one-half its initial value. The half-life of a first-order reaction is a constant that is related to the rate constant for the reaction: $t_{1/2} = 0.693/k$.

Radioactive decay reactions are first-order reactions. The **rate of decay**, or **activity**, of a sample of a radioactive substance is the decrease in the number of radioactive nuclei per unit time.

14.6 Reaction Rates—A Microscopic View (p. 649)

A **reaction mechanism** is the microscopic path by which reactants are transformed into products. Each step is an **elementary reaction**. Species that are formed in one step and consumed in another are **intermediates**. Each elementary step can be described in terms of its **molecularity**, the number of molecules that collide in that step.

The slowest step in a reaction mechanism is the **rate-determining step**. **Chain reactions** consist of three kinds of reactions: initiation, propagation, and termination. Intermediates in chain reactions are often **radicals**, species that have an unpaired valence electron.

14.7 The Collision Model of Chemical Kinetics (p. 654)

A minimum energy (**activation energy, E_a**) is required for a collision between molecules to result in a chemical reaction. Plots of potential energy for a system versus the reaction coordinate show an energy barrier that must be overcome for the reaction to occur. The arrangement of atoms at the highest point of this barrier is the **activated complex**, or **transition state**, of the reaction. At a given temperature, the higher the E_a, the slower the reaction. The fraction of orientations that result in a reaction is the **steric factor**. The **frequency factor**, steric factor, and activation energy are related to the rate constant in the **Arrhenius equation**: $k = Ae^{-E_a/RT}$. A plot of the natural logarithm of k versus $1/T$ is a straight line with a slope of $-E_a/R$.

14.8 Catalysis (p. 660)

Catalysts participate in a chemical reaction and increase its rate. They do not appear in the reaction's net equation, and they are not consumed during the reaction. Catalysts allow a reaction to proceed via a pathway that has a lower activation energy than the uncatalyzed reaction. **Heterogeneous catalysts** provide a surface to which reactants bind in a process of **adsorption**. **Homogeneous catalysts** are in the same phase as the reactants. **Enzymes** are biological catalysts that produce large increases in reaction rates and tend to be specific for certain reactants and products. The reactant in an enzyme-catalyzed reaction is called a **substrate**. **Enzyme inhibitors** cause a decrease in the rate of an enzyme-catalyzed reaction.

KEY EQUATIONS

General definition of rate for A ⟶ B	$\text{rate} = \dfrac{\Delta[B]}{\Delta t} = -\dfrac{\Delta[A]}{\Delta t}$	(14.4)
General form of rate law when A and B are reactants	$\text{rate} = k[A]^m[B]^n$	(14.10)
Zeroth-order reaction	$\text{rate} = -\dfrac{\Delta[A]}{\Delta t} = k$	(14.16)
	$[A] = [A]_0 - kt$	(14.17)
First-order reaction	$\text{rate} = -\dfrac{\Delta[A]}{\Delta t} = k[A]$	(14.21)
	$[A] = [A]_0e^{-kt}$	(14.22)
	$\ln[A] = \ln[A]_0 - kt$	(14.23)
Second-order reaction	$\text{rate} = -\dfrac{\Delta[A]}{\Delta t} = k[A]^2$	(14.25)
	$\dfrac{1}{[A]} = \dfrac{1}{[A]_0} + kt$	(14.26)
Half-life of first-order reaction	$t_{1/2} = \dfrac{0.693}{k}$	(14.32)
Radioactive decay	$A = kN$	(14.34)
Arrhenius equation	$k = Ae^{-E_a/RT}$	(14.43)

QUESTIONS AND PROBLEMS

MGC For instructor-assigned homework, go to **www.masteringgeneralchemistry.com**

Questions and Problems with colored numbers have answers in the Appendix and complete solutions in the Student Solutions Manual.

CONCEPTUAL

14.1 Factors That Affect Reaction Rates

1. What information can you obtain by studying the chemical kinetics of a reaction? Does a balanced chemical equation provide the same information? Why or why not?
2. If you were given the task of determining whether to proceed with a particular reaction in an industrial facility, why would studying the chemical kinetics of the reaction be important to you?
3. What is the relationship between each of these factors and the rate of a reaction: reactant concentration, temperature of the reaction, physical properties of the reactants, physical and chemical properties of the solvent, and the presence of a catalyst?
4. A slurry is a mixture of a substance that is only sparingly soluble in a liquid. As you prepare a reaction, you notice that one of your reactants forms a slurry with the solvent. What effect will this have on the reaction rate? What steps can you take to try to solve the problem?
5. Why does the rate of virtually all reactions increase with an increase in temperature? If you were to make a glass of sweetened iced tea the old-fashioned way, by adding sugar and ice cubes to a glass of hot tea, which would you add first?
6. In a typical laboratory setting, a reaction is carried out in a ventilated hood with air circulation provided by outside air. A student noticed that a reaction that gave a high yield of a

product during the winter months gave a low yield of that same product in the summer, even though his technique did not change and the reagents and concentrations used were identical. What is a plausible explanation for the different yields?

7. A very active area of chemical research involves the development of solubilized catalysts that are not made inactive during the reaction process. Such catalysts are expected to increase reaction rates significantly relative to the same reaction run in the presence of a heterogeneous catalyst. What is the reason for anticipating that the relative rate will increase?
8. Water has a dielectric constant more than two times greater than that of methanol (dielectric constant = 80.1 for H_2O and 33.0 for CH_3OH). Which would be your solvent of choice for a substitution reaction between an ionic compound and a polar reagent, both of which are soluble in either methanol or water? Why?

14.2 Reaction Rates and Rate Laws

9. Explain why the rate of a reaction is generally fastest at early time intervals. For the second-order $A + B \rightarrow C$, what would the plot of the concentration of C versus time look like during the course of the reaction?
10. Explain the differences between a differential rate law and an integrated rate law. What two components do they have in common? Which form is preferred for obtaining a reaction order and a rate constant? Why?
11. Diffusion-controlled reactions have rates that are determined only by the rate at which two reactant molecules can diffuse together. These reactions are rapid, with second order rate constants typically on the order of 10^{10} L/mol · s. Would you expect the reactions to be faster or slower in solvents that have a low viscosity? Why? Consider the reactions $H_3O^+ + OH^- \rightarrow 2H_2O$ and $H_3O^+ + N(CH_3)_3 \rightarrow H_2O + HN(CH_3)_3{}^+$ in aqueous solution. Which would have the higher rate constant? Why?
12. What information can you get from the reaction order? What correlation does the reaction order have with the stoichiometry of the overall equation?
13. During the hydrolysis $A + H_2O \rightarrow B + C$, the concentration of A decreases much more rapidly in a polar solvent than in one that is not polar. How do you expect this effect to be reflected in the overall reaction order?

14.3 Methods of Determining Reaction Orders

14. What are the characteristics of a zeroth-order reaction? Experimentally, how would you determine whether a reaction is zeroth order?
15. Predict whether the following reactions are zeroth order: (a) a substitution reaction of an alcohol with HCl to form an alkyl halide and water; (b) catalytic hydrogenation of an alkene; (c) hydrolysis of an alkyl halide to an alcohol; and (d) enzymatic conversion of nitrate to nitrite in a soil bacterium. Explain your reasoning.
16. In a first-order reaction, what is the advantage of using the integrated rate law expressed in natural logarithms over the rate law expressed in exponential form?
17. If the rate of a reaction is directly proportional to the concentration of a reactant, what does this tell you about (a) the order of the reaction with respect to the reactant, and (b) the overall order of the reaction?
18. The reaction of NO with O_2 is found to be second order with respect to NO and first order with respect to O_2. What is the overall reaction order? What is the effect of doubling the concentration of each reagent on the rate?

14.4 Using Graphs to Determine Rate Laws, Rate Constants, and Reaction Orders

19. Compare first-order differential and integrated rate laws with respect to (a) the magnitude of the rate constant, (b) the information needed to determine the order, and (c) the shape of the graphs. Is there any information that can be obtained from the integrated rate law that cannot be obtained from the differential rate law?
20. In the single step second order reaction $2A \rightarrow$ products, how would a graph of concentration of A vs. time compare to a plot of 1/A vs. time? Which of these would be the most similar to the same set of graphs for A during the single step second order reaction $A + B \rightarrow$ products? Explain.
21. For reactions of the same order, what is the relationship between the magnitude of the rate constant and reaction rate? If you were comparing reactions with different orders, could the same arguments be made? Why?

14.5 Half-Lives and Radioactive Decay Kinetics

22. What do chemists mean when they refer to the *half-life* of a reaction?
23. If a sample of one isotope undergoes more disintegrations per second than the same number of atoms of another isotope, how do their half-lives compare?

14.6 Reaction Rates—A Microscopic View

24. How does the term *molecularity* relate to the elementary reactions? How does it relate to the overall balanced chemical equation?
25. What is the relationship between the order of a rate law and the molecularity of a reaction? What is the relationship between the rate law and the balanced chemical equation?
26. When you determine the rate law for a given reaction, why is it valid to assume that the concentration of an intermediate does not change with time during the course of the reaction?
27. If you know the rate law for an overall reaction, how would you determine which elementary step is rate determining? If an intermediate is contained in the rate-determining step, how can the experimentally determined rate law for the reaction be derived from this step?
28. Give the rate-determining step for each case: (a) traffic is backed up on a highway because two lanes merge into one; (b) gas flows from a pressurized cylinder fitted with a gas regulator and then is bubbled through a solution; and (c) a document containing text and graphics is downloaded from the Internet.
29. Before being sent on an assignment, an aging James Bond was sent off to a health farm where part of the program's focus was to purge his body of radicals. Why was this goal considered important to his health?

14.7 The Collision Model of Chemical Kinetics

30. Although an increase in temperature results in an increase in kinetic energy, this increase in kinetic energy is not sufficient to explain the relationship between temperature and reaction rates. How does the activation energy relate to the chemical kinetics of a reaction? Why does an increase in temperature increase the reaction rate despite the fact that the average kinetic energy is still lower than the activation energy?
31. For any given reaction, what is the relationship between the energy of activation and (a) electrostatic repulsions, (b) bond formation, and (c) the nature of the activated complex?

32. If you are concerned with whether a reaction will occur rapidly, why would you be more interested in knowing the magnitude of the activation energy than the change in potential energy for the reaction?
33. The product C in the reaction A + B → C + D can be separated easily from the reaction mixture. You have been given pure A and pure B and are told to determine the activation energy for this reaction to determine whether the reaction is suitable for the industrial synthesis of C. How would you do this? Why do you need to know the magnitude of the activation energy to make a decision about feasibility?
34. Above E_a, molecules collide with enough energy to overcome the energy barrier for a reaction. Is it possible for a reaction to occur at a temperature lower than that needed to reach E_a? Explain your answer.
35. What is the relationship between *A*, activation energy, and temperature? How does an increase in the frequency factor affect the rate of reaction?
36. Of two highly exothermic reactions with different values of E_a, which would need to be monitored more carefully: the one with the smaller value or the one with the higher value? Why?

14.8 Catalysis

37. What effect does a catalyst have on the energy of activation of a reaction? What effect does it have on the frequency factor, *A*? What effect does it have on the change in potential energy for the reaction?
38. How is it possible to affect the product distribution of a reaction by using a catalyst?
39. A heterogeneous catalyst works by interacting with a reactant in a process called *adsorption*. What occurs during the process of adsorption? Explain how this can lower the activation energy.
40. What effect does increasing the surface area of a heterogeneous catalyst have on a reaction? Does increasing the surface area affect the energy of activation? Explain your answer.
41. What are the differences between a heterogeneous catalyst and a homogeneous catalyst in terms of (a) ease of recovery, (b) collision frequency, (c) temperature sensitivity, and (d) cost?
42. An area of intensive chemical research involves the development of homogeneous catalysts, even though homogeneous catalysts generally have a number of operational difficulties. Propose one or two reasons why a homogenous catalyst may be preferred.
43. Consider the reaction between cerium(IV) and thallium(I) ions

$$2Ce^{4+} + Tl^{+} \longrightarrow 2Ce^{3+} + Tl^{3+}$$

This reaction is slow, but it is catalyzed by Mn^{2+}, as shown in the following mechanism:

$$Ce^{4+} + Mn^{2+} \longrightarrow Ce^{3+} + Mn^{3+}$$
$$Ce^{4+} + Mn^{3+} \longrightarrow Ce^{3+} + Mn^{4+}$$
$$Mn^{4+} + Tl^{+} \longrightarrow Tl^{3+} + Mn^{2+}$$

In what way does Mn^{2+} increase the reaction rate?
44. The text lists several factors that limit the industrial applications of enzymes. Still, there is keen interest in understanding how enzymes work in order to design catalysts for industrial applications. Why?
45. Most enzymes have an optimal pH range; however, care must be taken when determining pH effects on enzyme activity. A decrease in activity could be due to the effects of changes in pH on groups at the catalytic center or to the effects on groups located elsewhere in the enzyme. Both examples are observed in chymotrypsin, a digestive enzyme that is a protease which hydrolyzes polypeptide chains. Explain how a change in pH could affect the catalytic activity due to (a) effects at the catalytic center and (b) effects elsewhere in the enzyme. (*Hint:* Remember that enzymes are composed of functional amino acids.)

NUMERICAL

This section includes "paired problems" (marked by brackets) that require similar problem-solving skills.

14.2 Reaction Rates and Rate Laws

46. The rate of a particular reaction in which A and B react to make C is rate $= -\Delta[A]/\Delta t = \frac{1}{2}\,\Delta[C]/\Delta t$. Write a reaction equation that is consistent with this rate law. What is the rate expression with respect to time if 2A is converted to 3C?
47. While commuting to work, a person drove for 12 min at 35 mi/h, then stopped at an intersection for 2 min, continued the commute at 50 mi/h for 28 min, drove slowly through traffic at 38 mi/h for 18 min, and then spent 1 min pulling into a parking space at 3 mi/h. What was the average rate of the commute? What was the instantaneous rate at 13 min? At 28 min?
48. Why do most studies of chemical reactions use the initial rates of reaction to generate a rate law? How is this initial rate determined? Given the following instantaneous rate data, what is the reaction order? Estimate.

Time, s	[A], (*M*)
120	0.158
240	0.089
360	0.062

49. Predict how the rate of each reaction will be affected by doubling the concentration of the first species in each equation.
(a) $C_2H_5I \longrightarrow C_2H_4 + HI$ Rate $= k[C_2H_5I]$
(b) $SO + O_2 \longrightarrow SO_2 + O$ Rate $= k[SO][O_2]$
(c) $2CH_3 \longrightarrow C_2H_6$ Rate $= k[CH_3]^2$
(d) $ClOO \longrightarrow Cl + O_2$ Rate $= k$
50. Three chemical processes occur at an altitude of approximately 100 km in Earth's atmosphere.

$$N_2^{+} + O_2 \xrightarrow{k_1} N_2 + O_2^{+}$$
$$O_2^{+} + O \xrightarrow{k_2} O_2 + O^{+}$$
$$O^{+} + N_2 \xrightarrow{k_3} NO^{+} + N$$

Write a rate law for each of the elementary steps. If the rate law for the overall reaction were found to be rate $= k[N_2^{+}][O_2]$, which one of the steps is rate limiting?
51. Cleavage of C_2H_6 to produce two $CH_3\cdot$ radicals is a gas-phase reaction that occurs at 700°C. This reaction is first order, with $k = 5.46 \times 10^{-4}\ s^{-1}$. How long will it take for the reaction to go to 15% completion? To 50% completion?
52. The oxidation of aqueous iodide by arsenic acid to give I_3^- and arsenious acid proceeds via the reaction

$$H_3AsO_4(aq) + 3I^-(aq) + 2H^+(aq) \underset{k_r}{\overset{k_f}{\rightleftharpoons}} H_3AsO_3(aq) + I_3^-(aq) + H_2O(l)$$

When the rate of the forward reaction is equal to the rate of the reverse reaction, what is $k_f/k_r = [H_3AsO_3][I_3^-]/[H_3AsO_4][I^-]^3[H^+]^2$? Then write an expression for $\Delta[I_3^-]/\Delta t$.

14.3 Methods of Determining Reaction Orders

53. Iodide reduces Fe(III) according to the reaction

$$2Fe^{3+}(soln) + 2I^-(soln) \longrightarrow 2Fe^{2+}(soln) + I_2(soln)$$

Experimentally, it was found that doubling the concentration of Fe(III) doubled the rate, and doubling the iodide concentration increased the reaction rate by a factor of 4. What is the order of the reaction with respect to each species? What is the overall rate law? What is the overall reaction order?

54. Benzoyl peroxide is a medication used to treat acne. Its rate of thermal decomposition at several concentrations was determined experimentally, and the data were tabulated as follows:

Experiment	[Benzoyl Peroxide]$_0$, M	Initial Rate, M/s
1	1.00	2.22×10^{-4}
2	0.70	1.64×10^{-4}
3	0.50	1.12×10^{-4}
4	0.25	0.59×10^{-4}

What is the order of this reaction with respect to benzoyl peroxide? What is the rate law for this reaction?

55. 1-Bromopropane is a colorless liquid that reacts with $S_2O_3^{2-}$ according to the equation

$$C_3H_7Br + S_2O_3^{2-} \longrightarrow C_3H_7S_2O_3^- + Br^-$$

The reaction is first order in 1-bromopropane and first order in $S_2O_3^{2-}$, with a rate constant of $8.05 \times 10^{-4}\ M^{-1} \cdot s^{-1}$. If you began a reaction with 40 mmol/100 mL of C_3H_7Br and an equivalent concentration of $S_2O_3^{2-}$, what would the initial rate of reaction be? If you were to decrease the concentration of each reactant to 20 mmol/100 mL, what would the initial rate of reaction be?

56. The experimental rate law for the reaction $3A + 2B \rightarrow C + D$ was found to be $\Delta[C]/\Delta t = k[A]^2[B]$ for an overall reaction that is third order. Because graphical analysis is difficult beyond second-order reactions, explain the procedure for determining the rate law experimentally.

14.4 Using Graphs to Determine Rate Laws, Rate Constants, and Reaction Orders

57. One method of using graphs to determine reaction order is to use relative rate information. Plotting the log of the relative rate vs. log of relative concentration provides information about the reaction. Here is an example of data from a zeroth-order reaction:

Relative [A], M	Relative Rate, M/s
1	1
2	1
3	1

Varying [A] does not alter the rate of the reaction. Using relative rates as shown in the table, generate plots of log (rate) vs. log (concentration) for zeroth-, first- and second order reactions. What does the slope of each line represent?

58. The table below follows the decomposition of N_2O_5 gas by examining the partial pressure of the gas as a function of time at 45°C. What is the order of the reaction? What is the rate constant? How long would it take for the pressure to reach 105 mm Hg at 45°C?

Time, s	Pressure, mm Hg
0	348
400	276
1600	156
3200	69
4800	33

14.5 Half-Lives and Radioactive Decay Kinetics

59. Half-lives for the reaction $A + B \rightarrow C$ were calculated at three values of $[A]_0$, and the concentration of B was the same in all cases. The data are listed in the table.

$[A]_0$, M	$t_{1/2}$, s
0.50	420
0.75	280
1.0	210

Does this reaction follow first-order kinetics? On what do you base your answer?

60. Ethyl-2-nitrobenzoate ($NO_2C_6H_4CO_2C_2H_5$) hydrolyzes under basic conditions. A plot of [ethyl-2-nitrobenzoate] versus t was used to calculate $t_{1/2}$, with the following results:

[Ethyl-2-nitrobenzoate], mol/cm^3	$t_{1/2}$, s
0.050	240
0.040	300
0.030	400

Is this a first-order reaction? Explain your reasoning.

61. Azomethane($CH_3N_2CH_3$) decomposes at 600 K to C_2H_6 and N_2. The decomposition is first order in azomethane. Calculate $t_{1/2}$ from the data in the table.

Time, s	$P_{CH_3N_2CH_3}$, atm
0	8.2×10^{-2}
2000	3.99×10^{-2}
4000	1.94×10^{-2}

How long will it take for the decomposition to be 99.9% complete?

62. The first-order decomposition of hydrogen peroxide has a half-life of 10.7 hr at 20°C. What is the rate constant (expressed in s^{-1}) for this reaction? If you started with a solution that was $7.5 \times 10^{-3}\ M$ H_2O_2, what would be the initial rate of decomposition (M/s)? What would be the concentration of H_2O_2 after 3.3 hr?

14.6 Reaction Rates—A Microscopic View

63. Cyclopropane, a mild anesthetic, rearranges to propylene via a collision that produces and destroys an energized species. Here are the important steps in this rearrangement.

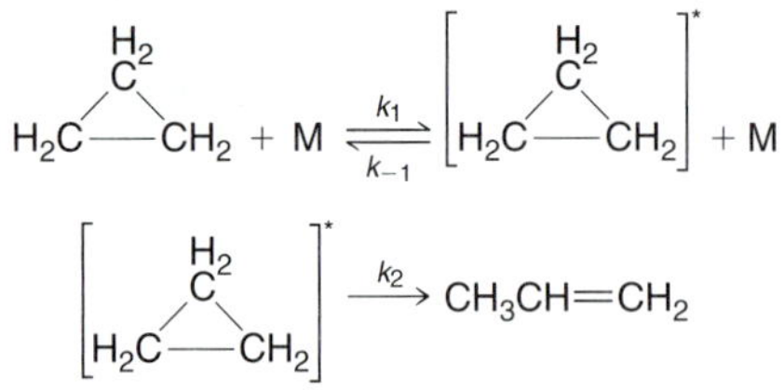

where M is any molecule, including cyclopropane. Only those cyclopropane molecules with sufficient energy (denoted with an asterisk) can rearrange to propylene. Which step determines the rate constant of the overall reaction?

64. Above approximately 500 K, the reaction between NO_2 and CO to produce CO_2 and NO follows the second-order rate law $\Delta[CO_2]/\Delta t = k[NO_2][CO]$. At lower temperatures, however, the rate law is $\Delta[CO_2]/\Delta t = k'[NO_2]^2$, for which it is known that NO_3 is an intermediate in the mechanism. Propose a complete low-temperature mechanism for the reaction based on this rate law. Which step is the slowest?

65. Nitramide (O_2NNH_2) decomposes in aqueous solution to N_2O and H_2O. What is the experimental rate law, $\Delta[N_2O]/\Delta t$, for the decomposition of nitramide if the mechanism for the decomposition is

$$O_2NNH_2 \underset{k_{-1}}{\overset{k_1}{\rightleftharpoons}} O_2NNH^- + H^+ \quad \text{(fast)}$$

$$O_2NNH^- \xrightarrow{k_2} N_2O + OH^- \quad \text{(slow)}$$

$$H^+ + OH^- \xrightarrow{k_3} H_2O \quad \text{(fast)}$$

66. Given the reactions

$$A + B \underset{k_{-1}}{\overset{k_1}{\rightleftharpoons}} C + D$$

$$D + E \xrightarrow{k_2} F$$

what is the relationship between the relative magnitudes of k_{-1} and k_2 if these reactions have the rate law $\Delta[F]/\Delta t = k[A][B][E]/[C]$? How does the magnitude of k_1 compare to that of k_2? Under what conditions would you expect the rate law to be $\Delta[F]/\Delta t = k'[A][B]$?

14.7 The Collision Model of Chemical Kinetics

67. What happens to the approximate rate of a reaction when the temperature of the reaction is increased from 20°C to 30°C? What happens to the reaction rate when the temperature is raised to 70°C? For a given reaction at room temperature (20°C), what is the shape of a plot of reaction rate versus temperature as the temperature is increased to 70°C?

68. Acetaldehyde, used in silvering mirrors and in some perfumes, undergoes a second-order decomposition between 700 and 840 K. From the data in the table, would you say that acetaldehyde follows the general rule that each 10 K increase in temperature doubles the reaction rate?

T, K	k_2, *M*/s
720	0.024
740	0.051
760	0.105
800	0.519

69. Bromoethane reacts with hydroxide ion in water to produce ethanol. The energy of activation for this reaction is 90 kJ/mol. If the rate of reaction is 3.6×10^{-5} *M*/s at 25°C, what would the rate of reaction be at (a) 15°C, (b) 30°C, (c) 45°C.

70. An enzyme-catalyzed reaction has an activation energy of 15 kcal/mol. How would the value of the rate constant differ between 20°C and 30°C? If the enzyme reduced the value of E_a from 25 kcal/mol to 15 kcal/mol, by what factor has the enzyme increased the reaction rate at each temperature?

71. The data in the table are the rate constants as a function of temperature for the dimerization of 1,3-butadiene. What is the energy of activation for this reaction?

T, K	*k*, $M^{-1} \cdot min^{-1}$
529	1.4
560	3.7
600	25
645	82

72. The rate of a reaction at 25°C is 1.0×10^{-4} *M*/s. Increasing the temperature to 75°C causes the rate to increase to 7.0×10^{-2} *M*/s. Estimate the activation energy for this process. If E_a were 25 kJ/mol and the rate at 25°C is 1.0×10^{-4} *M*/s, what would be the rate at 75°C?

14.8 Catalysis

73. At some point during an enzymatic reaction, the concentration of the activated complex, called an enzyme–substrate complex (ES), and other intermediates involved in the reaction is nearly constant. When a single substrate is involved, the reaction can be represented by the following sequence of equations:

enzyme (E) + substrate (S) $\rightleftharpoons$
enzyme–substrate complex (ES) $\rightleftharpoons$
enzyme + product (P)

This can also be shown as

$$E + S \underset{k_{-1}}{\overset{k_1}{\rightleftharpoons}} ES \underset{k_{-2}}{\overset{k_2}{\rightleftharpoons}} E + P$$

Using molar concentrations and rate constants, write an expression for the rate of disappearance of the enzyme–substrate complex. Typically, enzyme concentrations are small and substrate concentrations are high. If you were determining the rate law by varying the substrate concentrations under these conditions, what would be your apparent reaction order?

74. A particular reaction was found to proceed via the mechanism

$$A + B \longrightarrow C + D$$

$$2C \longrightarrow E$$

$$E + A \longrightarrow 3B + F$$

What is the overall reaction? Is this reaction catalytic, and if so, what species is the catalyst? Identify the intermediates.

75. A particular reaction has two accessible pathways (A and B), each of which favors conversion of X to a different product (Y and Z, respectively). Under uncatalyzed conditions pathway A is favored, but in the presence of a catalyst pathway B is favored. Pathway B is reversible, whereas pathway A is not. Which product is favored in the presence of a catalyst? Without a catalyst? Draw a diagram illustrating what is occurring with and without the catalyst.

76. The kinetics of an enzyme-catalyzed reaction can be analyzed by plotting the rate of reaction versus the substrate concentration. This type of analysis is referred to as a Michaelis–Menten treatment. At low substrate concentrations the plot shows behavior characteristic of first-order kinetics, but at very high substrate concentrations the behavior shows zeroth-order kinetics. Explain this phenomenon.

APPLICATIONS

77. Atmospheric chemistry in the region below the clouds of Venus appears to be dominated by reactions of sulfur and carbon-containing compounds. Included in representative elementary reaction steps are the following:

$$SO_2 + CO \longrightarrow SO + CO_2$$

$$SO + CO \longrightarrow S + CO_2$$

$$SO + SO_2 \longrightarrow S + SO_3$$

For each elementary reaction, write an expression for the reaction rate in terms of changes in the concentrations of each reactant and product with time.

78. In acid, nitriles hydrolyze to produce a carboxylic acid and ammonium ion. For example, acetonitrile, a substance used to extract fatty acids from fish liver oils, is hydrolyzed to acetic acid via the reaction

$$CH_3C{\equiv}N(l) + 2H_2O(l) + H^+(aq) \longrightarrow CH_3\overset{\overset{\displaystyle O}{\|}}{C}OH(aq) + NH_4{}^+(aq)$$

Express the rate of the reaction in terms of changes in the concentrations of each reactant and each product with time.

79. Ozone production occurs at lower altitudes according to the elementary reaction $O + O_2 \longrightarrow O_3$, with an estimated rate of ozone production of 4.86×10^{31} molecules $\cdot$ s^{-1} worldwide. What is the overall order of this reaction? If the rate of loss of O is 0.89×10^{31} molecules $\cdot$ s^{-1}, and 0.06×10^{31} molecules $\cdot$ s^{-1} of ozone is transported to other atmospheric regions, is ozone being produced faster than it is being destroyed? Measurements show that ozone concentrations are not increasing rapidly. What conclusion can you draw from these data?

80. The water in a fishery became polluted when toxic waste was dumped into its pond, causing the fish population to substantially decline. The percents of fish that survived are recorded in the table.

Day	1	2	3	4	5
% Survival	79	55	38	31	19

What is the order of the reaction live fish $\longrightarrow$ dead fish? What is the rate constant? If the fish continue to die at this rate, how many fish will be alive after 10 days?

81. Until 200 years ago, manufactured iron contained charcoal produced from freshly cut wood that was added during the smelting process. As a result of this practice, older samples of iron can be dated accurately using the carbon-14 method. An archaeologist found a cast iron specimen that she believed dated to the period between 480 and 221 B.C. in Hunan, China. Radiocarbon dating of the sample indicated a 24% reduction in carbon-14 content. Was the archaeologist correct?

82. Because of its short half-life, ^{32}P-labeled compounds must be shipped as quickly as possible so that they can be used as radioactive tags in biological studies. A 50-g sample that contained 0.60 % ^{32}P by mass was shipped at 11 A.M. on Monday morning. The package was delivered to a chemist via an overnight delivery service such that it arrived the next day.
   (a) What would be the mass of ^{32}P remaining in the sample if he received the package on Tuesday afternoon but was unable to use it until 9 A.M. on Wednesday?
   (b) What would be the mass of ^{32}P present in the sample if the shipper had not delivered the sample until Friday afternoon and then it sat on a loading dock until 9 A.M. Monday morning?
   (c) The late shipment was used immediately on Monday morning, but the biological samples were not analyzed until Thursday at 5 P.M. What percentage of the sample still consists of ^{32}P?

83. Tritium (^{3}H) is a radioactive isotope that is commonly used to follow biochemical reactions. (a) Using the data in Table 14.7, calculate the radioactive decay constant, k, for tritium. (b) Use the value of k to determine mass of tritium that is still present in a 5.00 g sample of NaB^3H_4 that is 17.57 years old.

84. L-Aspartic acid is an amino acid found in fossil bone. It can convert to a geometrically different form (D-aspartic acid) at 20°C, with a half-life corresponding to the conversion of L $\rightarrow$ D of 14,000–20,000 years. If the temperature of an archaeological site is constant, then the extent of the conversion can be used to date fossils. In one such case, archaeologists dated the arrival of humans on the North American continent to be 20,000 years ago, but the conversion of L-aspartic acid to D-aspartic acid in human fossils indicated that Paleo-Indians were living in California at least 48,000 years ago. What would be the relative concentrations of the L- and D-forms that produced this result? Carbon-14 has a half-life of approximately 5700 years. What percent of the carbon-14 originally present would have been found in the bones?

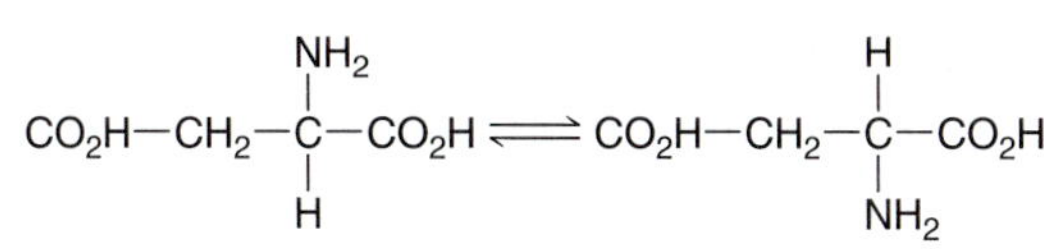

L-Aspartic acid D-Aspartic acid

The technique described is frequently used in conjunction with radiocarbon dating. In cases where the results from the two techniques are in gross disagreement, what information can you get by comparing the two results?

85. Peroxides are able to initiate the radical polymerization of alkenes. Polyethylene, for example, is a high-molecular-weight polymer used as a film in packaging, as kitchenware, and as tubing. It is produced by heating ethylene at high pressure in the presence of oxygen or peroxide. It is formed by this radical process:

$$RO{:}OR \xrightarrow{\Delta} 2RO\cdot$$

$$RO\cdot + CH_2{=}CH_2 \longrightarrow RO{-}CH_2{-}CH_2\cdot$$

$$RO{-}CH_2{-}CH_2\cdot + CH_2{=}CH_2 \longrightarrow RO{-}CH_2{-}CH_2{-}CH_2{-}CH_2\cdot$$

(a) Label the steps that correspond to initiation and propagation.
(b) Show all available chain-terminating steps.
(c) The polymerization of styrene ($C_6H_5CH{=}CH_2$) occurs by a similar process to produce polystyrene, also used as a packaging material. Draw the structure of the polymer that results from five propagation cycles.

86. Lucite and Plexiglas are transparent polymers used as a glass substitute when a plastic material is preferred for safety. The compound used to synthesize Lucite and Plexiglas is methyl methacrylate, shown here. During the polymerization reaction, light produces a radical initiator from hydrogen peroxide ($H_2O_2 \rightarrow HO\cdot$). Show the mechanism for the polymerization, being sure to include the initiation and propagation steps.

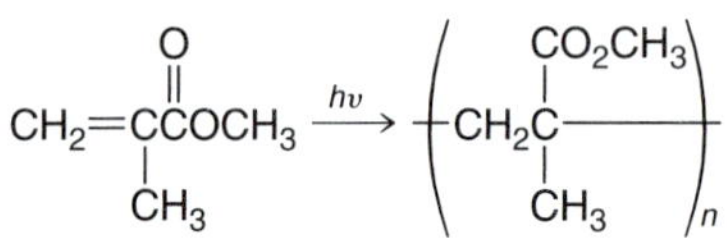

Methyl methacrylate **Plexiglas**

87. At higher altitudes ozone is converted to O_2 by the reaction $O + O_3 \rightarrow 2O_2$, with a rate constant at 220 K of 6.8×10^{-16} $cm^3 \cdot molecule^{-1} \cdot s^{-1}$.
(a) What is the overall reaction order?
(b) What is E_a for this reaction if $A = 8 \times 10^{-12}$ $cm^3 \cdot molecule^{-1} \cdot s^{-1}$?
If Cl is present, the rate constant at 220 K becomes 3.7×10^{-11} $cm^3 \cdot molecule^{-1} \cdot s^{-1}$, with $A = 4.7 \times 10^{-11}$ $cm^3 \cdot molecule^{-1} \cdot s^{-1}$.
(c) Calculate E_a for the depletion of ozone in the presence of Cl.
(d) Show an energy-level diagram for these two processes, clearly labeling reactants, products, and energies of activation.
(e) If you were an environmental scientist using these data to explain the effects of Cl on ozone concentration, what would be your conclusions?

88. Nitric acid is produced commercially by the catalytic oxidation of ammonia by air over platinum gauze at approximately 900°C. The following reactions occur:

$$NH_3(g) + \tfrac{5}{4}O_2(g) \longrightarrow NO(g) + \tfrac{3}{2}H_2O(g) \qquad \Delta H^\circ = -226.3 \text{ kJ/mol}$$

$$NO(g) + \tfrac{1}{2}O_2(g) \longrightarrow NO_2(g) \qquad \Delta H^\circ = -57.1 \text{ kJ/mol}$$

$$3NO_2(g) + H_2O(l) \longrightarrow 2HNO_3(l) + NO(g) \qquad \Delta H^\circ = -71.7 \text{ kJ/mol}$$

Why is platinum gauze rather than platinum wire used for the initial reaction? The reaction $4NH_3(g) + 3O_2(g) \longrightarrow 2N_2(g) + 6H_2O(g)$ has $\Delta H^\circ = -316.6$ kJ/mol. What would occur if the catalyst were not present? If the gas leaving the catalyst is not free of NH_3, the following reaction takes place: $6NO(g) + 4NH_3(g) \longrightarrow 5N_2(g) + 6H_2O(g)$. If this occurs, what will be the overall reaction?

89. Figure 14.22 illustrates the mechanism for the reduction of ethylene on a platinum surface to produce ethane. Industrially important silanes are synthesized using a related mechanism and are used to increase adhesion between layers of glass fiber and between layers of silicone rubber. Predict the products of these reactions:

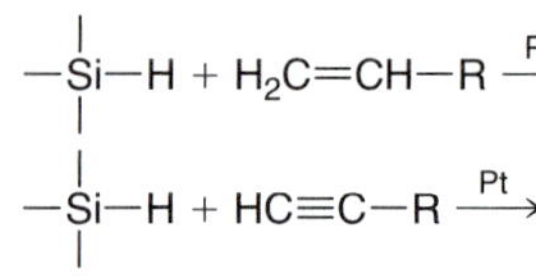

90. In catalysis, if a molecule forms strong bonds to the catalyst, then the catalyst may become poisoned. Experiments on various catalysts showed these results.
i. Fe, Ru, and Os form weak bonds with N_2; however, O_2, alkynes, alkenes, CO, H_2, and CO_2 interact more strongly.
ii. CO_2 and H_2 form weak bonds with a Co or Ni surface.
iii. Rh, Pd, Ir, and Pt form weak bonds with H_2 but do not bond at all with CO_2.
iv. Cu, Ag, and Au form weak bonds with CO and ethylene.
(a) Explain why Fe was chosen as a catalyst to convert nitrogen and hydrogen to ammonia. Why is Fe more suitable than Ru or Os?
(b) Because alkenes generally interact more strongly with metal surfaces than does H_2, what catalyst would you choose for hydrogenation of an alkene such as ethylene?
(c) Although platinum is used in catalytic converters for automobile exhaust, it was not found to be a particularly effective catalyst for the reaction of H_2 with a mixture of carbon monoxide and carbon dioxide to produce methane. Why?
(d) If you were interested in developing a catalyst to reversibly bind ethylene, which of the catalysts listed here would you choose?

91. Nonstoichiometric metal oxides can be effective catalysts for oxidation–reduction reactions. One such catalyst is $Ni_{1-x}O$, found to be effective for converting CO to CO_2 when oxygen is present. Why is it so effective?

92. The chemical reactions in an organism can be controlled by regulating the activity of certain enzymes. Efficient regulation results in an enzyme being active only when it is needed. For example, if a cell needed histidine, the nine enzymes needed to synthesize histidine would all be active. If the cell had adequate histidine, however, those enzymes would be inactive. The following diagram illustrates a situation in which three amino acids (D, F, H) are all synthesized from a common species, A. The numbers above the arrows refer to the enzymes that catalyze each step. Which enzymes would need to be regulated to produce (a) D, (b) F, (c) H?

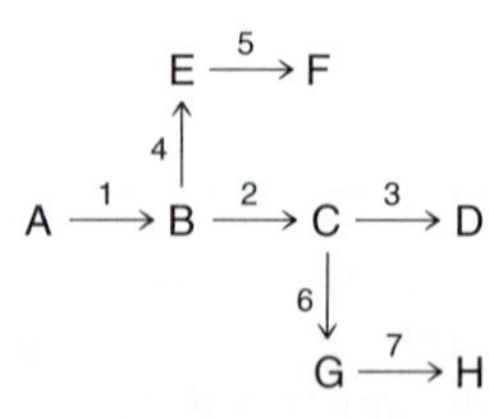

93. Because phosphorus-32 is incorporated into DNA, it can be used to detect DNA fragments. Consequently, it is used extensively in biological research, including the Human Genome Project, whose goal is to determine the complete sequence of human DNA. If you were to start with a 20 g sample of phosphorus that contained 10% ^{32}P by mass, converted it into DNA via several chemical steps that had an overall yield of 75% and took 25 days, and then incorporated it into bacteria and allowed them to grow for 5 more days, what mass of ^{32}P would be available for analysis at the end of this time?

94. The enzyme urease contains two atoms of nickel and catalyzes the hydrolysis of urea by the reaction

$$H_2NC(O)NH_2 + H_2O \longrightarrow 2NH_3 + CO_2$$

Urease is one of the most powerful catalysts known. It lowers the activation energy for the hydrolysis of urea from 137 kJ/mol to only 37 kJ/mol. Calculate the ratio of the rate of the catalyzed reaction to the rate of the uncatalyzed reaction at 37°C. Assume that the frequency factor is the same for both reactions.

95. As noted in Section 14.8, the rate for the hydrogenation of ethylene to give ethane can be increased by heterogeneous catalysts such as Pt or Ni:

$$H_2(g) + H_2C{=}CH_2(g) \xrightarrow{\text{Pt,Ni}} H_3C{-}CH_3(g)$$

The activation energy for the uncatalyzed reaction is large (188 kJ/mol), so the reaction is very slow at room temperature. In the presence of finely divided metallic Ni, the activation energy is only 84 kJ/mol. Calculate the ratio of the rate of the catalyzed reaction to the rate of the uncatalyzed reaction at 75°C.

15 Chemical Equilibrium

LEARNING OBJECTIVES

- To understand what is meant by chemical equilibrium
- To know the relationship between the equilibrium constant and the rate constants for the forward and reverse reactions
- To be able to write an equilibrium constant expression for any reaction
- To be able to solve quantitative problems involving chemical equilibrium
- To predict the direction of reaction
- To predict the effects of stresses on a system at equilibrium
- To understand different ways to control the products of a reaction

CHAPTER OUTLINE

15.1 The Concept of Chemical Equilibrium
15.2 The Equilibrium Constant
15.3 Solving Equilibrium Problems
15.4 Nonequilibrium Conditions
15.5 Factors That Affect Equilibrium
15.6 Controlling the Products of Reactions

ESSENTIAL SKILLS 7

The Quadratic Formula

In Chapter 14, we discussed the principles of chemical kinetics, which deal with the *rate of change*, or how quickly a given chemical reaction occurs. We now turn our attention to the *extent* to which a reaction occurs and how reaction conditions affect the final concentrations of reactants and products. For most of the reactions that we have discussed so far, you may have assumed that once reactants are converted to products, they are likely to remain that way. In fact, however, virtually all chemical reactions are *reversible* to some extent. That is, an opposing reaction occurs in which the products react, to a greater or lesser degree, to re-form the reactants. Eventually, the rates of the forward and reverse reactions become the same, and the system reaches **chemical equilibrium**, the point at which the composition of the system no longer changes with time.

A smoggy sunset in Shenzhen, China. The reaction of O_2 with N_2 at high temperature in an internal combustion engine produces small amounts of NO, which in turn reacts with atmospheric O_2 to form NO_2, an important component of smog. The reddish-brown color of NO_2 is responsible for the characteristic color of smog, as shown in this true-color photo.

We introduced the concept of equilibrium in Chapter 11, where you learned that a liquid and a vapor are in equilibrium when the number of molecules evaporating from the surface of the liquid per unit time is the same as the number of molecules condensing out of the vapor phase. Vapor pressure is an example of a *physical equilibrium* because only the physical form of the substance changes. Similarly, in Chapter 13, we discussed saturated solutions, another example of a physical equilibrium, in which the rate of dissolution of a solute is the same as the rate at which it crystallizes from solution.

In this chapter, we describe the methods chemists use to quantitatively describe the composition of chemical systems at equilibrium, and we discuss how factors such as temperature and pressure influence the equilibrium composition. As you study these concepts, you will also learn how urban smog forms, and how reaction conditions can be altered to produce H_2 rather than the combustion products CO_2 and H_2O from the methane in natural gas. You will discover how to control the composition of the gases emitted in automobile exhaust, and how synthetic polymers such as the polyacrylonitrile used in sweaters and carpets are produced on an industrial scale.

15.1 The Concept of Chemical Equilibrium

Chemical equilibrium is a dynamic process that consists of a forward reaction, in which reactants are converted to products, and a reverse reaction, in which products are converted to reactants. At equilibrium, the forward and reverse reactions proceed at equal rates. Consider, for example, a simple system that contains only one reactant and one product, the reversible dissociation of dinitrogen tetroxide, N_2O_4, to nitrogen dioxide, NO_2. You may recall from Chapter 14 that NO_2 is responsible for the brown color we associate with smog. When a sealed tube containing solid N_2O_4 (mp = −9.3°C, bp = 21.2°C) is heated from −78.4°C to 25°C, the red-brown color of NO_2 appears (Figure 15.1). The reaction can be followed visually because the product, NO_2, is colored, whereas the reactant, N_2O_4, is colorless:

$$\underset{\text{Colorless}}{N_2O_4(g)} \rightleftharpoons \underset{\text{Red-brown}}{2NO_2(g)} \tag{15.1}$$

The double arrow indicates that both the forward and reverse reactions are occurring simultaneously; it is read "is in equilibrium with."

Figure 15.1 The $N_2O_4(g) \rightleftharpoons 2NO_2$ system at different temperatures. (*left*) At dry ice temperature (−78.4°C), the system contains essentially pure solid N_2O_4, which is colorless. (*center*) As the system is warmed above the melting point of N_2O_4 (−9.3°C), the N_2O_4 melts and then evaporates, and some of the vapor dissociates to red-brown NO_2. (*right*) Eventually the sample reaches room temperature, and a mixture of gaseous N_2O_4 and NO_2 is present. The composition of the mixture and hence the color do not change further with time: the system has reached equilibrium at the new temperature.

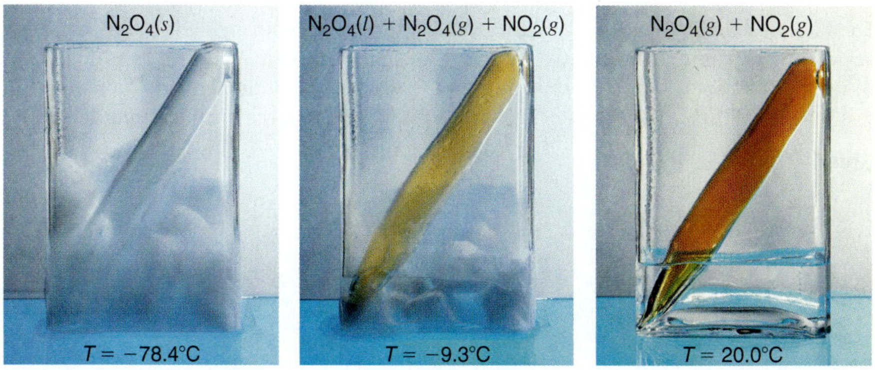

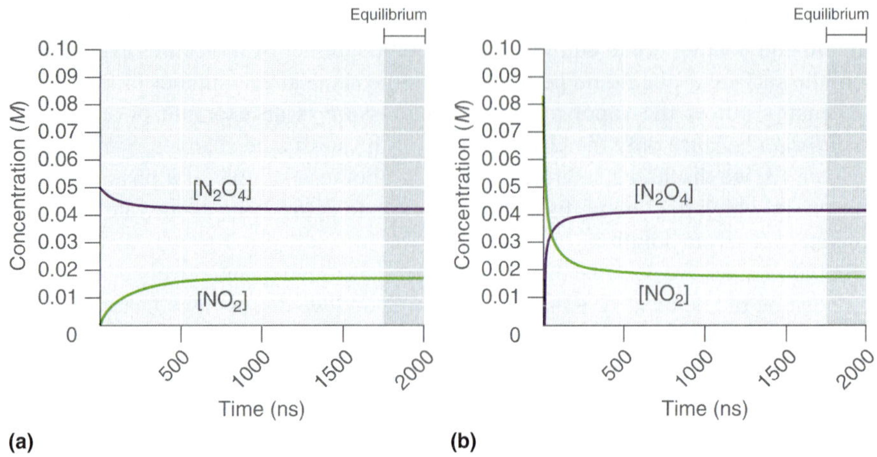

Figure 15.2 The composition of N_2O_4/NO_2 mixtures as a function of time at room temperature. (a) Initially, this idealized system contains 0.0500 *M* gaseous N_2O_4 and no gaseous NO_2. The concentration of N_2O_4 decreases with time as the concentration of NO_2 increases. **(b)** Initially, this system contains 0.1000 *M* NO_2 and no N_2O_4. The concentration of NO_2 decreases with time as the concentration of N_2O_4 increases. In both cases, the final concentrations of the substances are the same: $[N_2O_4] = 0.0422\ M$ and $[NO_2] = 0.0156\ M$ at equilibrium. Interactive Graph

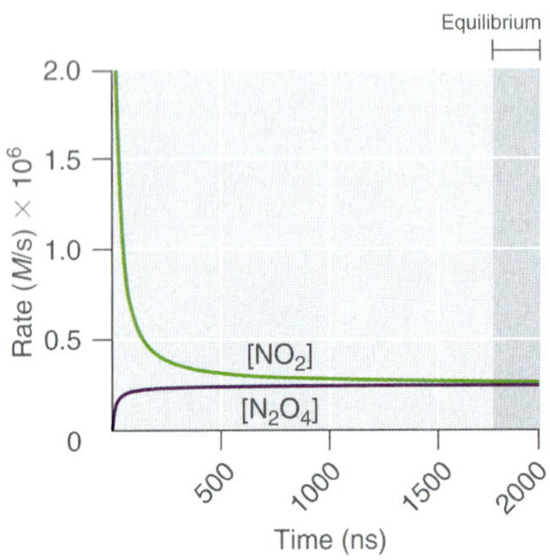

Figure 15.3 The rates of the forward and reverse reactions as a function of time for the $N_2O_4(g) \rightleftharpoons 2NO_2(g)$ system shown in Figure 15.2b. The rate of dimerization of NO_2 (reverse reaction) decreases rapidly with time, as expected for a second-order reaction. Because the initial concentration of N_2O_4 is zero, the rate of the dissociation reaction (forward reaction) at $t = 0$ is also zero. As the dimerization reaction proceeds, the N_2O_4 concentration increases, and its rate of dissociation also increases. Eventually the rates of the two reactions are equal: chemical equilibrium has been reached, and the concentrations of N_2O_4 and NO_2 no longer change. Interactive Graph

Figure 15.2 shows how the composition of this system would vary as a function of time at a constant temperature. If the initial concentration of NO_2 were zero, then it increases as the concentration of N_2O_4 decreases. Eventually the composition of the system stops changing with time; thus, chemical equilibrium has been achieved. Conversely, if we start with a sample that contains no N_2O_4 but an initial NO_2 concentration twice the initial concentration of N_2O_4 in Figure 15.2a, in accordance with the stoichiometry of the reaction, we reach exactly the same equilibrium composition, as shown in Figure 15.2b. Thus, equilibrium can be approached from *either direction* in a chemical reaction.

Figure 15.3 shows the rates of the forward and reverse reactions for a sample that initially contains pure NO_2. Because the initial concentration of N_2O_4 is zero, the rate of the forward reaction (dissociation of N_2O_4) is initially zero as well. In contrast, the rate of the reverse reaction (dimerization of NO_2) is initially very high (2.0×10^6 *M*/s), but it decreases rapidly as the concentration of NO_2 decreases. (Recall from Chapter 14 that the rate of the dimerization reaction is expected to decrease rapidly because the reaction is second order in NO_2: rate $= k_r[NO_2]^2$, where k_r is the rate constant for the reverse reaction shown in Equation 15.1.) As the concentration of N_2O_4 increases, the rate of dissociation of N_2O_4 increases, but more slowly than the dimerization of NO_2 because the reaction is only first order in N_2O_4 (rate $= k_f[N_2O_4]$, where k_f is the rate constant for the forward reaction in Equation 15.1). Eventually, the rates of the forward and reverse reactions become identical, and the system has reached chemical equilibrium. If the forward and reverse reactions occur at *different* rates, then the system is *not* at equilibrium.

Note the pattern

At equilibrium, the rate of the forward reaction is equal to the rate of the reverse reaction.

EXAMPLE 15.1

The three reaction systems (1, 2, and 3) depicted in the accompanying illustration can all be described by the equation $2A \rightleftharpoons B$, where the blue circles are A and the purple ovals are B. Each set of panels shows the changing composition of one of the three reaction mixtures as a function of time. Which system has taken the longest to reach chemical equilbrium?

Given Three reaction systems

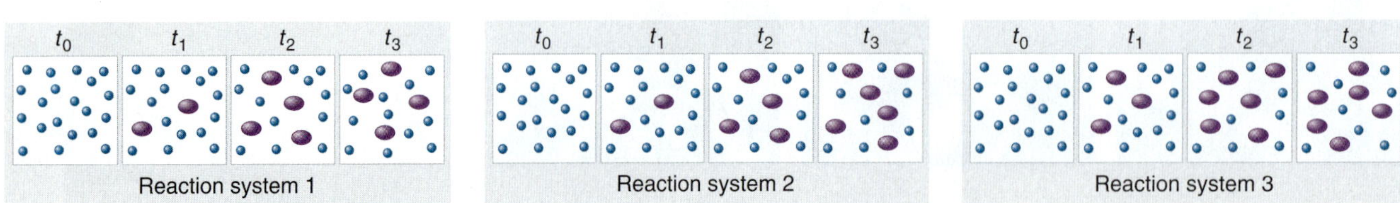

Asked for Relative time to reach chemical equilibrium

Strategy

Compare the concentrations of A and B at different times. The system whose composition takes the longest to stabilize has taken the longest to reach chemical equilibrium.

Solution

In both systems 1 and 3, the concentration of A decreases from t_0 through t_2 but is the same at both t_2 and t_3. Thus, systems 1 and 3 are at equilibrium by t_3. In system 2, the concentrations of A and B are still changing between t_2 and t_3, so system 2 may not yet have reached equilibrium by t_3. Thus, system 2 has taken the longest to reach chemical equilibrium.

EXERCISE 15.1

In the illustration, A is represented by blue circles, B by purple squares, and compound C by orange ovals; the equation for the reaction is $A + B \rightleftharpoons C$. The sets of panels represent the compositions of three reaction mixtures as a function of time. Which, if any, of the systems shown has reached equilibrium?

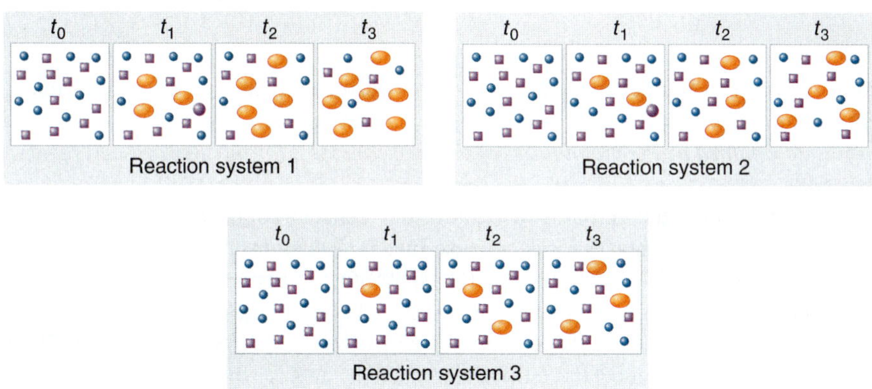

Answer System 2

15.2 • The Equilibrium Constant

Because an equilibrium state is achieved when the rate of the forward reaction equals the rate of the reverse reaction, under a given set of conditions there must be a relationship between the composition of the system at equilibrium and the kinetics of a reaction (represented by rate constants). We can show this relationship using the system described in Equation 15.1, the decomposition of N_2O_4 to NO_2. Both the forward and reverse reactions for this system consist of a single elementary step, so the rates are

MGC The Equilibrium Constant Expression

$$\text{forward rate} = k_f[N_2O_4] \tag{15.2a}$$

$$\text{reverse rate} = k_r[NO_2]^2 \tag{15.2b}$$

At equilibrium, the forward rate is equal to the reverse rate:

$$k_f[N_2O_4] = k_r[NO_2]^2 \tag{15.2c}$$

TABLE 15.1 Initial and equilibrium concentrations for NO_2/N_2O_4 mixtures at 25°C

	Initial Concentrations		Concentrations at Equilibrium		
Experiment	**$[N_2O_4]$, *M***	**$[NO_2]$, *M***	**$[N_2O_4]$, *M***	**$[NO_2]$, *M***	**$K = [NO_2]^2/[N_2O_4]$**
1	0.0500	0.0000	0.0417	0.0165	6.54×10^{-3}
2	0.0000	0.1000	0.0417	0.0165	6.54×10^{-3}
3	0.0750	0.0000	0.0647	0.0206	6.56×10^{-3}
4	0.0000	0.0750	0.0304	0.0141	6.54×10^{-3}
5	0.0250	0.0750	0.0532	0.0186	6.50×10^{-3}

so

$$\frac{k_f}{k_r} = \frac{[NO_2]^2}{[N_2O_4]} \tag{15.2d}$$

The ratio of the rate constants gives us a new constant, the **equilibrium constant (*K*)**, which is defined as

$$K = \frac{k_f}{k_r} \tag{15.3}$$

Note the pattern

The equilibrium constant K is equal to the rate constant for the forward reaction divided by the rate constant for the reverse reaction.

Hence, there is a fundamental relationship between chemical kinetics and chemical equilibrium: *Under a given set of conditions, the composition of the equilibrium mixture is determined by the magnitudes of the rate constants for the forward and reverse reactions.*

Table 15.1 lists the initial and equilibrium concentrations from five different experiments using the reaction system described by Equation 15.1. Note that at equilibrium the magnitude of the quantity $[NO_2]^2/[N_2O_4]$ is essentially the same for all five experiments. In fact, no matter what the initial concentrations of NO_2 and N_2O_4, at equilibrium the quantity $[NO_2]^2/[N_2O_4]$ will *always* be $6.53 \pm 0.03 \times 10^{-3}$ at 25°C, which corresponds to the ratio of the rate constants for the forward and reverse reactions. That is, at a given temperature, the equilibrium constant K for a reaction always has the same value, even though the specific concentrations of the reactants and products vary depending on their initial concentrations.

Developing an Equilibrium Constant Expression

In 1864, the Norwegian chemists Cato Guldberg (1836–1902) and Peter Waage (1833–1900) carefully measured the compositions of many reaction systems at equilibrium. They discovered that for *any* reversible reaction of the general form

$$aA + bB \rightleftharpoons cC + dD \tag{15.4}$$

where A and B are reactants, C and D are products, and a, b, c, and d are the stoichiometric coefficients in the balanced equation for the reaction, the ratio of the product of the equilibrium concentrations of the products (raised to their coefficients in the balanced equation) to the product of the equilibrium concentrations of the reactants (raised to their coefficients in the balanced equation) is always a constant under a given set of conditions. This relationship is known as the **law of mass action** and can be stated as

$$K = \frac{[C]^c[D]^d}{[A]^a[B]^b} \tag{15.5}$$

TABLE 15.2 Equilibrium constants for selected reactions[a]

Reaction	Temperature, K	Equilibrium Constant (K)
$S(s) + O_2(g) \rightleftharpoons SO_2(g)$	300	4.4×10^{53}
$2H_2(g) + O_2(g) \rightleftharpoons 2H_2O(g)$	500	2.4×10^{47}
$H_2(g) + Cl_2(g) \rightleftharpoons 2HCl(g)$	300	1.6×10^{33}
$H_2(g) + Br_2(g) \rightleftharpoons 2HBr(g)$	300	4.1×10^{18}
$2NO(g) + O_2(g) \rightleftharpoons 2NO_2(g)$	300	4.2×10^{13}
$3H_2(g) + N_2(g) \rightleftharpoons 2NH_3(g)$	300	2.7×10^{8}
$H_2(g) + D_2(g) \rightleftharpoons 2HD(g)$	100	1.92
$H_2(g) + I_2(g) \rightleftharpoons 2HI(g)$	300	2.9×10^{-1}
$Br_2(g) \rightleftharpoons 2Br(g)$	1000	4.0×10^{-7}
$I_2(g) \rightleftharpoons 2I(g)$	800	4.6×10^{-7}
$Cl_2(g) \rightleftharpoons 2Cl(g)$	1000	1.8×10^{-9}
$F_2(g) \rightleftharpoons 2F(g)$	500	7.4×10^{-13}

[a]Equilibrium constants vary with temperature. The K values shown are for systems at the indicated temperatures.

where K is the equilibrium constant for the reaction. Equation 15.4 is called the **equilibrium equation**, and the right side of Equation 15.5 is called the **equilibrium constant expression**. The relationship shown in Equation 15.5 is true for *any* pair of opposing reactions regardless of the mechanism of the reaction or of the number of steps in the mechanism.

The equilibrium constant can vary over a wide range of values. The values of K shown in Table 15.2, for example, vary by 60 orders of magnitude. Because products are in the numerator of the equilibrium constant expression and reactants are in the denominator, values of K greater than 10^3 indicate a strong tendency for reactants to form products. In this case, chemists say that equilibrium lies to the right as written, favoring the formation of products. An example is the reaction between H_2 and Cl_2 to produce HCl, which has an equilibrium constant of 1.6×10^{33} at 300 K. Because H_2 is a good reductant and Cl_2 is a good oxidant, the reaction proceeds essentially to completion. In contrast, values of K less than 10^{-3} indicate that the ratio of products to reactants at equilibrium is very small. That is, reactants do not tend to form products readily, and the equilibrium lies to the left as written, favoring the formation of reactants.

You will also notice in Table 15.2 that equilibrium constants have no units, even though Equation 15.5 suggests that the units of concentration might not always cancel because the exponents may vary. In fact, equilibrium constants are calculated using "effective concentrations", or *activities*, of reactants and products, which are the ratios of the measured concentrations to a standard state of 1 M.* As shown in Equation 15.6, the units of concentration cancel, which makes K unitless as well:

$$\frac{[\mathrm{A}]_{\text{measured}}}{[\mathrm{A}]_{\text{standard state}}} = \frac{\cancel{M}}{\cancel{M}} = \frac{\cancel{\text{mol/L}}}{\cancel{\text{mol/L}}} \tag{15.6}$$

Many reactions have equilibrium constants between 1000 and 0.001 ($10^3 \geq K \geq 10^{-3}$), neither very large nor very small. At equilibrium, these systems tend to

* As you will discover in more advanced chemistry courses, the effective concentration (the activity) is unitless. The standard state is assigned an activity of 1.

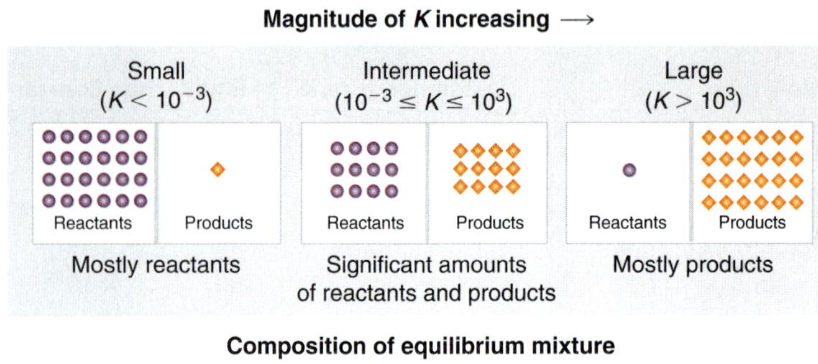

Figure 15.4 The relationship between the composition of the mixture at equilibrium and the magnitude of the equilibrium constant. The larger the value of K, the farther the reaction proceeds to the right before equilibrium is reached, and the higher the ratio of products to reactants at equilibrium.

contain significant amounts of both products and reactants, indicating that there is not a strong tendency to form either products from reactants or reactants from products. An example of this type of system is the reaction of gaseous hydrogen and deuterium, a component of high-stability fiber-optic light sources used in ocean studies, to form HD:

$$H_2(g) + D_2(g) \rightleftharpoons 2HD(g) \tag{15.7}$$

The equilibrium constant expression for this reaction is $[HD]^2/[H_2][D_2]$, and the value of K is between 1.9 and 4 over a wide temperature range (100–1000 K). Thus, an equilibrium mixture of H_2, D_2, and HD contains significant concentrations of both product and reactants.

Figure 15.4 summarizes the relationship between the magnitude of K and the relative concentrations of reactants and products at equilibrium for a general reaction, written as reactants $\rightleftharpoons$ products. Because there is a direct relationship between the kinetics of a reaction and the equilibrium concentrations of products and reactants (Equations 15.3 and 15.5), when $k_f \gg k_r$, K is a large number, and the concentration of products at equilibrium predominate. This corresponds to an essentially irreversible reaction. Conversely, when $k_f \ll k_r$, K is a very small number, and the reaction produces almost no products as written. Systems for which $k_f \approx k_r$ have significant concentrations of both reactants and products at equilibrium.

Note the pattern

A large value of the equilibrium constant K means that products predominate at equilibrium; a small value means that reactants predominate at equilibrium.

EXAMPLE 15.2

Write equilibrium constant expressions for these reactions:

(a) $N_2(g) + 3H_2(g) \rightleftharpoons 2NH_3(g)$
(b) $CO(g) + \frac{1}{2}O_2(g) \rightleftharpoons CO_2(g)$
(c) $2CO_2(g) \rightleftharpoons 2CO(g) + O_2(g)$

Given Balanced chemical equations

Asked for Equilibrium constant expressions

Strategy

Refer to Equation 15.5. Place the arithmetic product of the concentrations of the products (raised to their stoichiometric coefficients) in the numerator, and the product of the concentrations of the reactants (raised to their stoichiometric coefficients) in the denominator.

Solution

(a) The only product is ammonia, which has a coefficient of 2. For the reactants, N_2 has a coefficient of 1 and H_2 has a coefficient of 3. The equilibrium constant expression is therefore

$$\frac{[NH_3]^2}{[N_2][H_2]^3}$$

(b) The only product is carbon dioxide, which has a coefficient of 1. The reactants are CO, with a coefficient of 1, and O_2, with a coefficient of $\frac{1}{2}$. Thus, the equilibrium constant expression is

$$\frac{[CO_2]}{[CO][O_2]^{1/2}}$$

(c) This reaction is the reverse of the reaction in part b, with all coefficients multiplied by 2 to remove the fractional coefficient for O_2. The equilibrium constant expression is therefore the inverse of the expression in part b, with all exponents multiplied by 2:

$$\frac{[CO]^2[O_2]}{[CO_2]^2}$$

EXERCISE 15.2

Write equilibrium constant expressions for these reactions:

(a) $N_2O(g) \rightleftharpoons N_2(g) + \frac{1}{2}O_2(g)$
(b) $2C_8H_{18}(g) + 25O_2(g) \rightleftharpoons 16CO_2(g) + 18H_2O(g)$

Answer **(a)** $\dfrac{[N_2][O_2]^{1/2}}{[N_2O]}$; **(b)** $\dfrac{[CO_2]^{16}[H_2O]^{18}}{[C_8H_{18}]^2[O_2]^{25}}$

EXAMPLE 15.3

Predict which of the following systems at equilibrium will **(a)** contain essentially only products; **(b)** contain essentially only reactants; and **(c)** contain appreciable amounts of both products and reactants:

1. $H_2(g) + I_2(g) \rightleftharpoons 2HI(g)$, $K_{(700K)} = 54$
2. $2CO_2(g) \rightleftharpoons 2CO(g) + O_2(g)$, $K_{(1200K)} = 3.1 \times 10^{-18}$
3. $PCl_5(g) \rightleftharpoons PCl_3(g) + Cl_2(g)$, $K_{(613K)} = 97$
4. $2O_3(g) \rightleftharpoons 3O_2(g)$, $K_{(298K)} = 5.9 \times 10^{55}$

Given Systems and values of K

Asked for Composition of systems at equilibrium

Strategy

Use the value of the equilibrium constant to determine whether the equilibrium mixture will contain essentially only products, essentially only reactants, or significant amounts of both.

Solution

(a) Only system 4 has $K \gg 10^3$, so at equilibrium it will consist of essentially only products.
(b) System 2 has $K \ll 10^{-3}$, so the reactants have little tendency to form products under the conditions specified; thus, at equilibrium the system will contain essentially only reactants.
(c) Both systems 1 and 3 have equilibrium constants in the range $10^3 \geqslant K \geqslant 10^{-3}$, indicating that the equilibrium mixtures will contain appreciable amounts of both products and reactants.

EXERCISE 15.3

Hydrogen and nitrogen react to form ammonia according to this balanced chemical equation:

$$3H_2(g) + N_2(g) \rightleftharpoons 2NH_3(g)$$

Values of the equilibrium constant at various temperatures were reported as $K_{25°C} = 3.3 \times 10^8$, $K_{177°C} = 2.6 \times 10^3$, and $K_{327°C} = 4.1$. **(a)** At which temperature would you expect to find the highest proportion of H_2 and N_2 in the equilibrium mixture? **(b)** Assuming that the reaction rates are fast enough so that equilibrium is reached quickly, at what temperature would you design a commercial reactor to operate to maximize the yield of ammonia?

Answer **(a)** 327°C, where *K* is smallest; **(b)** 25°C

Variations in the Form of the Equilibrium Constant Expression

Because equilibrium can be approached from either direction in a chemical reaction, the equilibrium constant expression and thus the magnitude of the equilibrium constant depend on the form in which the chemical reaction is written. For example, if we write the reaction described in Equation 15.4 in reverse, we obtain

$$cC + dD \rightleftharpoons aA + bB \tag{15.8}$$

The corresponding equilibrium constant K' is

$$K' = \frac{[A]^a[B]^b}{[C]^c[D]^d} \tag{15.9}$$

This expression is the inverse of the expression for the original equilibrium constant *K*, so $K' = 1/K$. That is, when we write a reaction in the reverse direction, the equilibrium constant expression is inverted. For instance, the equilibrium constant *K* for the reaction $N_2O_4 \rightleftharpoons 2NO_2$ is

$$K = \frac{[NO_2]^2}{[N_2O_4]} \tag{15.10}$$

but for the opposite reaction, $2NO_2 \rightleftharpoons N_2O_4$, the equilibrium constant K' is given by the inverse expression:

$$K' = \frac{[N_2O_4]}{[NO_2]^2} \tag{15.11}$$

Consider another example, the formation of water: $2H_2(g) + O_2(g) \rightleftharpoons 2H_2O(g)$. Because H_2 is a good reductant and O_2 is a good oxidant, this reaction has a very large equilibrium constant ($K = 2.4 \times 10^{47}$at 500 K). Consequently, the equilibrium constant for the reverse reaction, the decomposition of water to form O_2 and H_2, is very small: $K' = 1/K = 1/(2.4 \times 10^{47} = 4.2 \times 10^{-48}$. As suggested by the very small equilibrium constant, and fortunately for life as we know it, a substantial amount of energy is indeed needed to dissociate water into H_2 and O_2.

Note the pattern

The equilibrium constant for a reaction written in reverse is the inverse of the original equilibrium constant.

Writing an equation in different but chemically equivalent forms also causes both the equilibrium constant expression and the magnitude of the equilibrium constant to be different. For example, we could write the equation for the reaction $2NO_2 \rightleftharpoons N_2O_4$ as $NO_2 \rightleftharpoons \frac{1}{2}N_2O_4$, for which the equilibrium constant K'' is

$$K'' = \frac{[N_2O_4]^{1/2}}{[NO_2]} \tag{15.12}$$

The values for K' (Equation 15.11) and K'' are related as follows:

$$K'' = (K')^{1/2} = \sqrt{K'} \tag{15.13}$$

In general, if all the coefficients in a balanced chemical equation are subsequently multiplied by *n*, then the new equilibrium constant is the original equilibrium constant raised to the *n*th power.

EXAMPLE 15.4

At 745 K, the value of K is 0.118 for the reaction

$$N_2(g) + 3H_2(g) \rightleftharpoons 2NH_3(g)$$

What are the equilibrium constants for these related reactions at 745 K?

(a) $2NH_3(g) \rightleftharpoons N_2(g) + 3H_2(g)$

(b) $\frac{1}{2}N_2(g) + \frac{3}{2}H_2(g) \rightleftharpoons NH_3(g)$

Given Balanced equilibrium equation, value of K at a given temperature, and equations of related reactions

Asked for Values of K for related reactions

Strategy

Write the equilibrium constant expression for the given reaction and for each related reaction. From these expressions, calculate K for each reaction.

Solution

The equilibrium constant expression for the given reaction of $N_2(g)$ with $H_2(g)$ to produce $NH_3(g)$ at 745 K is

$$K = \frac{[NH_3]^2}{[N_2][H_2]^3} = 0.118$$

(a) This reaction is the reverse of the one given, so its equilibrium constant expression is

$$K' = \frac{1}{K} = \frac{[N_2][H_2]^3}{[NH_3]^2} = \frac{1}{0.118} = 8.47$$

(b) In this reaction, all the stoichiometric coefficients of the given reaction are divided by 2, so the equilibrium constant is calculated as

$$K'' = \frac{[NH_3]}{[N_2]^{1/2}[H_2]^{3/2}} = K^{1/2} = \sqrt{K} = \sqrt{0.118} = 0.344$$

EXERCISE 15.4

At 527°C, the equilibrium constant for the reaction

$$2SO_2(g) + O_2(g) \rightleftharpoons 2SO_3(g)$$

is 7.9×10^4. Calculate the equilibrium constant for the following reaction at that same temperature:

$$SO_3(g) \rightleftharpoons SO_2(g) + \tfrac{1}{2}O_2(g)$$

Answer 3.6×10^{-3}

Equilibrium Constant Expressions for Systems That Contain Gases

For reactions that involve species in solution, the concentrations used in equilibrium calculations are usually expressed in moles/liter. For gases, however, the concentrations are usually expressed in terms of partial pressures rather than molarity, where the standard state is 1 atm of pressure. The symbol $\boldsymbol{K_p}$ is used to denote equilibrium constants calculated from partial pressures. For the general reaction

$aA + bB \rightleftharpoons cC + dD$ in which all the components are gases, we can write the equilibrium constant expression as the ratio of the partial pressures of the products and reactants (each raised to its coefficient in the chemical equation):

$$K_p = \frac{(P_A)^c (P_C)^d}{(P_B)^a (P_D)^b} \quad (15.14)$$

Thus, K_p for the decomposition of N_2O_4 (Equation 15.1) is

$$K_p = \frac{(P_{NO_2})^2}{P_{N_2O_4}} \quad (15.15)$$

Like K, K_p is a unitless quantity because the quantity that is actually used to calculate it is an "effective pressure," the ratio of the measured pressure to a standard state of 1 bar (approximately 1 atm) which produces a unitless quantity.*

Because partial pressures are usually expressed in atmospheres or mmHg, the molar concentration of a gas and its partial pressure do not have the same numerical value. Consequently, the numerical values of K and K_p are usually different. They are, however, related by the ideal gas constant R and the temperature T:

$$K_p = K(RT)^{\Delta n} \quad (15.16)$$

where K is the equilibrium constant expressed in units of concentration and Δn is the difference between the numbers of moles of gaseous products and gaseous reactants $(n_p - n_r)$. The temperature is expressed as the absolute temperature in kelvins. According to Equation 15.16, $K_p = K$ only if the numbers of moles of gaseous products and gaseous reactants are the same (that is, $\Delta n = 0$). For the decomposition of N_2O_4, there are 2 mol of gaseous product and 1 mol of gaseous reactant, so $\Delta n = 1$. Thus, for this reaction, $K_p = K(RT)^1 = KRT$.

EXAMPLE 15.5

The equilibrium constant K for the reaction of nitrogen and hydrogen to give ammonia is 0.118 at 745 K:

$$N_2(g) + 3H_2(g) \rightleftharpoons 2NH_3(g)$$

What is K_p for this reaction at the same temperature?

Given Equilibrium equation, equilibrium constant, and temperature

Asked for Value of K_p

Strategy

Use the coefficients in the balanced equation to calculate Δn. Then use Equation 15.16 to calculate K from K_p.

Solution

This reaction has 2 mol of gaseous product and 4 mol of gaseous reactants, so $\Delta n = (2 - 4) = -2$. We know the value of K, and $T = 745$ K. Thus, from Equation 15.16, we have

$$K_p = K(RT)^{-2} = \frac{K}{(RT)^2} = \frac{0.118}{\{[0.08206\ (\text{L} \cdot \text{atm})/(\text{mol} \cdot \text{K})][745\ \text{K}]\}^2} = 3.16 \times 10^{-5}$$

Because K_p is a unitless quantity, the answer is $K_p = 3.16 \times 10^{-5}$.

---

* The "effective pressure" is called the *fugacity*, just as activity is the effective concentration.

EXERCISE 15.5

Calculate K_p for the reaction $2SO_2(g) + O_2(g) \rightleftharpoons 2SO_3(g)$ at 527°C, if $K = 7.9 \times 10^4$ at this temperature.

Answer $K_p = 1.2 \times 10^3$

Homogeneous and Heterogeneous Equilibria

When the products and reactants of an equilibrium reaction form a single phase, whether gas or liquid, the system is a **homogeneous equilibrium**. In such situations, the concentrations of the reactants and products can vary over a wide range. In contrast, a system whose reactants, products, or both are in more than one phase is a **heterogeneous equilibrium**, such as the reaction of a gas with a solid or liquid.

Because the molar concentrations of pure liquids and solids normally do not vary greatly with temperature, their concentrations are treated as constants, which allows us to simplify equilibrium constant expressions that involve pure solids or liquids.* (Recall from Chapter 11, for example, that the density of water, and thus its volume, changes by only a few percent between 0°C and 100°C.)

Consider the following reaction, which is used in the final firing of some types of pottery to produce brilliant metallic glazes:

$$CO_2(g) + C(s) \rightleftharpoons 2CO(g) \tag{15.17}$$

The glaze is created when metal oxides are reduced to metals by the product, carbon monoxide. The equilibrium constant expression for this reaction is

$$K = \frac{[CO]^2}{[CO_2][C]} \tag{15.18}$$

Because graphite is a solid, however, its molar concentration, determined from its density and molar mass, is essentially constant and has the following value:

$$[C] = \frac{2.26\ \cancel{g}/\cancel{cm^3}}{12.01\ \cancel{g}/mol} \times 1000\ \cancel{cm^3}/L = 188\ mol/L = 188\ M \tag{15.19}$$

We can rearrange Equation 15.18 so that the constant terms are on one side:

$$K[C] = K(188) = \frac{[CO]^2}{[CO_2]} \tag{15.20}$$

Incorporating the constant value of [C] into the equilibrium equation for the reaction in Equation 15.17 gives

$$K' = \frac{[CO]^2}{[CO_2]} \tag{15.21}$$

The equilibrium constant for this reaction can also be written in terms of the partial pressures of the gases:

$$K_p = \frac{(P_{CO})^2}{P_{CO_2}} \tag{15.22}$$

Incorporating all the constant values into K' or K_p allows us to focus on the substances whose concentrations change during the reaction.

---

* The reference states for pure solids and liquids are those forms stable at 1 bar (approximately 1 atm), which are assigned an activity of 1.

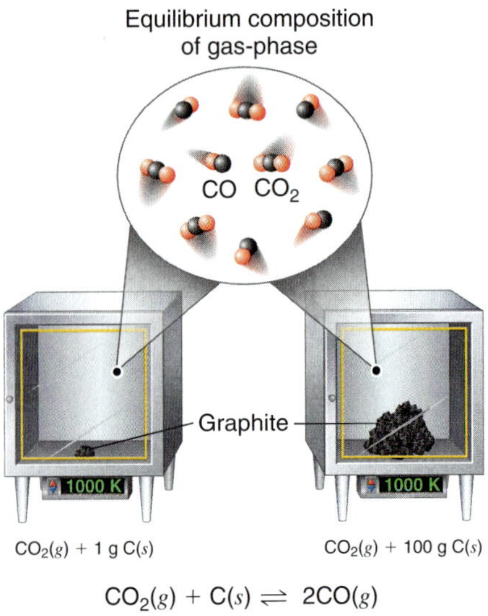

Figure 15.5 Effect of the amount of solid present on equilibrium in a heterogeneous solid–gas system. In the $CO_2(g) + C(s) \rightleftharpoons 2CO(g)$ system, the equilibrium composition of the gas phase at a given temperature, 1000 K in this case, is the same whether a small amount of solid carbon is present (*left*) or a large amount (*right*).

Although the concentrations of pure liquids or solids are not written explicitly in the equilibrium constant expression, these substances must be present in the reaction mixture for chemical equilibrium to occur. Whatever the concentrations of CO and CO_2, the system described in Equation 15.17 will reach chemical equilibrium only if a stoichiometric amount of solid carbon or excess solid carbon has been added so that some is still present once the system has reached equilibrium. As shown in Figure 15.5, it does not matter whether 1 g or 100 g of solid carbon is present; in either case, the composition of the gaseous components of the system will be the same at equilibrium.

EXAMPLE 15.6

Write expressions for K, incorporating all constants, and K_p for these equilibrium reactions:

(a) $PCl_3(l) + Cl_2(g) \rightleftharpoons PCl_5(s)$
(b) $Fe_3O_4(s) + 4H_2(g) \rightleftharpoons 3Fe(s) + 4H_2O(g)$

Given Balanced equilibrium equations

Asked for Expressions for K and K_p

Strategy

Find K by writing each equilibrium constant expression as the ratio of the concentrations of the products and reactants, each raised to its coefficient in the chemical equation. Then express K_p as the ratio of the partial pressures of the products and reactants, each also raised to its coefficient in the chemical equation.

Solution

(a) This reaction contains a pure solid (PCl_5) and a pure liquid (PCl_3). Their concentrations do not appear in the equilibrium constant expression because they do not change significantly. Thus,

$$K = \frac{1}{[Cl_2]} \quad \text{and} \quad K_p = \frac{1}{P_{Cl_2}}$$

(b) This reaction contains two pure solids (Fe_3O_4 and Fe), which do not appear in the equilibrium constant expressions. The two gases do, however, appear in the expressions:

$$K = \frac{[H_2O]^4}{[H_2]^4} \quad \text{and} \quad K_p = \frac{(P_{H_2O})^4}{(P_{H_2})^4}$$

EXERCISE 15.6

Write expressions for K and K_p for these reactions:

(a) $CaCO_3(s) \rightleftharpoons CaO(s) + CO_2(g)$
(b) $\underset{\text{Glucose}}{C_6H_{12}O_6(s)} + 6O_2(g) \rightleftharpoons 6CO_2(g) + 6H_2O(g)$

Answer **(a)** $K = [CO_2]$, $K_p = P_{CO_2}$; **(b)** $K = \dfrac{[CO_2]^6[H_2O]^6}{[O_2]^6}$, $K_p = \dfrac{(P_{CO_2})^6(P_{H_2O})^6}{(P_{O_2})^6}$

Note the pattern

The concentrations of pure solids, pure liquids, and solvents are omitted from equilibrium constant expressions because they don't change significantly during reactions when enough is present to reach equilibrium.

For reactions carried out in solution, the concentration of the solvent is omitted from the equilibrium constant expression even when the solvent appears in the balanced equation for the reaction. The concentration of the solvent is also typically much higher than the concentration of the reactants or products (recall that pure water is about 55.5 *M*, and pure ethanol is about 17 *M*). Consequently, the solvent concentration is essentially constant during chemical reactions, and the solvent is therefore treated as a pure liquid. The equilibrium constant expression for a reaction contains only those species whose concentrations could change significantly during the reaction.

Equilibrium Constant Expressions for the Sums of Reactions

Chemists frequently need to know the equilibrium constant for a reaction that has not been studied before. In such cases, the desired reaction can often be written as the sum of other reactions for which the equilibrium constants are known. The equilibrium constant for the unknown reaction can then be calculated from the tabulated values for the other reactions.

To illustrate this procedure, let's consider the reaction of N_2 with O_2 to give NO_2. As we stated in Section 15.1, this reaction is an important source of the NO_2 that gives urban smog its typical brown color. The reaction normally occurs in two distinct steps. In the first reaction [shown below as (1)], N_2 reacts with O_2 at the high temperatures inside an internal combustion engine to give NO. The released NO then reacts with additional O_2 to give NO_2 [shown below as (2)]. The equilibrium constant for each reaction at 100°C is also given.

$$(1)\ N_2(g) + O_2(g) \rightleftharpoons 2NO(g) \qquad K_1 = 2.0 \times 10^{-25}$$
$$(2)\ 2NO(g) + O_2(g) \rightleftharpoons 2NO_2(g) \qquad K_2 = 6.4 \times 10^{9}$$

Summing reactions (1) and (2) gives the overall reaction of N_2 with O_2:

$$(3)\ N_2(g) + 2O_2(g) \rightleftharpoons 2NO_2(g) \qquad K_3 = ?$$

The equilibrium constant expressions for the reactions are

$$K_1 = \frac{[NO]^2}{[N_2][O_2]} \qquad K_2 = \frac{[NO_2]^2}{[NO]^2[O_2]} \qquad K_3 = \frac{[NO_2]^2}{[N_2][O_2]^2}$$

What is the relationship between K_1, K_2, and K_3, all at 100°C? Note that the expression for K_1 has $[NO]^2$ in the numerator, the expression for K_2 has $[NO]^2$ in the denominator, and $[NO]^2$ does not appear at all in the expression for K_3. Multiplying K_1 by K_2 and canceling the $[NO]^2$ terms give

$$K_1K_2 = \frac{\cancel{[NO]^2}}{[N_2][O_2]} \times \frac{[NO_2]^2}{\cancel{[NO]^2}[O_2]} = \frac{[NO_2]^2}{[N_2][O_2]^2} = K_3$$

Thus, the product of the equilibrium constant expressions for K_1 and K_2 is the same as the equilibrium constant expression for K_3:

$$K_3 = K_1K_2 = (2.0 \times 10^{-25})(6.4 \times 10^{9}) = 1.3 \times 10^{-15}$$

The equilibrium constant for a reaction that is the sum of two or more reactions is equal to the product *of the equilibrium constants for the individual reactions.* In contrast, recall that according to Hess's Law, ΔH for the sum of two or more reactions is the *sum* of the ΔH values for the individual reactions.

Note the pattern

To determine K for a reaction that is the sum of two or more reactions, add reactions, but multiply equilibrium constants.

EXAMPLE 15.7

The following reactions occur at 1200°C:

$$(1)\ CO(g) + 3H_2(g) \rightleftharpoons CH_4(g) + H_2O(g) \qquad K_1 = 9.17 \times 10^{-2}$$
$$(2)\ CH_4(g) + 2H_2S(g) \rightleftharpoons CS_2(g) + 4H_2(g) \qquad K_2 = 3.3 \times 10^{4}$$

Calculate the equilibrium constant for this reaction at the same temperature:

$$(3)\ CO(g) + 2H_2S(g) \rightleftharpoons CS_2(g) + H_2O(g) + H_2(g) \qquad K_3 = ?$$

Given Two balanced equilibrium equations, values of K, and an equilibrium equation for the overall reaction

Asked for Equilibrium constant for the overall reaction

Strategy

Arrange the equations so that their sum produces the overall equation. If an equation had to be reversed, invert the value of K for that equation. Calculate K for the overall equation by multiplying the equilibrium constants for the individual equations.

Solution

The key to solving this problem is to recognize that reaction (3) is the sum of reactions (1) and (2):

$$CO(g) + \cancel{3H_2(g)} \rightleftharpoons \cancel{CH_4(g)} + H_2O(g)$$
$$\cancel{CH_4(g)} + 2H_2S(g) \rightleftharpoons CS_2(g) + \cancel{3H_2(g)} + H_2(g)$$
$$CO(g) + 2H_2S(g) \rightleftharpoons CS_2(g) + H_2O(g) + H_2(g)$$

The values for K_1 and K_2 are given, so it is straightforward to calculate K_3:

$$K_3 = K_1K_2 = (9.17 \times 10^{-2})(3.3 \times 10^4) = 3.03 \times 10^3$$

EXERCISE 15.7

In the first of two steps in the industrial synthesis of sulfuric acid, elemental sulfur reacts with oxygen to produce sulfur dioxide. In the second step, sulfur dioxide reacts with additional oxygen to form sulfur trioxide. The reaction for each step is shown, as is the value of the corresponding equilibrium constant at 25°C. Calculate the equilibrium constant for the overall reaction at this same temperature.

$$(1)\ \frac{1}{8}S_8(s) + O_2(g) \rightleftharpoons SO_2(g) \qquad K_1 = 4.4 \times 10^{53}$$

$$(2)\ SO_2(g) + \frac{1}{2}O_2(g) \rightleftharpoons SO_3(g) \qquad K_2 = 2.6 \times 10^{12}$$

$$(3)\ \frac{1}{8}S_8(s) + \frac{3}{2}O_2(g) \rightleftharpoons SO_3(g) \qquad K_3 = ?$$

Answer $K_3 = 1.1 \times 10^{66}$

MGC Calculating Equilibrium Constants

15.3 Solving Equilibrium Problems

There are two fundamental kinds of equilibrium problems: (1) those in which we are given the concentrations of the reactants and products at equilibrium (or, more often, information that allows us to calculate these concentrations) and we are asked to calculate the equilibrium constant for the reaction; and (2) those in which we are given the equilibrium constant and the initial concentrations of reactants and we are asked to calculate the concentration of one or more substances at equilibrium. In this section, we describe methods for solving both kinds of calculations.

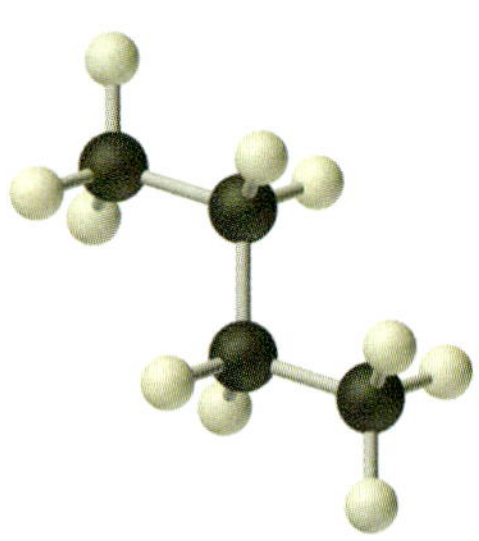
n-Butane

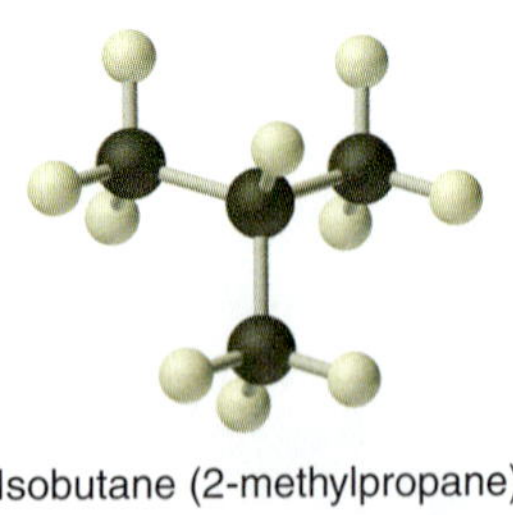
Isobutane (2-methylpropane)

Calculating an Equilibrium Constant from Equilibrium Concentrations

We saw in Exercise 15.6 that the equilibrium constant for the decomposition of $CaCO_3(s)$ to $CaO(s)$ and $CO_2(g)$ is $K = [CO_2]$. At 800°C, the concentration of CO_2 in equilibrium with solid $CaCO_3$ and CaO is 2.5×10^{-3} *M*. Thus, the value of K at 800°C is 2.5×10^{-3} (remember that equilibrium constants are unitless).

A more complex example of this type of problem is the conversion of *n*-butane, an additive used to increase the volatility of gasoline, to isobutane (2-methylpropane). This reaction can be written as

$$n\text{-butane}(g) \rightleftharpoons \text{isobutane}(g) \qquad (15.23)$$

for which the equilibrium constant $K = [i\text{-but}]/[n\text{-but}]$. At equilibrium, a mixture of n-butane and isobutane at room temperature was found to contain 0.041 M isobutane and 0.016 M n-butane. Substituting these concentrations into the equilibrium constant expression gives

$$K = \frac{[i\text{-but}]}{[n\text{-but}]} = \frac{0.041\ \cancel{M}}{0.016\ \cancel{M}} = 2.6 \qquad (15.24)$$

Thus, the equilibrium constant K for the reaction as written is 2.6.

EXAMPLE 15.8

The reaction between gaseous sulfur dioxide and oxygen is a key step in the industrial synthesis of sulfuric acid:

$$2SO_2(g) + O_2(g) \rightleftharpoons 2SO_3(g)$$

A mixture of SO_2 and O_2 was maintained at a temperature of 800 K until the system reached equilibrium. The equilibrium mixture contained 5.0×10^{-2} M SO_3, 3.5×10^{-3} M O_2, and 3.0×10^{-3} M SO_2. Calculate the equilibrium constant, K and K_p, at this temperature.

Given Balanced equilibrium equation and composition of equilibrium mixture

Asked for Equilibrium constant

Strategy

Write the equilibrium constant expression for the reaction. Then substitute the appropriate equilibrium concentrations into this equation to obtain K.

Solution

Substituting the appropriate equilibrium concentrations into the equilibrium constant expression gives

$$K = \frac{[SO_3]^2}{[SO_2]^2[O_2]} = \frac{(5.0 \times 10^{-2})^2}{(3.0 \times 10^{-3})^2(3.5 \times 10^{-3})} = 7.9 \times 10^4$$

To solve for K_p we use Equation 16.16, where $\Delta n = 2 - 3 = -1$:

$$K_p = K(RT)^{\Delta n}$$
$$K_p = 7.9 \times 10^4\ [(0.08206\ \text{L} \cdot \text{atm/mol} \cdot \text{K})(800\ \cancel{K})]^{-1}$$
$$= 1.2 \times 10^3$$

EXERCISE 15.8

Hydrogen gas and iodine react to form hydrogen iodide via the reaction

$$H_2(g) + I_2(g) \rightleftharpoons 2HI(g)$$

A mixture of H_2 and I_2 was maintained at a temperature of 740 K until the system reached equilibrium. The equilibrium mixture contained 0.0137 M HI, 6.47×10^{-3} M H_2, and 5.94×10^{-4} M I_2. Calculate K and K_p for this reaction.

Answer $K = 48.8$, $K_p = 48.8$

Often chemists are not given the concentrations of all the substances or they are not likely to measure the equilibrium concentrations of all the relevant substances for a particular system. In such cases, we can obtain the equilibrium concentrations from the initial concentrations of the reactants and the balanced equation for the reaction, as long as the equilibrium concentration of one of the substances is known. The next example shows one way to do this.

EXAMPLE 15.9

A 1.00-mol sample of NOCl was placed in a 2.00-L reactor and heated to 227°C until the system reached equilibrium. The contents of the reactor were then analyzed and found to contain 0.056 mol of Cl_2. Calculate K at this temperature. The equation for the decomposition of NOCl to NO and Cl_2 is

$$2NOCl(g) \rightleftharpoons 2NO(g) + Cl_2(g)$$

Given Balanced equilibrium equation, amount of reactant, volume, and amount of one product at equilibrium

Asked for Value of K

Strategy

Ⓐ Write the equilibrium constant expression for the reaction. Construct a table showing the initial concentrations, the changes in concentrations, and the final concentrations (as initial concentrations plus changes in concentrations).

Ⓑ Calculate all possible initial concentrations from the data given and insert them in the table.

Ⓒ Use the coefficients in the balanced equation to obtain the changes in concentration of all other substances in the reaction. Insert those concentration changes in the table.

Ⓓ Obtain the final concentrations by summing the columns. Calculate the equilibrium constant for the reaction.

Solution

Ⓐ The first step in any such problem is to balance the equation for the reaction (if not already balanced) and use it to derive the equilibrium constant expression. In this case, the equation is already balanced, and the equilibrium constant expression is thus

$$K = \frac{[NO]^2[Cl_2]}{[NOCl]^2}$$

To obtain the concentrations of NOCl, NO, and Cl_2 at equilibrium, we construct a table showing what is known and what needs to be calculated. We begin by writing the balanced equation at the top of the table, followed by three lines corresponding to the initial concentrations, the changes in concentrations required to get from the initial to the final state, and the final concentrations.

	$2NOCl(g) \rightleftharpoons$	$2NO(g)$ +	$Cl_2(g)$
	[NOCl]	**[NO]**	**[Cl_2]**
Initial	____	____	____
Change	____	____	____
Final	____	____	____

Ⓑ Initially, the system contains 1.00 mol of NOCl in a 2.00-L container. Thus, $[NOCl]_i$ = 1.00 mol/2.00 L = 0.500 M. The initial concentrations of NO and Cl_2 are 0 M because initially no products are present. Moreover, we are told that at equilibrium the system contains 0.056 mol of Cl_2 in a 2.00-L container, so $[Cl_2]_f$ = 0.056 mol/2.00 L = 0.028 M. We insert these values into the table.

	$2NOCl(g) \rightleftharpoons$	$2NO(g)$ +	$Cl_2(g)$
	[NOCl]	**[NO]**	**[Cl_2]**
Initial	0.500	0	0
Change	____	____	____
Final	____	____	0.028

C We use the stoichiometric relationships given in the balanced equation to find the change in the concentration of Cl_2, the substance for which initial and final concentrations are known:

$$\Delta[Cl_2] = [0.028\ M\ (\text{final}) - 0.00\ M\ (\text{initial})] = +0.028\ M$$

According to the coefficients in the balanced equation, 2 mol of NO is produced for every 1 mol of Cl_2, so the change in the NO concentration is

$$\Delta[NO] = \left(\frac{0.028\ \cancel{\text{mol Cl}_2}}{\text{L}}\right)\left(\frac{2\ \text{mol NO}}{1\ \cancel{\text{mol Cl}_2}}\right) = 0.056\ M$$

Similarly, 2 mol of NOCl are consumed for every 1 mol of Cl_2 produced, so the change in the NOCl concentration is

$$\Delta[NOCl] = \left(\frac{0.028\ \cancel{\text{mol Cl}_2}}{\text{L}}\right)\left(\frac{-2\ \text{mol NOCl}}{1\ \cancel{\text{mol Cl}_2}}\right) = -0.056\ M$$

We insert these values into the table.

	$2NOCl(g)$ $\rightleftharpoons$	$2NO(g)$ +	$Cl_2(g)$
	[NOCl]	**[NO]**	**[Cl_2]**
Initial	0.500	0	0
Change	−0.056	+0.056	+0.028
Final	_____	_____	0.028

D We sum the numbers in the [NOCl] and [NO] columns to obtain the final concentrations of NO and NOCl:

$$[NO]_f = 0.000\ M + 0.056\ M = 0.056\ M$$
$$[NOCl]_f = 0.500\ M + (-0.056\ M) = 0.444\ M$$

We can now complete the table.

	$2NOCl(g)$ $\rightleftharpoons$	$2NO(g)$ +	$Cl_2(g)$
	[NOCl]	**[NO]**	**[Cl_2]**
Initial	0.500	0	0
Change	−0.056	+0.056	+0.028
Final	0.444	0.056	0.028

We can now calculate the equilibrium constant for the reaction:

$$K = \frac{[NO]^2[Cl_2]}{[NOCl]^2} = \frac{(0.056)^2(0.028)}{(0.444)^2} = 4.5 \times 10^{-4}$$

EXERCISE 15.9

The German chemist Fritz Haber (1868–1934, Nobel prize 1918) was able to synthesize ammonia, NH_3, by the reaction of 0.1248 M H_2 and 0.0416 M N_2 at about 500°C. At equilibrium, the mixture contained 0.00272 M NH_3. What is the equilibrium constant K for the reaction $N_2 + 3H_2 \rightleftharpoons 2NH_3$ at this temperature? What is K_p?

Answer $K = 0.105$, $K_p = 2.61 \times 10^{-5}$

The original laboratory apparatus designed by Fritz Haber and Robert Le Rossignol in 1908 for synthesizing ammonia from its elements. A metal catalyst bed, where ammonia was produced, is in the large cylinder at the left. The Haber–Bosch process used for the industrial production of ammonia uses essentially the same process and components, but on a much larger scale. Unfortunately, Haber's process enabled Germany to prolong World War I when German supplies of nitrogen compounds, which were used for explosives, had been exhausted in 1914.

Calculating Equilibrium Concentrations from the Equilibrium Constant

To describe how to calculate equilibrium concentrations from an equilibrium constant, we first consider a system that contains only a single product and a single reactant, the conversion of *n*-butane to isobutane (Equation 15.23), for which $K = 2.6$ at 25°C. If we begin with a 1.00 *M* sample of *n*-butane, we can determine the concentration of *n*-butane and isobutane at equilibrium by constructing a table showing what is known and what needs to be calculated, just as we did in Example 15.9.

$$n\text{-butane}(g) \rightleftharpoons \text{isobutane}(g)$$

	[*n*-Butane]	**[Isobutane]**
Initial	______	______
Change	______	______
Final	______	______

The initial concentrations of the reactant and product are both known: $[n\text{-but}]_i = 1.00\ M$ and $[i\text{-but}]_i = 0\ M$. We need to calculate the equilibrium concentrations of both *n*-butane and isobutane. Because it is generally difficult to calculate final concentrations directly, we focus on the *change* in the concentrations of the substances between the initial and final (equilibrium) conditions. If, for example, we define the change in the concentration of isobutane, $\Delta[i\text{-but}]$, as $+x$, then the change in the concentration of *n*-butane is $\Delta[n\text{-but}] = -x$. This is because the balanced equation for the reaction tells us that 1 mol of *n*-butane is consumed for every 1 mol of isobutane produced. We can then express the final concentrations in terms of the initial concentrations and the changes they have undergone.

$$n\text{-butane}(g) \rightleftharpoons \text{isobutane}(g)$$

	[*n*-Butane]	**[Isobutane]**
Initial	1.00	0
Change	$-x$	$+x$
Final	$(1.00 - x)$	$(0 + x) = x$

Substituting the expressions for the final concentrations of *n*-butane and isobutane from the table into the equilibrium equation gives

$$K = \frac{[i\text{-but}]}{[n\text{-but}]} = \frac{x}{1.00 - x} = 2.6$$

Rearranging and solving for x give

$$x = 2.6(1.00 - x) = 2.6 - 2.6x$$
$$x + 2.6x = 2.6$$
$$x = 0.72$$

We obtain the final concentrations by substituting this value of x into the expressions for the final concentrations of *n*-butane and isobutane listed in the table:

$$[n\text{-but}]_f = (1.00 - x)\ M = (1.00 - 0.72)\ M = 0.28\ M$$
$$[i\text{-but}]_f = (0.00 + x)\ M = (0.00 + 0.72)\ M = 0.72\ M$$

We can check the results by substituting them back into the equilibrium constant expression to see whether they give the same value of K that we used in the calculation:

$$K = \frac{[i\text{-but}]}{[n\text{-but}]} = \frac{0.72\ \cancel{M}}{0.28\ \cancel{M}} = 2.6 \quad \text{(to two significant figures)}$$

This is the same value of K we were given, so we can be confident of our results.

The next examples illustrate some common types of equilibrium problems you are likely to encounter.

EXAMPLE 15.10

The *water–gas shift reaction* is important in a number of chemical processes, such as the production of H_2 for fuel cells. This reaction can be written as

$$H_2(g) + CO_2(g) \rightleftharpoons H_2O(g) + CO(g) \qquad K = 0.106 \text{ at } 700 \text{ K}$$

If a mixture of gases that initially contains 0.0150 M H_2 and 0.0150 M CO_2 is allowed to equilibrate at 700 K, what are the final concentrations of all substances present?

Given Balanced equilibrium equation, value of K, and initial concentrations

Asked for Final concentrations

Strategy

Ⓐ Construct a table showing what is known and what needs to be calculated. Define x as the change in the concentration of one substance. Then use the reaction stoichiometry to express the changes in the concentrations of the other substances in terms of x. From the values in the table, calculate the final concentrations.

Ⓑ Write the equilibrium equation for the reaction. Substitute appropriate values from the table to obtain the value of x.

Ⓒ Calculate the final concentrations of all species present. Check your answers by substituting these values into the equilibrium constant expression to obtain K.

Solution

Ⓐ The initial concentrations of the reactants are $[H_2]_i = [CO_2]_i = 0.0150\ M$. Just as before, we will focus on the *change* in the concentrations of the various substances between the initial and final states. If we define the change in the concentration of H_2O as x, then $\Delta[H_2O] = +x$. We can use the stoichiometry of the reaction to express the changes in the concentrations of the other substances in terms of x. For example, 1 mol of CO is produced for every 1 mol of H_2O, so the change in the CO concentration can be expressed as $\Delta[CO] = +x$. Similarly, for every 1 mol of H_2O produced, 1 mol each of H_2 and CO_2 is consumed, so the change in the concentration of the reactants is $\Delta[H_2] = \Delta[CO_2] = -x$. We enter the values in the table and calculate the final concentrations.

	$H_2(g)$	+	$CO_2(g)$	$\rightleftharpoons$	$H_2O(g)$	+	$CO(g)$
	$[H_2]$		**$[CO_2]$**		**$[H_2O]$**		**$[CO]$**
Initial	0.0150		0.0150		0		0
Change	$-x$		$-x$		$+x$		$+x$
Final	$(0.0150 - x)$		$(0.0150 - x)$		x		x

Ⓑ We can now use the equilibrium equation and the value of K given to solve for x:

$$K = \frac{[H_2O][CO]}{[H_2][CO_2]} = \frac{(x)(x)}{(0.0150 - x)(0.0150 - x)} = \frac{x^2}{(0.0150 - x)^2} = 0.106$$

We could solve this equation using the quadratic formula (see Essential Skills 7 at the end of this chapter), but it is far easier to solve for x by recognizing that the left side of the equation is a perfect square; that is,

$$\frac{x^2}{(0.0150 - x)^2} = \left(\frac{x}{0.0150 - x}\right)^2 = 0.106$$

Taking the square root of the middle and right terms gives

$$\frac{x}{(0.0150 - x)} = (0.106)^{1/2} = 0.326$$

$$x = (0.326)(0.0150) - 0.326x$$

$$1.326x = 0.00489$$

$$x = 0.00369 = 3.69 \times 10^{-3} \quad \text{(to three significant figures)}$$

C The final concentrations of all species in the reaction mixture are

$$[H_2]_f = [H_2]_i + \Delta[H_2] = (0.0150 - 0.00369)\ M = 0.0113\ M$$

$$[CO_2]_f = [CO_2]_i + \Delta[CO_2] = (0.0150 - 0.00369)\ M = 0.0113\ M$$

$$[H_2O]_f = [H_2O]_i + \Delta[H_2O] = (0 + 0.00369)\ M = 0.00369\ M$$

$$[CO]_f = [CO]_i + \Delta[CO] = (0 + 0.00369)\ M = 0.00369\ M$$

We can check our work by inserting the calculated values back into the equilibrium constant expression:

$$K = \frac{[H_2O][CO]}{[H_2][CO_2]} = \frac{(0.00369)^2}{(0.0113)^2} = 0.107$$

To 2 significant figures, this value of K is the same as the value given in the problem, so our answer is confirmed.

EXERCISE 15.10

Hydrogen gas reacts with iodine vapor to give hydrogen iodide according to the chemical equation

$$H_2(g) + I_2(g) \rightleftharpoons 2HI(g) \qquad K = 54 \text{ at } 425°C$$

If 0.172 M H_2 and I_2 are injected into a reactor and maintained at 425°C until the system equilibrates, what is the final concentration of each substance in the reaction mixture?

Answer $[HI]_f = 0.270\ M$; $[H_2]_f = [I_2]_f = 0.037\ M$

In Example 15.10, the initial concentrations of the reactants were the same, which gave us an equation that was a perfect square and simplified our calculations. Often, however, the initial concentrations of the reactants are not the same, or one or more of the products may be present when the reaction starts. Under these conditions, there is usually no way to simplify the problem, and we must determine the equilibrium concentrations using other means. Such a case is described in the next example.

EXAMPLE 15.11

In the water–gas shift reaction shown in Example 15.10, a sample containing 0.632 M CO_2 and 0.570 M H_2 is allowed to equilibrate at 700 K. At this temperature $K = 0.106$. What is the composition of the reaction mixture at equilibrium?

Given Balanced equilibrium equation, concentrations of reactants, and value of K

Asked for Composition of reaction mixture at equilibrium

Strategy

A Write the equilibrium equation. Construct a table showing the initial concentrations of all substances in the mixture. Complete the table showing the changes in the concentrations (x) and the final concentrations.

B Write the equilibrium constant expression for the reaction. Substitute the known value of K and the final concentrations to solve for x.

C Calculate the final concentration of each substance in the reaction mixture. Check your answers by substituting these values into the equilibrium constant expression to obtain K.

Solution

A The initial concentrations of CO_2 and H_2 are given: $[CO_2]_i = 0.632\ M$ and $[H_2]_i = 0.570\ M$. Again, x is defined as the change in the concentration of H_2O: $\Delta[H_2O] = +x$. Because 1 mol of CO is produced for every 1 mol of H_2O, the change in the concentration of CO is the same as the change in the concentration of H_2O, so $\Delta[CO] = +x$. Similarly, because 1 mol of both H_2 and CO_2 is consumed for every 1 mol of H_2O produced, $\Delta[H_2] = \Delta[CO_2] = -x$. The final concentrations are the sums of the initial concentrations and the changes in concentrations at equilibrium.

	$H_2(g)$	+	$CO_2(g)$	$\rightleftharpoons$	$H_2O(g)$	+	$CO(g)$
	$[H_2]$		**$[CO_2]$**		**$[H_2O]$**		**[CO]**
Initial	0.570		0.632		0		0
Change	$-x$		$-x$		$+x$		$+x$
Final	$(0.570 - x)$		$(0.632 - x)$		x		x

B We can now use the equilibrium equation and the known value of K to solve for x:

$$K = \frac{[H_2O][CO]}{[H_2][CO_2]} = \frac{x^2}{(0.570 - x)(0.632 - x)} = 0.106$$

In contrast to Example 15.10, however, there is no obvious way to simplify this expression. Thus, we must expand the expression and multiply both sides by the denominator:

$$x^2 = 0.106(0.360 - 1.202x + x^2)$$

Collecting terms on one side of the equation gives

$$0.894x^2 + 0.127x - 0.0382 = 0$$

This equation can be solved using the quadratic formula (see Essential Skills 7):

$$x = \frac{-b \pm \sqrt{b^2 - 4ac}}{2a} = \frac{-0.127 \pm \sqrt{(0.127)^2 - 4(0.894)(-0.0382)}}{2(0.894)} = 0.148, -0.290$$

Only the answer with the positive value has any physical significance, so $\Delta[H_2O] = \Delta[CO] = +0.148\ M$, and $\Delta[H_2] = \Delta[CO_2] = -0.148\ M$. **C** The final concentrations of all species in the reaction mixture are

$$[H_2]_f = [H_2]_i + \Delta[H_2] = 0.570\ M - 0.148\ M = 0.422\ M$$
$$[CO_2]_f = [CO_2]_i + \Delta[CO_2] = 0.632\ M - 0.148\ M = 0.484\ M$$
$$[H_2O]_f = [H_2O]_i + \Delta[H_2O] = 0\ M + 0.148\ M = 0.148\ M$$
$$[CO]_f = [CO]_i + \Delta[CO] = 0\ M + 0.148\ M = 0.148\ M$$

We can check our work by substituting these values into the equilibrium constant expression:

$$K = \frac{[H_2O][CO]}{[H_2][CO_2]} = \frac{(0.148)^2}{(0.422)(0.484)} = 0.107$$

Because the value of K is essentially the same as the value given in the problem, our calculations are confirmed.

EXERCISE 15.11

Exercise 15.8 showed the reaction of hydrogen and iodine vapor to form hydrogen iodide, for which $K = 54$ at 425°C. If a sample containing 0.200 M H_2 and 0.0450 M I_2 is allowed to equilibrate at 425°C, what is the final concentration of each substance in the reaction mixture?

Answer $[HI]_f = 0.0882\ M$; $[H_2]_f = 0.156\ M$; $[I_2]_f = 9.2 \times 10^{-4}\ M$

In many situations it is not necessary to solve a quadratic (or higher-order) equation. Most of these cases involve reactions for which the equilibrium constant is either very small ($K \leqslant 10^{-3}$) or very large ($K \geqslant 10^3$), which means that the change in the concentration (defined as x) is essentially negligible compared with the initial concentration of a substance. Knowing this simplifies the calculations dramatically, as illustrated in the next example.

EXAMPLE 15.12

Atmospheric nitrogen and oxygen react to form nitric oxide:

$$N_2(g) + O_2(g) \rightleftharpoons 2NO(g) \qquad K_p = 2.0 \times 10^{-31} \text{ at } 25^\circ C$$

What is the partial pressure of NO in equilibrium with N_2 and O_2 in the atmosphere (at 1 atm, $P_{N_2} = 0.78$ atm and $P_{O_2} = 0.21$ atm)?

Given Balanced equilibrium equation and values of K_p, P_{O_2}, and P_{N_2}

Asked for Partial pressure of NO

Strategy

A Construct a table and enter the initial partial pressures, the changes in the partial pressures that occur during the course of the reaction, and the final partial pressures of all substances.

B Write the equilibrium equation for the reaction. Then substitute values from the table to solve for the change in concentration, x.

C Calculate the partial pressure of NO. Check your answer by substituting values into the equilibrium equation and solving for K.

Solution

A Because we are given K_p and partial pressures are reported in atmospheres, we will use partial pressures. The initial partial pressure of O_2 is 0.21 atm, and that of N_2 is 0.78 atm. If we define the change in the partial pressure of NO as $2x$, then the change in the partial pressure of O_2 and of N_2 is $-x$ because 1 mol each of N_2 and of O_2 is consumed for every 2 mol of NO produced. Each substance has a final partial pressure equal to the sum of the initial pressure and the change in that pressure at equilibrium.

	$N_2(g)$	+	$O_2(g)$	$\rightleftharpoons$	$2NO(g)$
	P_{N_2}, atm		P_{O_2}, atm		P_{NO}, atm
Initial *P*	0.78		0.21		0
Change in *P*	$-x$		$-x$		$+2x$
Final *P*	$(0.78 - x)$		$(0.21 - x)$		$2x$

B Substituting these values into the equation for the equilibrium constant gives

$$K_p = \frac{(P_{NO})^2}{(P_{N_2})(P_{O_2})} = \frac{(2x)^2}{(0.78 - x)(0.21 - x)} = 2.0 \times 10^{-31}$$

In principle, we could multiply out the terms in the denominator, rearrange, and solve the resulting quadratic equation. In practice, it is far easier to recognize that an equilibrium constant of this magnitude means that the extent of the reaction will be very small; therefore, the value of x will be negligible compared with the initial concentrations. If this assumption is correct, then to two significant figures, $(0.78 - x) = 0.78$ and $(0.21 - x) = 0.21$. Substituting these expressions into our original equation gives

$$\frac{(2x)^2}{(0.78)(0.21)} = 2.0 \times 10^{-31}$$

$$\frac{4x^2}{0.16} = 2.0 \times 10^{-31}$$

$$x^2 = \frac{0.33 \times 10^{-31}}{4}$$
$$x = 9.1 \times 10^{-17}$$

C Substituting this value of x into our expressions for the final partial pressures of the substances gives

$$P_{NO} = 2x \text{ atm} = 1.8 \times 10^{-16} \text{ atm}$$
$$P_{N_2} = (0.78 - x) \text{ atm} = 0.78 \text{ atm}$$
$$P_{O_2} = (0.21 - x) \text{ atm} = 0.21 \text{ atm}$$

From these calculations, we see that our initial assumption regarding x was correct: given two significant figures, 2.0×10^{-16} is certainly negligible compared with 0.78 and 0.21. When can we make such an assumption? As a general rule, if x is less than about 5% of the total, or $10^{-3} > K > 10^3$, then the assumption is justified. Otherwise, we must use the quadratic formula or some other approach. The results we have obtained agree with the general observation that toxic NO, an ingredient of smog, does not form from atmospheric concentrations of N_2 and O_2 to a substantial degree at 25°C. We can verify our results by substituting them into the original equilibrium equation:

$$K_p = \frac{(P_{NO})^2}{(P_{N_2})(P_{O_2})} = \frac{(1.8 \times 10^{-16})^2}{(0.78)(0.21)} = 2.0 \times 10^{-31}$$

The final value of K_p agrees with the value given at the beginning of this example.

EXERCISE 15.12

Under certain conditions, oxygen will react to form ozone, as shown in the equation

$$3O_2(g) \rightleftharpoons 2O_3(g) \qquad K_p = 2.5 \times 10^{-59} \text{ at } 25^\circ\text{C}$$

What ozone partial pressure is in equilibrium with oxygen in the atmosphere ($P_{O_2} = 0.21$ atm)?

Answer 4.8×10^{-31} atm

Another type of problem that can be simplified by assuming that changes in concentration are negligible is one in which the equilibrium constant is very large ($K \geq 10^3$). A large equilibrium constant implies that reactants are converted almost entirely to products, so we can assume that the reaction proceeds 100% to completion. When we solve this type of problem, we view the system as equilibrating from the *products* side of the reaction rather than the reactants side. This approach is illustrated in the next example.

EXAMPLE 15.13

The chemical equation for the reaction of hydrogen with ethylene (C_2H_4) to give ethane (C_2H_6) is

$$H_2(g) + C_2H_4(g) \underset{}{\overset{\text{Ni}}{\rightleftharpoons}} C_2H_6(g) \qquad K = 9.6 \times 10^{18} \text{ at } 25^\circ\text{C}$$

If a mixture of 0.200 M H_2 and 0.155 M C_2H_4 is maintained at 25°C in the presence of a powdered nickel catalyst, what is the equilibrium concentration of each substance in the mixture?

Given Balanced chemical equation, value of K, and initial concentrations of reactants

Asked for Equilibrium concentrations

Strategy

A Construct a table showing initial concentrations, concentrations that would be present if the reaction were to go to completion, changes in concentrations, and final concentrations.

B Write the equilibrium constant expression for the reaction. Then substitute values from the table into the expression to solve for x, the change in concentration.

C Calculate the equilibrium concentrations. Check your answers by substituting these values into the equilibrium equation.

Solution

A From the magnitude of the equilibrium constant, we see that the reaction goes essentially to completion. Because the initial concentration of ethylene (0.155 M) is less than the concentration of hydrogen (0.200 M), ethylene is the limiting reactant; that is, no more than 0.155 M ethane can be formed from 0.155 M ethylene. If the reaction were to go to completion, the concentration of ethane would be 0.155 M and the concentration of ethylene would be 0 M. Because the concentration of hydrogen is greater than is needed for complete reaction, the concentration of unreacted hydrogen in the reaction mixture would be 0.200 M − 0.155 M = 0.045 M. The equilibrium constant for the forward reaction is very large, so the equilibrium constant for the reverse reaction must be very small. The problem then is identical to the ones described in Example 15.12 and Exercise 15.12. If we define $-x$ as the change in the ethane concentration for the reverse reaction, then the change in the ethylene and hydrogen concentrations is $+x$. The final equilibrium concentrations are the sums of the concentrations for the forward and reverse reactions.

	$H_2(g)$ +	$C_2H_4(g)$ $\xrightleftharpoons{\text{Ni}}$	$C_2H_6(g)$
	$[H_2]$	$[C_2H_4]$	$[C_2H_6]$
Initial	0.200	0.155	0
Assuming 100% reaction	0.045	0	0.155
Change	$+x$	$+x$	$-x$
Final	$(0.045 + x)$	$(0 + x)$	$(0.155 - x)$

B Substituting values into the equilibrium constant expression gives:

$$K = \frac{[C_2H_6]}{[H_2][C_2H_4]} = \frac{0.155 - x}{(0.045 + x)x} = 9.6 \times 10^{18}$$

Once again, the magnitude of the equilibrium constant tells us that the equilibrium will lie far to the right as written, so the reverse reaction is negligible. Thus, x is likely to be very small compared with either 0.155 M or 0.045 M, and the equation can be simplified $[(0.045 + x) = 0.045$ and $(0.155 - x) = 0.155]$ to give

$$K = \frac{0.155}{0.045x} = 9.6 \times 10^{18}$$

$$x = 3.6 \times 10^{-19}$$

C The small value of x indicates that our assumption concerning the reverse reaction is correct, and we can therefore calculate the final concentrations by evaluating the expressions from the last line of the table:

$$[C_2H_6]_f = (0.155 - x)\,M = 0.155\,M$$

$$[C_2H_4]_f = x\,M = 3.6 \times 10^{-19}\,M$$

$$[H_2]_f = (0.045 + x)\,M = 0.045\,M$$

We can verify our calculations by substituting the final concentrations into the equilibrium constant expression:

$$K = \frac{[C_2H_6]}{[H_2][C_2H_4]} = \frac{0.155}{(0.045)(3.6 \times 10^{-19})} = 9.6 \times 10^{18}$$

This value of K agrees with our initial value at the beginning of the example.

EXERCISE 15.13

Hydrogen reacts with chlorine gas to form hydrogen chloride:

$$H_2(g) + Cl_2(g) \rightleftharpoons 2HCl(g) \qquad K_p = 4.0 \times 10^{31} \text{ at } 47^\circ C$$

If a mixture of 0.257 M H_2 and 0.392 M Cl_2 is allowed to equilibrate at 47°C, what is the equilibrium composition of the mixture?

Answer $[H_2]_f = 4.8 \times 10^{-32}$ M; $[Cl_2]_f = 0.135$ M; $[HCl]_f = 0.514$ M

15.4 • Nonequilibrium Conditions

MGC The Reaction Quotient

In Section 15.3, we saw that knowing the magnitude of the equilibrium constant under a given set of conditions allows chemists to predict the extent of a reaction. Often, however, chemists must decide whether a system has reached equilibrium or the composition of the mixture will continue to change with time. In this section, we describe how to analyze the composition of a reaction mixture quantitatively in order to make this determination.

The Reaction Quotient (Q)

To determine whether a system has reached equilibrium, chemists use a quantity called the **reaction quotient (Q)**. The expression for the reaction quotient has precisely the same form as the equilibrium constant expression, except that Q may be derived from a set of values measured at *any* time during the reaction of *any* mixture of the reactants and products, regardless of whether the system is at equilibrium. Therefore, for the general reaction

$$aA + bB \rightleftharpoons cC + dD$$

the reaction quotient is defined as

$$Q = \frac{[C]^c[D]^d}{[A]^a[B]^b} \tag{15.25}$$

A related quantity Q_p, analogous to K_p, can be written for any reaction that involves gases by using the partial pressures of the components.

To understand how information is obtained using a reaction quotient, consider the dissociation of dinitrogen tetroxide to nitrogen dioxide, $N_2O_4(g) \rightleftharpoons 2NO_2(g)$, for which $K = 4.65 \times 10^{-3}$ at 298 K. We can write Q for this reaction as

$$Q = \frac{[NO_2]^2}{[N_2O_4]} \tag{15.26}$$

The table lists data from three experiments in which samples of the reaction mixture were obtained and analyzed at equivalent time intervals, and the corresponding values of Q calculated for each. Note that each experiment begins with different proportions of product and reactant:

Experiment	$[NO_2]$, M	$[N_2O_4]$, M	$Q = [NO_2]^2/[N_2O_4]$
1	0	0.0400	$\frac{0^2}{0.0400} = 0$
2	0.0600	0	$\frac{(0.0600)^2}{0} =$ undefined
3	0.0200	0.0600	$\frac{(0.0200)^2}{0.0600} = 6.67 \times 10^{-3}$

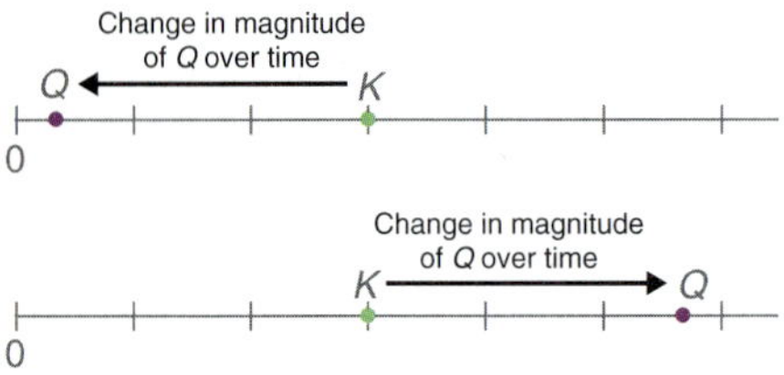

(a) Q versus K, t_n

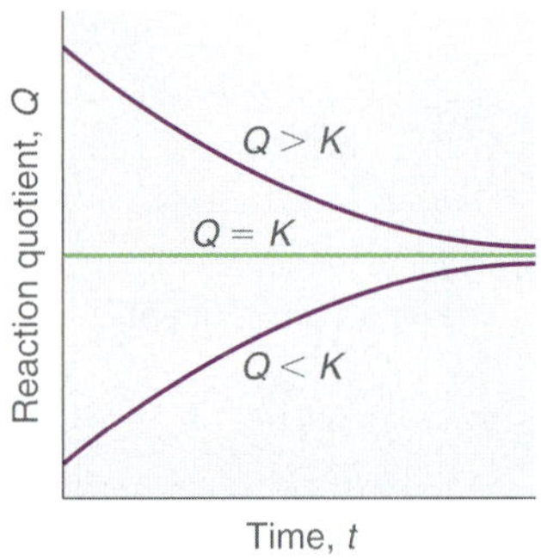

(b) Q versus K, t_0 to t_∞

Figure 15.6 Two different ways of illustrating how the composition of a system will change depending on the relative values of Q and K. (a) Both Q and K are plotted as points along a number line: the system will always react in the way that causes Q to approach K. **(b)** The change in the composition of a system with time is illustrated for systems with initial values of $Q > K$, $Q < K$, and $Q = K$. Interactive Graph

As these calculations demonstrate, Q can have any numerical value between 0 and infinity (undefined); that is, the value of Q can be greater than, less than, or equal to K.

Comparing the magnitudes of Q and K enables us to determine whether a reaction mixture is already at equilibrium and, if it is not, to predict how its composition will change with time in order to reach equilibrium (that is, whether the reaction will proceed to the right or to the left as written). All you need to remember is that the composition of a system that is not at equilibrium will change in a way that makes the value of Q approach the value of K. If $Q = K$, for example, then the system is already at equilibrium, and no further change in the composition of the system will occur unless the conditions are changed. If $Q < K$, then the ratio of the concentrations of products to the concentrations of reactants is *less* than the ratio at equilibrium. Therefore, the reaction will proceed to the right as written, forming products at the expense of reactants. Conversely, if $Q > K$, then the ratio of the concentrations of products to the concentrations of reactants is *greater* than at equilibrium, so the reaction will proceed to the left as written, forming reactants at the expense of products. These points are illustrated graphically in Figure 15.6.

> **Note the pattern**
>
> *If $Q < K$, the reaction will proceed to the right as written. If $Q > K$, the reaction will proceed to the left as written. If $Q = K$, then the system is at equilibrium.*

EXAMPLE 15.14

At elevated temperatures, methane (CH_4) reacts with water to produce hydrogen and carbon monoxide in what is known as a *steam-reforming* reaction:

$$CH_4(g) + H_2O(g) \rightleftharpoons CO(g) + 3H_2(g) \qquad K = 2.4 \times 10^{-4} \text{ at } 900 \text{ K}$$

Huge amounts of hydrogen are produced from natural gas in this way and are then used for the industrial synthesis of ammonia. If 1.2×10^{-2} mol of CH_4, 8.0×10^{-3} mol of H_2O, 1.6×10^{-2} mol of CO, and 6.0×10^{-3} mol of H_2 are placed in a 2.0-L steel reactor and heated to 900 K, will the reaction be at equilibrium or will it proceed to the right to produce CO and H_2 or to the left to form CH_4 and H_2O?

Given Balanced equation, value of K, amounts of reactants and products, and volume

Asked for Direction of reaction

Strategy

A Calculate the molar concentrations of the reactants and products.
B Use Equation 15.25 to determine Q. Compare the values of Q and K to decide in which direction the reaction will proceed.

Solution

A We must first determine the initial concentrations of the substances present. For example, we have 1.2×10^{-2} mol of CH_4 in a 2.0-L container, so

$$[CH_4] = \frac{1.2 \times 10^{-2} \text{ mol}}{2.0 \text{ L}} = 6.0 \times 10^{-3}\ M$$

We can calculate the other concentrations in a similar way: $[H_2O] = 4.0 \times 10^{-3}\ M$, $[CO] = 8.0 \times 10^{-3}\ M$, and $[H_2] = 3.0 \times 10^{-3}\ M$. **B** We now compute the value of Q and compare it with the value of K:

$$Q = \frac{[CO][H_2]^3}{[CH_4][H_2O]} = \frac{(8.0 \times 10^{-3})(3.0 \times 10^{-3})^3}{(6.0 \times 10^{-3})(4.0 \times 10^{-3})} = 9.0 \times 10^{-6}$$

Because $K = 2.4 \times 10^{-4}$, we see that $Q < K$. Thus, the ratio of the concentrations of products to the concentrations of reactants is less than the ratio for an equilibrium mixture. The reaction will therefore proceed to the right as written, forming H_2 and CO at the expense of H_2O and CH_4.

EXERCISE 15.14

In the water–gas shift reaction introduced in Example 15.10, CO produced by steam reforming of methane (Example 15.14) reacts with steam at elevated temperatures to produce more hydrogen:

$$CO(g) + H_2O(g) \rightleftharpoons CO_2(g) + H_2(g) \qquad K = 0.64 \text{ at } 900 \text{ K}$$

If 0.010 mol of both CO and H_2O, 0.0080 mol of CO_2, and 0.012 mol of H_2 are injected into a 4.0-L reactor and heated to 900 K, will the reaction proceed to the left or to the right as written?

Answer $Q = 0.96$ ($Q > K$), so the reaction will proceed to the left and CO and H_2O will form.

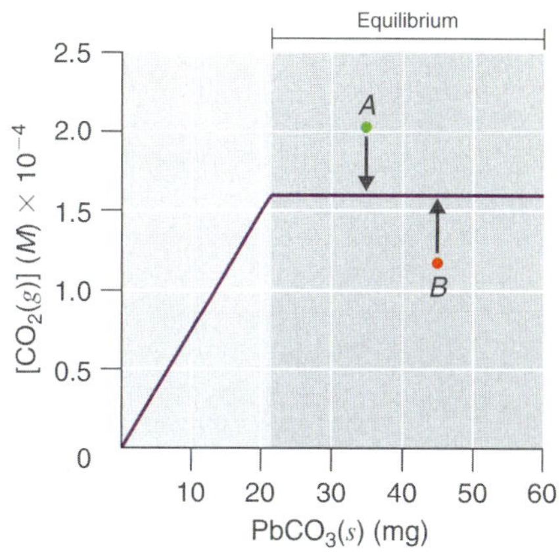

Figure 15.7 The concentration of gaseous CO_2 in a closed system at equilibrium as a function of the amount of solid $PbCO_3$ added. Initially the concentration of $CO_2(g)$ increases linearly with the amount of solid $PbCO_3$ added, as $PbCO_3$ decomposes to $CO_2(g)$ and solid PbO. Once the CO_2 concentration reaches the value that corresponds to the equilibrium concentration, however, adding more solid $PbCO_3$ has no effect on $[CO_2]$, as long as the temperature remains constant. Interactive Graph

Predicting the Direction of Reaction Using a Graph

By graphing a few equilibrium concentrations for a system at a given temperature and pressure, we can readily see the range of reactant and product concentrations that correspond to equilibrium conditions, for which $Q = K$. Such a graph allows us to predict what will happen to a reaction when conditions change so that Q no longer equals K, such as when a reactant or product concentration is increased or decreased.

Lead carbonate decomposes to lead oxide and carbon dioxide according to the equation

$$PbCO_3(s) \rightleftharpoons PbO(s) + CO_2(g) \qquad (15.27)$$

Because $PbCO_3$ and PbO are solids, the equilibrium constant is simply $K = [CO_2]$. At a given temperature, therefore, any system that contains solid $PbCO_3$ and solid PbO will have exactly the same concentration of CO_2 at equilibrium, regardless of the ratio or the amounts of the solids present. This situation is represented in Figure 15.7, which shows a plot of $[CO_2]$ versus the amount of $PbCO_3$ added. Initially, the added $PbCO_3$ decomposes completely to CO_2 because the amount of $PbCO_3$ is not sufficient to give a CO_2 concentration equal to K. Thus, the left portion of the graph represents a system that is *not* at equilibrium because it contains only $CO_2(g)$ and $PbO(s)$. In contrast, when just enough $PbCO_3$ has been added to give $[CO_2] = K$, the system has reached equilibrium, and adding more $PbCO_3$ has no effect on the CO_2 concentration: the graph is a horizontal line. Thus, any CO_2 concentration that is not on the horizontal line represents a nonequilibrium state, and the system will adjust its composition to achieve equilibrium provided enough $PbCO_3$ and PbO are present. For example, the point labeled A in Figure 15.7 lies above the horizontal line, so it corresponds to a $[CO_2]$ that is greater than the equilibrium concentration of CO_2 ($Q > K$). To reach equilibrium, the system must decrease $[CO_2]$, which it can do only by the reaction of CO_2 with solid PbO to form solid $PbCO_3$. Thus, the reaction in Equation 15.27 will proceed to the left as written, until $[CO_2] = K$. Conversely, the point labeled B in Figure 15.7 lies below the horizontal line, so it corresponds to a $[CO_2]$ that is less than the equilibrium concentration of CO_2 ($Q < K$). To reach equilibrium, the system must increase $[CO_2]$, which it can do only by the decomposition of solid $PbCO_3$ to form CO_2 and solid PbO. The reaction in Equation 15.27 will therefore proceed to the right as written, until $[CO_2] = K$.

In contrast, for the reduction of cadmium oxide by hydrogen to give metallic cadmium and water vapor:

$$CdO(s) + H_2(g) \rightleftharpoons Cd(s) + H_2O(g) \qquad (15.28)$$

the equilibrium constant K is $[H_2O]/[H_2]$. Thus, if $[H_2O]$ is doubled at equilibrium, then $[H_2]$ must also be doubled for the system to remain at equilibrium. A plot of $[H_2O]$ versus $[H_2]$ at equilibrium is a straight line with a slope of K (Figure 15.8). Again, only those pairs of concentrations of H_2O and H_2 that lie on the line correspond to equilibrium states. Any point representing a pair of concentrations that does

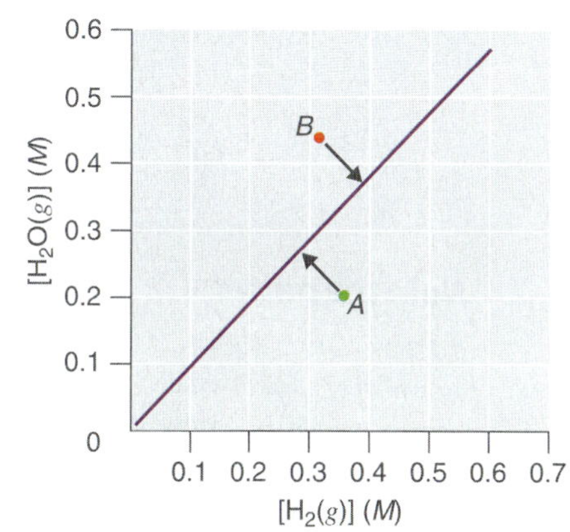

Figure 15.8 The concentration of water vapor versus the concentration of hydrogen for the $CdO(s) + H_2(g) \rightleftharpoons Cd(s) + H_2O(g)$ system at equilibrium. For any equilibrium concentration of $H_2O(g)$, there is only one equilibrium concentration of $H_2(g)$. Because the magnitudes of the two concentrations are directly proportional, a large $[H_2O]$ at equilibrium requires a large $[H_2]$, and vice versa. In this case, the slope of the line is equal to the value of the equilibrium constant K. Interactive Graph

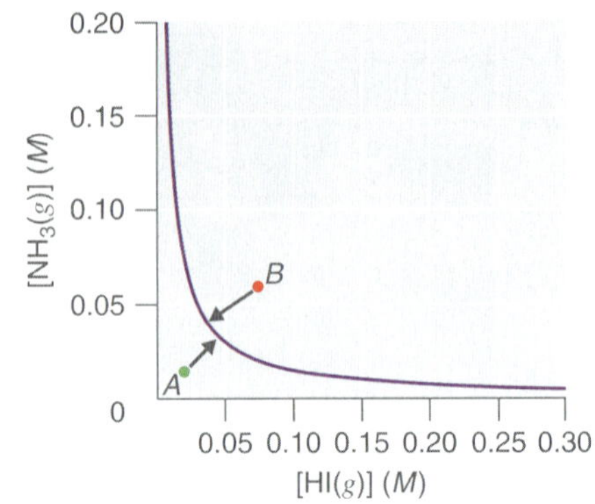

Figure 15.9 The concentration of $NH_3(g)$ versus the concentration of HI(g) for the $NH_4I(s) \rightleftharpoons NH_3(g) + HI(g)$ system at equilibrium. Only one equilibrium concentration of $NH_3(g)$ is possible for any given equilibrium concentration of HI(g). In this case, the two are inversely proportional. Thus, a large [HI] at equilibrium requires a small $[NH_3]$ at equilibrium, and vice versa.

Interactive Graph

not lie on the line corresponds to a nonequilibrium state. In such cases, the reaction in Equation 15.28 will proceed in whichever direction causes the composition of the system to move toward the equilibrium line. For example, point A in Figure 15.8 lies below the line, indicating that the $[H_2O]/[H_2]$ ratio is lower than the ratio of an equilibrium mixture ($Q < K$). Thus, the reaction in Equation 15.28 will proceed to the right as written, consuming H_2 and producing H_2O, which causes the concentration ratio to move up and to the left toward the equilibrium line. Conversely, point B in Figure 15.8 lies above the line, indicating that the $[H_2O]/[H_2]$ ratio is higher than the ratio of an equilibrium mixture ($Q > K$). Thus, the reaction in Equation 15.28 will proceed to the left as written, consuming H_2O and producing H_2, which causes the concentration ratio to move down and to the right toward the equilibrium line.

In another example, solid ammonium iodide dissociates to gaseous ammonia and hydrogen iodide at elevated temperatures:

$$NH_4I(s) \rightleftharpoons NH_3(g) + HI(g) \tag{15.29}$$

For this system, the equilibrium constant K is equal to the product of the concentrations of the two products, $[NH_3] \times [HI]$. Thus, if we double the concentration of NH_3, the concentration of HI must decrease by approximately a factor of 2 to maintain equilibrium, as shown in Figure 15.9. As a result, for a given concentration of either HI or NH_3, only a *single* equilibrium composition that contains equal concentrations of both NH_3 and HI is possible, for which $[NH_3] = [HI] = K^{1/2}$. Any point that lies below and to the left of the equilibrium curve (such as point A in Figure 15.9) corresponds to $Q < K$, and the reaction in Equation 15.29 will therefore proceed to the right as written, causing the composition of the system to move toward the equilibrium line. Conversely, any point that lies above and to the right of the equilibrium curve (such as point B in Figure 15.9) corresponds to $Q > K$, and the reaction in Equation 15.29 will therefore proceed to the left as written, again causing the composition of the system to move toward the equilibrium line. Thus, by graphing equilibrium concentrations for a given system at a given temperature and pressure, we can predict the direction of reaction of that mixture when the system is not at equilibrium.

Le Châtelier's Principle

When a system at equilibrium is perturbed in some way, the effects of the perturbation can be predicted qualitatively using **Le Châtelier's principle** (named after the French chemist Henri Louis Le Châtelier, 1850–1936).* This principle can be stated as follows: *If a stress is applied to a system at equilibrium, the composition of the system will change to counteract the applied stress.* Stress occurs when any change in the system affects the magnitude of Q or K. In Equation 15.29, for example, increasing $[NH_3]$ produces a stress on the system that requires a decrease in [HI] for the system to return to equilibrium. Consider esters, which are one of the products of an equilibrium reaction between a carboxylic acid and an alcohol. Esters are responsible for the scents we associate with fruits (such as oranges and bananas), and they are also used as scents in perfumes. Applying a stress to the reaction of a carboxylic acid and an alcohol will change the composition of the system, leading to an increase or a decrease in the amount of ester produced. In the next section, we explore factors that affect equilibrium concentrations.

Note the pattern

If a stress is applied to a system at equilibrium, the composition of the system will change to counteract the applied stress (Le Châtelier's principle).

EXAMPLE 15.15

Write an equilibrium constant expression for each of the following reactions, and use this expression to predict what will happen to the concentration of the substance in red type when the indicated change is made if the system is to maintain equilibrium.

* The name is pronounced "Luh SHOT-lee-ay."

(a) $2HgO(s) \rightleftharpoons 2Hg(l) + O_2(g)$; the amount of HgO is doubled.
(b) $NH_4HS(s) \rightleftharpoons NH_3(g) + H_2S(g)$; the concentration of H_2S is tripled.
(c) n-butane(g) $\rightleftharpoons$ isobutane(g); the concentration of isobutane is halved.

Given Equilibrium systems and changes

Asked for Equilibrium constant expressions and effects of changes

Strategy

Write the equilibrium constant expression, remembering that pure liquids and solids do not appear in the expression. From this expression, predict the change that must occur to maintain equilibrium when the indicated changes is made.

Solution

(a) Because $HgO(s)$ and $Hg(l)$ are pure substances, they do not appear in the equilibrium constant expression. Thus, for this reaction, $K = [O_2]$. The equilibrium concentration of O_2 is a constant and does not depend on the amount of HgO present. Hence, adding more HgO will not affect the equilibrium concentration of O_2, so no compensatory change is necessary.
(b) The reactant, NH_4HS, does not appear in the equilibrium constant expression because it is a solid. Thus, $K = [NH_3][H_2S]$, which means that the concentrations of the products are inversely proportional. If the H_2S concentration is tripled by adding H_2S, for example, then the NH_3 concentration must decrease by about a factor of 3 for the system to remain at equilibrium so that the product of the concentrations is equal to K.
(c) For this reaction, $K = [i\text{-bu}]/[n\text{-bu}]$, so halving the concentration of isobutane means that the n-butane concentration must also decrease by about half if the system is to maintain equilibrium.

EXERCISE 15.15

Write an equilibrium constant expression for each reaction. What must happen to the concentration of the substance in red type when the indicated change occurs if the system is to maintain equilibrium?

(a) $HBr(g) + NaH(s) \rightleftharpoons NaBr(s) + H_2(g)$; the concentration of HBr is decreased by a factor of 3.
(b) $6Li(s) + N_2(g) \rightleftharpoons 2Li_3N(s)$; the amount of Li is tripled.
(c) $SO_2(g) + Cl_2(g) \rightleftharpoons SO_2Cl_2(l)$; the concentration of Cl_2 is doubled.

Answer **(a)** $[H_2]$ must decrease by about a factor of 3; **(b)** solid lithium does not appear in the equilibrium constant expression, so no compensatory change is necessary; **(c)** $[SO_2]$ must decrease by about half.

15.5 ○ Factors That Affect Equilibrium

MGC Le Châtelier's Principle

Chemists use various strategies to increase the yield of the desired products of reactions. When synthesizing an ester, for example, how can a chemist control the reaction conditions to obtain the maximum amount of the desired product? Only three types of stresses can change the composition of an equilibrium mixture: (1) a change in the concentrations (or partial pressures) of the components by the addition or removal of reactants or products, (2) a change in the total pressure or volume, and (3) a change in the temperature of the system. In this section, we explore how changes in reaction conditions can affect the equilibrium composition of a system and will explore each of these possibilities in turn.

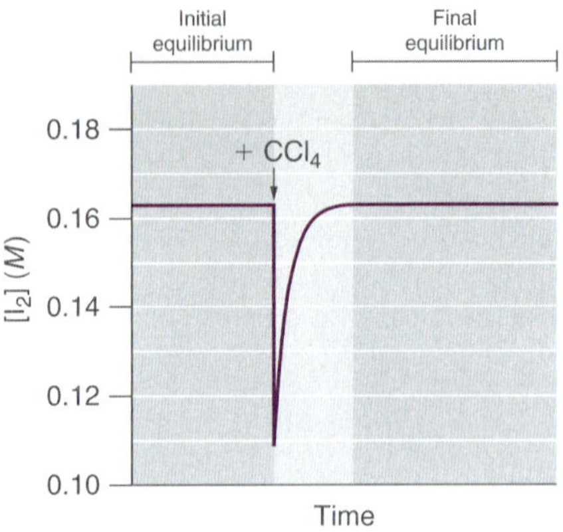

Figure 15.10 The concentration of dissolved I_2 as a function of time following the addition of more solvent to a saturated solution in contact with excess solid I_2. The concentration of I_2 decreases initially due to dilution but returns to its original value as long as solid I_2 is present. Interactive Graph

Changes in Concentration

If we add a small volume of carbon tetrachloride (CCl_4) solvent to a flask containing crystals of iodine, we obtain a saturated solution of I_2 in CCl_4, along with undissolved crystals:

$$I_2(s) \overset{\text{solvent}}{\rightleftharpoons} I_2\,(soln) \tag{15.30}$$

The system reaches equilibrium, with $K = [I_2]$. If we add more CCl_4, thereby diluting the solution, the equilibrium quotient Q is now less than K. Le Châtelier's principle tells us that the system will react to relieve the stress, but how? Adding solvent stressed the system by decreasing the concentration of dissolved I_2. Hence, more crystals will dissolve, thereby increasing the concentration of dissolved I_2 until the system again reaches equilibrium if enough solid I_2 is available (Figure 15.10). By adding solvent, we drove the reaction shown in Equation 15.30 to the right as written.

We encounter a more complex system in the reaction of hydrogen and nitrogen to form ammonia:

$$N_2(g) + 3H_2(g) \rightleftharpoons 2NH_3(g) \tag{15.31}$$

The K_p for this reaction is 2.14×10^{-2} at about 540 K. Under one set of equilibrium conditions, the partial pressure of ammonia is $P_{NH_3} = 0.454$ atm, that of hydrogen is $P_{H_2} = 2.319$ atm, and that of nitrogen is $P_{N_2} = 0.773$ atm. If an additional 1 atm of hydrogen is added to the reactor to give $P_{H_2} = 3.319$ atm, how will the system respond? Because the stress is an increase in P_{H_2}, the system must respond in some way that decreases the partial pressure of hydrogen to counteract the stress. The reaction will therefore proceed to the right as written, consuming H_2 and N_2 and forming additional NH_3. Initially, the partial pressures of H_2 and N_2 will decrease, and the partial pressure of NH_3 will increase until the system eventually reaches a new equilibrium composition, which will have a net increase in P_{H_2}.

We can confirm that this is indeed what will happen by evaluating the reaction quotient Q_p under the new conditions and comparing its value with the value of K_p. The equations used to evaluate K_p and Q_p have the same form: substituting the values after the addition of hydrogen into the expression for Q_p gives

$$Q_p = \frac{(P_{NH_3})^2}{(P_{N_2})(P_{H_2})^3} = \frac{(0.454)^2}{(0.773)(2.319 + 1.00)^3} = 7.29 \times 10^{-3}$$

Thus, $Q_p < K_p$, which tells us that the ratio of products to reactants is lower than at equilibrium. To reach equilibrium, the reaction must proceed to the right as written: the partial pressures of the products will increase, and the partial pressures of the reactants will decrease. Q_p will thereby increase until it equals K_p, and the system will once again be at equilibrium. Changes in the partial pressures of the various substances in the reaction mixture (Equation 15.31) as a function of time are shown in Figure 15.11.

We can force a reaction to go essentially to completion, regardless of the magnitude of K, by continually removing one of the products from the reaction mixture. Consider, for example, the *methanation* reaction, in which hydrogen reacts with carbon monoxide to form methane and water:

$$CO(g) + 3H_2(g) \rightleftharpoons CH_4(g) + H_2O(g) \tag{15.32}$$

This reaction is used for the industrial production of methane, whereas the reverse reaction is used for the production of H_2 (Example 15.14). The reaction quotient Q has the form

$$Q = \frac{[CH_4][H_2O]}{[CO][H_2]^3} \tag{15.33}$$

Regardless of the magnitude of K, if either H_2O or CH_4 can be removed from the reaction mixture so that $[H_2O]$ or $[CH_4]$ is approximately zero, then $Q \approx 0$. In other words, when product is removed, the system is stressed ($Q << K$), and more product

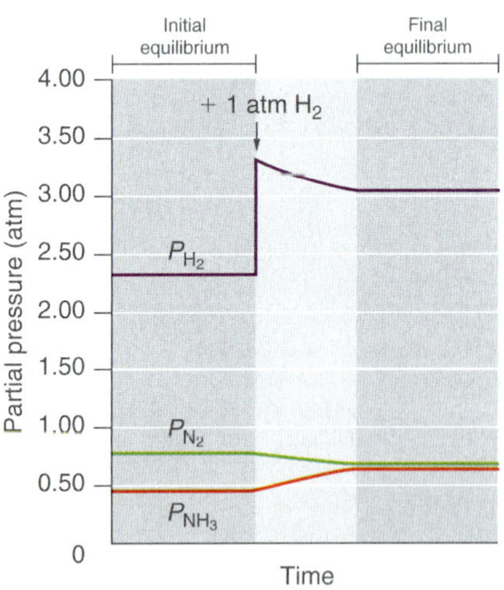

Figure 15.11 The partial pressures of H_2, N_2, and NH_3 as a function of time following the addition of more H_2 to an equilibrium mixture. Some of the added hydrogen is consumed by reaction with nitrogen to produce more ammonia, allowing the system to reach a new equilibrium composition. Interactive Graph

will form to counter the stress. Because water (bp = 100°C) is much less volatile than methane, hydrogen, or carbon monoxide (all of which have boiling points below −100°C), passing the gaseous reaction mixture through a cold coil will cause the water vapor to condense to a liquid that can be drawn off. Continuing to remove water from the system forces the reaction to the right as the system attempts to equilibrate, thus enriching the reaction mixture in methane. This technique, referred to as *driving a reaction to completion*, can be used to force a reaction to completion even if K is relatively small. For example, esters are usually synthesized by removing water, the other product of the condensation reaction. In Chapter 19, we will describe the thermodynamic basis for the change in the equilibrium position caused by changes in the concentrations of reaction components.

EXAMPLE 15.16

For each equilibrium system, predict the effect of the indicated stress on the specified quantity:

(a) $2SO_2(g) + O_2(g) \rightleftharpoons 2SO_3(g)$: (1) the effect of removing O_2 on P_{SO_2}, (2) the effect of removing O_2 on P_{SO_3}
(b) $CaCO_3(s) \rightleftharpoons CaO(s) + CO_2(g)$: (1) the effect of removing CO_2 on the amount of $CaCO_3$, (2) the effect of adding $CaCO_3$ on P_{CO_2}

Given Balanced chemical equations and changes

Asked for Effects of indicated stresses

Strategy

Use Q and K to predict the effect of the stress on each reaction.

Solution

(a) (1) Removing O_2 will decrease P_{O_2}, thereby decreasing the denominator in the reaction quotient and making $Q_p > K_p$. The reaction will proceed to the left as written, increasing the partial pressures of SO_2 and O_2 until Q_p once again equals K_p. (2) Removing O_2 will decrease P_{O_2} and thus increase Q_p, so the reaction will proceed to the left. The partial pressure of SO_3 will decrease.
(b) K_p and Q_p are both equal to P_{CO_2}. (1) Removing CO_2 from the system causes more $CaCO_3$ to react to produce CO_2, which increases P_{CO_2} to the partial pressure required by K_p. (2) Adding (or removing) solid $CaCO_3$ has no effect on P_{CO_2} because it does not appear in the expression for K_p (or Q_p).

EXERCISE 15.16

For each equilibrium system, predict the effect that the indicated stress will have upon the specified quantity:

(a) $H_2(g) + CO_2(g) \rightleftharpoons H_2O(g) + CO(g)$: (1) the effect of adding CO on $[H_2]$, (2) the effect of adding CO_2 on $[H_2]$
(b) $CuO(s) + CO(g) \rightleftharpoons Cu(s) + CO_2(g)$: (1) the effect of adding CO on the amount of Cu, (2) the effect of adding CO_2 on [CO]

Answer **(a)** (1) $[H_2]$ increases, (2) $[H_2]$ decreases; **(b)** (1) the amount of Cu increases, (2) [CO] increases

Changes in Total Pressure or Volume

Because liquids are relatively incompressible, changing the pressure above a liquid solution has little effect on the concentrations of dissolved substances. Consequently, changes in external pressure have very little effect on equilibrium systems

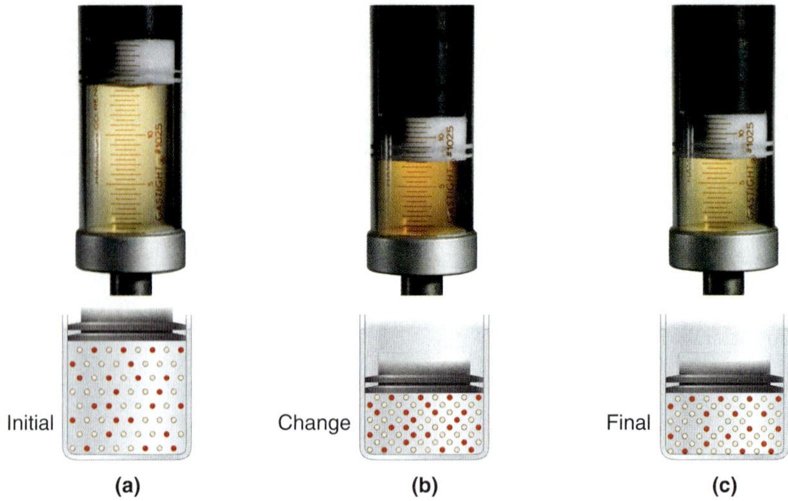

Figure 15.12 The effect of changing the volume (and thus the pressure) of an equilibrium mixture of N_2O_4 and NO_2 at constant temperature. **(a)** The syringe with a total volume of 12 mL contains an equilibrium mixture of N_2O_4 and NO_2; the red-brown color is proportional to the NO_2 concentration. **(b)** If the volume is rapidly decreased by a factor of 2 to 6 mL, the initial effect is to double the concentrations of all species present, including NO_2. Hence, the color becomes more intense. **(c)** With time, the system adjusts its composition in response to the stress as predicted by Le Châtelier's principle, forming colorless N_2O_4 at the expense of red-brown NO_2, which decreases the intensity of the color of the mixture.

that contain only solids or liquids. In contrast, because gases are highly compressible, their concentrations vary dramatically with pressure. From the ideal gas law, $PV = nRT$, described in Chapter 11, the concentration C of a gas is related to its pressure as follows:

$$C = \frac{n}{V} = \frac{P}{RT} \tag{15.34}$$

Hence, the concentration of any gaseous reactant or product is directly proportional to the applied pressure P and inversely proportional to the total volume V. Consequently, the equilibrium compositions of systems that contain gaseous substances are quite sensitive to changes in pressure, volume, and T.

These principles can be illustrated using the reversible dissociation of gaseous N_2O_4 to gaseous NO_2 (Equation 15.1). The syringe shown in Figure 15.12 initially contains an equilibrium mixture of colorless N_2O_4 and red-brown NO_2. Decreasing the volume by 50% causes the mixture to become darker because all concentrations have doubled. Decreasing the volume also constitutes a stress, however, as we can see by examining the effect of a change in volume on the reaction quotient Q. At equilibrium, $Q = K = [NO_2]^2/[N_2O_4]$ (Equation 15.10). If the volume is decreased by half, the concentrations of the substances in the mixture are doubled, so the new reaction quotient is

$$Q = \frac{[NO_2]_i^2}{[N_2O_4]_i} = \frac{2([NO_2]_i)^2}{2[N_2O_4]_i} = \frac{4([NO_2]_i)^2}{2[N_2O_4]_i} = 2K \tag{15.35}$$

> **Note the pattern**
>
> *Increasing the pressure of a system (or decreasing the volume) favors the side of the reaction that has fewer gaseous molecules, and vice versa.*

Because Q is now greater than K, the system is no longer at equilibrium. The stress can be relieved if the reaction proceeds to the left, consuming 2 mol of NO_2 for every 1 mol of N_2O_4 produced. This will decrease the concentration of NO_2 and increase the concentration of N_2O_4, causing Q to decrease until it is once again equal to K. Thus, as shown in Figure 15.12c, the intensity of the brown color due to NO_2 decreases with time following the change in volume.

In general, *if a balanced reaction contains different numbers of gaseous reactant and product molecules, the equilibrium will be sensitive to changes in volume or pressure.* Increasing the pressure on a system (or decreasing the volume) will favor the side of the reaction that has fewer gaseous molecules, and vice versa.

EXAMPLE 15.17

For each equilibrium system, predict the direction of the reaction if the pressure is decreased by a factor of 2 (for example, if the volume is doubled) at constant temperature:

(a) $N_2(g) + 3H_2(g) \rightleftharpoons 2NH_3(g)$
(b) $C_2H_2(g) + C_2H_6(g) \rightleftharpoons 2C_2H_4(g)$
(c) $2NO_2(g) \rightleftharpoons 2NO(g) + O_2(g)$

Given Balanced chemical equations

Asked for Direction of reaction if pressure is halved

Strategy

Use Le Châtelier's principle to predict the effect of the stress.

Solution

(a) Two moles of gaseous products are formed from 4 mol of gaseous reactants. Decreasing the pressure will cause the reaction to shift to the left because that side contains the larger number of moles of gas. Thus, the pressure increases, counteracting the stress.
(b) Two moles of gaseous products form from 2 mol of gaseous reactants. Decreasing the pressure will have no effect on the equilibrium composition because both sides of the balanced equation have the same number of moles of gas.
(c) Three moles of gaseous products are formed from 2 mol of gaseous reactants. Decreasing the pressure will favor the side that contains more moles of gas, so the reaction will shift toward the products to increase the pressure.

EXERCISE 15.17

For each equilibrium system, predict the direction in which the system will shift if the pressure is increased by a factor of 2 (for example, if the volume is halved) at constant temperature:

(a) $H_2O(g) + CO(g) \rightleftharpoons H_2(g) + CO_2(g)$
(b) $H_2(g) + C_2H_4(g) \rightleftharpoons C_2H_6(g)$
(c) $2SO_2(g) + O_2(g) \rightleftharpoons 2SO_3(g)$

Answer **(a)** no effect; **(b)** to the right; **(c)** to the right

Changes in Temperature

In all the cases we have considered so far, the magnitude of the equilibrium constant, K or K_p, was constant. Changes in temperature can, however, change the value of the equilibrium constant without immediately affecting the reaction quotient ($Q \neq K$). In this case, the system is no longer at equilibrium; hence, the composition of the system will change until Q equals K at the new temperature.

To predict how an equilibrium system will respond to a change in temperature, we must know something about the enthalpy change of the reaction, ΔH_{rxn}. As you learned in Chapter 5, heat is released to the surroundings in an exothermic reaction ($\Delta H_{rxn} < 0$), and heat is absorbed from the surroundings in an endothermic reaction ($\Delta H_{rxn} > 0$). We can express these changes in the following way:

$$\text{Exothermic:}\quad \text{reactants} \rightleftharpoons \text{products} + \text{heat} \quad (\Delta H < 0) \tag{15.36}$$

$$\text{Endothermic:}\quad \text{reactants} + \text{heat} \rightleftharpoons \text{products} \quad (\Delta H > 0) \tag{15.37}$$

Figure 15.13 The effect of temperature on the equilibrium between gaseous N_2O_4 and NO_2. (*center*) A tube containing a mixture of N_2O_4 and NO_2 in the same proportion at room temperature is red-brown due to the NO_2 present. (*left*) Immersing the tube in ice water causes the mixture to become lighter in color due to a shift in the equilibrium composition toward colorless N_2O_4. (*right*) In contrast, immersing the same tube in boiling water causes the mixture to become darker due to a shift in the equilibrium composition toward the highly colored NO_2.

Thus, heat can be thought of as a product in an exothermic reaction and as a reactant in an endothermic reaction. Increasing the temperature of a system corresponds to adding heat. Le Châtelier's principle predicts that an exothermic reaction will shift to the left (toward the reactants) if the temperature of the system is increased (heat is added). Conversely, an endothermic reaction will shift to the right (toward the products) if the temperature of the system is increased. If a reaction is thermochemically neutral ($\Delta H_{rxn} = 0$), then a change in temperature will not affect the equilibrium composition.

We can examine the effects of temperature on the dissociation of N_2O_4 to NO_2, for which $\Delta H = +58$ kJ/mol. This reaction can be written as

$$58 \text{ kJ} + N_2O_4(g) \rightleftharpoons 2NO_2(g) \tag{15.38}$$

Increasing the temperature (adding heat to the system) is a stress that will drive the reaction to the right, as illustrated in Figure 15.13. Thus, increasing the temperature increases the ratio of NO_2 to N_2O_4 at equilibrium, which increases the value of K.

The effect of increasing the temperature on a system at equilibrium can be summarized in the following way: *Increasing the temperature increases the magnitude of the equilibrium constant for an endothermic reaction, decreases the equilibrium constant for an exothermic reaction, and has no effect on the equilibrium constant for a thermally neutral reaction.* Table 15.3 shows the temperature dependence of the equilibrium constants for the synthesis of ammonia from hydrogen and nitrogen, which is an exothermic reaction with $\Delta H^0 = -91.8$ kJ/mol. Note that the values of both K and K_p decrease dramatically with increasing temperature, as predicted for an exothermic reaction.

Note the pattern

Increasing the temperature causes endothermic reactions to favor products and exothermic reactions to favor reactants.

TABLE 15.3 Temperature dependence of the equilibrium constants, K and K_p, for the reaction $N_2(g) + 3H_2(g) \rightleftharpoons 2NH_3(g)$

Temperature, K	K	K_p
298	3.3×10^8	5.6×10^5
400	3.9×10^4	3.6×10^1
450	2.6×10^3	1.9
500	1.7×10^2	1.0×10^{-1}
550	2.6×10^1	1.3×10^{-2}
600	4.1	1.7×10^{-3}

EXAMPLE 15.18

For each equilibrium reaction, predict the effect of decreasing the temperature:

(a) $N_2(g) + 3H_2(g) \rightleftharpoons 2NH_3(g)$, $\Delta H_{rxn} = -91.8$ kJ/mol
(b) $CaCO_3(s) \rightleftharpoons CaO(s) + CO_2(g)$, $\Delta H_{rxn} = 178$ kJ/mol

Given Balanced chemical equations and values of ΔH_{rxn}

Asked for Effects of decreasing temperature

Strategy

Use Le Châtelier's principle to predict the effect of decreasing the temperature on each reaction.

Solution

(a) The formation of NH_3 is exothermic, so we can view heat as one of the products:

$$N_2(g) + 3H_2(g) \rightleftharpoons 2NH_3(g) + 91.8 \text{ kJ}$$

If the temperature of the mixture is decreased, heat (one of the products) is being removed from the system, which causes the equilibrium to shift to the right. Hence, the formation of ammonia is favored at lower temperatures.

(b) The decomposition of calcium carbonate is endothermic, so heat can be viewed as one of the reactants:

$$CaCO_3(s) + 178 \text{ kJ} \rightleftharpoons CaO(s) + CO_2(g)$$

If the temperature of the mixture is decreased, heat (one of the reactants) is being removed from the system, which causes the equilibrium to shift to the left. Hence, the thermal decomposition of calcium carbonate is less favored at lower temperatures.

EXERCISE 15.18

For each equilibrium system, predict the effect of increasing the temperature on the reaction mixture:

(a) $2SO_2(g) + O_2(g) \rightleftharpoons 2SO_3(g)$, $\Delta H_{rxn} = -198$ kJ/mol
(b) $N_2(g) + O_2(g) \rightleftharpoons 2NO(g)$, $\Delta H_{rxn} = +181$ kJ/mol

Answer **(a)** reaction shifts to the *left*; **(b)** reaction shifts to the *right*

15.6 • Controlling the Products of Reactions

MGC Effects of Temperature and Pressure on Gaseous Reactions

Whether in the synthetic laboratory or in industrial settings, one of the primary goals of modern chemistry is to control the identity and quantity of the products of chemical reactions. For example, a process aimed at synthesizing ammonia is designed to maximize the amount of ammonia produced using a given amount of energy. Alternatively, other processes may be designed to minimize the creation of undesired products, such as pollutants emitted from an internal combustion engine. To achieve these goals, chemists must consider the competing effects of the reaction conditions that they can control.

One way to get a high yield of a desired compound is to make the rate of the desired reaction much faster than the rates of any other possible reactions that might occur in the system. Altering reaction conditions to control reaction rates, thereby obtaining a single product or set of products, is called **kinetic control**. A second approach, called **thermodynamic control**, consists of adjusting conditions so that at equilibrium only the desired products are present in significant quantities.

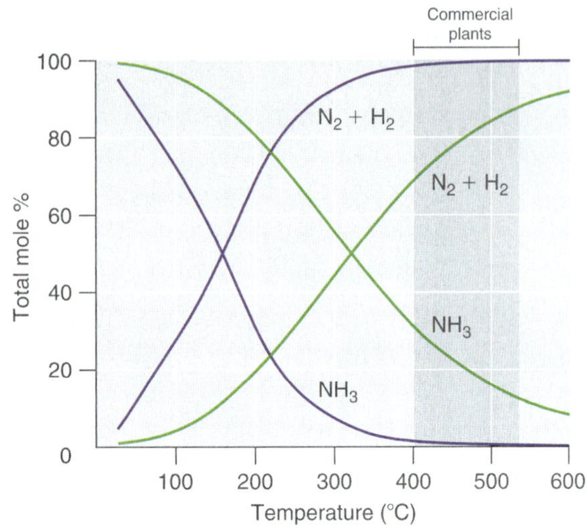

Figure 15.14 Effect of temperature and pressure on the equilibrium composition of two systems that originally contained a 3:1 mixture of hydrogen and nitrogen. At all temperatures, the total pressure in the systems was initially either 4 atm (purple curves) or 200 atm (green curves). Note the dramatic *decrease* in the proportion of NH_3 at equilibrium at higher temperatures in both cases, as well as the large *increase* in the proportion of NH_3 at equilibrium at any temperature for the system at higher pressure (green) versus lower pressure (purple). Commercial plants that use the Haber–Bosch process to synthesize ammonia on an industrial scale operate at temperatures of 400–530°C (indicated by the darker gray band) and total pressures of 130–330 atm. Interactive Graph

An example of thermodynamic control is the Haber–Bosch process* used to synthesize ammonia via the reaction

$$N_2(g) + 3H_2(g) \rightleftharpoons 2NH_3(g) \qquad \Delta H_{rxn} = -91.8 \text{ kJ/mol} \qquad (15.39)$$

Because the reaction converts 4 mol of gaseous reactants to only 2 mol of gaseous product, Le Châtelier's principle predicts that the formation of NH_3 will be favored when the pressure is increased. The reaction is exothermic, however, ($\Delta H_{rxn} = -91.8$ kJ/mol), so the equilibrium constant decreases with increasing temperature, which causes an equilibrium mixture to contain only relatively small amounts of ammonia at high temperatures (Figure 15.14). Taken together, these considerations suggest that the maximum yield of NH_3 will be obtained if the reaction is carried out at as low a temperature and as high a pressure as possible. Unfortunately, at temperatures lower than approximately 300°C, where the equilibrium yield of ammonia would be relatively high, the reaction is too *slow* to be of any commercial use. The industrial process therefore uses a mixed oxide (Fe_2O_3/K_2O) catalyst that enables the reaction to proceed at a significant rate at temperatures of 400–530°C, where the formation of ammonia is less unfavorable than at higher temperatures.

Because of the low value of the equilibrium constant at high temperatures (for example, $K = 0.039$ at 800 K), there is no way to produce an equilibrium mixture that contains large proportions of ammonia at high temperature. We can, however, control the temperature and the pressure while using a catalyst to convert a fraction of the N_2 and H_2 in the reaction mixture to NH_3, as is done in the Haber–Bosch process. This process also makes use of the fact that the product, ammonia, is less volatile than the reactants. Because NH_3 is a liquid at room temperature at pressures higher than 10 atm, cooling the reaction mixture causes NH_3 to condense from the vapor as liquid ammonia, which is easily separated from unreacted N_2 and H_2. The unreacted gases are recycled until complete conversion of hydrogen and nitrogen to ammonia is eventually achieved. Figure 15.15 is a simplified layout of a Haber–Bosch process plant.

The Sohio acrylonitrile process, in which propene and ammonia react with oxygen to form acrylonitrile, is an example of a kinetically controlled reaction:

$$CH_2{=}CHCH_3(g) + NH_3(g) + \frac{3}{2}O_2(g) \rightleftharpoons CH_2{=}CHC{\equiv}N(g) + 3H_2O(g) \qquad (15.40)$$

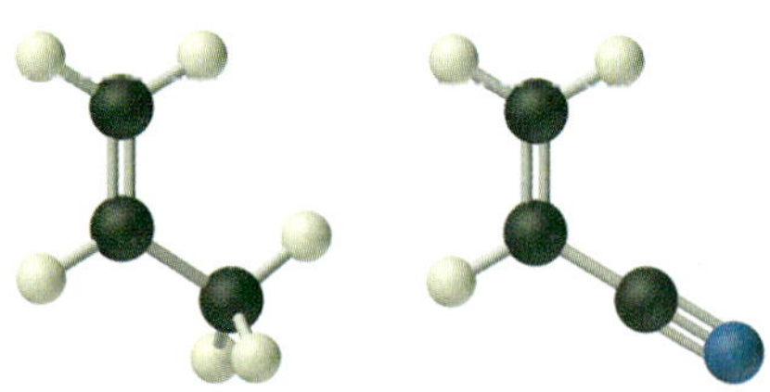

Propene Acrylonitrile

Like most oxidation reactions of organic compounds, this reaction is highly exothermic ($\Delta H^0 = -519$ kJ/mol) and has a very large equilibrium constant ($K = 1.2 \times 10^{94}$). Nonetheless, the reaction shown in Equation 15.40 is not the reaction a chemist would expect to occur when propene or ammonia is heated in the presence of oxygen. Competing combustion reactions that produce CO_2 and N_2 from the reactants, such as those shown below, are even more exothermic and have even larger equilibrium constants, thereby reducing the yield of the desired product, acrylonitrile:

$$CH_2{=}CHCH_3(g) + \frac{9}{2}O_2(g) \rightleftharpoons 3CO_2(g) + 3H_2O(g) \qquad \Delta H^0 = -1926.1 \text{ kJ/mol}, K = 4.5 \times 10^{338} \qquad (15.41)$$

$$2NH_3(g) + 3O_2(g) \rightleftharpoons N_2(g) + 6H_2O(g) \qquad \Delta H^0 = -1359.2 \text{ kJ/mol}, K = 4.4 \times 10^{234} \qquad (15.42)$$

---

* Karl Bosch (1874–1940) was a German chemical engineer who was responsible for designing the process that took advantage of Fritz Haber's discoveries regarding the $N_2 + H_2/NH_3$ equilibrium to make ammonia synthesis via this route cost effective. He received the Nobel Prize in chemistry in 1931 for his work. The industrial process is called either the Haber process or the Haber–Bosch process.

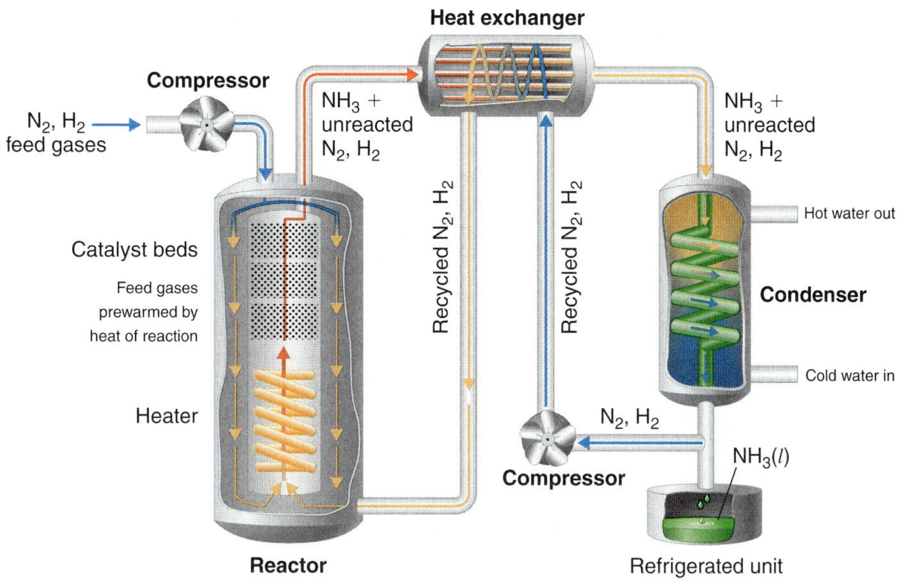

Figure 15.15 A schematic diagram of an industrial plant for the production of ammonia via the Haber–Bosch process. A 3:1 mixture of gaseous H_2 and N_2 is compressed to 130–330 atm, heated to 400–530°C, and passed over an Fe_2O_3/K_2O catalyst, which results in partial conversion to gaseous NH_3. The resulting mixture of gaseous NH_3, H_2, and N_2 is passed through a heat exchanger, which uses the hot gases to prewarm recycled N_2 and H_2, and a condensor to cool the NH_3, giving a liquid that is readily separated from unreacted N_2 and H_2. (Although the normal boiling point of NH_3 is −33°C, the boiling point increases rapidly with increasing pressure, to 20°C at 8.5 atm and 126°C at 100 atm.) The unreacted N_2 and H_2 are recycled to form more NH_3.

In fact, the formation of acrylonitrile in Equation 15.40 is accompanied by the release of approximately 760 kJ/mol of heat due to partial combustion of propene during the reaction.

The Sohio process uses a catalyst that selectively accelerates the rate of formation of acrylonitrile without significantly affecting the rates of the competing combustion reactions. Consequently, acrylonitrile is formed more rapidly than CO_2 and N_2 under the optimized reaction conditions (approximately 1.5 atm and 450°C). The reaction mixture is rapidly cooled to prevent further oxidation or combustion of acrylonitrile, which is then washed out of the vapor with a liquid water spray. Thus, controlling the kinetics of the reaction causes the desired product to be formed under conditions where equilibrium is not established. In industry, this reaction is carried out on an enormous scale. Acrylonitrile is the building block of the polymer called *polyacrylonitrile*, found in all the products referred to collectively as *acrylics*, whose wide range of uses includes the synthesis of fibers woven into clothing and carpets.

EXAMPLE 15.19

Recall that methanation is the reaction of hydrogen with carbon monoxide to form methane and water:

$$CO(g) + 3H_2(g) \rightleftharpoons CH_4(g) + H_2O(g)$$

Notice that this reaction is the reverse of the steam reforming of methane described in Example 15.14. The reaction is exothermic ($\Delta H^0 = -206$ kJ/mol), with an equilibrium constant at room temperature of $K_p = 7.6 \times 10^{24}$. Unfortunately, however, CO and H_2 do not react at an appreciable rate at room temperature. What conditions would you select to maximize the amount of methane formed per unit time by this reaction?

Given Balanced chemical equation and values of ΔH^0 and K

Asked for Conditions to maximize yield of product

Strategy

Consider the effect of changes in temperature and pressure and the addition of an effective catalyst on the rate and equilibrium of the reaction. Determine which combination of reaction conditions will result in the maximum production of methane.

Solution

The products are highly favored at equilibrium, but the rate at which equilibrium is reached is too slow to be useful. You learned in Chapter 14 that the rate of a reaction can often be increased dramatically by increasing the temperature of the reactants. Unfortunately, however, because the reaction is quite exothermic, an increase in temperature will shift the equilibrium to the left, causing more reactants to form and relieving the stress on the system by absorbing the added heat. Thus, if we increase the temperature too much, the equilibrium will no longer favor methane formation. (In fact, the equilibrium constant for this reaction is very temperature sensitive, decreasing to only 1.9×10^{-3} at 1000°C.) To increase the rate, we can try to find a catalyst that will operate at lower temperatures where equilibrium favors the formation of products. Higher pressures will also favor the formation of products because 4 mol of gaseous reactant is converted to only 2 mol of gaseous product. Very high pressures should not be needed, however, because the equilibrium constant favors the formation of products. Thus, optimal conditions for the reaction include carrying it out at temperatures higher than room temperature (but not too high), adding a catalyst, and using pressures higher than atmospheric pressure.

Industrially, catalytic methanation is typically carried out at pressures of 1–100 atm and temperatures of 250–450°C in the presence of a nickel catalyst. (At 425°C, K_p is 3.7×10^3, so the formation of products is still favored.) The synthesis of methane can also be favored by the removal of either H_2O or CH_4 from the reaction mixture by condensation as they form.

EXERCISE 15.19

As you learned in Example 15.10, the water–gas shift reaction is

$$H_2(g) + CO_2(g) \rightleftharpoons H_2O(g) + CO(g) \qquad K_p = 0.106 \text{ at } 700 \text{ K}$$

At 700 K, $\Delta H = 41.2$ kJ/mol. What reaction conditions would you use to maximize the yield of carbon monoxide?

Answer high temperatures to increase the reaction rate and favor product formation, a catalyst to increase the reaction rate, and atmospheric pressure because the equilibrium will not be greatly affected by pressure

SUMMARY AND KEY TERMS

15.1 The Concept of Chemical Equilibrium (p. 673)

Chemical equilibrium is a dynamic process consisting of forward and reverse reactions that proceed at equal rates. At equilibrium, the composition of the system no longer changes with time. The composition of an equilibrium mixture is independent of the direction from which equilibrium is approached.

15.2 The Equilibrium Constant (p. 675)

The ratio of the rate constants for the forward and reverse reactions at equilibrium is the **equilibrium constant (K)**, a unitless quantity. The composition of the equilibrium mixture is therefore determined by the magnitudes of the forward and reverse rate constants at equilibrium. Under a given set of conditions, a reaction will always have the same value of K. For a system at equilibrium, the **law of mass action** relates K to the ratio of the equilibrium concentrations of the products to the concentrations of the reactants raised to their respective exponents to match the coefficients in the **equilibrium equation**. The ratio is called the **equilibrium constant expression**. When a reaction is written in the reverse direction, K and the equilibrium constant expression are inverted. For gases, the equilibrium constant expression can be written as the ratio of the partial pressures of the products to the partial pressures of the reactants, each raised to its coefficient in the chemical equation. An equilibrium constant calculated from partial pressures, K_p, is related to K by the ideal gas constant R, the temperature T, and the change in the number of moles of gas during the reaction. An equilibrium system that contains products and reactants in a single phase is a **homogeneous equilibrium**, whereas a system whose reactants, products, or both are in more than one phase is a **heterogeneous equilibrium**. When a reaction can be expressed as the sum of two or more reactions, its equilibrium constant is equal to the product of the equilibrium constants for the individual reactions.

15.3 Solving Equilibrium Problems (p. 686)

When an equilibrium constant is calculated from equilibrium concentrations, molar concentrations or partial pressures are substituted into the equilibrium constant expression for the reaction. Equilibrium constants can be used to calculate the equilibrium concentrations of reactants and products by using the quantities or concentrations of the reactants, the stoichiometry of the balanced equation for the reaction, and a tabular format to obtain the final concentrations of all species at equilibrium.

15.4 Nonequilibrium Conditions (p. 697)

The **reaction quotient (Q or Q_p)** has the same form as the equilibrium constant expression, but it is derived from concentrations obtained at any time. When a reaction system is at equilibrium, $Q = K$. Graphs derived by plotting a few equilibrium concentrations for a system at a given temperature and pressure can be used to predict the direction in which a reaction will proceed. Points that do not lie on the line or curve represent nonequilibrium states, and the system will adjust, if it can, to achieve equilibrium. **Le Châtelier's principle** states that if a stress is applied to a system at equilibrium, the composition of the system will adjust to counteract the stress.

15.5 Factors That Affect Equilibrium (p. 701)

Three types of stresses can alter the composition of an equilibrium system: adding or removing reactants or products, changing the total pressure or volume, and changing the temperature of the system. A reaction with an unfavorable equilibrium constant can be driven to completion by continually removing one of the products of the reaction. Equilibria that contain different numbers of gaseous reactant and product molecules are sensitive to changes in volume or pressure; higher pressures favor the side with fewer gaseous molecules. Removing heat from an exothermic reaction favors the formation of products, whereas removing heat from an endothermic reaction favors the formation of reactants.

15.6 Controlling the Products of Reactions (p. 707)

Changing conditions to affect the *rates* of reactions in order to obtain a single product is called **kinetic control** of the system. In contrast, **thermodynamic control** is adjusting conditions to ensure that at equilibrium only the desired product or products are present in significant concentrations.

KEY EQUATIONS

Definition of equilibrium constant in terms of forward and reverse rate constants	$K = \dfrac{k_f}{k_r}$	(15.3)
Equilibrium constant expression (law of mass action)	$K = \dfrac{[C]^c\,[D]^d}{[A]^a[B]^b}$	(15.5)
Equilibrium constant expression for reactions involving gases using partial pressures	$K_p = \dfrac{(P_C)^c\,(P_D)^d}{(P_A)^a\,(P_B)^b}$	(15.14)
Relationship between K_p and K	$K_p = K(RT)^{\Delta n}$	(15.16)
Reaction quotient	$Q = \dfrac{[C]^c[D]^d}{[A]^a[B]^b}$	(15.25)

QUESTIONS AND PROBLEMS

MGC For instructor-assigned homework, go to **www.masteringgeneralchemistry.com**

Please be sure you are familiar with the topics discussed in Essential Skills 7 at the end of this chapter before proceeding to the Questions and Problems.

Questions and Problems with colored numbers have answers in the Appendix and complete solutions in the Student Solutions Manual.

CONCEPTUAL

15.1 The Concept of Chemical Equilibrium

1. What is meant when a reaction is described as "having reached equilibrium"? What does this statement mean regarding the rates of the forward and reverse reactions? What does this statement mean regarding the concentrations or amounts of the reactants and products?
2. Is it correct to say that the reaction has "stopped" when it has reached equilibrium? Explain your answer and support it with a specific example.
3. Why is chemical equilibrium described as a dynamic process? Describe this process in the context of a saturated solution of NaCl in water. What is occurring on a microscopic level? What is happening on a macroscopic level?
4. Which of these systems exist(s) in a state of chemical equilibrium? (a) oxygen and hemoglobin in the human circulatory system; (b) iodine crystals in an open beaker; (c) the combustion of wood; (d) the amount of ^{14}C in a decomposing organism

15.2 The Equilibrium Constant

5. For an equilibrium reaction, what effect does reversing the reactants and products have on the value of the equilibrium constant?

6. Which of the following equilibria are homogeneous and which are heterogeneous?
(a) $2HF(g) \rightleftharpoons H_2(g) + F_2(g)$
(b) $C(s) + 2H_2(g) \rightleftharpoons CH_4(g)$
(c) $H_2C{=}CH_2(g) + H_2(g) \rightleftharpoons C_2H_6(g)$
(d) $2Hg(l) + O_2(g) \rightleftharpoons 2HgO(s)$

7. Classify each equilibrium system as being either homogeneous or heterogeneous:
(a) $NH_4NH_2CO_2(s) \rightleftharpoons 2NH_3(g) + CO_2(g)$
(b) $C(s) + O_2(g) \rightleftharpoons CO_2(g)$
(c) $2Mg(s) + O_2(g) \rightleftharpoons 2MgO(s)$
(d) $AgCl(s) \rightleftharpoons Ag^+(aq) + Cl^-(aq)$

8. If an equilibrium reaction is endothermic, what happens to the equilibrium constant if the temperature of the reaction is increased? If the temperature is decreased?

9. Industrial production of NO by the reaction $N_2(g) + O_2(g) \rightleftharpoons 2NO(g)$ is carried out at elevated temperatures in order to drive the reaction toward the formation of product. After sufficient product has formed, the reaction mixture is quickly cooled. Why?

10. How would you differentiate between a system that has reached chemical equilibrium and one that is reacting so slowly that changes in concentration are difficult to observe?

11. What is the relationship among the equilibrium constant K, the concentration of each component of the system, and the rate constants for the forward and reverse reactions?

12. Write expressions for K and K_p for each of these reactions:
(a) $CO(g) + H_2O(g) \rightleftharpoons CO_2(g) + H_2(g)$
(b) $PCl_3(g) + Cl_2(g) \rightleftharpoons PCl_5(g)$
(c) $2O_3(g) \rightleftharpoons 3O_2(g)$

13. Write expressions for K and K_p for each reaction:
(a) $2NO(g) + O_2(g) \rightleftharpoons 2NO_2(g)$
(b) $\frac{1}{2}H_2(g) + \frac{1}{2}I_2(g) \rightleftharpoons HI(g)$
(c) *cis*-stilbene $\rightleftharpoons$ *trans*-stilbene

14. Why is it incorrect to state that pure liquids, pure solids, and solvents are not part of an equilibrium constant expression?

15. Write expressions for K and K_p for each equilibrium reaction:
(a) $2S(s) + 3O_2(g) \rightleftharpoons 2SO_3(g)$
(b) $C(s) + CO_2(g) \rightleftharpoons 2CO(g)$
(c) $2ZnS(s) + 3O_2(g) \rightleftharpoons 2ZnO(s) + 2SO_2(g)$

16. Write expressions for K and K_p for each equilibrium reaction:
(a) $2HgO(s) \rightleftharpoons 2Hg(l) + O_2(g)$
(b) $H_2(g) + I_2(s) \rightleftharpoons 2HI(g)$
(c) $NH_4CO_2NH_2(s) \rightleftharpoons 2NH_3(g) + CO_2(g)$

17. At room temperature, the equilibrium constant of the reaction $2A(g) \rightleftharpoons B(g)$ is 1. What does this indicate about the concentrations of A and B at equilibrium? Would you expect K and K_p to vary significantly from each other? If so, how would their difference be affected by temperature?

18. For a certain series of reactions, if $[OH^-][HCO_3^-]/[CO_3^{2-}] = K_1$ and $[OH^-][H_2CO_3]/[HCO_3^-] = K_2$, what is the equilibrium expression for the overall reaction? Write the overall equilibrium equation.

19. In the equation for an enzymatic reaction, ES represents the complex formed between the substrate S and the enzyme protein E. In the final step of the oxidation reaction shown below, the product P dissociates from the ESO2 complex, which regenerates the active enzyme:

$$E + S \rightleftharpoons ES \qquad K_1$$
$$ES + O_2 \rightleftharpoons ESO_2 \qquad K_2$$
$$ESO_2 \rightleftharpoons E + P \qquad K_3$$

Give the overall reaction equation, and show that $K = K_1 \times K_2 \times K_3$.

15.3 Solving Equilibrium Problems

20. Calculations involving systems with very small or very large equilibrium constants can be dramatically simplified by making certain assumptions about the concentrations of products and reactants. What are these assumptions (a) when K is very large and (b) when K is very small? Illustrate this technique using the system $A + 2B \rightleftharpoons C$ for which you are to calculate the concentration of the product at equilibrium starting with only A and B. Under what circumstances should simplifying assumptions not be used?

15.4 Nonequilibrium Conditions

21. During a set of experiments, graphs were drawn of [reactants] versus [products] at equilibrium. Using Figures 15.8 and 15.9 as your guide, sketch the shape of each graph using appropriate labels.
(a) $H_2O(l) \rightleftharpoons H_2O(g)$
(b) $2Mg(s) + O_2(g) \rightleftharpoons 2MgO(s)$
(c) $2O_3(g) \rightleftharpoons 3O_2(g)$
(d) $2PbS(s) + 3O_2(g) \rightleftharpoons 2PbO(s) + 2SO_2(g)$

22. Write an equilibrium constant expression for each reaction system. Given the indicated changes, how must the concentration of the species in red type change if the system is to maintain equilibrium?
(a) $2NaHCO_3(s) \rightleftharpoons Na_2CO_3(s) + CO_2(g) + H_2O(g)$; $[CO_2]$ is doubled.
(b) $N_2F_4(g) \rightleftharpoons 2NF_2(g)$; $[NF_2]$ is decreased by a factor of 2.
(c) $H_2(g) + I_2(g) \rightleftharpoons 2HI(g)$; $[I_2]$ is doubled.

23. Write an equilibrium constant expression for each reaction system. Given the indicated changes, how must the concentration of the species in red type change if the system is to maintain equilibrium?
(a) $CS_2(g) + 4H_2(g) \rightleftharpoons CH_4(g) + 2H_2S(g)$; $[CS_2]$ is doubled.
(b) $PCl_5(g) \rightleftharpoons PCl_3(g) + Cl_2(g)$; $[Cl_2]$ is decreased by a factor of 2.
(c) $4NH_3(g) + 5O_2(g) \rightleftharpoons 4NO(g) + 6H_2O(g)$; [NO] is doubled.

15.5 Factors That Affect Equilibrium

24. If an equilibrium reaction is endothermic in the forward direction, what is the expected change in the concentration of each component of the system if the temperature of the reaction is increased? If the temperature is decreased?

25. Write the equilibrium equation for the system $4NH_3(g) + 5O_2(g) \rightleftharpoons 4NO(g) + 6H_2O(g)$. Would you expect equilibrium to shift toward the products or reactants with an increase in pressure? Why?

26. The rate of a reaction approximately doubles for every 10°C rise in temperature. What happens to K?

27. The formation of $A_2B_2(g)$ via the equilibrium reaction $2AB(g) \rightleftharpoons A_2B_2(g)$ is exothermic. What happens to the ratio k_f/k_r if the temperature is increased? If both temperature and pressure are increased?

28. In each system, predict the effect that the indicated change will have on the specified quantity at equilibrium:
   (a) $H_2(g) + I_2(g) \rightleftharpoons 2HI(g)$
   H_2 is removed; what is the effect on P_{I_2}?
   (b) $2NOBr(g) \rightleftharpoons 2NO(g) + Br_2(g)$
   Br_2 is removed; what is the effect on P_{NOBr}?
   (c) $2NaHCO_3(s) \rightleftharpoons Na_2CO_3(s) + CO_2(g) + H_2O(g)$
   CO_2 is removed; what is the effect on P_{NaHCO_3}?

29. What effect will the indicated change have on the specified quantity at equilibrium?
   (a) $NH_4Cl(s) \rightleftharpoons NH_3(g) + HCl(g)$
   NH_4Cl is increased; what is the effect on P_{HCl}?
   (b) $H_2O(g) \rightleftharpoons 2H_2(g) + O_2(g)$
   O_2 is added; what is the effect on P_{H_2}?
   (c) $PCl_3(g) + Cl_2(g) \rightleftharpoons PCl_5(g)$
   Cl_2 is removed; what is the effect on P_{PCl_5}?

15.6 Controlling the Products of Reactions

30. A reaction mixture will produce either product A or B depending on the reaction pathway. In the absence of a catalyst, product A is formed; in the presence of a catalyst, product B is formed. What conclusions can you draw about the forward and reverse rates of the reaction that produces A versus the reaction that produces B in (a) the absence of a catalyst and (b) the presence of a catalyst?

31. Describe how you would design an experiment to determine the equilibrium constant for the synthesis of ammonia: $N_2(g) + 3H_2(g) \rightleftharpoons 2NH_3(g)$. The forward reaction is exothermic ($\Delta H^\circ = -91.8$ kJ). What effect would an increase in temperature have on the equilibrium constant?

32. What effect does a catalyst have on the equilibrium position of a reaction? What effect does a catalyst have on the rate at which equilibrium is reached? What effect does a catalyst have on the equilibrium constant?

33. How can the ratio Q/K be used to determine in which direction a reaction will proceed to reach equilibrium?

34. Industrial reactions are frequently run under conditions in which competing reactions can occur. Explain how a catalyst can be used to achieve reaction selectivity. Does the ratio Q/K for the selected reaction change in the presence of a catalyst?

NUMERICAL

This section includes "paired problems" (marked by brackets) that require similar problem-solving skills.

15.2 The Equilibrium Constant

35. Explain what each of the following values for K tells you about the relative concentrations of reactants versus products in a given equilibrium reaction: $K = 0.892$; $K = 3.25 \times 10^8$; $K = 5.26 \times 10^{-11}$. Are products or reactants favored at equilibrium?

36. Write the equilibrium constant expression for each reaction:
   (a) $N_2O_4(g) \rightleftharpoons 2NO_2(g)$
   (b) $\frac{1}{2}N_2O_4(g) \rightleftharpoons NO_2(g)$
   Are these equilibrium constant expressions equivalent? Explain.

37. Write the equilibrium constant expression for each reaction:
   (a) $\frac{1}{2}N_2(g) + \frac{3}{2}H_2(g) \rightleftharpoons NH_3(g)$
   (b) $\frac{1}{3}N_2(g) + H_2(g) \rightleftharpoons \frac{2}{3}NH_3(g)$
   How are these two expressions mathematically related to the equilibrium constant expression for $N_2(g) + 3H_2(g) \rightleftharpoons 2NH_3(g)$?

38. Write an equilibrium constant expression for each reaction:
   (a) $C(s) + 2H_2O(g) \rightleftharpoons CO_2(g) + 2H_2(g)$
   (b) $SbCl_3(g) + Cl_2(g) \rightleftharpoons SbCl_5(g)$
   (c) $2O_3(g) \rightleftharpoons 3O_2(g)$

39. Give an equilibrium constant expression for each reaction:
   (a) $2NO(g) + O_2(g) \rightleftharpoons 2NO_2(g)$
   (b) $\frac{1}{2}H_2(g) + \frac{1}{2}I_2(g) \rightleftharpoons HI(g)$
   (c) $CaCO_3(s) + 2HOCl(aq) \rightleftharpoons Ca^{2+}(aq) + 2OCl^-(aq) + 2H_2O(l) + CO_2(g)$

40. Calculate K and K_p for each reaction:
   (a) $2NOBr(g) \rightleftharpoons NO(g) + Br_2(g)$; at 727°C, the equilibrium concentration of NO is 1.29 M, Br_2 is 10.52 M, and NOBr is 0.423 M.
   (b) $C(s) + CO_2(g) \rightleftharpoons 2CO(g)$; at 1200 K, a 2.00-L vessel at equilibrium has partial pressures of 93.5 atm CO_2 and 76.8 atm CO, and the vessel contains 3.55 g of carbon.

41. Calculate K and K_p for each reaction:
   (a) $N_2O_4(g) \rightleftharpoons 2NO_2(g)$; at the equilibrium temperature of −40°C, a 0.150 M sample of N_2O_4 undergoes a 0.456% decomposition.
   (b) $CO(g) + 2H_2(g) \rightleftharpoons CH_3OH(g)$; an equilibrium is reached at 227°C in a 15.5-L reaction vessel with a total pressure of 6.71×10^2 atm and is found to contain 37.8 g of hydrogen gas, 457.7 g of carbon monoxide, and 7193 g of methanol.

42. Determine the equilibrium constant expression, K, and K_p (where applicable) for each reaction:
   (a) $2H_2S(g) \rightleftharpoons 2H_2(g) + S_2(g)$; at 1065°C, an equilibrium mixture consists of 1.00×10^{-3} M H_2, 1.20×10^{-3} M S_2, and 3.32×10^{-3} M H_2S.
   (b) $Ba(OH)_2(s) \rightleftharpoons 2OH^-(aq) + Ba^{2+}(aq)$; at 25°C, a 250-mL beaker contains 0.330 mol of barium hydroxide in equilibrium with 0.0267 mol of barium ions and 0.0537 mol of hydroxide ions.

43. Determine the equilibrium constant expression, K, and K_p for each reaction:
   (a) $2NOCl(g) \rightleftharpoons 2NO(g) + Cl_2(g)$; at 500 K, a 24.3-m$M$ sample of NOCl has decomposed, leaving an equilibrium mixture that contains 72.7% of the original amount of NOCl.
   (b) $Cl_2(g) + PCl_3(g) \rightleftharpoons PCl_5(g)$; at 250°C, a 500-mL reaction vessel contains 16.9 g of chlorine gas, 0.500 g of PCl_3, and 10.2 g of PCl_5 at equilibrium.

44. The equilibrium constant expression for a reaction is $[CO_2]^2/[SO_2]^2[O_2]$. What is the balanced chemical equation for the overall reaction if one of the reactants is $Na_2CO_3(s)$?

45. The equilibrium constant expression for a reaction is $[NO][H_2O]^{3/2}/[NH_3][O_2]^{5/4}$. What is the balanced chemical equation for the overall reaction?

46. Given $K = k_f/k_r$, what happens to the magnitude of the equilibrium constant if the rate of the forward reaction is doubled? If the rate of the reverse reaction for the overall reaction is decreased by a factor of 3?

47. The value of the equilibrium constant for the reaction $2H_2(g) + S_2(g) \rightleftharpoons 2H_2S(g)$ is 1.08×10^7 at 700°C.

What is the value of the equilibrium constant for the following related reactions?
(a) $H_2(g) + \frac{1}{2}S_2(g) \rightleftharpoons H_2S(g)$
(b) $4H_2(g) + 2S_2(g) \rightleftharpoons 4H_2S(g)$
(c) $H_2S(g) \rightleftharpoons H_2(g) + \frac{1}{2}S_2(g)$

15.3 Solving Equilibrium Problems

48. In the equilibrium reaction $A + B \rightleftharpoons C$, what happens to the value of the equilibrium constant K if the concentrations of the reactants are doubled? If they are tripled? Can the same be said about the equilibrium reaction $2A \rightleftharpoons B + C$?

49. The table shows reported values of the equilibrium P_{O_2} at three temperatures for the reaction $Ag_2O(s) \rightleftharpoons 2Ag(s) + \frac{1}{2}O_2(g)$, for which $\Delta H^0 = 31$ kJ/mol. Are these data consistent with what you would expect to occur? Why or why not?

T, °C	P_{O_2}, mmHg
150	182
184	143
191	126

50. Given the equilibrium system $N_2O_4(g) \rightleftharpoons 2NO_2(g)$, what happens to K_p if the initial concentration of N_2O_4 is doubled? If K_p is 1.7×10^{-1} at 2300°C, and the system initially contains 100% N_2O_4 at a pressure of 2.6×10^2 atm, what is the equilibrium pressure of each component?

51. At 430°C, 4.20 mol of HI in a 9.60-L reaction vessel reaches equilibrium according to the equation $H_2(g) + I_2(g) \rightleftharpoons 2HI(g)$. At equilibrium, $[H_2] = 0.047$ M and $[HI] = 0.345$ M. What are the values of K and K_p for this reaction?

52. Methanol, a liquid used as an automobile fuel additive, is commercially produced from carbon monoxide and hydrogen at 300°C according to the reaction $CO(g) + 2H_2(g) \rightleftharpoons CH_3OH(g)$, for which $K_p = 1.3 \times 10^{-4}$. If 56.0 g of CO is mixed with excess hydrogen in a 250-mL flask at this temperature, and the hydrogen pressure is continuously maintained at 100 atm, what would be the maximum percent yield of methanol? What pressure of hydrogen would be required to obtain a minimum yield of methanol of 95% under these conditions?

53. If the total equilibrium pressure is 0.969 atm for the reaction $A(s) \rightleftharpoons 2B(g) + C(g)$, what is the value of K_p?

54. The decomposition of ammonium carbamate to ammonia and CO_2 at 40°C is written as $NH_4CO_2NH_2(s) \rightleftharpoons 2NH_3(g) + CO_2(g)$. If the partial pressure of NH_3 at equilibrium is 0.242 atm, what is the equilibrium partial pressure of CO_2? What is the total gas pressure of the system? What is K_p?

55. At 375 K, K_p for the reaction $SO_2Cl_2(g) \rightleftharpoons SO_2(g) + Cl_2(g)$ is 2.4, with pressures expressed in atmospheres. At 303 K, K_p is 2.9×10^{-2}.
(a) What is the equilibrium constant, K, for the reaction at each temperature?
(b) If a sample at 375 K has 0.100 M Cl_2 and 0.200 M SO_2 at equilibrium, what is the concentration of SO_2Cl_2?
(c) If the sample given in part b is cooled to 303 K, what is the pressure inside the bulb?

56. For the gas-phase reaction $aA \rightleftharpoons bB$, show that $K_p = K(RT)^{\Delta n}$ assuming ideal gas behavior.

57. For the gas-phase reaction $I_2 \rightleftharpoons 2I$, show that the total pressure is related to the equilibrium concentration by the equation $P_T = \sqrt{K_p P_{I_2}} + P_{I_2}$.

58. Experimental data on the system $Br_2(l) \rightleftharpoons Br_2(aq)$ are given in the table. Graph the moles of $Br_2(l)$ present versus $[Br_2]$, and then write the equilibrium equation and determine the value of K.

Grams Br_2 in 100 mL water	$[Br_2]$, M
1.0	0.0626
2.5	0.156
3.0	0.188
4.0	0.219
4.5	0.219

59. Data accumulated for the reaction n-butane $\rightleftharpoons$ isobutane at equilibrium are shown in the table. What is the equilibrium constant for this conversion? If 1 mol of n-butane is allowed to equilibrate under the same reaction conditions, what is the final number of moles each of n-butane and isobutane?

Moles n-butane	Moles isobutane
0.5	1.25
1.0	2.5
1.50	3.75

60. Solid ammonium carbamate, $NH_4CO_2NH_2$, dissociates completely to ammonia and carbon dioxide when it vaporizes:

$$NH_4CO_2NH_2(s) \rightleftharpoons 2NH_3(g) + CO_2(g)$$

At 25°C, the total pressure of the gases in equilibrium with the solid is 0.116 atm. What is the equilibrium partial pressure of each gas? What is K_p? If the concentration of CO_2 is doubled, and then equilibrates to its initial equilibrium partial pressure $+x$ atm, what change in the NH_3 concentration is necessary for the system to restore equilibrium?

61. The equilibrium constant for the reaction $COCl_2(g) \rightleftharpoons CO(g) + Cl_2(g)$ is $K_p = 2.2 \times 10^{-10}$ at 100°C. If the initial concentration of $COCl_2$ is 3.05×10^{-3} M, what is the partial pressure of each gas at equilibrium at 100°C? What assumption can be made to simplify your calculations?

62. Aqueous dilution of IO_4^- results in the reaction $IO_4^-(aq) + 2H_2O(l) \rightleftharpoons H_4IO_6^-(aq)$, for which $K = 3.5 \times 10^{-2}$. If you begin with 50 mL of a 0.896 M solution of IO_4^- that is diluted to 250 mL with water, how many moles of $H_4IO_6^-$ are formed at equilibrium?

63. Iodine and bromine react to form IBr, which then sublimes. At 184.4°C, the overall reaction proceeds according to the equation

$$I_2(g) + Br_2(g) \rightleftharpoons 2IBr(g) \qquad K_p = 1.2 \times 10^2$$

If you begin the reaction with 7.4 g of I_2 vapor and 6.3 g of Br_2 vapor in a 1.00 L container, what is the concentration of IBr(g) at equilibrium? What is the partial pressure of each gas at equilibrium? What is the total pressure of the system?

64. For the reaction $C(s) + \frac{1}{2}N_2(g) + \frac{5}{2}H_2(g) \rightleftharpoons CH_3NH_2(g)$, $K = 1.8 \times 10^{-1}$. If you begin the reaction with 1.0 mol of N_2, 2.0 mol of H_2, and sufficient $C(s)$ in a 2.00-L container, what are the concentrations of N_2 and CH_3NH_2 at equilibrium? What happens to K if the concentration of H_2 is doubled?

15.4 Nonequilibrium Conditions

65. The data listed in the table were collected at 450°C for the reaction $N_2(g) + 3H_2(g) \rightleftharpoons 2NH_3(g)$:

	Equilibrium Partial Pressure, atm		
P, atm	**NH_3**	**N_2**	**H_2**
30 (equilibrium)	1.740	6.588	21.58
100	15.20	19.17	65.13
600	321.6	56.74	220.8

The reaction equilibrates at 30 atm pressure. The pressure on the system is first increased to 100 atm and then to 600 atm. Is the system at equilibrium at each of these higher pressures? If not, in which direction will the reaction proceed in order to reach equilibrium?

66. For the reaction $2A \rightleftharpoons B + 3C$, K at 200°C is 2.0. A 6.00-L flask was used to carry out the reaction at this temperature. Given the experimental data in the table, all at 200°C, when the data for each experiment were collected, was the reaction at equilibrium? If it was not at equilibrium, in which direction will the reaction proceed?

Experiment	A	B	C
1	2.50 *M*	2.50 *M*	2.50 *M*
2	1.30 atm	1.75 atm	14.15 atm
3	12.61 mol	18.72 mol	6.51 mol

67. The following two reactions are carried out at 823 K:

$$CoO(s) + H_2(g) \rightleftharpoons Co(s) + H_2O(g) \qquad K = 67$$
$$CoO(s) + CO(g) \rightleftharpoons Co(s) + CO_2(g) \qquad K = 490$$

(a) Write the equilibrium expression for each reaction.
(b) Calculate the partial pressure of both gaseous components at equilibrium in each of the two reactions, if a 1.00 L reaction vessel initially contains 0.316 mol of H_2 or CO plus 0.500 mol CoO.
(c) Using the information provided, calculate K_p for the reaction $H_2(g) + CO_2(g) \rightleftharpoons CO(g) + H_2O(g)$
(d) Describe the shape of the graphs of [reactants] versus [products] as the amount of CoO changes.

68. Hydrogen iodide, HI, is synthesized via the reaction $H_2(g) + I_2(g) \rightleftharpoons 2HI(g)$, for which $K_p = 54.5$ at 425°C. Given a 2.0-L vessel containing 1.12×10^{-2} mol of H_2 and 1.8×10^{-3} mol of I_2 at equilibrium, what is the concentration of HI? Excess hydrogen is added to the vessel, so that the vessel now contains 3.64×10^{-1} mol of H_2. Calculate Q and then predict the direction in which the reaction will proceed. What are the new equilibrium concentrations?

15.5 Factors That Affect Equilibrium

69. For each equilibrium reaction, describe how Q and K change when (1) the pressure is increased; (2) the temperature is increased; (3) the volume of the system is increased; and (4) the concentration(s) of the reactant(s) is increased.
(a) $A(g) \rightleftharpoons B(g)$ $\qquad \Delta H = -20.6$ kJ/mol
(b) $2A(g) \rightleftharpoons B(g)$ $\qquad \Delta H = 0.3$ kJ/mol
(c) $A(g) + B(g) \rightleftharpoons 2C(g)$ $\qquad \Delta H = 46$ kJ/mol

70. For each reaction, describe how Q and K change when (1) the pressure in decreased; (2) the temperature is increased; (3) the volume of the system is decreased; (d) the concentration(s) of the reactant(s) is increased.
(a) $2A(g) \rightleftharpoons B(g)$ $\qquad \Delta H = -80$ kJ/mol
(b) $A(g) \rightleftharpoons 2B(g)$ $\qquad \Delta H = -18$ kJ/mol
(c) $2A(g) \rightleftharpoons 2B(g) + C(g)$ $\qquad \Delta H = 1.2$ kJ/mol

71. Le Châtelier's principle states that a system will change its composition to counteract stress. For the system $CO(g) + Cl_2(g) \rightleftharpoons COCl_2(g)$, write the equilibrium constant expression, K_p. What changes in the values of Q and K would you anticipate when (a) the volume is doubled; (b) the pressure is increased by a factor of 2; and (c) $COCl_2$ is removed from the system?

72. For the equilibrium system $3O_2(g) \rightleftharpoons 2O_3(g)$, $\Delta H^0 = 284$ kJ, write the equilibrium constant expression, K_p. What happens to the values of Q and K if the reaction temperature is increased? What happens to these values if both the temperature and pressure are increased?

73. Carbon and oxygen react to form CO_2 gas via the reaction $C(s) + O_2(g) \rightleftharpoons CO_2(g)$, for which $K = 1.2 \times 10^{69}$. Would you expect K to increase or decrease if the volume of the system were tripled? Why?

74. The reaction $COCl_2(g) \rightleftharpoons CO(g) + Cl_2(g)$ has $K = 2.2 \times 10^{-10}$ at 100°C. Starting with an initial P_{COCl_2} of 1.0 atm, you determine the following values of P_{CO} at three successive time intervals: 6.32×10^{-6} atm, 1.78×10^{-6} atm, and 1.02×10^{-5} atm. Based on these data, in which direction will the reaction proceed after each measurement? If chlorine gas is added to the system, what will be the effect on Q?

75. The table lists experimentally determined partial pressures at three temperatures for the reaction $Br_2(g) \rightleftharpoons 2Br(g)$.

T, K	1123	1173	1273
P_{Br_2} (atm)	3.000	0.3333	6.755×10^{-2}
P_{Br} (atm)	3.477×10^{-2}	2.159×10^{-2}	2.191×10^{-2}

Is this an endothermic or an exothermic reaction? Explain your reasoning.

76. The dissociation of water vapor proceeds according to the reaction $H_2O(g) \rightleftharpoons \frac{1}{2}O_2(g) + H_2(g)$. At 1300 K, there is 0.0027% dissociation, whereas at 2155 K, the dissociation is 1.18%. Calculate K and K_p. Is this an endothermic or an exothermic reaction? How do the magnitudes of the two equilibria compare? Would increasing the pressure improve the yield of H_2 gas at either temperature? (*Hint:* Assume that the system initially contains 1.00 mol of H_2O in a 1.00-L container.)

77. When 1.33 mol of CO_2 and 1.33 mol of H_2 are mixed in a 0.750-L container and are heated to 395°C, they react according to the equation $CO_2(g) + H_2(g) \rightleftharpoons CO(g) + H_2O(g)$. If $K = 0.802$, what are the equilibrium concentrations of each component of the equilibrium mixture? What happens to the value of K if H_2O is removed during the course of the reaction?

78. The equilibrium reaction $H_2(g) + Br_2(g) \rightleftharpoons 2HBr(g)$ has an equilibrium constant of 2.2×10^9. If you begin with 2.0 mol of

Br_2 and 2.0 mol of H_2 in a 5.0-L container, what is the partial pressure of HBr at equilibrium? What is the partial pressure of H_2 at equilibrium? If H_2 is removed from the system, what is the effect on the partial pressure of Br_2?

79. Iron(II) oxide reacts with carbon monoxide according to the equation $FeO(s) + CO(g) \rightleftharpoons Fe(s) + CO_2(g)$. At 800°C, $K = 0.34$; at 1000°C, $K = 0.40$.
   (a) A 20.0-L container is charged with 800.0 g of carbon dioxide, 1436 g of FeO, and 1120 g of iron. What are the equilibrium concentrations of all components of the mixture at each temperature?
   (b) What are the partial pressures of the gases at each temperature?
   (c) If CO were removed, what would be the effect on P_{CO_2} at each temperature?

80. The equilibrium constant K for the reaction $C(s) + CO_2(g) \rightleftharpoons 2CO(g)$ is 1.9 at 1000 K and 0.133 at 298 K.
   (a) If excess C is allowed to react with 25.0 g of CO_2 in a 3.00-L flask, how many grams of CO are produced at each temperature?
   (b) What are the partial pressures of each gas at 298 K? At 1000 K?
   (c) Would you expect K to increase or decrease if the pressure were increased at constant temperature and volume?

81. Data for the oxidation of methane, $CH_4(g) + 2O_2(g) \rightleftharpoons CO_2(g) + 2H_2O(g)$, in a closed 5.0 L vessel are listed in the table. Fill in the blanks in the following table, and determine the missing values of Q and K (indicated by ?) as the reaction is driven to completion.

	CH_4	O_2	CO_2	H_2O	Q	K
Initial (moles)	0.45	0.90	0	0	?	
At equilibrium	___	___	___	___		1.29
Add 0.50 mol of methane	0.95	___	___	___	?	
New equilibrium	___	___	___	___		?
Remove water	___	___	___	0	?	
New equilibrium	___	___	___	___		1.29

15.6 Controlling the Products of Reactions

82. The oxidation of carbon monoxide via the reaction $CO(g) + \frac{1}{2}O_2(g) \rightleftharpoons CO_2(g)$ has $\Delta H^0 = -283$ kJ. If you were interested in maximizing the yield of CO_2, what general conditions would you select with regard to temperature, pressure, and volume?

83. The oxidation of acetylene via the reaction $2C_2H_2(g) + 5O_2(g) \rightleftharpoons 4CO_2(g) + 2H_2O(l)$ has $\Delta H^0 = -2600$ kJ. What strategy would you use with regard to temperature, volume, and pressure to maximize the yield of product?

84. You are interested in maximizing the product yield of the system

$$2SO_2(g) + O_2(g) \rightleftharpoons 2SO_3(g) \qquad K = 280,\ \Delta H^0 = -158\ \text{kJ}$$

What general conditions would you select with regard to temperature, pressure, and volume? If SO_2 has an initial concentration of 0.200 M and the amount of O_2 is stoichiometric, what amount of SO_3 is produced at equilibrium?

APPLICATIONS

85. The total concentrations of dissolved Al in a soil sample represent the sum of "free" Al^{3+} and bound forms of Al that are stable enough to be considered definite chemical species. The distribution of aluminum among its possible chemical forms can be described using equilibrium constants such as these:

$$K_1 = [AlOH^{2+}]/[Al^{3+}][OH^-] = 1.0 \times 10^9$$
$$K_2 = [AlSO_4^{\ +}]/[Al^{3+}][SO_4^{\ 2-}] = 1.0 \times 10^3$$
$$K_3 = [AlF^{2+}]/[Al^{3+}][F^-] = 1.0 \times 10^7$$

Write an equilibrium equation for each expression. Which anion has the highest affinity for Al^{3+}: OH^-, $SO_4^{\ 2-}$, or F^-? Explain your reasoning. A 1.0 M solution of Al^{3+} is mixed with a 1.0 M solution of each of the anions. Which mixture has the lowest Al^{3+} concentration?

86. Many hydroxy acids undergo reactions forming lactones (cyclic esters) that contain a five- or six-member ring. Common hydroxy acids found in nature are glycolic acid, a constituent of cane-sugar juice; lactic acid, which has the characteristic odor and taste of sour milk; and citric acid, found in fruit juices. The general reaction is written as hydroxy acid $\rightleftharpoons$ lactone + H_2O:

$HOCH_2CH_2COOH \rightleftharpoons$ 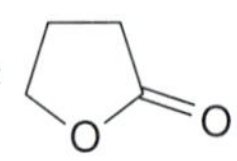 $+ H_2O$

Hydroxyacid Lactone

Given the information in the table, calculate the equilibrium constant for this reaction for each of the hydroxy acids indicated, and determine which ring size is most stable.

		At Equilibrium	
Hydroxy Acid Formula	**Size of Lactone Ring (atoms)**	**Hydroxy Acid, M**	**Lactone, M**
$HOCH_2CH_2COOH$	4	4.99×10^{-3}	5.00×10^{-5}
$HOCH_2CH_2CH_2COOH$	5	8.10×10^{-5}	2.19×10^{-4}
$HOCH_2CH_2CH_2CH_2COOH$	6	5.46×10^{-2}	5.40×10^{-9}
$HOCH_2CH_2CH_2CH_2CH_2COOH$	7	9.90×10^{-3}	1.00×10^{-4}

87. Phosphorus pentachloride, an important reagent in organic chemistry for converting alcohols to alkyl chlorides ($ROH \rightarrow RCl$), is hydrolyzed in water to form phosphoric acid and hydrogen chloride. In the gaseous state, however, PCl_5 can decompose at 250°C according to the reaction $PCl_5(g) \rightleftharpoons PCl_3(g) + Cl_2(g)$, for which $K = 0.0420$.
   (a) Are products or reactants favored in the decomposition of $PCl_5(g)$?
   (b) If a 2.00-L flask containing 104.1 g PCl_5 is heated to 250°C, what is the equilibrium concentration of each species in this reaction?
   (c) What effect would an increase in pressure have on the equilibrium position? Why?
   (d) If a 1.00×10^3 L vessel containing 2.00×10^3 kg of PCl_3 with a constant chlorine pressure of 2.00 atm is allowed to reach equilibrium, how many kilograms of PCl_5 are produced? What is the percent yield of PCl_5?

88. Carbon disulfide is used in the manufacture of rayon and in electronic vacuum tubes. However, extreme caution must be used when handling CS_2 in its gaseous state because it is extremely toxic and can cause fatal convulsions. Chronic toxicity is marked by psychic disturbances and tremors. Carbon disulfide is used to synthesize H_2S at elevated T via the reaction
$CS_2(g) + 4H_2(g) \rightleftharpoons CH_4(g) + 2H_2S(g)$; $K = 3.3 \times 10^4$.

(a) If the equilibrium concentration of methane in this reaction is $2.5 \times 10^{-2} M$ and the initial concentration of each reactant is $0.1635\ M$, what is the concentration of H_2S at equilibrium?

(b) Exposure to CS_2 concentrations greater than 300 ppm for several hours can start to produce adverse effects. After working for several hours in a laboratory that contains large quantities of carbon disulfide, you notice that the fume hoods were off and there was not enough ventilation to remove any carbon disulfide vapor. Given the equilibrium $CS_2(l) \rightleftharpoons CS_2(g)$, where $T = 20°C$ and $K_p = 0.391$, determine whether you are in any danger.

89. Chloral hydrate, a sedative commonly referred to as "knockout drops," is in equilibrium with trichloroacetaldehyde in aqueous solution:

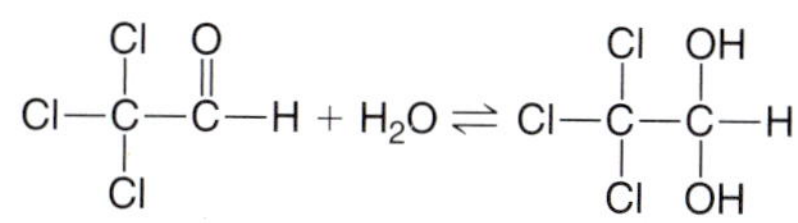

Trichloroacetaldehyde **Chloral hydrate**

The equilibrium constant for this reaction as written is 3×10^4. Are the products or the reactants favored? Write an equilibrium expression for this reaction. How could you drive this reaction to completion?

90. Hydrogen cyanide is commercially produced in the United States by the reaction $CH_4(g) + NH_3(g) + \frac{3}{2}O_2(g) \rightleftharpoons HCN(g) + 3H_2O(g)$, where HCN is continuously removed from the system. This reaction is carried out at approximately 1100°C in the presence of a catalyst; however, the high temperature causes other reactions to occur. Why is it necessary to run this reaction at such an elevated temperature? Does the presence of the catalyst affect the equilibrium position?

91. Hemoglobin (Hb) transports oxygen from the lungs to the capillaries, and consists of four subunits, each capable of binding a single molecule of O_2. In the lungs, P_{O_2} is relatively high (100 mmHg), so Hb becomes nearly saturated with O_2. In the tissues, however, P_{O_2} is relatively low (40 mmHg), and Hb releases about half of its bound oxygen. Myoglobin, a protein in muscle, resembles a single subunit of hemoglobin. The plots

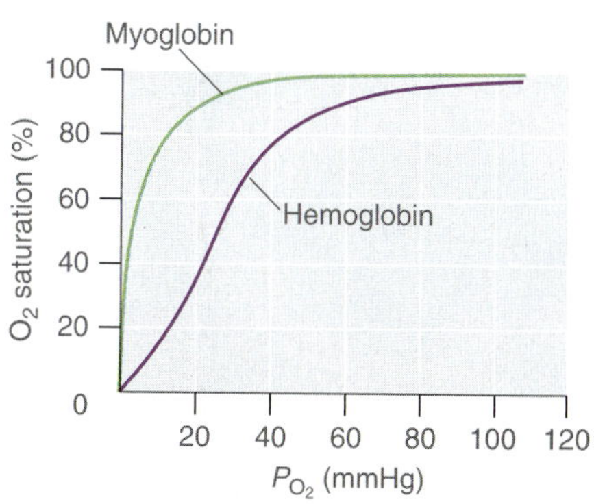

show the percent O_2 saturation versus P_{O_2} for Hb and myoglobin. Based on these plots, which molecule has the higher affinity for oxygen? What advantage does Hb have over myoglobin as the oxygen transporter? Why is it advantageous to have myoglobin in muscle tissue? Use equilibrium to explain why it is more difficult to exercise at high altitudes where the partial pressure of oxygen is lower.

92. Sodium sulfate is widely used in the recycling industry as well as in the detergent and glass industries. This compound combines with H_2SO_4 via the reaction $Na_2SO_4(s) + H_2SO_4(g) \rightleftharpoons 2NaHSO_4(s)$. Sodium hydrogen sulfate is used as a cleaning agent because it is water soluble and acidic.

(a) Write an expression for the equilibrium constant K for this reaction.

(b) Relate this equilibrium constant to the equilibrium constant for the related reaction $2Na_2SO_4(s) + 2H_2SO_4(g) \rightleftharpoons 4NaHSO_4(s)$.

(c) The dissolution of Na_2SO_4 in water produces the equilibrium reaction $SO_4^{2-}(aq) + H_2O(l) \rightleftharpoons HSO_4^-(aq) + OH^-(aq)$ with $K = 8.33 \times 10^{-13}$. What is the concentration of OH^- in a solution formed from the dissolution of 1.00 g of sodium sulfate to make 150.0 mL of aqueous solution? Neglect the autoionization of water in your answer.

93. One of the Venera orbiter satellites measured S_2 concentrations at the surface of Venus. The resulting thermochemical data suggest that S_2 formation at the planet's surface occurs via the equilibrium reaction $4CO(g) + 2SO_2(g) \rightleftharpoons 4CO_2(g) + S_2(g)$. Write an expression for the equilibrium constant K for this reaction, and then relate this expression to those for the following reactions:

(a) $2CO(g) + SO_2(g) \rightleftharpoons 2CO_2(g) + \frac{1}{2}S_2(g)$

(b) $CO(g) + \frac{1}{2}SO_2(g) \rightleftharpoons CO_2(g) + \frac{1}{4}S_2\ (g)$

(c) At 450°C, the equilibrium pressure of CO_2 is 85.0 atm, SO_2 is 1.0 atm, CO is 1.0 atm, and S_2 is 3.0×10^{-8} atm. What are K and K_p at this temperature? What is the concentration of S_2?

94. Until the early part of the 20th century, commercial production of sulfuric acid was carried out by the "lead-chamber" process, in which SO_2 was oxidized to H_2SO_4 in a lead-lined room. This process may be summarized by the following sequence of reactions:

(1) $NO(g) + NO_2(g) + 2H_2SO_4(l) \rightleftharpoons 2NOHSO_4(s) + H_2O(l)$ K_1

(2) $2NOHSO_4(s) + SO_2(g) + 2H_2O(l) \rightleftharpoons 3H_2SO_4(l) + 2NO(g)$ K_2

(a) Write the equilibrium constant expressions for reactions 1 and 2 and the sum of the reactions (reaction 3).

(b) Show that $K_3 = K_1 \times K_2$.

(c) If insufficient water is added in reaction 2 such that the reaction becomes $NOHSO_4(s) + \frac{1}{2}SO_2(g) + H_2O(l) \rightleftharpoons \frac{3}{2}H_2SO_4(l) + NO(g)$, does K'_3 still equal $K_1 \times K_2$?

(d) Based on part c, write the equilibrium constant expression for K_2.

95. Phosgene (carbonic dichloride, $COCl_2$) is a colorless, highly toxic gas with an odor similar to that of moldy hay. Used as a lethal gas in war, phosgene can be immediately fatal; inhalation can cause either pneumonia or pulmonary edema. For the equilibrium reaction $COCl_2(g) \rightleftharpoons CO(g) + Cl_2(g)$, K_p is 0.680 at $-10°C$. If the initial pressure of $COCl_2$ is 0.681 atm what is the partial pressure of each component of this equilibrium system? Is the formation of products or reactant favored in this reaction?

96. British bituminous coal has a high sulfur content and produces much smoke when burned. In 1952, burning of this coal in London led to elevated levels of smog containing high concentrations of sulfur dioxide, a lung irritant, and more than 4000 people died. Sulfur dioxide emissions can be converted to SO_3 and ultimately to H_2SO_4, the cause of acid rain. The initial reaction is $2SO_2(g) + O_2(g) \rightleftharpoons 2SO_3(g)$, for which $K_p = 44$.

(a) Given this value of K_p, are products or reactant favored in this reaction?

(b) What is the partial pressure of each species under equilibrium conditions if the initial pressure of each reactant is 0.50 atm?

(c) Would an increase in pressure favor the formation of products or reactant? Why?

97. Oxyhemoglobin is the oxygenated form of hemoglobin, the oxygen-carrying pigment of red blood cells. Hemoglobin is built from α and β protein chains. Assembly of the oxygenated (oxy) and deoxygenated (deoxy) β-chains has been studied with the following results:

$$4\beta\ (\text{oxy}) \rightleftharpoons \beta_4\,(\text{oxy}) \qquad K = 9.07 \times 10^{15}$$

$$4\beta\ (\text{deoxy}) \rightleftharpoons \beta_4\,(\text{deoxy}) \qquad K = 9.20 \times 10^{13}$$

Is it more likely that hemoglobin β chains assemble in an oxygenated or deoxygenated state? Explain your answer.

98. Inorganic weathering reactions can turn silicate rocks, such as diopside ($CaMgSi_2O_6$), to carbonate via the reaction

$$CaMgSi_2O_6(s) + CO_2(g) \rightleftharpoons MgSiO_3(s) + CaCO_3(s) + SiO_2(s)$$

Write an expression for the equilibrium constant. Although this reaction occurs on both Earth and Venus, the high surface temperature of Venus causes the reaction to be driven in one direction on that planet. Predict whether the reaction will be driven to the right or the left by high temperatures, and then justify your answer. The estimated partial pressure of carbon dioxide on Venus is 85 atm due to the dense Venusian atmosphere. How does this pressure influence the reaction?

99. Silicon and its inorganic compounds are widely used to manufacture textile glass fibers, cement, ceramic products, and synthetic fillers. Two of the most important industrially utilized silicon halides are $SiCl_4$ and $SiHCl_3$, formed by reaction of elemental silicon with HCl at temperatures higher than 300°C:

$$Si(s) + 4HCl(g) \rightleftharpoons SiCl_4(g) + 2H_2(g)$$

$$Si(s) + 3HCl(g) \rightleftharpoons SiHCl_3(g) + H_2(g)$$

Which of these two reactions is favored by increasing the HCl concentration? By decreasing the volume of the system?

100. The first step in the utilization of glucose in humans is the conversion of glucose to glucose-6-phosphate via the transfer of a phosphate group from ATP (adenosine triphosphate), which produces glucose-6-phosphate and ADP (adenosine diphosphate):

$$\text{glucose} + \text{ATP} \rightleftharpoons \text{glucose-6-phosphate} + \text{ADP} \qquad K = 680 \text{ at } 25^\circ\text{C}$$

(a) Is the formation of products or reactant favored in this reaction?

(b) Would the value of K increase, decrease, or remain the same if the glucose concentration were doubled?

(c) If $-RT \ln K = -RT' \ln K$, what would the value of K be if the temperature were decreased to 0°C?

(d) Is the formation of products favored by an increase or a decrease in the temperature of the system?

101. In the presence of O_2, the carbon atoms of glucose can be fully oxidized to CO_2 with a release of free energy almost 20 times greater than that possible under conditions in which O_2 is not present. In many animal cells, the TCA cycle (*Tri*C*arboxylic* *A*cid cycle) is the second stage in the complete oxidation of glucose. One reaction in the TCA cycle is the conversion of citrate to isocitrate, for which $K = 0.08$ in the forward direction. Speculate as to why the cycle continues despite this unfavorable value of K. What happens if the citrate concentration increases?

102. Soil is an open system, subject to natural inputs and outputs that may change its chemical composition. Aqueous-phase, adsorbed, and solid-phase forms of Al(III) are of critical importance in controlling the acidity of soils, although industrial effluents, such as sulfur and nitrogen oxide gases, and fertilizers containing nitrogen can also have a large effect. Dissolution of the mineral gibbsite, which contains Al^{3+} in the form $Al(OH)_3(s)$, occurs in soil according to the reaction

$$Al(OH)_3(s) + 3H^+(aq) \rightleftharpoons Al^{3+}(aq) + 3H_2O(l)$$

When gibbsite is in a highly crystalline state, $K = 9.35$ for this reaction at 298 K. In the microcrystalline state, $K = 8.11$. Is this change consistent with the increased surface area of the microcrystalline state?

ESSENTIAL SKILLS

7

TOPICS

- The Quadratic Formula

Previous Essential Skills sections introduced many of the mathematical operations you need to solve chemical problems. We now introduce the quadratic formula, a mathematical relationship involving sums of powers in a single variable that you will need to apply to solve some of the problems in this chapter.

The Quadratic Formula

Mathematical expressions that involve a sum of powers in one or more variables (x, for example) multiplied by coefficients (such as a) are called *polynomials*. Polynomials of a single variable have the general form

$$a_n x^n + \cdots + a_2 x^2 + a_1 x + a_0$$

The highest power to which the variable in a polynomial is raised is called its *order*. Thus, the polynomial shown above is of the *nth* order. For example, if n were 3, the polynomial would be third order.

A *quadratic equation* is a second-order polynomial equation in a single variable x:

$$ax^2 + bx + c = 0$$

According to the fundamental theorem of algebra, a second-order polynomial equation has two solutions, called *roots*, which can be found using a method called *completing the square*. In this method, we first add $-c$ to both sides of the quadratic equation and then divide both sides by a:

$$x^2 + \frac{b}{a}x = -\frac{c}{a}$$

We can convert the left side of this equation to a perfect square by adding $b^2/4a^2$, which is equal to $(b/2a)^2$:

Left side: $$x^2 + \frac{b}{a}x + \frac{b^2}{4a^2} = \left(x + \frac{b}{2a}\right)^2$$

Having added a value to the left side, we must now add that same value to the right side:

$$\left(x + \frac{b}{2a}\right)^2 = -\frac{c}{a} + \frac{b^2}{4a^2}$$

The common denominator on the right side is $4a^2$. Rearranging the right side, we obtain

$$\left(x + \frac{b}{2a}\right)^2 = \frac{b^2 - 4ac}{4a^2}$$

Taking the square root of both sides and solving for x give us

$$x + \frac{b}{2a} = \frac{\pm\sqrt{b^2 - 4ac}}{2a}$$

$$x = \frac{-b \pm \sqrt{b^2 - 4ac}}{2a}$$

This equation, known as the *quadratic formula*, has two roots:

$$x = \frac{-b + \sqrt{b^2 - 4ac}}{2a} \quad \text{and} \quad x = \frac{-b - \sqrt{b^2 - 4ac}}{2a}$$

Thus, we can obtain the solutions to a quadratic equation by substituting the values of the coefficients (a, b, c) into the quadratic formula.

When you apply the quadratic formula to obtain solutions to a quadratic equation, it is important to remember that one of the two solutions may not make sense, or neither may make sense. There may be times, for example, when a negative solution is not reasonable, or when both solutions require that a square root be taken of a negative number. In such cases, we simply discard any solution that is unreasonable, and only report a solution that is reasonable. The next exercise gives you practice using the quadratic formula.

SKILL BUILDER ES15.1

Use the quadratic formula to solve for x in each equation. Report your answers to three significant figures.

(a) $x^2 + 8x - 5 = 0$
(b) $2x^2 - 6x + 3 = 0$
(c) $3x^2 - 5x - 4 = 6$
(d) $2x(-x + 2) + 1 = 0$
(e) $3x(2x + 1) - 4 = 5$

Solution

(a) $$x = \frac{-8 + \sqrt{8^2 - 4(1)(-5)}}{2(1)} = 0.583 \text{ and}$$

$$x = \frac{-8 - \sqrt{8^2 - 4(1)(-5)}}{2(1)} = -8.58$$

(b) $$x = \frac{-(-6) + \sqrt{(-6)^2 - 4(2)(3)}}{2(2)} = 2.37 \text{ and}$$

$$x = \frac{-(-6) - \sqrt{(-6)^2 - 4(2)(3)}}{2(2)} = 0.634$$

(c) $$x = \frac{-(-5) + \sqrt{(-5)^2 - 4(3)(-10)}}{2(3)} = 2.84 \text{ and}$$

$$x = \frac{-(-5) - \sqrt{(-5)^2 - 4(3)(-10)}}{2(3)} = -1.17$$

(d) $$x = \frac{-4 + \sqrt{4^2 - 4(-2)(1)}}{2(-2)} = -0.225 \text{ and}$$

$$x = \frac{-4 - \sqrt{4^2 - 4(-2)(1)}}{2(-2)} = 2.22$$

(e) $$x = \frac{-3 + \sqrt{3^2 - 4(6)(-9)}}{2(6)} = 1.00 \text{ and}$$

$$x = \frac{-3 - \sqrt{3^2 - 4(6)(-9)}}{2(6)} = -1.50$$

16 Aqueous Acid–Base Equilibria

LEARNING OBJECTIVES

- To understand the autoionization reaction of liquid water
- To know the relationship among pH, pOH, and pK_w
- To understand the concept of conjugate acid–base pairs
- To know the relationship between acid or base strength and the magnitude of K_a, K_b, pK_a, and pK_b
- To understand the leveling effect
- To be able to predict whether reactants or products are favored in an acid–base equilibrium
- To understand how molecular structure determines acid and base strengths
- To be able to use K_a and K_b values to calculate the percent ionization and pH of a solution of an acid or a base
- To be able to calculate the pH at any point in an acid–base titration
- To understand how the addition of a common ion affects the position of an acid–base equilibrium
- To understand how a buffer works and how to use the Henderson–Hasselbalch equation to calculate the pH of a buffer

Many vital chemical and physical processes take place exclusively in aqueous solution, including the complex biochemical reactions that occur in living organisms and the reactions that rust and corrode steel objects, such as bridges, ships, and automobiles. Among the most important reactions in aqueous solution are those that can be categorized as acid–base, precipitation, and complexation reactions. So far, our discussions of these reactions have been largely qualitative. In this chapter and the next, however, we take a more quantitative approach to understanding such reactions, using the concept of chemical equilibrium that we developed in Chapter 15 for simple gas-phase reactions. We will begin by revisiting acid–base reactions in a qualitative fashion and then develop quantitative methods to describe acid–base equilibria. In Chapter 17, we will use the same approach to describe the equilibria involved in the dissolution of sparingly soluble solids and the formation of metal complexes.

Indicaters are used to monitor changes in pH. The pH of a solution can be monitored using an acid–base indicator, a substance that undergoes a color change within a specific pH range that is characteristic of that indicator. The color changes for seven commonly used indicators over a pH range of 1–10 are shown here.

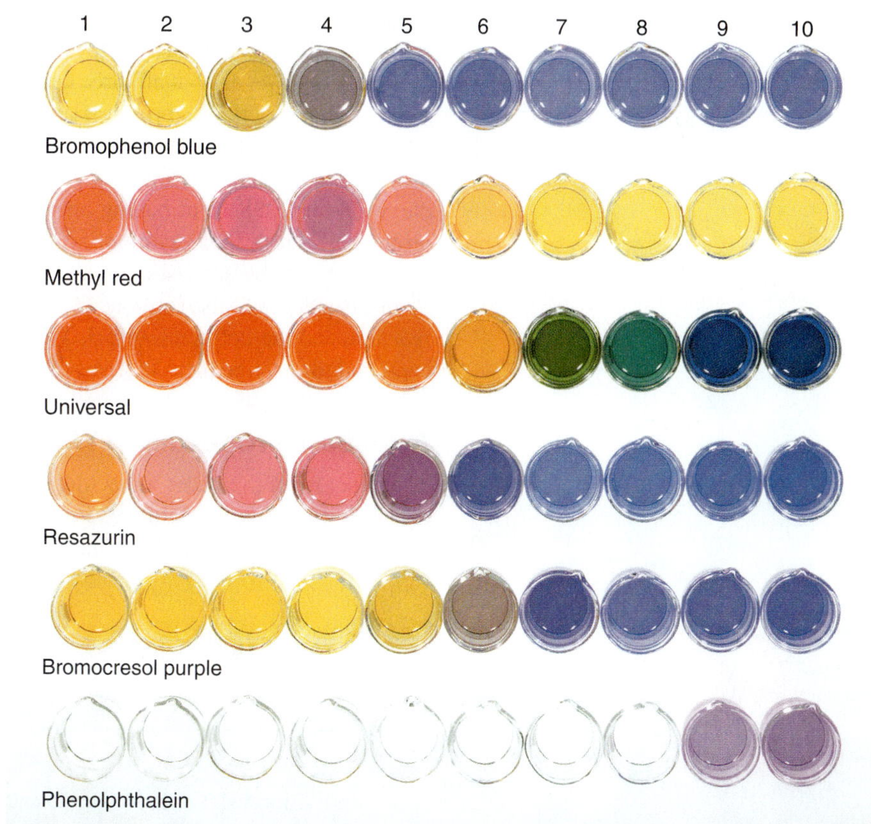

CHAPTER OUTLINE

In Chapter 4, we described how acid rain can adversely affect the survival of marine life and plant growth. Many significant phenomena such as acid rain can be understood only in terms of the acid–base behavior of chemical species. As we expand our discussion of acid–base behavior in this chapter, you will learn why lemon slices are served with fish, why the strengths of acids and bases can vary over many orders of magnitude, and why rhubarb leaves are toxic to humans. You will also understand how the pH of your blood is kept constant, even though you produce large amounts of acid when you exercise.

16.1 • The Autoionization of Water

As you learned in Chapters 4 and 8, acids and bases can be defined in several different ways. Recall that the Arrhenius definition of an acid is a substance that dissociates in water to produce H^+ ions (protons), and an Arrhenius base is a substance that dissociates in water to produce OH^- ions (hydroxide). According to this view, an acid–base reaction involves the reaction of a proton with a hydroxide ion to form water. Although Brønsted and Lowry defined an acid similarly to Arrhenius, as any substance that can *donate* a proton, the Brønsted–Lowry definition of a base is much more general than the Arrhenius definition. In Brønsted–Lowry terms, a base is any substance that can *accept* a proton and therefore is not limited to just a hydroxide ion. Recall from Chapter 4 that for every Brønsted–Lowry acid, there exists a corresponding conjugate base that contains one fewer proton. Consequently, all Brønsted–Lowry acid–base reactions actually involve *two* conjugate acid–base pairs and the transfer of a *proton* from one substance (the acid) to another (the base). In contrast, the Lewis definition of acids and bases, discussed in Chapter 8, focuses on accepting or donating *pairs of electrons* rather than protons. A Lewis base is an electron-pair donor, and a Lewis acid is an electron-pair acceptor. The definitions of acids and bases are summarized in Table 16.1.

Because this chapter deals with acid–base equilibria in *aqueous solution*, our discussion will use primarily the Brønsted–Lowry definitions and nomenclature. Remember, however, that all three definitions are just different ways of looking at the same kind of reaction: a proton is an acid and the hydroxide ion is a base, no matter which definition you use. In practice, chemists tend to use whichever definition is most helpful to make a particular point or to understand a given system. If, for example, we refer to a base as having one or more lone pairs of electrons that can accept a proton, we are simply combining the Lewis and Brønsted–Lowry definitions to emphasize the characteristic properties of a base.

In Chapter 4, we also introduced the acid–base properties of water, its *autoionization reaction,* and the definition of pH. The purpose of this section is to review those concepts and to describe them in terms of chemical equilibrium.

Acid–Base Properties of Water

The structure of the water molecule, with its polar O—H bonds and two lone pairs of electrons on the oxygen atom, was described in Chapters 4 and 8, and the structure of liquid water was discussed in Chapter 13. Recall that its highly polar structure allows

TABLE 16.1 Definitions of acids and bases

	Acids	Bases
Arrhenius	H^+ donor	OH^- donor
Brønsted–Lowry	H^+ donor	H^+ acceptor
Lewis	Electron-pair acceptor	Electron-pair donor

liquid water to act as either an acid (by donating a proton to a base) or a base (by using a lone pair of electrons to accept a proton). For example, when a strong acid such as HCl dissolves in water, it dissociates into chloride ions (Cl^-) and protons (H^+). As you learned in Chapter 4, the proton, in turn, reacts with a water molecule to form the **hydronium ion**, H_3O^+:

$$\underset{\text{Acid}}{HCl(aq)} + \underset{\text{Base}}{H_2O(l)} \longrightarrow \underset{\text{Acid}}{H_3O^+(aq)} + \underset{\text{Base}}{Cl^-(aq)} \quad (16.1)$$

In this reaction, HCl is the acid and water is acting as a base by accepting an H^+ ion. The reaction in Equation 16.1 is often written in a simpler form, with the hydronium ion represented by H^+ (remember that free H^+ ions do not exist in liquid water):

$$HCl(aq) \longrightarrow H^+(aq) + Cl^-(aq) \quad (16.2)$$

Conversely, in Equation 16.3, water acts as an acid by donating a proton to NH_3, which acts as a base:

$$\underset{\text{Acid}}{H_2O(l)} + \underset{\text{Base}}{NH_3(aq)} \rightleftharpoons \underset{\text{Acid}}{NH_4^+(aq)} + \underset{\text{Base}}{OH^-(aq)} \quad (16.3)$$

Thus, water is **amphiprotic**, meaning that it can behave as either an acid or a base, depending on the nature of the other reactant. Notice in Equation 16.3 that the reaction is not complete, as indicated by the double arrow.

The Ion-Product Constant of Liquid Water

Because water is amphiprotic, one water molecule can react with another to form an OH^- ion and an H_3O^+ ion in an autoionization process:

$$2H_2O(l) \rightleftharpoons H_3O^+(aq) + OH^-(aq) \quad (16.4)$$

The equilibrium constant K for this reaction can be written as

$$K = \frac{[H_3O^+][OH^-]}{[H_2O]^2} \quad (16.5)$$

When pure liquid water is in equilibrium with hydronium and hydroxide ions at 25°C, the concentrations of hydronium ion and hydroxide ion are equal: $[H_3O^+] = [OH^-] = 1.003 \times 10^{-7}\ M$ (to three decimal places). At 25°C, the density of liquid water is 0.997 g/mL. Hence, the concentration of liquid water at 25°C is

$$[H_2O] = \frac{\text{mol}}{\text{L}} = \left(\frac{0.997\ \cancel{\text{g}}}{\cancel{\text{mL}}}\right)\left(\frac{1\ \text{mol}}{18.02\ \cancel{\text{g}}}\right)\left(\frac{1000\ \cancel{\text{mL}}}{\text{L}}\right) = 55.3\ M \quad (16.6)$$

Because the number of dissociated water molecules is very small (approximately 2 ppb), the equilibrium of the autoionization reaction (Equation 16.4) lies far to the left. Consequently, the concentration of water is essentially unchanged by the autoionization reaction [$55.3\ M - (1.00 \times 10^{-7}\ M)$] and can be treated as a constant.

By treating $[H_2O]$ as a constant, we can rearrange Equation 16.5 to define a new equilibrium constant, the **ion-product constant of liquid water (K_w)**:

$$\begin{aligned} K[H_2O]^2 &= [H_3O^+][OH^-] \\ K_w &= [H_3O^+][OH^-] \end{aligned} \quad (16.7)$$

Substituting the values for $[H_3O^+]$ and $[OH^-]$ at 25°C into this expression gives

$$K_w = (1.00 \times 10^{-7})(1.00 \times 10^{-7}) = 1.01 \times 10^{-14} \quad (16.8)$$

Like any other equilibrium constant, the value of K_w varies with temperature, ranging from 1.15×10^{-15} at 0°C to 4.99×10^{-13} at 100°C.

In pure water, the concentrations of the hydronium ion and the hydroxide ion are the same; hence, the solution is neutral. If $[H_3O^+] > [OH^-]$, however, the solution is acidic. Conversely, if $[H_3O^+] < [OH^-]$, the solution is basic. For an aqueous solution, the H_3O^+ concentration is a quantitative measure of acidity: the higher the H_3O^+ concentration, the more acidic the solution. Conversely, the higher the OH^- concentration, the more basic the solution. In most situations that you will encounter, the H_3O^+ and OH^- concentrations from the dissociation of water are so small ($\leq 10^{-7}$ *M*) that they can be ignored in calculating the H_3O^+ or OH^- concentrations of solutions of acids and bases, but this is not always the case.

The Relationship Among pH, pOH, and pK_w

Note the pattern

For any neutral solution, pH + pOH = 14.00 (at 25°C) and pH = $\frac{1}{2}$pK_w.

The *pH scale* is a concise way of describing the H_3O^+ concentration, and hence the acidity or basicity of a solution. Recall from Chapter 4 that pH and the H^+ (actually H_3O^+) concentration are related as follows:

$$\text{pH} = -\log_{10}[H^+] \qquad (16.9a)$$

$$[H^+] = 10^{-\text{pH}} \qquad (16.9b)$$

Recall also that the pH of a neutral solution ($[H_3O^+] = 1.00 \times 10^{-7}$ *M*)* is 7.00, whereas acidic solutions have pH < 7 (corresponding to $[H_3O^+] > 1.00 \times 10^{-7}$) and basic solutions have pH > 7 (corresponding to $[H_3O^+] < 1.00 \times 10^{-7}$). Because the scale is logarithmic, a pH difference of *one* between two solutions corresponds to a difference of a factor of *ten* in their hydronium ion concentrations. (Refer to Essential Skills 3 at the end of Chapter 4 if you need to refresh your memory about how to use logarithms.)

Similar notation systems are used to describe many other chemical quantities that contain a large negative exponent. For example, chemists use an analogous *pOH scale* to describe the hydroxide ion concentration of a solution. The pOH and $[OH^-]$ are related as follows:

$$\text{pOH} = -\log_{10}[OH^-] \qquad (16.10a)$$

$$[OH^-] = 10^{-\text{pOH}} \qquad (16.10b)$$

The constant K_w can also be expressed using this notation, where p$K_w = -\log K_w = 14$.

Because a neutral solution has $[OH^-] = 1.00 \times 10^{-7}$, the pOH of a neutral solution is 7.00. Consequently, the sum of the pH and the pOH for a neutral solution at 25°C is 7.00 + 7.00 = 14.00.

We can show that the sum of pH and pOH is equal to 14.00 for *any* aqueous solution at 25°C by taking the negative logarithm of both sides of Equation 16.7:

$$\begin{aligned} -\log K_w = \text{p}K_w &= -\log([H_3O^+][OH^-]) \\ &= (-\log[H_3O^+]) + (-\log[OH^-]) \\ &= \text{pH} + \text{pOH} \end{aligned} \qquad (16.11)$$

At any temperature, pH + pOH = pK_w. Thus, at 25°C, where $K_w = 1.01 \times 10^{-14}$, pH + pOH = 14.00. More generally, the pH of any neutral solution is just half the value of pK_w at that temperature. The relationship among pH, pOH, and the acidity or basicity of a solution is summarized graphically in Figure 16.1 over the common pH range of 0 to 14. Notice the inverse relationship between the pH and pOH scales.

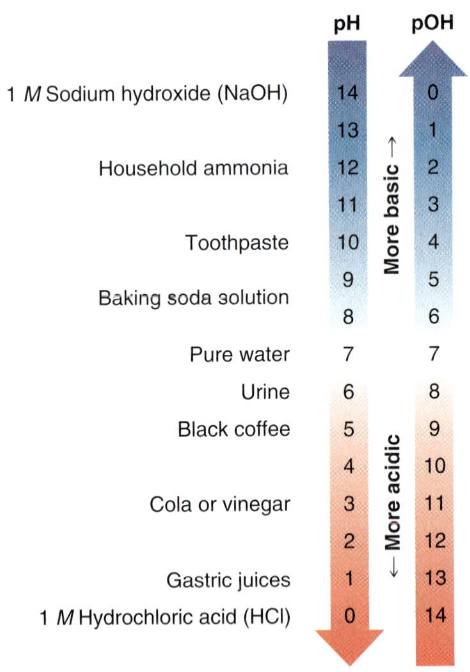

Figure 16.1 The inverse relationship between the pH and pOH scales. As pH decreases, $[H^+]$ and the acidity increase. As pOH increases, $[OH^-]$ and the basicity decrease. Common substances have pH values that range from extremely acidic to extremely basic.

* Because most pH meters used in the laboratory read to only two decimal places, we will use only two decimal places for hydronium and hydroxide concentrations and K_w values in the rest of the chapter.

EXAMPLE 16.1

The value of K_w for water at 100°C is 4.99×10^{-13}. Calculate **(a)** the value of pK_w for water at this temperature and **(b)** the values of pH and pOH for a neutral aqueous solution at 100°C.

Given K_w

Asked for pK_w, pH, and pOH

Strategy

Ⓐ Calculate the value of pK_w by taking the negative logarithm of K_w.
Ⓑ For a neutral aqueous solution, $[H_3O^+] = [OH^-]$. Use this relationship and Equation 16.7 to calculate $[H_3O^+]$ and $[OH^-]$. Then determine the values of pH and pOH for the solution.

Solution

(a) Ⓐ Because pK_w is the negative logarithm of K_w, we can write

$$pK_w = -\log K_w = -\log(4.99 \times 10^{-13}) = 12.302$$

The answer is reasonable: K_w is between 10^{-13} and 10^{-12}, so pK_w must be between 12 and 13.

(b) Ⓑ Equation 16.7 shows that $K_w = [H_3O^+][OH^-]$. Because $[H_3O^+] = [OH^-]$ in a neutral solution, we can let $x = [H_3O^+] = [OH^-]$ to give

$$K_w = [H_3O^+][OH^-] = (x)(x) = x^2$$
$$x = \sqrt{K_w} = \sqrt{4.99 \times 10^{-13}} = 7.06 \times 10^{-7}\ M$$

Because x is equal to both $[H_3O^+]$ and $[OH^-]$,

$$pH = pOH = -\log(7.06 \times 10^{-7}) = 6.15 \text{ (to two decimal places)}$$

We could obtain the same answer more easily (without using logarithms) by using the value of pK_w. In this case, we know that $pK_w = 12.302$, and from Equation 16.11, we know that $pK_w = pH + pOH$. Because pH = pOH in a neutral solution, we can use Equation 16.11 directly, setting pH = pOH = y. Solving to two decimal places we obtain:

$$pK_w = pH + pOH = y + y = 2y$$
$$y = \frac{pK_w}{2} = \frac{12.302}{2} = 6.15 = pH = pOH$$

EXERCISE 16.1

Humans maintain an internal temperature of about 37°C. At this temperature, $K_w = 3.55 \times 10^{-14}$. Calculate pK_w and the pH and pOH of a neutral solution at 37°C.

Answer $pK_w = 13.49$; pH = pOH = 6.73

16.2 ○ A Qualitative Description of Acid–Base Equilibria

We now turn our attention to acid–base reactions to see how the concepts of chemical equilibrium and equilibrium constants can deepen our understanding of this kind of chemical behavior. We begin with a qualitative description of acid–base equilibria in terms of the Brønsted–Lowry model, and then we proceed to a quantitative description in Section 16.4.

Conjugate Acid–Base Pairs

We discussed the concept of conjugate acid–base pairs in Chapter 4, using the reaction of ammonia, the base, with water, the acid, as an example. In aqueous solutions, acids and bases can be defined in terms of the transfer of a proton from an acid to a base. Thus, for every acidic species in an aqueous solution, there exists a species derived from the acid by the loss of a proton. These two species that differ by only a proton constitute a **conjugate acid–base pair**. For example, in the reaction of HCl with water (Equation 16.1), HCl, the *parent acid*, donates a proton to a water molecule, the *parent base*, thereby forming Cl^-. Thus, HCl and Cl^- constitute a conjugate acid–base pair. (By convention, we always write a conjugate acid–base pair as the acid followed by its conjugate base.) In the reverse reaction, the Cl^- ion in solution acts as a base to accept a proton from H_3O^+, forming H_2O and HCl. Thus, H_3O^+ and H_2O constitute a second conjugate acid–base pair. In general, any acid–base reaction must contain *two* conjugate acid–base pairs, which in this case are HCl/Cl^- and H_3O^+/H_2O:

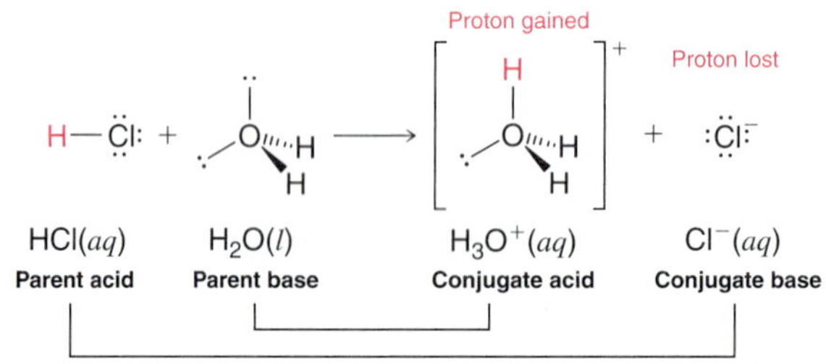

Similarly, in the reaction of acetic acid with water, acetic acid donates a proton to water, which acts as the base. In the reverse reaction, H_3O^+ is the acid that donates a proton to the acetate ion, which acts as the base. Once again, we have two conjugate acid–base pairs: the parent acid and its conjugate base, $CH_3CO_2H/CH_3CO_2^-$, and the parent base and its conjugate acid, written as H_3O^+/H_2O.

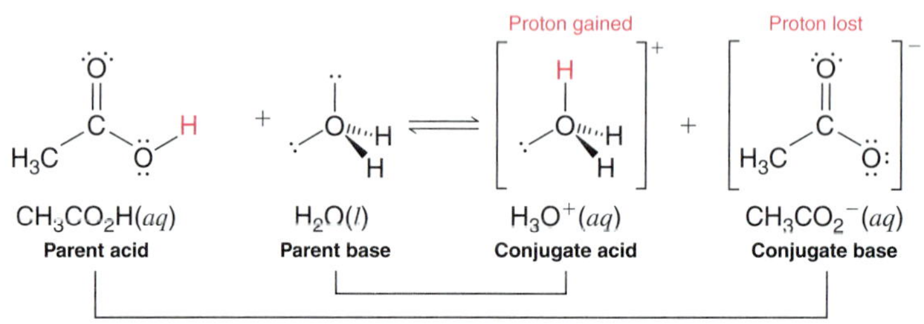

Note the pattern

All acid-base reactions contain two conjugate acid-base pairs.

In the reaction of ammonia with water to give ammonium ions and hydroxide ions (Equation 16.3), ammonia acts as a base by accepting a proton from a water molecule, which in this case is acting as an acid. In the reverse reaction, an ammonium ion acts as an acid by donating a proton to a hydroxide ion, which acts as a base. The conjugate acid–base pairs for this reaction are NH_4^+/NH_3 and H_2O/OH^-:

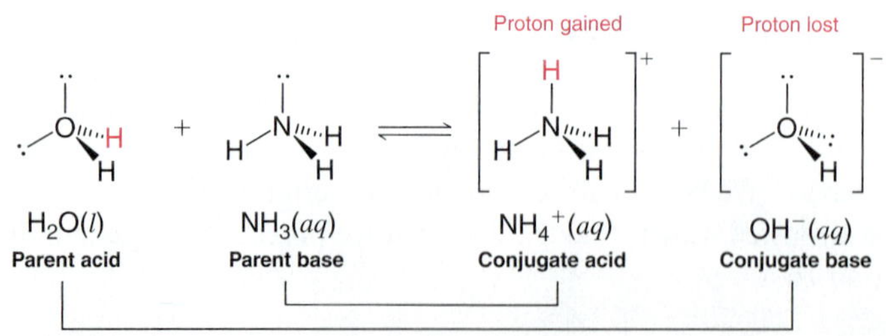

Acid–Base Equilibrium Constants: K_a, K_b, pK_a, and pK_b

The magnitude of the equilibrium constant for an ionization reaction can be used to determine the relative strengths of acids and bases. For example, the general equation for the ionization of a weak acid in water, where HA is the parent acid and A^- is its conjugate base, is

$$HA(aq) + H_2O(l) \rightleftharpoons H_3O^+(aq) + A^-(aq) \tag{16.12}$$

The equilibrium constant for this dissociation is

$$K = \frac{[H_3O^+][A^-]}{[H_2O][HA]} \tag{16.13}$$

As we noted earlier, the concentration of water is essentially constant for all reactions in aqueous solution, so $[H_2O]$ in Equation 16.13 can be incorporated into a new quantity, the **acid ionization constant (K_a)**, also called the *acid dissociation constant*:

$$K_a = K[H_2O] = \frac{[H_3O^+][A^-]}{[HA]} \tag{16.14}$$

Thus, the numerical values of K and K_a differ by the concentration of water (55.3 *M*). Again, for simplicity, H_3O^+ can be written as H^+ in Equation 16.14. Keep in mind, though, that free H^+ does not exist in aqueous solutions and that a proton is transferred to H_2O in all acid ionization reactions to form H_3O^+. Note that the larger the value of K_a, the stronger the acid and the higher the H^+ concentration at equilibrium.* Values of K_a for a number of common acids are given in Table 16.2.

TABLE 16.2 Values of K_a, pK_a, K_b, and pK_b for selected acids (HA) and their conjugate bases (A^-)

Acid	HA	K_a	pK_a	A^-	K_b	pK_b
Hydroiodic acid	HI	2×10^9	−9.3	I^-	5.0×10^{-24}	23.3
Sulfuric acid (1)[a]	H_2SO_4	1×10^2	−2	HSO_4^-	1.0^{-16}	16
Nitric acid	HNO_3	2.3×10^1	−1.37	NO_3^-	4.3×10^{-16}	15.37
Hydronium ion	H_3O^+	1.00	0.00	H_2O	1.0×10^{-14}	14.00
Sulfuric acid (2)[a]	HSO_4^-	1.0×10^{-2}	1.99	SO_4^{2-}	9.8×10^{-13}	12.01
Hydrofluoric acid	HF	6.3×10^{-4}	3.20	F^-	1.6×10^{-11}	10.80
Nitrous acid	HNO_2	5.6×10^{-4}	3.25	NO_2^-	1.8×10^{-11}	10.75
Formic acid	HCO_2H	1.78×10^{-4}	3.75	HCO_2^-	5.6×10^{-11}	10.25
Benzoic acid	$C_6H_5CO_2H$	6.3×10^{-5}	4.20	$C_6H_5CO_2^-$	1.6×10^{-10}	9.80
Acetic acid	CH_3CO_2H	1.7×10^{-5}	4.76	$CH_3CO_2^-$	5.8×10^{-10}	9.24
Pyridinium ion	$C_5H_5NH^+$	5.9×10^{-6}	5.23	C_5H_5N	1.7×10^{-9}	8.77
Hypochlorous acid	HOCl	4.0×10^{-8}	7.40	OCl^-	2.5×10^{-7}	6.60
Hydrocyanic acid	HCN	6.2×10^{-10}	9.21	CN^-	1.6×10^{-5}	4.79
Ammonium ion	NH_4^+	5.6×10^{-10}	9.25	NH_3	1.8×10^{-5}	4.75
Water	H_2O	1.0×10^{-14}	14.00	OH^-	1.00	0.00
Acetylene	C_2H_2	1×10^{-26}	26	HC_2^-	1×10^{12}	−12
Ammonia	NH_3	1×10^{-35}	35	NH_2^-	1×10^{21}	−21

[a] The number in parentheses indicates the ionization step referred to for a polyprotic acid.

* Like all equilibrium constants, acid–base ionization constants are actually measured in terms of the *activities* of H^+ or OH^-, making them unitless.

Weak bases react with water to produce the hydroxide ion, as shown in the following general equation, where B is the parent base and BH^+ is its conjugate acid:

$$B(aq) + H_2O(l) \rightleftharpoons BH^+(aq) + OH^-(aq) \quad (16.15)$$

The equilibrium constant for this reaction is the **base ionization constant (K_b)**, also called the *base dissociation constant*:

$$K_b = K[H_2O] = \frac{[BH^+][OH^-]}{[B]} \quad (16.16)$$

Once again, the concentration of water is constant, so it does not appear in the equilibrium constant expression; instead, it is included in the value of K_b. The larger the value of K_b, the stronger the base and the higher the OH^- concentration at equilibrium. Values of K_b for a number of common weak bases are given in Table 16.3.

There is a simple relationship between the magnitude of K_a for an acid and K_b for its conjugate base. Consider, for example, the ionization of hydrocyanic acid (HCN) in water to produce an acidic solution, and the reaction of CN^- with water to produce a basic solution:

$$HCN(aq) \rightleftharpoons H^+(aq) + CN^-(aq) \quad (16.17)$$

$$CN^-(aq) + H_2O(l) \rightleftharpoons OH^-(aq) + HCN(aq) \quad (16.18)$$

The equilibrium constant expression for the ionization of HCN is

$$K_a = \frac{[H^+][CN^-]}{[HCN]} \quad (16.19a)$$

and the corresponding expression for the reaction of cyanide with water is

$$K_b = \frac{[OH^-][HCN]}{[CN^-]} \quad (16.19b)$$

If we add Equations 16.17 and 16.18, we obtain the following (recall from Chapter 15 that the equilibrium constant for the *sum* of two reactions is the *product* of the equilibrium constants for the individual reactions):

$$\begin{array}{rlr} \cancel{HCN(aq)} \rightleftharpoons H^+(aq) + \cancel{CN^-(aq)} & & K_a = [H^+]\cancel{[CN^-]}/\cancel{[HCN]} \\ \cancel{CN^-(aq)} + H_2O(l) \rightleftharpoons OH^-(aq) + \cancel{HCN(aq)} & & K_b = [OH^-]\cancel{[HCN]}/\cancel{[CN^-]} \\ \hline H_2O(l) \rightleftharpoons H^+(aq) + OH^-(aq) & & K = K_a \times K_b = [H^+][OH^-] \end{array}$$

TABLE 16.3 Values of K_b, pK_b, K_a, and pK_a for selected weak bases (B) and their conjugate acids (BH^+)

Base	B	K_b	pK_b	BH^+	K_a	pK_a
Hydroxide ion	OH^-	1.00	0.00	H_2O	1.0×10^{-14}	14.00
Phosphate ion	PO_4^{3-}	2.1×10^{-2}	1.68	HPO_4^{2-}	4.8×10^{-13}	12.32
Dimethylamine	$(CH_3)_2NH$	5.4×10^{-4}	3.27	$(CH_3)_2NH_2^+$	1.9×10^{-11}	10.73
Methylamine	CH_3NH_2	4.6×10^{-4}	3.34	$CH_3NH_3^+$	2.2×10^{-11}	10.66
Trimethylamine	$(CH_3)_3N$	6.3×10^{-5}	4.20	$(CH_3)_3NH^+$	1.6×10^{-10}	9.80
Ammonia	NH_3	1.8×10^{-5}	4.75	NH_4^+	5.6×10^{-10}	9.25
Pyridine	C_5H_5N	1.7×10^{-9}	8.77	$C_5H_5NH^+$	5.9×10^{-6}	5.23
Aniline	$C_6H_5NH_2$	7.4×10^{-10}	9.13	$C_6H_5NH_3^+$	1.3×10^{-5}	4.87
Water	H_2O	1.0×10^{-14}	14.00	H_3O^+	1.00	0.00

In this case, the sum of the reactions described by K_a and K_b is the equation for the autoionization of water, and the product of the two equilibrium constants is K_w:

$$K_a K_b = K_w \tag{16.20}$$

Thus, if we know the value of either K_a for an acid or K_b for its conjugate base, we can calculate the magnitude of the other equilibrium constant for any conjugate acid–base pair.

Just as with pH, pOH, and pK_w, we can use negative logarithms to avoid exponential notation in writing acid and base ionization constants by defining pK_a as

$$pK_a = -\log_{10} K_a \tag{16.21a}$$

$$K_a = 10^{-pK_a} \tag{16.21b}$$

and pK_b as

$$pK_b = -\log_{10} K_b \tag{16.22a}$$

$$K_b = 10^{-pK_b} \tag{16.22b}$$

Values of pK_a and pK_b are given for several common acids and bases in Tables 16.2 and 16.3, respectively, and a more extensive set of data is provided in Appendixes C and D. Because of the use of negative logarithms, *smaller values of* pK_a *correspond to larger acid ionization constants and hence stronger acids.* For example, nitrous acid (HNO_2), with a pK_a of 3.25, is about a thousand times stronger an acid than hydrocyanic acid (HCN), with a pK_a of 9.21. Conversely, *smaller values of* pK_b *correspond to larger base ionization constants and hence stronger bases.* Similarly, Equation 16.20, which expresses the relationship between K_a and K_b, can be written in logarithmic form as

$$pK_a + pK_b = pK_w \tag{16.23a}$$

At 25°C, this becomes

$$pK_a + pK_b = 14.00 \tag{16.23b}$$

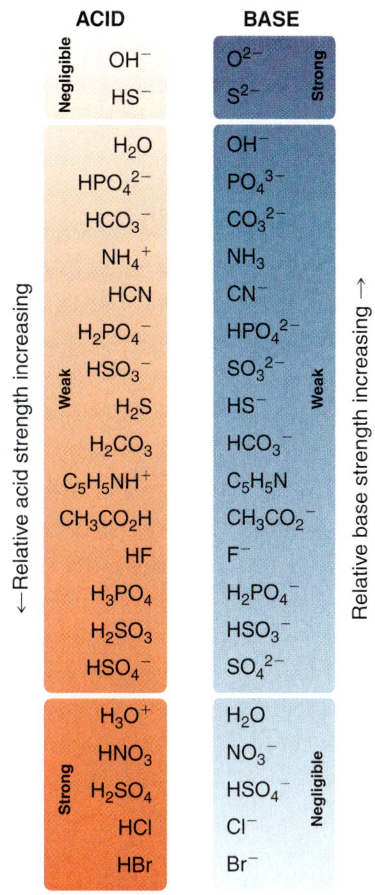

Figure 16.2 The relative strengths of some common conjugate acid–base pairs. The strongest acids are at the bottom left, and the strongest bases are at the top right. Note that the conjugate base of a strong acid is a very weak base and, conversely, the conjugate acid of a strong base is a very weak acid.

The relative strengths of some common acids and their conjugate bases are shown graphically in Figure 16.2. The conjugate acid–base pairs are listed in order (from top to bottom) of increasing acid strength, which corresponds to decreasing values of pK_a. This order corresponds to decreasing strength of the conjugate base, or increasing values of pK_b. At the bottom left of Figure 16.2 are the common strong acids, and at the top right are the most common strong bases. Notice the inverse relationship between the strength of the parent acid and the strength of the conjugate base. Thus, the conjugate base of a strong acid is a very weak base, and the conjugate base of a very weak acid is a strong base.

We can use the relative strengths of acids and bases to predict the direction of an acid–base reaction by following a single rule: *An acid-base equilibrium always favors the side with the weaker acid and base*:

$$\text{stronger acid} + \text{stronger base} \rightleftharpoons \text{weaker acid} + \text{weaker base}$$

Thus, in an acid–base reaction, the proton always reacts with the stronger base.

For example, hydrochloric acid is a strong acid that ionizes essentially completely in dilute aqueous solution to produce H_3O^+ and Cl^-; only negligible amounts of HCl molecules remain undissociated. Hence, the ionization equilibrium lies virtually all the way to the right, as represented by a single arrow:

$$HCl(aq) + H_2O(l) \longrightarrow H_3O^+(aq) + Cl^-(aq) \tag{16.24}$$

Note the pattern

The conjugate base of a strong acid is a weak base, and vice versa.

Note the pattern

All acid–base equilibria favor the side with the weaker acid and base. Thus, the proton is bound to the stronger base.

In contrast, acetic acid is a weak acid and water is a weak base. Consequently, aqueous solutions of acetic acid contain mostly acetic acid molecules in equilibrium with a small concentration of H_3O^+ and acetate ions, and the ionization equilibrium lies far to the left, as represented by these arrows:

$$CH_3CO_2H(aq) + H_2O(l) \rightleftharpoons H_3O^+(aq) + CH_3CO_2^-(aq) \qquad (16.25)$$

Similarly, in the reaction of ammonia with water, the hydroxide ion is a strong base and ammonia is a weak base, whereas the ammonium ion is a stronger acid than water. Hence, this equilibrium also lies to the left:

$$H_2O(l) + NH_3(aq) \rightleftharpoons NH_4^+(aq) + OH^-(aq) \qquad (16.26)$$

EXAMPLE 16.2

(a) Calculate the base ionization constant K_b and the pK_b of the butyrate ion, $CH_3CH_2CH_2CO_2^-$. The pK_a of butyric acid at 25°C is 4.83; butyric acid is responsible for the foul smell of rancid butter. **(b)** Calculate the acid ionization constant K_a and the pK_a of the dimethylammonium ion, $(CH_3)_2NH_2^+$. The base ionization constant K_b of dimethylamine, $(CH_3)_2NH$, is 5.4×10^{-4} at 25°C.

Given pK_a and K_b

Asked for Corresponding K_b and pK_b, K_a and pK_a

Strategy

The constants K_a and K_b are related as shown in Equation 16.20. The pK_a and pK_b for an acid and its conjugate base are related as shown in Equations 16.23a and b. Use the relationships $pK = -\log K$ and $K = 10^{-pK}$ (Equations 16.21a, b and 16.22a, b) to convert between K_a and pK_a or K_b and pK_b.

Solution

(a) We are given the value of pK_a for butyric acid and asked to calculate K_b and pK_b for its conjugate base, the butyrate ion. Because the pK_a value cited is for a temperature of 25°C, we can use Equation 16.23b: $pK_a + pK_b = pK_w = 14.00$. Substituting the value of pK_a and solving for pK_b give

$$4.83 + pK_b = 14.00$$
$$pK_b = 14.00 - 4.83 = 9.17$$

Because $pK_b = -\log K_b$, K_b is $10^{-9.17} = 6.8 \times 10^{-10}$.

(b) In this case, we are given K_b for a base (dimethylamine) and asked to calculate K_a and pK_a for its conjugate acid, the dimethylammonium ion. Because the initial quantity given is K_b rather than pK_b, we can use Equation 16.20: $K_aK_b = K_w$. Substituting the values of K_b and K_w at 25°C and solving for K_a give

$$K_a(5.4 \times 10^{-4}) = 1.01 \times 10^{-14}$$
$$K_a = 1.9 \times 10^{-11}$$

Because $pK_a = -\log K_a$, we have $pK_a = -\log(1.9 \times 10^{-11}) = 10.72$. We could also have converted K_b to pK_b to obtain the same answer:

$$pK_b = -\log(5.4 \times 10^{-4}) = 3.27$$
$$pK_a + pK_b = 14.00$$
$$pK_a = 10.73$$
$$K_a = 10^{-pKa} = 10^{-10.73} = 1.9 \times 10^{-11}$$

Notice that, if we are given any one of these four quantities for an acid or a base (K_a, pK_a, K_b, or pK_b), we can calculate the other three.

EXERCISE 16.2

Lactic acid, $CH_3CH(OH)CO_2H$, is responsible for the pungent taste and smell of sour milk; it is also thought to produce soreness in fatigued muscles. Its pK_a is 3.86 at 25°C. Calculate the value of K_a for lactic acid, and pK_b and K_b for the lactate ion.

Answer $K_a = 1.4 \times 10^{-4}$ for lactic acid; $pK_b = 10.14$ and $K_b = 7.2 \times 10^{-11}$ for the lactate ion.

Solutions of Strong Acids and Bases: The Leveling Effect

You will notice in Table 16.2 that acids like H_2SO_4 and HNO_3 lie *above* the hydronium ion, meaning that they have pK_a values less than zero and are stronger acids than the H_3O^+ ion.* In fact, all six of the common strong acids that we first encountered in Chapter 4 have pK_a values less than zero, which means that they have a greater tendency to lose a proton than does the H_3O^+ ion. Conversely, the conjugate bases of these strong acids are weaker bases than water. Consequently, the proton-transfer equilibria for these strong acids lie far to the right, and the addition of any of the common strong acids to water results in an essentially stoichiometric reaction of the acid with water to form a solution of the H_3O^+ ion and the conjugate base of the acid. Although, for example, K_a for HI is about 10^8 greater than K_a for HNO_3, the reaction of either HI or HNO_3 with water gives an essentially stoichiometric solution of H_3O^+ and I^- or NO_3^-. In fact, a 0.1 *M* aqueous solution of any strong acid actually contains 0.1 *M* H_3O^+, *regardless of the identity of the strong acid.* This phenomenon is called the **leveling effect**: Any species that is a stronger acid than the conjugate acid of water (H_3O^+) is *leveled* to the strength of H_3O^+ in aqueous solution because H_3O^+ is the strongest acid that can exist in equilibrium with water. Consequently, it is impossible to distinguish between the strengths of acids such as HI and HNO_3 in aqueous solution, and an alternative approach must be used to determine their relative acid strengths.

One method is to use a solvent such as anhydrous acetic acid. Since acetic acid is a stronger acid than water, it must also be a weaker base, with a lesser tendency to accept a proton than H_2O. Measurements of the conductivity of 0.1 *M* solutions of both HI and HNO_3 in acetic acid show that HI is completely dissociated but HNO_3 is only partially dissociated, behaving like a weak acid *in this solvent*. This result clearly tells us that HI is a stronger acid than HNO_3. The relative order of acid strengths and approximate K_a and pK_a values for the strong acids at the top of Table 16.2 were determined using measurements like this and different nonaqueous solvents.

The leveling effect applies to solutions of strong bases as well: In aqueous solution, any base stronger than OH^- is *leveled* to the strength of OH^- because OH^- is the strongest base that can exist in equilibrium with water. Salts such as K_2O, $NaOCH_3$ (sodium methoxide), and $NaNH_2$ (sodamide, or sodium amide), whose anions are the conjugate bases of species that would lie below water in Table 16.3, are all strong bases that react essentially completely (and often violently) with water, accepting a proton to give a solution of OH^- and the corresponding cation:

$$K_2O(s) + H_2O(l) \longrightarrow 2OH^-(aq) + 2K^+(aq) \tag{16.27a}$$

$$NaOCH_3(s) + H_2O(l) \longrightarrow OH^-(aq) + Na^+(aq) + CH_3OH(aq) \tag{16.27b}$$

$$NaNH_2(s) + H_2O(l) \longrightarrow OH^-(aq) + Na^+(aq) + NH_3(aq) \tag{16.27c}$$

* Recall from Chapter 4 that the acidic proton(s) in virtually all oxoacids is bonded to one of the oxygen atoms of the oxoanion. Thus, nitric acid should properly be written as $HONO_2$. Unfortunately, however, the formulas of oxoacids are almost always written with hydrogen on the left and oxygen on the right, giving HNO_3 instead.

In general, any substance whose anion is the conjugate base of a compound that is a weaker acid than water is a strong base that reacts quantitatively with water to form hydroxide ion. Other examples that you may encounter are potassium hydride (KH) and organometallic compounds such as methyl lithium (CH_3Li).

Polyprotic Acids and Bases

As you learned in Chapter 4, *polyprotic acids* such as H_2SO_4, H_3PO_4, and H_2CO_3 contain more than one ionizable proton, and the protons are lost in a stepwise manner. The fully protonated species is always the strongest acid because it is easier to remove a proton from a neutral molecule than from a negatively charged ion. Thus, acid strength decreases with the loss of subsequent protons, and, correspondingly, the pK_a increases. Consider H_2SO_4, for example:

$$H_2SO_4(aq) \rightleftharpoons HSO_4^-(aq) + H^+(aq) \qquad pK_a \approx -2 \qquad (16.28a)$$

$$HSO_4^-(aq) \rightleftharpoons SO_4^{2-}(aq) + H^+(aq) \qquad pK_a = 1.99 \qquad (16.28b)$$

The equilibrium in the first reaction lies far to the right, consistent with H_2SO_4 being a strong acid. In contrast, in the second reaction, appreciable quantities of both HSO_4^- and SO_4^{2-} are present at equilibrium.

Notice that HSO_4^- is both the conjugate base of H_2SO_4 and the conjugate acid of SO_4^{2-}. Just like water, HSO_4^- can therefore act as *either* an acid or a base, depending on whether the other reactant is a stronger acid or a stronger base. Conversely, the sulfate ion, SO_4^{2-}, is a *polyprotic base* that is capable of accepting two protons in a stepwise manner:

$$SO_4^{2-}(aq) + H_2O(aq) \rightleftharpoons HSO_4^-(aq) + OH^-(aq) \qquad (16.29a)$$

$$HSO_4^-(aq) + H_2O(aq) \rightleftharpoons H_2SO_4(aq) + OH^-(aq) \qquad (16.29b)$$

Like any other conjugate acid–base pair, the strengths of the conjugate acids and bases are related by $pK_a + pK_b = pK_w$. Consider, for example, the HSO_4^-/SO_4^{2-} conjugate acid–base pair. From Table 16.2, we see that the pK_a of HSO_4^- is 1.99. Hence, the pK_b of SO_4^{2-} is $14.00 - 1.99 = 12.01$. Thus, sulfate is a rather weak base, whereas OH^- is a strong base, so the equilibrium shown in Equation 16.29a lies to the left. The HSO_4^- ion is also a very weak base [pK_a of $H_2SO_4 = -2$, pK_b of $HSO_4^- = 14 - (-2) = 16$], consistent with what we expect for the conjugate base of a strong acid. Thus, the equilibrium shown in Equation 16.29b also lies almost completely to the left. Once again, equilibrium favors formation of the weaker acid–base pair.

EXAMPLE 16.3

Predict whether the equilibrium for each reaction lies to the left or the right as written:

(a) $NH_4^+(aq) + PO_4^{3-}(aq) \rightleftharpoons NH_3(aq) + HPO_4^{2-}(aq)$
(b) $CH_3CH_2CO_2H(aq) + CN^-(aq) \rightleftharpoons CH_3CH_2CO_2^-(aq) + HCN(aq)$

Given Balanced chemical equation

Asked for Equilibrium position

Strategy

Identify the conjugate acid–base pairs in each reaction. Then refer to Tables 16.2 and 16.3 to determine which is the stronger acid and base. Equilibrium always favors the formation of the weaker acid–base pair.

Solution

(a) The conjugate acid–base pairs are NH_4^+/NH_3 and HPO_4^{2-}/PO_4^{3-}. According to Tables 16.2 and 16.3, NH_4^+ is a stronger acid ($pK_a = 9.25$) than HPO_4^{2-} ($pK_a = 12.32$), and PO_4^{3-} is a stronger base ($pK_b = 1.68$) than NH_3 ($pK_b = 4.75$). The equilibrium will therefore lie to the right, favoring the formation of the weaker acid–base pair:

$$\underset{\text{Stronger acid}}{NH_4^+(aq)} + \underset{\text{Stronger base}}{PO_4^{3-}(aq)} \rightleftharpoons \underset{\text{Weaker base}}{NH_3(aq)} + \underset{\text{Weaker acid}}{HPO_4^{2-}(aq)}$$

(b) The conjugate acid–base pairs are $CH_3CH_2CO_2H/CH_3CH_2CO_2^-$ and HCN/CN^-. According to Table 16.2, HCN is a weak acid ($pK_a = 9.21$) and CN^- is a moderately weak base ($pK_b = 4.79$). Propionic acid, $CH_3CH_2CO_2H$, is not listed in Table 16.2, however. In a situation like this, the best approach is to look for a similar compound whose acid–base properties *are* listed. For example, propionic acid and acetic acid are identical except for the groups attached to the carbon atom of the carboxylic acid ($-CH_2CH_3$ versus $-CH_3$), so we might expect the two compounds to have similar acid–base properties. In particular, we would expect the pK_a of propionic acid to be similar in magnitude to the pK_a of acetic acid. (In fact, the pK_a of propionic acid is 4.87, compared to 4.76 for acetic acid, which makes propionic acid a slightly weaker acid than acetic acid.) Thus, propionic acid should be a significantly stronger acid than HCN. Because the stronger acid forms the weaker conjugate base, we predict that cyanide will be a stronger base than propionate. The equilibrium will therefore lie to the right, favoring the formation of the weaker acid–base pair:

$$\underset{\text{Stronger acid}}{CH_3CH_2CO_2H(aq)} + \underset{\text{Stronger base}}{CN^-(aq)} \rightleftharpoons \underset{\text{Weaker base}}{CH_3CH_2CO_2^-(aq)} + \underset{\text{Weaker acid}}{HCN(aq)}$$

EXERCISE 16.3

Predict whether the equilibrium for each reaction lies to the left or the right as written:

(a) $H_2O(l) + HS^-(aq) \rightleftharpoons OH^-(aq) + H_2S(aq)$
(b) $HCO_2^-(aq) + HSO_4^-(aq) \rightleftharpoons HCO_2H(aq) + SO_4^{2-}(aq)$

Answer **(a)** left; **(b)** left, $K_{a(HSO_4^-)} > K_{a(HCO_2H)}$

Acid–Base Properties of Solutions of Salts

We can also use the relative strengths of conjugate acid–base pairs to understand the acid–base properties of solutions of salts. In Chapter 4, you learned that a *neutralization reaction* can be defined as the reaction of an acid and a base to produce a salt and water. That is, another cation, such as Na^+, replaces the proton on the acid. An example is the reaction of CH_3CO_2H, a weak acid, with NaOH, a strong base:

$$\underset{\text{Acid}}{CH_3CO_2H(l)} + \underset{\text{Base}}{NaOH(s)} \xrightarrow{H_2O} \underset{\text{Salt}}{CH_3CO_2Na(aq)} + \underset{\text{Water}}{H_2O(l)} \qquad (16.30)$$

Depending on the acid–base properties of its component ions, however, a salt can dissolve in water to produce a neutral solution, a basic solution, or an acidic solution.

When a salt such as NaCl dissolves in water, it produces $Na^+(aq)$ and $Cl^-(aq)$ ions. Using a Lewis approach, the Na^+ ion can be viewed as an acid because it is an electron pair acceptor, although its low charge and relatively large radius make it a very weak acid. The Cl^- ion is the conjugate base of the strong acid HCl, so it has essentially no basic character. Consequently, dissolving NaCl in water has no effect on the pH of a solution, and the solution remains neutral.

In contrast, aqueous solutions of potassium cyanide and sodium acetate are basic. Again, the cations (K^+ and Na^+) have essentially no acidic character, but the anions

(CN^- and $CH_3CO_2^-$) are the conjugate bases of the weak acids HCN and acetic acid, respectively. Hence, they are weak bases that can react with water:

$$CN^-(aq) + H_2O(l) \rightleftharpoons HCN(aq) + OH^-(aq) \tag{16.31}$$
$$CH_3CO_2^-(aq) + H_2O(l) \rightleftharpoons CH_3CO_2H(aq) + OH^-(aq) \tag{16.32}$$

Neither reaction proceeds very far to the right as written because HCN and acetic acid are stronger acids than water, and hydroxide is a stronger base than either acetate or cyanide. Thus, in both cases, the equilibrium lies to the left, favoring the weaker acid–base combination. Nonetheless, each of these reactions generates enough hydroxide ions to produce a basic solution. For example, the pH of a 0.1 *M* solution of sodium acetate or potassium cyanide at 25°C is 8.8 or 11.1, respectively. From Table 16.2 and Figure 16.2, we can see that CN^- is a stronger base ($pK_b = 4.79$) than acetate ($pK_b = 9.24$), which is consistent with KCN producing a more basic solution than sodium acetate at the same concentration.

In contrast, the conjugate acid of a weak base should be a weak acid. For example, ammonium chloride and pyridinium chloride are salts produced by the reaction of ammonia and pyridine, respectively, with HCl. As you already know, the chloride ion is such a weak base that it does not react with water. In contrast, the cations of the two salts are weak acids that react with water as follows:

$$NH_4^+(aq) + H_2O(l) \rightleftharpoons NH_3(aq) + H_3O^+(aq) \tag{16.33}$$
$$C_5H_5NH^+(aq) + H_2O(l) \rightleftharpoons C_5H_5N(aq) + H_3O^+(aq) \tag{16.34}$$

Figure 16.2 shows that H_3O^+ is a stronger acid than either NH_4^+ or $C_5H_5NH^+$. Conversely, ammonia and pyridine are both stronger bases than water. The equilibrium will therefore lie far to the left in both cases, favoring the weaker acid–base pair. The H_3O^+ concentration produced by the reactions is great enough, however, to decrease the pH of the solution significantly: the pH of a 0.1 *M* solution of ammonium chloride or pyridinium chloride at 25°C is 5.13 or 3.12, respectively. This is consistent with the information shown in Figure 16.2, indicating that the pyridinium ion is more acidic than the ammonium ion.

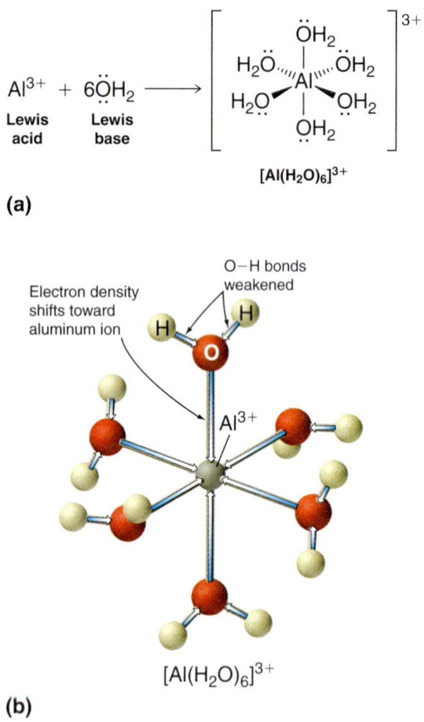

Figure 16.3 Effect of a metal ion on the acidity of water. (a) Reaction of the metal ion Al^{3+} with water to form the hydrated metal ion is an example of a Lewis acid–base reaction. **(b)** The positive charge on the aluminum ion attracts electron density from the oxygen atoms, which in turn shifts electron density away from the O—H bonds. The decrease in electron density weakens the O—H bonds in the water molecules and makes it easier for them to lose a proton.

What happens with aqueous solutions of a salt such as ammonium acetate, where both the cation and the anion can react separately with water to produce an acid and a base, respectively? According to Equation 16.33, the ammonium ion will lower the pH, while according to Equation 16.32, the acetate ion will raise the pH. This particular case is unusual, in that the cation is as strong an acid as the anion is a base ($pK_a \approx pK_b$). Consequently, the two effects cancel, and the solution remains neutral. With salts in which the cation is a stronger acid than the anion is a base, the final solution has pH < 7. Conversely, if the cation is a weaker acid than the anion is a base, the final solution has pH > 7.

Solutions of simple salts of metal ions can also be acidic, even though a metal ion cannot donate a proton directly to water to produce H_3O^+. Instead, a metal ion can act as a Lewis acid and interact with water, a Lewis base, by coordinating to a lone pair of electrons on the oxygen atom to form a hydrated metal ion (Figure 16.3a), as discussed in Chapter 4. A water molecule coordinated to a metal ion is more acidic than a free water molecule for two reasons. First, repulsive electrostatic interactions between the positively charged metal ion and the partially positively charged hydrogen atoms of the coordinated water molecule make it easier for the coordinated water to lose a proton:

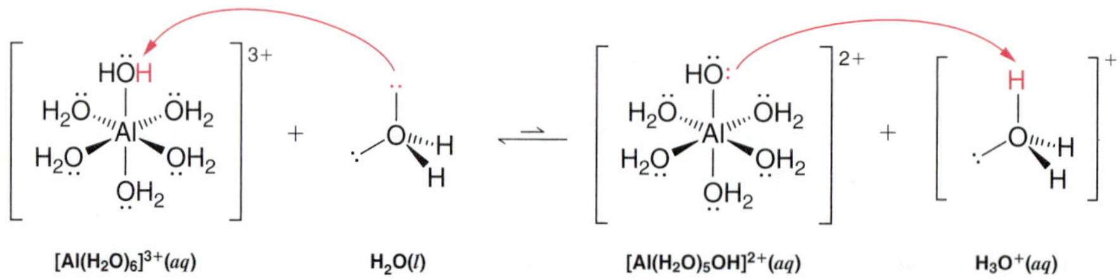

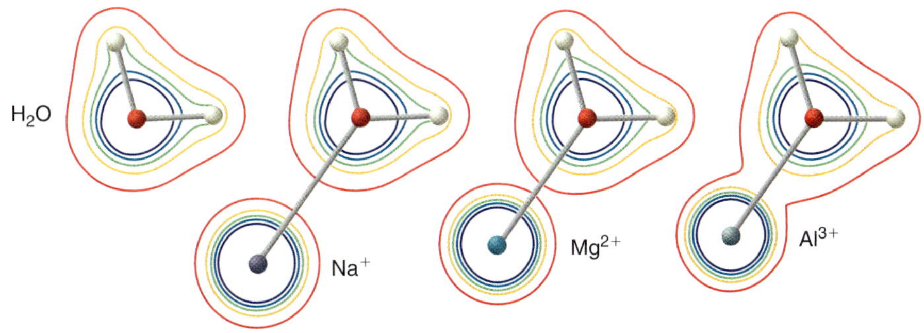

Figure 16.4 The effect of the charge and radius of a metal ion on the acidity of a coordinated water molecule. The contours show the electron density on the O and H atoms in both a free water molecule (*left*) and water molecules coordinated to Na^+, Mg^{2+}, and Al^{3+} ions. These contour maps demonstrate that the smallest, most highly charged metal ion (Al^{3+}) causes the greatest decrease in electron density of the O—H bonds of the water molecule. Due to this effect, the acidity of hydrated metal ions increases as the charge on the metal ion increases and its radius decreases.

Second, the positive charge on the Al^{3+} ion attracts electron density from the oxygen atoms of the water molecules, which in turn decreases the electron density in the O—H bonds, as shown in Figure 16.3b. With less electron density between the O and the H atoms, the O—H bonds are weaker than in a free H_2O molecule, making it easier to lose a H^+ ion.

The magnitude of this effect depends on these two factors (Figure 16.4):

- The *charge* on the metal ion: A divalent ion (M^{2+}) has approximately twice as strong an effect on the electron density in a coordinated water molecule as a monovalent ion (M^+) of the same radius.
- The *radius* of the metal ion: For metal ions with the same charge, the smaller the ion, the shorter the internuclear distance to the oxygen atom of the water molecule, and the greater the effect of the metal on the electron density distribution in the water molecule.

Thus, aqueous solutions of small, highly charged metal ions, such as Al^{3+} and Fe^{3+}, are acidic:

$$[Al(H_2O)_6]^{3+}(aq) \rightleftharpoons [Al(H_2O)_5(OH)]^{2+}(aq) + H^+(aq) \qquad (16.35)$$

The $[Al(H_2O)_6]^{3+}$ ion has a pK_a of 5.0, making it almost as strong an acid as acetic acid. The most important parameter for predicting the effect of a metal ion on the acidity of coordinated water molecules is actually the *charge-to-radius ratio* of the metal ion. A number of pairs of metal ions that lie on a diagonal line in the periodic table, such as Li^+ and Mg^{2+} or Ca^{2+} and Y^{3+}, have different sizes and charges but similar charge-to-radius ratios. As a result, these pairs of metal ions have similar effects on the acidity of coordinated water molecules, and they often exhibit other significant similarities in chemistry as well.

Reactions such as those discussed in this section, in which a salt reacts with water to give an acidic or basic solution, are often called **hydrolysis reactions**. The use of a separate name for this type of reaction is unfortunate because it suggests that they are somehow different. In fact, hydrolysis reactions are just acid–base reactions in which the acid is a cation or the base is an anion; they obey the same principles and rules as all other acid–base reactions.

EXAMPLE 16.4

Predict whether aqueous solutions of these compounds are acidic, basic, or neutral: **(a)** KNO_3; **(b)** $CrBr_3·6H_2O$; **(c)** Na_2SO_4.

Given Compound

Asked for Acidity or basicity of aqueous solution

Strategy

Ⓐ Assess the acid–base properties of the cation and anion. If the cation is a weak Lewis acid, it will not affect the pH of the solution. If the cation is the conjugate acid of a weak base or a relatively highly charged metal cation, however, it will react with water to produce an acidic solution.
Ⓑ If the anion is the conjugate base of a strong acid, it will not affect the pH of the solution. If, however, the anion is the conjugate base of a weak acid, the solution will be basic.

Solution

(a) Ⓐ✓ The K^+ cation has a small positive charge (+1) and a relatively large radius (because it is in the fourth row of the periodic table), so it is a very weak Lewis acid. Ⓑ✓ The NO_3^- anion is the conjugate base of a strong acid, so it has essentially no basic character (Table 16.1). Hence, neither the cation nor the anion will react with water to produce H^+ or OH^-, and the solution will be neutral.

(b) Ⓐ✓ The Cr^{3+} ion is a relatively highly charged metal cation that should behave similarly to the Al^{3+} ion and form the $[Cr(H_2O)_6]^{3+}$ complex, which will behave as a weak acid:

$$[Cr(H_2O)_6]^{3+}(aq) \rightleftharpoons [Cr(H_2O)_5(OH)]^{2+}(aq) + H^+(aq)$$

Ⓑ✓ The Br^- anion is a very weak base (it is the conjugate base of the strong acid HBr), so it does not affect the pH of the solution. Hence, the solution will be acidic.

(c) Ⓐ✓ The Na^+ ion, like K^+, is a very weak acid, so it should not affect the acidity of the solution. Ⓑ✓ In contrast, SO_4^{2-} is the conjugate base of HSO_4^-, which is a weak acid. Hence, the SO_4^{2-} ion will react with water as shown in Equation 16.29a to give a slightly basic solution.

EXERCISE 16.4

Predict whether aqueous solutions of the following are acidic, basic, or neutral: **(a)** KI; **(b)** $Mg(ClO_4)_2$; **(c)** NaHS.

Answer **(a)** neutral; **(b)** acidic; **(c)** basic (due to the reaction of HS^- with water to form H_2S and OH^-)

16.3 • Molecular Structure and Acid–Base Strength

We have seen that the strengths of acids and bases vary over many orders of magnitude. In this section, we explore some of the structural and electronic factors that control the acidity or basicity of a molecule.

Bond Strengths

In general, the stronger the A—H or B—H^+ bond, the less likely the bond is to break to form H^+ ions, and thus the less acidic the substance. This effect can be illustrated using the hydrogen halides:

Relative Acid Strength	HF	<	HCl	<	HBr	<	HI
H—X Bond Energy, kJ/mol	570		432		366		298
pK_a	3.20		−6.1		−8		−9

The trend in bond energies is due to a steady decrease in overlap between the 1*s* orbital of hydrogen and the valence orbital of the halogen atom as the size of the halogen increases. The larger the atom to which H is bonded, the weaker the bond. Thus, the bond between

H and a large atom in a given family, such as I or Te, is weaker than the bond between H and a smaller atom in the same family, such as F or O. As a result, *acid strengths of binary hydrides increase as we go down a column of the periodic table*. For example, the order of acidity for the binary hydrides of Group 16 is the following, with pK_a values in parentheses: H_2O (14.00 = pK_w) < H_2S (7.05) < H_2Se (3.89) < H_2Te (2.6).

Note the pattern

Acid strengths of binary hydrides increase as we go down a column or from left to right across a row of the periodic table.

Stability of the Conjugate Base

Whether we write an acid–base reaction as $AH \rightleftharpoons A^- + H^+$ or as $BH^+ \rightleftharpoons B + H^+$, the conjugate base ($A^-$ or B) contains one more lone pair of electrons than the parent acid (AH or BH^+). *Any factor that stabilizes the lone pair on the conjugate base favors dissociation of H^+ and makes the parent acid a stronger acid.* Let's see how this explains the relative acidity of the binary hydrides of the elements in the second row of the periodic table. The observed order of increasing acidity is the following, with pK_a values in parentheses: CH_4 (~50) ≪ NH_3 (~36) < H_2O (14.00) < HF (3.20). Consider, for example, the compounds at both ends of this series: methane and hydrogen fluoride. The conjugate base of CH_4 is CH_3^-, and the conjugate base of HF is F^-. Because fluorine is much more electronegative than carbon, fluorine can better stabilize the negative charge in the F^- ion than carbon can stabilize the negative charge in the CH_3^- ion. Consequently, HF has a greater tendency to dissociate to form H^+ and F^- than does methane to form H^+ and CH_3^-, making HF a much stronger acid than CH_4.

The same trend is predicted by analysis of the properties of the conjugate acids. For a series of compounds of the general formula HE, as the electronegativity of E increases, the E—H bond becomes more polar, favoring dissociation to form E^- and H^+. Thus, due to both the increasing stability of the conjugate base and the increasing polarization of the E—H bond in the conjugate acid, *acid strengths of binary hydrides increase as we go from left to right across a row of the periodic table.*

Inductive Effects

Atoms or groups of atoms in a molecule other than those to which H is bonded can induce a change in the distribution of electrons within the molecule. This is called an *inductive effect*, and, much like the coordination of water to a metal ion, it can have a major effect on the acidity or basicity of the molecule. For example, the hypohalous acids (general formula HOX, with X representing a halogen) all have a hydrogen atom bonded to an oxygen atom. In aqueous solution, they all produce the following equilibrium:

$$HOX(aq) \rightleftharpoons H^+(aq) + OX^-(aq) \qquad (16.36)$$

The acidities of these acids vary by about three orders of magnitude, however, due to the difference in electronegativity of the halogen atoms:

HOX	Electronegativity of X	pK_a
HOCl	3.0	7.40
HOBr	2.8	8.55
HOI	2.5	10.5

As the electronegativity of X increases, the distribution of electron density within the molecule changes: the electrons are drawn more strongly toward the halogen atom, and in turn away from the H in the O—H bond, thus weakening the O—H bond and allowing dissociation of hydrogen as H^+.

The acidity of oxoacids, with the general formula $HOXO_n$ (n = 0–3), depends strongly on the number of terminal oxygen atoms attached to the central atom X. As shown in Figure 16.5, the K_a values of the oxoacids of chlorine increase by a factor

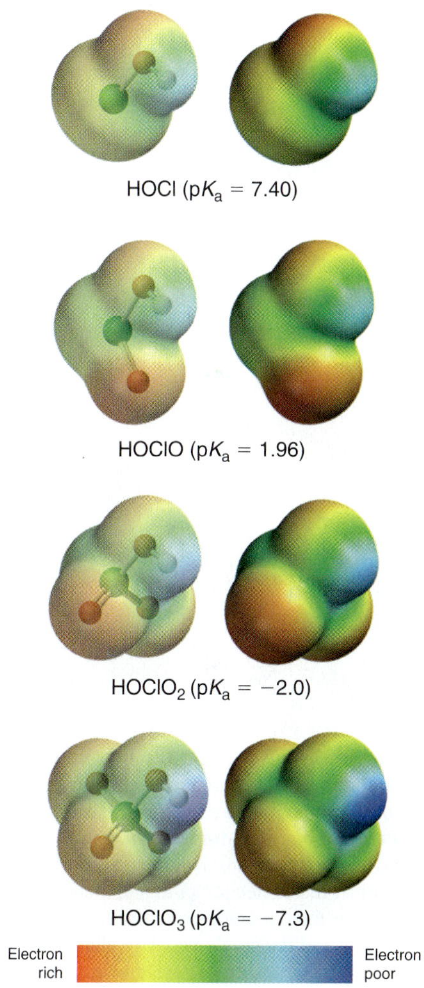

Figure 16.5 The relationship between the acid strengths of the oxoacids of chlorine and the electron density on the O—H unit. These electrostatic potential maps show how the electron density on the O—H unit decreases as the number of terminal oxygen atoms increases. Blue corresponds to low electron densities, whereas red corresponds to high electron densities. The oxygen atom in the O—H unit becomes steadily less red from HClO to $HClO_4$ (also written as $HOClO_3$), while the H atom becomes steadily bluer, indicating that the electron density on the O—H unit is decreasing. (Chlorine oxoacids pK_a values from J. R. Bowser, *Inorganic Chemistry*, 1993.)

of about 10^4 to 10^6 with each oxygen as successive oxygen atoms are added. The increase in acid strength with increasing number of terminal oxygen atoms is due to both an inductive effect and increased stabilization of the conjugate base.

Because oxygen is the second most electronegative element, adding terminal oxygen atoms causes electrons to be drawn away from the O—H bond, making it weaker and thereby increasing the strength of the acid. The colors in Figure 16.5 show how the electrostatic potential, a measure of the strength of the interaction of a point charge at any place on the surface of the molecule, changes as the number of terminal oxygen atoms increases. In Figures 16.5 and 16.6, blue corresponds to low electron densities, while red corresponds to high electron densities. Note that the oxygen atom in the O—H unit becomes steadily less red from HClO to $HClO_4$ (also written as $HOClO_3$), while the H atom becomes steadily bluer, indicating that the electron density on the O—H unit decreases as the number of terminal oxygen atoms increases. The decrease in electron density in the O—H bond weakens it, making it easier to lose hydrogen as H^+ ions and increasing the strength of the acid.

At least as important, however, is the effect of delocalization of the negative charge in the conjugate base. As shown in Figure 16.6, the number of resonance structures that can be written for the oxoanions of chlorine increases as the number of terminal oxygen atoms increases, allowing the single negative charge to be delocalized over successively more oxygen atoms. The electrostatic potential plots in Figure 16.6 demonstrate that the electron density on the terminal oxygen atoms decreases steadily as their number increases. Note that the oxygen atom in ClO^- is red, indicating that it is electron rich, and that the color of oxygen progressively changes to green in ClO_4^-, indicating that the oxygen atoms are becoming steadily less electron rich through the series. For example, in the perchlorate ion (ClO_4^-), the single negative charge is delocalized over all four oxygen atoms, whereas in the hypochlorite ion (OCl^-), the negative charge is largely localized on a single oxygen atom (Figure 16.6). As a result, the perchlorate ion has no localized negative charge to which a proton can bind. Consequently, the perchlorate anion has a much lower affinity for a proton than does the hypochlorite ion, and perchloric acid is one of the strongest acids known.

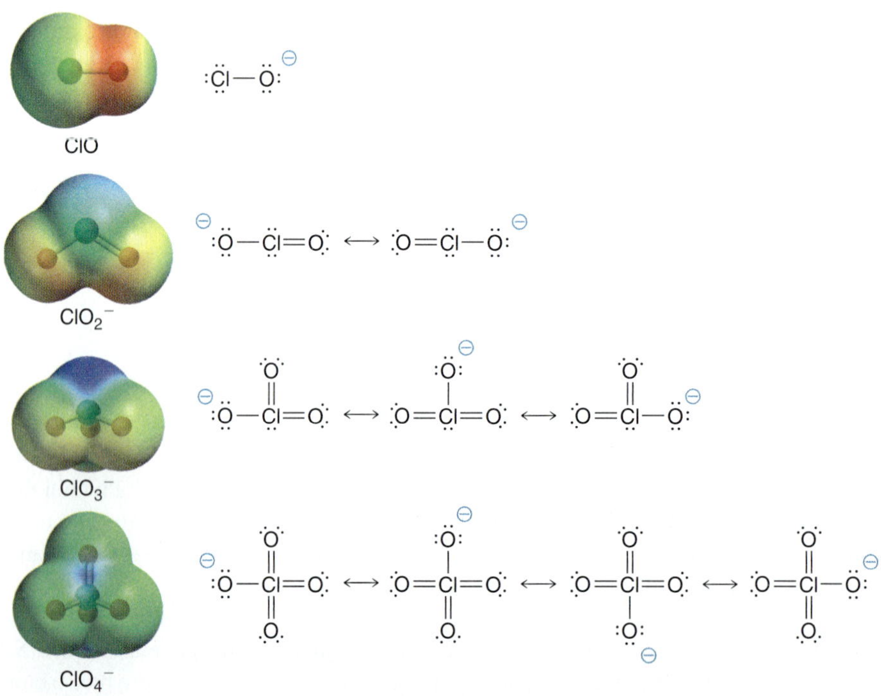

Figure 16.6 The relationship between delocalization of the negative charge in the oxoanions of chlorine and the number of terminal oxygen atoms. As the number of terminal oxygen atoms increases, the number of resonance structures that can be written for the oxoanions of chlorine also increases, and the single negative charge is delocalized over more oxygen atoms. As these electrostatic potential plots demonstrate, the electron density on the terminal oxygen atoms decreases steadily as their number increases. As the electron density on the oxygen atoms decreases, so does their affinity for a proton, making the anion less basic. As a result, the parent oxoacid is more acidic.

Similar inductive effects are also responsible for the trend in the acidities of oxoacids that have the same number of oxygen atoms as we go across a row of the periodic table from left to right. For example, H_3PO_4 is a weak acid, H_2SO_4 is a strong acid, and $HClO_4$ is one of the strongest acids known. The number of terminal oxygen atoms increases steadily across the row, consistent with the observed increase in acidity. In addition, the electronegativity of the central atom increases steadily from P to S to Cl, which causes electrons to be drawn from oxygen to the central atom, weakening the O—H bond and increasing the strength of the oxoacid.

Careful inspection of the data in Table 16.4 shows two apparent anomalies: carbonic acid and phosphorous acid. If carbonic acid (H_2CO_3) were really a discrete molecule with the structure $(HO)_2C{=}O$, it would have a single terminal oxygen atom and should be comparable in acid strength to phosphoric acid (H_3PO_4), for which $pK_{a1} = 2.16$. Instead, the tabulated value of pK_{a1} for carbonic acid is 6.35, making it about 10,000 times weaker than expected. As we shall see in Section 16.6, however, H_2CO_3 is only a minor component of the aqueous solutions of CO_2 that are referred to as carbonic acid. Similarly, if phosphorous acid (H_3PO_3) actually had the structure $(HO)_3P$, it would have no terminal oxygen atoms attached to phosphorous. It would therefore be expected to be about as strong an acid as HOCl ($pK_a = 7.40$). In fact, the value of pK_{a1} for phosphorous acid is 1.30, and the structure of phosphorous acid is $(HO)_2P({=}O)H$ with one H atom directly bonded to P and one P═O bond. Thus, the value of pK_{a1} for phosphorous acid is similar to that of other oxoacids with one terminal oxygen atom, such as H_3PO_4. Fortunately, phosphorous acid is the only common oxoacid in which a hydrogen atom is bonded to the central atom rather than oxygen.

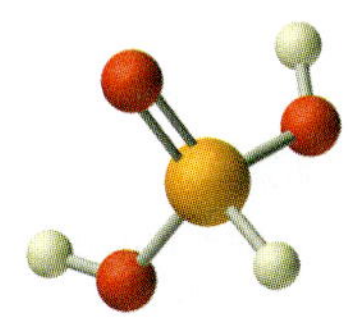
Phosphorous acid, H_3PO_3

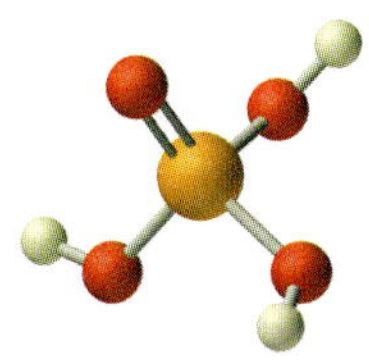
Phosphoric acid, H_3PO_4

Inductive effects are also observed in organic molecules that contain electronegative substituents. The magnitude of the electron-withdrawing effect depends on both

TABLE 16.4 Values of pK_a for selected polyprotic acids and bases

	Formula	pK_{a1}	pK_{a2}	pK_{a3}
Polyprotic Acids				
Carbonic acid[a]	"H_2CO_3"	6.35	10.33	
Citric acid	$HO_2CCH_2C(HOCCO_2H)CH_2CO_2H$	3.13	4.76	6.40
Malonic acid	$HO_2CCH_2CO_2H$	2.85	5.70	
Oxalic acid	HO_2CCO_2H	1.25	3.81	
Phosphoric acid	H_3PO_4	2.16	7.21	12.32
Phosphorous acid	H_3PO_3	1.3	6.70	
Succinic acid	$HO_2CCH_2CH_2CO_2H$	4.21	5.64	
Sulfuric acid	H_2SO_4	−2	1.99	
Sulfurous acid[a]	"H_2SO_3"	1.85	7.21	
Polyprotic Bases		pK_{b1}	pK_{b2}	
Ethylenediamine	$H_2N(CH_2)_2NH_2$	4.08	7.14	
Piperazine	$HN(CH_2)_4NH$	4.27	8.67	
Propylenediamine	$H_2N(CH_2)_3NH_2$	3.45	5.12	

[a] H_2CO_3 and H_2SO_3 are at best minor components of aqueous solutions of $CO_2(g)$ and $SO_2(g)$, respectively, but such solutions are commonly referred to as containing carbonic acid and sulfurous acid, respectively.

Note the pattern

Any inductive effect that withdraws electron density from an O—H bond increases the acidity of a compound.

the nature and the number of halogen substituents, as shown by the pK_a values for several acetic acid derivatives:

$$\underset{4.76}{CH_3CO_2H} < \underset{2.87}{CH_2ClCO_2H} < \underset{1.35}{CHCl_2CO_2H} < \underset{0.66}{CCl_3CO_2H} < \underset{0.52}{CF_3CO_2H} \quad (\text{p}K_a)$$

As you might expect, fluorine, which is more electronegative than chlorine, causes a larger effect than chlorine, and the effect of three halogens is greater than the effect of two or one. Notice from these data that inductive effects can be quite large. For instance, replacing the —CH_3 group of acetic acid by a —CF_3 group results in about a 10,000-fold increase in acidity!

EXAMPLE 16.5

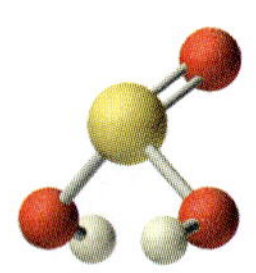

Sulfurous acid, H_2SO_3

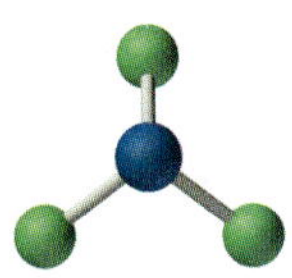

Trifluoramine, NF_3

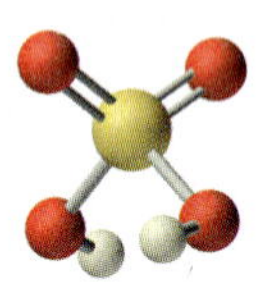

Sulfuric acid, H_2SO_4

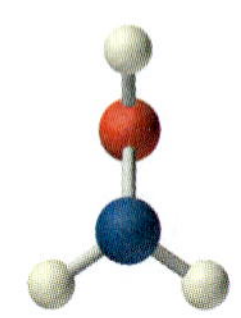

Hydroxylamine, NH_2OH

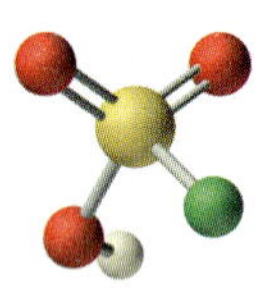

Fluorosulfonic acid, FSO_3H

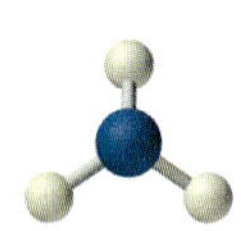

Ammonia, NH_3

Arrange the compounds of each series in order of increasing acid or base strength: (a) sulfuric acid [H_2SO_4, or $(HO)_2SO_2$], fluorosulfonic acid (FSO_3H, or FSO_2OH), sulfurous acid [H_2SO_3, or $(HO)_2SO$]; (b) ammonia (NH_3), trifluoramine (NF_3), hydroxylamine (NH_2OH). The structures are shown in the margin.

Given Series of compounds

Asked for Relative acid or base strengths

Strategy

Use relative bond strengths, stability of the conjugate base, and inductive effects to arrange the compounds in order of increasing tendency to ionize in aqueous solution.

Solution

(a) Although both sulfuric acid and sulfurous acid have two —OH groups, the sulfur atom in sulfuric acid is bonded to two terminal oxygen atoms versus one in sulfurous acid. Because oxygen is highly electronegative, sulfuric acid is the stronger acid because the negative charge on the anion is stabilized by the additional oxygen atom. In comparing sulfuric acid and fluorosulfonic acid, we note that fluorine is more electronegative than oxygen. Thus, replacing an —OH by —F will remove *more* electron density from the central S atom, which will in turn remove electron density from the S—OH bond and the O—H bond. Because its O—H bond is weaker, FSO_3H is a stronger acid than sulfuric acid. The predicted order of acid strengths given here is confirmed by the measured pK_a values for these acids:

$$\underset{1.85}{H_2SO_3} < \underset{-2}{H_2SO_4} < \underset{-10}{FSO_3H} \quad (\text{p}K_a)$$

(b) The structures of both trifluoramine and hydroxylamine are similar to that of ammonia. In trifluoramine, all of the hydrogen atoms in NH_3 are replaced by fluorine atoms, whereas in hydroxylamine, one hydrogen atom is replaced by OH. Replacing the three hydrogen atoms by fluorine will withdraw electron density from N, making the lone electron pair on N less available to bond to an H^+ ion. Thus, NF_3 is predicted to be a much weaker base than NH_3. Similarly, since oxygen is more electronegative than hydrogen, replacing one hydrogen atom in NH_3 by OH will make the amine less basic. Because oxygen is less electronegative than fluorine and only one hydrogen atom is replaced, however, the effect will be smaller. The predicted order of increasing base strength shown here is confirmed by the measured pK_b values:

$$\underset{—}{NF_3} \ll \underset{8.06}{NH_2OH} < \underset{4.75}{NH_3} \quad (\text{p}K_b)$$

Trifluoramine is such a weak base that it does not react with aqueous solutions of strong acids. Hence, its base ionization constant has not been measured.

EXERCISE 16.5

Arrange the following in order of **(a)** decreasing acid strength: H_3PO_4, $CH_3PO_3H_2$, and $HClO_3$; and **(b)** increasing base strength: CH_3S^-, OH^-, and CF_3S^-.

Answer **(a)** $HClO_3 > CH_3PO_3H_2 > H_3PO_4$; **(b)** $CF_3S^- < CH_3S^- < OH^-$

16.4 Quantitative Aspects of Acid–Base Equilibria

This section presents a quantitative approach to analyzing acid–base equilibria. You will learn how to determine the values of K_a and K_b, how to use K_a or K_b to calculate the percent ionization and pH of an aqueous solution of an acid or a base, and how to calculate the equilibrium constant for the reaction of an acid with a base from the K_a and K_b of the reactants.

Determining K_a and K_b

The ionization constants K_a and K_b are equilibrium constants that are calculated from experimentally measured concentrations, just like the equilibrium constants discussed in Chapter 15. Before proceeding further, it is important to understand exactly what is meant when we talk about the concentration of an aqueous solution of a weak acid or base. Suppose, for example, we have a bottle labeled 1 *M* acetic acid or 1 *M* ammonia. As you learned in Chapter 4, such a solution is usually prepared by dissolving 1 mol of acetic acid or ammonia in water and adding enough water to give a final volume of exactly 1 L. If, however, we were to list the actual concentrations of all the species present in either solution, we would find that *none* of the values is exactly 1 *M* because a weak acid such as acetic acid or a weak base such as ammonia always reacts with water to some extent. The extent of the reaction depends on the value of K_a or K_b, the concentration of the acid or base, and the temperature. Consequently, only the *total concentration* of both the ionized and unionized species is equal to 1 *M*.

The *analytical concentration* (C), is defined as the *total* concentration of all forms of an acid or base that are present in solution, regardless of their state of protonation. Thus, a "1 *M*" solution of acetic acid has an analytical concentration of 1 *M*, which is the sum of the *actual* concentrations of unionized acetic acid (CH_3CO_2H) and the ionized form ($CH_3CO_2^-$):

$$C_{CH_3CO_2H} = [CH_3CO_2H] + [CH_3CO_2^-] \tag{16.37}$$

As we shall see shortly, if we know the analytical concentration and the value of K_a, we can calculate the actual values of $[CH_3CO_2H]$ and $[CH_3CO_2^-]$.

The equilibrium equations for the reaction of acetic acid and ammonia with water are

$$K_a = \frac{[H^+][CH_3CO_2^-]}{[CH_3CO_2H]} \tag{16.38}$$

$$K_b = \frac{[NH_4^+][OH^-]}{[NH_3]} \tag{16.39}$$

where K_a and K_b are the ionization constants for acetic acid and ammonia, respectively. In addition to the analytical concentration of the acid (or base), we must have a way to measure the concentration of at least *one* of the species in the equilibrium constant expression in order to determine the value of K_a (or K_b). There are two common ways to obtain the concentrations: by measuring the electrical conductivity of the solution, which is related to the total concentration of ions present, and by measuring the pH of the solution, which gives $[H^+]$ or $[OH^-]$.

The following examples illustrate the procedure for determining K_a for a weak acid and K_b for a weak base. In both cases, we will follow the procedure developed in Chapter 15: the analytical concentration of the acid or base is the *initial* concentration, and the stoichiometry of the reaction with water determines the *change* in concentrations. The *final* concentrations of all species are calculated from the initial concentrations and the changes in the concentrations. Inserting the final concentrations into the equilibrium constant expression enables us to calculate the value of K_a or K_b.

EXAMPLE 16.6

Electrical conductivity measurements indicate that 0.42% of the acetic acid molecules in a 1.000 M solution are ionized at 25°C. Calculate the values of K_a and pK_a for acetic acid at this temperature.

Given Analytical concentration and percent ionization

Asked for K_a and pK_a

Strategy

A Write the balanced equilibrium equation for the reaction, and derive the equilibrium constant expression.

B Use the data given and the stoichiometry of the reaction to construct a table showing initial concentrations, changes in concentrations, and final concentrations for all species in the equilibrium constant expression.

C Substitute the final concentrations into the equilibrium constant expression, and calculate the value of K_a. Take the negative logarithm of K_a to obtain pK_a.

Solution

A The balanced equilibrium equation for the dissociation of acetic acid is

$$CH_3CO_2H(aq) \rightleftharpoons H^+(aq) + CH_3CO_2^-(aq)$$

and the equilibrium constant expression is

$$K_a = \frac{[H^+][CH_3CO_2^-]}{[CH_3CO_2H]}$$

B To calculate the value of K_a, we need to know the equilibrium concentrations of CH_3CO_2H, $CH_3CO_2^-$, and H^+. The most direct way to do this is to construct a table that lists the initial concentrations and the changes in concentrations that occur during the reaction to give the final concentrations, using the procedure introduced in Chapter 15. The initial concentration of unionized acetic acid, $[CH_3CO_2H]_i$, is the analytical concentration, 1.000 M, and the initial acetate concentration, $[CH_3CO_2^-]_i$, is zero. The initial concentration of H^+ is not zero, however; $[H^+]_i$ is 1.00×10^{-7} M due to the autoionization of water. The measured percent ionization tells us that 0.42% of the acetic acid molecules are ionized at equilibrium. Consequently, the change in the concentration of acetic acid is $\Delta[CH_3CO_2H] = -(4.2 \times 10^{-3})(1.000\ M) = -0.0042\ M$. Conversely, the change in the acetate concentration is $\Delta[CH_3CO_2^-] = +0.0042\ M$ because every 1 mol of acetic acid that ionizes gives 1 mol of acetate. Because one proton is produced for each acetate ion formed, $\Delta[H^+] = +0.0042\ M$ as well. These results are summarized in the table.

	$CH_3CO_2H(aq)$	$\rightleftharpoons$ $H^+(aq)$	+ $CH_3CO_2^-(aq)$
	$[CH_3CO_2H]$	$[H^+]$	$[CH_3CO_2^-]$
Initial	1.000	1.00×10^{-7}	0
Change	−0.0042	+0.0042	+0.0042
Final	0.9958	0.0042	0.0042

The final concentrations of all species are therefore

$$[CH_3CO_2H]_f = [CH_3CO_2H]_i + \Delta[CH_3CO_2H] = 1.000\ M + (-0.0042\ M) = 0.9958\ M$$

$$[CH_3CO_2^-]_f = [CH_3CO_2^-]_i + \Delta[CH_3CO_2^-] = 0\ M + (+0.0042\ M) = 0.0042\ M$$

$$[H^+]_f = [H^+]_i + \Delta[H^+] = 1.00 \times 10^{-7}\ M + (+0.0042\ M) = 0.0042\ M$$

C We can now calculate K_a by inserting the final concentrations into the equilibrium constant expression:

$$K_a = \frac{[H^+][CH_3CO_2^-]}{[CH_3CO_2H]} = \frac{(0.0042)(0.0042)}{0.9958} = 1.8 \times 10^{-5}$$

The pK_a is the negative logarithm of K_a: p$K_a = -\log K_a = -\log(1.8 \times 10^{-5}) = 4.74$.

EXERCISE 16.6

Picric acid is the common name for 2,4,6-trinitrophenol, a derivative of phenol (C_6H_5OH) in which three H atoms are replaced by nitro (—NO_2) groups. The presence of the nitro groups removes electron density from the phenyl ring, making picric acid a much stronger acid than phenol (pK_a = 9.99). The nitro groups also make picric acid potentially explosive, as you might expect based on its chemical similarity to 2,4,6-trinitrotoluene, better known as TNT. A 0.20 *M* solution of picric acid is 73% ionized at 25°C. Calculate K_a and pK_a for picric acid.

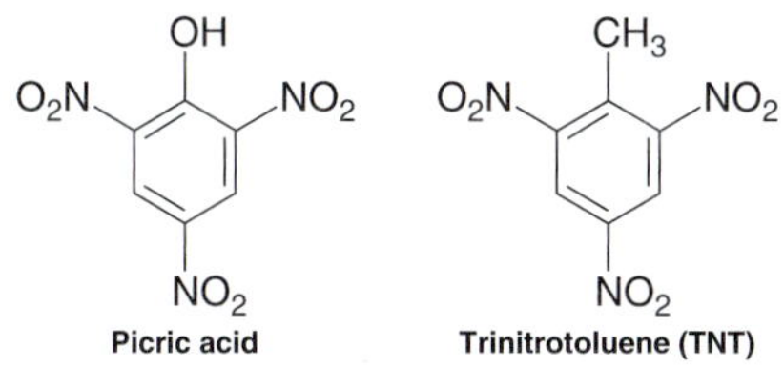

Answer $K_a = 0.39$; p$K_a = 0.41$

EXAMPLE 16.7

A 1.000 *M* aqueous solution of ammonia has a pH of 11.63 at 25°C. Calculate K_b and pK_b for ammonia.

Given Analytical concentration and pH

Asked for K_b and pK_b

Strategy

A Write the balanced equilibrium equation for the reaction, and derive the equilibrium constant expression.

B Use the data given and the stoichiometry of the reaction to construct a table showing initial concentrations, changes in concentrations, and final concentrations for all species in the equilibrium constant expression.

C Substitute the final concentrations into the equilibrium constant expression and calculate the value of K_b. Take the negative logarithm of K_b to obtain pK_b.

Solution

A The balanced equilibrium equation for the reaction of ammonia with water is

$$NH_3(aq) + H_2O(l) \rightleftharpoons NH_4^+(aq) + OH^-(aq)$$

for which the equilibrium constant expression is

$$K_b = \frac{[NH_4^+][OH^-]}{[NH_3]}$$

(Remember that water does not appear in the equilibrium constant expression for K_b.) **B** To calculate K_b, we need to know the equilibrium concentrations of NH_3, NH_4^+, and OH^-. The initial concentration of NH_3 is the analytical concentration, 1.000 *M*, and the initial concentrations of NH_4^+ and OH^- are 0 *M* and 1.00×10^{-7} *M*, respectively. In this case, we are given the pH of the solution, which allows us to calculate the final concentration of one species (OH^-) directly, rather than the change in concentration. Recall that pK_w = pH + pOH = 14.00

at 25°C. Thus, pOH = 14.00 − pH = 14.00 − 11.63 = 2.37, and $[OH^-]_f = 10^{-2.37} = 4.27 \times 10^{-3}$ *M*. Our data thus far are listed in the table.

$$NH_3(aq) \rightleftharpoons NH_4^+(aq) + OH^-(aq)$$

	$[NH_3]$	$[NH_4^+]$	$[OH^-]$
Initial	1.000	0	1.00×10^{-7}
Change			
Final			4.25×10^{-3}

The final $[OH^-]$ is much greater than the initial $[H^+]$, so the change in $[OH^-]$ is

$$\Delta[OH^-] = (4.25 \times 10^{-3}\,M) - (1.00 \times 10^{-7}\,M) \approx 4.25 \times 10^{-3}\,M$$

The stoichiometry of the reaction tells us that 1 mol of NH_3 is converted to NH_4^+ for each 1 mol of OH^- formed, so

$$\Delta[NH_4^+] = +4.25 \times 10^{-3}\,M \quad \text{and} \quad \Delta[NH_3] = -4.25 \times 10^{-3}\,M$$

Inserting these values for the changes in concentrations in the table enables us to complete the table

$$H_2O(l) + NH_3(aq) \rightleftharpoons NH_4^+(aq) + OH^-(aq)$$

	$[NH_3]$	$[NH_4^+]$	$[OH^-]$
Initial	1.000	0	1.00×10^{-7}
Change	-4.25×10^{-3}	$+4.25 \times 10^{-3}$	$+4.27 \times 10^{-3}$
Final	0.996	4.25×10^{-3}	4.27×10^{-3}

Thus, the final concentrations of all species in the equilibrium constant expression are:

$$[NH_4^+]_f = [NH_4^+]_i + \Delta[NH_4^+] = 0\,M + 4.27 \times 10^{-3}\,M = 4.27 \times 10^{-3}\,M$$
$$[NH_3]_f = [NH_3]_i + \Delta[NH_3] = 1.000\,M + (-4.27 \times 10^{-3}\,M) \approx 0.996\,M$$

Inserting the final concentrations into the equilibrium constant expression gives

$$K_b = \frac{[NH_4^+][OH^-]}{[NH_3]} = \frac{(4.27 \times 10^{-3})^2}{0.996} = 1.83 \times 10^{-5}$$

and $pK_b = -\log K_b = 4.737$.

Note that the values of K_b and pK_b for ammonia are almost exactly the same as the values of K_a and pK_a for acetic acid at 25°C. In other words, ammonia is almost exactly as strong a base as acetic acid is an acid. Consequently, the extent of the ionization reaction in an aqueous solution of ammonia at a given concentration is the same as in an aqueous solution of acetic acid at the same concentration.

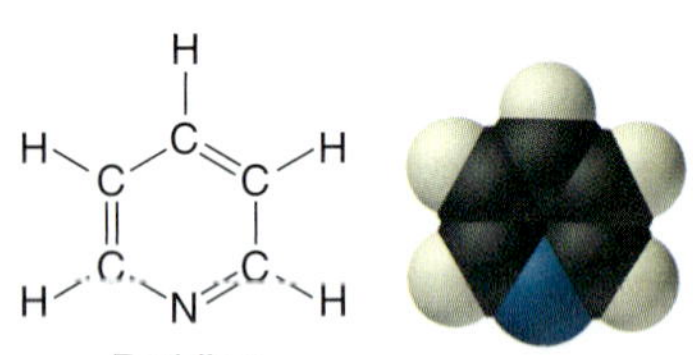

Pyridine

EXERCISE 16.7

The pH of a 0.0500 *M* solution of pyridine (C_6H_5N) is 8.96 at 25°C. Calculate the values of K_b and pK_b for pyridine.

Answer $K_b = 1.7 \times 10^{-9}$; $pK_b = 8.77$

Calculating Percent Ionization from K_a or K_b

When carrying out a laboratory analysis, chemists frequently need to know the concentrations of all species in solution. Because the reactivity of a weak acid or a weak base is usually very different from the reactivity of its conjugate base or acid, we often need to know the percent ionization of a solution of an acid or a base in order to understand a chemical reaction. The percent ionization is defined as

$$\text{percent ionization of acid} = \frac{[H^+]}{C_{HA}} \times 100 \qquad (16.40)$$

$$\text{percent ionization of base} = \frac{[OH^-]}{C_B} \times 100 \qquad (16.41)$$

One way to determine the concentrations of species in solutions of weak acids and bases is a variation of the tabular method we used above to determine K_a and K_b values. As a demonstration, let's calculate the concentrations of all species and the percent ionization in a 0.150 *M* solution of formic acid at 25°C. The data in Table 16.2 show that formic acid ($K_a = 1.8 \times 10^{-4}$ at 25°C) is a slightly stronger acid than acetic acid. The equilibrium equation for the ionization of formic acid in water is

$$HCO_2H(aq) \rightleftharpoons H^+(aq) + HCO_2^-(aq) \qquad (16.42)$$

and the equilibrium constant expression for this reaction is

$$K_a = \frac{[H^+][HCO_2^-]}{[HCO_2H]} \qquad (16.43)$$

We set the initial concentration of HCO_2H equal to 0.150 *M*, and that of HCO_2^- is 0 *M*. The initial concentration of H^+ is 1.00×10^{-7} *M* due to the autoionization of water. Because the equilibrium constant for the ionization reaction is small, the equilibrium will lie to the left, favoring the unionized form of the acid. Hence, we can define *x* as the amount of formic acid that dissociates.

If the change in $[HCO_2H]$ is $-x$, then the change in $[H^+]$ and $[HCO_2^-]$ is $+x$. The final concentration of each species is the sum of its initial concentration and the change in concentration, as summarized in the table.

	$HCO_2H(aq)$	$\rightleftharpoons$	$H^+(aq)$	+	$HCO_2^-(aq)$
	$[HCO_2H]$		**$[H^+]$**		**$[HCO_2^-]$**
Initial	0.150		1.00×10^{-7}		0
Change	$-x$		$+x$		$+x$
Final	$(0.150 - x)$		$(1.00 \times 10^{-7} + x)$		x

We can calculate the value of *x* by substituting the final concentrations from the table into the equilibrium constant expression:

$$K_a = \frac{[H^+][HCO_2^-]}{[HCO_2H]} = \frac{(1.00 \times 10^{-7} + x)x}{0.150 - x}$$

Because the ionization constant K_a is small, *x* is likely to be small compared with the initial concentration of formic acid: $(0.150 - x)\ M \approx 0.150\ M$. Moreover, $[H^+]$ due to the autoionization of water (1.00×10^{-7} *M*) is likely to be negligible compared with $[H^+]$ due to the dissociation of formic acid: $(1.00 \times 10^{-7} + x)\ M \approx x\ M$. Inserting these values into the equilibrium constant expression and solving for *x* give

$$K_a = \frac{x^2}{0.150} = 1.8 \times 10^{-4}$$

$$x = 5.2 \times 10^{-3}$$

We can now calculate the concentrations of the species present in a 0.150 *M* formic acid solution by inserting this value of *x* into the expressions in the last line of the table:

$$[HCO_2H] = (0.150 - x)\ M = 0.145\ M$$

$$[HCO_2] = x = 5.2 \times 10^{-3}\ M$$

$$[H^+] = (1.00 \times 10^{-7} + x)\ M = 5.2 \times 10^{-3}\ M$$

Thus, the pH of the solution is $-\log(5.2 \times 10^{-3}) = 2.28$. We can also use these concentrations to calculate the fraction of the original acid that is ionized. In this case, the percent ionization is the ratio of $[H^+]$ (or $[HCO_2^-]$) to the analytical concentration, multiplied by 100 to give a percentage:

$$\text{percent ionization} = \frac{[H^+]}{C_{HA}} \times 100 = \frac{5.2 \times 10^{-3}\ M}{0.150} \times 100 = 3.5\%$$

Always check to make sure that any simplifying assumption was valid. As a general rule of thumb, approximations such as those used above are valid *only* if the quantity being neglected is no more than about 5% of the quantity to which it is being added or from which it is being subtracted. If the quantity that was neglected is much greater than about 5%, then the approximation is probably not valid, and you should go back and solve the problem using the quadratic formula. In the example above, both simplifying assumptions were justified: the percent ionization is only 3.5%, which is well below the approximately 5% limit, and the $1.00 \times 10^{-7}\ M\ [H^+]$ due to the autoionization of water is much, much less than the $0.0052\ M\ [H^+]$ due to the ionization of formic acid.

As a general rule, the $[H^+]$ contribution due to the autoionization of water can be ignored as long as the product of the acid (or base) ionization constant and the analytical concentration of the acid (or base) is at least 10 times greater than the $[H^+]$ or $[OH^-]$ from the autoionization of water—that is, if

$$K_aC_{HA} \geq 10(1.00 \times 10^{-7}) = 1.0 \times 10^{-6} \tag{16.44}$$

or

$$K_bC_B \geq 10(1.00 \times 10^{-7}) = 1.0 \times 10^{-6} \tag{16.45}$$

By substituting the appropriate values for the formic acid solution considered above into Equation 16.44, we see that the simplifying assumption is valid in this case:

$$K_aC_{HA} = (1.8 \times 10^{-4})(0.150) = 2.7 \times 10^{-5} > 1.0 \times 10^{-6} \tag{16.46}$$

Doing this simple calculation before solving this type of problem saves time and allows you to write simplified expressions for the final concentrations of species. In practice, it is necessary to include the $[H^+]$ contribution due to the autoionization of water only for extremely dilute solutions of very weak acids or bases.

The next example illustrates how the procedure outlined above can be used to calculate the pH of a solution of a weak base.

EXAMPLE 16.8

Calculate the pH and percent ionization of a 0.225 M solution of ethylamine ($CH_3CH_2NH_2$), which is used in the synthesis of some dyes and medicines. The pK_b of ethylamine is 3.19 at 20°C.

Given Concentration and pK_b

Asked for pH and percent ionization

Strategy

A Write the balanced equilibrium equation for the reaction and the equilibrium constant expression. Calculate the value of K_b from pK_b.

B Use Equation 16.44 to see whether you can ignore $[H^+]$ due to the autoionization of water. Then use a tabular format to write expressions for the final concentrations of all species in solution. Substitute these values into the equilibrium equation, and solve for $[OH^-]$. Use Equation 16.41 to calculate the percent ionization.

C Use the relationship $K_w = [OH^-][H^+]$ to obtain $[H^+]$. Then calculate the pH of the solution.

Solution

A We begin by writing the balanced equilibrium equation for the reaction:

$$CH_3CH_2NH_2(aq) + H_2O(l) \rightleftharpoons CH_3CH_2NH_3^+(aq) + OH^-(aq)$$

The corresponding equilibrium constant expression is

$$K_b = \frac{[CH_3CH_2NH_3^+][OH^-]}{[CH_3CH_2NH_2]}$$

From the value of pK_b, we have $K_b = 10^{-3.19} = 6.46 \times 10^{-4}$. **B** To calculate the pH, we need to determine the H^+ concentration. Unfortunately, H^+ does not appear in either the chemical equation or the equilibrium constant expression. However, $[H^+]$ and $[OH^-]$ in an aqueous solution are related by the equation $K_w = [H^+][OH^-]$. Hence, if we can determine $[OH^-]$, we can calculate $[H^+]$ and then the pH. The initial concentration of $CH_3CH_2NH_2$ is 0.225 *M*, and the initial $[OH^-]$ is 1.00×10^{-7} *M*. Because ethylamine is a weak base, the extent of the reaction will be small, and it makes sense to let x equal the amount of $CH_3CH_2NH_2$ that reacts with water. The change in $[CH_3CH_2NH_2]$ is therefore $-x$, and the change in both $[CH_3CH_2NH_3^+]$ and $[OH^-]$ is $+x$. To see whether the autoionization of water can safely be ignored, we substitute the values of K_b and C_B into Equation 16.45 to give

$$K_bC_B = (6.46 \times 10^{-4})(0.225) = 1.45 \times 10^{-4} > 1.0 \times 10^{-6}$$

Thus, the simplifying assumption is valid, and we will not include $[OH^-]$ due to the autoionization of water in our calculations.

$$H_2O(l) + CH_3CH_2NH_2(aq) \rightleftharpoons CH_3CH_2NH_3^+(aq) + OH^-(aq)$$

	$[CH_3CH_2NH_2]$	$[CH_3CH_2NH_3^+]$	$[OH^-]$
Initial	0.225	0	1.00×10^{-7}
Change	$-x$	$+x$	$+x$
Final	$(0.225 - x)$	x	x

Substituting the quantities from the last line of the table into the equilibrium constant expression gives

$$K_b = \frac{[CH_3CH_2NH_3^+][OH^-]}{[CH_3CH_2NH_2]} = \frac{(x)(x)}{0.225 - x} = 6.46 \times 10^{-4}$$

As before, we assume the amount of $CH_3CH_2NH_2$ that ionizes is small compared with the initial concentration, so $[CH_3CH_2NH_2]_f = 0.225 - x \approx 0.225$. With this assumption, we can simplify the equilibrium equation and solve for x:

$$K_b = \frac{x^2}{0.225} = 6.46 \times 10^{-4}$$

$$x = 0.0121 = [CH_3CH_2NH_3^+]_f = [OH^-]_f$$

The percent ionization is therefore

$$\text{percent ionization} = \frac{[OH^-]}{C_B} \times 100 = \frac{0.0121\ M}{0.225\ M} \times 100 = 5.4\%$$

which is at the upper limit of the approximately 5% range that can be ignored. The final hydroxide concentration is thus 0.0121 *M*. **C** We can now determine the value of $[H^+]$ using the expression for K_w:

$$K_w = [OH^-][H^+]$$
$$1.01 \times 10^{-14} = (0.0121\ M)[H^+]$$
$$[H^+] = 8.35 \times 10^{-13}\ M$$

The pH of the solution is $-\log(8.35 \times 10^{-13}) = 12.08$. Alternatively, we could have calculated pOH as $-\log(0.0121) = 1.917$ and determined the pH as follows:

$$pH + pOH = pK_w = 14.00$$
$$pH = 14.00 - 1.917 = 12.08$$

The two methods are equivalent.

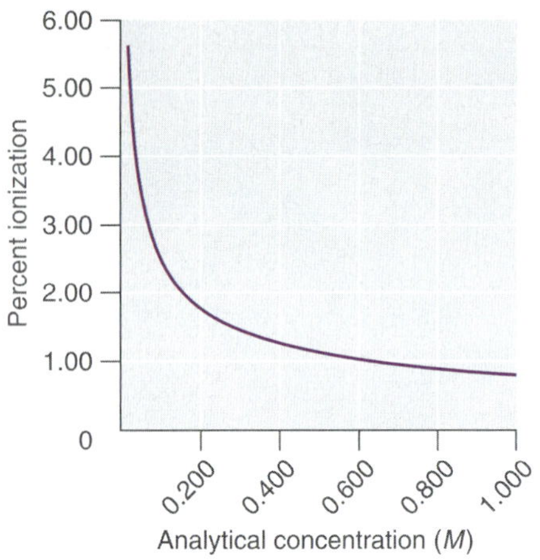

Figure 16.7 The relationship between the analytical concentration of a weak acid and the percent ionization. As shown here for benzoic acid ($C_6H_5CO_2H$), the percent ionization decreases as the analytical concentration of a weak acid increases.

EXERCISE 16.8

Aromatic amines, in which the nitrogen atom is bonded directly to a phenyl ring ($C_6H_5^-$) tend to be much weaker bases than simple alkylamines. For example, aniline ($C_6H_5NH_2$) has a pK_b of 9.13 at 25°C. What is the pH of a 0.050 *M* solution of aniline?

Answer 8.78

The examples worked above illustrate a key difference between solutions of strong acids and bases and solutions of weak acids and bases. Because strong acids and bases ionize essentially completely in water, the percent ionization is always approximately 100%, regardless of the concentration. In contrast, the percent ionization in solutions of weak acids and bases is small and depends on the analytical concentration of the weak acid or base. As illustrated for benzoic acid in Figure 16.7, the percent ionization of a weak acid or a weak base actually *increases* as its analytical concentration decreases. The percent ionization also increases as the magnitude of the ionization constants K_a and K_b increases.

Thus, unlike K_a or K_b, the percent ionization is not a constant for weak acids and bases but depends on *both* the acid or base ionization constant *and* the analytical concentration. Consequently, the procedure outlined above must be used to calculate the percent ionization and pH for solutions of weak acids and bases. The following example and exercise demonstrate that the combination of a dilute solution and a relatively large value of K_a or K_b can give a percent ionization much greater than 5%, making it necessary to use the quadratic equation to determine the concentrations of species in solution.

Note the pattern

The percent ionization in a solution of a weak acid or a weak base increases as the analytical concentration decreases and as the value of K_a or K_b increases.

EXAMPLE 16.9

Benzoic acid ($C_6H_5CO_2H$) is used in the food industry as a preservative and medically as an antifungal agent. Its pK_a at 25°C is 4.20, making it a somewhat stronger acid than acetic acid. Calculate the percentage of benzoic acid molecules that are ionized in **(a)** a 0.0500 *M* solution and **(b)** a 0.00500 *M* solution.

Given Concentrations and pK_a

Asked for Percent ionization

Strategy

A Write both the balanced equilibrium equation for the ionization reaction and the equilibrium equation (Equation 16.13). Use Equation 16.21b to calculate the value of K_a from pK_a.

B For both the concentrated and dilute solutions, use a tabular format to write expressions for the final concentrations of all species in solution. Substitute these values into the equilibrium equation, and solve for $[C_6H_5CO_2^-]_f$ for each solution.

C Use the values of $[C_6H_5CO_2^-]_f$ and Equation 16.40 to calculate the percent ionization.

Solution

A If we abbreviate benzoic acid as $PhCO_2H$ where Ph = C_6H_5—, the balanced equilibrium equation for the ionization reaction and the equilibrium equation can be written as

$$PhCO_2H(aq) \rightleftharpoons H^+(aq) + PhCO_2^-(aq)$$

$$K_a = \frac{[H^+][PhCO_2^-]}{[PhCO_2H]}$$

From the value of pK_a, we have $K_a = 10^{-4.20} = 6.31 \times 10^{-5}$.

(a) B For the more concentrated solution, we set up our table of initial concentrations, changes in concentrations, and final concentrations:

	$PhCO_2H(aq)$	$\rightleftharpoons$	$H^+(aq)$	+	$PhCO_2^-(aq)$
	$[PhCO_2H]$		$[H^+]$		$[PhCO_2^-]$
Initial	0.0500		1.00×10^{-7}		0
Change	$-x$		$+x$		$+x$
Final	$(0.0500 - x)$		$(1.00 \times 10^{-7} + x)$		x

Inserting the expressions for the final concentrations into the equilibrium equation and making our usual assumptions, that $[PhCO_2^-]_i$ and $[H^+]$ due to the autoionization of water are negligible, give

$$K_a = \frac{[H^+][PhCO_2^-]}{[PhCO_2H]} = \frac{(x)(x)}{0.0500 - x} = \frac{x^2}{0.0500} = 6.31 \times 10^{-5}$$
$$x = 1.78 \times 10^{-3}$$

This value is less than 5% of 0.0500, so our simplifying assumption is justified, and $[PhCO_2^-]$ at equilibrium is 1.78×10^{-3} *M*. We reach the same conclusion using C_{HA}: $K_aC_{HA} = (6.31 \times 10^{-5})(0.0500) = 3.16 \times 10^{-6} > 1.0 \times 10^{-6}$. C The percent ionized is the ratio of the concentration of $PhCO_2^-$ to the analytical concentration, multiplied by 100:

$$\text{percent ionized} = \frac{[PhCO_2^-]}{C_{PhCO_2H}} \times 100 = \frac{1.78 \times 10^{-3}}{0.0500} \times 100 = 3.6\%$$

Because only 3.6% of the benzoic acid molecules are ionized in a 0.0500 *M* solution, our simplifying assumptions are confirmed.

(b) B For the more dilute solution, we proceed in exactly the same manner. Our table of concentrations is therefore

	$PhCO_2H(aq)$	$\rightleftharpoons$	$H^+(aq)$	+	$PhCO_2^-(aq)$
	$[PhCO_2H]$		$[H^+]$		$[PhCO_2^-]$
Initial	0.00500		1.00×10^{-7}		0
Change	$-x$		$+x$		$+x$
Final	$(0.00500 - x)$		$(1.00 \times 10^{-7} + x)$		x

Inserting the expressions for the final concentrations into the equilibrium equation and making our usual simplifying assumptions give

$$K_a = \frac{[H^+][PhCO_2^-]}{[PhCO_2H]} = \frac{(x)(x)}{0.00500 - x} = \frac{x^2}{0.00500} = 6.31 \times 10^{-5}$$
$$x = 5.62 \times 10^{-4}$$

Unfortunately, this number is greater than 10% of 0.00500, so our assumption that the fraction of benzoic acid that is ionized in this solution could be neglected and that $(0.00500 - x) \approx x$ is not valid. Furthermore, we see that $K_a\,C_{HA} = (6.31 \times 10^{-5})(0.00500) = 3.16 \times 10^{-7} < 1.0 \times 10^{-6}$. Thus, the relevant equation is

$$\frac{x^2}{0.00500 - x} = 6.31 \times 10^{-5}$$

which must be solved using the quadratic formula. Multiplying out the quantities gives

$$x^2 = (6.31 \times 10^{-5})(0.00500 - x) = 3.16 \times 10^{-7} - (6.31 \times 10^{-5})x$$

Rearranging the equation to fit the standard quadratic equation format gives

$$x^2 + (6.31 \times 10^{-5})x - (3.16 \times 10^{-7}) = 0$$

This equation can be solved using the quadratic formula:

$$x = \frac{-b \pm \sqrt{b^2 - 4ac}}{2a}$$

$$= \frac{-(6.31 \times 10^{-5}) \pm \sqrt{(6.31 \times 10^{-5})^2 - 4(1)(-3.16 \times 10^{-7})}}{2(1)}$$

$$= \frac{-(6.31 \times 10^{-5}) \pm (1.12 \times 10^{-3})}{2} = 5.28 \times 10^{-4} \text{ or } -5.92 \times 10^{-4}$$

Because a negative value of x corresponds to a negative $[PhCO_2^-]$, which is not physically meaningful, we use the positive solution: $x = 5.28 \times 10^{-4}$. Thus, $[PhCO_2^-] = 5.28 \times 10^{-4}\ M$. The percent ionized is therefore

$$\text{percent ionized} = \frac{[PhCO_2^-]}{C_{PhCO_2H}} \times 100 = \frac{5.28 \times 10^{-4}}{0.00500} \times 100 = 10.6\%$$

Thus, in the more dilute solution ($C = 0.00500\ M$), 10.6% of the benzoic acid molecules are ionized versus only 3.6% in the more concentrated solution ($C = 0.0500\ M$). Decreasing the analytical concentration by a factor of 10 results in an approximately threefold increase in the percentage of benzoic acid molecules that are ionized.

EXERCISE 16.9

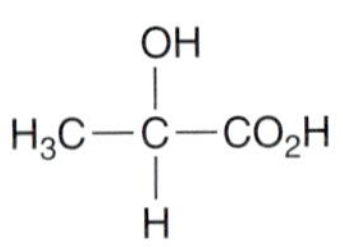

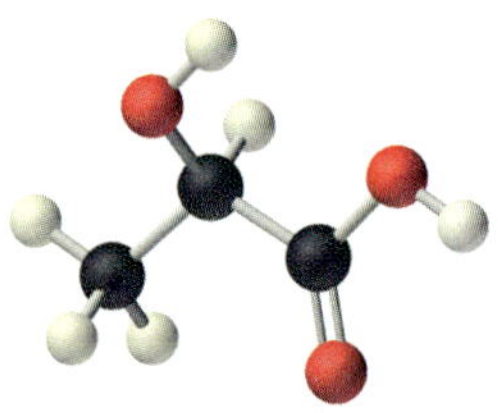

Lactic acid

Lactic acid, $CH_3CH(OH)CO_2H$, is a weak acid with a pK_a of 3.86 at 25°C. What percentage of the lactic acid is ionized in **(a)** a 0.10 M solution and **(b)** a 0.0020 M solution?

Answer **(a)** 3.7%; **(b)** 26%

Note the pattern

Simplifying assumptions in equilibrium calculations for solutions of weak acids and bases are most likely to be valid for concentrated solutions.

Determining K_{eq} from K_a and K_b

In Section 16.2, you learned how to use K_a and K_b values to qualitatively predict whether reactants or products are favored in an acid–base reaction. Tabulated values of K_a (or pK_a) and K_b (or pK_b), plus the value of K_w, enable us to quantitatively determine the direction and extent of reaction for a weak acid and a weak base by calculating the value of K for the reaction. To illustrate how to do this, we begin by writing the dissociation equilibria for a weak acid and a weak base and then summing them:

Acid	$HA \rightleftharpoons H^+ + A^-$	K_a
Base	$B + H_2O \rightleftharpoons HB^+ + OH^-$	K_b
Sum	$HA + B + H_2O \rightleftharpoons H^+ + A^- + HB^+ + OH^-$	$K_{sum} = K_aK_b$

(16.47)

Note that the overall reaction has H_2O on the left and H^+ and OH^- on the right, which means it involves the autoionization of water ($H_2O \rightleftharpoons H^+ + OH^-$) in addition to the acid–base equilibrium in which we are interested. We can obtain an equation that includes only the acid–base equilibrium by simply adding the equation for the reverse of the autoionization of water ($H^+ + OH^- \rightleftharpoons H_2O$), for which $K = 1/K_w$, to the overall equilibrium in Equation 16.47 and canceling:

$HA + B + \cancel{H_2O} \rightleftharpoons \cancel{H^+} + A^- + HB^+ + \cancel{OH^-}$	$K_{sum} = K_aK_b$
$\cancel{H^+} + \cancel{OH^-} \rightleftharpoons \cancel{H_2O}$	$1/K_w$
$HA + B \rightleftharpoons A^- + HB^+$	$K = (K_aK_b)/K_w$

(16.48)

Thus, the equilibrium constant for the reaction of a weak acid with a weak base is the product of the ionization constants of the acid and the base divided by K_w. The next example illustrates how to calculate the equilibrium constant for the reaction of a weak acid with a weak base.

CH_2CO_2H
|
$HOCCO_2H$
|
CH_2CO_2H

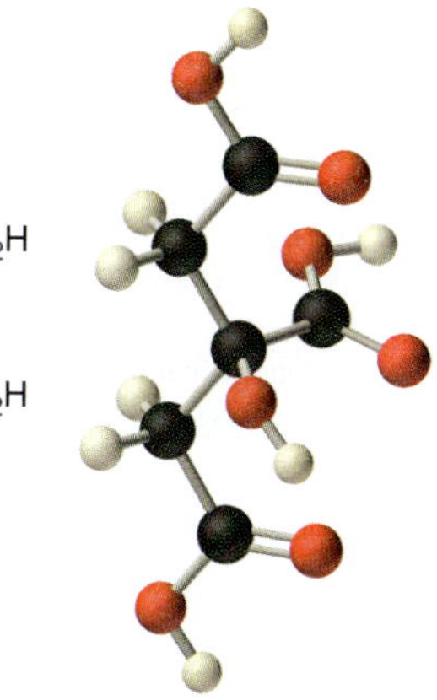

Citric acid

EXAMPLE 16.10

Fish tend to spoil rapidly even when refrigerated. The cause of the resulting "fishy" odor is a mixture of amines, particularly methylamine (CH_3NH_2), a volatile weak base ($pK_b = 3.34$). Fish is often served with a wedge of lemon because lemon juice contains citric acid, a triprotic acid with pK_a's of 3.13, 4.76, and 6.40 that can neutralize amines. Calculate the equilibrium constant for the reaction of excess citric acid with methylamine, assuming that only the first dissociation constant of citric acid is important.

Given pK_b for base and pK_a for acid

Asked for K

Strategy

A Write the balanced equilibrium equation and the equilibrium constant expression for the reaction.
B Convert pK_a and pK_b to K_a and K_b, and then use Equation 16.48 to calculate the value of K.

Solution

A If we abbreviate citric acid as H_3citrate, the equilibrium equation for its reaction with methylamine is

$$CH_3NH_2(aq) + H_3\text{citrate}(aq) \rightleftharpoons CH_3NH_3^+(aq) + H_2\text{citrate}^-(aq)$$

The equilibrium constant expression for this reaction is

$$K = \frac{[CH_3NH_3^+][H_2\text{citrate}^-]}{[CH_3NH_2][H_3\text{citrate}]}$$

B Equation 16.48 is $K = (K_aK_b)/K_w$. Converting pK_a and pK_b to K_a and K_b gives $K_a = 10^{-3.13} = 7.41 \times 10^{-4}$ for citric acid and $K_b = 10^{-3.34} = 4.57 \times 10^{-4}$ for methylamine. Substituting these values into the equilibrium equation gives

$$K = \frac{K_aK_b}{K_w} = \frac{(7.41 \times 10^{-4})(4.57 \times 10^{-4})}{1.01 \times 10^{-14}} = 3.35 \times 10^7 = 3.4 \times 10^7$$

The value of pK can also be calculated directly by taking the negative logarithm of both sides of Equation 16.48, which gives

$$pK = pK_a + pK_b - pK_w = 3.13 + 3.34 - 14.00 = -7.53$$

Thus, $K = 10^{-(-7.53)} = 3.4 \times 10^7$, in agreement with the earlier value. In either case, the K values show that the reaction of citric acid with the volatile, foul-smelling methylamine lies very far to the right, favoring the formation of a much less volatile salt with no odor. This is one reason a little lemon juice helps make less-than-fresh fish more appetizing.

EXERCISE 16.10

Dilute aqueous ammonia solution, often used as a cleaning agent, is also effective as a deodorizing agent. To see why, calculate the equilibrium constant for the reaction of aqueous ammonia with butyric acid ($CH_3CH_2CH_2CO_2H$), a particularly foul-smelling substance associated with the odor of rancid butter and smelly socks. The pK_b of ammonia is 4.75, and the pK_a of butyric acid is 4.83.

Answer 2.6×10^4

16.5 Acid–Base Titrations

In Chapter 4, you learned that in an acid–base titration, a buret is used to deliver measured volumes of an acid or base solution of known concentration (the *titrant*) to a flask that contains a solution of a base or an acid, respectively, of unknown concentration

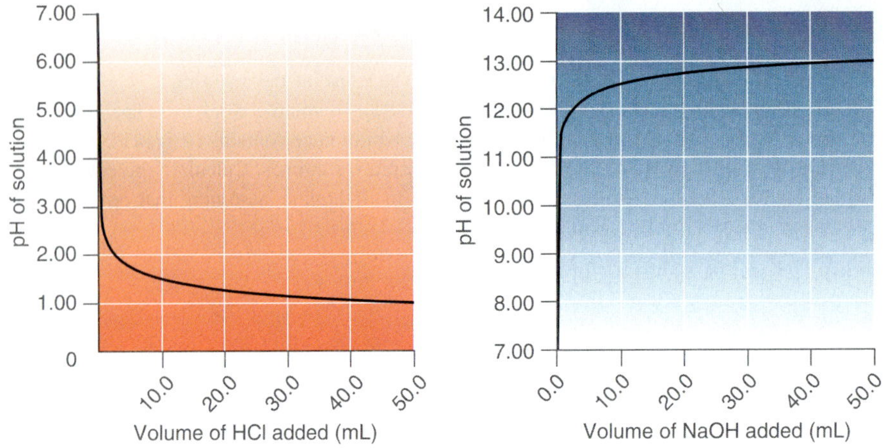

(a) Water with strong acid added **(b) Water with strong base added**

Figure 16.8 Solution pH as a function of the volume of a strong acid or base added to distilled water. **(a)** When 0.200 *M* HCl is added to 50.0 mL of distilled water, the pH rapidly decreases from 7.0 to 0.70 at a point beyond 50.0 mL (the pH of 0.200 *M* HCl). **(b)** Conversely, when 0.200 *M* NaOH is added to 50.0 mL of distilled water, the pH rapidly increases from 7.0 to 13.30 at a point beyond 50.0 mL (the pH of 0.200 *M* NaOH).

(the *unknown*). If the concentration of the titrant is known, then the concentration of the unknown can be determined. The following discussion focuses on the pH changes that occur during an acid–base titration. Plotting the pH of the solution in the flask against the amount of acid or base added produces a **titration curve**. The shape of the curve provides important information about what is occurring in solution during the titration.

Titrations of Strong Acids and Bases

Figure 16.8a shows a plot of the pH as 0.200 *M* HCl is gradually added to 50.00 mL of pure water. The pH of the sample in the flask is initially 7.00 (as expected for pure water), but it drops very rapidly as HCl is added. Eventually the pH becomes constant at a value of 0.70 at a point well beyond its value of 1.00 with the addition of 50.0 mL of HCl, the pH of 0.200 *M* HCl. In contrast, when 0.200 *M* NaOH is added to 50.00 mL of distilled water, the pH (initially 7.00) climbs very rapidly at first, but then more gradually, eventually approaching a limit of 13.30 (the pH of 0.200 *M* NaOH), again well beyond its value of 13.00 with the addition of 50.0 mL of NaOH as shown in Figure 16.8b. As you can see from these plots, the titration curve for the addition of a base is the mirror image of the curve for the addition of an acid.

Suppose that we now add 0.200 *M* NaOH to 50.00 mL of a 0.100 *M* solution of HCl. Because HCl is a strong acid that is completely ionized in water, the initial $[H^+]$ is 0.100 *M*, and the initial pH is 1.00. Adding NaOH decreases the concentration of H^+ because of the neutralization reaction: $OH^- + H^+ \longrightarrow H_2O$ (Figure 16.9a). Thus, the pH of the

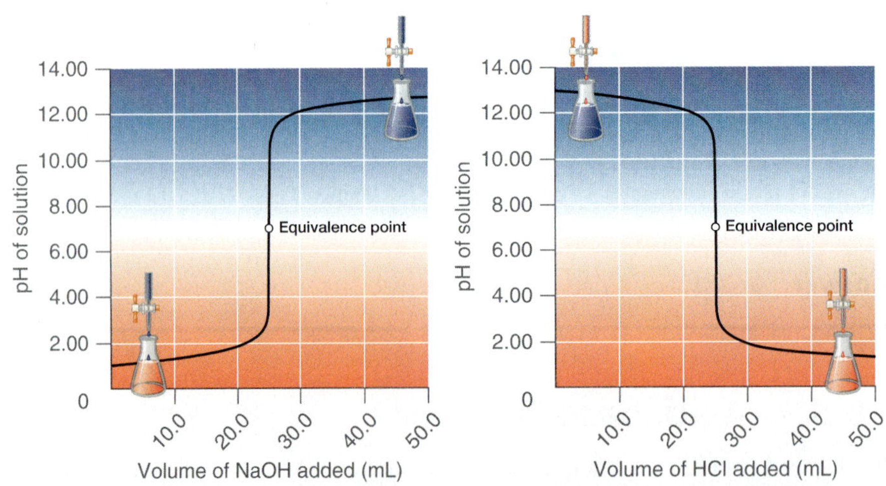

(a) Strong acid titrated with strong base **(b) Strong base titrated with strong acid**

Figure 16.9 The titration of (a) a strong acid with a strong base and (b) a strong base with a strong acid. **(a)** As 0.200 *M* NaOH is slowly added to 50.0 mL of 0.100 *M* HCl, the pH increases slowly at first, then increases very rapidly as the equivalence point is approached, and finally increases slowly once more. **(b)** Conversely, as 0.200 *M* HCl is slowly added to 50.0 mL of 0.100 *M* NaOH, the pH decreases slowly at first, then decreases very rapidly as the equivalence point is approached, and finally decreases slowly once more.

solution increases gradually. Near the **equivalence point**, however, the point at which the number of moles of base (or acid) added equals the number of moles of acid (or base) originally present in the solution (see Chapter 4) the pH increases much more rapidly because most of the H^+ ions originally present have been consumed. For the titration of a monoprotic strong acid (HCl) with a *monobasic* strong base (NaOH), we can calculate the volume of base needed to reach the equivalence point from the following relationship:

$$\begin{aligned} \text{moles of base} &= \text{moles of acid} \\ (\text{volume})_b(\text{molarity})_b &= (\text{volume})_a(\text{molarity})_a \\ V_bM_b &= V_aM_a \end{aligned} \qquad (16.49)$$

If 0.200 *M* NaOH is added to 50.00 mL of a 0.100 *M* solution of HCl, we solve for V_b to obtain

$$\begin{aligned} V_b(0.200\ \cancel{M}) &= (0.0500\ \text{L})(0.100\ \cancel{M}) \\ V_b &= 0.0250\ \text{L} = 25.0\ \text{mL} \end{aligned}$$

At the equivalence point (when 25.00 mL of NaOH solution has been added), the neutralization is complete: only a salt remains in solution (NaCl), and the pH of the solution is 7.00. Adding more NaOH produces a rapid increase in pH, but eventually the pH levels off at a value of about 13.30, the pH of 0.200 *M* NaOH.

As shown in Figure 16.9b, the titration of 50.0 mL of a 0.100 *M* solution of NaOH with 0.200 *M* HCl produces a titration curve that is nearly the mirror image of the titration curve in Figure 16.9a. The pH is initially 13.00, and it slowly decreases as HCl is added. As the equivalence point is approached, the pH drops rapidly before leveling off at a value of about 0.70, the pH of 0.2 *M* HCl.

Notice that the titration of either a strong acid with a strong base or a strong base with a strong acid produces an S-shaped curve. The curve is somewhat asymmetrical because the steady increase in the volume of the solution during the titration causes the solution to become more dilute. Due to the leveling effect, the shape of the curve for a titration involving a strong acid and a strong base depends *only* on the concentrations of the acid and base, *not* on their identity.

EXAMPLE 16.11

Calculate the pH of the solution after 24.90 mL of 0.200 *M* NaOH has been added to 50.00 mL of 0.100 *M* HCl.

Given Volumes and concentrations of strong base and acid

Asked for pH

Strategy

A Calculate the number of millimoles of H^+ and OH^- to determine which, if either, is in excess after the neutralization reaction has occurred. If one species is in excess, calculate the amount that remains after the neutralization reaction.

B Determine the final volume of the solution. Calculate the concentration of the species in excess, and convert this value to pH.

Solution

A Because 0.100 mol/L is equivalent to 0.100 mmol/mL, the number of millimoles of H^+ in 50.00 mL of 0.100 *M* HCl can be calculated as

$$50.00\ \cancel{\text{mL}}\left(\frac{0.100\ \text{mmol HCl}}{\cancel{\text{mL}}}\right) = 5.00\ \text{mmol HCl} = 5.00\ \text{mmol H}^+$$

The number of millimoles of NaOH added is

$$24.90\ \cancel{\text{mL}}\left(\frac{0.200\ \text{mmol NaOH}}{\cancel{\text{mL}}}\right) = 4.98\ \text{mmol NaOH} = 4.98\ \text{mmol OH}^-$$

Thus, H^+ is in excess. To completely neutralize the acid requires the addition of 5.00 mmol of OH^- to the HCl solution. Because only 4.98 mmol of OH^- has been added, the amount of excess H^+ is 5.00 mmol − 4.98 mmol = 0.02 mmol of H^+. The final volume of the solution is 50.00 mL + 24.90 mL = 74.90 mL, so the final concentration of H^+ is

$$[H^+] = \frac{0.02 \text{ mmol } H^+}{74.90 \text{ mL}} = 3 \times 10^{-4}\,M$$

The pH is $-\log[H^+] = -\log(3 \times 10^{-4}) = 3.5$, which is significantly lower than the pH of 7.00 for a neutral solution.

EXERCISE 16.11

Calculate the pH of a solution prepared by adding 40.00 mL of 0.237 *M* HCl to 75.00 mL of a 0.133 *M* solution of NaOH.

Answer 11.6

Titrations of Weak Acids and Bases

In contrast to strong acids and bases, the shape of the titration curve for a weak acid or a weak base depends dramatically on the identity of the acid or base and the corresponding value of K_a or K_b. As we shall see, the pH also changes much more gradually around the equivalence point in the titration of a weak acid or a weak base. As you learned in the preceding section, $[H^+]$ of a solution of a weak acid (HA) is *not* equal to the concentration of the acid, but depends on both its pK_a and its concentration. Because only a fraction of a weak acid dissociates, $[H^+]$ is less than [HA]. Thus, the pH of a solution of a weak acid is higher than the pH of a solution of a strong acid of the same concentration. Figure 16.10a shows the titration curve for 50.0 mL of a 0.100 *M* solution of acetic acid with 0.200 *M* NaOH, superimposed on the curve for the titration of 0.100 *M* HCl shown in Figure 16.9a. Below the equivalence point, the two curves are very different. Notice that before any base is added, the pH of the acetic acid solution is higher than the pH of the HCl solution, and the pH changes more rapidly during the first part of the titration. Note also that the pH of the acetic acid solution at the equivalence point is greater than 7.00. That is, at the equivalence point, the solution is basic. In addition, the change in pH around the equivalence point is only about half as large as for the HCl titration; the magnitude of the pH change at the equivalence point depends on the pK_a of the acid being titrated. Above the equivalence point, however, the two curves are identical. Once the acid

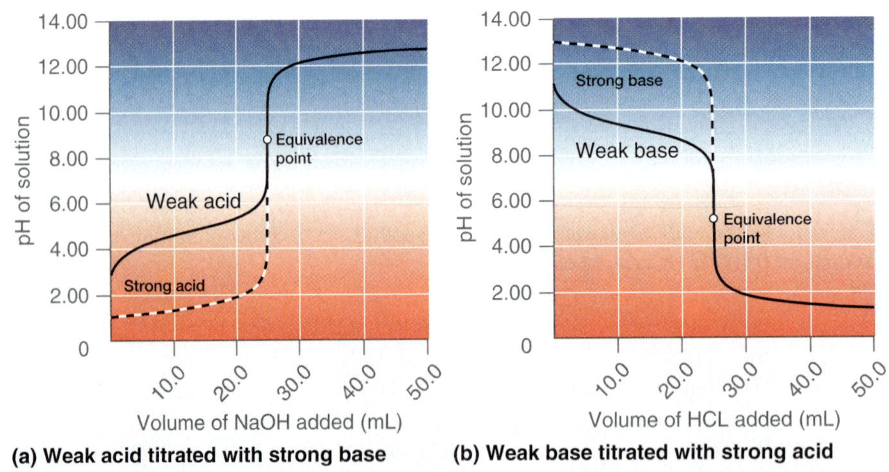

Figure 16.10 The titration of (a) a weak acid with a strong base and (b) a weak base with a strong acid. (a) As 0.200 *M* NaOH is slowly added to 50.0 mL of 0.100 *M* acetic acid, the pH increases slowly at first, then increases rapidly as the equivalence point is approached, and then again increases more slowly. The corresponding curve for the titration of 50.0 mL of 0.100 *M* HCl with 0.200 *M* NaOH is shown in yellow. **(b)** As 0.200 *M* HCl is slowly added to 50.0 mL of 0.100 *M* NH_3, the pH decreases slowly at first, then decreases rapidly as the equivalence point is approached, and then again decreases more slowly. The corresponding curve for the titration of 50.0 mL of 0.100 *M* NaOH with 0.200 *M* HCl is shown in yellow.

has been neutralized, the pH of the solution is controlled only by the amount of excess NaOH present, regardless of whether the acid is weak or strong.

The titration curve in Figure 16.10a was created by calculating the starting pH of the acetic acid solution before any NaOH is added, and then calculating the pH of the solution after the addition of increasing volumes of NaOH. The procedure is illustrated below and in Example 16.12 for three points on the titration curve, using the pK_a of acetic acid (4.76 at 25°C; $K_a = 1.7 \times 10^{-5}$).

Calculating the pH of a Solution of a Weak Acid or Base

As explained in the preceding section, if we know K_a or K_b and the initial concentration of a weak acid or base, we can calculate the pH of a solution of a weak acid or base by setting up a table of initial concentrations, changes in concentrations, and final concentrations. In this example, the initial concentration of acetic acid is 0.100 *M*. If we define x as $[H^+]$ due to the dissociation of the acid, then the table of concentrations for the ionization of 0.100 *M* acetic acid is shown here.

	$CH_3CO_2H(aq)$	$\rightleftharpoons$	$H^+(aq)$	+	$CH_3CO_2^-(aq)$
	$[CH_3CO_2H]$		$[H^+]$		$[CH_3CO_2^-]$
Initial	0.100		1.00×10^{-7}		0
Change	$-x$		$+x$		$+x$
Final	$(0.100 - x)$		x		x

In this and all subsequent examples, we will ignore $[H^+]$ and $[OH^-]$ due to the autoionization of water when calculating the final concentration. However, you should use Equations 16.44 and 16.45 to check that this assumption is justified.

Inserting the expressions for the final concentrations into the equilibrium equation (and using approximations) gives

$$K_a = \frac{[H^+][CH_3CO_2^-]}{[CH_3CO_2H]} = \frac{(x)(x)}{0.100 - x} \approx \frac{x^2}{0.100} = 1.74 \times 10^{-5}$$

Solving this equation gives $x = [H^+] = 1.32 \times 10^{-3}$ *M*. Thus, the pH of a 0.100 *M* solution of acetic acid is

$$\text{pH} = -\log(1.32 \times 10^{-3}) = 2.88$$

Calculating the pH During Titration of a Weak Acid or Base

Now consider what happens when we add 5.00 mL of 0.200 *M* NaOH to 50.00 mL of 0.100 *M* CH_3CO_2H (Figure 16.10a). Because the neutralization reaction proceeds to completion, all of the OH^- ions added will react with the acetic acid to generate acetate ion and water:

$$CH_3CO_2H(aq) + OH^-(aq) \rightarrow CH_3CO_2^-(aq) + H_2O(l) \qquad (16.50)$$

All problems of this type must be solved in two steps: a stoichiometric calculation followed by an equilibrium calculation. In the first step, we use the stoichiometry of the neutralization reaction to calculate the *amounts* of acid and conjugate base present in solution after the neutralization reaction has occurred. In the second step, we use the equilibrium equation (Equation 16.13) to determine $[H^+]$ of the resulting solution.

Step 1 To determine the amount of acid and conjugate base in solution after the neutralization reaction, we calculate the amount of CH_3CO_2H in the original solution and the amount of OH^- in the NaOH solution that was added. The acetic acid solution contained

$$50.00\ \cancel{\text{mL}}\left(\frac{0.100\ \text{mmol}\ CH_3CO_2H}{\cancel{\text{mL}}}\right) = 5.00\ \text{mmol}\ CH_3CO_2H$$

The NaOH solution contained

$$50.00\ \cancel{\text{mL}}\left(\frac{0.200\ \text{mmol NaOH}}{\cancel{\text{mL}}}\right) = 1.00\ \text{mmol NaOH}$$

Comparing the amounts shows that CH_3CO_2H is in excess. Since OH^- reacts with CH_3CO_2H in a 1:1 stoichiometry, the amount of excess CH_3CO_2H is

$$5.00\ \text{mmol}\ CH_3CO_2H - 1.00\ \text{mmol}\ OH^- = 4.00\ \text{mmol}\ CH_3CO_2H$$

Each 1 mmol of OH^- reacts to produce 1 mmol of acetate ion, so the final amount of $CH_3CO_2^-$ is 1.00 mmol.

The stoichiometry of the reaction is summarized in the table, which shows the numbers of moles of the various species, *not* their concentrations.

	$CH_3CO_2H(aq)$	+ $OH^-(aq)$	$\longrightarrow$ $CH_3CO_2^-(aq) + H_2O(l)$
Initial	5.00 mmol	1.00 mmol	0 mmol
Change	− 1.00 mmol	− 1.00 mmol	+ 1.00 mmol
Final	4.00 mmol	0 mmol	1.00 mmol

Note that the table gives the initial amount of acetate and the final amount of OH^- ions as 0. Because an aqueous solution of acetic acid always contains at least a small amount of acetate ion in equilibrium with acetic acid, however, the initial acetate concentration is not actually 0. The value can be ignored in this calculation because the amount of $CH_3CO_2^-$ in equilibrium is insignificant compared to the amount of OH^- added. Moreover, due to the autoionization of water, no aqueous solution can contain 0 mmol of OH^-, but the amount of OH^- due to the autoionization of water is insignificant compared to the amount of OH^- added. We use the initial amounts of the reactants to determine the stoichiometry of the reaction, and defer a consideration of the equilibrium until the second half of the problem.

Step 2 To calculate $[H^+]$ at equilibrium following the addition of NaOH, we must first calculate $[CH_3CO_2H]$ and $[CH_3CO_2^-]$ using the number of millimoles of each and the total volume of the solution at this point in the titration:

$$\text{final volume} = 50.00\ \text{mL} + 5.00\ \text{mL} = 55.00\ \text{mL}$$

$$[CH_3CO_2H] = \frac{4.00\ \text{mmol}\ CH_3CO_2H}{55.00\ \text{mL}} = 7.27 \times 10^{-2}\ M$$

$$[CH_3CO_2^-] = \frac{1.00\ \text{mmol}\ CH_3CO_2^-}{55.00\ \text{mL}} = 1.82 \times 10^{-2}\ M$$

Knowing the concentrations of acetic acid and acetate ion at equilibrium and K_a for acetic acid (1.74×10^{-5}), we can use Equation 16.13 to calculate $[H^+]$ at equilibrium:

$$K_a = \frac{[CH_3CO_2^-][H^+]}{[CH_3CO_2H]}$$

$$[H^+] = \frac{K_a[CH_3CO_2H]}{[CH_3CO_2^-]} = \frac{(1.74 \times 10^{-5})(7.27 \times 10^{-2}\ M)}{1.82 \times 10^{-2}} = 6.95 \times 10^{-5}\ M$$

Calculating $-\log[H^+]$ gives pH $= -\log(6.95 \times 10^{-5}) = 4.16$.

Comparing the titration curves for HCl and acetic acid in Figure 16.10a, we see that adding the same amount (5.00 mL) of 0.200 *M* NaOH to 50 mL of a 0.10 *M* solution of both acids causes a much smaller pH change for HCl (from 1.00 to 1.14) than for acetic acid (2.88 to 4.16). This is consistent with the qualitative description of the shapes of the titration curves at the beginning of this section. In the next example, we calculate another point for constructing the titration curve of acetic acid.

EXAMPLE 16.12

What is the pH of the solution after 25.00 mL of 0.200 *M* NaOH is added to 50.00 mL of 0.100 *M* acetic acid?

Given Volume and molarity of base and acid

Asked for pH

Strategy

Ⓐ Write the balanced chemical equation for the reaction. Then calculate the initial numbers of millimoles of OH^- and CH_3CO_2H. Determine which species, if either, is present in excess.
Ⓑ Tabulate the results showing initial numbers, changes, and final numbers of millimoles.
Ⓒ If excess acetate is present after the reaction with OH^-, write the equation for the reaction of acetate with water. Use a tabular format to obtain the concentrations of all species present.
Ⓓ Calculate K_b using the relationship $K_w = K_a K_b$ (Equation 16.20). Calculate $[OH^-]$, and use this to calculate the pH of the solution.

Solution

Ⓐ Ignoring the spectator ion, Na^+, the equation for this reaction is

$$CH_3CO_2H(aq) + OH^-(aq) \rightarrow CH_3CO_2^-(aq) + H_2O(l)$$

The initial numbers of millimoles of OH^- and CH_3CO_2H are

$$25.00\ \cancel{mL}\left(\frac{0.200\text{ mmol }OH^-}{\cancel{mL}}\right) = 5.00\text{ mmol }OH^-$$

$$50.00\ \cancel{mL}\left(\frac{0.100\text{ mmol }CH_3CO_2H}{\cancel{mL}}\right) = 5.00\text{ mmol }CH_3CO_2H$$

The number of millimoles of OH^- equals the number of millimoles of CH_3CO_2H, so neither species is present in excess.
Ⓑ Because the number of millimoles of OH^- added corresponds to the number of millimoles of acetic acid in solution, this is the equivalence point. The results of the neutralization reaction can be summarized in tabular form.

	$CH_3CO_2H(aq)$	+	$OH^-(aq)$	⟶	$CH_3CO_2^-(aq) + H_2O(l)$
Initial	5.00 mmol		5.00 mmol		0 mmol
Change	− 5.00 mmol		− 5.00 mmol		+ 5.00 mmol
Final	0 mmol		0 mmol		5.00 mmol

Ⓒ Because the product of the neutralization reaction is a weak base, we must consider the reaction of the weak base *with water* in order to calculate $[H^+]$ at equilibrium and thus the final pH of the solution. The initial concentration of acetate is obtained from the neutralization reaction:

$$[CH_3CO_2] = \frac{5.00\text{ mmol }CH_3CO_2}{(50.00 + 25.00)\text{ mL}} = 6.67 \times 10^{-2}\,M$$

The equilibrium reaction of acetate with water is

$$CH_3CO_2^-(aq) + H_2O(l) \rightleftharpoons CH_3CO_2H(aq) + OH^-(aq)$$

The equilibrium constant for this reaction is $K_b = K_w/K_a$, where K_a is the acid ionization constant of acetic acid. We therefore define x as $[OH^-]$ produced by the reaction of acetate with water. Here is the completed table of concentrations:

$H_2O(l)$ +	$CH_3CO_2^-(aq)$ $\rightleftharpoons$	$CH_3CO_2H(aq)$ +	$OH^-(aq)$
	$[CH_3CO_2^-]$	$[CH_3CO_2H]$	$[OH^-]$
Initial	0.0667	0	1.00×10^{-7}
Change	$-x$	$+x$	$+x$
Final	$(0.0667 - x)$	x	x

Substituting the expressions for the final values from this table into Equation 16.16 gives

$$K_b = \frac{[CH_3CO_2H][OH^-]}{[CH_3CO_2^-]} = \frac{(x)(x)}{0.0667 - x} \approx \frac{x^2}{0.0667}$$

We can obtain the value of K_b by rearranging Equation 16.20 and substituting the known values:

$$K_b = \frac{K_w}{K_a} = \frac{1.01 \times 10^{-14}}{1.74 \times 10^{-5}} = 5.80 \times 10^{-10} = \frac{x^2}{0.0667}$$

which we can solve to get $x = 6.22 \times 10^{-6}$. Thus, $[OH^-] = 6.22 \times 10^{-6}$ M, and the pH of the final solution is 8.79 (Figure 16.10a). As expected for the titration of a weak acid, the pH at the equivalence point is greater than 7 because the product of the titration is a base, the acetate ion, which then reacts with water to produce OH^-.

EXERCISE 16.12

Calculate the pH of a solution prepared by adding 45.0 mL of a 0.213 *M* HCl solution to 125.0 mL of a 0.150 *M* solution of ammonia. The pK_b of ammonia is 4.75 at 25°C.

Answer 9.23

As shown in Figure 16.10b, the titration curve for NH_3, a weak base, is the reverse of the titration curve for acetic acid. In particular, note that the pH at the equivalence point in the titration of a weak base is *less* than 7.00 because the titration produces an acid.

The identity of the weak acid or base being titrated strongly affects the shape of the titration curve. Figure 16.11 illustrates the shape of titration curves as a function of the pK_a or pK_b. As the acid or base being titrated becomes weaker (its pK_a or pK_b becomes larger), the pH change around the equivalence point decreases significantly.

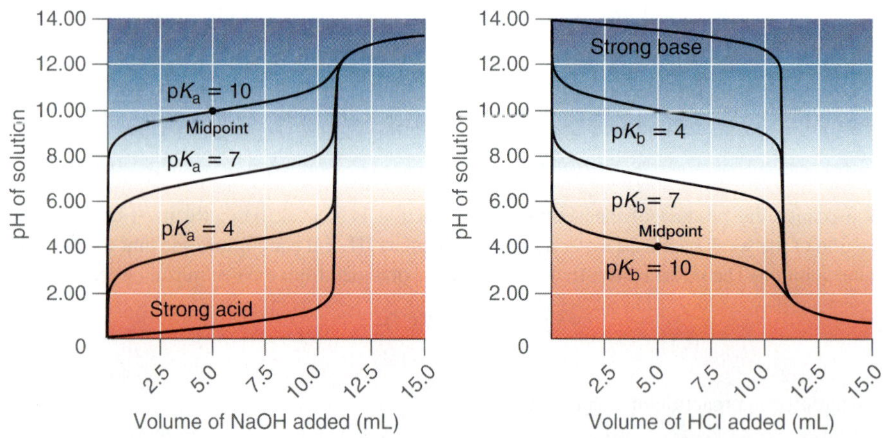

(a) Weak acids titrated with strong base **(b) Weak bases titrated with strong acid**

Figure 16.11 Effect of acid or base strength on the shape of titration curves. Unlike strong acids or bases, the shape of the titration curve for a weak acid or base depends on the pK_a or pK_b of the weak acid or base being titrated. **(a)** Solution pH as function of the volume of 1.00 *M* NaOH added to 10.00 mL of 1.00 *M* solutions of weak acids with the indicated pK_a values. **(b)** Solution pH as a function of the volume of 1.00 *M* HCl added to 10.00 mL of 1.00 *M* solutions of weak bases with the indicated pK_b values. Note that the shapes of the two sets of curves are essentially identical, but one is flipped vertically in relation to the other. Midpoints are indicated for the titration curves corresponding to $pK_a = 10$ and $pK_b = 10$.

With very dilute solutions, the curve becomes so shallow that it can no longer be used to determine the equivalence point.

One point in the titration of a weak acid or a weak base is particularly important: The **midpoint** of a titration is defined as the point at which exactly enough acid (or base) has been added to neutralize one-half of the acid (or base) originally present, and occurs halfway to the equivalence point. The midpoint is indicated in Figure 16.11a and b for the two shallowest curves. By definition, at the midpoint of the titration of an acid, [HA] = [A^-]. Recall from Equation 16.13 that the ionization constant for a weak acid is

$$K_a = \frac{[H_3O^+][A^-]}{[HA]}$$

If [HA] = [A^-], this reduces to $K_a = [H_3O^+]$. Taking the negative logarithm of both sides gives

$$-\log K_a = -\log[H_3O^+]$$

From the definitions of pK_a and pH, we see that this is identical to

$$pK_a = pH \qquad (16.51)$$

Thus, *the pH at the midpoint of the titration of a weak acid is equal to the* pK_a *of the weak acid*, as indicated in Figure 16.11a for the weakest acid where we see that the midpoint for $pK_a = 10$ occurs at pH = 10. Titration methods can therefore be used to determine both the concentration *and* the pK_a (or pK_b) of a weak acid (or base).

Note the pattern

The pH at the midpoint of the titration of a weak acid is equal to the pK_a *of the weak acid.*

Titrations of Polyprotic Acids or Bases

When a strong base is added to a solution of a polyprotic acid, the neutralization reaction occurs in stages. The most acidic group is titrated first, followed by the next most acidic, and so forth. If the pK_a values are separated by at least three pK_a units, then the overall titration curve shows well-resolved "steps" corresponding to the titration of each proton. A titration of H_3PO_4, a triprotic acid, with NaOH is illustrated in Figure 16.12, which shows two well-defined steps: the first midpoint corresponds to pK_{a1}, and the second midpoint to pK_{a2}. Because HPO_4^{2-} is such a weak acid, pK_{a3} has such a high value that the third step cannot be resolved using 0.10 *M* NaOH as the titrant.

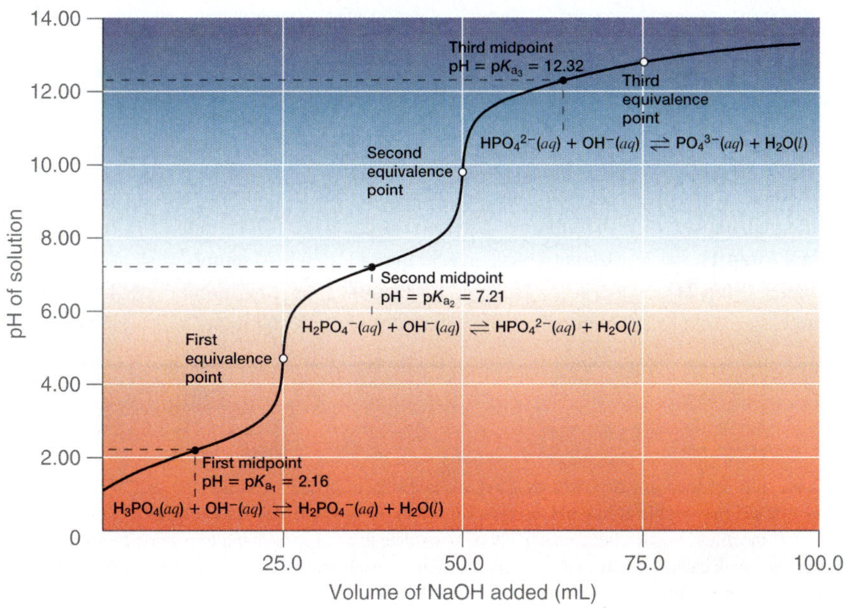

Figure 16.12 Titration curve for phosphoric acid (H_3PO_4), a typical polyprotic acid. Shown here is the curve for the titration of 25.0 mL of a 0.100 *M* H_3PO_4 solution with 0.100 *M* NaOH along with the species in solution at each K_a. Note the two distinct equivalence points corresponding to deprotonation of H_3PO_4 at pH ≈ 4.6 and $H_2PO_4^-$ at pH ≈ 9.8. Because HPO_4^{2-} is a very weak acid, the third equivalence point, at pH ≈13, is not well defined.

The titration curve for the reaction of a polyprotic base with a strong acid is the mirror image of the curve shown in Figure 16.12. The initial pH is high, but as acid is added, the pH decreases in steps if the successive pK_b values are well separated. Table 16.4 lists the ionization constants and pK_a values for some common polyprotic acids and bases.

EXAMPLE 16.13

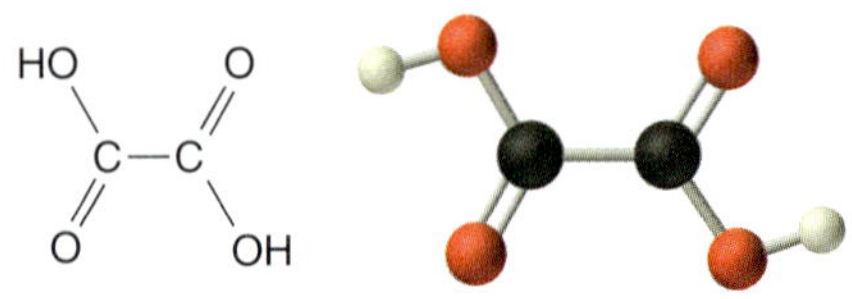

Oxalic acid

Calculate the pH of a solution prepared by adding 55.0 mL of a 0.120 *M* NaOH solution to 100.0 mL of a 0.0510 *M* solution of oxalic acid (HO_2CCO_2H), a diprotic acid (abbreviated as H_2ox). Oxalic acid, the simplest dicarboxylic acid, is found in rhubarb and many other plants. Rhubarb leaves are toxic because they contain the calcium salt of the fully deprotonated form of oxalic acid, the oxalate ion ($^-O_2CCO_2^-$, abbreviated ox^{2-}).*

Given Volume and concentration of acid and base

Asked for pH

Strategy

Ⓐ Calculate the initial numbers of millimoles of the acid and base. Use a tabular format to determine the amounts of all species in solution.

Ⓑ Calculate the concentrations of all species in the final solution. Use Equation 16.14 to determine $[H^+]$, and convert this value to pH.

Solution

Ⓐ Table 16.4 gives the pK_a values of oxalic acid as 1.25 and 3.81. Again we proceed by determining the numbers of millimoles of acid and base initially present:

$$100.0\ \text{mL}\left(\frac{0.0510\ \text{mmol H}_2\text{ox}}{\text{mL}}\right) = 5.10\ \text{mmol H}_2\text{ox}$$

$$55.0\ \text{mL}\left(\frac{0.120\ \text{mmol NaOH}}{\text{mL}}\right) = 6.60\ \text{mmol NaOH}$$

The strongest acid (H_2ox) reacts with the base first. This leaves $(6.60 - 5.10) = 1.50$ mmol of OH^- to react with Hox^-, forming ox^{2-} and H_2O. The reactions can be written as

$$\underset{5.10\ \text{mmol}}{H_2ox} + \underset{6.60\ \text{mmol}}{OH^-} \longrightarrow \underset{5.10\ \text{mmol}}{Hox^-} + \underset{5.10\ \text{mmol}}{H_2O}$$

$$\underset{5.10\ \text{mmol}}{Hox^-} + \underset{1.50\ \text{mmol}}{OH^-} \longrightarrow \underset{1.50\ \text{mmol}}{ox^{2-}} + \underset{1.50\ \text{mmol}}{H_2O}$$

In tabular form,

	H_2ox	OH^-	Hox^-	ox^{2-}
Initial	5.10 mmol	6.60 mmol	0 mmol	0 mmol
Change (Step 1)	− 5.10 mmol	− 5.10 mmol	+ 5.10 mmol	0 mmol
Final (Step 1)	0 mmol	1.50 mmol	5.10 mmol	0 mmol
Change (Step 2)	—	− 1.50 mmol	− 1.50 mmol	+ 1.50 mmol
Final	0 mmol	0 mmol	3.60 mmol	1.50 mmol

* Oxalate salts are toxic for two reasons. First, oxalate salts of divalent cations such as Ca^{2+} are insoluble at neutral pH but soluble at low pH, as we shall see in the next chapter. As a result, calcium oxalate dissolves in the dilute acid of the stomach, allowing oxalate to be absorbed and transported into cells, where it can react with calcium to form tiny calcium oxalate crystals that damage tissues. Second, oxalate forms stable complexes with metal ions, which can alter their normal distribution in biological fluids.

The equilibrium between the weak acid (Hox^-) and its conjugate base (ox^{2-}) in the final solution is determined by the magnitude of the second ionization constant, $K_{a2} = 10^{-3.81} = 1.55 \times 10^{-4}$. To calculate the pH of the solution, we need to know $[H^+]$, which is determined using exactly the same method as in the acetic acid titration in Example 16.12:

$$\text{final volume of solution} = 100.0 \text{ mL} + 55.0 \text{ mL} = 155.0 \text{ mL}$$

Thus, the concentrations of Hox^- and ox^{2-} are

$$[Hox^-] = \frac{3.60 \text{ mmol Hox}^-}{155.0 \text{ mL}} = 2.32 \times 10^{-2}\,M$$

$$[ox^{2-}] = \frac{1.50 \text{ mmol ox}^{2-}}{155.0 \text{ mL}} = 9.68 \times 10^{-3}\,M$$

We can now calculate $[H^+]$ at equilibrium using the equation

$$K_{a2} = \frac{[ox^{2-}][H^+]}{[Hox^-]}$$

Rearranging this equation and substituting the values for the concentrations of Hox^- and ox^{2-} give

$$[H^+] = \frac{K_{a2}\,[Hox^-]}{[ox^{2-}]} = \frac{(1.55 \times 10^{-4})(2.32 \times 10^{-2})}{9.68 \times 10^{-3}} = 3.71 \times 10^{-4}\,M$$

Thus,

$$pH = -\log[H^+] = -\log(3.71 \times 10^{-4}) = 3.43$$

This answer makes chemical sense because the pH is between the first and second pK_a values of oxalic acid, as it must be. We added enough hydroxide ion to completely titrate the first, more acidic proton (which should give us a pH greater than pK_{a1}), but only enough to titrate less than half of the second, less acidic proton, with pK_{a2}. If we had added exactly enough hydroxide to completely titrate the first proton plus half of the second, we would be at the midpoint of the second step in the titration, and the pH would be 3.81, equal to pK_{a2}.

EXERCISE 16.13

Piperazine is a diprotic base used to control intestinal parasites ("worms") in pets and humans. A dog is given 500 mg (5.80 mmol) of piperazine ($pK_{b1} = 4.27$, $pK_{b2} = 8.67$). If the dog's stomach initially contains 100 mL of 0.10 *M* HCl (pH = 1.00), calculate the pH of the stomach contents after ingestion of the piperazine.

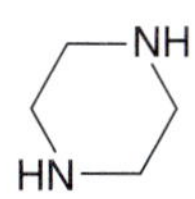

Piperazine

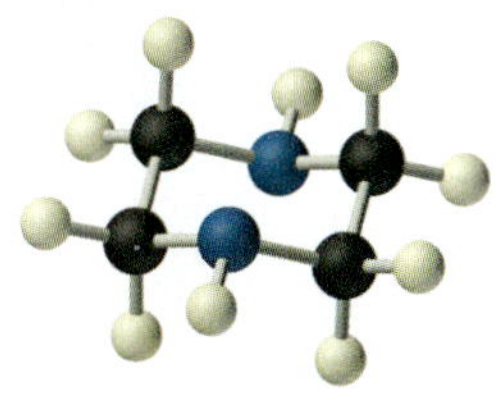

Answer 4.91

Indicators

In practice, most acid–base titrations are not monitored by recording the pH as a function of the amount of the strong acid or base solution used as the titrant. Instead, an **acid–base indicator** is often used that, if carefully selected, undergoes a dramatic color change at the pH corresponding to the equivalence point of the titration. Indicators are weak acids or bases that exhibit intense colors that vary with pH. The conjugate acid and conjugate base of a good indicator have very different colors so that they can be distinguished easily. Some indicators are colorless in the conjugate acid form but intensely colored when deprotonated (phenolphthalein, for example), which makes them particularly useful.

We can describe the chemistry of indicators by the general equation

$$HIn(aq) \rightleftharpoons H^+(aq) + In^-(aq)$$

Figure 16.13 Naturally occurring pH indicators in red cabbage juice. Red cabbage juice contains a mixture of substances whose color depends on the pH. Each test tube contains a solution of red cabbage juice in water, but the pH of the solutions vary from pH = 2.0 (far left) to pH = 11.0 (far right). At pH = 7.0, the solution is blue.

where the protonated form is designated by HIn and the conjugate base by In^-. The ionization constant for the deprotonation of indicator HIn is

$$K_{in} = \frac{[H^+][In^-]}{[HIn]} \tag{16.52}$$

The value of pK_{in} (its pK_a) determines the pH at which the indicator changes color.

Many different substances can be used as indicators, depending on the particular reaction to be monitored. For example, red cabbage juice contains a mixture of colored substances that change from deep red at low pH to light blue at intermediate pH's to yellow at high pH (Figure 16.13). In all cases, though, a good indicator must have these properties:

1. The color change must be easily detected.
2. The color change must be rapid.
3. The indicator molecule must not react with the substance being titrated.
4. To minimize errors, the indicator should have a pK_{in} that is within one pH unit of the expected pH at the equivalence point of the titration.

Synthetic indicators have been developed that meet these criteria and cover virtually the entire pH range. Figure 16.14 shows the approximate pH range over which some common indicators change color and their change in color. Some indicators are colorless when protonated but intensely colored when deprotonated, which makes the color change particularly easy to detect. One common example is phenolphthalein, which changes from colorless to pink at a pH of 9.5. In addition, some indicators (such as thymol blue) are polyprotic acids or bases, which change color twice at widely separated pH values.

It is important to be aware that an indicator does not change color abruptly at a particular pH value; instead, it actually undergoes a pH titration just like any other acid or base. As the concentration of HIn decreases and the concentration of In^- increases, the color of the solution slowly changes from the characteristic color of HIn to that of In^-. As we will see in the next section, the $[In^-]/[HIn]$ ratio changes from a value of 0.1 at a pH one unit *below* pK_{in} to a value of 10 at a pH one unit *above* pK_{in}. Thus, most indicators change color over a pH range of about two pH units.

How do you select the right indicator for the titration of a particular acid or base? As we stated above, the indicator should have a pK_{in} value that is close to the expected pH at the equivalence point. For a strong acid–strong base titration, the choice of indicator is not especially critical due to the very large change in pH that occurs around the equivalence point. In contrast, using the wrong indicator for a titration of a weak acid or a weak base can result in relatively large errors, as illustrated in

Indicator	pK_{in}	Color change
Alizarin yellow R	11.0	yellow to red
Thymolphthalein	9.9	colorless to blue
Phenolphthalein	9.5	colorless to pink
Thymol blue	9.2	yellow to blue
Phenol red	7.4	yellow to red
Bromothymol blue	7.3	yellow to blue
Bromocresol purple	6.4	yellow to purple
Methyl red	5.0	red to yellow
Resorcin blue	5.3	red to blue
Ethyl red	5.4	colorless to red
Resazurin	5.1	orange to violet
Bromophenol blue	4.1	yellow to blue
Thymol blue	1.7	red to yellow
Malachite green	1.3	yellow to turquoise
Methyl violet	0.8	yellow to blue

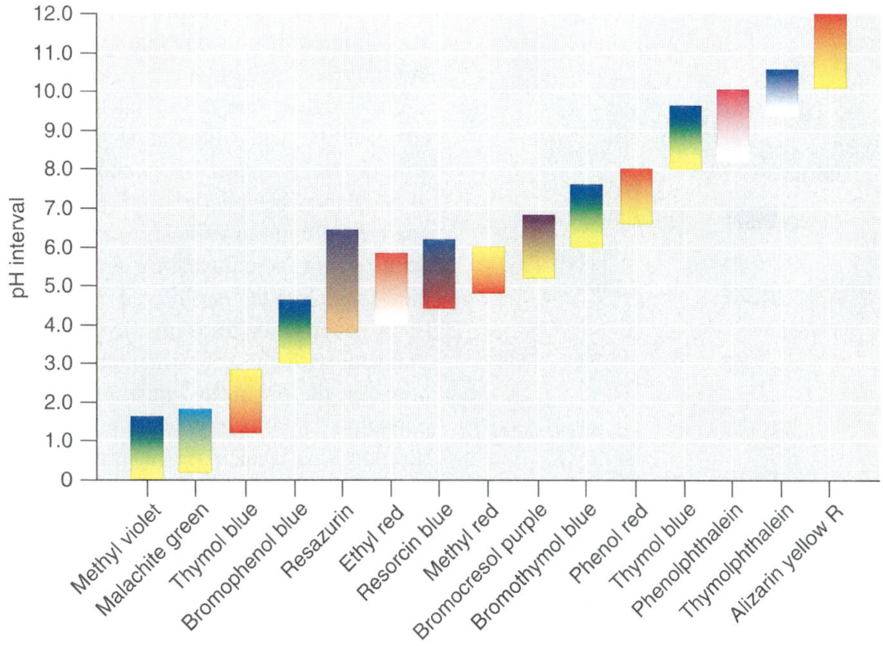

Figure 16.14 Some common acid–base indicators. Approximate colors are shown, along with values of pK_{in} and the pH range over which the color changes.

Figure 16.15. This figure shows plots of pH versus volume of base added for the titration of 50.0 mL of a 0.100 M solution of a strong acid (HCl) and a weak acid (acetic acid) with 0.100 M NaOH. The pH ranges over which two common indicators (methyl red, pK_{in} = 5.0, and phenolphthalein, pK_{in} = 9.5) change color are also shown. The horizontal bars indicating the pH ranges over which both indicators change color cross the HCl titration curve where it is almost vertical. Hence, both indicators change color when essentially the same volume of NaOH has been added (about 50 mL), which

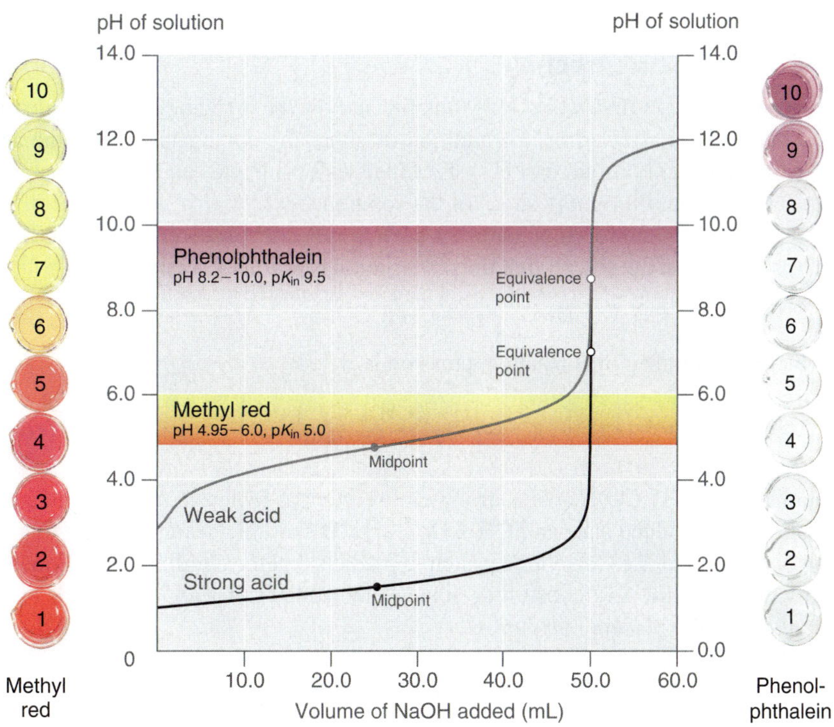

Figure 16.15 Choosing the correct indicator for an acid–base titration. The graph shows the results obtained using two indicators (methyl red and phenolphthalein) for the titration of 0.100 M solutions of a strong acid (HCl) and a weak acid (acetic acid) with 0.100 M NaOH. Due to the steepness of the titration curve of a strong acid around the equivalence point, either indicator will rapidly change color at the equivalence point for the titration of the strong acid. In contrast, the pK_{in} for methyl red (5.0) is very close to the pK_a of acetic acid (4.76); hence, the midpoint of the color change for methyl red occurs near the midpoint of the titration, rather than at the equivalence point.

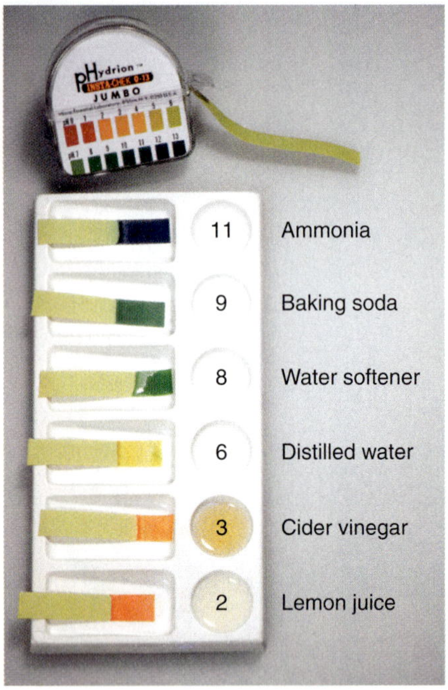

Figure 16.16 pH paper. pH paper contains a set of indicators that change color at different pH values. The approximate pH of a solution can be determined by simply dipping a paper strip into the solution and comparing the color to the standards provided.

corresponds to the equivalence point. In contrast, the titration of acetic acid will give very different results depending on whether methyl red or phenolphthalein is used as the indicator. Although the pH range over which phenolphthalein changes color is slightly higher than the pH at the equivalence point of the strong acid titration, the error will be negligible due to the slope of this portion of the titration curve. Just as with the HCl titration, the phenolphthalein indicator will turn pink when about 50 mL of NaOH has been added to the acetic acid solution. In contrast, methyl red begins to change from red to yellow around pH 5, which is near the *midpoint* of the acetic acid titration, not the equivalence point. Addition of only about 25–30 mL of NaOH will therefore cause the methyl red indicator to change color, resulting in a huge error.

In general, for titrations of strong acids with strong bases (and vice versa), any indicator with a pK_{in} between about 4 and 10 will do. For the titration of a weak acid, however, the pH at the equivalence point is greater than 7, and an indicator such as phenolphthalein or thymol blue, with $pK_{in} > 7$, should be used. Conversely, for the titration of a weak base, where the pH at the equivalence point is less than 7, an indicator such as methyl red or bromocresol blue, with $pK_{in} < 7$, should be used.

The existence of many different indicators with different colors and pK_{in} values also provides a convenient way to estimate the pH of a solution without using an expensive electronic pH meter and a fragile pH electrode. Paper or plastic strips that contain combinations of indicators are used as "pH paper," which allows you to estimate the pH of a solution by simply dipping a piece of pH paper into it and comparing the resulting color with the standards printed on the container (Figure 16.16).

16.6 Buffers

Buffers are solutions that maintain a relatively constant pH when an acid or a base is added. They therefore protect, or "buffer," other molecules in solution from the effects of the added acid or base. Buffers contain either a weak acid (HA) and its conjugate base (A^-) or a weak base (B) and its conjugate acid (BH^+), and they are critically important for the proper functioning of biological systems. In fact, every biological fluid is buffered to maintain its physiological pH.

The Common Ion Effect

To understand how buffers work, let's look first at how the ionization equilibrium of a weak acid (HA) is affected by the addition of either the conjugate base of the acid (A^-) or a strong acid (a source of H^+). Le Châtelier's principle can be used to predict the effect on the equilibrium position of the solution.

A typical buffer used in biochemistry laboratories contains acetic acid and a salt such as sodium acetate. Recall that the dissociation reaction of acetic acid is

$$CH_3CO_2H(aq) \rightleftharpoons CH_3CO_2^-(aq) + H^+(aq) \tag{16.53}$$

for which the equilibrium constant expression is

$$K_a = \frac{[H^+][CH_3CO_2^-]}{[CH_3CO_2H]} \tag{16.54}$$

Sodium acetate (CH_3CO_2Na) is a strong electrolyte that ionizes completely in aqueous solution to produce Na^+ and $CH_3CO_2^-$ ions. If sodium acetate is added to a solution of acetic acid, Le Châtelier's principle predicts that the equilibrium in Equation 16.53 will shift to the left, consuming some of the added $CH_3CO_2^-$ and some of the H^+ ions originally present in solution:

$$\underleftarrow{CH_3CO_2H(aq) \rightleftharpoons CH_3CO_2^-(aq) + H^+(aq)}$$
$$+\ CH_3CO_2^-$$

(Because Na^+ is a spectator ion, it has no effect on the position of the equilibrium and can be ignored.) The addition of sodium acetate therefore produces a new equilibrium composition, in which the value of $[H^+]$ is lower than the initial value. Since $[H^+]$ has decreased, the pH will be higher. Thus, adding a salt of the conjugate base to a solution of a weak acid increases the pH. This makes sense because sodium acetate is a base, and adding *any* base to a solution of a weak acid should increase the pH.

If we instead add a strong acid such as HCl to the system, $[H^+]$ increases. Once again the equilibrium is temporarily disturbed, but the excess H^+ ions react with the conjugate base ($CH_3CO_2^-$), whether from the parent acid or sodium acetate, to drive the equilibrium to the left. The net result is a new equilibrium composition that has a lower $[CH_3CO_2^-]$ than before. In both cases, *only the equilibrium composition has changed; the ionization constant* K_a *for acetic acid remains the same.* Adding a strong electrolyte that contains one ion in common with a reaction system that is at equilibrium, in this case $CH_3CO_2^-$, will therefore shift the equilibrium in the direction that reduces the concentration of the common ion. The shift in equilibrium is called the **common ion effect**.

EXAMPLE 16.14

In Section 16.4, we calculated that a 0.150 *M* solution of formic acid at 25°C ($pK_a = 3.75$) has a pH of 2.28 and is 3.5% ionized. **(a)** Is there a change to the pH of the solution if enough solid sodium formate is added to make the final formate concentration 0.100 *M* (assume that the formic acid concentration does not change)? **(b)** What percentage of the formic acid is ionized if 0.200 *M* HCl is added to the system?

Given Solution concentration and pH, pK_a, and percent ionization of acid; final concentration of conjugate base or strong acid added

Asked for pH and percent ionization of formic acid

Strategy

A Write a balanced equilibrium equation for the ionization equilibrium of formic acid. Tabulate the initial concentrations, changes, and final concentrations.

B Substitute the expressions for the final concentrations into the expression for K_a. Calculate $[H^+]$ and the pH of the solution.

C Construct a table of concentrations for the dissociation of the formic acid. To determine the percent ionization, determine the anion concentration, divide it by the initial concentration of formic acid, and multiply the result by 100.

Solution

(a) **A** Because sodium formate is a strong electrolyte, it ionizes completely in solution to give formate and sodium ions. The Na^+ ions are spectator ions, so they can be ignored in the equilibrium equation. Because water is both a much weaker acid than formic acid and a much weaker base than formate, the acid–base properties of the solution are determined solely by the formic acid ionization equilibrium:

$$HCO_2H(aq) \rightleftharpoons HCO_2^-(aq) + H^+(aq)$$

The initial concentrations, the changes in concentration that occur as equilibrium is reached, and the final concentrations can be tabulated.

	$HCO_2H(aq)$	$\rightleftharpoons$	$H^+(aq)$	+	$HCO_2^-(aq)$
	$[HCO_2H]$		$[H^+]$		$[HCO_2^-]$
Initial	0.150		1.00×10^{-7}		0.100
Change	$-x$		$+x$		$+x$
Final	$(0.150 - x)$		x		$(0.100 + x)$

Substituting the expressions for the final concentrations into the equilibrium constant expression and making our usual simplifying assumptions give

$$K_a = \frac{[H^+][HCO_2^-]}{[HCO_2H]} = \frac{(x)(0.100 + x)}{0.150 - x} = \frac{x(0.100)}{0.150} = 10^{-3.75} = 1.8 \times 10^{-4}$$

Rearranging and solving for x give

$$x = (1.8 \times 10^{-4}) \times \frac{0.150\ M}{0.100\ M} = 2.7 \times 10^{-4} = [H^+]$$

The value of x is small compared with 0.150 or 0.100 M, so our assumption about the extent of ionization is justified. Moreover, $K_aC_{HA} = (1.8 \times 10^{-4})(0.150) = 2.7 \times 10^{-5}$, which is greater than 1.0×10^{-6}, so again, our assumption is justified. The final pH is $-\log(2.7 \times 10^{-4}) =$ 3.57, compared with the initial value of 2.29. Thus, adding a salt containing the conjugate base of the acid has increased the pH of the solution, as we expect based on Le Châtelier's principle; the stress on the system has been relieved by the consumption of H^+ ions, driving the equilibrium to the left.

(b) Because HCl is a strong acid, it ionizes completely, and chloride is a spectator ion that can be neglected. Thus, the only relevant acid–base equilibrium is again the dissociation of formic acid, and initially the concentration of formate is zero. We can construct a table of initial concentrations, changes in concentration, and final concentrations.

	$HCO_2H(aq)$	$\rightleftharpoons$	$H^+(aq)$	+	$HCO_2^-(aq)$
	$[HCO_2H]$		$[H^+]$		$[HCO_2^-]$
Initial	0.150		0.200		0
Change	$-x$		$+x$		$+x$
Final	$(0.150 - x)$		$(0.200 + x)$		x

To calculate the percentage of formic acid that is ionized under these conditions, we have to determine the final $[HCO_2^-]$. Substituting final concentrations into the equilibrium constant expression and making the usual simplifying assumptions give

$$K_a = \frac{[H^+][HCO_2^-]}{[HCO_2H]} = \frac{(0.200 + x)(x)}{0.150 - x} = \frac{x(0.200)}{0.150} = 1.80 \times 10^{-4}$$

Rearranging and solving for x give

$$x = (1.80 \times 10^{-4}) \times \frac{0.150\ M}{0.200\ M} = 1.35 \times 10^{-4} = [HCO_2^-]$$

Once again, our simplifying assumptions are justified. The percent ionization of formic acid is

$$\text{percent ionization} = \frac{1.35 \times 10^{-4}\ M}{0.150\ M} \times 100 = 0.090\%$$

Adding the strong acid to the solution, as shown in the table, decreased the percent ionization of formic acid by a factor of approximately 38 (3.45%/0.090%). Again, this is consistent with Le Châtelier's principle: adding H^+ ions drives the dissociation equlibrium to the left.

EXERCISE 16.14

As you learned in Example 16.8, a 0.225 M solution of ethylamine ($CH_3CH_2NH_2$, $pK_b = 3.19$) has a pH of 12.08 and a percent ionization of 5.4% at 20°C. Calculate **(a)** the pH of the solution if enough solid ethylamine hydrochloride ($EtNH_3Cl$) is added to make the solution 0.100 M in $EtNH_3^+$ and **(b)** the percentage of ethylamine that is ionized if enough solid NaOH is added to the original solution to give a final concentration of 0.050 M NaOH.

Answer **(a)** 11.16; **(b)** 1.3%

Now let's suppose we have a buffer solution that contains equimolar concentrations of a weak base (B) and its conjugate acid (BH^+). The general equation for the ionization of a weak base is

$$B(aq) + H_2O(l) \rightleftharpoons BH^+(aq) + OH^-(aq) \tag{16.55}$$

If the equilibrium constant for the reaction as written in Equation 16.55 is small, for example $K_b = 10^{-5}$, then the equilibrium constant for the reverse reaction is very large: $K = 1/K_b = 10^5$. Adding a strong base such as OH^- to the solution therefore causes the equilibrium in Equation 16.55 to shift to the left, consuming the added OH^-. As a result, the hydroxide ion concentration in solution remains relatively constant, and the pH of the solution changes very little. Le Châtelier's principle predicts the same outcome: when the system is stressed by an increase in the OH^- concentration, the reaction will proceed to the left to counteract the stress.

If the pK_b of the base (B) is 5, the pK_a of its conjugate acid (BH^+) is $pK_a = pK_w - pK_b = 14 - 5 = 9$. Thus, the equilibrium constant for ionization of the conjugate acid (BH^+) is even smaller than that for ionization of the base (B). The ionization reaction for the conjugate acid (BH^+) of a weak base is written

$$BH^+(aq) + H_2O(l) \rightleftharpoons B(aq) + H_3O^+(aq) \tag{16.56}$$

Again, the equilibrium constant for the reverse of this reaction is very large: $K = 1/K_a = 10^9$. Thus, if a strong acid is added, it is neutralized by reaction with B as the reaction in Equation 16.56 shifts to the left. As a result, the H^+ ion concentration does not increase very much, and the pH changes only slightly. In effect, a buffer solution behaves somewhat like a sponge that can absorb H^+ and OH^- ions, thereby preventing large changes in pH when appreciable amounts of strong acid or base are added to a solution.

Buffers are characterized by the *pH range* over which they can maintain a more or less constant pH, and by their *buffer capacity*, the amount of strong acid or base that can be absorbed before the pH changes significantly. Although the useful pH range of a buffer depends strongly on the chemical properties of the weak acid and weak base used to prepare the buffer (that is, on K), its buffer capacity depends solely on the concentrations of the species in the buffered solution. The more concentrated the buffer solution, the greater its buffer capacity. As illustrated in Figure 16.17, when NaOH is added to solutions that contain different concentrations of an acetic acid/sodium acetate buffer, the observed change in the pH of the buffer is inversely proportional to the concentration of the buffer. If the buffer capacity is ten times larger, then the buffer solution can absorb ten times more strong acid or base before undergoing a significant change in pH.

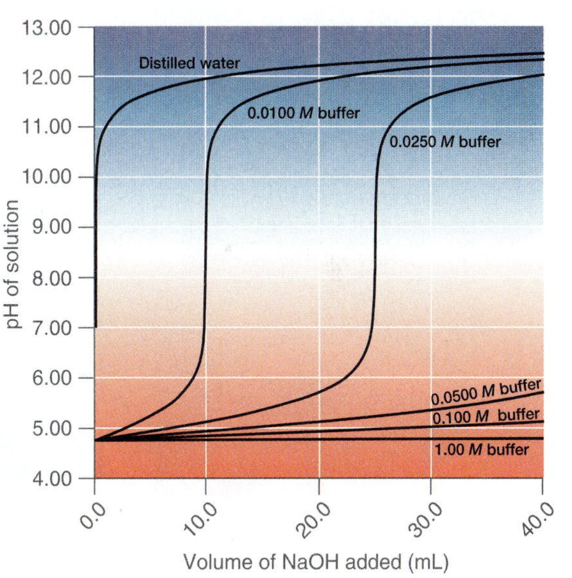

Figure 16.17 Effect of buffer concentration on the capacity of a buffer. A buffer maintains a relatively constant pH when acid or base is added to a solution. The addition of even tiny volumes of 0.10 *M* NaOH to 100.0 mL of distilled water results in a very large change in pH. As the concentration of a 50:50 mixture of sodium acetate/acetic acid buffer in the solution is increased from 0.010 *M* to 1.00 *M*, the change in the pH produced by the addition of the same volume of NaOH solution decreases steadily. For buffer concentrations of at least 0.500 *M*, the addition of even 25 mL of the NaOH solution results in only a relatively small change in pH.

Calculating the pH of a Buffer

The pH of a buffer can be calculated from the concentrations of the weak acid and the weak base used to prepare it, the concentration of the conjugate base and acid, and the pK_a or pK_b of the weak acid or base. The procedure is analogous to that used in Example 16.14 to calculate the pH of a solution containing known concentrations of formic acid and formate.

An alternative method frequently used to calculate the pH of a buffer solution is based on a rearrangement of the equilibrium equation for the dissociation of a weak acid. The simplified ionization reaction is $HA \rightleftharpoons H^+ + A^-$, for which the equilibrium constant expression is

$$K_a = \frac{[H^+][A^-]}{[HA]} \tag{16.57}$$

This equation can be rearranged to give

$$[H^+] = \frac{K_a[HA]}{[A^-]} \tag{16.58}$$

Taking the logarithm of both sides and multiplying both sides by −1 give

$$-\log[H^+] = -\log K_a - \log\left(\frac{[HA]}{[A^-]}\right) = -\log K_a + \log\left(\frac{[A^-]}{[HA]}\right) \tag{16.59}$$

Replacing the negative logarithms in Equation 16.59 gives

$$pH = pK_a + \log\left(\frac{[A^-]}{[HA]}\right) \tag{16.60}$$

or, more generally,

$$pH = pK_a + \log\left(\frac{[base]}{[acid]}\right) \tag{16.61}$$

Equations 16.60 and 16.61 are both forms of the **Henderson–Hasselbalch equation**, named after the two early-20th-century chemists who first noticed that this rearranged version of the equilibrium constant expression provides an easy way to calculate the pH of a buffer solution. In general, the validity of the Henderson–Hasselbalch equation may be limited to solutions whose concentrations are at least a hundred times greater than the value of their K_a's.

There are three special cases where the Henderson–Hasselbalch equation is easily interpreted without the need for calculations:

Note the pattern

If [base] = [acid] for a buffer, then pH = pK_a. Changing this ratio by a factor of 10 either way changes the pH by ±1.

- **[base] = [acid].** Under these conditions, [base]/[acid] = 1 in Equation 16.61. Because log 1 = 0, pH = pK_a, *regardless of the actual concentrations of the acid and base*. Recall from Section 16.5 that this corresponds to the midpoint in the titration of a weak acid or base.
- **[base]/[acid] = 10.** In Equation 16.61, because log 10 = 1, pH = pK_a + 1.
- **[base]/[acid] = 100.** In Equation 16.61, because log 100 = 2, pH = pK_a + 2.

Thus, each time we increase the [base]/[acid] ratio by 10, the pH of the solution increases by one pH unit. Conversely, if the [base]/[acid] ratio is 0.1, then pH = pK_a −1. Each additional factor-of-10 decrease in the [base]/[acid] ratio causes the pH to decrease by one pH unit.

EXAMPLE 16.15

What is the pH of a solution that contains **(a)** 0.135 *M* HCO_2H and 0.215 *M* HCO_2Na (the pK_a of formic acid is 3.75); **(b)** 0.0135 *M* HCO_2H and 0.0215 *M* HCO_2Na; **(c)** 0.119 *M* pyridine and 0.234 *M* pyridine hydrochloride (the pK_b of pyridine is 8.77)?

Given Concentration of acid, conjugate base, and pK_a; concentration of base, conjugate acid, and pK_b

Asked for pH

Strategy

Substitute values into either form of the Henderson–Hasselbalch equation (Equation 16.60 or 16.61) to calculate the pH.

Solution

(a) According to the Henderson–Hasselbalch equation, the pH of a solution that contains both a weak acid and its conjugate base is pH = pK_a + log($[A^-]$/[HA]). Inserting the given values into the equation gives

$$pH = 3.75 + \log\left(\frac{0.215}{0.135}\right) = 3.75 + \log 1.593 = 3.95$$

This result makes sense because the $[A^-]/[HA]$ ratio is between 1 and 10, so the pH of the buffer must be between the pK_a (3.75) and $pK_a + 1$, or 4.75.

(b) This is identical to part (a), except for the concentrations of the acid and the conjugate base, which are ten times lower. Inserting the concentrations into the Henderson–Hasselbalch equation gives

$$pH = 3.75 + \log\left(\frac{0.0215}{0.0135}\right) = 3.75 + \log 1.593 = 3.95$$

This result is identical to the result in part (a), which emphasizes the point that the pH of a buffer depends *only* on the *ratio* of the concentrations of the conjugate base and the acid, *not* on the magnitude of the concentrations. Because the $[A^-]/[HA]$ ratio is the same as in part (a), the pH of the buffer must also be the same (3.95).

(c) In this case, we have a weak base, pyridine (Py), and its conjugate acid, the pyridinium ion (HPy^+). We will therefore use Equation 16.61, the more general form of the Henderson–Hasselbalch equation, in which "base" and "acid" refer to the appropriate members of the conjugate acid–base pair. We are given [base] = [Py] = 0.119 *M* and [acid] = $[HPy^+]$ = 0.234 *M*. We also are given $pK_b = 8.77$ for pyridine, but we need pK_a for the pyridinium ion. Recall from Equation 16.23 that the pK_b of a weak base and the pK_a of its conjugate acid are related: $pK_a + pK_b = pK_w$. Thus, pK_a for the pyridinium ion is $pK_w - pK_b = 14.00 - 8.77 = 5.23$. Substituting this pK_a value into the Henderson–Hasselbalch equation gives

$$pH = pK_a + \log\left(\frac{[\text{base}]}{[\text{acid}]}\right) = 5.23 + \log\left(\frac{0.119}{0.234}\right) = 5.23 + \log 0.509 = 4.94$$

Once again, this result makes sense: the $[B]/[BH^+]$ ratio is about 1/2, which is between 1 and 0.1, so the final pH must be between the pK_a (5.23) and $pK_a - 1$, or 4.23.

EXERCISE 16.15

Calculate the pH of a solution that contains **(a)** 0.333 *M* benzoic acid and 0.252 *M* sodium benzoate, and **(b)** 0.050 *M* trimethylamine and 0.066 *M* trimethylamine hydrochloride. The pK_a of benzoic acid is 4.20, and the pK_b of trimethylamine is also 4.20.

Answer **(a)** 4.08; **(b)** 9.68

The Henderson–Hasselbalch equation can also be used to calculate the pH of a buffer solution after the addition of a given amount of strong acid or base, as demonstrated in the next example.

EXAMPLE 16.16

The buffer solution in Example 16.15 contained 0.135 *M* HCO_2H and 0.215 *M* HCO_2Na and had a pH of 3.95. **(a)** What is the final pH if 5.00 mL of 1.00 *M* HCl is added to 100 mL of this solution? **(b)** What is the final pH if 5.00 mL of 1.00 *M* NaOH is added?

Given Composition and pH of buffer; concentration and volume of added acid or base

Asked for Final pH

Strategy

A Calculate the amounts of formic acid and formate present in the buffer solution using the procedure from Example 16.14. Then calculate the amount of acid or base added.

B Construct a table showing the amounts of all species after the neutralization reaction. Use the final volume of the solution to calculate the concentrations of all species. Finally, substitute the appropriate values into the Henderson–Hasselbalch equation (Equation 16.61) to obtain the pH.

Solution

The added HCl (a strong acid) or NaOH (a strong base) will react completely with formate (a weak base) or formic acid (a weak acid), respectively, to give formic acid or formate and water. We must therefore calculate the amounts of formic acid and formate present after the neutralization reaction.

(a) **A** We begin by calculating the numbers of millimoles of formic acid and formate present in 100 mL of the initial pH 3.95 buffer:

$$100\ \cancel{\text{mL}}\left(\frac{0.135\ \text{mmol HCO}_2\text{H}}{\cancel{\text{mL}}}\right) = 13.5\ \text{mmol HCO}_2\text{H}$$

$$100\ \cancel{\text{mL}}\left(\frac{0.215\ \text{mmol HCO}_2^-}{\cancel{\text{mL}}}\right) = 21.5\ \text{mmol HCO}_2^-$$

The number of millimoles of H^+ in 5.00 mL of 1.00 M HCl is

$$5.00\ \cancel{\text{mL}}\left(\frac{1.00\ \text{mmol H}^+}{\cancel{\text{mL}}}\right) = 5.00\ \text{mmol H}^+$$

B Next, we construct a table of initial amounts, changes in amounts, and final amounts:

	HCO_2^-	+	H^+	→	HCO_2H
Initial	21.5 mmol		5.00 mmol		13.5 mmol
Change	− 5.00 mmol		− 5.00 mmol		+ 5.00 mmol
Final	16.5 mmol		~0 mmol		18.5 mmol

The final amount of H^+ in solution is given as "~0 mmol." For the purposes of the stoichiometry calculation, this is essentially true, but remember that the point of the problem is to calculate the final $[H^+]$ and thus the pH. We now have all the information we need to calculate the pH. We can use either the lengthy procedure of Example 16.14 or the Henderson–Hasselbach equation. Since we have performed *many* equilibrium calculations in this chapter, we'll take the latter approach. The Henderson–Hasselbalch equation requires the concentrations of HCO_2^- and HCO_2H, which can be calculated using the number of millimoles (n) of each and the total volume (V_T). Substituting these values into the Henderson–Hasselbalch equation gives

$$\text{pH} = \text{p}K_a + \log\left(\frac{[\text{HCO}_2^-]}{[\text{HCO}_2\text{H}]}\right) = \text{p}K_a + \log\left(\frac{n_{\text{HCO}_2^-}/V_f}{n_{\text{HCO}_2\text{H}}/V_f}\right) = \text{p}K_a + \log\left(\frac{n_{\text{HCO}_2^-}}{n_{\text{HCO}_2\text{H}}}\right)$$

Notice that because the total volume appears in both the numerator and denominator, it cancels. We therefore need to use only the ratio of the number of millimoles of the conjugate base to the number of millimoles of the weak acid. Thus,

$$\text{pH} = \text{p}K_a + \log\left(\frac{n_{\text{HCO}_2^-}}{n_{\text{HCO}_2\text{H}}}\right) = 3.75 + \log\left(\frac{16.5\ \text{mmol}}{18.5\ \text{mmol}}\right) = 3.75 + \log(0.892) = 3.70$$

Note the pattern

Only the amounts (in moles or millimoles) of the acidic and basic components of the buffer are needed to use the Henderson–Hasselbalch equation, not their concentrations.

Once again, this result makes sense on two levels. First, the addition of HCl has decreased the pH from 3.95, as expected. Second, the ratio of HCO_2^- to HCO_2H is slightly less than 1, so the pH should be between the pK_a and pK_a − 1.

(b) **A** The procedure for solving this part of the problem is exactly the same as that used in part (a). We have already calculated the numbers of millimoles of formic acid and formate in 100 mL of the initial pH 3.95 buffer: 13.5 mmol of HCO_2H and 21.5 mmol of HCO_2^-. The number of millimoles of OH^- in 5.00 mL of 1.00 M NaOH is

$$5.00\ \cancel{\text{mL}}\left(\frac{1.00\ \text{mmol OH}^-}{\cancel{\text{mL}}}\right) = 5.00\ \text{mmol OH}^-$$

B With this information, we can construct a table of initial amounts, changes in amounts, and final amounts.

	HCO_2H	+	OH^-	⟶	$HCO_2^- + H_2O$
Initial	13.5 mmol		5.00 mmol		21.5 mmol
Change	− 5.00 mmol		− 5.00 mmol		+ 5.00 mmol
Final	8.5 mmol		~0 mmol		26.5 mmol

The final amount of OH^- in solution is not actually zero; this is only approximately true based on the stoichiometric calculation. We can calculate the final pH by inserting the numbers of millimoles of both HCO_2^- and HCO_2H into the simplified Henderson–Hasselbalch expression used in part (a):

$$\mathrm{pH} = \mathrm{p}K_a + \log\left(\frac{n_{HCO_2^-}}{n_{HCO_2H}}\right) = 3.75 + \log\left(\frac{26.5\ \text{mmol}}{8.5\ \text{mmol}}\right) = 3.75 + \log(3.12) = 4.24$$

Once again, this result makes chemical sense: the pH has increased, as would be expected upon the addition of a strong base, and the final pH is between the $\mathrm{p}K_a$ and $\mathrm{p}K_a + 1$, as expected for a solution with a HCO_2^-/HCO_2H ratio between 1 and 10.

EXERCISE 16.16

The buffer solution from Example 16.15 contained 0.119 *M* pyridine and 0.234 *M* pyridine hydrochloride and had a pH of 4.94. **(a)** What is the final pH if 12.0 mL of 1.5 *M* NaOH is added to 250 mL of this solution? **(b)** What is the final pH if 12.0 mL of 1.5 *M* HCl is added?

Answer **(a)** 5.30; **(b)** 4.42

Note the pattern

The most effective buffers contain equal concentrations of an acid and its conjugate base.

The results obtained in Example 16.16 and Exercise 16.16 demonstrate how little the pH of a well-chosen buffer solution changes despite the addition of a significant quantity of strong acid or base. Suppose we had added the same amount of HCl or NaOH solution to 100 mL of an *un*buffered solution at pH 3.95 (corresponding to 1.12×10^{-4} *M* HCl). In this case, adding 5.00 mL of 1.00 *M* HCl would lower the final pH to 1.32 instead of 3.70, whereas adding 5.00 mL of 1.00 *M* NaOH would raise the final pH to 12.68 rather than 4.24. (Try verifying these values by doing the calculations yourself.) Thus, the presence of a buffer significantly increases the ability of a solution to maintain an almost constant pH.

A buffer that contains approximately equal amounts of a weak acid and its conjugate base in solution is equally effective at neutralizing either added base or added acid. This is shown in Figure 16.18 for an acetic acid/sodium acetate buffer. Adding a given amount of strong acid shifts the system along the horizontal axis to the left, whereas adding the same amount of strong base shifts the system the same distance to the right. In either case, the change in the ratio of $CH_3CO_2^-$ to CH_3CO_2H from 1:1 reduces the buffer capacity of the solution.

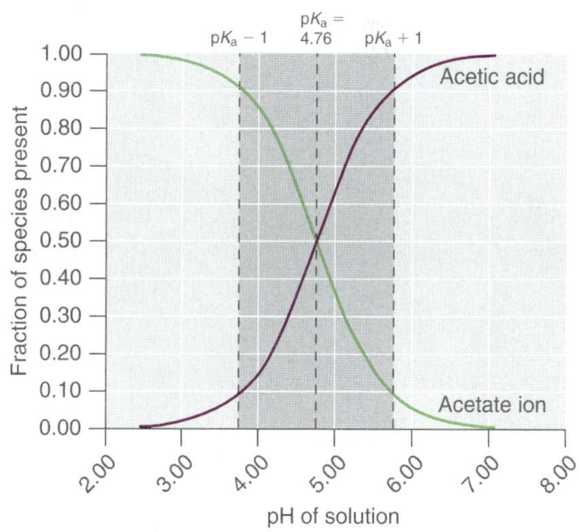

Figure 16.18 Distribution curve showing the fraction of acetic acid molecules and acetate ions as a function of pH in a solution of acetic acid. Note that the pH range over which the acetic acid/sodium acetate system is an effective buffer (the darker shaded region) corresponds to the region in which appreciable concentrations of both species are present (pH 3.76–5.76, corresponding to pH = $\mathrm{p}K_a \pm 1$).

The Relationship Between Titrations and Buffers

There is a strong correlation between the effectiveness of a buffer solution and the titration curves discussed in Section 16.5. Consider the schematic titration curve of a weak acid with a strong base shown in Figure 16.19. As indicated by the labels, the region around $\mathrm{p}K_a$ corresponds to the midpoint of the titration, when approximately half the weak acid has been neutralized. This portion of the titration curve corresponds to a buffer: it exhibits the smallest change in pH per increment of added strong base, as shown by the nearly horizontal nature of the curve in this region. Note that the nearly flat portion of the curve extends only from approximately a pH value of one unit less than the $\mathrm{p}K_a$ to approximately a pH value of one unit greater than the

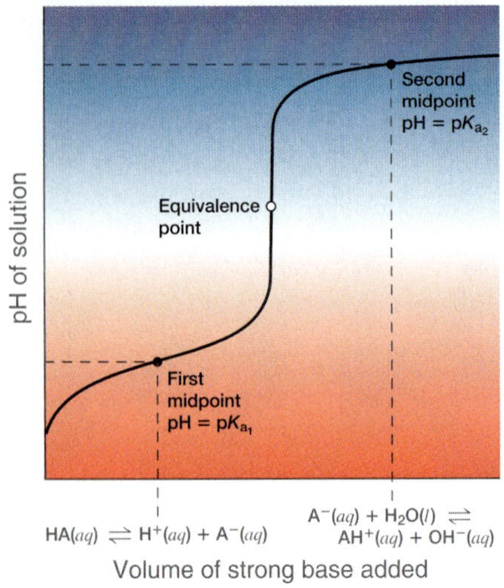

Figure 16.19 The relationship between titration curves and buffers. This schematic plot of pH for the titration of a weak acid with a strong base shows the nearly flat region of the titration curve around the midpoint, which corresponds to the formation of a buffer. At the lower left, the pH of the solution is determined by the equilibrium for dissociation of the weak acid; at the upper right, the pH is determined by the equilibrium for reaction of the conjugate base with water.

pK_a, which is why buffer solutions usually have a pH that is within ±1 pH units of the pK_a of the acid component of the buffer.

In the region of the titration curve at the lower left, before the midpoint, the acid–base properties of the solution are dominated by the equilibrium for dissociation of the weak acid, corresponding to K_a. In the region of the titration curve at the upper right, after the midpoint, the acid–base properties of the solution are dominated by the equilibrium for reaction of the conjugate base of the weak acid with water, corresponding to K_b. Note, however, that we can calculate either K_a or K_b from the other because they are related by K_w.

Blood: A Most Important Buffer

Metabolic processes produce large amounts of acids and bases, yet organisms are able to maintain an almost constant internal pH because their fluids contain buffers. This is not to say that the pH is uniform throughout all cells and tissues of a mammal. The internal pH of a red blood cell is about 7.2, but the pH of most other kinds of cells is lower, around 7.0. Even within a single cell, different compartments can have very different pH values. For example, one intracellular compartment in white blood cells has a pH of around 5.

Because no single buffer system can effectively maintain a constant pH value over the entire physiological range of approximately pH 5 to 7.4, biochemical systems use a set of buffers with overlapping ranges. The most important of these is the CO_2/HCO_3^- system, which dominates the buffering action of blood plasma.

The acid–base equilibrium in the CO_2/HCO_3^- buffer system is usually written as

$$H_2CO_3\,(aq) \rightleftharpoons H^+(aq) + HCO_3^-(aq) \qquad (16.62)$$

with $K_a = 4.5 \times 10^{-7}$ and $pK_a = 6.35$ at 25°C. In fact, Equation 16.62 is a grossly oversimplified version of the CO_2/HCO_3^- system because a solution of CO_2 in water contains only rather small amounts of $H_2CO_3(aq)$. Thus, Equation 16.62 does not allow us to understand how blood is actually buffered, particularly at a physiological temperature of 37°C. As shown in Equation 16.63, $CO_2(aq)$ is in equilibrium with $H_2CO_3(aq)$, but the equilibrium lies far to the left, with an $H_2CO_3(aq)/CO_2(aq)$ ratio less than 0.01 under most conditions:

$$CO_2(aq) + H_2O(l) \rightleftharpoons H_2CO_3(aq) \qquad (16.63)$$

with $K' = 4.0 \times 10^{-3}$ at 37°C. The true pK_a of carbonic acid at 37°C is 3.70, corresponding to a K_a value of 2.0×10^{-4}, which makes it a much stronger acid than Equation 16.62 suggests. Adding Equations 16.62 and 16.63 and canceling H_2CO_3 from both sides give the following overall equation for the reaction of CO_2 with water to give a proton and the bicarbonate ion:

$$\begin{array}{lll} CO_2(aq) + H_2O(l) \rightleftharpoons \cancel{H_2CO_3(aq)} & & K' = 4.0 \times 10^{-3}\,(37°C) \\ \cancel{H_2CO_3(aq)} \rightleftharpoons H^+(aq) + HCO_3^-(aq) & & K_a = 2.0 \times 10^{-4}\,(37°C) \\ \hline CO_2(aq) + H_2O(l) \rightleftharpoons H^+(aq) + HCO_3^-(aq) & & K = 8.0 \times 10^{-7}\,(37°C) \end{array} \qquad (16.64)$$

Note that the equilibrium constant K for the reaction in Equation 16.64 is the product of the *true* ionization constant for carbonic acid K_a and the equilibrium constant K' for the reaction of $CO_2(aq)$ with water to give carbonic acid. The equilibrium equation for the reaction of CO_2 with water to give bicarbonate and a proton is therefore

$$K = \frac{[H^+][HCO_3^-]}{[CO_2]} = 8.0 \times 10^{-7} \qquad (16.65)$$

The presence of a gas in the equilibrium constant expression for a buffer is unusual. According to Henry's law (described in Chapter 13), $[CO_2] = kP_{CO_2}$, where k is the Henry's law constant for CO_2, 0.030 mM/mmHg at 37°C. Substituting this expression for $[CO_2]$ in Equation 16.65 gives

$$K = \frac{[H^+][HCO_3^-]}{(3.0 \times 10^{-5}\ M/\text{mmHg})(P_{CO_2})} \quad (16.66)$$

where P_{CO_2} is in mmHg. Taking the negative logarithm of both sides and rearranging give

$$pH = 6.10 + \log \frac{[HCO_3^-]}{(3.0 \times 10^{-5}\ M/\text{mm Hg})(P_{CO_2})} \quad (16.67)$$

Thus, the pH of the solution depends on both the CO_2 pressure over the solution and $[HCO_3^-]$. Figure 16.20 plots the relationship between pH and $[HCO_3^-]$ under physiological conditions for several different values of P_{CO_2}, with normal pH and $[HCO_3^-]$ values indicated by the dashed lines.

According to Equation 16.64, adding a strong acid to the CO_2/HCO_3^- system causes $[HCO_3^-]$ to decrease as HCO_3^- is converted to CO_2. Excess CO_2 is released in the lungs and exhaled into the atmosphere, however, so there is essentially no change in P_{CO_2}. Because the change in $[HCO_3^-]/P_{CO_2}$ is small, Equation 16.67 predicts that the change in pH will also be rather small. Conversely, if a strong base is added, the OH^- reacts with CO_2 to form HCO_3^-, but CO_2 is replenished by the body, again limiting the change in both $[HCO_3^-]/P_{CO_2}$ and pH. The CO_2/HCO_3^- buffer system is an example of an *open* system, in which the total concentration of the components of the buffer change in order to keep the pH at a nearly constant value.

If a passenger steps out of an airplane in Denver, Colorado, for example, the lower P_{CO_2} at higher elevations (typically 31 mmHg at an elevation of 2000 m versus 40 mmHg at sea level) causes a shift to a new pH and $[HCO_3^-]$. The increase in pH and decrease in $[HCO_3^-]$ in response to the decrease in P_{CO_2} are responsible for the general malaise that many people experience at high altitudes. If their blood pH does not adjust rapidly, the condition can develop into the life-threatening phenomenon known as altitude sickness.

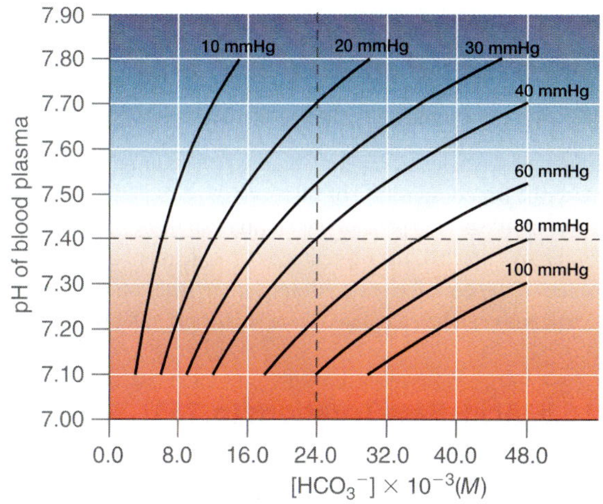

Figure 16.20 Buffering in blood: pH versus $[HCO_3^-]$ curves for buffers with different values of P_{CO_2}. Only those combinations of pH and $[HCO_3^-]$ that lie on a given line are allowed for the particular value of P_{CO_2} indicated. Normal values of blood plasma pH and $[HCO_3^-]$ are indicated by the dashed lines.

SUMMARY AND KEY TERMS

16.1 The Autoionization of Water (p. 722)

Water is **amphiprotic**: it can act as an acid by donating a proton to a base to form the hydroxide ion, or as a base by accepting a proton from an acid to form the **hydronium ion**, H_3O^+. The *autoionization* of liquid water produces OH^- and H_3O^+ ions. The equilibrium constant for this reaction is called the **ion-product constant of liquid water (K_w)** and is defined as $K_w = [H_3O^+][OH^-]$. At 25°C, K_w has a value of 1.01×10^{-14}; hence, pH + pOH = pK_w = 14.00.

16.2 A Qualitative Description of Acid–Base Equilibria (p. 725)

Two species that differ by only a proton constitute a **conjugate acid–base pair**. The magnitude of the equilibrium constant for an ionization reaction can be used to determine the relative strengths of acids and bases. For an aqueous solution of a weak acid, the dissociation constant is called the **acid ionization constant (K_a)**. Similarly, the equilibrium constant for the reaction of a weak base with water is the **base ionization constant (K_b)**. For any conjugate acid–base pair, $K_aK_b = K_w$. Smaller values of pK_a correspond to larger acid ionization constants and hence stronger acids. Conversely, smaller values of pK_b correspond to larger base ionization constants and hence stronger bases. At 25°C, $pK_a + pK_b = 14.00$. Acid–base reactions always proceed in the direction that produces the weaker acid–base pair. No acid stronger than H_3O^+ and no base stronger than OH^- can exist in aqueous solution, leading to the phenomenon known as the **leveling effect**. Polyprotic acids (and bases) lose (and gain) protons in a stepwise manner, with the fully protonated species being the strongest acid and the fully deprotonated species the strongest base. A salt can dissolve in water to produce a neutral, basic, or acidic solution, depending on whether it contains the conjugate base of a weak acid as the anion (A^-) or the conjugate acid of a weak base as the cation (BH^+), or both. Salts that contain small, highly charged metal ions produce acidic solutions in water. The reaction of a salt with water to produce an acidic or basic solution is called a **hydrolysis reaction**.

16.3 Molecular Structure and Acid–Base Strength (p. 736)

The acid–base strength of a molecule depends strongly on its structure. The weaker the A—H or B—H^+ bond, the more likely it is to dissociate to form an H^+ ion. In addition, any factor that stabilizes the lone pair on the conjugate base favors the dissociation of H^+, making the conjugate acid a stronger acid. Atoms or groups of atoms elsewhere in a molecule can also be important in determining acid or base strength through an *inductive effect*, which can weaken an O—H bond and allow hydrogen to be more easily lost as H^+ ions.

16.4 Quantitative Aspects of Acid–Base Equilibria (p. 741)

If the concentration of one or more of the species in a solution of an acid or a base is determined experimentally, the values of K_a and K_b can be calculated. Values of K_a, pK_a, K_b, and pK_b can be used to quantitatively describe the composition of solutions of acids and bases. The concentrations of all species present in solution can be determined, as can the pH of the solution and the percentage of the acid or base that is ionized. The value of the equilibrium constant for the reaction of a weak acid with a weak base can be calculated from K_a (or pK_a), K_b (or pK_b), and K_w.

16.5 Acid–Base Titrations (p. 751)

The shape of a **titration curve**, a plot of pH versus the amount of acid or base added, provides important information about what is occurring in solution during the titration. The shapes of titration curves for weak acids and bases depend dramatically on the identity of the compound. The **equivalence point** of an acid–base titration is the point at which exactly enough acid or base has been added to react completely with the other component. The equivalence point in the titration of a strong acid or base occurs at pH 7.00. In titrations of weak acids or bases, however, the pH at the equivalence point is greater or less than 7, respectively. The pH tends to change more slowly before the equivalence point is reached in titrations of weak acids and weak bases than in titrations of strong acids and strong bases. The pH at the **midpoint**, the point halfway on the titration curve to the equivalence point, is equal to the pK_a of the weak acid or the pK_b of the weak base. Thus, titration methods can be used to determine both the concentration *and* the pK_a (or pK_b) of a weak acid (or base). **Acid–base indicators** are compounds that change color at a particular pH. They are typically weak acids or bases whose changes in color correspond to deprotonation or protonation of the indicator itself.

16.6 Buffers (p. 764)

Buffers are solutions that resist a change in pH upon the addition of an acid or a base. Buffers contain a weak acid (HA) and its conjugate weak base (A^-). Adding a strong electrolyte that contains one ion in common with a reaction system that is at equilibrium shifts the equilibrium in such a way as to reduce the concentration of the common ion. The shift in equilibrium is called the **common ion effect**. Buffers are characterized by their pH range and their buffer capacity. The useful pH range of a buffer depends strongly on the chemical properties of the conjugate weak acid–base pair used to prepare the buffer (the K_a or K_b), whereas its buffer capacity depends solely on the concentrations of the species in the solution. The pH of a buffer can be calculated using the **Henderson–Hasselbalch equation**, which is valid for solutions whose concentrations are at least a hundred times greater than the value of their K_a's. Because no single buffer system can effectively maintain a constant pH value over the physiological range of approximately 5 to 7.4, biochemical systems use a set of buffers with overlapping ranges. The most important of these is the CO_2/HCO_3^- system, which dominates the buffering action of blood plasma.

KEY EQUATIONS

Ion-product constant of liquid water	$K_w = [H_3O^+][OH^-]$	(16.7)
Definition of pH	$pH = -\log_{10}[H^+]$ $[H^+] = 10^{-pH}$	(16.9a) (16.9b)
Definition of pOH	$pOH = -\log_{10}[OH^-]$ $[OH^-] = 10^{-pOH}$	(16.10a) (16.10b)
Relationship among pH, pOH, and pK_w	$pK_w = pH + pOH$	(16.11)
Acid ionization constant	$K_a = \dfrac{[H_3O^+][A^-]}{[HA]}$	(16.14)
Base ionization constant	$K_b = \dfrac{[BH^+][OH^-]}{[B]}$	(16.16)
Relationship between K_a and K_b of a conjugate acid–base pair	$K_aK_b = K_w$	(16.20)
Definition of pK_a	$pK_a = -\log_{10}K_a$ $K_a = 10^{-pK_a}$	(16.21a) (16.21b)
Definition of pK_b	$pK_b = -\log_{10}K_b$ $K_b = 10^{-pK_b}$	(16.22a) (16.22b)
Relationship between pK_a and pK_b of a conjugate acid–base pair	$pK_a + pK_b = pK_w$ $pK_a + pK_b = 14.00$ (at 25°C)	(16.23a) (16.23b)
Percent ionization of acid	$\dfrac{[H^+]}{C_{HA}} \times 100$	(16.40)
Percent ionization of base	$\dfrac{[OH^-]}{C_B} \times 100$	(16.41)
Equilibrium constant for reaction of a weak acid with a weak base	$K = \dfrac{K_aK_b}{K_w}$	(16.48)
Henderson–Hasselbalch equation	$pH = pK_a + \log\left(\dfrac{[A^-]}{[HA]}\right)$	(16.60)
	$pH = pK_a + \log\left(\dfrac{[base]}{[acid]}\right)$	(16.61)

QUESTIONS AND PROBLEMS

For instructor-assigned homework, go to **www.masteringgeneralchemistry.com**

Questions and Problems with colored numbers have answers in the Appendix and complete solutions in the Student Solutions Manual.

CONCEPTUAL

16.1 The Autoionization of Water

1. What is the relationship between the value of the equilibrium constant for the autoionization of liquid water and the tabulated value of the ion-product constant of liquid water, K_w?
2. The density of liquid water decreases as the temperature increases from 25°C to 50°C. Will this effect cause the value of K_w to increase or decrease? Why?
3. Show that water is amphiprotic by writing balanced equations for the reactions of water with HNO_3 and NH_3. In which reaction does water act as the acid? In which does it act as the base?
4. Write each equation:
   (a) nitric acid is added to water
   (b) potassium hydroxide is added to water
   (c) calcium hydroxide is added to water
   (d) sulfuric acid is added to water
5. Show that K for the sum of the following reactions is equal to K_w:
   (a) $HMnO_4(aq) \rightleftharpoons H^+(aq) + MnO_4^-(aq)$
   (b) $MnO_4^-(aq) + H_2O(l) \rightleftharpoons HMnO_4(aq) + OH^-(aq)$

16.2 A Qualitative Description of Acid–Base Equilibria

6. Identify the conjugate acid–base pairs in each of these equilibria:
   (a) $HF(aq) + H_2O(l) \rightleftharpoons H_3O^+(aq) + F^-(aq)$
   (b) $CH_3CH_2NH_2(aq) + H_2O(l) \rightleftharpoons CH_3CH_2NH_3^+(aq) + OH^-(aq)$
   (c) $C_3H_7NO_2(aq) + OH^-(aq) \rightleftharpoons C_3H_6NO_2^-(aq) + H_2O(l)$
   (d) $CH_3CO_2H(aq) + 2HF(aq) \rightleftharpoons CH_3C(OH)_2^+(aq) + HF_2^-(aq)$
7. Identify the conjugate acid–base pairs in each equilibrium:
   (a) $HSO_4^-(aq) + H_2O(l) \rightleftharpoons SO_4^{2-}(aq) + H_3O^+(aq)$
   (b) $C_3H_7NO_2(aq) + H_3O^+(aq) \rightleftharpoons C_3H_8NO_2^+(aq) + H_2O(l)$
   (c) $CH_3CO_2H(aq) + NH_3(aq) \rightleftharpoons CH_3CO_2^-(aq) + NH_4^+(aq)$
   (d) $SbF_5(aq) + 2HF(aq) \rightleftharpoons H_2F^+(aq) + SbF_6^-(aq)$
8. Salts such as NaH contain the hydride ion, H^-. When sodium hydride is added to water, it produces hydrogen gas in a highly vigorous reaction. Write a balanced equation for this reaction, and identify the conjugate acid–base pairs.
9. Write the expression for K_a for each reaction:
   (a) $HCO_3^-(aq) + H_2O(l) \rightleftharpoons CO_3^{2-}(aq) + H_3O^+(aq)$
   (b) formic acid$(aq) + H_2O(l) \rightleftharpoons$ formate$(aq) + H_3O^+(aq)$
   (c) $H_3PO_4(aq) + H_2O(l) \rightleftharpoons H_2PO_4^-(aq) + H_3O^+(aq)$
10. Write an expression for the ionization constant K_b for each reaction:
   (a) $OCH_3^-(aq) + H_2O(l) \rightleftharpoons HOCH_3(aq) + OH^-(aq)$
   (b) $NH_2^-(aq) + H_2O(l) \rightleftharpoons NH_3(aq) + OH^-(aq)$
   (c) $S^{2-}(aq) + H_2O(l) \rightleftharpoons HS^-(aq) + OH^-(aq)$
11. Predict whether each equilibrium lies primarily to the left or to the right:
   (a) $HBr(aq) + H_2O(l) \rightleftharpoons H_3O^+(aq) + Br^-(aq)$
   (b) $NaH(soln) + NH_3(l) \rightleftharpoons H_2(soln) + NaNH_2(soln)$
   (c) $OCH_3^-(aq) + NH_3(aq) \rightleftharpoons CH_3OH(aq) + NH_2^-(aq)$
   (d) $NH_3(aq) + HCl(aq) \rightleftharpoons NH_4^+(aq) + Cl^-(aq)$
12. Species that are strong bases in water, such as CH_3^-, NH_2^-, and S^{2-}, are leveled to the strength of OH^-, the conjugate base of H_2O. Because their relative base strengths are indistinguishable in water, suggest a method for identifying which is the strongest base. How would you distinguish between the strong acids HIO_3, H_2SO_4 and $HClO_4$?
13. Is it accurate to say that a 2.0 *M* solution of H_2SO_4, which contains two acidic protons per molecule, is 4.0 *M* in H^+? Explain your answer.
14. The alkalinity of soil is defined by the equation: alkalinity = $[HCO_3^-] + 2[CO_3^{2-}] + [OH^-] - [H^+]$. The source of both HCO_3^- and CO_3^{2-} is H_2CO_3. Explain why the basicity of soil is defined in this way.
15. Why are aqueous solutions of salts such as $CaCl_2$ neutral? Why is an aqueous solution of $NaNH_2$ basic?
16. Predict whether aqueous solutions of the following are acidic, basic, or neutral: (a) Li_3N; (b) NaH; (c) KBr; (d) $C_2H_5NH_3^+Cl^-$.
17. When each of these compounds is added to water, would you expect the pH of the solution to increase, decrease, or remain the same? (a) $LiCH_3$; (b) $MgCl_2$; (c) K_2O; (d) $(CH_3)_2NH_2^+Br^-$
18. Which complex ion would you expect to be more acidic: $Pb(H_2O)_4^{2+}$ or $Sn(H_2O)_4^{2+}$? Why?
19. Would you expect $Sn(H_2O)_4^{2+}$ or $Sn(H_2O)_6^{4+}$ to be more acidic? Why?
20. Is it possible to arrange the hydrides LiH, RbH, KH, CsH, and NaH in order of increasing base strength in aqueous solution? Why or why not?

16.3 Molecular Structure and Acid–Base Strength

21. Section 16.3 presented several factors that affect the relative strengths of acids and bases. For each of the following pairs, identify the most important factor in determining which is the stronger acid or base in aqueous solution:
   (a) $CH_3CCl_2CH_2CO_2H$ vs. $CH_3CH_2CH_2CO_2H$
   (b) CH_3CO_2H vs. CH_3CH_2OH
   (c) HClO vs. HBrO
   (d) $CH_3C(=O)NH_2$ vs. $CH_3CH_2NH_2$
   (e) H_3AsO_4 vs. H_3AsO_3
22. The stability of the conjugate base is an important factor in determining the strength of an acid. Which would you expect to be the stronger acid in aqueous solution: $C_6H_5NH_3^+$ or NH_4^+? Justify your reasoning.
23. Explain why H_2Se is a weaker acid than HBr.
24. Arrange the following in order of decreasing acid strength in aqueous solution: H_3PO_4, $CH_3PO_3H_2$, and $HClO_3$.
25. Arrange the following in order of increasing base strength in aqueous solution: CH_3S^-, OH^-, and CF_3S^-.
26. Arrange the following in order of increasing acid strength in aqueous solution: $HClO_2$, HNO_2, HNO_3.

27. Do you expect H_2SO_3 or H_2SeO_3 to be the stronger acid? Why?
28. Give a plausible explanation for why CF_3OH is a stronger acid than CH_3OH in aqueous solution. Do you expect $CHCl_2CH_2OH$ to be a stronger or weaker acid than CH_3OH? Why?
29. Do you expect Cl_2NH or NH_3 to be the stronger base in aqueous solution? Why?

16.4 Quantitative Aspects of Acid–Base Equilibria

30. Explain why the analytical concentration C of H_2SO_4 is equal to $[H_2SO_4] + [HSO_4^-] + [SO_4^{2-}]$.
31. Write an expression for the analytical concentration C of H_3PO_4 in terms of the concentrations of the species actually present in solution.
32. For a dilute solution of a weak acid such as acetic acid, CH_3CO_2H, the concentration of undissociated acetic acid in solution is often assumed to be the same as the analytical concentration. Explain why this is a valid practice.
33. How does dilution affect the percent ionization of a weak acid or weak base?
34. What is the relationship between the K_a of a weak acid and its percent ionization? Does a compound with a large pK_a value have a higher or lower percent ionization than a compound with a small pK_a value (assuming the same analytical concentration in both cases)? Explain.
35. For a dilute solution of a weak acid, HA, show that the pH of the solution can be approximated using the following equation (where C_{HA} is the analytical concentration of the weak acid):

$$\text{pH} = -\log\sqrt{K_a \cdot C_{HA}}$$

Under what conditions is this approximation valid?

16.5 Acid–Base Titrations

36. Why is the portion of the titration curve that lies below the equivalence point of a solution of a weak acid displaced upward relative to the titration curve of a strong acid? How are the slopes of the curves different at the equivalence point? Why?
37. Predict whether each solution will be neutral, basic, or acidic at the equivalence point of the following titrations: (a) an aqueous solution of NaOH is titrated with 0.100 *M* HCl; (b) an aqueous solution of ethylamine, $CH_3CH_2NH_2$, is titrated with 0.150 *M* HNO_3; (c) an aqueous solution of aniline hydrochloride, $C_6H_5NH_3^+Cl^-$, is titrated with 0.050 *M* KOH.
38. The pK_a values of phenol red, bromophenol blue, and phenolphthalein are 7.4, 4.1, and 9.5, respectively. Which indicator is best suited for each of the following acid–base titrations? (a) titrating a solution of $Ba(OH)_2$ with 0.100 *M* HCl; (b) titrating a solution of trimethylamine, Me_3N, with 0.150 *M* HNO_3; (c) titrating a solution of aniline hydrochloride, $C_6H_5NH_3^+Cl^-$, with 0.050 *M* KOH
39. For the titration of any strong acid with any strong base, the pH at the equivalence point is 7. Why is this not usually the case in titrations of weak acids or bases?
40. Why are the titration curves for a strong acid with a strong base and a weak acid with a strong base identical in shape above the equivalence points but not below?
41. Describe what is occurring on a molecular level during the titration of a weak acid, such as acetic acid, with a strong base, such as NaOH, at the following points along the titration curve: (a) at the beginning of the titration, (b) at the midpoint of the titration, (c) at the equivalence point, and (d) when excess titrant has been added. (e) Which of these points corresponds to the pH = pK_a?
42. On a molecular level, describe what is happening during the titration of a weak base, such as ammonia, with a strong acid, such as HCl, at the following points along the titration curve: (a) at the beginning of the titration, (b) at the midpoint of the titration, (c) at the equivalence point, and (d) when excess titrant has been added. (e) Which of these points corresponds to the pOH = pK_b?
43. For the titration of a weak acid with a strong base, use the K_a expression to show that pH = pK_a at the midpoint of the titration.
44. Chemical indicators can be used to monitor pH rapidly and inexpensively. Nevertheless, electronic methods are generally preferred. Why?
45. Why does adding ammonium chloride to a solution of ammonia in water decrease the pH of the solution?
46. Given the equilibrium system $CH_3NH_2(aq) + H_2O(l) \rightleftharpoons CH_3NH_3^+(aq) + OH^-(aq)$, explain what happens to the position of the equilibrium and the pH in each case: (a) dilute HCl is added; (b) dilute NaOH is added; (c) solid $CH_3NH_3^+Cl^-$ is added.
47. Given the equilibrium system $CH_3CO_2H(aq) \rightleftharpoons CH_3CO_2^-(aq) + H^+(aq)$, explain what happens to the position of the equilibrium and the pH in each case: (a) dilute HCl is added; (b) dilute NaOH is added; (c) solid sodium acetate is added.

16.6 Buffers

48. Explain why buffers are crucial for the proper functioning of biological systems.
49. What is the role of a buffer in chemistry and biology? Is it correct to say that buffers prevent a change in $[H_3O^+]$? Explain your reasoning.
50. Explain why the most effective buffers are those that contain approximately equal amounts of the weak acid and its conjugate base.
51. Which region of the titration curve of a weak acid or base corresponds to the region of the smallest change in pH per amount of added strong acid or base?
52. If you were given a solution of sodium acetate, describe two ways you could convert the solution to a buffer.
53. Why are buffers usually used only within approximately one pH unit of the pK_a or pK_b of the parent weak acid or base?
54. The titration curve for a monoprotic acid can be divided into four regions: the starting point, the region around the midpoint of the titration, the equivalence point, and the region after the equivalence point. For which region would you use each of the following approaches to describe the behavior of the solution? (a) a buffer; (b) a solution of a salt of a weak base; (c) a solution of a weak acid; (d) diluting a strong base
55. Which of the following will produce a buffer solution? Explain your reasoning in each case. (a) mixing 100 mL of 0.1 *M* HCl and 100 mL of 0.1 *M* sodium acetate; (b) mixing 50 mL of a 0.1 *M* solution of HCl and 100 mL of a 0.1 *M* solution of sodium acetate; (c) mixing 100 mL of a 0.1 *M* acetic acid solution and 100 mL of a 0.1 *M* NaOH solution; (d) mixing 100 mL of a 0.1 *M* acetic acid solution and 50 mL of a 0.1 *M* NaOH solution; (e) mixing 100 mL of a 0.1 *M* sodium acetate solution and 50 mL of a 0.1 *M* acetic acid solution
56. Which of the following will produce a buffer solution? Explain your reasoning in each case. (a) mixing 100 mL of 0.1 *M* HCl

and 100 mL of 0.1 *M* sodium fluoride; (b) mixing 50 mL of a 0.1 *M* solution of HCl and 100 mL of a 0.1 *M* solution of sodium fluoride; (c) mixing 100 mL of a 0.1 *M* hydrofluoric acid solution and 100 mL of a 0.1 *M* HCl solution; (d) mixing 100 mL of a 0.1 *M* hydrofluoric acid solution and 50 mL of a 0.1 *M* NaOH solution; (e) mixing 100 mL of a 0.1 *M* sodium fluoride solution and 50 mL of a 0.1 *M* NaOH solution.

57. Use the definition of K_b for a weak base to derive the expression shown here, which is analogous to the Henderson–Hasselbalch equation, but for a weak base, B, rather than a weak acid, HA:

$$pOH = pK_b - \log\left(\frac{[base]}{[acid]}\right)$$

58. Why do biological systems use overlapping buffer systems to maintain a constant pH?

59. The CO_2/HCO_3^- buffer system of blood has an effective pK_a of approximately 6.1, yet the normal pH of blood is 7.4. Why is CO_2/HCO_3^- an effective buffer when the pK_a is more than one unit below the pH of blood? What happens to the pH of blood when the CO_2 pressure increases? When the O_2 pressure increases?

60. Carbon dioxide produced during respiration is converted to carbonic acid, H_2CO_3. The pK_{a1} of carbonic acid is 6.35, and its pK_{a2} is 10.33. Write the equations corresponding to each p*K* value, and predict the equilibrium position for each reaction.

NUMERICAL

This section includes paired problems (marked by brackets) that require similar problem-solving skills.

16.1 The Autoionization of Water

61. The autoionization of sulfuric acid can be described by the equation

$$H_2SO_4(l) + H_2SO_4(l) \rightleftharpoons H_3SO_4^+(soln) + HSO_4^-(soln)$$

At 25°C, $K = 3 \times 10^{-4}$. Write an equilibrium constant expression for $K_{H_2SO_4}$ that is analogous to K_w. The density of H_2SO_4 is 1.8 g/cm^3 at 25°C. What is the concentration of $H_3SO_4^+$? What fraction of H_2SO_4 is ionized?

62. An aqueous solution of a substance is found to have $[H_3O]^+ = 2.48 \times 10^{-8}$ *M*. Is the solution acidic, neutral, or basic?

63. The pH of a solution is 5.63. What is the pOH? What is the value of $[OH^-]$? Is the solution acidic or basic?

64. State whether each solution is acidic, neutral, or basic:
(a) solution A: $[H_3O^+] = 8.6 \times 10^{-3}$ *M*
(b) solution B: $[H_3O^+] = 3.7 \times 10^{-9}$ *M*
(c) solution C: $[H_3O^+] = 2.1 \times 10^{-7}$ *M*
(d) solution D: $[H_3O^+] = 1.4 \times 10^{-6}$ *M*

65. Calculate the pH and the pOH of each solution: (a) 0.15 *M* HBr; (b) 0.03 *M* KOH; (c) 2.3×10^{-3} *M* HNO_3; (d) 9.78×10^{-2} *M* NaOH; (e) 0.00017 *M* HCl; (f) 5.78 *M* HI.

66. Calculate the pH and the pOH of each solution: (a) 25.0 mL of 2.3×10^{-2} *M* HCl, diluted to 100 mL; (b) 5.0 mL of 1.87 *M* NaOH, diluted to 125 mL; (c) 5.0 mL of 5.98 *M* HCl added to 100 mL of water; (d) 25.0 mL of 3.7 *M* HNO_3 added to 250 mL of water; (e) 35.0 mL of 0.046 *M* HI added to 500 mL of water; (f) 15.0 mL of 0.0087 *M* KOH added to 250 mL of water.

67. The pH of stomach acid is approximately 1.5. What is the H^+ concentration?

68. Given the pH values in parentheses, what is the H^+ concentration of each solution? (a) household bleach (11.4); (b) milk (6.5); (c) orange juice (3.5); (d) seawater (8.5); (e) tomato juice (4.2)

69. A reaction requires the addition of 250.0 mL of a solution with a pH of 3.50. What mass of HCl(*g*) must be dissolved in 250 mL of water to produce a solution with this pH?

70. If you require 333 mL of a pH 12.50 solution, how would you prepare it using a 0.500 *M* sodium hydroxide stock solution?

16.2 A Qualitative Description of Acid–Base Equilibria

71. Arrange the acids in order of increasing strength:
acid A: $pK_a = 1.52$
acid B: $pK_a = 6.93$
acid C: $pK_a = 3.86$
Given solutions with the same initial concentration of each acid, which would have the highest percent ionization?

72. Arrange the bases in order of increasing strength:
base A: $pK_b = 13.10$
base B: $pK_b = 8.74$
base C: $pK_b = 11.87$
Given solutions with the same initial concentration of each base, which would have the highest percent ionization?

73. Calculate the K_a and pK_a of the conjugate acid of a base with each of these pK_b values: (a) 3.8; (b) 7.9; (c) 13.7; (d) 1.4; (e) −2.5.

74. Determine K_b and pK_b for benzoic acid, a food preservative, which has a pK_a of 4.20.

75. Determine K_a and pK_a of boric acid, $B(OH)_3$, solutions of which are occasionally used as an eyewash, if the pK_b of its conjugate base is 4.8.

16.4 Quantitative Aspects of Acid–Base Equilibria

76. The pK_a of NH_3 is estimated to be 9.25. Its conjugate base, amide ion, NH_2^-, can be isolated as an alkali metal salt, such as sodium amide, $NaNH_2$. Calculate the pH of a solution prepared by adding 0.100 mol of sodium amide to 1.00 L of water. Does the pH differ appreciably from the pH of a NaOH solution of the same concentration? Why or why not?

77. Phenol is a topical anesthetic that has been used in throat lozenges to relieve sore throat pain. Describe in detail how you would prepare a 2.00 *M* solution of phenol, C_6H_5OH, in water; then write equations to show all species present in the solution. What is the equilibrium constant expression for the reaction of phenol with water? Use the information in Table 16.2 to calculate the pH of the phenol solution.

78. Describe in detail how you would prepare a 1.50 *M* solution of methylamine in water; then write equations to show all species present in the solution. What is the equilibrium constant expression for the reaction of methylamine with water? Use the information in Table 16.3 to calculate the pH of the solution.

79. A 0.200 *M* solution of diethylamine, a substance used in insecticides and fungicides, is only 3.9% ionized at 25°C. Write an equation showing the equilibrium reaction, and then calculate the pK_b of diethylamine. What is the pK_a of its conjugate acid, the diethylammonium ion? What is the equilibrium constant expression for the reaction of diethylammonium chloride with water?

80. A 1.00 *M* solution of fluoroacetic acid, FCH_2CO_2H, is 5% dissociated in water. What is the equilibrium constant expression for the dissociation reaction? Calculate the concentration of each species in solution, and then calculate the pK_a of FCH_2CO_2H.

81. The pK_a of 3-chlorobutanoic acid, $CH_3CHClCH_2CO_2H$, is 4.05. What percentage is dissociated in a 1.0 *M* solution? Do you expect the pK_a of butanoic acid to be higher or lower than the pK_a of 3-chlorobutanoic acid? Why?

82. The pK_a of the ethylammonium ion, $C_2H_5NH_3^+$, is 10.64. What percentage of ethylamine is ionized in a 1.00 *M* solution of ethylamine?

83. The pH of a 0.150 *M* solution of aniline hydrochloride, $C_6H_5NH_3^+Cl^-$, is 2.70. What is the pK_b of the conjugate base, aniline, $C_6H_5NH_2$? Do you expect the pK_b of $(CH_3)_2CHNH_2$ to be higher or lower than the pK_b of $C_6H_5NH_2$? Why?

84. The pK_a of Cl_3CCO_2H is 0.64. What is the pH of a 0.580 *M* solution? What percentage of the Cl_3CCO_2H is dissociated?

85. What is the pH of a 0.620 *M* solution of $CH_3NH_3^+Br^-$ if the pK_b of CH_3NH_2 is 10.62?

86. The pK_b of 4-hydroxypyridine is 10.80 at 25°C. What is the pH of a 0.0250 *M* solution?

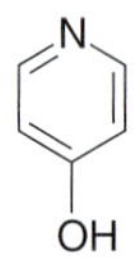

87. The pK_a's of butanoic acid and ammonium ion are 4.82 and 9.24, respectively. Calculate the value of *K* for the reaction

$$CH_3CH_2CH_2CO_2^-(aq) + NH_4^+(aq) \rightleftharpoons CH_3CH_2CH_2CO_2H(aq) + NH_3(aq)$$

88. The pK_a's of formic acid and the methylammonium ion are 3.75 and 10.62, respectively. Calculate the value of *K* for the reaction

$$HCO_2^-(aq) + CH_3NH_3^+(aq) \rightleftharpoons HCO_2H(aq) + CH_3NH_2(aq)$$

89. Use the information in Table 16.2 to calculate the pH of a 0.0968 *M* solution of calcium formate.

90. Calculate the pH of a 0.24 *M* solution of sodium lactate. The pK_a of lactic acid is 3.9.

91. Use the information in Table 16.2 to determine the pH of a solution prepared by dissolving 855 mg of sodium nitrite, $NaNO_2$, in enough water to make 100.0 mL of solution.

92. Use the information in Table 16.3 to determine the pH of a solution prepared by dissolving 750.0 mg of methylammonium chloride, $CH_3NH_3^+Cl^-$, in enough water to make 150.0 mL of solution.

16.5 Acid–Base Titrations

93. Calculate the pH of each solution: (a) 25.0 mL of 6.09 *M* HCl is added to 100.0 mL of distilled water; (b) 5.0 mL of 2.55 *M* NaOH is added to 75.0 mL of distilled water.

94. What is the pH of a solution prepared by mixing 50.0 mL of 0.225 *M* HCl with 100.0 mL of a 0.184 *M* solution of NaOH?

95. What volume of 0.50 *M* HCl is needed to completely neutralize 25.00 mL of 0.86 *M* NaOH?

96. Calculate the final pH when each of the following pairs of solutions is mixed: (a) 100 mL of 0.105 *M* HCl and 100 mL of 0.115 *M* sodium acetate; (b) 50 mL of a 0.10 *M* solution of HCl and 100 mL of a 0.15 *M* solution of sodium acetate; (c) 100 mL of a 0.109 *M* acetic acid solution and 100 mL of a 0.118 *M* NaOH solution; (d) 100 mL of a 0.998 *M* acetic acid solution and 50.0 mL of a 0.110 *M* NaOH solution; (e) 100 mL of a 0.107 *M* sodium acetate solution and 50.0 mL of a 0.987 *M* acetic acid solution.

97. Calculate the final pH when each of the following pairs of solutions is mixed: (a) 100 mL of 0.983 *M* HCl and 100 mL of 0.102 *M* sodium fluoride; (b) 50.0 mL of a 0.115 *M* solution of HCl and 100 mL of a 0.109 *M* solution of sodium fluoride; (c) 100 mL of a 0.106 *M* hydrofluoric acid solution and 50.0 mL of a 0.996 *M* NaOH solution.

98. Calcium carbonate is a major contributor to the "hardness" of water. The amount of $CaCO_3$ in a water sample can be determined by titrating the sample with an acid, such as HCl, which produces water and CO_2. Write a balanced equation for this reaction. Generate a plot of solution pH versus volume of 0.100 *M* HCl added for the titration of a solution of 250 mg of $CaCO_3$ in 200.0 mL of water with 0.100 *M* HCl; assume that the HCl solution is added in 5.00-mL increments. What volume of HCl corresponds to the equivalence point?

99. For a titration of 50.0 mL of 0.288 *M* NaOH, you would like to prepare a 0.200 *M* HCl solution. The only HCl solution available to you, however, is 12.0 *M*.
(a) How would you prepare 500 mL of a 0.200 *M* HCl solution?
(b) Approximately what volume of your 0.200 *M* HCl solution is needed to neutralize the NaOH solution?
(c) After completing the titration, you find that your "0.200 *M*" HCl solution is actually 0.187 *M*. What was the exact volume of titrant used in the neutralization?

100. While titrating 50.0 mL of a 0.582 *M* solution of HCl with a solution labeled "0.500 *M* KOH," you overshoot the endpoint. To correct the problem, you add 10.00 mL of the HCl solution to your flask and then carefully continue the titration. The total volume of titrant needed for neutralization is 71.9 mL.
(a) What is the actual molarity of your KOH solution?
(b) What volume of titrant was needed to neutralize 50.0 mL of the acid?

101. Complete the table and generate a titration curve showing the pH versus volume of added base for the titration of 50.0 mL of 0.288 *M* HCl with 0.321 *M* NaOH. Clearly indicate the equivalence point.

Base Added, mL	10.0	30.0	40.0	45.0	50.0	55.0	65.0	75.0
pH								

102. The following data were obtained while titrating 25.0 mL of 0.156 *M* NaOH with a solution labeled "0.202 *M* HCl." Plot the pH versus volume of titrant added. Then determine the equivalence point from your graph, and calculate the exact concentration of your HCl solution.

Volume of HCl, mL	5	10	15	20	25	30	35
pH	11.46	11.29	10.98	4.40	2.99	2.70	2.52

103. Fill in the data for the titration of 50.0 mL of 0.241 *M* formic acid with 0.0982 *M* KOH. The pK_a of formic acid is 3.75. What is the pH of the solution at the equivalence point?

Volume of Base Added, mL	0	5	10	15	20	25
pH						

104. Glycine hydrochloride, which contains the fully protonated form of the amino acid glycine, has this structure:

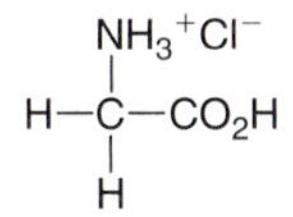

Glycine hydrochloride

It is a strong electrolyte that completely dissociates in water. Titration with base gives two equivalence points: the first corresponds to deprotonation of the carboxylic acid group, and the second to loss of the proton from the ammonium group. The corresponding equilibrium equations are

$$^{+}NH_3—CH_2—CO_2H(aq) \rightleftharpoons {}^{+}NH_3—CH_2—CO_2^{-}(aq) + H^{+} \quad pK_{a1} = 2.3$$

$$^{+}NH_3—CH_2—CO_2^{-}(aq) \rightleftharpoons NH_2—CH_2—COO^{-}(aq) + H^{+} \quad pK_{a2} = 9.6$$

(a) Given 50.0 mL of solution that is 0.430 M in glycine hydrochloride, how many milliliters of 0.150 M KOH are needed to fully deprotonate the carboxylic acid group?
(b) How many additional milliliters of KOH are needed to deprotonate the ammonium group?
(c) What is the pH of the solution at each of the equivalence points?
(d) How many milliliters of titrant are needed to obtain a solution in which glycine has no net electrical charge ? The pH at which a molecule such as glycine has no net charge is its *isoelectric point*. What is the isoelectric point of glycine?

105. What is the pH of a solution prepared by adding 38.2 mL of a 0.197 M HCl solution to 150.0 mL of a 0.242 M solution of pyridine? The pK_b of pyridine is 8.77.

106. Calculate the pH of a solution prepared by adding 40.3 mL of 0.289 M NaOH to 150.0 mL of a 0.564 M succinic acid, $HO_2CCH_2CH_2CO_2H$, solution ($pK_{a1} = 4.21$, $pK_{a2} = 5.64$).

107. Calculate the pH of a 0.150 M solution of malonic acid, $HO_2CCH_2CO_2H$ ($pK_{a1} = 2.85$, $pK_{a2} = 5.70$).

16.6 Buffers

108. Benzenesulfonic acid ($pK_a = 2.55$) is synthesized by treating benzene with concentrated sulfuric acid. Calculate (a) the pH of a 0.286 M solution of benzenesulfonic acid, and (b) the pH after adding enough sodium benzenesulfonate to increase the benzenesulfonate concentration to 0.100 M.

109. Phenol has a pK_a of 9.99. Calculate (a) the pH of a 0.195 M solution, and (b) the percent increase in the concentration of phenol after adding enough solid sodium phenoxide (the sodium salt of the conjugate base) to give a total phenoxide concentration of 0.100 M.

110. An intermediate used in the synthesis of perfumes is valeric acid, also called pentanoic acid. The pK_a of pentanoic acid is 4.84 at 25°C.
(a) What is the pH of a 0.259 M solution of pentanoic acid?
(b) Sodium pentanoate is added to make a buffered solution. What is the pH of the solution if it is 0.210 M in sodium pentanoate?
(c) What is the final pH if 8.00 mL of 0.100 M HCl is added to 75.0 mL of the buffered solution?
(d) What is the final pH if 8.00 mL of 0.100 M NaOH is added to 75.0 mL of the buffered solution?

111. Salicylic acid is used in the synthesis of acetylsalicylic acid, or aspirin. One gram dissolves in 460 mL of water to create a saturated solution with a pH of 2.40.
(a) What is the K_a of salicylic acid?
(b) What is the final pH of a saturated solution that is also 0.238 M in sodium salicylate?
(c) What is the final pH if 10.00 mL of 0.100 M HCl is added to 150.0 mL of the buffered solution?
(d) What is the final pH if 10.00 mL of 0.100 M NaOH is added to 150.0 mL of the buffered solution?

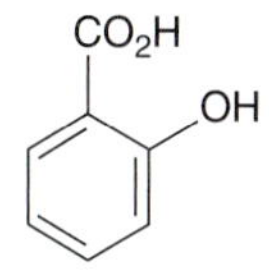

Salicylic acid

APPLICATIONS

112. The analytical concentration of lactic acid in blood is generally less than 1.2×10^{-3} M, corresponding to the sum of [lactate] and [lactic acid]. During strenuous exercise, however, oxygen in the muscle tissue is depleted, and overproduction of lactic acid occurs. This leads to a condition known as lactic acidosis, which is characterized by elevated blood lactic acid levels (approximately 5×10^{-3} M). The pK_a of lactic acid is 3.9.
(a) What is the actual lactic acid concentration under normal physiological conditions?
(b) What is the actual lactic acid concentration during lactic acidosis?

113. When the internal temperature of a human reaches 105°F, immediate steps must be taken to prevent the person from having convulsions. At this temperature, K_w has a value of approximately 2.94×10^{-14}. (a) Calculate the value of pK_w and the pH and pOH of a neutral solution at 105°F. (b) Is the pH higher or lower than that calculated in Exercise 16.1 for a neutral solution at a normal body temperature of 98.6°F?

114. The compound diphenhydramine is the active ingredient in a number of over-the-counter antihistamine medications used to treat the runny nose and watery eyes associated with hay fever and other allergies. Diphenhydramine is a derivative of trimethylamine (one methyl group is replaced by a more complex organic "arm" containing two phenyl rings) $(C_6H_5)_2\,CHOCH_2CH_2N(CH_3)_2$. The compound is sold as the water-soluble hydrochloride salt, $DPH^{+}Cl^{-}$. A tablet of diphenhydramine hydrochloride contains 25.0 mg of the active ingredient. Calculate the pH of the solution if two tablets are dissolved in 100 mL of water. The pK_b of diphenhydramine is 5.47, and the formula mass of diphenhydramine hydrochloride is 291.81 amu.

115. Fluoroacetic acid is a poison that has been used by ranchers in the western United States. The ranchers place the poison in the carcasses of dead animals to kill coyotes that feed on them; unfortunately, however, eagles and hawks are also killed in the process. How many mL of 0.0953 M $Ca(OH)_2$ are needed to completely neutralize 50.0 mL of 0.262 M fluoroacetic acid solution ($pK_a = 2.59$)? What is the initial pH of the solution? What is the pH of the solution at the equivalence point?

116. Epinephrine, a secondary amine, is used to counter allergic reactions as well as to bring patients out of anesthesia and cardiac arrest. The pK_b of epinephrine is 4.31. What is the percent ionization in a 0.280 *M* solution? What is the percent ionization after enough solid epinephrine hydrochloride is added to make the final epinephrineH^+ concentration 0.982 *M*? What is the final pH of the solution?

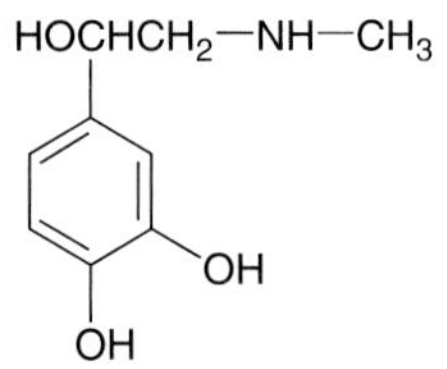

Epinephrine

$$\text{epinephrine}(aq) + H_2O(l) \rightleftharpoons \text{epinephrineH}^+(aq) + OH^-(aq)$$

117. Accidental ingestion of aspirin (acetylsalicylic acid) is probably the most common cause of childhood poisoning. Initially, salicylates stimulate the portion of the brain that controls breathing, resulting in hyperventilation (excessively intense breathing that lowers the P_{CO_2} in the lungs). Subsequently, a potentially serious rebound effect occurs, as the salicylates are converted to a weak acid, salicylic acid, in the body. Starting with the normal values of P_{CO_2} = 40 mmHg, pH = 7.40, and $[HCO_3^-]$ = 24 m*M*, show what happens during (a) the initial phase of respiratory stimulation, and (b) the subsequent phase of acid production. Why is the rebound effect dangerous?

118. Emphysema is a disease that reduces the efficiency of breathing, and as a result less CO_2 is exchanged with the atmosphere. What effect will this have on blood pH, P_{CO_2}, and $[HCO_3^-]$?

17 Solubility and Complexation Equilibria

LEARNING OBJECTIVES

- To be able to calculate the solubility of an ionic compound from its K_{sp}
- To understand the factors that determine the solubility of ionic compounds
- To be able to describe complex-ion formation quantitatively
- To understand why the solubility of many compounds depends on pH
- To know how to separate metal ions by selective precipitation

CHAPTER OUTLINE

Although Chapter 16 focused exclusively on acid–base equilibria in aqueous solutions, equilibrium concepts can also be applied to many other kinds of reactions that occur in aqueous solution. In this chapter, we describe the equilibria involved in the solubility of ionic compounds and the formation of complex ions.

Solubility equilibria involving ionic compounds are important in fields as diverse as medicine, biology, geology, and industrial chemistry. Carefully controlled precipitation reactions of calcium salts, for example, are used by many organisms to produce structural materials, such as bone and the shells that surround mollusks and bird eggs. In contrast, uncontrolled precipitation

Scanning electron micrograph of kettle scale. Hard water is a solution that consists largely of calcium and magnesium carbonate in CO_2-rich water. When the water is heated, CO_2 gas is released and the carbonate salts precipitate from solution producing a solid called *scale*.

reactions of calcium salts are partially or wholly responsible for the formation of scale in coffee makers and boilers, "bathtub rings," and kidney stones, which can be excruciatingly painful. The principles discussed in this chapter will enable you to understand how these apparently diverse phenomena are related. Solubility equilibria are also responsible for the formation of caves and their striking features, such as stalactites and stalagmites, through a long process involving the repeated dissolution and precipitation of calcium carbonate. By the end of this chapter, you will understand why barium sulfate is ideally suited for X-ray imaging of the digestive tract, and why soluble complexes of gadolinium can be used for imaging soft tissue and blood vessels using magnetic resonance imaging (MRI), even though most simple salts of both metals are toxic to humans.

17.1 Determining the Solubility of Ionic Compounds

We begin our discussion of solubility and *complexation equilibria*, the equilibria associated with the formation of complex ions, by developing quantitative methods for describing dissolution and precipitation reactions of ionic compounds in aqueous solution. Just as with acid–base equilibria, we can describe the concentrations of ions in equilibrium with an ionic solid using an equilibrium constant expression.

The Solubility Product, K_{sp}

When a slightly soluble ionic compound is added to water, some of it dissolves to form a solution, establishing an equilibrium between the pure solid and a solution of its ions. For the dissolution of calcium phosphate, one of the two main components of kidney stones, the equilibrium can be written as follows, with the solid salt on the left:*

$$Ca_3(PO_4)_2(s) \rightleftharpoons 3Ca^{2+}(aq) + 2PO_4^{3-}(aq) \tag{17.1}$$

Note the pattern

As with K, the concentration of a pure solid does not appear explicitly in K_{sp}.

The equilibrium constant for the dissolution of a sparingly soluble salt is the **solubility product** of the salt, abbreviated $\boldsymbol{K_{sp}}$. Because the concentration of a pure solid such as $Ca_3(PO_4)_2$ is a constant, it does not appear explicitly in the equilibrium constant expression (see Chapter 15). The equilibrium constant expression for the above reaction is therefore

$$K = \frac{[Ca^{2+}]^3[PO_4^{3-}]^2}{[Ca_3(PO_4)_2]} \tag{17.2}$$

$$[Ca_3(PO_4)_2]K = K_{sp} = [Ca^{2+}]^3[PO_4^{3-}]^2$$

At 25°C and pH 7.00, K_{sp} for calcium phosphate is 2.07×10^{-33}, indicating that the concentrations of $Ca^{2+}(aq)$ and $PO_4^{3-}(aq)$ ions in equilibrium with solid calcium phosphate are very low. The values of K_{sp} for some common salts are listed in Table 17.1; they show that the magnitude of K_{sp} varies dramatically for different compounds. Although in Equation 17.1, the K_{sp} is not a function of pH, changes in pH can affect the solubility of a compound, as you will discover in Section 17.4.

Solubility products are determined experimentally by either directly measuring the concentration of one of the component ions or by measuring the solubility of the compound in a given amount of water. Note, however, that whereas solubility is usually expressed in terms of *mass* of solute per 100 mL of solvent, K_{sp}, like K, is defined in terms of *molar* concentrations of the component ions.

---

* As you will discover in Section 17.4 and in more advanced chemistry courses, basic anions such as S^{2-}, PO_4^{3-}, and CO_3^{2-} react with water to produce OH^- and the corresponding protonated anion. Consequently, their calculated molarities assuming no protonation in aqueous solution are approximate.

TABLE 17.1 Solubility products, K_{sp}, for selected ionic substances at 25°C

Solid	Color	K_{sp}	Solid	Color	K_{sp}
	Acetates			**Iodides**	
$Ca(O_2CCH_3)_2 \cdot 3H_2O$	White	4×10^{-3}	Hg_2I_2[a]	Yellow	5.2×10^{-29}
	Bromides		PbI_2	Yellow	9.8×10^{-9}
AgBr	Off-white	5.35×10^{-13}		**Oxalates**	
Hg_2Br_2[a]	Yellow	6.40×10^{-23}	$Ag_2C_2O_4$	White	5.40×10^{-12}
	Carbonates		$MgC_2O_4 \cdot 2H_2O$	White	4.83×10^{-6}
$CaCO_3$	White	3.36×10^{-9}	PbC_2O_4	White	4.8×10^{-10}
$PbCO_3$	White	7.40×10^{-14}		**Phosphates**	
	Chlorides		Ag_3PO_4	White	8.89×10^{-17}
AgCl	White	1.77×10^{-10}	$Sr_3(PO_4)_2$	White	4.0×10^{-28}
Hg_2Cl_2[a]	White	1.43×10^{-18}	$FePO_4 \cdot 2H_2O$	Pink	9.91×10^{-16}
$PbCl_2$	White	1.70×10^{-5}		**Sulfates**	
	Chromates		Ag_2SO_4	White	1.20×10^{-5}
$CaCrO_4$	Yellow	7.1×10^{-4}	$BaSO_4$	White	1.08×10^{-10}
$PbCrO_4$	Yellow	2.8×10^{-13}	$PbSO_4$	White	2.53×10^{-8}
	Fluorides			**Sulfides**	
BaF_2	White	1.84×10^{-7}	Ag_2S	Black	6.3×10^{-50}
PbF_2	White	3.3×10^{-8}	CdS	Yellow	8.0×10^{-27}
	Hydroxides		PbS	Black	8.0×10^{-28}
$Ca(OH)_2$	White	5.02×10^{-6}	ZnS	White	1.6×10^{-24}
$Cu(OH)_2$	Pale blue	1×10^{-14}			
$Mn(OH)_2$	Light pink	1.9×10^{-13}			
$Cr(OH)_3$	Gray-green	6.3×10^{-31}			
$Fe(OH)_3$	Rust red	2.79×10^{-39}			

[a] These contain the Hg_2^{2+} ion.

EXAMPLE 17.1

Calcium oxalate monohydrate, $Ca(O_2CCO_2) \cdot H_2O$ (also written as $CaC_2O_4 \cdot H_2O$), is a sparingly soluble salt that is the other major component of kidney stones [along with $Ca_3(PO_4)_2$]. Its solubility in water at 25°C is 7.36×10^{-4} g/100 mL. Calculate its K_{sp}.

Given Solubility in g/100 mL

Asked for K_{sp}

Strategy

Ⓐ Write the balanced dissolution equilibrium and the corresponding solubility product expression.
Ⓑ Convert the solubility of the salt to moles per liter. From the balanced dissolution equilibrium, determine the equilibrium concentrations of the dissolved solute ions. Substitute these values into the solubility product expression to calculate K_{sp}.

Solution

Ⓐ We need to write the solubility product expression in terms of the concentrations of the component ions. For calcium oxalate monohydrate, the balanced dissolution equilibrium and the solubility product expression (abbreviating oxalate as ox^{2-}) are

$$Ca(O_2CCO_2)\cdot H_2O(s) \rightleftharpoons Ca^{2+}(aq) + {}^-O_2CCO_2{}^-(aq) + H_2O(l) \qquad K_{sp} = [Ca^{2+}][ox^{2-}]$$

Neither solid calcium oxalate monohydrate nor water appears in the solubility product expression because their concentrations are essentially constant. Ⓑ Next, we need to determine $[Ca^{2+}]$ and $[ox^{2-}]$ at equilibrium. We can use the mass of calcium oxalate monohydrate that dissolves in 100 mL of water to calculate the number of moles that dissolve in 100 mL of water. From this we can determine the number of moles that dissolve in 1.00 L of water. For dilute solutions, the density of the solution is nearly the same as that of water, so dissolving the salt in 1.00 L of water gives essentially 1.00 L of solution. Because each 1 mol of dissolved calcium oxalate monohydrate dissociates to produce 1 mol of calcium ions and 1 mol of oxalate ions, we can obtain the equilibrium concentrations that must be inserted into the solubility product expression. The number of moles of calcium oxalate monohydrate that dissolve in 100 mL of water is

$$\frac{7.36\times10^{-4}\ \cancel{g}}{146.1\ \cancel{g}/mol} = 5.04\times10^{-6}\ mol\ Ca(O_2CCO_2)\cdot H_2O$$

The number of moles of calcium oxalate monohydrate that dissolve in 1.00 L of the saturated solution is

$$\left(\frac{5.04\times10^{-6}\ mol\ Ca(O_2CCO_2)\cdot H_2O}{100\ \cancel{mL}}\right)\left(\frac{1000\ \cancel{mL}}{1.00\ L}\right) = 5.04\times10^{-5}\ mol/L = 5.04\times10^{-5}\ M$$

Because of the stoichiometry of the reaction, the concentration of Ca^{2+} and ox^{2-} ions are both $5.04\times10^{-5}\ M$. Inserting these values into the solubility product expression gives

$$K_{sp} = [Ca^{2+}][ox^{2-}] = (5.04\times10^{-5})(5.04\times10^{-5}) = 2.54\times10^{-9}$$

In our calculation in Example 17.1 we have ignored the reaction of the weakly basic anion with water, which tends to make the actual solubility of salts greater than the calculated value.

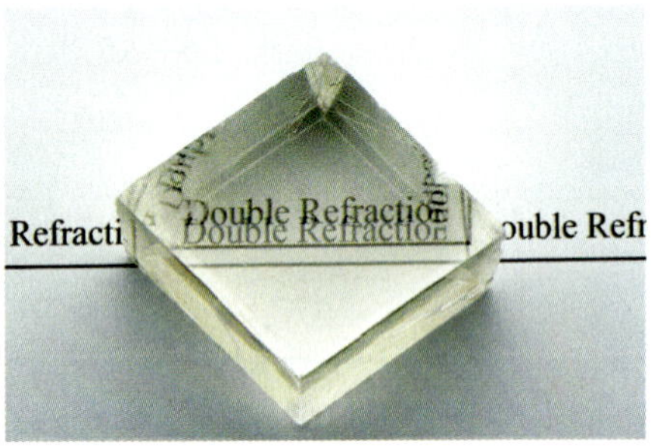

A crystal of calcite ($CaCO_3$), illustrating the phenomenon of double refraction. When a transparent crystal of calcite is placed over a page, we see *two* images of the letters.

EXERCISE 17.1

One crystalline form of calcium carbonate, $CaCO_3$, is the mineral sold as "calcite" in mineral and gem shops. The solubility of calcite in water is 0.67 mg/100 mL. Calculate its K_{sp}.

Answer 4.5×10^{-9}

Tabulated values of K_{sp} can also be used to estimate the solubility of a salt with a procedure that is essentially the reverse of the one used in Example 17.1. In this case, we treat the problem as a typical equilibrium problem and set up a table of initial concentrations, changes in concentration, and final concentrations as we did in Chapter 15, remembering that the concentration of the pure solid is essentially constant.

EXAMPLE 17.2

We saw that the value of K_{sp} for $Ca_3(PO_4)_2$ is 2.07×10^{-33} at 25°C. Calculate the aqueous solubility of $Ca_3(PO_4)_2$ in terms of (a) the molarity of ions produced in solution, and (b) the mass of salt that dissolves in 100 mL of water at 25°C.

Given K_{sp}

Asked for Molar concentration, mass of salt that dissolves in 100 mL of water

Strategy

Ⓐ Write the balanced equilibrium equation for the dissolution reaction, and construct a table showing the concentrations of the species produced in solution. Insert the appropriate values into the solubility product expression, and calculate the molar solubility at 25°C.

Ⓑ Calculate the mass of solute in 100 mL of solution from the molar solubility of the salt. Assume that the volume of the solution is the same as the volume of the solvent.

Solution

(a) The dissolution equilibrium for $Ca_3(PO_4)_2$ (Equation 17.1) is shown in the table. Because we are starting with distilled water, the initial concentration of both calcium and phosphate ions is zero. For every 1 mol of $Ca_3(PO_4)_2$ that dissolves, 3 mol of Ca^{2+} and 2 mol of PO_4^{3-} ions are produced in solution. If we let x equal the solubility of $Ca_3(PO_4)_2$ in moles per liter, then the change in $[Ca^{2+}]$ will be $+3x$, and the change in $[PO_4^{3-}]$ will be $+2x$. We can insert these values into the table.

	$Ca_3(PO_4)_2(s)$	$\rightleftharpoons$ $3Ca^{2+}(aq)$	+ $2PO_4^{3-}(aq)$
	$Ca_3(PO_4)_2$	**$[Ca^{2+}]$**	**$[PO_4^{3-}]$**
Initial	Pure solid	0	0
Change	—	$+3x$	$+2x$
Final	Pure solid	$3x$	$2x$

Although the *amount* of solid $Ca_3(PO_4)_2$ changes as some of it dissolves, its *molar concentration* does not change. We now insert the expressions for the equilibrium concentrations of the ions into the solubility product expression (Equation 17.2):

$$K_{sp} = [Ca^{2+}]^3[PO_4^{3-}]^2 = (3x)^3(2x)^2$$
$$108x^5 = 2.1 \times 10^{-33}$$
$$x^5 = 1.9 \times 10^{-35}$$
$$x = 1.1 \times 10^{-7}\ M$$

This is the molar solubility of calcium phosphate at 25°C. Notice, however, that the molarity of the ions is $2x$ and $3x$, which means that $[PO_4^{3-}] = 2.2 \times 10^{-7}$ and $[Ca^{2+}] = 3.3 \times 10^{-7}$.

(b) To find the mass of solute in 100 mL of solution, we assume that the density of this dilute solution is the same as the density of water because of the low solubility of the salt, so that 100 mL of water gives 100 mL of solution. We can then determine the amount of salt that dissolves in 100 mL of water:

$$\left(\frac{1.1 \times 10^{-7}\ \text{mol}}{1\ \text{L}}\right) 100\ \text{mL}\left(\frac{1\ \text{L}}{1000\ \text{mL}}\right)\left(\frac{310.18\ \text{g}\ Ca_3(PO_4)_2}{1\ \text{mol}}\right) = 3.4 \times 10^{-6}\ \text{g}\ Ca_3(PO_4)_2 \text{ in 100 mL water}$$

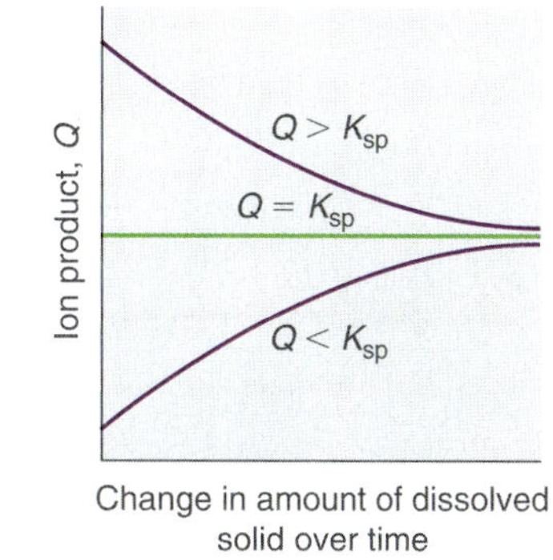

Figure 17.1 The relationship between Q and K_{sp}. If Q is less than K_{sp}, the solution is unsaturated and more solid will dissolve until the system reaches equilibrium ($Q = K_{sp}$). If Q is greater than K_{sp}, the solution is supersaturated and solid will precipitate until $Q = K_{sp}$. If $Q = K_{sp}$, the rate of dissolution is equal to the rate of precipitation; the solution is saturated and no net change in the amount of dissolved solid will occur.

EXERCISE 17.2

The solubility product of silver carbonate, Ag_2CO_3, is 8.46×10^{-12} at 25°C. Calculate **(a)** the molarity of a saturated solution, and **(b)** the mass of silver carbonate that will dissolve in 100 mL of water at this temperature.

Answer **(a)** $1.28 \times 10^{-4}\ M$; **(b)** 3.54 mg

The Ion Product

The **ion product (Q)** of a salt is the product of the concentrations of the ions in solution raised to the same powers as in the solubility product expression. It is analogous to the reaction quotient (Q) discussed for gaseous equilibria in Chapter 15. Whereas K_{sp} describes equilibrium concentrations, the ion product describes concentrations that are *not* necessarily equilibrium concentrations.

Note the pattern

The ion product Q is analogous to the reaction quotient Q for gaseous equilibria.

As summarized in Figure 17.1, there are three possible conditions for an aqueous solution of an ionic solid:

1. $Q < K_{sp}$: the solution is unsaturated, and more of the ionic solid, if available, will dissolve.

2. $Q = K_{sp}$: the solution is saturated and at equilibrium.
3. $Q > K_{sp}$: the solution is supersaturated, and ionic solid will precipitate.

The process of calculating the value of the ion product and comparing it with the magnitude of the solubility product is a straightforward way to determine whether a solution is unsaturated, saturated, or supersaturated. More important, the ion product tells chemists whether a precipitate will form when solutions of two soluble salts are mixed.

EXAMPLE 17.3

We mentioned that barium sulfate is used in medical imaging of the gastrointestinal tract. Its solubility product is 1.08×10^{-10} at 25°C, so it is ideally suited for this purpose because of its low solubility when a "barium milkshake" is consumed by a patient. The pathway of the sparingly soluble salt can be easily monitored by X-rays. Will barium sulfate precipitate if 10.0 mL of 0.0020 *M* Na_2SO_4 is added to 100 mL of 3.2×10^{-4} *M* $BaCl_2$? Recall that NaCl is highly soluble in water.

Given K_{sp}, and volumes and concentrations of reactants

Asked for Whether precipitate will form

Strategy

A Write the balanced equilibrium equation for the precipitation reaction and the expression for K_{sp}.

B Determine the concentrations of all ions in solution when the solutions are mixed, and use them to calculate the value of the ion product Q.

C Compare the values of Q and K_{sp} to decide whether a precipitate will form.

Solution

A The only slightly soluble salt that can be formed when these two solutions are mixed is $BaSO_4$ because NaCl is highly soluble. The equation for the precipitation of $BaSO_4$ is

$$BaSO_4(s) \rightleftharpoons Ba^{2+}(aq) + SO_4{}^{2-}(aq)$$

The solubility product expression is

$$K_{sp} = [Ba^{2+}][SO_4{}^{2-}] = 1.08 \times 10^{-10}$$

B To solve this problem we must first calculate the value of the ion product, $Q = [Ba^{2+}][SO_4^{2-}]$, using the concentrations of the ions that are present after the solutions are mixed and before any reaction occurs. The concentration of Ba^{2+} when the solutions are mixed is the total number of moles of Ba^{2+} in the original 100 mL of $BaCl_2$ solution divided by the final volume (100 mL + 10 mL = 110 mL):

$$\text{moles } Ba^{2+} = 100 \cancel{\text{mL}}\left(\frac{1 \cancel{\text{L}}}{1000 \cancel{\text{mL}}}\right)\left(\frac{3.2 \times 10^{-4} \text{ mol}}{1 \cancel{\text{L}}}\right) = 3.2 \times 10^{-5} \text{ mol } Ba^{2+}$$

$$[Ba^{2+}] = \left(\frac{3.2 \times 10^{-5} \text{ mol } Ba^{2+}}{110 \cancel{\text{mL}}}\right)\left(\frac{1000 \cancel{\text{mL}}}{1 \text{ L}}\right) = 2.9 \times 10^{-4}\ M\ Ba^{2+}$$

Similarly, the concentration of $SO_4{}^{2-}$ after mixing is the total number of moles of SO_4^{2-} in the original 10.0 mL of Na_2SO_4 solution divided by the final volume (110 mL):

$$\text{moles } SO_4{}^{2-} = 10.0 \cancel{\text{mL}}\left(\frac{1 \cancel{\text{L}}}{1000 \cancel{\text{mL}}}\right)\left(\frac{0.0020 \text{ mol}}{1 \cancel{\text{L}}}\right) = 2.0 \times 10^{-5} \text{ mol } SO_4{}^{2-}$$

$$[SO_4{}^{2}] = \left(\frac{2.0 \times 10^{-5} \text{ mol } SO_4{}^{2-}}{110 \cancel{\text{mL}}}\right)\left(\frac{1000 \cancel{\text{mL}}}{1 \text{ L}}\right) = 1.8 \times 10^{-4}\ M\ SO_4{}^{2-}$$

We can now calculate the ion product Q:

$$Q = [Ba^{2+}][SO_4^{2-}] = (2.9 \times 10^{-4})(1.8 \times 10^{-4}) = 5.2 \times 10^{-8}$$

C We now compare Q, the ion product, with the solubility product K_{sp}. If $Q > K_{sp}$, then $BaSO_4$ will precipitate, but if $Q < K_{sp}$, it will not. Since $Q > K_{sp}$, we predict that $BaSO_4$ will precipitate when the two solutions are mixed. In fact, $BaSO_4$ will continue to precipitate until the system reaches equilibrium, which occurs when $[Ba^{2+}][SO_4^{2-}] = K_{sp} = 1.08 \times 10^{-10}$.

EXERCISE 17.3

The solubility product of calcium fluoride, CaF_2, is 3.45×10^{-11}. If 2.0 mL of a 0.10 M solution of NaF is added to 128 mL of a 2.0×10^{-5} M solution of $Ca(NO_3)_2$, will CaF_2 precipitate?

Answer yes ($Q = 4.7 \times 10^{-11} > K_{sp}$)

The Common Ion Effect and Solubility

The solubility product expression tells us that the equilibrium concentrations of cation and anion are inversely related. That is, as the concentration of the anion increases, the maximum concentration of the cation needed for precipitation to occur decreases, and vice versa, so that K_{sp} is constant. Consequently, the solubility of an ionic compound depends on the concentrations of other salts that contain the same ions. This dependency is another example of the common ion effect discussed in Section 16.6: adding a common cation or anion shifts a solubility equilibrium in the direction predicted by Le Châtelier's principle. As a result, *the solubility of any sparingly soluble salt is almost always decreased by the presence of a soluble salt that contains a common ion.**

Consider, for example, the effect of adding a soluble salt such as $CaCl_2$ to a saturated solution of calcium phosphate. We have seen that the solubility of $Ca_3(PO_4)_2$ in water at 25°C is 1.14×10^{-7} M ($K_{sp} = 2.07 \times 10^{-33}$). Thus, a saturated solution of calcium phosphate in water contains $3 \times 1.14 \times 10^{-7}$ M $= 3.42 \times 10^{-7}$ M Ca^{2+} and $2 \times 1.14 \times 10^{-7}$ M $= 2.28 \times 10^{-7}$ M PO_4^{3-} (neglecting hydrolysis to HPO_4^{2-} as described in Chapter 16), according to the stoichiometry shown in Equation 17.1. If $CaCl_2$ is added to a saturated solution of $Ca_3(PO_4)_2$, the Ca^{2+} ion concentration will increase such that $[Ca^{2+}] > 3.42 \times 10^{-7}$ M, making $Q > K_{sp}$. The only way the system can return to equilibrium is for the reaction in Equation 17.1 to proceed to the left, resulting in precipitation of calcium phosphate. This will decrease the concentration of both Ca^{2+} and PO_4^{3-} until $Q = K_{sp}$.

Note the pattern

The common ion effect usually decreases the solubility of a sparingly soluble salt.

EXAMPLE 17.4

Calculate the solubility of calcium phosphate in 0.20 M $CaCl_2$.

Given Concentration of $CaCl_2$ solution

Asked for Solubility of $Ca_3(PO_4)_2$ in $CaCl_2$ solution

Strategy

A Write the balanced equilibrium equation for the dissolution of $Ca_3(PO_4)_2$. Tabulate the concentrations of all species produced in solution.

B Substitute the appropriate values into the expression for the solubility product, and calculate the solubility of $Ca_3(PO_4)_2$.

* The exceptions generally involve the formation of complex ions, which is discussed in Section 17.3.

Solution

A The balanced equilibrium equation is given in the table. If we let x equal the solubility of $Ca_3(PO_4)_2$ in moles per liter, then the change in $[Ca^{2+}]$ is once again $+3x$, and the change in $[PO_4^{3-}]$ is $+2x$. We can insert these values into the table.

	$Ca_3(PO_4)_2(s)$ $\rightleftharpoons$	$3Ca^{2+}(aq)$ +	$2PO_4^{3-}(aq)$
	$Ca_3(PO_4)_2$	$[Ca^{2+}]$	$[PO_4^{3-}]$
Initial	Pure solid	0.20	0
Change	—	$+ 3x$	$+ 2x$
Final	Pure solid	$0.20 + 3x$	$2x$

B The K_{sp} expression is

$$K_{sp} = [Ca^{2+}]^3[PO_4^{3-}]^2 = (0.20 + 3x)^3(2x)^2 = 2.07 \times 10^{-33}$$

Because $Ca_3(PO_4)_2$ is a sparingly soluble salt, we can reasonably expect that $x << 0.20$. Thus, $(0.20 + 3x)$ M is approximately 0.20 M, which simplifies the K_{sp} expression to

$$K_{sp} = (0.20)^3(2x)^2 = 2.07 \times 10^{-33}$$
$$x^2 = 6.5 \times 10^{-32}$$
$$x = 2.5 \times 10^{-16}\ M$$

This value is the solubility of calcium phosphate in 0.20 M $CaCl_2$ at 25°C. It is approximately nine orders of magnitude less than its solubility in pure water, as we would expect based on Le Châtelier's principle. Note that, with one exception, this example was identical to Example 17.2—here the initial $[Ca^{2+}]$ was 0.20 M rather than 0.

EXERCISE 17.4

Calculate the solubility of silver carbonate in a 0.25 M solution of sodium carbonate. The solubility of silver carbonate in pure water is 8.45×10^{-12} at 25°C.

Answer 2.9×10^{-6} M (versus 1.3×10^{-4} M in pure water)

17.2 • Factors That Affect Solubility

The solubility product of an ionic compound describes the concentrations of *ions* in equilibrium with a solid, but what happens if some of the cations become associated with anions rather than being completely surrounded by solvent? Then predictions of the total solubility of the compound based either on the assumption that the solute exists solely as ions or on the measured concentrations of those ions would substantially underestimate the actual solubility. In general, there are four reasons that the solubility of a compound may be other than expected: ion-pair formation, incomplete dissociation of molecular solutes, formation of complex ions, and changes in pH. The first two situations are described in this section, the third is discussed in Section 17.3, and changes in pH are discussed in Section 17.4.

Ion-Pair Formation

An **ion pair** consists of a cation and anion that are in intimate contact in solution, rather than separated by solvent (Figure 17.2). The ions in an ion pair are held together by the same attractive electrostatic forces that we discussed in Chapter 8 for ionic solids. As a result, the ions in an ion pair migrate as a single unit, whose net charge is the sum of the charges on the ions. In many ways, we can view an ion pair as a species intermediate between the ionic solid (in which each ion participates in many cation–anion interactions that hold the ions in a rigid array) and the completely dissociated ions in solution (where each is fully surrounded by water molecules and free to migrate independently).

As illustrated for calcium sulfate in the following equation, a second equilibrium must be included to describe the solubility of salts that form ion pairs:

$$CaSO_4(s) \rightleftharpoons \underset{\text{Ion pair}}{Ca^{2+} \cdot SO_4{}^{2-}(aq)} \rightleftharpoons Ca^{2+}(aq) + SO_4{}^{2-}(aq) \quad (17.3)$$

The ion pair is represented by the symbols of the individual ions separated by a dot, to indicate that they are associated in solution. The formation of an ion pair is a dynamic process, just like any other equilibrium, so a particular ion pair may exist only briefly before dissociating into the free ions, each of which may later associate briefly with other ions.

Note the pattern

The formation of ion pairs increases the solubility of a salt.

Ion-pair formation can have a major effect on the measured solubility of a salt. For example, the measured value of K_{sp} for calcium sulfate is 4.93×10^{-5} at 25°C. The solubility of $CaSO_4$ should be 7.02×10^{-3} *M* if the only equilibrium involved were

$$CaSO_4(s) \rightleftharpoons Ca^{2+}(aq) + SO_4{}^{2-}(aq) \quad (17.4)$$

In fact, the experimentally measured solubility of calcium sulfate at 25°C is 1.6×10^{-2} *M*, almost twice the value predicted from its K_{sp}. The reason for the discrepancy is that the concentration of ion pairs in a saturated $CaSO_4$ solution is almost as high as the concentration of hydrated ions. Recall that the magnitude of attractive electrostatic interactions is greatest for small, highly charged ions. Hence, ion-pair formation is most important for salts that contain M^{2+} and M^{3+} ions, such as Ca^{2+} and La^{3+}, and is relatively unimportant for salts that contain monopositive cations, except for the smallest, Li^+. We therefore expect a saturated solution of $CaSO_4$ to contain a high concentration of ion pairs and its solubility to be greater than predicted from its K_{sp}.

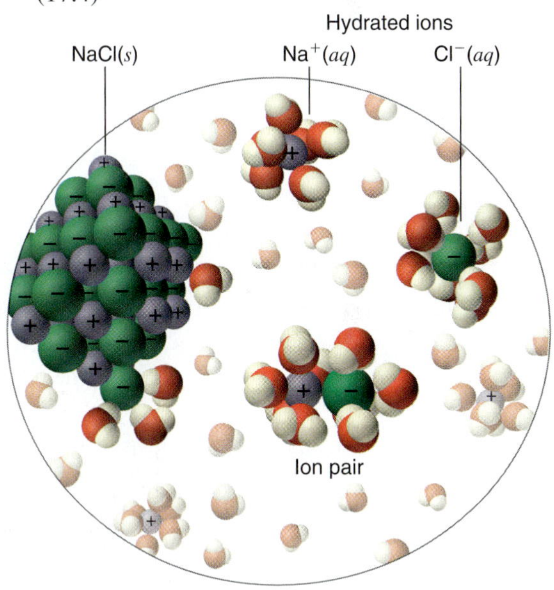

Figure 17.2 Ion-pair formation. In an ion pair, the cation and anion are in intimate contact in solution and migrate as a single unit. They are not completely dissociated and individually surrounded by solvent molecules, as are the hydrated ions, which are free to migrate independently.

Incomplete Dissociation

A molecular solute may also be more soluble than predicted by the measured concentrations of ions in solution due to *incomplete dissociation*. This is particularly common with weak organic acids (see Chapter 16). Although strong acids (HA) dissociate completely into their constituent ions (H^+ and A^-) in water, weak acids such as carboxylic acids do not ($K_a = 1.5 \times 10^{-5}$). However, the molecular (undissociated) form of a weak acid (HA) is often quite soluble in water; for example, acetic acid (CH_3CO_2H) is completely miscible with water. Many carboxylic acids, however, have only limited solubility in water, such as benzoic acid ($C_6H_5CO_2H$), with $K_a = 6.25 \times 10^{-5}$. Just as with calcium sulfate, we need to include an additional equilibrium to describe the solubility of benzoic acid:

$$C_6H_5CO_2H(s) \rightleftharpoons C_6H_5CO_2H(aq) \rightleftharpoons C_6H_5CO_2{}^-(aq) + H^+(aq) \quad (17.5)$$

Note the pattern

Incomplete dissociation of a molecular solute can increase the solubility of the solute.

In a case like this, measuring only the concentration of the ions grossly underestimates the total concentration of the organic acid in solution. In the case of benzoic acid, for example, the pH of a saturated solution at 25°C is 2.85, corresponding to $[H^+] = [C_6H_5CO_2{}^-] = 1.4 \times 10^{-3}$ *M*. The total concentration of benzoic acid in the solution, however, is 2.8×10^{-2} *M*. Thus, approximately

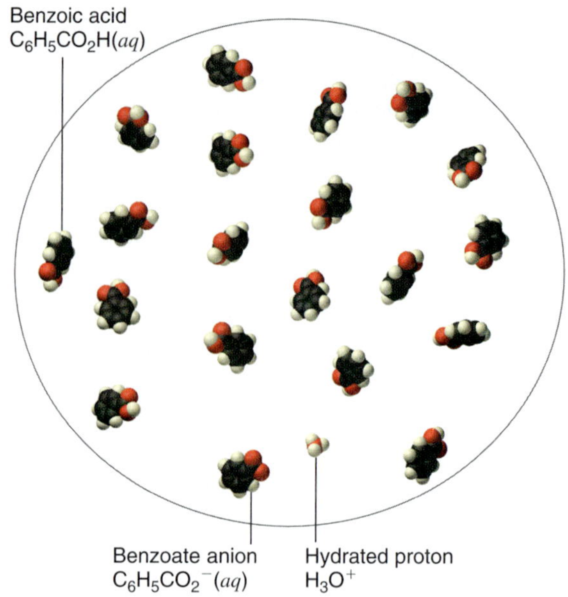

Figure 17.3 Incomplete dissociation of a molecular solute. In a saturated solution of benzoic acid in water at 25°C, only about 5% of the dissolved benzoic acid molecules are dissociated to form benzoate anions and hydrated protons. The remaining 95% exists in solution in the form of hydrated neutral molecules. (H_2O molecules are omitted for clarity.)

95% of the benzoic acid in solution is in the form of hydrated neutral molecules, $C_6H_5CO_2H(aq)$, and only about 5% is present as the dissociated ions (Figure 17.3).

Although ion pairs such as $Ca^{2+} \cdot SO_4^{2-}$, and undissociated electrolytes, such as $C_6H_5CO_2H$, are both electrically neutral, there is a major difference in the forces responsible for their formation. The ion pair is held together by simple electrostatic attractive forces between the cation and the anion, whereas the undissociated acid is held together by a polar covalent O—H bond.

17.3 Complex-Ion Formation

In Chapter 4, you learned that metal ions in aqueous solution are *hydrated*—that is, surrounded by a shell of usually four or six water molecules. A hydrated ion is one kind of **complex ion** (or, simply, *complex*), a species formed between a central metal ion and one or more surrounding **ligands**, molecules or ions that contain at least one lone pair of electrons.

A complex ion forms from a metal ion and a ligand because of a Lewis acid–base interaction. The positively charged metal ion acts as a Lewis acid, and the ligand, with one or more lone pairs of electrons, acts as a Lewis base. Small, highly charged metal ions such as Cu^{2+} or Ru^{3+}, have the greatest tendency to act as Lewis acids, and consequently they have the greatest tendency to form complex ions.

As an example of complex-ion formation, consider the addition of ammonia to an aqueous solution of the hydrated Cu^{2+} ion, $[Cu(H_2O)_6]^{2+}$. Because it is a stronger base than H_2O, ammonia replaces the water molecules in the hydrated ion to form the $[Cu(NH_3)_4(H_2O)_2]^{2+}$ ion. Formation of the $[Cu(NH_3)_4(H_2O)_2]^{2+}$ complex is accompanied by a dramatic color change, as shown in Figure 17.4. The solution changes from the light blue of $[Cu(H_2O)_6]^{2+}$ to the blue-violet characteristic of the $[Cu(NH_3)_4(H_2O)_2]^{2+}$ ion.

Figure 17.4 Complex-ion formation. An aqueous solution of $CuSO_4$ consists of hydrated Cu^{2+} ions in the form of pale blue $[Cu(H_2O)_6]^{2+}$ *(left)*. Addition of aqueous ammonia to the solution results in the formation of the intensely blue-violet $[Cu(NH_3)_4(H_2O)_2]^{2+}$, usually written as $[Cu(NH_3)_4^{2+}]$ ion *(right)* because ammonia, a stronger base than H_2O, replaces water molecules from the hydrated Cu^{2+} ion.

The Formation Constant

The replacement of water molecules from $[Cu(H_2O)_6]^{2+}$ by ammonia occurs in sequential steps. Omitting the water molecules bound to Cu^{2+} for simplicity, we can write the equilibrium reactions as

$$\begin{aligned} Cu^{2+}(aq) + NH_3(aq) &\rightleftharpoons [Cu(NH_3)]^{2+}(aq) \quad K_1 \\ [Cu(NH_3)]^{2+}(aq) + NH_3(aq) &\rightleftharpoons [Cu(NH_3)_2]^{2+}(aq) \quad K_2 \end{aligned} \tag{17.6}$$

and so on. The sum of the stepwise reactions is the overall equation for the formation of the complex ion:*

$$Cu^{2+}(aq) + 4NH_3(aq) \rightleftharpoons [Cu(NH_3)_4]^{2+}(aq) \tag{17.7}$$

The equilibrium constant for the formation of the complex ion from the hydrated ion is called the **formation constant (K_f)**. The equilibrium constant expression for K_f has the same general form as any other equilibrium constant expression. In this case, the expression is

Note the pattern

The formation constant, K_f, has the same general form as any other equilibrium constant expression.

$$K_f = \frac{[[Cu(NH_3)_4]^{2+}]}{[Cu^{2+}][NH_3]^4} = 2.1 \times 10^{13} = K_1K_2K_3K_4 \tag{17.8}$$

Notice that water a pure liquid, does not appear explicitly in the equilibrium constant expression, and that the hydrated $Cu^{2+}(aq)$ ion is represented as Cu^{2+} for simplicity. As for any equilibrium, the larger the value of the equilibrium constant (in this case, K_f), the more stable the product. With $K_f = 2.1 \times 10^{13}$, the $[Cu(NH_3)_4(H_2O)_2]^{2+}$ complex ion is very stable. The formation constants for some common complex ions are listed in Table 17.2.

EXAMPLE 17.5

If 12.5 g of $Cu(NO_3)_2{\cdot}6H_2O$ is added to 500 mL of 1.00 *M* aqueous ammonia, what is the equilibrium concentration of $Cu^{2+}(aq)$?

Given Mass of Cu^{2+} salt, volume and concentration of ammonia solution

Asked for Equilibrium concentration of $Cu^{2+}(aq)$

Strategy

Ⓐ Calculate the initial concentration of Cu^{2+}due to the addition of copper(II) nitrate hexahydrate. Use the stoichiometry of the reaction shown in Equation 17.7 to construct a table

* Note that the hydrated Cu^{2+} ion contains six H_2O ligands, but the complex ion that is produced contains only four NH_3 ligands, not six. The reasons for this apparently unusual behavior will be discussed in Chapter 23.

TABLE 17.2 Formation constants, K_f, for selected complex ions in aqueous solution*

	Complex Ion	Equilibrium Equation	K_f
Ammonia Complexes	$[Ag(NH_3)_2]^+$	$Ag^+ + 2NH_3 \rightleftharpoons [Ag(NH_3)_2]^+$	1.1×10^7
	$[Cu(NH_3)_4]^{2+}$	$Cu^{2+} + 4NH_3 \rightleftharpoons [Cu(NH_3)_4]^{2+}$	2.1×10^{13}
	$[Ni(NH_3)_6]^{2+}$	$Ni^{2+} + 6NH_3 \rightleftharpoons [Ni(NH_3)_6]^{2+}$	5.5×10^8
Cyanide Complexes	$[Ag(CN)_2]^-$	$Ag^+ + 2CN^- \rightleftharpoons [Ag(CN)_2]^-$	1.1×10^{18}
	$[Ni(CN)_4]^{2-}$	$Ni^{2+} + 4CN^- \rightleftharpoons [Ni(CN)_4]^{2-}$	2.2×10^{31}
	$[Fe(CN)_6]^{3-}$	$Fe^{3+} + 6CN^- \rightleftharpoons [Fe(CN)_6]^{3-}$	1×10^{42}
Hydroxide Complexes	$[Zn(OH)_4]^{2-}$	$Zn^{2+} + 4OH^- \rightleftharpoons [Zn(OH)_4]^{2-}$	4.6×10^{17}
	$[Cr(OH)_4]^-$	$Cr^{3+} + 4OH^- \rightleftharpoons [Cr(OH)_4]^-$	8.0×10^{29}
Halide Complexes	$[HgCl_4]^{2-}$	$Hg^{2+} + 4Cl^- \rightleftharpoons [HgCl_4]^{2-}$	1.2×10^{15}
	$[CdI_4]^{2-}$	$Cd^{2+} + 4I^- \rightleftharpoons [CdI_4]^{2-}$	2.6×10^5
	$[AlF_6]^{3-}$	$Al^{3+} + 6F^- \rightleftharpoons [AlF_6]^{3-}$	6.9×10^{19}
Other Complexes	$[Ag(S_2O_3)_2]^{3-}$	$Ag^+ + 2S_2O_3^{2-} \rightleftharpoons [Ag(S_2O_3)_2]^{3-}$	2.9×10^{13}
	$[Fe(C_2O_4)_3]^{3-}$	$Fe^{3+} + 3C_2O_4^{2-} \rightleftharpoons [Fe(C_2O_4)_3]^{3-}$	2.0×10^{20}

Source of data: *Lange's Handbook of Chemistry*, 15th Edition (1999).

*Reported values are overall formation constants.

showing the initial concentrations, changes in concentrations, and final concentrations of all species in solution.

B Substitute the final concentrations into the expression for the formation constant (Equation 17.8) to calculate the equilibrium concentration of $Cu^{2+}(aq)$.

Solution

Adding an ionic compound that contains Cu^{2+} to an aqueous ammonia solution will result in the formation of $[Cu(NH_3)_4]^{2+}(aq)$, as shown in Equation 17.7. We assume that the volume change caused by adding solid copper(II) nitrate to aqueous ammonia is negligible. **A** The initial concentration of Cu^{2+} from the amount of added copper nitrate prior to any reaction is

$$12.5\ \cancel{g}\ Cu(NO_3)_2 \cdot 6H_2O \left(\frac{1\ mol}{295.65\ \cancel{g}}\right)\left(\frac{1}{500\ \cancel{mL}}\right)\left(\frac{1000\ \cancel{mL}}{1\ L}\right) = 0.0846\ M$$

Because the stoichiometry of the reaction is four NH_3 to one Cu^{2+}, the amount of NH_3 required to react completely with the Cu^{2+} is $4(0.0846) = 0.338\ M$. The concentration of ammonia after complete reaction is $1.00\ M - 0.338\ M = 0.66\ M$. These results are summarized in the first two lines of the table below. Because the equilibrium constant for the reaction is large (2.1×10^{13}) the equilibrium will lie far to the right. Thus, we will assume that the formation of $[Cu(NH_3)_4]^{2+}$ in the first step is complete and then allow some of it to dissociate into Cu^{2+} and NH_3 until equilibrium has been reached. If we define x as the amount of Cu^{2+} produced by the dissociation reaction, then the stoichiometry of the reaction tells us that the change in the concentration of $[Cu(NH_3)_4]^{2+}$ is $-x$, and the change in the concentration of ammonia is $+4x$, as indicated in the table. The final concentrations of all species (in the bottom row of the table) are the sums of the concentrations after complete reaction and the changes in concentrations.

	Cu^{2+}	+	$4NH_3$	$\rightleftharpoons$	$[Cu(NH_3)_4]^{2+}$
	$[Cu^{2+}]$		$[NH_3]$		$[[Cu(NH_3)_4]^{2+}]$
Initial	0.0846		1.00		0
After complete reaction	0		0.66		0.0846
Change	$+x$		$+4x$		$-x$
Final	x		$0.66 + 4x$		$0.0846 - x$

B Substituting the final concentrations into the expression for the formation constant (Equation 17.8) and assuming that $x << 0.0846$, which allows us to remove x from the sum and difference, give

$$K_f = \frac{[[Cu(NH_3)_4]^{2+}]}{[Cu^{2+}][NH_3]^4} = \frac{0.0846 - x}{x(0.66 + 4x)^4} \approx \frac{0.0846}{x(0.66)^4} = 2.1 \times 10^{13}$$

$$x = 2.1 \times 10^{-14}$$

The value of x indicates that our assumption was justified. The equilibrium concentration of $Cu^{2+}(aq)$ in a 1.00 M ammonia solution is therefore 2.1×10^{-14} M.

EXERCISE 17.5

The ferrocyanide ion, $[Fe(CN)_6]^{4-}$, is very stable, with a K_f of 1×10^{35}. Calculate the concentration of cyanide ion in equilibrium with a 0.65 M solution of $K_4[Fe(CN)_6]$.

Answer 2×10^{-6} M

The Effect of Complex-Ion Formation on Solubility

What happens to the solubility of a sparingly soluble salt if a ligand that forms a stable complex ion is added to the solution? One such example occurs in black-and-white photography, which was discussed briefly in Chapter 4.

Recall that black-and-white photographic film contains light-sensitive microcrystals of AgBr, or mixtures of AgBr and other silver halides. AgBr is a sparingly soluble salt, with a K_{sp} of 5.35×10^{-13} at 25°C. When the shutter of the camera opens, the light from the object being photographed strikes some of the crystals on the film and initiates a photochemical reaction that converts AgBr to black Ag metal. Well-formed, stable negative images appear in tones of gray, corresponding to the number of grains of AgBr converted, with the areas exposed to the most light being darkest. To fix the image and prevent more AgBr crystals from being converted to Ag metal during processing of the film, the unreacted AgBr on the film is removed using a complexation reaction to dissolve the sparingly soluble salt.

The reaction for the dissolution of silver bromide is

$$AgBr(s) \rightleftharpoons Ag^+(aq) + Br^-(aq) \qquad K_{sp} = 5.35 \times 10^{-13} \text{ at } 25°C \qquad (17.9)$$

The equilibrium lies far to the left, and the equilibrium concentrations of Ag^+ and Br^- ions are very low (7.31×10^{-7} M). As a result, removing unreacted AgBr from even a single roll of film using pure water would require tens of thousands of liters of water and a great deal of time. Le Châtelier's principle tells us, however, that we can drive the reaction to the right by removing one of the products, which will cause more AgBr to dissolve. Bromide ion is difficult to remove chemically, but silver ion forms a variety of stable two-coordinate complexes with neutral ligands such as ammonia or with anionic ligands such as cyanide or thiosulfate ($S_2O_3^{2-}$). In photographic processing, excess AgBr is dissolved using a concentrated solution of sodium thiosulfate.

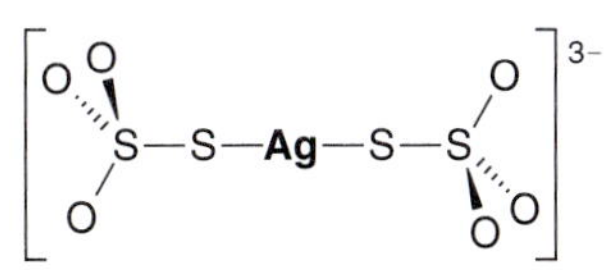

Thiosulfate complex of Ag^+

The reaction of Ag^+ with thiosulfate is

$$Ag^+(aq) + 2S_2O_3^{2-}(aq) \rightleftharpoons [Ag(S_2O_3)_2]^{3-}(aq) \qquad K_f = 2.9 \times 10^{13} \qquad (17.10)$$

The magnitude of the equilibrium constant indicates that almost all Ag^+ ions in solution will be immediately complexed by thiosulfate to form $[Ag(S_2O_3)_2]^{3-}$. We can see the effect of thiosulfate on the solubility of AgBr by writing the appropriate reactions and adding them:

$$\begin{array}{rll} AgBr(s) \rightleftharpoons \cancel{Ag^+(aq)} + Br^-(aq) & & K_{sp} = 5.35 \times 10^{-13} \\ \cancel{Ag^+(aq)} + 2S_2O_3^{2-}(aq) \rightleftharpoons [Ag(S_2O_3)_2]^{3-}(aq) & & K_f = 2.9 \times 10^{13} \\ \hline AgBr(s) + 2S_2O_3^{2-}(aq) \rightleftharpoons [Ag(S_2O_3)_2]^{3-}(aq) + Br^-(aq) & & K = K_{sp}K_f = 15 \end{array} \tag{17.11}$$

Comparing K with K_{sp} shows that the formation of the complex ion increases the solubility of AgBr by approximately 3×10^{13}. The dramatic increase in solubility combined with the low cost and low toxicity explains why sodium thiosulfate is almost universally used for developing black-and-white film. If desired, the silver can be recovered from the thiosulfate solution using any of several methods and recycled.

Note the pattern

If a complex ion has a large K_f, complex-ion formation can dramatically increase the solubility of sparingly soluble salts.

EXAMPLE 17.6

Due to the common ion effect, we might expect a salt such as AgCl to be much less soluble in a concentrated solution of KCl than in water. Such an assumption would be incorrect, however, because it ignores the fact that silver ion tends to form a two-coordinate complex with chloride ions, $AgCl_2^-$. Calculate the solubility of AgCl in (a) pure water; (b) a 1.0 *M* KCl solution, ignoring complex-ion formation; and (c) the same solution taking complex-ion formation into account, assuming that $AgCl_2^-$ is the only Ag^+ complex that forms in significant concentrations. At 25°C, $K_{sp} = 1.77 \times 10^{-10}$ for AgCl and $K_f = 1.1 \times 10^5$ for $AgCl_2^-$.

Given K_{sp} of AgCl, K_f of $AgCl_2^-$, and KCl concentration

Asked for Solubility of AgCl in water and in KCl solution with and without complex-ion formation

Strategy

A Write the solubility product expression for AgCl, and calculate the concentration of Ag^+ and Cl^- in water.

B Calculate the concentration of Ag^+ in the KCl solution.

C Write balanced equations for the dissolution of AgCl and for the formation of the $AgCl_2^-$ complex. Add the two equations, and calculate the equilibrium constant for the overall equilibrium.

D Write the equilibrium constant expression for the overall reaction. Solve for the concentration of the complex ion.

Solution

(a) **A** If we let x equal the solubility of AgCl, then at equilibrium $[Ag^+] = [Cl^-] = x\ M$. Substituting this value into the solubility product expression gives

$$K_{sp} = [Ag^+][Cl^-] = (x)(x) = x^2 = 1.77 \times 10^{-10}$$
$$x = 1.33 \times 10^{-5}$$

Thus, the solubility of AgCl in pure water at 25°C is 1.33×10^{-5} *M*.

(b) **B** If x equals the solubility of AgCl in the KCl solution, then at equilibrium $[Ag^+] = x\ M$ and $[Cl^-] = (1.0 + x)\ M$. Substituting these values into the solubility product expression and assuming that $x << 1.0$ give

$$K_{sp} = [Ag^+][Cl^-] = (x)(1.0 + x) \approx x(1.0) = 1.77 \times 10^{-10} = x$$

Thus, if the common ion effect were the only important factor, we would predict that AgCl is approximately five orders of magnitude less soluble in a 1.0 *M* KCl solution than in water.

(c) To account for the effects of complex-ion formation, we must first write the equation for both the dissolution and complex-ion formation equilibria. Adding the equations corresponding to K_{sp} and K_f gives us an equation that describes the dissolution of AgCl in a KCl solution. The equilibrium constant for the reaction is therefore the product of K_{sp} and K_f:

$$
\begin{array}{lll}
AgCl(s) \rightleftharpoons \cancel{Ag^{+}(aq)} + \cancel{Cl^{-}(aq)} & & K_{sp} = 1.77 \times 10^{-10} \\
\cancel{Ag^{+}(aq)} + \cancel{2}Cl^{-} \rightleftharpoons [AgCl_2]^{-} & & K_f = 1.1 \times 10^{5} \\
\hline
AgCl(s) + Cl^{-} \rightleftharpoons [AgCl_2]^{-} & & K = K_{sp}K_f = 1.9 \times 10^{-5}
\end{array}
$$

If we let x equal the solubility of AgCl in the KCl solution, then at equilibrium $[AgCl_2^-] = x$ and $[Cl^-] = 1.0 - x$. Substituting these quantities into the equilibrium constant expression for the net reaction and assuming that $x << 1.0$ give

$$K = \frac{[AgCl_2^-]}{[Cl^-]} = \frac{x}{1.0 - x} \approx \frac{x}{1.0} = 1.9 \times 10^{-5} = x$$

That is, AgCl dissolves in 1.0 *M* KCl to produce a 1.9×10^{-5} *M* solution of the $AgCl_2^-$ complex ion. Thus, we predict that AgCl has approximately the same solubility in a 1.0 *M* KCl solution as it does in pure water, which is 10^5 times greater than that predicted based on the common ion effect. (In fact, the measured solubility of AgCl in 1.0 *M* KCl is almost a factor of ten *greater* than that in pure water, largely due to the formation of other chloride-containing complexes.)

EXERCISE 17.6

Calculate the solubility of mercury(II) iodide, HgI_2, in (a) pure water, and (b) a 3.0 *M* solution of NaI, assuming $[HgI_4]^{2-}$ is the only Hg-containing species present in significant amounts. $K_{sp} = 2.9 \times 10^{-29}$ for HgI_2 and $K_f = 6.8 \times 10^{29}$ for $[HgI_4]^{2-}$.

Answer **(a)** 1.9×10^{-10} *M*; **(b)** 1.4 *M*

Complexing agents, which are molecules or ions that increase the solubility of metal salts by forming soluble metal complexes, are common components of laundry detergents. Long-chain carboxylic acids, the major components of soaps, form insoluble salts with Ca^{2+} and Mg^{2+}, which are present in high concentrations in "hard" water. Precipitation of these salts produces a bathtub ring and gives a gray tinge to clothing. Adding a complexing agent such as pyrophosphate ($O_3POPO_3^{4-}$, or $P_2O_7^{4-}$) or triphosphate to detergents prevents the magnesium and calcium salts from precipitating because the equilibrium constant for complex-ion formation is large:

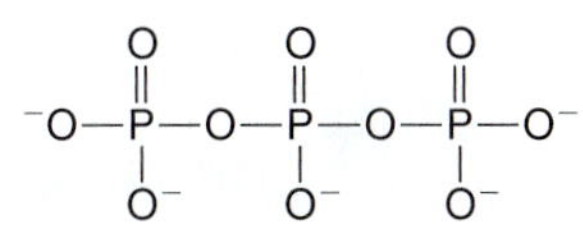

Triphosphate

$$Ca^{2+}(aq) + O_3POPO_3^{4-}(aq) \rightleftharpoons [Ca(O_3POPO_3)]^{2-}(aq)^2 \qquad K_f = 4 \times 10^4 \qquad (17.12)$$

However, phosphates can cause environmental damage by promoting *eutrophication*, the growth of excessive amounts of algae in a body of water, which can eventually lead to large decreases in levels of dissolved oxygen that kill fish and other aquatic organisms. Consequently, many states in the United States have banned the use of phosphate-containing detergents, and France has banned their use beginning in 2007. "Phosphate-free" detergents contain different kinds of complexing agents, such as triacetic acid derivatives or hydroxy polycarboxylic acids. The development of phosphate substitutes is an area of intense research.

Commercial water softeners also use complexing agent to treat hard water by passing the water over ion-exchange resins, which are complex sodium salts. When water

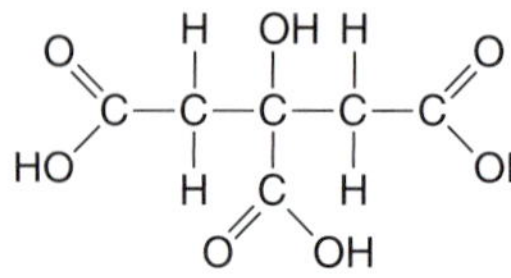

Citric acid
A hydroxy polycarboxylic acid

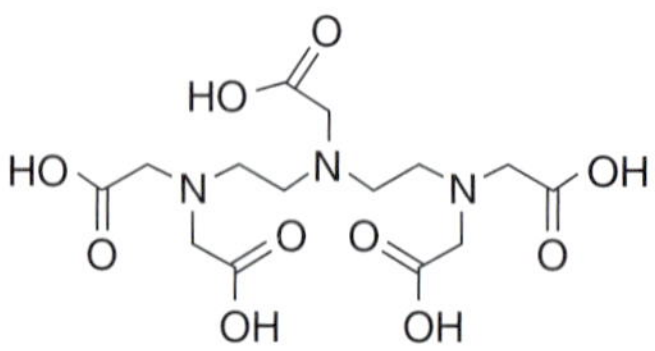

DTPA (diethylenetriaminepentaacetic acid)

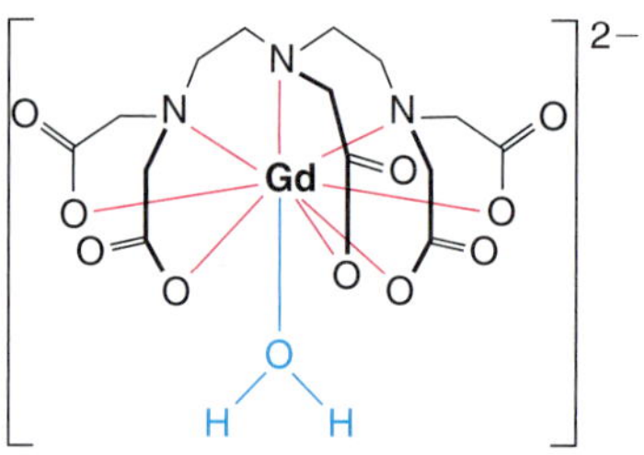

Gadolinium-DTPA complex, $[Gd(DTPA{\cdot}H_2O)]^{2-}$

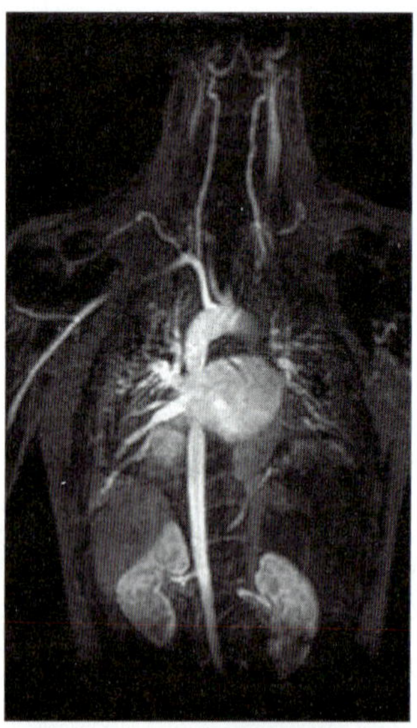

Figure 17.5 An MRI image of heart, arteries, and veins. When a patient is injected with a paramagnetic metal cation in the form of a stable complex known as an MRI contrast agent, the magnetic properties of water in cells are altered. Because the different environments in different types of cells respond differently, a physician can obtain detailed images of soft tissues.

flows over the resin, sodium ion is dissolved and insoluble salts precipitate onto the resin surface. Water treated in this way has a saltier taste due to the presence of Na^+, but it contains fewer dissolved minerals.

Another application of complexing agents is found in medicine. Unlike X rays, magnetic resonance imaging (MRI) can give relatively good images of soft tissues such as internal organs. MRI is based on the magnetic properties of the ^1H nucleus of hydrogen atoms in water, which is a major component of soft tissues. Because the properties of water don't depend very much on whether it is inside a cell or in the blood, it is hard to get detailed images of these tissues that have good contrast. To solve this problem, scientists have developed a class of metal complexes known as "MRI contrast agents." Injecting an MRI contrast agent into a patient selectively affects the magnetic properties of water in cells of normal tissues, in tumors, or in blood vessels, and allows doctors to "see" each of these separately (Figure 17.5). One of the most important metal ions for this application is Gd^{3+}, which with seven unpaired electrons is highly paramagnetic. Because $Gd^{3+}(aq)$ is quite toxic, it must be administered as a very stable complex that does not dissociate in the body and can be excreted intact by the kidneys. The complexing agents used for gadolinium are ligands such as $DTPA^{5-}$.

17.4 Solubility and pH

The solubility of many compounds depends strongly on the pH of the solution. For example, the anion in many sparingly soluble salts is the conjugate base of a weak acid that may become protonated in solution. In addition, the solubility of simple binary compounds such as oxides and sulfides, both strong bases, is often dependent on pH. In this section, we discuss the relationship between the solubility of these classes of compounds and pH.

The Effect of Acid–Base Equilibria on the Solubility of Salts

We begin our discussion by examining the effect of pH on the solubility of a representative salt, M^+A^-, where A^- is the conjugate base of the weak acid HA. When the salt dissolves in water, this reaction occurs:

$$MA(s) \rightleftharpoons M^+(aq) + A^-(aq) \qquad K_{sp} = [M^+][A^-] \tag{17.13}$$

The anion can also react with water in a hydrolysis reaction:

$$A^-(aq) + H_2O(l) \rightleftharpoons OH^-(aq) + HA(aq) \tag{17.14}$$

Because of the reaction described in Equation 17.14, the predicted solubility of a sparingly soluble salt that has a basic anion such as S^{2-}, PO_4^{3-}, or CO_3^{2-} is increased, as described in Section 17.1. If instead a strong acid is added to the solution, the added H^+ will react essentially completely with A^- to form HA. This reaction decreases $[A^-]$, which in turn decreases the magnitude of the ion product, $Q = [M^+][A^-]$. According to Le Châtelier's principle, more MA will dissolve until $Q = K_{sp}$. Hence, *an acidic pH dramatically increases the solubility of virtually all sparingly soluble salts whose anion is the conjugate base of a weak acid*. In contrast, the pH has little to no effect on the solubility of salts whose anion is the conjugate base of a stronger weak acid or a *strong* acid, respectively (chlorides, bromides, iodides, and sulfates, for example).

For example, the hydroxide salt $Mg(OH)_2$ is relatively insoluble in water:

$$Mg(OH)_2(s) \rightleftharpoons Mg^{2+}(aq) + 2OH^-(aq) \qquad K_{sp} = 5.61 \times 10^{-12} \tag{17.15}$$

When acid is added to a saturated solution that contains excess solid $Mg(OH)_2$, the following reaction occurs, removing OH^- from solution:

$$H^+(aq) + OH^-(aq) \longrightarrow H_2O(l) \tag{17.16}$$

The overall equation for the reaction of $Mg(OH)_2$ with acid is thus

$$Mg(OH)_2(s) + 2H^+(aq) \rightleftharpoons Mg^{2+}(aq) + 2H_2O(l) \qquad (17.17)$$

As more acid is added to a suspension of $Mg(OH)_2$, the equilibrium shown in Equation 17.17 is driven to the right, and more $Mg(OH)_2$ dissolves.

Such pH-dependent solubility is not restricted to salts that contain anions derived from water. For example, CaF_2 is a sparingly soluble salt:

$$CaF_2(s) \rightleftharpoons Ca^{2+}(aq) + 2F^-(aq) \qquad K_{sp} = 3.45 \times 10^{-11} \qquad (17.18)$$

When strong acid is added to a saturated solution of CaF_2, this reaction occurs:

$$H^+(aq) + F^-(aq) \rightleftharpoons HF(aq) \qquad (17.19)$$

Because the forward reaction decreases the fluoride ion concentration, more CaF_2 dissolves in order to relieve the stress on the system. The net reaction of CaF_2 with strong acid is thus

$$CaF_2(s) + 2H^+(aq) \longrightarrow Ca^{2+}(aq) + 2HF(aq) \qquad (17.20)$$

The next example shows how to calculate the effect of adding a strong acid to a solution of a sparingly soluble salt on its solubility.

Note the pattern

Sparingly soluble salts derived from weak acids tend to be more soluble in an acidic solution.

EXAMPLE 17.7

Lead oxalate (PbC_2O_4), lead iodide (PbI_2), and lead sulfate ($PbSO_4$) are all rather insoluble, with K_{sp} values of 4.8×10^{-10}, 9.8×10^{-9}, and 2.53×10^{-8}, respectively. What effect does adding a strong acid, such as perchloric acid, have on their relative solubilities?

Given K_{sp} values for three compounds

Asked for Relative solubilities in acid solution

Strategy

Write the balanced equation for the dissolution of each salt. Because the strongest conjugate base will be most affected by the addition of strong acid, determine the relative solubilities from the relative basicity of the anions.

Solution

The solubility equilibria for the three salts are

$$PbC_2O_4(s) \rightleftharpoons Pb^{2+}(aq) + C_2O_4{}^{2-}(aq)$$
$$PbI_2(s) \rightleftharpoons Pb^{2+}(aq) + 2I^-(aq)$$
$$PbSO_4(s) \rightleftharpoons Pb^{2+}(aq) + SO_4{}^{2-}(aq)$$

The addition of a strong acid will have the greatest effect on the solubility of a salt that contains the conjugate base of a weak acid as the anion. Because HI is a strong acid we predict that adding a strong acid to a saturated solution of PbI_2 will not greatly affect its solubility: the acid will simply dissociate to form $H^+(aq)$ and the corresponding anion. In contrast, oxalate is the fully deprotonated form of oxalic acid, HO_2CCO_2H, which is a weak diprotic acid (pK_a = 1.23 and 4.19). Consequently, the oxalate ion has a significant affinity for one proton, and a lower affinity for a second proton. Adding a strong acid to a saturated solution of lead oxalate will result in these reactions:

$$C_2O_4{}^{2-}(aq) + H^+(aq) \longrightarrow HO_2CCO_2{}^-(aq)$$
$$HO_2CCO_2{}^-(aq) + H^+(aq) \longrightarrow HO_2CCO_2H(aq)$$

These reactions will decrease $[C_2O_4{}^{2-}]$, causing more lead oxalate to dissolve to relieve the stress on the system.

The pK_a of HSO_4^- (1.99) is similar in magnitude to pK_{a1} of oxalic acid, so adding a strong acid to a saturated solution of $PbSO_4$ will result in this reaction:

$$SO_4^{2-}(aq) + H^+(aq) \rightleftharpoons HSO_4^-(aq)$$

Because HSO_4^- has a pK_a of 1.99, this reaction will lie largely to the left as written. Consequently, we predict that the effect of added strong acid on the solubility of $PbSO_4$ will be significantly less than for PbC_2O_4.

EXERCISE 17.7

Which of the following insoluble salts will be substantially more soluble in 1.0 *M* HNO_3 than in pure water? AgCl, Ag_2CO_3, Ag_3PO_4, AgBr

Answer Ag_2CO_3 and Ag_3PO_4

Caves and their associated pinnacles and spires of stone provide one of the most impressive examples of pH-dependent solubility equilibria (Figure 17.6a). Perhaps the most familiar caves are formed from limestone, such as Carlsbad Caverns in New Mexico, Mammoth Cave in Kentucky, and Luray Caverns in Virginia. The primary reactions that are responsible for the formation of limestone caves are

$$CO_2(aq) + H_2O(l) \rightleftharpoons H^+(aq) + HCO_3^-(aq) \quad (17.21)$$

$$HCO_3^-(aq) \rightleftharpoons H^+(aq) + CO_3^{2-}(aq) \quad (17.22)$$

$$Ca^{2+}(aq) + CO_3^{2-}(aq) \rightleftharpoons CaCO_3(s) \quad (17.23)$$

Limestone deposits that form caves consist primarily of $CaCO_3$ from the remains of living creatures such as clams and corals, which used it for making structures such as

Figure 17.6 The chemistry of cave formation. (a) This cave in Campanet, Mallorca, Spain, and its associated formations are examples of pH-dependent solubility equilibria. (b) A cave forms when groundwater containing atmospheric CO_2, forming an acidic solution, dissolves limestone, $CaCO_3$, in a process that may take tens of thousands of years. As groundwater seeps into a cave, water evaporates from the solution of $CaCO_3$ in CO_2-rich water, producing a supersaturated solution and a shift in equilibrium that causes precipitation of $CaCO_3$. The deposited limestone eventually forms stalactites and stalagmites.

(a) (b)

shells. When a saturated solution of $CaCO_3$ in CO_2-rich water rises toward the earth's surface or is otherwise heated, CO_2 gas is released as the water warms. Calcium carbonate then precipitates from the solution according to this equation (Figure 17.6b):

$$Ca^{2+}(aq) + 2HCO_3^-(aq) \rightleftharpoons CaCO_3(s) + CO_2(g) + H_2O(l) \quad (17.24)$$

The forward direction is the same reaction that produces the solid called *scale* in teapots, coffee makers, water heaters, boilers, and other places where hard water is repeatedly heated.

When groundwater containing atmospheric CO_2 (Equations 17.21 and 17.22) finds its way into microscopic cracks in the limestone deposits, $CaCO_3$ dissolves in the acidic solution in the reverse direction of Equation 17.24. The cracks gradually enlarge from 10–50 μm to 5–10 mm, a process that can take as long as ten thousand years. Eventually, after about another ten thousand years, a cave forms. Groundwater from the surface seeps into the cave and clings to the ceiling, where the water evaporates and causes the equilibrium in Equation 17.24 to shift to the right. A circular layer of solid $CaCO_3$ is deposited, which eventually produces a long, hollow spire of limestone called a *stalactite* that grows down from the ceiling. Below, where the droplets land when they fall from the ceiling, a similar process causes another spire, called a *stalagmite*, to grow up. The same processes that carve out hollows below ground are also at work above ground, in some cases producing fantastically convoluted landscapes like that of Yunnan Province in China (Figure 17.7).

Figure 17.7 Solubility equilibria in the formation of karst landscapes. Landscapes such as the steep limestone pinnacles of the Stone Forest in Yunnan Province, China, are formed from the same process that produces caves and their associated formations.

Acidic, Basic, and Amphoteric Oxides and Hydroxides

One of the earliest classifications of substances was based on their solubility in acidic versus basic solution, which led to the classification of oxides and hydroxides as being either basic or acidic. **Basic oxides** and hydroxides either react with water to produce a basic solution or dissolve readily in aqueous acid. **Acidic oxides** or hydroxides either react with water to produce an acidic solution or are soluble in aqueous base. As shown in Figure 17.8, there is a clear correlation between the acidic or basic character of an oxide and the position of the element combined with oxygen in the periodic table. *Oxides of metallic elements are generally basic oxides, and oxides of nonmetallic elements are acidic oxides.* Compare, for example, the reactions of a typical metal oxide, cesium oxide, and a typical nonmetal oxide, sulfur trioxide, with water:

Note the pattern

Metal oxides generally react with water to produce basic solutions, whereas nonmetal oxides produce acidic solutions.

$$Cs_2O(s) + H_2O(l) \longrightarrow 2Cs^+(aq) + 2OH^-(aq) \quad (17.25)$$

$$SO_3(g) + H_2O(l) \longrightarrow H_2SO_4(aq) \quad (17.26)$$

Cesium oxide reacts with water to produce a basic solution of cesium hydroxide, whereas sulfur trioxide reacts with water to produce a solution of sulfuric acid—very different behavior indeed!

The difference in reactivity is due to the difference in bonding in the two kinds of oxides. Because of the low electronegativity of the metals at the far left of the periodic table, their oxides are best viewed as containing discrete M^{n+} cations and O^{2-} anions. At the other end of the spectrum are nonmetal oxides; due to their higher electronegativities, nonmetals form oxides that contain covalent bonds to oxygen. Because of the high electronegativity of oxygen, however, the covalent bond between oxygen and the other atom, E, is usually polarized $E^{\delta+}$–$O^{\delta-}$. These oxides act as Lewis acids that react with water to produce an oxoacid. Oxides of metals in high oxidation states also tend to be acidic oxides for the same reason: they contain covalent bonds to oxygen. An example of an acidic metal oxide is MoO_3, which is insoluble in both water and acid but dissolves in strong base to give solutions of the molybdate ion, MoO_4^{2-}:

$$MoO_3(s) + 2OH^-(aq) \longrightarrow MoO_4^{2-}(aq) + H_2O(l) \quad (17.27)$$

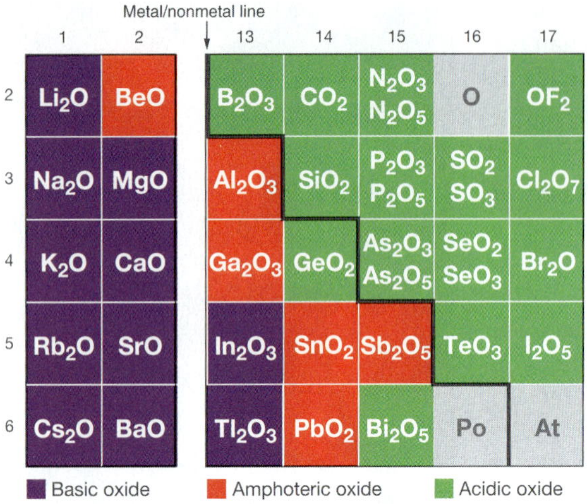

Figure 17.8 Classification of the oxides of the main group elements according to their acidic or basic character. There is a gradual transition from basic oxides to acidic oxides from the lower left to the upper right of the periodic table. Oxides of metallic elements are generally basic oxides, which either react with water to form a basic solution or dissolve in aqueous acid. In contrast, oxides of nonmetallic elements are acidic oxides, which either react with water to form an acidic solution or are soluble in aqueous base. Oxides of intermediate character, called amphoteric oxides, are located along a diagonal line between the two extremes. Amphoteric oxides either dissolve in acid to produce water or dissolve in base to produce a soluble complex ion. (Radioactive elements are not classified.)

As shown in Figure 17.8, there is a gradual transition from basic metal oxides to acidic nonmetal oxides as we go from the lower left to the upper right of the periodic table, with a broad diagonal band of oxides of intermediate character separating the two extremes. Many of the oxides of the elements in this diagonal region of the periodic table are soluble in both acidic *and* basic solutions; consequently, they are called **amphoteric** oxides (from the Greek *ampho*, "both," as in *amphiprotic*, defined in Section 16.1). Amphoteric oxides either dissolve in acid to produce water or dissolve in base to produce a soluble complex. As shown in Figure 17.9, for example, mixing $Cr(OH)_3$ with water gives a muddy, purple-brown suspension. Adding acid causes the $Cr(OH)_3$ to dissolve to give a bright violet solution of $Cr^{3+}(aq)$, which is in the form $[Cr(H_2O)_6]^{3+}$, whereas adding strong base gives a green solution of the $[Cr(OH)_4]^-$ ion. The chemical equations for the reactions are

$$Cr(OH)_3(s) + 3H^+(aq) \longrightarrow \underset{\text{Violet}}{Cr^{3+}(aq)} + 3H_2O(l) \tag{17.28}$$

$$Cr(OH)_3(s) + OH^-(aq) \longrightarrow \underset{\text{Green}}{[Cr(OH)_4]^-(aq)} \tag{17.29}$$

EXAMPLE 17.8

Aluminum hydroxide, written as either $Al(OH)_3$ or $Al_2O_3 \cdot 3H_2O$, is amphoteric. Write chemical equations to describe the dissolution of aluminum hydroxide in **(a)** acid and **(b)** base.

Given Amphoteric compound

Asked for Dissolution reactions in acid and base

Strategy

Using Equations 17.28 and 17.29 as a guide, write the dissolution reactions in acid and base solutions.

Solution

(a) An acid donates protons to hydroxide to give water and the hydrated metal ion, so aluminum hydroxide, which contains three OH^- ions per Al, needs three H^+ ions:

$$Al(OH)_3(s) + 3H^+(aq) \longrightarrow Al^{3+}(aq) + 3H_2O(l)$$

In aqueous solution, Al^{3+} forms the complex ion $[Al(H_2O)_6]^{3+}$.

(b) In basic solution, OH^- is added to the compound to produce a soluble and stable poly(hydroxo) complex:

$$Al(OH)_3(s) + OH^-(aq) \longrightarrow [Al(OH)_4]^-(aq)$$

Figure 17.9 Chromium(III) hydroxide, $Cr(OH)_3$, is an example of an amphoteric hydroxide. All three beakers originally contained a suspension of brownish purple $Cr(OH)_3(s)$ (*center*). When concentrated acid (6 *M* H_2SO_4) was added to the beaker on the left, $Cr(OH)_3$ dissolved to produce violet $[Cr(H_2O)_6]^{3+}$ ions and water. The addition of concentrated base (6 *M* NaOH) to the beaker on the right caused $Cr(OH)_3$ to dissolve, producing green $[Cr(OH)_4]^-$ ions.

EXERCISE 17.8

Copper(II) hydroxide, written as either $Cu(OH)_2$ or $CuO \cdot H_2O$, is amphoteric. Write chemical equations that describe the dissolution of cupric hydroxide both in an acid and in a base.

Answer

$$Cu(OH)_2(s) + 2H^+(aq) \longrightarrow Cu^{2+}(aq) + 2H_2O(l)$$
$$Cu(OH)_2(s) + 2OH^-(aq) \longrightarrow [Cu(OH)_4]^{2-}(aq)$$

Selective Precipitation Using pH

Many dissolved metal ions can be separated by selective precipitation of the cations from solution under specific conditions. In this technique, pH is often used to control the concentration of the anion in solution, which in turn controls which cations precipitate.

Suppose, for example, we have a solution that contains 1.0 m*M* Zn^{2+} and 1.0 m*M* Cd^{2+}, and we would like to separate the two metals by selective precipitation as the insoluble sulfide salts, ZnS and CdS. The relevant solubility equilibria can be written as:

$$ZnS(s) \rightleftharpoons Zn^{2+}(aq) + S^{2-}(aq) \qquad K_{sp} = 1.6 \times 10^{-24} \qquad (17.30)$$
$$CdS(s) \rightleftharpoons Cd^{2+}(aq) + S^{2-}(aq) \qquad K_{sp} = 8.0 \times 10^{-27} \qquad (17.31)$$

Because the S^{2-} ion is quite basic and reacts extensively with water to give HS^- and OH^-, the solubility equilibria are more accurately written as $MS(s) \rightleftharpoons M^{2+}(aq) + HS^-(aq) + OH^-$ rather than $MS(s) \rightleftharpoons M^{2+}(aq) + S^{2-}(aq)$. Here we use the simpler form involving S^{2-}, which is justified because we take the reaction of S^{2-} with water into account later in the solution, arriving at the same answer using either equilibrium equation.

The sulfide concentrations needed to cause ZnS and CdS to precipitate are

$$K_{sp} = [Zn^{2+}][S^{2-}]$$
$$1.6 \times 10^{-24} = (0.0010\ M)[S^{2-}]$$
$$[S^{2-}] = 1.6 \times 10^{-21}\ M \qquad (17.32)$$
$$K_{sp} = [Cd^{2+}][S^{2-}]$$
$$8.0 \times 10^{-27} = (0.0010\ M)[S^{2-}] \qquad (17.33)$$
$$[S^{2-}] = 8.0 \times 10^{-24}\ M$$

Thus, sulfide concentrations between 1.6×10^{-21} *M* and 8.0×10^{-24} *M* will precipitate CdS from solution but not ZnS. How do we obtain such low concentrations of sulfide? A saturated aqueous solution of H_2S contains 0.10 *M* H_2S at 20°C. The value of pK_{a1} for H_2S is 6.97, and pK_{a2} corresponding to the formation of $[S^{2-}]$ is 12.90. The equations for these reactions are

$$H_2S(aq) \rightleftharpoons H^+(aq) + HS^-(aq) \qquad pK_{a1} = 6.97, K_{a1} = 1.1 \times 10^{-7} \qquad (17.34)$$
$$HS^-(aq) \rightleftharpoons H^+(aq) + S^{2-}(aq) \qquad pK_{a2} = 12.90, K_{a2} = 1.3 \times 10^{-13}$$

We can show that the concentration of S^{2-} is 1.3×10^{-13} by comparing K_{a1} and K_{a2} and recognizing that the contribution to $[H^+]$ from the dissociation of HS^- is negligible compared with $[H^+]$ from the dissociation of H_2S. Thus, substituting 0.10 *M* in the equation for K_{a1} for the concentration of H_2S, which is essentially constant regardless of the pH, gives

$$K_{a1} = 1.1 \times 10^{-7} = \frac{[H^+][HS^-]}{[H_2S]} = \frac{x^2}{0.10\ M} \qquad (17.35)$$
$$x = 1.1 \times 10^{-4}\ M = [H^+] = [HS^-]$$

Substituting this value for $[H^+]$ and $[HS^-]$ into the equation for K_{a2} gives

$$K_{a2} = 1.3 \times 10^{-13} = \frac{[H^+][S^{2-}]}{[HS^-]} = \frac{(1.1 \times 10^{-4}\,M)x}{1.1 \times 10^{-4}\,M} = x = [S^{2-}]$$

Although $[S^{2-}]$ in an H_2S solution is very low (1.3×10^{-13} *M*), bubbling H_2S through the solution until it is saturated would precipitate both metal ions because the concentration of S^{2-} would then be much higher than 1.6×10^{-21} *M*. Thus, we must adjust $[S^{2-}]$ to stay within the desired range. The most direct way to do this is to adjust $[H^+]$ by adding acid to the H_2S solution, thereby driving the equilibrium in Equation 17.34 to the left. The overall equation for the dissociation of H_2S is

$$H_2S(aq) \rightleftharpoons 2H^+(aq) + S^{2-}(aq) \tag{17.36}$$

Now we can use the equilibrium constant *K* for the overall reaction, which is the product of K_{a1} and K_{a2}, and the concentration of H_2S in a saturated solution to calculate the H^+ concentration needed to produce $[S^{2-}]$ of 1.6×10^{-21} *M*:

$$K = K_{a1}K_{a2} = (1.1 \times 10^{-7})(1.3 \times 10^{-13}) = 1.4 \times 10^{-20} = \frac{[H^+]^2[S^{2-}]}{[H_2S]} \tag{17.37}$$

$$[H^+]^2 = \frac{K[H_2S]}{[S^{2-}]} = \frac{(1.4 \times 10^{-20})(0.10\,M)}{1.6 \times 10^{-21}\,M} = 0.88 \tag{17.38}$$

$$[H^+] = 0.94\,M$$

Thus, adding a strong acid such as HCl to make the solution 0.94 *M* in H^+ will prevent ZnS from precipitating, while ensuring that the less soluble CdS will precipitate when the solution is saturated with H_2S.

EXAMPLE 17.9

A solution contains 0.010 *M* Ca^{2+} and 0.010 *M* La^{3+}. What concentration of HCl is needed to precipitate $La_2(C_2O_4)_3 \cdot 9H_2O$ but not $Ca(C_2O_4) \cdot H_2O$ if the concentration of oxalic acid is 1.0 *M*? K_{sp} values are 2.32×10^{-9} for $Ca(C_2O_4)$ and 2.5×10^{-27} for $La_2(C_2O_4)_3$; $pK_{a1} = 1.25$ and $pK_{a2} = 3.81$ for oxalic acid.

Given Concentrations of cations, K_{sp} values, concentration and pK_a values for oxalic acid

Asked for Concentration of HCl needed for selective precipitation of $La_2(C_2O_4)_3$

Strategy

A Write each solubility product expression, and calculate the oxalate concentration needed for precipitation to occur. Determine the concentration range needed for selective precipitation of $La_2(C_2O_4)_3 \cdot 9H_2O$.

B Add the equations for the first and second dissociations of oxalic acid to get an overall equation for the dissociation of oxalic acid to oxalate. Substitute the $[ox^{2-}]$ needed to precipitate $La_2(C_2O_4)_3 \cdot 9H_2O$ into the overall equation for the dissociation of oxalic acid to calculate the required $[H^+]$.

Solution

A Because the salts have different stoichiometries, we cannot directly compare the magnitudes of the solubility products. Instead, we must use the equilibrium constant expression for each solubility product to calculate the concentration of oxalate needed for precipitation to occur. Using ox^{2-} for oxalate, we write the solubility product expression for calcium oxalate as

$$K_{sp} = [Ca^{2+}][ox^{2-}] = (0.010)[ox^{2-}] = 2.32 \times 10^{-9}$$

$$[ox^{2-}] = 2.32 \times 10^{-7}\,M$$

The expression for lanthanum oxalate is

$$K_{sp} = [La^{3+}]^2[ox^{2-}]^3 = (0.010)^2[ox^{2-}]^3 = 2.5 \times 10^{-27}$$
$$[ox^{2-}] = 2.9 \times 10^{-8}\ M$$

Thus, lanthanum oxalate is less soluble and will selectively precipitate when the oxalate concentration is between $2.9 \times 10^{-8}\ M$ and $2.32 \times 10^{-7}\ M$. B To prevent Ca^{2+} from precipitating as calcium oxalate, we must add enough H^+ to give a maximum oxalate concentration of $2.32 \times 10^{-7}\ M$. We can calculate the required $[H^+]$ by using the overall equation for the dissociation of oxalic acid to oxalate:

$$HO_2CCO_2H(aq) \rightleftharpoons 2H^+(aq) + C_2O_4^{2-}(aq)$$
$$K = K_{a1}K_{a2} = (10^{-1.25})(10^{-3.81}) = 10^{-5.06} = 8.71 \times 10^{-6}$$

Substituting the desired oxalate concentration into the equilibrium constant expression gives

$$8.71 \times 10^{-6} = \frac{[H^+]^2[ox^{2-}]}{[HO_2CCO_2H]} \approx \frac{[H^+]^2(2.32 \times 10^{-7})}{1.0}$$
$$[H^+] = 6.1\ M$$

Thus, adding enough HCl to give $[H^+] = 6.1\ M$ will cause only $La_2(C_2O_4)_3 \cdot 9H_2O$ to precipitate from the solution.

EXERCISE 17.9

A solution contains $0.015\ M\ Fe^{2+}$ and $0.015\ M\ Pb^{2+}$. What concentration of acid is needed to ensure that Pb^{2+} precipitates as PbS in a saturated solution of H_2S but Fe^{2+} does not precipitate as FeS? K_{sp} values are 6.3×10^{-18} for FeS and 8.0×10^{-28} for PbS.

Answer $0.018\ M\ H^+$

17.5 Qualitative Analysis Using Selective Precipitation

The composition of relatively complex mixtures of metal ions can be determined using **qualitative analysis**, a procedure for discovering the *identity* of metal ions present in the mixture (rather than quantitative information about their amounts).

The procedure used to separate and identify more than 20 common metal cations from a single solution consists of selectively precipitating only a few kinds of metal ions at a time under given sets of conditions. Consecutive precipitation steps become progressively *less* selective until almost all of the metal ions are precipitated, as illustrated in Figure 17.10.

Group 1: Insoluble Chlorides

Most metal chloride salts are soluble in water; only Ag^+, Pb^{2+}, and Hg_2^{2+} form chlorides that precipitate from water. Thus, the first step in a qualitative analysis is to add about 6 M HCl, thereby causing AgCl, $PbCl_2$, and/or Hg_2Cl_2 to precipitate. If no precipitate forms, then these cations are not present in significant amounts. The precipitate can be collected by filtration or centrifugation.

Group 2: Acid-Insoluble Sulfides

Next, the acidic solution is saturated with H_2S gas. Only those metal ions that form very insoluble sulfides, such as As^{3+}, Bi^{3+}, Cd^{2+}, Cu^{2+}, Hg^{2+}, Sb^{3+}, and Sn^{2+}, precipitate as their sulfide salts under these acidic conditions. All others, such as Fe^{2+} and Zn^{2+}, remain in solution. Once again, the precipitates are collected by filtration or centrifugation.

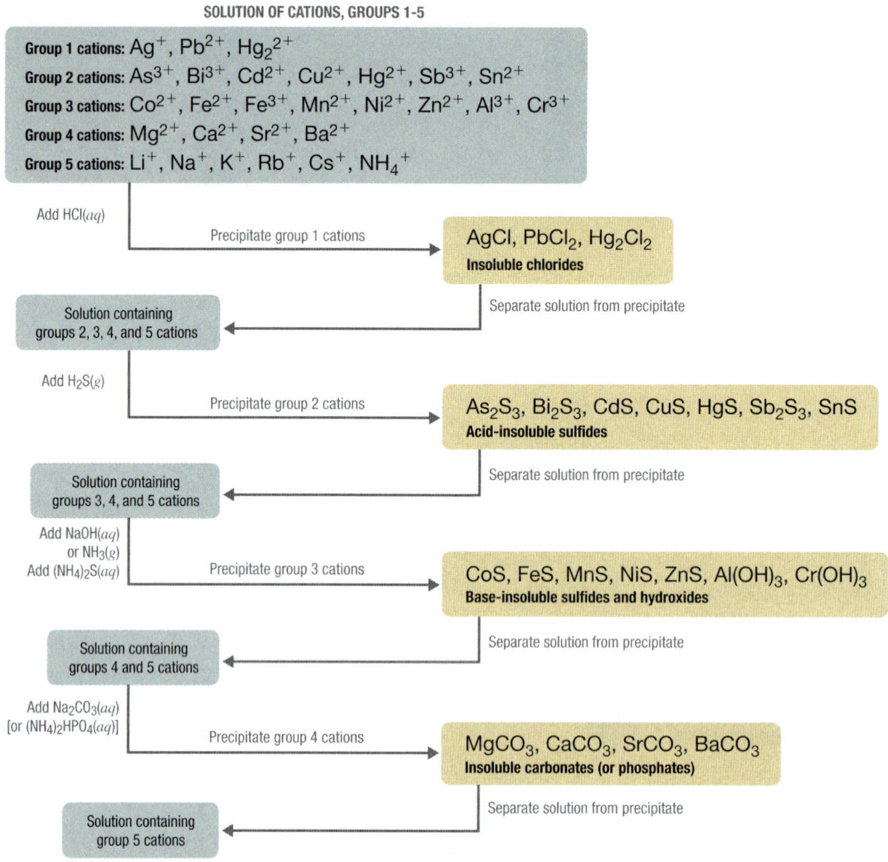

Figure 17.10 **Steps in a typical qualitative analysis scheme for a solution that contains several metal ions.**

Group 3: Base-Insoluble Sulfides (and Hydroxides)

Ammonia or NaOH is now added to the solution until it is basic, and then $(NH_4)_2S$ is added. This treatment removes any remaining cations that form insoluble hydroxides or sulfides. The divalent metal ions Co^{2+}, Fe^{2+}, Mn^{2+}, Ni^{2+}, and Zn^{2+} precipitate as their sulfides, and the trivalent metal ions Al^{3+} and Cr^{3+} precipitate as their hydroxides, $Al(OH)_3$ and $Cr(OH)_3$. If the mixture contains Fe^{3+}, sulfide reduces the cation to Fe^{2+}, which precipitates as FeS.

Group 4: Insoluble Carbonates or Phosphates

The next metal ions to be removed from solution are those that form insoluble carbonates and phosphates. When Na_2CO_3 is added to the basic solution that remains after the precipitated metal ions are removed, insoluble carbonates precipitate and are collected. Alternatively, adding $(NH_4)_2HPO_4$ causes the same metal ions to precipitate as insoluble phosphates.

Group 5: Alkali Metals

At this point, we have removed all the metal ions that form water-insoluble chlorides, sulfides, carbonates, or phosphates. The only common ions that might remain are any alkali metals (Li^+, Na^+, K^+, Rb^+, and Cs^+) and ammonium (NH_4^+). We now take a second sample from the *original* solution and add a small amount of NaOH to neutralize the ammonium ion and produce NH_3. (We can't use the same sample we used for the first four groups because we added ammonium to that sample in earlier steps.) Any ammonia produced can be detected either by its odor or by a litmus paper test. A flame test on another original sample is used to detect sodium, which produces a characteristic bright yellow color. As discussed in Chapter 6, the other alkali metal ions also give characteristic colors in flame tests, which allows them to be identified if only one is present.

Metal ions that precipitate together are separated by various additional techniques, such as forming complex ions, changing the pH of the solution, or increasing the temperature to redissolve some of the solids. For example, the precipitated metal chlorides of group 1 cations, containing Ag^+, Pb^{2+}, and Hg_2^{2+}, are all quite insoluble in water. Because $PbCl_2$ is much more soluble in hot water than are the other two chloride salts, however, adding water to the precipitate and heating the resulting slurry will dissolve any $PbCl_2$ present. Isolating the solution and adding a small amount of Na_2CrO_4 solution to it will produce a bright yellow precipitate of $PbCrO_4$ if Pb^{2+} was in the original sample (Figure 17.11).

As another example, treating the precipitates from group 1 cations with aqueous ammonia will dissolve any AgCl because Ag^+ forms a stable complex with ammonia, $[Ag(NH_3)_2]^+$. In addition, Hg_2Cl_2 *disproportionates* in ammonia ($2Hg_2^{2+} \longrightarrow Hg + Hg^{2+}$) to form a black solid that is a mixture of finely divided metallic mercury and an insoluble mercury(II) compound, which is separated from solution:

$$Hg_2Cl_2(s) + 2NH_3(aq) \longrightarrow Hg(l) + Hg(NH_2)Cl(s) + NH_4^+(aq) + Cl^-(aq) \quad (17.39)$$

Any silver ion in the solution is then detected by adding hydrochloric acid, which reverses the reaction and gives a precipitate of white AgCl that slowly darkens when exposed to light:

$$[Ag(NH_3)_2]^+(aq) + 2H^+(aq) + Cl^-(aq) \longrightarrow AgCl(s) + 2NH_4^+(aq) \quad (17.40)$$

Similar but slightly more complex reactions are used to separate and identify the individual components of the other groups as well.

Figure 17.11 Separation of metal ions from Group 1 using qualitative analysis. In (a), the cations of group 1 precipitate when $HCl(aq)$ is added to a solution containing a mixture of cations. (b) When a small amount of Na_2CrO_4 solution is added to sample containing Pb^{2+} ions in water, a bright yellow precipitate of $PbCrO_4$ forms. (c) Addition of aqueous ammonia to a second portion of the solid sample produces a black solid that is a mixture of finely divided metallic mercury, an insoluble mercury(II) compound, $Hg(NH_2)Cl$, and a stable $[Ag(NH_3)_2]^+(aq)$ complex. (d) The presence of Ag^+ is detected by decanting the solution from the precipitated mercury and mercury complex and adding hydrochloric acid to the decanted solution, which causes AgCl to precipitate.

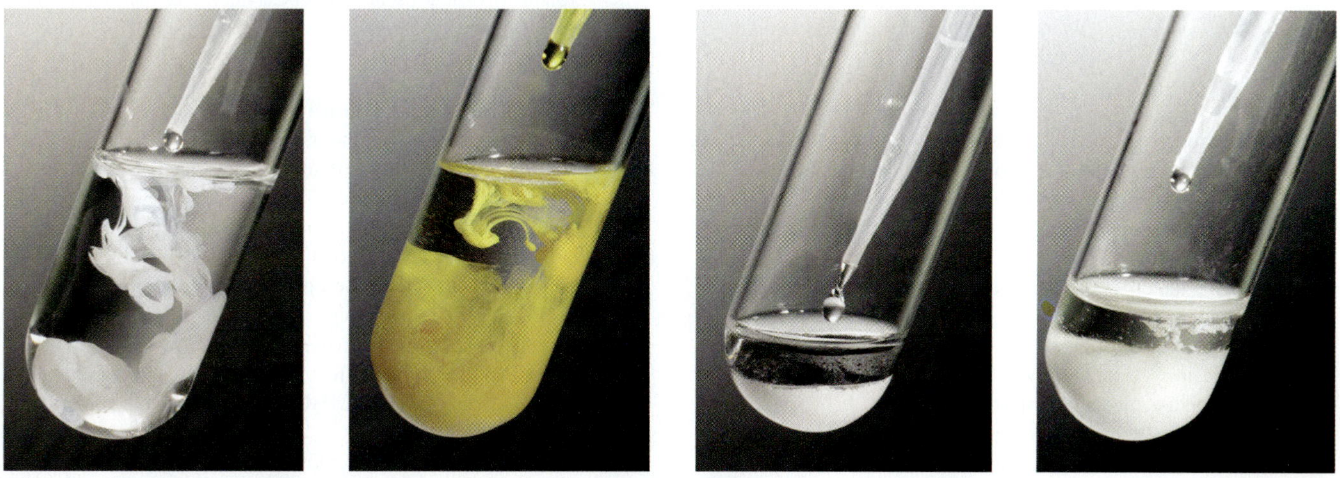

(a) Precipitation of group 1 cations, $AgCl(s) + Hg_2Cl_2(s) + PbCl_2(s)$

(b) Confirmation test for lead: $PbCrO_4(s)$

(c) Confirmation test for mercury: $Hg(l) + Hg(NH_2)Cl(s)$

(d) Confirmation test for silver: $AgCl(s)$

SUMMARY AND KEY TERMS

17.1 Determining the Solubility of Ionic Compounds (p. 782)

The equilibrium constant for a dissolution reaction, called the **solubility product** (**K_{sp}**), is a measure of the solubility of a compound. Whereas solubility is usually expressed in terms of *mass* of solute per 100 mL of solvent, K_{sp} is defined in terms of *molar* concentrations of the component ions. In contrast, the **ion product** (**Q**) describes concentrations that are not necessarily equilibrium concentrations. Comparing Q and K_{sp} enables us to determine whether a precipitate will form when solutions of two soluble salts are mixed. Adding a common cation or anion to a solution of a sparingly soluble salt shifts the solubility equilibrium in the direction predicted by Le Châtelier's principle. The solubility of the salt is almost always decreased by the presence of a common ion.

17.2 Factors That Affect Solubility (p. 788)

The solubility of a compound can differ from the solubility indicated by the concentrations of ions for four reasons: (1) **ion-pair** formation, in which an anion and cation are in intimate contact in solution and not separated by solvent, (2) incomplete dissociation of molecular solutes, (3) formation of complex ions, and (4) changes in pH. An ion pair is held together by electrostatic attractive forces between the cation and the anion, whereas incomplete dissociation results from intramolecular forces, such as polar covalent O—H bonds.

17.3 Complex-Ion Formation (p. 790)

A **complex ion** is a species formed between a central metal ion and one or more surrounding **ligands**, molecules or ions that contain at least one lone pair of electrons. Small, highly charged metal ions have the greatest tendency to act as Lewis acids and have the greatest tendency to form complex ions. The equilibrium constant for the formation of the complex ion is the **formation constant** (**K_f**). The formation of a complex ion by the addition of a *complexing agent* increases the solubility of a compound.

17.4 Solubility and pH (p. 796)

The anion in many sparingly soluble salts is the conjugate base of a weak acid. At low pH, protonation of the anion can dramatically increase the solubility of the salt. Oxides can be classified as acidic oxides or basic oxides. **Acidic oxides** either react with water to give an acidic solution or dissolve in strong base; most acidic oxides are nonmetal oxides or oxides of metals in high oxidation states. **Basic oxides** either react with water to give a basic solution or dissolve in strong acid; most basic oxides are oxides of metallic elements. Oxides or hydroxides that are soluble in both acidic and basic solutions are called **amphoteric**. Most elements whose oxides exhibit amphoteric behavior are located along the diagonal line separating metals and nonmetals in the periodic table. In solutions that contain mixtures of dissolved metal ions, the pH can be used to control the anion concentration needed to selectively precipitate the desired cation.

17.5 Qualitative Analysis Using Selective Precipitation (p. 803)

In **qualitative analysis**, the identity of metal ions present in a mixture is determined, not their amounts. The technique consists of selectively precipitating only a few kinds of metal ions at a time under given sets of conditions. Consecutive precipitation steps become progressively *less* selective until almost all the metal ions are precipitated. Other additional steps are needed to separate metal ions that precipitate together.

QUESTIONS AND PROBLEMS

Questions and Problems with colored numbers have answers in the Appendix and complete solutions in the Student Solutions Manual.

CONCEPTUAL

17.1 Determining the Solubility of Ionic Compounds

1. Write an expression for K_{sp} for each of these salts: (a) AgI; (b) CaF_2; (c) $PbCl_2$; (d) Ag_2CrO_4.
2. Some species are not represented in a solubility product expression. Why?
3. Describe the differences between Q and K_{sp}.
4. How can an ion product be used to determine whether a solution is saturated?
5. When using K_{sp} to directly compare the solubilities of compounds, why is it important to compare only the K_{sp} values of salts that have the same stoichiometry?
6. Describe the effect of a common ion on the solubility of a salt. Is this effect similar to the common ion effect found in buffers? Explain your answer.
7. Explain why the presence of $MgCl_2$ decreases the molar solubility of the sparingly soluble salt $MgCO_3$.

17.2 Factors That Affect Solubility

8. Do you expect the actual molar solubility of $LaPO_4$ to be higher than, the same as, or lower than the value calculated from its K_{sp}? Explain your reasoning.
9. Do you expect the difference between the calculated molar solubility and the actual molar solubility of $Ca_3(PO_4)_2$ to be greater or less than the difference in the solubilities of $Mg_3(PO_4)_2$? Why?
10. Write equations to describe the interactions in a solution that contains $Mg(OH)_2$, which forms ion pairs, and in one that contains propanoic acid, $CH_3CH_2CO_2H$, which forms a hydrated neutral molecule.
11. Draw circular atom representations of $Ca(IO_3)_2$ in solution (a) as an ionic solid, (b) in the form of ion pairs, and (c) as discrete ions.

17.3 Complex-Ion Formation

12. What is the difference between an equilibrium constant K_{eq} and a formation constant K_f?

13. Which would you expect to have the greater tendency to form a complex ion: Mg^{2+} or Ba^{2+}? Why?

14. How can a ligand be used to affect the concentration of hydrated metal ions in solution? How is K_{sp} affected? Explain your answer.

15. Co(II) forms a complex ion with pyridine, C_5H_5N. Which is the Lewis acid and which is the Lewis base? Use Lewis electron structures to justify your answer.

17.4 Solubility and pH

16. Of the following compounds, which do you expect to be most soluble in 1.0 *M* HNO_3 and which least soluble? (a) $CuCl_2$; (b) $K[Pb(OH)_3]$; (c) $Ba(CH_3CO_2)_2$; (d) $CaCO_3$. Explain your reasoning.

17. Of the compounds $Sn(CH_3CO_2)_2$ and SnS, one is soluble in dilute HCl and the other is soluble only in hot, concentrated HCl. Which is which? Provide a reasonable explanation.

18. Where in the periodic table do you expect to find elements that form basic oxides? Where do you expect to find elements that form acidic oxides?

19. Because water can autoionize, it reacts with oxides either as a base (as OH^-) or as an acid (as H_3O^+). Do you expect oxides of elements in high oxidation states to be more acidic (reacting with OH^-) or more basic (reacting with H_3O^+) than the corresponding oxides in low oxidation states? Why?

20. Given solid samples of CrO, Cr_2O_3, and CrO_3, which would you expect to be the most acidic (reacts most readily with OH^-)? Which would be the most basic (reacts most readily with H_3O^+)? Why?

21. Which of these metals do you expect to form an amphoteric oxide, and why? Be, B, Al, N, Se, In, Tl, Pb

17.5 Qualitative Analysis Using Selective Precipitation

22. Given a solution that contains a mixture of NaCl, $CuCl_2$, and $ZnCl_2$, propose a method for separating the metal ions.

NUMERICAL

17.1 Determining the Solubility of Ionic Compounds

23. Predict the molar solubility of each compound using the K_{sp} values given in Appendix B: (a) $Cd(IO_3)_2$, (b) AgCN, (c) HgI_2.

24. Predict the molar solubility of each compound using the K_{sp} values given: (a) Li_3PO_4, 2.37×10^{-11}; (b) $Ca(IO_3)_2$, 6.47×10^{-6}; (c) $Y(IO_3)_3$, 1.12×10^{-10}.

25. A student prepared 750 mL of a saturated solution of silver sulfate. How many grams of Ag_2SO_4 does the solution contain? $K_{sp} = 1.20 \times 10^{-5}$.

26. Given the K_{sp} values in Table 17.1 and Appendix B, predict the molar concentration of each species in a saturated aqueous solution of the following: (a) silver bromide; (b) lead oxalate; (c) iron(II) carbonate; (d) silver phosphate; (e) copper(I) cyanide.

27. Use the K_{sp} values in Table 17.1 and Appendix B to predict the molar concentration of each species in a saturated aqueous solution of each of the following: (a) copper(I) chloride; (b) lanthanum(III) iodate; (c) magnesium phosphate; (d) silver chromate; (e) strontium sulfate.

28. Silicon dioxide, the most common binary compound of silicon and oxygen, constitutes approximately 60% of the earth's crust. Under certain conditions, this amphoteric compound can react with water to form silicic acid, which is written as either H_4SiO_4 or $Si(OH)_4$. Write balanced chemical equations for the dissolution of SiO_2 under acidic and basic conditions.

29. The K_{sp} of $Mg(OH)_2$ is 5.61×10^{-12}. If you tried to dissolve 24 mg of $Mg(OH)_2$ in 250 mL of water and then filtered the solution and dried the remaining solid, what would you predict to be the mass of the undissolved solid? You discover that only 1.0 mg remains undissolved. Explain the difference between your expected value and the actual value.

30. The K_{sp} of lithium carbonate is 8.15×10^{-4}. If 2.34 g of the salt is stirred with 500 mL of water and any undissolved solid is filtered from the solution and dried, what do you predict to be the mass of the solid? You discover that all of your sample dissolves. Explain the difference between your predicted value and the actual value.

31. You have calculated that 24.6 mg of $BaSO_4$ will dissolve in 1.0 L of water at 25°C. After adding your calculated amount to 1.0 L of water and stirring for several hours, you notice that the solution contains undissolved solid. After carefully filtering the solution and drying the solid, you find that 22.1 mg did not dissolve. According to your measurements, what is the K_{sp} of barium sulfate?

32. In a saturated silver chromate solution, the molar solubility of chromate is 6.54×10^{-5}. What is the K_{sp}?

33. A saturated lead(II) chloride solution contains a chloride concentration of 3.24×10^{-2} mol/L. What is the K_{sp}?

34. From the solubility data given, predict the value of K_{sp} for each compound: (a) AgI, 2.89×10^{-7} g/100 mL; (b) SrF_2, 1.22×10^{-2} g/100 mL; (c) $Pb(OH)_2$, 78 mg/500 mL; (d) $BiAsO_4$, 14.4 mg/2.0 L.

35. From the solubility data given, predict the value of K_{sp} for each compound: (a) $BaCO_3$, 10.0 mg/500 mL; (b) CaF_2, 3.50 mg/200 mL; (c) $Mn(OH)_2$, 6.30×10^{-4} g/300 mL; (d) Ag_2S, 1.60×10^{-13} mg/100 mL.

36. Given the following solubilities, predict the K_{sp} for each compound: (a) $BaCO_3$, 7.00×10^{-5} mol/L; (b) CaF_2, 1.70 mg/100 mL; (c) $Pb(IO_3)_2$, 2.30 mg/100 mL; (d) SrC_2O_4, 1.58×10^{-7}mol/L.

37. From the solubility data given, predict the K_{sp} of each compound? (a) Ag_2SO_4, 4.2×10^{-1} g/100 mL; (b) $SrSO_4$, 1.5×10^{-3} g/100 mL; (c) CdC_2O_4, 6.0×10^{-3} g/100 mL; (d) $Ba(IO_3)_2$, 3.96×10^{-2} g/100 mL

38. The K_{sp} of the phosphate fertilizer $CaHPO_4 \cdot 2H_2O$ is 2.7×10^{-7} at 25°C. What is the molar concentration of a saturated solution? What mass of this compound will dissolve in 3.0 L of water at this temperature?

39. The K_{sp} of zinc carbonate monohydrate is $5.5. \times 10^{-11}$ at 25°C. What is the molar concentration of a saturated solution? What mass of compound will dissolve in 2.0 L of water at this temperature?

40. Silver nitrate eye drops were at one time administered to newborn infants to guard against eye infections contracted during birth. Although silver nitrate is highly water soluble, silver sulfate has a K_{sp} of 1.20×10^{-5} at 25°C. If you add 25.0 mL of 0.015 *M* $AgNO_3$ to 150 mL of 2.8×10^{-3} *M* Na_2SO_4, will you get a precipitate? If so, what will its mass be?

41. Use the data in Appendix B to predict whether precipitation will occur when these pairs of solutions are mixed: (a) 150 mL of 0.142 *M* $Ba(NO_3)_2$ with 200 mL of 0.089 *M* NaF; (b) 250 mL of 0.079 *M* K_2CrO_4 with 175 mL of 0.087 *M* $CaCl_2$; (c) 300 mL of 0.109 *M* $MgCl_2$ with 230 mL of 0.073 *M* $Na_2(C_2O_4)$.

42. Given a 0.048 *M* solution of $Pb(NO_3)_2$, what is the maximum volume that can be added to 250 mL of 0.10 *M* NaSCN before precipitation occurs? $K_{sp} = 2.0 \times 10^{-5}$ for $Pb(SCN)_2$.

43. Given 300 mL of a solution that is 0.056 *M* in lithium nitrate, what volume of 0.53 *M* sodium carbonate can be added before precipitation occurs? $K_{sp} = 8.15 \times 10^{-4}$ for Li_2CO_3.

44. Given the information in the table, calculate the molar solubility of each sparingly soluble salt in 0.95 *M* $MgCl_2$.

Saturated Solution	K_{sp}
$MgCO_3 \cdot 3H_2O$	2.4×10^{-6}
$Mg(OH)_2$	5.6×10^{-12}
$Mg_3(PO_4)_2$	1.04×10^{-24}

45. Ferric phosphate has a molar solubility of 5.44×10^{-16} in 1.82 *M* Na_3PO_4. Predict its K_{sp}. The actual K_{sp} is 9.9×10^{-16}. Explain this discrepancy.

17.3 Complex-Ion Formation

46. Zn(II) forms the complex ion $[Zn(NH_3)_4]^{2+}$ through equilibrium reactions in which ammonia replaces coordinated water molecules in a stepwise manner. If $\log K_1 = 2.37$, $\log K_2 = 2.44$, $\log K_3 = 2.50$, and $\log K_4 = 2.15$, what is the overall K_f? Write the equilibrium equation that corresponds to each stepwise equilibrium constant. Do you expect the $[Zn(NH_3)_4]^{2+}$ complex to be stable? Explain your reasoning.

47. Fe(II) forms the complex ion $[Fe(OH)_4]^{2-}$ through equilibrium reactions in which hydroxide replaces water in a stepwise manner. If $\log K_1 = 5.56$, $\log K_2 = 9.77$, $\log K_3 = 9.67$, and $\log K_4 = 8.58$, what is K_f? Write the equilibrium equation that corresponds to each stepwise equilibrium constant. Do you expect the $[Fe(OH)_4]^{2-}$ complex to be stable? Explain your reasoning.

48. Although thallium has limited commercial applications because it is toxic to humans (10 mg/kg body weight is fatal to children), it is used as a substitute for mercury in industrial switches. The complex ion $[TlBr_6]^{3-}$ is highly stable, with $\log K_f = 31.6$. What is the concentration of Tl(III)(*aq*) in equilibrium with a 1.12 *M* solution of $Na_3[TlBr_6]$?

17.4 Solubility and pH

49. A 1.0-L solution contains 1.98 *M* $Al(NO_3)_3$. What are $[OH^-]$ and $[H^+]$? What pH is required to precipitate the cation as $Al(OH)_3$? $K_{sp} = 1.3 \times 10^{-33}$, $K_a = 1.05 \times 10^{-5}$.

50. A 1.0-L solution contains 2.03 *M* $CoCl_2$. What is $[H^+]$? What pH is required to precipitate the cation as $Co(OH)_2$? $K_{sp} = 5.92 \times 10^{-15}$, $K_a = 1.26 \times 10^{-9}$.

51. Given 100 mL of a solution that contains 0.80 m*M* Ag^+ and 0.80 m*M* Cu^+, can the two metals be separated by selective precipitation as the insoluble bromide salts by adding 10 mL of an 8.0 m*M* solution of KBr? K_{sp} values are 6.27×10^{-9} for CuBr and 5.35×10^{-13} for AgBr. What maximum $[Br^-]$ will separate the ions?

52. Given 100 mL of a solution that is 1.5 m*M* in Tl^+, Zn^{2+}, and Ni^{2+}, which ions can be separated from solution by adding 5.0 mL of a 12.0 m*M* solution of $Na_2C_2O_4$?

Precipitate	K_{sp}
$Tl_2C_2O_4$	2×10^{-4}
$ZnC_2O_4 \cdot 2H_2O$	1.38×10^{-9}
NiC_2O_4	4×10^{-10}

How many milliliters of 12.0 m*M* $Na_2C_2O_4$ should be added to separate Tl^+ and Zn^{2+} from Ni^{2+}?

APPLICATIONS

53. Gypsum ($CaSO_4 \cdot 2H_2O$) is added to soil to enhance plant growth. It dissolves according to the equation: $CaSO_4 \cdot 2H_2O(s) \rightleftharpoons Ca^{2+}(aq) + SO_4^{2-}(aq) + 2H_2O(l)$.
   (a) The K_{sp} of gypsum is 3.14×10^{-5}. How much gypsum should you add to your 5.0-L watering can to produce a solution that is 0.0050 *M* in Ca^{2+}?
   (b) Gibbsite $[Al(OH)_3]$ is a component of clay, found in places such as Molokai, Hawaii. It dissolves according to the equation: $Al(OH)_3(s) \rightleftharpoons Al^{3+}(aq) + 3OH^-(aq)$, with a K_{sp} of 1.3×10^{-33}. You are interested in using gypsum to counteract harmful growth effects of Al^{3+} on plants from dissolved gibbsite, and you have found the pH of your soil to be 8.7. What is the apparent concentration of OH^- in your soil?

54. Egyptian blue is a synthetic pigment developed about 4500 years ago that is difficult to prepare. It was the only blue pigment identified in a study of stocks of dry pigments found in color merchants' shops in Pompeii. The pigment contains 12.9% calcium carbonate (calcite). A major source of $CaCO_3$ is limestone, which also contains $MgCO_3$. Assuming that the masses of $CaCO_3$ and $MgCO_3$ are equal, and that a sample of limestone is dissolved in acidified water to give $[Ca^{2+}] = [Mg^{2+}] = 0.010$ *M* in 5.0 L of solution, would selective precipitation be a viable method for purifying enough $CaCO_3$ from limestone to produce 1.0 g of pigment? Why? K_{sp} values are 3.36×10^{-9} for $CaCO_3$ and 6.8×10^{-6} for $MgCO_3$.

55. One method of mining gold is to extract it through the process of cyanidation. Mined ores are milled and treated with cyanide to produce a gold complex ion that is very stable, $[Au(CN)_2]^-$. Given a sample of AuCl, what is the solubility of AuCl in (a) pure water and (b) a 1.68 *M* solution of NaCN? K_{sp} of AuCl $= 2.0 \times 10^{-13}$; $\log K_f([Au(CN)_2]^-) = 38.3$.

56. Almost all barium carbonate is produced synthetically. The compound is used in manufacturing clay tiles and ceramic products as well as in CRT screens and special optical glasses. $BaCO_3$ is synthesized by allowing barium sulfate to react with coal at 1000–1200°C in a rotary kiln, followed by treatment of a solution of the product with either CO_2 or Na_2CO_3. The reactions are

$$BaSO_4(s) + 4C(s) \longrightarrow 4BaS(s) + 4CO(g)$$

$$(1)\quad BaS(s) + CO_2(g) + H_2O(l) \longrightarrow BaCO_3(aq) + H_2S(g)$$

$$(2)\quad BaS(s) + Na_2CO_3(aq) \rightleftharpoons BaCO_3(aq) + Na_2S(aq)$$

Barium carbonate has a K_{sp} of 2.58×10^{-9}. The pK_a for $H_2S \rightleftharpoons H^+ + HS^-$ is 6.97, and the pK_a for $HS^- \rightleftharpoons H^+ + S^{2-}$ is 12.90. Given this information, answer these questions:
   (a) If reaction 1 occurs, what is the pH of the solution if 80.0 g of BaS reacts with CO_2 in 1.00 L of water?

(b) If reaction 2 occurs, how many grams of $BaCO_3$ are produced by the reaction of 80.0 g of BaS with excess Na_2CO_3?

57. A person complaining of chronic indigestion continually consumed antacid tablets containing $Ca(OH)_2$ over a two week period. A blood test at the end of this period showed that the person had become anemic. Explain the reactions that caused this test result.

58. Although the commercial production of radium has virtually ceased since artificial radionuclides were discovered to have similar properties and lower costs, commercial radium is still isolated using essentially the same procedure developed by Marie Curie, as outlined in the figure. Explain what is happening chemically at each step of the purification process. What is precipitate A? What metal ions are present in solution A? What is precipitate B? What metal ions are present in solution B?

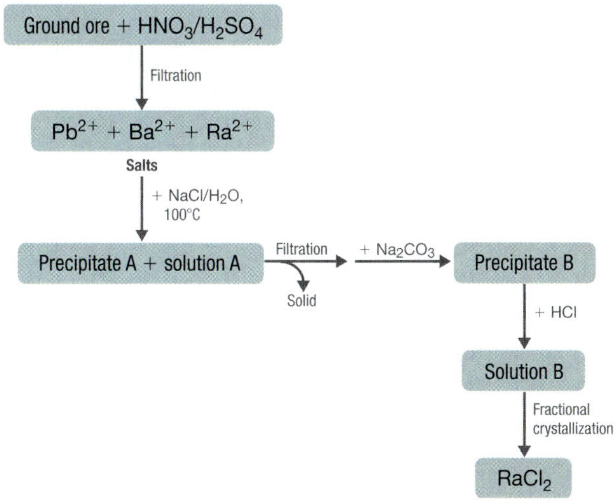

59. In a qualitative analysis laboratory, a student initially treated his sample of metal ions with 6 *M* HNO_3 instead of 6 *M* HCl, and recognized his mistake only after the acid-insoluble sulfides had been precipitated. He decided to simply add 6 *M* HCl to the filtrate from which the sulfides had been removed, but he obtained no precipitate. The student therefore concluded that there were no Ag^+, Hg_2^{2+}, or Pb^{2+} cations in his original sample. Is this conclusion valid?

60. Using qualitative analysis, a student decided to treat her sample with $(NH_4)_2S$ solution directly, skipping the HCl and acidic H_2S treatments because she was running out of time. In a sample that contained Ag^+, Hg_2^{2+}, Cd^{2+}, Sb^{3+}, and Zn^{2+}, which metal ions was she most likely to obtain in the resulting precipitate?

18 Chemical Thermodynamics

LEARNING OBJECTIVES

- To understand the connections among work, heat, and energy
- To be familiar with the concept of *PV* work
- To be able to calculate changes in internal energy
- To understand the relationship between internal energy and entropy
- To be able to use a thermodynamic cycle to calculate changes in entropy
- To understand the relationship between Gibbs free energy and work
- To know the difference between the information that thermodynamics and kinetics provide about a system
- To understand the importance of thermodynamics in biochemical systems

CHAPTER OUTLINE

Chemical reactions obey two fundamental laws. The first of these, the law of conservation of mass (see Chapter 1), states that matter can be neither created nor destroyed. The law of conservation of mass explains why equations must balance and is the basis for all the stoichiometry and equilibrium calculations you have learned to do thus far in chemistry. The second, the law of conservation of energy (see Chapter 5), states that energy can be neither created nor destroyed. Instead, energy takes various forms that can be converted from one to the other. For example, the energy stored in chemical bonds can be released as heat during a chemical reaction.

In Chapter 5, you learned about thermochemistry, the study of energy changes that occur during chemical reactions. Our goal in this chapter is to extend the

The melting of ice is a thermodynamic process. When a cube of ice melts, there is a spontaneous and irreversible transfer of heat from a warm substance, the surrounding air, to a cold substance, the ice cube. The direction of heat flow in this process and the resulting increase in entropy illustrate the second law of thermodynamics.

concepts of thermochemistry to an exploration of **thermodynamics** (from the Greek for "heat" and "power"), the study of the interrelationships among heat, work, and the energy content of a system at equilibrium. Thermodynamics tells chemists *whether* a particular reaction is energetically possible in the direction in which it is written and the composition of the reaction system at equilibrium. It does not, however, say anything about whether an energetically feasible reaction will *actually* occur as written, and it tells us nothing about the rate of the reaction or the pathway by which it will occur. The rate of a reaction and its pathway are described by chemical kinetics (see Chapter 14).

Chemical thermodynamics provides a bridge between the macroscopic properties of a substance and the individual properties of its constituent molecules and atoms. As you will see, thermodynamics explains why graphite can be converted to diamond; how chemical energy stored in molecules can be utilized to perform work; and why certain processes, such as iron rusting and organisms aging and dying, proceed spontaneously in only one direction, requiring no input of energy to occur.

18.1 ◦ Thermodynamics and Work

We begin our discussion of thermodynamics by reviewing some important terms introduced in Chapter 5. First, we need to distinguish between a system and its surroundings. A *system* is that part of the universe in which we are interested, such as a mixture of gases in a glass bulb or a solution of substances in a flask. The *surroundings* are everything else—the rest of the *universe*. We can therefore state

$$\text{system} + \text{surroundings} = \text{universe} \tag{18.1}$$

A *closed system*, such as the contents of a sealed jar, cannot exchange matter with its surroundings, whereas an *open system* can; in this case, we can convert a closed system (the jar) to an open system by removing the jar's lid.

In Chapter 5, we also introduced the concept of a **state function**, a property of a system that depends only on the present state of the system and not on its history. Thus, a change in a state function depends only on the difference between the initial and final states, not on the pathway used to go from one to the other. To help understand the concept of a state function, imagine a person hiking up a mountain (Figure 18.1). If the person is well trained and fit, he or she may be able to climb almost vertically to the top (path A), whereas another less athletic person may choose a path that winds gradually to the top (path B). If both hikers start from the same point at the base of the mountain and end up at the same point at the top, their *net change in altitude* will be the same regardless of the path chosen. Hence, altitude is a state function. On the other hand, a person may or may not carry a heavy pack, and may climb in hot weather or cold. These conditions would influence changes in the hiker's fatigue level, which depends on the path taken and the conditions experienced. Fatigue, therefore, is not a state function. Thermodynamics is generally concerned with state functions and does not deal with how the change between the initial and final state occurs.

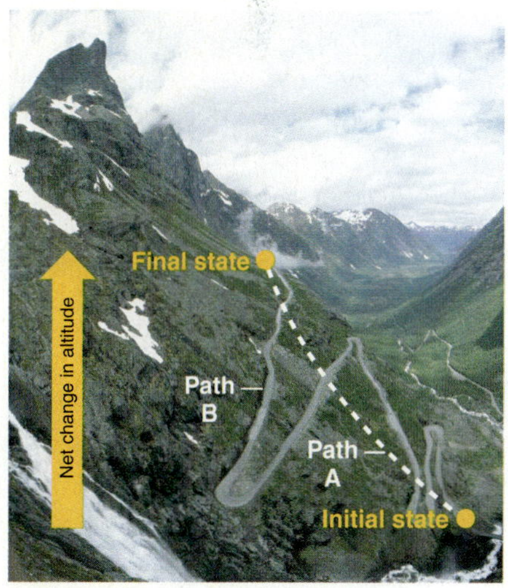

Figure 18.1 Illustration that altitude is a state function. When hiking up a mountain, a person may decide to take path A, which is almost vertical, or path B, which gradually winds up to the top. Regardless of the path taken, the net change in altitude going from the initial state (bottom of the climb) to the final state (top of the climb) is the same. Thus, altitude is a state function.

The Connections Among Work, Heat, and Energy

The **internal energy** (*E*), of a system is the sum of the potential energy and the kinetic energy of all the components; internal energy is a state function. Although a closed system cannot exchange matter with its surroundings, it can exchange energy with its surroundings in two ways: by doing work or by releasing or absorbing heat, the flow of thermal energy. Work and heat are therefore two distinct ways of changing the internal energy of a system. We defined *work* w in Chapter 5 as a force F acting through a distance d:

$$w = Fd \tag{18.2}$$

Because work occurs only when an object, such as a person, or a substance, such as water, moves against an opposing force, work requires that the system and its surroundings be physically connected. In contrast, the flow of *heat*, the transfer of energy due to differences in temperature between two objects, represents a thermal connection between the system and its surroundings. Thus, doing work causes a physical displacement, whereas the flow of heat causes a temperature change. The units of work and heat must be the same because both processes result in the transfer of energy. In the SI system, those units are joules (J), the same unit used for energy. There is no difference between an energy change brought about by doing work on a system and an equal energy change brought about by heating it.

The connections among work, heat, and energy were first described by Benjamin Thompson (1753–1814), an American-born scientist who was also known as Count Rumford. While supervising the manufacture of cannons, Rumford recognized the relationship between the amount of work required to drill out a cannon and the temperature of the water used to cool it during the drilling process (Figure 18.2). At that time, it was generally thought that heat and work were separate and unrelated phenomena. Hence, Rumford's ideas were not widely accepted until many years later, after his findings had been corroborated in other laboratories.

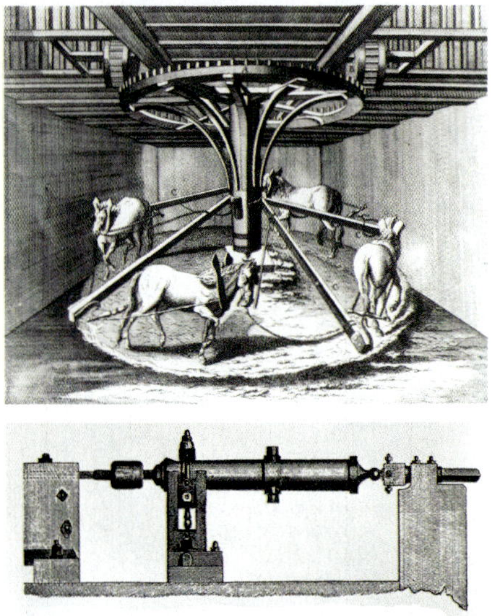

Figure 18.2 The relationship between heat and work. In the 1780s, an American scientist named Benjamin Thompson, also known as Count Rumford, was hired by the Elector of Bavaria to supervise the manufacture of cannons. During the manufacturing process, teams of horses harnessed to a large-toothed wheel supplied the power needed to drill a hole several inches in diameter straight down the center of a solid brass or bronze cylinder, which was cooled by water. Based on his observations, Rumford became convinced that heat and work are equivalent ways of transferring energy.

PV Work

As we saw in Chapter 5, there are many kinds of work, including mechanical work, electrical work, and work against a gravitational or magnetic field. In this section, we will consider only mechanical work, focusing on the work done during changes in the pressure or volume of a gas. To describe this *pressure–volume work* (*PV* work), we will use such imaginary oddities as frictionless pistons, which involve no component of resistance, and ideal gases, which have no attractive or repulsive interactions.

Imagine, for example, an ideal gas, confined by a frictionless piston, with internal pressure P_{int} and initial volume V_i (Figure 18.3). If $P_{ext} = P_{int}$, the system is at equilibrium; the piston does not move, and no work is done. If the external pressure on the piston, P_{ext}, is less than P_{int}, however, then the ideal gas inside the piston will expand, forcing the piston to perform work on the surroundings; that is, the final volume V_f will be greater than V_i. If $P_{ext} > P_{int}$, then the gas will be compressed and the surroundings will perform work on the system.

If the piston has cross-sectional area A, the external pressure exerted by the piston is by definition the force per unit area: $P_{ext} = F/A$. The volume of any three-dimensional object with parallel sides (such as a cylinder) is the cross-sectional area times the height ($V = Ah$). Rearranging to give $F = P_{ext}A$ and defining the distance the piston moves, d, as Δh, we can calculate the magnitude of the work performed by the piston by substituting into Equation 18.2:

$$w = Fd = P_{ext}A\Delta h \tag{18.3a}$$

The change in the volume of the cylinder, ΔV, as the piston moves a distance d is $\Delta V = A\Delta h$, as shown in Figure 18.4. The work performed is thus

$$w = P_{ext}\Delta V \tag{18.3b}$$

Note that the units of work obtained using this definition are correct: pressure is force per unit area (newton/m^2) and volume has units of cubic meters, so

$$w = \left(\frac{F}{A}\right)_{ext} \Delta V = \frac{\text{newton}}{\text{m}^2} \times \text{m}^3 = \text{newton} \cdot \text{m} = \text{joule}$$

If we use atmospheres for P and liters for V, we obtain units of L·atm for work. These units correspond to units of energy (see Chapter 10), as shown in the different values of the ideal gas constant R:

$$R = \frac{0.08206\ \text{L}\cdot\text{atm}}{\text{mol}\cdot\text{K}} = \frac{8.314\ \text{J}}{\text{mol}\cdot\text{K}}$$

Thus, 0.08206 L·atm = 8.314 J and 1 L·atm = 101.3 J.

Whether work is defined as having a positive or negative sign is a matter of convention. In Chapter 5, we defined heat flow *from* the system *to* the surroundings as negative. Using that same sign convention, we define work done *by* the system *on* the surroundings as having a negative sign because it too results in a transfer of energy from the system to the surroundings.* Because $\Delta V > 0$ for an expansion, Equation 18.3b must be written with a negative sign to describe *PV* work done *by* the system as negative:

$$w = -P_{ext}\Delta V \qquad (18.4)$$

The work done by a gas expanding against an external pressure is therefore negative, corresponding to work done *by* the system *on* the surroundings. Conversely, when a gas is compressed by an external pressure, $\Delta V < 0$ and the work is positive because work is being done *on* the system *by* the surroundings.

Suppose, for example, that the system under study is a mass of steam heated by the combustion of several hundred pounds of coal and enclosed within a cylinder housing a piston attached to the crankshaft of a large steam engine. The gas is not ideal, and the cylinder is not frictionless. Nonetheless, as steam enters the engine chamber and the expanding gas pushes against the piston, the piston moves and useful work is performed. In fact, *PV* work launched the Industrial Revolution of the 19th century and powers the internal combustion engine on which most of us still rely for transportation.

We can see that work is not a state function by examining Figure 18.5, in which two different, two-step pathways take a gaseous system from an initial state to a final state with corresponding changes in temperature. In pathway A, the volume of a gas is initially increased while its pressure stays constant (step 1), and then its pressure is decreased while the volume remains constant (step 2). In pathway B, the order of the steps is reversed. The temperatures, pressures, and volumes of the initial and final states are identical in both cases, but the amount of work done, indicated by the shaded areas in the figure, is substantially different. As we can see, the amount of work done depends on the pathway taken from (V_1, P_1) to (V_2, P_2), which means that work is not a state function.

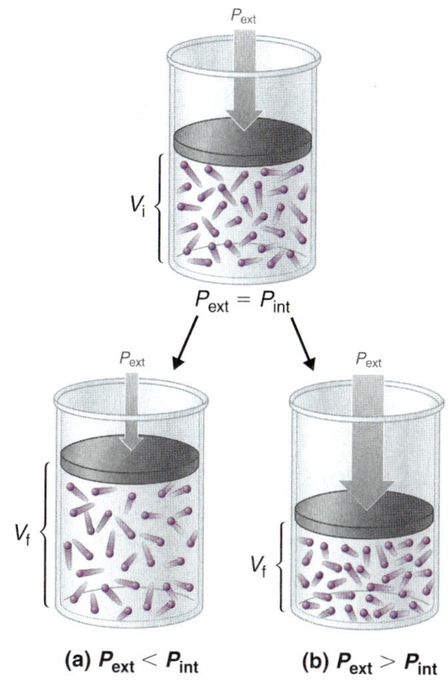

Figure 18.3 ***PV* work.** The frictionless piston is subject to external pressure P_{ext} and encloses an ideal gas with initial volume V_i and internal pressure P_{int}. If $P_{ext} = P_{int}$, the system is at equilibrium, the piston does not move, and no work is done. However, if the external pressure is less than the internal pressure **(a),** the ideal gas inside the piston will expand, forcing the piston to perform work on the surroundings. The final volume V_f will be greater than V_i. Alternatively, if the external pressure is greater than the internal pressure **(b),** the gas will be compressed and the surroundings will perform work on the system.

EXAMPLE 18.1

A small high-performance internal combustion engine has six cylinders with a total nominal displacement (volume) of 2.40 L and a 10:1 compression ratio (meaning that the volume of each cylinder decreases by a factor of 10 when the piston compresses the air–gas mixture inside the cylinder prior to ignition). How much work in joules is done when a gas in one cylinder of the engine expands at constant temperature against an opposing pressure of 40.0 atm during the engine cycle? Assume that the gas is ideal, the piston is frictionless, and no energy is lost as heat.

Given Final volume, compression ratio, external pressure

Asked for Work done

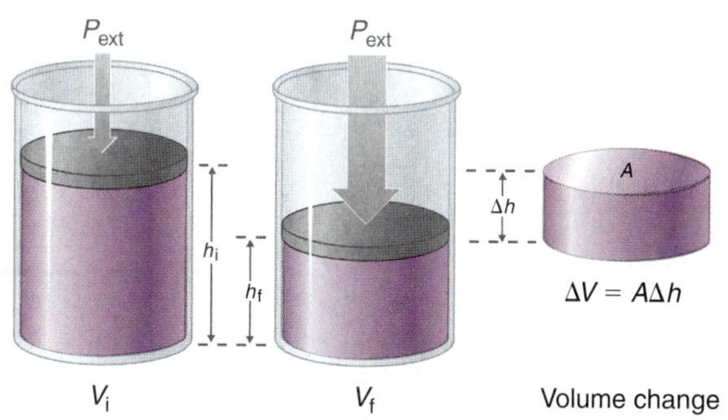

Figure 18.4 Work performed with a change in volume. The volume of a cylinder is equal to its cross-sectional area A times its height h. Thus, the change in the volume, ΔV, of the cylinder housing a piston is $\Delta V = A\ \Delta h$ as the piston moves. The work performed by the surroundings on the system as the piston moves inward is given by $w = P_{ext}\,\Delta V$.

* Note that this is an arbitrary convention and one that is not universally used. Some engineering disciplines are more interested in the work done *on* the surroundings than in the work done *by* the system, and therefore use the opposite convention.

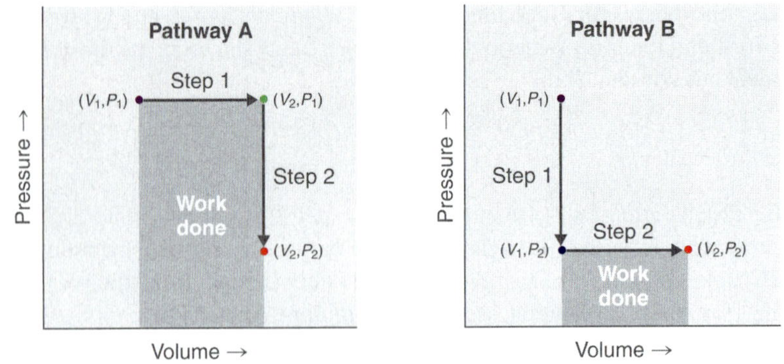

Figure 18.5 Illustration that work is not a state function. Two different pathways, A and B, are taken from an initial state (V_1, P_1) to a final state (V_2, P_2). In pathway A, the volume of a gas is initially increased while its pressure stays constant (step 1). Its pressure is then decreased while the volume remains constant (step 2). The order of the steps is reversed in pathway B. Although V_1, P_1 and V_2, P_2 are identical in both cases, the amount of work done (*shaded area*) depends on the pathway taken. Thus, work is not a state function.

Note the pattern

By convention, both heat flow and work have a negative sign when energy is transferred from the system to the surroundings, and vice versa.

Strategy

Ⓐ Calculate the final volume of gas in a single cylinder. Then compute the initial volume of gas in a single cylinder from the compression ratio.

Ⓑ Use Equation 18.4 to calculate the work done in liter-atmospheres. Convert from liter-atmospheres to joules.

Solution

Ⓐ To calculate the work done, we need to know the initial and final volumes. The final volume is the volume of one of the six cylinders with the piston all the way down: $V_f = 2.40 \text{ L}/6 = 0.400 \text{ L}$. With a 10:1 compression ratio, the volume of the same cylinder with the piston all the way up is $V_i = 0.400 \text{ L}/10 = 0.0400 \text{ L}$. Ⓑ To determine the work done, we substitute the values for P_{ext} and ΔV $(V_f - V_i)$ into Equation 18.4:

$$w = -P_{ext}\Delta V = -(40.0 \text{ atm})(0.400 \text{ L} - 0.0400 \text{ L}) = -14.4 \text{ L}\cdot\text{atm}$$

Converting from liter-atmospheres to joules gives

$$w = -(14.4 \cancel{\text{L}\cdot\text{atm}})[101.3 \text{ J}/(\cancel{\text{L}\cdot\text{atm}})] = -1.46 \times 10^3 \text{ J}$$

In Exercise 18.1, you will see that the concept of work is not confined to engines and pistons. It is found in other applications as well.

EXERCISE 18.1

While you may not be aware of it, breathing requires work. The lung volume of a 70-kg man at rest changed from 2200 mL to 2700 mL when he breathed in, while his lungs maintained a pressure of approximately 1.0 atm. How much work in liter-atmospheres and joules was required to take a single breath? During exercise, his lung volume changed from 2200 mL to 5200 mL on each in-breath. How much additional work in joules did he require to take a breath while exercising?

Answer $-0.500 \text{ L}\cdot\text{atm}$ or -50.7 J; -304 J. If he takes a breath every three seconds, this corresponds to 1.4 Calories per minute (1.4 kcal).

18.2 • The First Law of Thermodynamics

Note the pattern

The tendency of all systems, chemical or otherwise, is to move toward the state with the lowest possible energy.

The relationship between the energy change of the system and that of the surroundings is given by the **first law of thermodynamics**, which states that the *energy of the universe is constant.* Using Equation 18.1, we can express this law mathematically as

$$\begin{aligned}\Delta E_{univ} &= \Delta E_{sys} + \Delta E_{surr} = 0 \\ \Delta E_{sys} &= -\Delta E_{surr}\end{aligned} \tag{18.5}$$

where the subscripts refer to the universe, system, and surroundings. Thus, the change in energy of the system is identical in magnitude but opposite in sign to the change in energy of the surroundings.

One of the most important factors that determine the outcome of a chemical reaction is the tendency of all systems, chemical or otherwise, to move toward the lowest possible energy state. Thus, as a brick dropped from a rooftop falls, its potential energy is converted to kinetic energy, so that when it reaches ground level, it has achieved a state of lower potential energy. Anyone nearby will notice that energy is transferred to the surroundings as the noise of the impact reverberates and the dust rises when the brick hits the ground. Similarly, if a mixture of isooctane and oxygen in an internal combustion engine is ignited by a spark, carbon dioxide and water form spontaneously, while potential energy (in the form of the relative positions of atoms in the molecules) is released to the surroundings as heat and work. The internal energy content of the CO_2/H_2O product mixture is lower than that of the isooctane/O_2 reactant mixture. The two cases differ, however, in the form in which the energy is released to the surroundings. In the case of the falling brick, the energy is transferred as work done on whatever happens to be in the path of the brick; in the case of burning isooctane, the energy can be released as solely heat (if the reaction is carried out in an open container) or as a mixture of heat and work (if the reaction is carried out in the cylinder of an internal combustion engine).

Because heat and work are the only two ways in which energy can be transferred between a system and the surroundings, any change in the internal energy of the system is the sum of the heat transferred, q, and the work done, w:

$$\Delta E_{sys} = q + w \tag{18.6}$$

Although q and w are not state functions on their own, their sum, ΔE_{sys}, is independent of the path taken and is therefore a state function. A major task for the designers of any machine that converts energy to work is to maximize the amount of work obtained and to minimize the amount of energy released to the environment as heat. An example is the combustion of coal to produce electricity. Although the maximum amount of energy available from the process is fixed by the energy content of the reactants and products, the fraction of that energy that can be used to perform useful work is not fixed, as discussed in Section 18.5. Because we focus almost exclusively on the changes in the energy of the system, we will not use the "sys" subscript further unless we need to distinguish explicitly between the system and the surroundings.

Note the pattern

Although q and w are not state functions, their sum, ΔE_{sys}, is independent of the path taken and therefore is a state function.

EXAMPLE 18.2

A sample of an ideal gas in the cylinder of an engine is compressed from 400 mL to 50.0 mL during the compression stroke against a constant pressure of 8.00 atm. At the same time, 140 J of energy is transferred from the gas to the surroundings as heat. What is the total change in the internal energy, ΔE, of the gas in joules?

Given Initial volume, final volume, external pressure, quantity of energy transferred as heat

Asked for Total change in internal energy

Strategy

Ⓐ Determine the sign of q to use in Equation 18.6.

Ⓑ From Equation 18.4, calculate w from the values given. Substitute this value into Equation 18.6 to calculate ΔE.

Solution

Ⓐ From Equation 18.6, we know that $\Delta E = q + w$. We are given the magnitude of q (140 J) and need only determine its sign. Because energy is transferred from the system (the gas) to the surroundings, q is negative by convention. Ⓑ Because the gas is being compressed, we know that work is being done *on* the system, and hence w must be positive. From Equation 18.4,

$$w = -P_{ext}\Delta V = -8.00\ \cancel{\text{atm}}(0.0500\ \cancel{\text{L}} - 0.400\ \cancel{\text{L}})\left(\frac{101.3\ \text{J}}{\cancel{\text{L}}\cdot\cancel{\text{atm}}}\right) = 284\ \text{J}$$

Thus,

$$\Delta E = q + w = -140 \text{ J} + 284 \text{ J} = 144 \text{ J}$$

In this case, although work is done *on* the gas, increasing its internal energy, heat flows *from* the system *to* the surroundings, decreasing its internal energy by 144 J. Notice that the work done and the heat transferred can have opposite signs.

EXERCISE 18.2

A sample of an ideal gas is allowed to expand from an initial volume of 0.200 L to a final volume of 3.50 L against a constant external pressure of 0.995 atm. At the same time, 117 J of heat is transferred from the surroundings to the gas. What is the total change in the internal energy, ΔE, of the gas in joules?

Answer −216 J

Enthalpy

To further understand the relationship between heat flow q and the resulting change in internal energy ΔE, we can look at two sets of limiting conditions: reactions that occur at constant volume and reactions that occur at constant pressure. We will assume that PV work is the only kind of work possible for the system, so that we can substitute its definition from Equation 18.4 into Equation 18.6 to obtain

$$\Delta E = q - P\Delta V \tag{18.7}$$

where subscripts have been deleted.

If the reaction occurs in a closed vessel, the volume of the system is fixed and ΔV is zero. Under these conditions, the heat flow (often given the symbol q_v to indicate constant volume) must be equal to ΔE:

$$\underset{\text{Constant volume}}{q_v = \Delta E} \tag{18.7a}$$

No PV work can be done, and the change in the internal energy of the system is equal to the amount of heat transferred from the system to the surroundings, or vice versa.

Many chemical reactions are not, however, carried out in sealed containers at constant volume, but in open containers at a more or less constant pressure of about 1 atm. The heat flow under these conditions is given the symbol q_p to indicate constant pressure. Replacing q in Equation 18.7 by q_p and rearranging to solve for q_p give

$$\underset{\text{Constant pressure}}{q_p = \Delta E + P\Delta V} \tag{18.7b}$$

Note the pattern

At constant pressure, the change in the enthalpy of a system is equal to the heat flow: $\Delta H = q_p$.

Thus, at constant pressure, the heat flow for any process is equal to the change in the internal energy of the system plus the PV work done, as we stated in Chapter 5.

Because conditions of constant pressure are so important in chemistry, a new state function called **enthalpy (*H*)** is defined as $H = E + PV$. At constant pressure, the change in the enthalpy of a system is

$$\Delta H = \Delta E + \Delta(PV) = \Delta E + P\Delta V \tag{18.8}$$

Comparing the last two equations shows that at constant pressure, the change in the enthalpy of a system is equal to the heat flow: $\Delta H = q_p$. This expression is consistent with our definition of enthalpy in Chapter 5, where we stated that enthalpy is the heat absorbed or produced during any process that occurs at constant pressure.

EXAMPLE 18.3

The molar enthalpy of fusion for ice at 0.0°C and a pressure of 1.00 atm is 6.01 kJ (see Chapter 5), and the molar volumes of ice and water at 0°C are 0.0197 L and 0.0180 L, respectively. Calculate ΔH and ΔE for the melting of ice at 0.0°C.

Given Enthalpy of fusion for ice, pressure, molar volumes of ice and water

Asked for ΔH and ΔE for ice melting at 0.0°C

Strategy

Ⓐ Determine the sign of q, and set this value equal to ΔH.
Ⓑ Calculate $\Delta(PV)$ from the information given.
Ⓒ Determine ΔE by substituting the calculated values into Equation 18.8.

Solution

Ⓐ Because 6.01 kJ of heat is absorbed from the surroundings when 1 mol of ice melts, $q = +6.01$ kJ. When the process is carried out at constant pressure, $q = q_p = \Delta H = 6.01$ kJ. Ⓑ To find ΔE using Equation 18.8, we need to calculate $\Delta(PV)$. The process is carried out at a constant pressure of 1.00 atm, so

$$\Delta(PV) = P\Delta V = P(V_f - V_i) = (1.00 \text{ atm})(0.0180 \text{ L} - 0.0197 \text{ L})$$
$$= (-1.7 \times 10^{-3} \cancel{\text{L}\cdot\text{atm}})(101.3 \text{ J}/\cancel{\text{L}\cdot\text{atm}}) = -0.0017 \text{ J}$$

Ⓒ Substituting the calculated values of ΔH and $P\Delta V$ into Equation 18.8 gives

$$\Delta E = \Delta H - P\Delta V = 6010 \text{ J} - (-0.0017 \text{ J}) = 6010 \text{ J} = 6.01 \text{ kJ}$$

EXERCISE 18.3

At 298 K and 1 atm, the conversion of graphite to diamond requires the input of 1.850 kJ of heat per mole of carbon. The molar volumes of graphite and diamond are 0.00534 L and 0.00342 L, respectively. Calculate ΔH and ΔE for the conversion of C (graphite) to C (diamond) under these conditions.

Answer $\Delta H = 1.85$ kJ/mol; $\Delta E = 1.85$ kJ/mol

The Relationship Between ΔH and ΔE

If ΔH for a reaction is known, we can use the change in the enthalpy of the system (Equation 18.8) to calculate its change in internal energy. When a reaction involves only solids, liquids, liquid solutions, or any combination of these, the volume does not change appreciably ($\Delta V = 0$). Under these conditions, we can simplify Equation 18.8 to $\Delta H = \Delta E$. If gases are involved, however, ΔH and ΔE can differ significantly. We can calculate ΔE from the measured value of ΔH by using the right side of Equation 18.8 together with the ideal gas law, $PV = nRT$. Recognizing that $\Delta(PV) = \Delta(nRT)$, we can rewrite Equation 18.8 as

$$\Delta H = \Delta E + \Delta(PV) = \Delta E + \Delta(nRT) \tag{18.9}$$

At constant temperature, $\Delta(nRT) = RT\Delta n$, where Δn is the difference between the final and initial numbers of moles of gas. Thus,

$$\Delta E = \Delta H - RT\Delta n \tag{18.10}$$

For reactions that result in a net production of gas, $\Delta n > 0$ and therefore $\Delta E < \Delta H$. Conversely, endothermic reactions ($\Delta H > 0$) that result in a net consumption of gas

Note the pattern

For reactions that result in a net production of gas, $\Delta E < \Delta H$. For endothermic reactions that result in a net consumption of gas, $\Delta E > \Delta H$.

have $\Delta n < 0$ and $\Delta E > \Delta H$. The relationship between ΔH and ΔE for systems involving gases is illustrated in the next example.

EXAMPLE 18.4

The combustion of graphite to produce carbon dioxide is described by the equation C (graphite) + $O_2(g) \longrightarrow CO_2(g)$. At 298 K and 1.0 atm, $\Delta H = -393.5$ kJ/mol of graphite for this reaction, and the molar volume of graphite is 0.0053 L. What is the value of ΔE for the reaction?

Given Balanced chemical equation, temperature, pressure, ΔH, molar volume of reactant

Asked for ΔE

Strategy

Ⓐ Use the balanced chemical equation to calculate the change in the number of moles of gas during the reaction.
Ⓑ Substitute this value and the data given into Equation 18.10 to obtain ΔE.

Solution

Ⓐ In this reaction, 1 mol of gas (CO_2) is produced, and 1 mol of gas (O_2) is consumed. Thus, $\Delta n = 1 - 1 = 0$. Ⓑ Substituting this calculated value and the given values into Equation 18.10 gives

$$\Delta E = \Delta H - RT\Delta n = (-393.5 \text{ kJ/mol}) - [8.314 \text{ J/(mol} \cdot \text{K)}](298 \text{ K})(0)$$
$$= (-393.5 \text{ kJ/mol}) - (0 \text{ J/mol}) = -393.5 \text{ kJ/mol}$$

To understand why only the change in the volume of the gases needs to be considered, notice that the molar volume of graphite is only 0.0053 L. A change in the number of moles of gas corresponds to a volume change of 22.4 L/mol of gas at STP, so the volume of gas consumed or produced in this case is (1)(22.4 L) = 22.4 L, which is much, much greater than the volume of 1 mol of a solid such as graphite.

EXERCISE 18.4

Calculate ΔE for the conversion of oxygen gas to ozone at 298 K: $3O_2(g) \longrightarrow 2O_3(g)$. The value of ΔH for the reaction is 285.4 kJ.

Answer 288 kJ

As Exercise 18.4 illustrates, the magnitudes of ΔH and ΔE for reactions that involve gases are generally rather similar, even when there is a net production or consumption of gases.

18.3 ○ The Second Law of Thermodynamics

The first law of thermodynamics governs changes in the state function we have called internal energy E. As we have just seen, changes in the internal energy, ΔE, are closely related to changes in the enthalpy, ΔH, which is a measure of the heat flow between a system and its surroundings at constant pressure. You also learned in Chapter 5 that the enthalpy change for a chemical reaction can be calculated using tabulated values of enthalpies of formation. This information, however, does not tell us whether a particular process or reaction will occur spontaneously.

Let's consider a familiar example of spontaneous change. If a hot frying pan that has just been removed from the stove is allowed to come into contact with a cooler object, such as cold water in a sink, heat will flow from the hotter object to the cooler one. Eventually both objects will reach the same temperature, at a value between the initial temperatures of the two objects. This transfer of heat from a hot object to a cooler one obeys the first law of thermodynamics: energy is conserved.

Now consider the same process in reverse. Suppose that a hot frying pan in a sink of water were to become hotter while the water became cooler. As long as the same amount of thermal energy was gained by the frying pan and lost by the water, the first law of thermodynamics would be satisfied. Yet we all know that such a process cannot occur: heat always flows from a hot object to a cold one, and never in the reverse direction. That is, *by itself* the magnitude of the heat flow associated with a process does not predict whether the process will occur spontaneously.

For many years, chemists and physicists tried to identify a single measurable quantity that would enable them to predict whether a particular process or reaction would occur spontaneously. Initially, many of them focused on enthalpy changes and hypothesized that an exothermic process would always be spontaneous. But although it is true that many, if not most, spontaneous processes are exothermic, there are also many spontaneous processes that are *not* exothermic. For example, at a pressure of 1 atm, ice melts spontaneously at temperatures higher than 0°C, yet this is an endothermic process: heat is absorbed. Similarly, many salts (such as NH_4NO_3, NaCl, and KBr) dissolve spontaneously in water even though they absorb heat from the surroundings as they dissolve (that is, $\Delta H_{soln} > 0$). Reactions can also be both spontaneous and highly endothermic, like the reaction of barium hydroxide with ammonium thiocyanate shown in Figure 18.6.

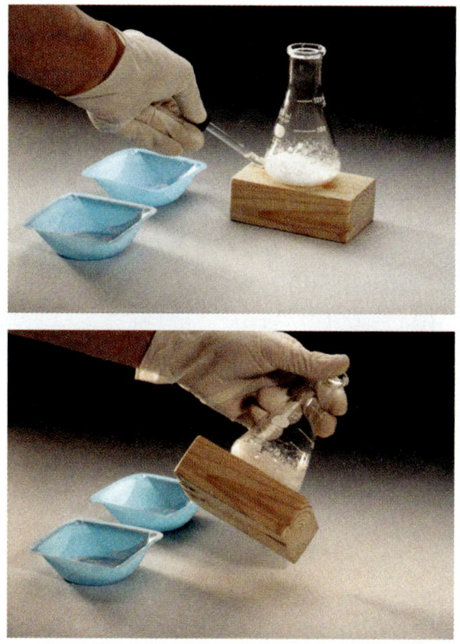

Figure 18.6 An endothermic reaction. The reaction of barium hydroxide with ammonium thiocyanate is spontaneous but highly endothermic, so water, one product of the reaction, quickly freezes into slush. When water is placed on a block of wood under the beaker, the highly endothermic reaction that takes place in the beaker freezes water that has been placed under the beaker, and the beaker becomes frozen to the wood.

Thus, enthalpy is not the only factor that determines whether a process is spontaneous. For example, once a cube of sugar has dissolved in a glass of water so that the sucrose molecules are uniformly dispersed in a dilute solution, they never spontaneously come back together in solution to form a sugar cube. Moreover, the molecules of a gas remain evenly distributed throughout the entire volume of a glass bulb and never spontaneously assemble in only one portion of the available volume. To help explain why these phenomena proceed spontaneously in only one direction requires an additional state function called **entropy (S)**, a thermodynamic property of all substances that is proportional to their degree of disorder. In Chapter 13, we introduced the concept of entropy in relation to solution formation. Here we further explore the nature of this state function and define it mathematically.

Entropy

Chemical and physical changes in a system may be accompanied by either an increase or a decrease in the disorder of the system, corresponding to an increase in entropy ($\Delta S > 0$) or a decrease in entropy ($\Delta S < 0$), respectively. As with any other state function, the change in entropy is defined as the difference between the entropies of the final and initial states: $\Delta S = S_f - S_i$.

When a gas expands into a vacuum, its entropy increases because the increased volume allows for greater atomic or molecular disorder. The greater the number of atoms or molecules in the gas, the greater the disorder. The magnitude of the entropy of a system depends on the number of microscopic states, or *microstates*, associated with it (in this case, the number of atoms or molecules); that is, the greater the number of microstates, the greater the entropy.

We can illustrate the concepts of microstates and entropy using a deck of playing cards, as shown in Figure 18.7. In any new deck, the 52 cards are arranged by suits, with each suit arranged in descending order. If the cards are shuffled, however, there are approximately 10^{68} different ways they might be arranged, which corresponds to 10^{68} different microscopic states. The entropy of an ordered new deck of cards is therefore low, whereas the entropy of a randomly shuffled deck is high. Card games assign a higher value to a hand that has a low degree of disorder. In games like five-card poker, only four of the 2,598,960 different possible hands, or microstates, contain the highly ordered and valued arrangement of cards called a royal flush, almost 1.1 million hands contain one pair, and more than 1.3 million hands are completely disordered and

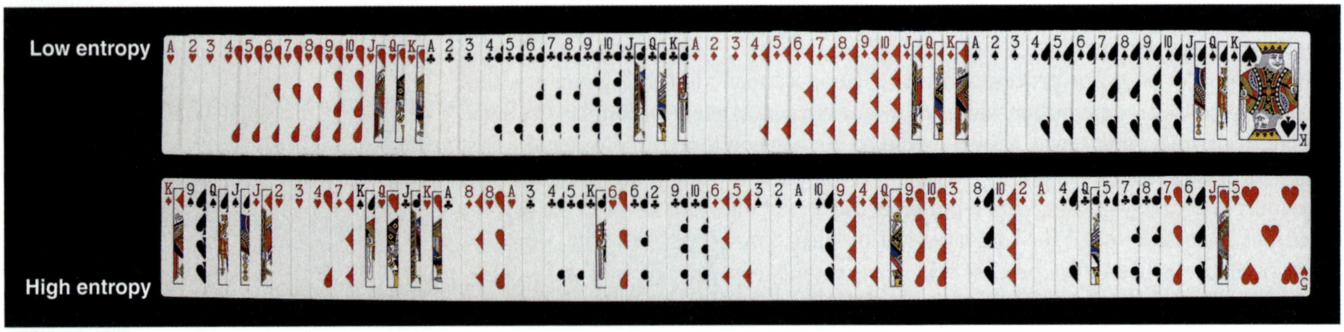

Figure 18.7 Illustrating low- and high-entropy states with a deck of playing cards. A new, unshuffled deck (*top*) contains 52 cards arranged by suits, with each suit arranged in descending order. With only a single arrangement possible, there is only one microstate. The entropy of the deck is therefore low. In contrast, a randomly shuffled deck (*bottom*) can have any one of approximately 10^{68} different arrangements, which correspond to 10^{68} different microstates. The entropy of a randomly shuffled deck is therefore high. One of the possible random arrangements is shown.

therefore have no value. Because the last two arrangements are far more probable than the first, the value of a poker hand is inversely proportional to its entropy.

We can see how to calculate these kinds of probabilities for a chemical system by considering the possible arrangements of a sample of four gas molecules in a two-bulb container (Figure 18.8). There are five possible arrangements: all four molecules in the left bulb (I); three molecules in the left bulb and one in the right bulb (II); two molecules in each bulb (III); one molecule in the left bulb and three molecules in the right bulb (IV); and four molecules in the right bulb (V). If we assign a different color to each molecule in order to keep track of it for this discussion (remember, however, that in reality molecules are indistinguishable from one another), we can see that there are 16 different ways the four molecules can be distributed in the bulbs, each corresponding to a particular microstate. As shown in Figure 18.8, arrangement I is associated with a single microstate, as is arrangement V. Hence each of these arrangements has a probability of $\frac{1}{16}$. Arrangements II and IV each have a probability of $\frac{4}{16}$ because each can exist in four microstates. Similarly, six different microstates can occur as arrangement III, making the probability of this arrangement $\frac{6}{16}$. Thus, the arrangement that we would expect to encounter, with half the gas molecules in each bulb, is the *most probable* arrangement. The others are not impossible, but simply less likely.

Figure 18.8 The possible microstates for a sample of four gas molecules in two bulbs of equal volume. There are 16 different ways to distribute four gas molecules between the bulbs, with each distribution corresponding to a particular microstate. The possible microstates can be grouped into five arrangements. Arrangements I and V each produce a single microstate with a probability of 1/16. This particular arrangement is so improbable that it is likely not observed. Arrangements II and IV each produce four microstates, with a probability of 4/16. Arrangement III, with half the gas molecules in each bulb, has a probability of 6/16. It is the one encompassing the most microstates, so it is the most probable.

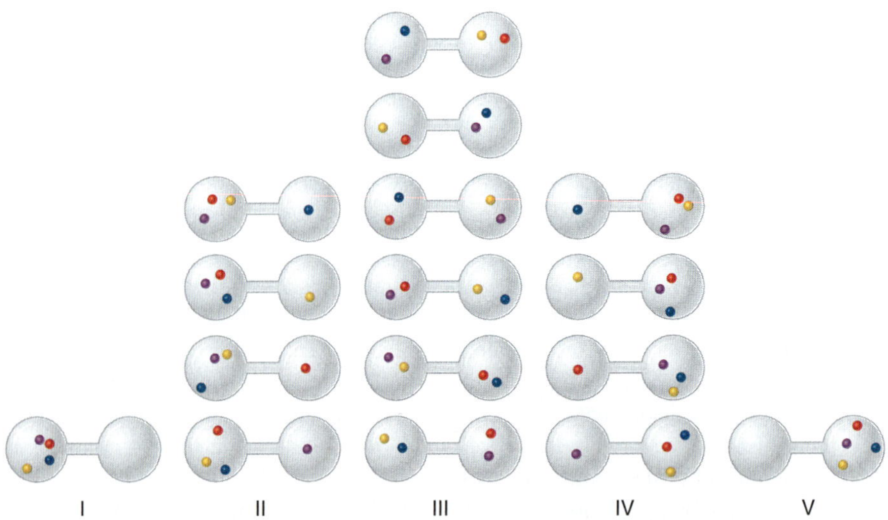

Instead of four molecules of gas, let's now consider 1 L of an ideal gas at STP, which contains 2.69×10^{22} molecules (6.022×10^{23} molecules/22.4 L). If we allow the sample of gas to expand into a second 1-L container, the probability of finding all 2.69×10^{22} molecules in one container and none in the other at any given time is infinitesimally small, approximately $1/(2.69 \times 10^{22})$. The denominator is larger than the total number of atoms in the universe, so the probability of such an occurrence is effectively zero. Thus, although nothing prevents the molecules in the gas sample from occupying only one of the two bulbs, that particular arrangement is so improbable that it is never actually observed. The probability of arrangements with essentially equal numbers of molecules in each bulb is quite high, however, because there are *many* equivalent microstates in which the molecules are distributed equally. Hence, a macroscopic sample of a gas occupies all of the space available to it, simply because this is the most probable arrangement.

A disordered system has a greater number of possible microstates than does an ordered system, so it has a higher entropy. This is most clearly seen in the entropy changes that accompany phase transitions, such as solid to liquid or liquid to gas. As you know from Chapters 11–13, a crystalline solid is composed of an ordered array of molecules that occupy fixed positions in a lattice, whereas the molecules in a liquid are free to move and tumble within the volume of the liquid; molecules in a gas have even more freedom to move than those in a liquid. Each degree of motion increases the number of available microstates, resulting in a higher entropy. Thus, the entropy of a system must increase during melting ($\Delta S_{fus} > 0$). Similarly, when a liquid is converted to a vapor, the greater freedom of motion of the molecules in the gas phase means that $\Delta S_{vap} > 0$. Conversely, the reverse processes (condensing a vapor to form a liquid or freezing a liquid to form a solid) must be accompanied by a decrease in the entropy of the system: $\Delta S < 0$.

Note the pattern

Entropy S is a thermodynamic property of all substances that is proportional to their degree of disorder. The greater the number of possible microstates for a system, the greater the disorder and the higher the entropy.

Experiments show that the magnitude of ΔS_{vap} is 80–90 J/ (mol · K) for a wide variety of liquids with different boiling points. However, liquids that have highly ordered structures due to hydrogen bonding or other intermolecular interactions tend to have significantly higher values of ΔS_{vap}. For instance, ΔS_{vap} for water is 102 J/(mol · K).

Another process that is accompanied by entropy changes is the formation of a solution. As illustrated in Figure 18.9, the formation of a liquid solution from a crystalline solid (the solute) and a liquid solvent is expected to result in an increase in the disorder of the system and hence in its entropy. Indeed, dissolving a substance such as NaCl in water disrupts both the ordered crystal lattice of NaCl and the ordered hydrogen-bonded structure of water, leading to an increase in the entropy of the system. At the same time, however, each dissolved Na^+ ion becomes hydrated by an ordered arrangement of at least six water molecules, and the Cl^- ions also cause the water to adopt a particular local structure. Both of these effects *increase* the order of the system, leading to a *decrease* in entropy. The overall entropy change for the formation of a solution therefore depends on the relative magnitudes of these opposing factors. In the case of an NaCl solution, disruption of the crystalline NaCl structure and the hydrogen-bonded interactions in water is quantitatively more important, and thus $\Delta S_{soln} > 0$.

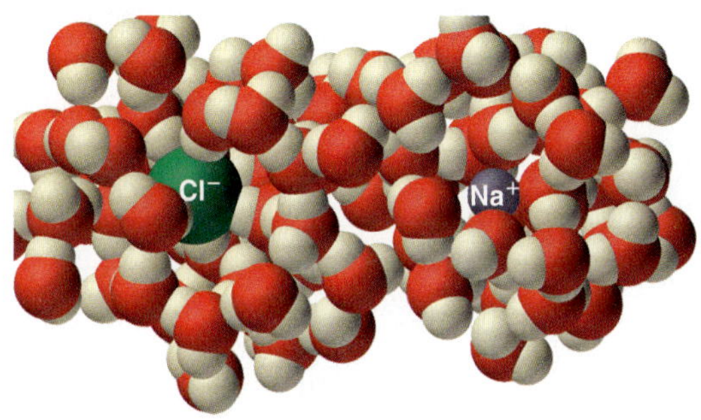

Figure 18.9 The effect of solution formation on entropy. Dissolving NaCl in water disrupts the highly ordered crystalline lattice structure of NaCl and the somewhat ordered hydrogen-bonded structure of liquid water, resulting in an increase in the entropy of the system. Each hydrated Na^+ and Cl^- ion, however, forms an ordered arrangement with at least six water molecules, thereby decreasing the entropy of the system. In the case of NaCl, the magnitude of the factors that increase entropy is greater than the magnitude of those that decrease entropy. Hence, the overall entropy change for the formation of an NaCl solution is positive.

EXAMPLE 18.5

Predict which substance in the following pairs has the higher entropy, and justify your answers: **(a)** 1 mol of $NH_3(g)$ or 1 mol of $He(g)$, both at 25°C; **(b)** 1 mol of $Pb(s)$ at 25°C or 1 mol of $Pb(l)$ at 800°C.

Given Amounts of substances and temperature

Asked for Higher entropy

Strategy

From the number of atoms present and the phase of each substance, predict which has the greater number of available microstates and hence the higher entropy.

Solution

(a) Both substances are gases at 25°C, but one consists of He atoms and the other consists of NH_3 molecules. With four atoms instead of one, the NH_3 molecules have more motions available, leading to a greater number of microstates. Hence, we predict that the NH_3 sample will have the higher entropy.

(b) The nature of the atomic species is the same in both cases, but the phase is different: one sample is a solid and one a liquid. Based on the greater freedom of motion available to atoms in a liquid, we predict that the liquid sample will have the higher entropy.

EXERCISE 18.5

Predict which substance in the following pairs has the higher entropy, and justify your answers: **(a)** 1 mol of He(g) at 10 K and 1 atm pressure or 1 mol of He(g) at 250°C and 0.2 atm; **(b)** a mixture of 3 mol of $H_2(g)$ and 1 mol of $N_2(g)$ at 25°C and 1 atm or a sample of 2 mol of $NH_3(g)$ at 25°C and 1 atm.

Answer **(a)** 1 mol of He(g) at 250°C and 0.2 atm (higher temperature and lower pressure indicate greater volume and more microstates); **(b)** a mixture of 3 mol of $H_2(g)$ and 1 mol of $N_2(g)$ at 25°C and 1 atm (more molecules of gas are present)

Reversible and Irreversible Changes

Changes in entropy, ΔS, together with changes in enthalpy, ΔH, enable us to predict in which direction a chemical or physical change will occur spontaneously. Before discussing how to do so, however, we must understand the difference between a reversible process and an irreversible process. In a **reversible process**, every intermediate state between the extremes is an equilibrium state, regardless of the direction of the change. In contrast, an **irreversible process** is one in which the intermediate states are not equilibrium states and change occurs spontaneously in only one direction. As a result, *a reversible process can change direction at any time, whereas an irreversible process cannot*. When a gas expands reversibly against an external pressure such as a piston, for example, the expansion can be reversed at any time by reversing the motion of the piston; once the gas is compressed, it can be allowed to expand again, and the process can continue indefinitely. In contrast, the expansion of a gas into a vacuum ($P_{ext} = 0$) is irreversible because the external pressure is measurably less than the internal pressure of the gas. No equilibrium states exist, and the gas expands irreversibly. When gas escapes from a microscopic hole in a balloon into a vacuum, for example, the process is irreversible; the direction of air flow cannot change.

Because work done during the expansion of a gas depends on the opposing external pressure ($w = P_{ext}\Delta V$), *work done in a reversible process is always equal to or greater than work done in a corresponding irreversible process*: $w_{rev} \geq w_{irrev}$. Whether a process is reversible or irreversible, $\Delta E = q + w$. Because E is a state function, the magnitude of ΔE does *not* depend on reversibility and is independent of the path taken. Thus,

> **Note the pattern**
>
> *Work done in a reversible process is always equal to or greater than work done in a corresponding irreversible process: $w_{rev} \geq w_{irrev}$.*

$$\Delta E = q_{rev} + w_{rev} = q_{irrev} + w_{irrev} \quad (18.11)$$

In other words, ΔE for a process is the same whether that process is carried out in a reversible or an irreversible manner. We now return to our earlier definition of entropy, using the magnitude of the heat flow for a reversible process, q_{rev}, to define entropy quantitatively.

The Relationship Between Internal Energy and Entropy

Because the quantity of heat transferred, q_{rev}, is directly proportional to the absolute temperature of an object, T ($q_{rev} \propto T$), the hotter the object, the greater the amount of heat transferred. Moreover, adding heat to a system increases the kinetic energy of the component atoms and molecules and hence their disorder ($\Delta S \propto q_{rev}$). Thus, for any reversible process,

$$q_{rev} = T\Delta S \quad \text{and} \quad \Delta S = \frac{q_{rev}}{T} \tag{18.12}$$

Because the numerator, q_{rev}, is expressed in units of energy (joules), the units of ΔS are joules/kelvin (J/K). Recognizing that the work done in a reversible process at constant pressure is $w_{rev} = -P\Delta V$, we can express Equation 18.11 as

$$\Delta E = q_{rev} + w_{rev} = T\Delta S - P\Delta V \tag{18.13}$$

Thus, the change in the internal energy of the system is related to the change in entropy, the absolute temperature, and the PV work done.

To illustrate the use of Equations 18.12 and 18.13, let's consider two reversible processes before turning to an irreversible process. When a sample of gas is allowed to expand reversibly at constant temperature, heat must be added to the gas during expansion in order to keep its T constant (Figure 18.10). The internal energy of the gas does not change because the temperature of the gas does not change; that is, $\Delta E = 0$ and $q_{rev} = -w_{rev}$. During an expansion, $\Delta V > 0$, so the gas performs work on its surroundings: $w_{rev} = -P\Delta V < 0$. According to Equation 18.13, this means that q_{rev} must increase during the expansion; that is, the gas must absorb heat from the surroundings during the expansion, and the surroundings must give up that same amount of heat. The entropy change of the system is therefore $\Delta S_{sys} = +q_{rev}/T$, and the entropy change of the surroundings is $\Delta S_{surr} = -q_{rev}/T$. The corresponding change in entropy of the universe is

$$\Delta S_{univ} = \Delta S_{sys} + \Delta S_{surr} = \frac{q_{rev}}{T} + \left(-\frac{q_{rev}}{T}\right) = 0 \tag{18.14}$$

Thus, no change in ΔS_{univ} has occurred.

Now consider the reversible melting of a sample of ice at 0°C and 1 atm. The enthalpy of fusion of ice is 6.01 kJ/mol, which means that 6.01 kJ of heat is absorbed reversibly from the surroundings when 1 mol of ice melts at 0°C, as illustrated in Figure 18.11. The surroundings is a sample of low-density carbon foam that is thermally conductive, and the system is the ice cube that has been placed on it. The direction of heat flow along the resulting temperature gradient is indicated with an arrow. From Equation 18.12, we see that the entropy of fusion of ice can be written as

$$\Delta S_{fus} = \frac{q_{rev}}{T} = \frac{\Delta H_{fus}}{T} \tag{18.15}$$

In this case, $\Delta S_{fus} = (6.01\ \text{kJ/mol})/(273\ \text{K}) = 22.0\ \text{J/(mol}\cdot\text{K)} = \Delta S_{sys}$. The amount of heat lost by the surroundings is the same as the amount gained by the ice, so $\Delta S_{surr} = q_{rev}/T = -(6.01\ \text{kJ/mol})/(273\ \text{K}) = -22.0\ \text{J/(mol}\cdot\text{K)}$. Once again, we see that the entropy of the universe does not change:

$$\Delta S_{univ} = \Delta S_{sys} + \Delta S_{surr} = 22.0\ \text{J/(mol}\cdot\text{K)} - 22.0\ \text{J/(mol}\cdot\text{K)} = 0$$

In both these examples of reversible processes, the entropy of the universe is unchanged. This is true of all reversible processes and constitutes part of the **second law of thermodynamics**: *The entropy of the universe remains constant in a reversible process, whereas the entropy of the universe increases in an irreversible (spontaneous) process.*

As an example of an irreversible process, consider the entropy changes that accompany the spontaneous and irreversible transfer of heat from a hot object to a

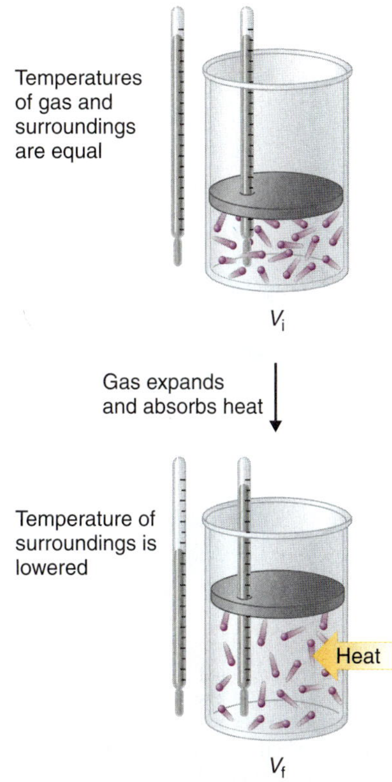

Figure 18.10 Expansion of gas at constant temperature. Internal energy is related to entropy, temperature, and the PV work done. In the initial state (*top*), the temperatures of the gas and the surroundings are the same. During the reversible expansion of the gas, heat must be added to the gas to maintain a constant temperature. Because the temperature of the gas does not change, the internal energy of the gas does not change, but work is performed on the surroundings. In the final state (*bottom*), the temperature of the surroundings is lower because the gas has absorbed heat from the surroundings during expansion.

Note the pattern

The entropy of the universe increases during a spontaneous process.

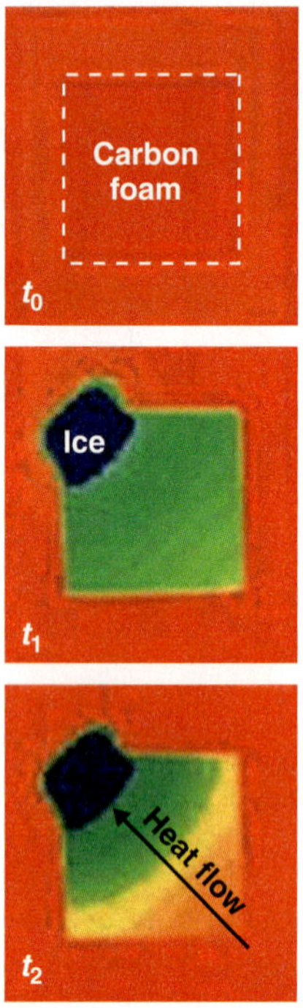

Figure 18.11 Thermograms showing that heat is absorbed from the surroundings when ice melts reversibly at 0°C. By convention, a thermogram shows cold regions in blue, warm regions in red, and thermally intermediate regions in green. When an ice cube (the system, dark blue) is placed on the corner of a square sample of low-density carbon foam with very high thermal conductivity, the temperature of the foam is lowered (going from red to green). As the ice melts, a temperature gradient appears, ranging from warm to very cold. The direction of heat flow from the surroundings (red and green) to the ice cube is indicated by the arrow. The amount of heat lost by the surroundings is the same as the amount gained by the ice, so the entropy of the universe does not change.

Figure 18.12 Spontaneous transfer of heat from a hot substance to a cold substance. When molten lava flows into cold oceanwater, so much heat is spontaneously transferred to the water that steam is produced.

cold one, as occurs when lava spewed from a volcano flows into cold ocean water. The cold substance, the water, gains heat ($q > 0$), so the change in the entropy of the water can be written as $\Delta S_{\text{cold}} = q/T_{\text{cold}}$. Similarly, the hot substance, the lava, loses heat ($q < 0$), so its entropy change can be written as $\Delta S_{\text{hot}} = -q/T_{\text{hot}}$. T_{cold} and T_{hot} are the temperatures of the cold and hot substances, respectively. The total entropy change of the universe accompanying this process is therefore

$$\Delta S_{\text{univ}} = \Delta S_{\text{cold}} + \Delta S_{\text{hot}} = \frac{q}{T_{\text{cold}}} + \left(-\frac{q}{T_{\text{hot}}}\right) \tag{18.16}$$

The numerators on the right side of Equation 18.16 are the same in magnitude but opposite in sign. Whether ΔS_{univ} is positive or negative depends on the relative magnitudes of the denominators. By definition $T_{\text{hot}} > T_{\text{cold}}$, so $-q/T_{\text{hot}}$ must be less than q/T_{cold}, and ΔS_{univ} must be positive. Thus, as predicted by the second law of thermodynamics, the entropy of the universe increases during this irreversible process. Any process for which ΔS_{univ} is positive is by definition a spontaneous one that will occur as written. Conversely, any process for which ΔS_{univ} is negative will not occur spontaneously as written, but will occur spontaneously in the opposite direction. We see, therefore, that heat is spontaneously transferred from a hot substance, the lava, to a cold substance, the ocean water. In fact, if the lava is hot enough (for example, if it is molten), so much heat can be transferred that the water is converted to steam (Figure 18.12).

EXAMPLE 18.6

Tin has two allotropes with different structures. Gray tin (α-tin) has a structure similar to that of diamond, whereas white tin (β-tin) is denser, with a unit cell structure that is based on a rectangular prism. At temperatures higher than 13.2°C, white tin is the more stable phase, but below that temperature, it slowly converts reversibly to the less dense, powdery gray phase. This phenomenon plagued Napoleon's army during his ill-fated invasion of Russia in 1812: the buttons on his soldiers' uniforms were made of tin and disintegrated during the Russian winter, adversely affecting the soldiers' health (and morale). The conversion of white tin to gray tin is exothermic, with $\Delta H = -2.1$ kJ/mol at 13.2°C. **(a)** What is ΔS for this process? **(b)** Which is the more highly ordered form of tin: white or gray?

Given ΔH, temperature

Asked for ΔS, relative degree of order

Strategy

Use Equation 18.12 to calculate the change in entropy for the reversible phase transition. From the calculated value of ΔS, predict which allotrope has the more highly ordered structure.

Solution

(a) We know from Equation 18.12 that the entropy change for any reversible process is the heat transferred (in joules) divided by the temperature at which the process occurs. Because the conversion occurs at constant pressure, and ΔH and ΔE are essentially equal for reactions that involve only solids, we can calculate the change in entropy for the reversible phase transition where $q_{\text{rev}} = \Delta H$. Substituting the given values for ΔH and temperature in kelvins (in this case, $T = 13.2°\text{C} = 286.4$ K) gives

$$\Delta S = \frac{q_{\text{rev}}}{T} = \frac{(-2.1\ \cancel{\text{kJ}}/\text{mol})(1000\ \text{J}/\cancel{\text{kJ}})}{286.4\ \text{K}} = -7.3\ \text{J}/(\text{mol}\cdot\text{K})$$

(b) The fact that $\Delta S < 0$ means that the entropy decreases when white tin is converted to gray tin. Thus, gray tin must be the more highly ordered structure.

EXERCISE 18.6

Elemental sulfur exists in two forms: an orthorhombic form (S_α), which is stable below 95.3°C, and a monoclinic form (S_β), which is stable above 95.3°C. The conversion of orthorhombic sulfur to monoclinic sulfur is endothermic, with $\Delta H = 0.401$ kJ/mol at 1 atm. **(a)** What is ΔS for this process? **(b)** Which is the more highly ordered form of sulfur: S_α or S_β?

Answer **(a)** 1.09 J/(mol · K); **(b)** orthorhombic, S_α

18.4 Entropy Changes and the Third Law of Thermodynamics

The atoms, molecules, or ions that make up a chemical system can undergo several types of molecular motion, including translation, rotation, and vibration (Figure 18.13). The greater the molecular motion of a system, the greater the number of possible microstates and hence the higher the entropy. A perfectly ordered system with only a single microstate available to it would have an entropy of zero. The only system that meets this criterion is a perfect crystal at a temperature of absolute zero (0 K), in which each component atom, molecule, or ion is fixed in place within a crystal lattice and thus exhibits no motion. Such a state of perfect order (or, conversely, zero disorder) corresponds to zero entropy. In practice, absolute zero is an ideal temperature that is unobtainable, and a perfect single crystal is also an ideal that cannot be achieved. Nonetheless, the combination of these two ideals constitutes the basis for the **third law of thermodynamics**: *The entropy of any perfectly ordered, crystalline substance at absolute zero is zero.*

The third law of thermodynamics has two important consequences: it defines the *sign* of the entropy of any substance at temperatures above absolute zero as positive, and it provides a fixed reference point that allows us to measure the *absolute entropy* of any substance at any temperature.* In contrast, other thermodynamic properties, such as internal energy and enthalpy, can be evaluated in only *relative* terms, not absolute terms. In this section, we examine two different ways to calculate ΔS for a reaction or physical change. The first, based on the definition of absolute entropy provided by the

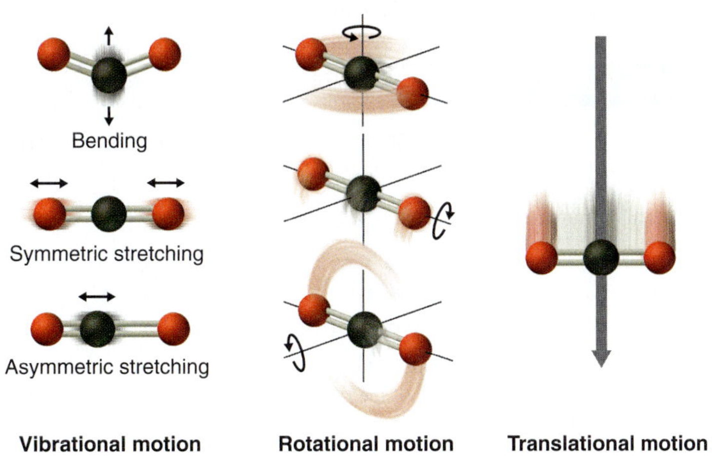

Figure 18.13 Molecular motions. Vibrational, rotational, and translational motions of a carbon dioxide molecule are illustrated here. The more molecular motions the atoms, molecules, or ions of a chemical system undergo, the higher the entropy of the system. Only a perfectly ordered, crystalline substance at absolute zero would exhibit no molecular motion and would thus have zero entropy. In practice, this is an unattainable ideal.

* In practice, chemists determine the value of the absolute entropy of a substance by measuring the molar heat capacity C_p as a function of temperature, and then plotting the quantity (C_p/T) versus T. The *area* under the curve between 0 K and any temperature T is the absolute entropy of the substance at temperature T.

TABLE 18.1 Standard molar entropy, $S°$, values of selected substances at 25°C

Substance	$S°$, J/(mol · K)
Gases	
He(g)	126.2
$H_2(g)$	130.7
Ne(g)	146.3
Ar(g)	154.8
Kr(g)	164.1
Xe(g)	169.7
$H_2O(g)$	188.8
$N_2(g)$	191.6
$O_2(g)$	205.2
$CO_2(g)$	213.8
$I_2(g)$	260.7
Liquids	
$H_2O(l)$	70.0
$CH_3OH(l)$	126.8
$Br_2(l)$	152.2
$CH_3CH_2OH(l)$	160.7
$C_6H_6(l)$	173.4
$CH_3COCl(l)$	200.8
$C_6H_{12}(l)$ (cyclohexane)	204.4
$C_8H_{18}(l)$ (isooctane)	329.3
Solids	
C(s) (diamond)	2.4
C(s) (graphite)	5.7
LiF(s)	35.7
$SiO_2(s)$ (quartz)	41.5
Ca(s)	41.6
Na(s)	51.3
$MgF_2(s)$	57.2
K(s)	64.7
NaCl(s)	72.1
KCl(s)	82.6
$I_2(s)$	116.1

third law, uses tabulated values of absolute entropies of substances. The second, based on the fact that entropy is a state function, uses a thermodynamic cycle similar to those we first encountered in Chapter 5.

Calculating ΔS from Standard Molar Entropy Values

One way of calculating ΔS for a reaction is to use tabulated values of the **standard molar entropy** (**$S°$**), which is the entropy of 1 mol of a substance at a standard temperature of 298 K; the units of $S°$ are J/(mol · K). Unlike enthalpy or internal energy, it is possible to obtain absolute entropy values by measuring the entropy change that occurs between the reference point of 0 K, corresponding to $S = 0$, and 298 K.

As shown in Table 18.1, for substances with approximately the same molar mass and number of atoms, $S°$ values fall in the order $S°(\text{gas}) > S°(\text{liquid}) > S°(\text{solid})$. For instance, $S°$ for liquid water is 70.0 J/(mol · K), whereas $S°$ for water vapor is 188.8 J/(mol · K). Likewise, $S°$ is 260.7 J/(mol · K) for gaseous I_2 and 116.1 J/(mol · K) for solid I_2. This order makes qualitative sense based on the kinds and extents of motion available to atoms and molecules in the three phases. The correlation between physical state and absolute entropy is illustrated in Figure 18.14, a generalized plot of the entropy of a substance versus temperature.

Closer examination of Table 18.1 also reveals that substances with similar molecular structures tend to have similar $S°$ values. Among crystalline materials, those with the lowest entropies tend to be rigid crystals composed of small atoms linked by strong, highly directional bonds, like diamond [$S° = 2.4$ J/(mol · K)]. In contrast, graphite, the softer, less rigid allotrope of carbon, has a higher $S°$ [5.7 J/(mol · K)] due to more disorder in the crystal. Soft crystalline substances and those that contain larger atoms tend to have higher entropies because of increased molecular motion and disorder. Similarly, the absolute entropy of a substance tends to increase with increasing molecular complexity because the number of available microstates increases with molecular complexity. For example, compare the $S°$ values for $CH_3OH(l)$ [126.8 J/(mol · K)] and $CH_3CH_2OH(l)$ [160.7 J/ (mol · K)]. Finally, substances with strong hydrogen bonds have lower values of $S°$, reflecting a more ordered structure.

To calculate $\Delta S°$ for a chemical reaction from standard molar entropies, we use the familiar "products minus reactants" rule, in which the absolute entropy of each reactant and product is multiplied by its stoichiometric coefficient in the balanced chemical equation. The next example illustrates this procedure for the combustion of the liquid hydrocarbon isooctane (2,2,4-trimethylpentane, C_8H_{18}).

EXAMPLE 18.7

Use the data in Table 18.1 to calculate $\Delta S°$ for the reaction of liquid isooctane with $O_2(g)$ to give $CO_2(g)$ and $H_2O(g)$ at 298 K.

Given Standard molar entropies, reactants and products

Asked for $\Delta S°$

Strategy

Write the balanced equation for the reaction, and identify the appropriate quantities in Table 18.1. Subtract the sum of the absolute entropies of the reactants from the sum of the absolute entropies of the products, each multiplied by their appropriate stoichiometric coefficients, to obtain $\Delta S°$ for the reaction.

Solution

The balanced equation for the complete combustion of isooctane (C_8H_{18}) is

$$C_8H_{18}(l) + \frac{25}{2}O_2(g) \rightarrow 8CO_2(g) + 9H_2O(g)$$

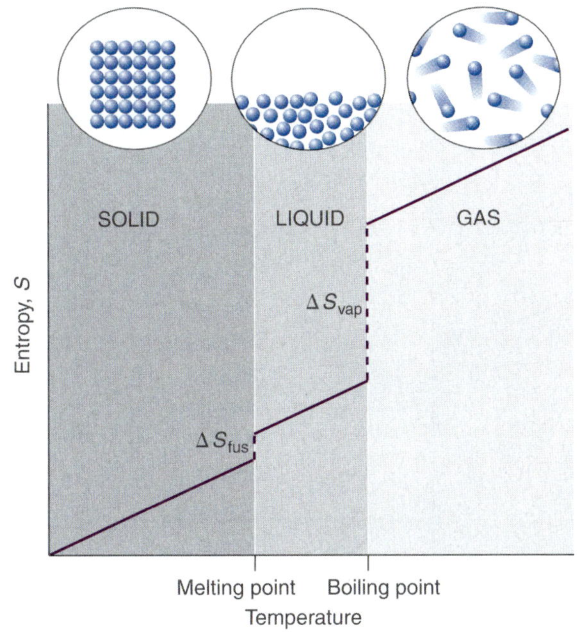

Figure 18.14 A generalized plot of entropy versus temperature for a single substance. According to the third law of thermodynamics, the entropy at 0 K is zero. The absolute entropy increases steadily with increasing temperature until the melting point is reached. At this point, the entropy jumps suddenly as the substance undergoes a phase change from a highly ordered solid to a disordered liquid (ΔS_{fus}). After this, the entropy again increases steadily with increasing temperature until the boiling point is reached and another phase change occurs. The entropy jumps suddenly as the liquid is converted to a highly disordered gas (ΔS_{vap}) and then increases again with increasing temperature. The increase in entropy with increasing temperature is approximately proportional to the heat capacity of the substance.

We calculate $\Delta S°$ for the reaction using the "products minus reactants" rule, where m and n are the stoichiometric coefficients of each product and each reactant:

$$\begin{aligned}\Delta S°_{rxn} &= \Sigma m S°(\text{products}) - \Sigma n S°(\text{reactants}) \\ &= [8S°(CO_2) + 9S°(H_2O)] - [S°(C_8H_{18}) + \frac{25}{2}S°(O_2)] \\ &= \{[8\ \cancel{\text{mol}}\ CO_2 \times 213.8\ \text{J/}(\cancel{\text{mol}}\cdot\text{K})] + [9\ \cancel{\text{mol}}\ H_2O \times 188.8\ \text{J/}(\cancel{\text{mol}}\cdot\text{K})]\} \\ &\quad - \left\{[1\ \cancel{\text{mol}}\ C_8H_{18} \times 329.3\ \text{J/}(\cancel{\text{mol}}\cdot\text{K})] + \left[\frac{25}{2}\ \cancel{\text{mol}}\ O_2 \times 205.2\ \text{J/}(\cancel{\text{mol}}\cdot\text{K})\right]\right\} \\ &= 515.3\ \text{J/K}\end{aligned}$$

Note that $\Delta S°$ is positive, as expected for a combustion reaction in which one large hydrocarbon molecule is converted to many molecules of gaseous products.

Note the pattern

Entropy increases with softer, less rigid solids, solids that contain larger atoms, and solids with complex molecular structures.

EXERCISE 18.7

Use the data in Table 18.1 to calculate $\Delta S°$ for the reaction of $H_2(g)$ with liquid benzene (C_6H_6) to give cyclohexane (C_6H_{12}).

Answer −361.1 J/K

Note the pattern

$\Delta S°$ for a reaction can be calculated from absolute entropy values using the same "products minus reactants" rule used to calculate $\Delta H°$.

Calculating ΔS from Thermodynamic Cycles

We can also calculate a change in entropy using a thermodynamic cycle. As you learned in Chapter 5, the molar heat capacity C_p is the amount of heat needed to raise the temperature of 1 mol of a substance by 1°C at constant pressure. Similarly, C_v is the amount of heat needed to raise the temperature of 1 mol of a substance by 1°C at constant volume. The increase in entropy with increasing temperature in Figure 18.14 is approximately proportional to the heat capacity of the substance.

Recall that the entropy change ΔS is related to the heat flow q_{rev} by $\Delta S = q_{rev}/T$. Because $q_{rev} = nC_p\,\Delta T$ at constant pressure or $nC_v\,\Delta T$ at constant volume, where n is

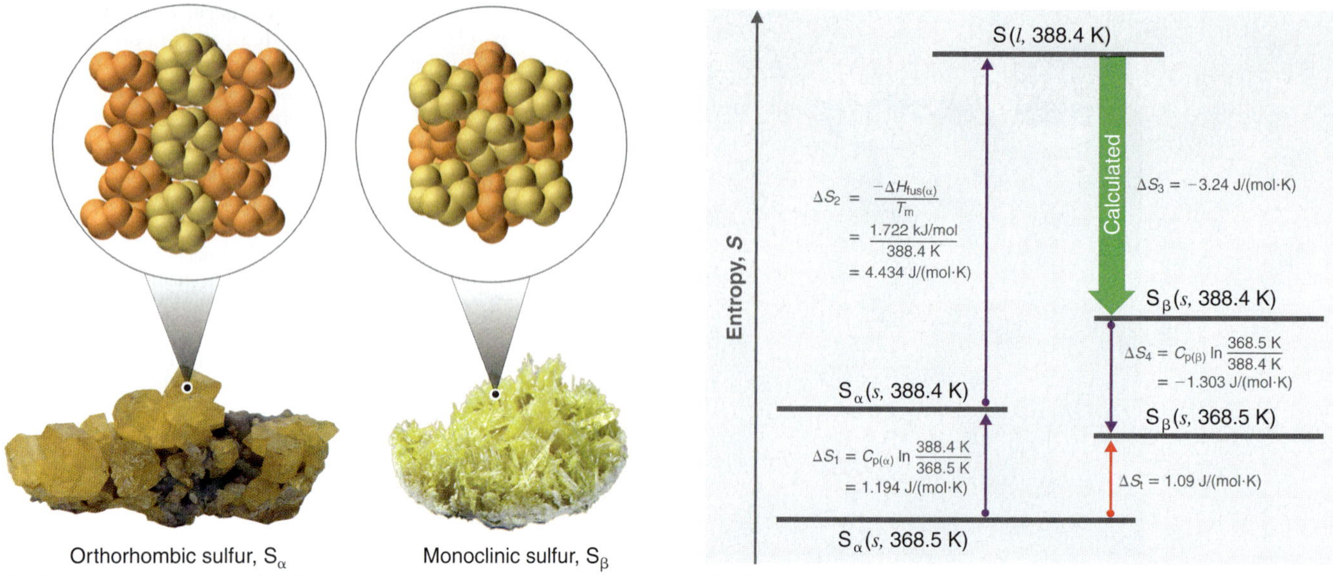

(a) Allotropes of elemental sulfur **(b) Thermodynamic cycle**

Figure 18.15 Two forms of elemental sulfur and a thermodynamic cycle showing the transition from one to the other. (a) Orthorhombic sulfur, S_α, has a highly ordered structure in which the S_8 rings are stacked in a "crankshaft" arrangement. Monoclinic sulfur, S_β, is also composed of S_8 rings but has a less-ordered structure. **(b)** At 368.5 K, S_α undergoes a phase transition to S_β. Although the change in entropy for the conversion of liquid sulfur to S_β, ΔS_3, cannot be measured directly, it can be calculated using the values shown in this thermodynamic cycle because $\Delta S_t = \Delta S_1 + \Delta S_2 + \Delta S_3 + \Delta S_4$, where the subscripts 1–4 refer to steps in the cycle.

the number of moles of substance present, the change in entropy for a substance whose temperature changes from T_1 to T_2 is

$$\Delta S = \frac{q_{rev}}{T} = nC_p \frac{\Delta T}{T} \quad \text{(constant pressure)}$$

As you will discover in more advanced math courses than is required here, it can be shown that this is equal to*

$$\Delta S = nC_p \ln \frac{T_2}{T_1} \quad \text{(constant pressure)} \tag{18.17}$$

Similarly,

$$\Delta S = nC_v \ln \frac{T_2}{T_1} \quad \text{(constant volume)} \tag{18.18}$$

Thus, we can use a combination of heat capacity measurements (Equations 18.17 or 18.18) and experimentally measured values of enthalpies of fusion or vaporization if a phase change is involved (Equation 18.15) to calculate the entropy change corresponding to a change in the temperature of a sample.

We can use a thermodynamic cycle to calculate the entropy change when the phase change for a substance such as sulfur cannot be measured directly. As noted in Exercise 18.6, elemental sulfur exists in two forms (Figure 18.15a): an orthorhombic form with a highly ordered structure (S_α) and a less-ordered monoclinic form (S_β). The orthorhombic (α) form is more stable at room temperature but undergoes a phase transition to the monoclinic (β) form at temperatures higher than 95.3°C (368.5 K). The transition from S_α to S_β can be described by the thermodynamic cycle shown in Figure 18.15b,

* For a review of natural logarithms, see Essential Skills 6.

in which liquid sulfur is an intermediate. The change in entropy that accompanies the conversion of liquid sulfur to S_β ($-\Delta S_{fus(\beta)} = \Delta S_3$ in the cycle) cannot be measured directly. Because entropy is a state function, however, ΔS_3 can be calculated from the overall entropy change, ΔS_t, for the S_α–S_β transition, which equals the sum of the ΔS values for the steps in the thermodynamic cycle, using Equation 18.17 and tabulated thermodynamic parameters (the heat capacities of S_α and S_β, $\Delta H_{fus(\alpha)}$, and the melting point of S_α.

If we know the melting point of S_α ($T_m = 115.2°C = 388.4$ K) and ΔS_t for the overall phase transition [(calculated to be 1.09 J/(mol·K) in Exercise 18.6], we can calculate ΔS_3 from the values given in Figure 18.15b where $C_{p(\alpha)} = 22.70$ J/mol·K and $C_{p(\beta)} = 24.77$ J/mol·K (subscripts on ΔS refer to steps in the cycle):

$$\Delta S_t = \Delta S_1 + \Delta S_2 + \Delta S_3 + \Delta S_4$$

$$1.09\ \text{J/(mol·K)} = C_{p(\alpha)}\ln\left(\frac{T_2}{T_1}\right) + \frac{\Delta H_{fus}}{T_m} + \Delta S_3 + C_{p(\beta)}\ln\left(\frac{T_4}{T_3}\right)$$

$$= 22.70\ \text{J/(mol·K)}\ \ln\left(\frac{388.4}{368.5}\right) + \left(\frac{1.722\ \text{kJ/mol}}{388.4\ \text{K}}\right) + \Delta S_3 + 24.77\ \text{J/(mol·K)}\ \ln\left(\frac{368.5}{388.4}\right)$$

$$= [1.194\ \text{J/(mol·K)}] + [4.434\ \text{J/(mol·K)}] + \Delta S_3 + [-1.303\ \text{J/(mol·K)}]$$

Solving for ΔS_3 gives a value of -3.24 J/(mol·K). As expected for the conversion of a less ordered state (a liquid) to a more ordered one (a crystal), ΔS_3 is negative.

18.5 • Free Energy

One of the major goals of chemical thermodynamics is to establish criteria for predicting whether a particular reaction or process will occur spontaneously. We have developed one such criterion, the change in entropy of the universe: if $\Delta S_{univ} > 0$ for a process or reaction, then the process will occur spontaneously as written. Conversely, if $\Delta S_{univ} < 0$, a process cannot occur spontaneously, whereas if $\Delta S_{univ} = 0$, the system is at equilibrium. The sign of ΔS_{univ} is a universally applicable and infallible indicator of the spontaneity of a reaction. Unfortunately, using ΔS_{univ} requires that we calculate ΔS for *both* the system *and* the surroundings. This is not particularly useful for two reasons: we are normally much more interested in the system than in the surroundings, and it is difficult to make quantitative measurements of the surroundings (that is, the rest of the universe). A criterion of spontaneity that is based solely on state functions of the *system* would be much more convenient and is provided by a new state function, the Gibbs free energy.

J. Willard Gibbs (1839–1903). Born in Connecticut, Josiah Willard Gibbs attended Yale, as did his father, a professor of sacred literature at Yale, who was involved in the *Amistad* trial. In 1863 Gibbs was awarded the first engineering doctorate granted in the United States. He was appointed Professor of Mathematical Physics at Yale in 1871, the first such professorship in the United States. His series of papers entitled "On the Equilibrium of Heterogeneous Substances" was the foundation of the field of physical chemistry and is considered one of the great achievements of the 19th century. Gibbs, whose work was translated into French by le Châtelier, lived with his sister and brother-in-law until his death in 1903, shortly before the inauguration of the Nobel Prizes.

Gibbs Free Energy and the Direction of Spontaneous Reactions

The **Gibbs free energy (*G*)**, often called simply *free energy*, was named in honor of J. Willard Gibbs (1838–1903), an American physicist who first developed the concept. It is defined in terms of three other state functions with which you are already familiar: enthalpy, temperature, and entropy:

$$G = H - TS \tag{18.19}$$

Because it is a combination of state functions, G is also a state function.

The criterion for predicting spontaneity is based on ΔG, the change in G, at constant temperature and pressure. Although very few chemical reactions actually occur under conditions of constant temperature and pressure, most systems can be brought back to the initial temperature and pressure without significantly affecting the value of thermodynamic state functions such as G. At constant temperature and pressure,

$$\Delta G = \Delta H - T\Delta S \tag{18.20}$$

where all thermodynamic quantities are those of the *system*. Recall that at constant pressure, $\Delta H = q$ whether a process is reversible or irreversible and $T\Delta S = q_{rev}$. Using these expressions, we can reduce Equation 18.20 to $\Delta G = q - q_{rev}$. Thus, ΔG is the difference between the heat released during a process (via a reversible or an irreversible path) and the heat released for the same process occurring in a reversible manner. Under the special condition in which a process occurs reversibly, $q = q_{rev}$ and hence $\Delta G = 0$. As we shall soon see, if ΔG is zero, the system is at equilibrium and there will be no net change.

What about processes for which $\Delta G \neq 0$? To understand how the sign of ΔG for the *system* determines the direction in which change is spontaneous, we can rewrite Equation 18.12 (where $q_p = \Delta H$, Equation 18.8) to give

$$\Delta S_{surr} = -\frac{\Delta H_{sys}}{T} \tag{18.21}$$

Thus, the entropy change of the *surroundings* is related to the enthalpy change of the *system*. We have stated that for a spontaneous reaction, $\Delta S_{univ} > 0$, so substituting we obtain

$$\begin{aligned} \Delta S_{univ} &= \Delta S_{sys} + \Delta S_{surr} > 0 \\ &= \Delta S_{sys} - \frac{\Delta H_{sys}}{T} > 0 \end{aligned}$$

Multiplying both sides of the inequality by $-T$ reverses its sign, and rearranging gives

$$\Delta H_{sys} - T\Delta S_{sys} < 0$$

which is equal to ΔG (Equation 18.20). We can therefore see that for a spontaneous process, $\Delta G < 0$.

The relationship between the entropy change of the *surroundings* and the heat gained or lost by the *system* provides the key connection between the thermodynamic properties of the system and the change in entropy of the universe. The relationship shown in Equation 18.20 allows us to predict spontaneity by focusing exclusively on the thermodynamic properties and temperature of the system. We predict that highly exothermic processes ($\Delta H << 0$) that increase the disorder of the *system* ($\Delta S_{sys} >> 0$) would therefore occur spontaneously. An example of such a process is the decomposition of ammonium nitrate fertilizer, which destroyed Texas City, Texas, in 1947; ammonium nitrate was also used to destroy the Murrah Federal Building in Oklahoma City, Oklahoma, in 1995.

For a system at constant temperature and pressure, we can summarize these results:

If $\Delta G < 0$, the process occurs spontaneously.

If $\Delta G = 0$, the system is at equilibrium.

If $\Delta G > 0$, the process is not spontaneous as written but occurs spontaneously in the reverse direction.

To further understand how the various components of ΔG dictate whether a process occurs spontaneously, let's look at a simple and familiar physical change: the conversion of liquid water to water vapor. If this process is carried out at 1 atm and the normal boiling point of 100.00°C (373.15 K), we can calculate ΔG from the experimentally measured value of ΔH_{vap} (40.657 kJ/mol). For vaporizing 1 mol of water, $\Delta H = 40{,}657$ J, so the process is highly endothermic. From the definition of ΔS (Equation 18.12), we know that for 1 mol of water,

$$\Delta S_{vap} = \frac{\Delta H_{vap}}{T_b} = \frac{40{,}657 \text{ J}}{373.15 \text{ K}} = 108.96 \text{ J/K}$$

Hence, there is an increase in the disorder of the system. At the normal boiling point of water,

$$\Delta G_{100°C} = \Delta H_{100°C} - T\Delta S_{100°C} = 40{,}657\ \text{J} - [(373.15\ \cancel{\text{K}})(108.96\ \text{J}/\cancel{\text{K}})] = 0\ \text{J}$$

The energy required for vaporization offsets the increase in disorder of the system. Thus, $\Delta G = 0$ and the liquid and vapor are in equilibrium, as is true of any liquid at its normal boiling point (see Chapter 11).

Now suppose we were to superheat 1 mol of liquid water to a temperature of 110°C. The value of ΔG for the vaporization of 1 mol of water at 110°C, assuming that ΔH and ΔS do not change significantly with temperature, becomes

$$\Delta G_{110°C} = \Delta H - T\Delta S = 40{,}657\ \text{J} - [(383.15\ \cancel{\text{K}})(108.96\ \text{J}/\cancel{\text{K}})] = -1091\ \text{J}$$

Thus, at a temperature of 110°C, $\Delta G < 0$ and vaporization is predicted to occur spontaneously and irreversibly.

We can also calculate ΔG for the vaporization of 1 mol of water at a temperature below its normal boiling point—for example, 90°C—making the same assumptions:

$$\Delta G_{90°C} = \Delta H - T\Delta S = 40{,}657\ \text{J} - [(363.15\ \cancel{\text{K}})(108.96\ \text{J}/\cancel{\text{K}})] = 1088\ \text{J}$$

At a temperature of 90°C, $\Delta G > 0$ and water does not spontaneously convert to water vapor. Notice that when using all digits in the calculator display in carrying out our calculations (see Essential Skills 1), $\Delta G_{110°C} = 1090\ \text{J} = -\Delta G_{90°C}$, as we would predict.

We can also calculate the temperature at which liquid water is in equilibrium with water vapor. Inserting the values of ΔH and ΔS into the definition of ΔG (Equation 18.20), setting $\Delta G = 0$, and solving for T give

$$0\ \text{J} = 40{,}657\ \text{J} - T(108.96\ \text{J/K})$$
$$T = 373.15\ \text{K}$$

Note the pattern

$\Delta G = 0$ only if $\Delta H = T\Delta S$.

Thus, $\Delta G = 0$ at $T = 373.15$ K and 1 atm, indicating that liquid water and water vapor are in equilibrium; this temperature is therefore the *normal boiling point* of water. At temperatures higher than 373.15 K, ΔG is negative, and water evaporates spontaneously and irreversibly. Below 373.15 K, ΔG is positive, and water does not evaporate spontaneously. Instead, water vapor at a temperature lower than 373.15 K and 1 atm will spontaneously and irreversibly condense to liquid water. Figure 18.16 shows how the ΔH and $T\Delta S$ terms vary with temperature for the vaporization of water. Note that $\Delta G = 0$ when the two lines cross, at which point $\Delta H = T\Delta S$.

A similar situation arises in the conversion of liquid egg white to a solid when an egg is boiled. The major component of egg white is a protein called albumin, which is held in a compact, ordered structure by a large number of hydrogen bonds. Breaking them requires an input of energy ($\Delta H > 0$), which converts the albumin to a highly disordered structure in which the molecules aggregate as a disorganized solid ($\Delta S > 0$). At temperatures higher than 373 K, the $T\Delta S$ term dominates and $\Delta G < 0$, so the conversion of a raw egg to a hard-boiled egg is an irreversible and spontaneous process above 373 K.

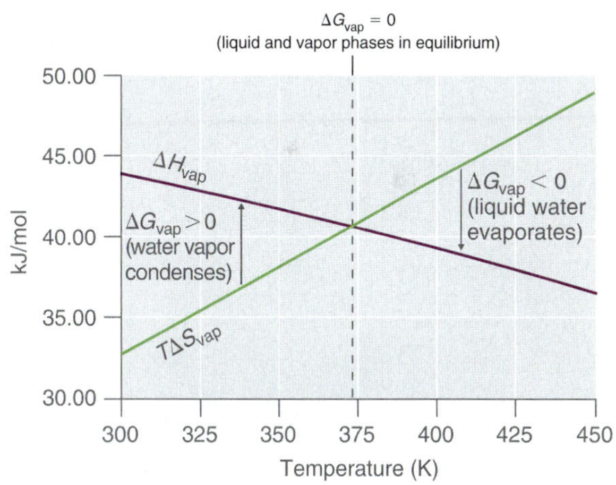

Figure 18.16 Temperature dependence of ΔH and $T\Delta S$ for the vaporization of water. Both ΔH and $T\Delta S$ are temperature dependent, but the lines have opposite slopes and cross at 373.15 K at 1 atm, where $\Delta H = T\Delta S$. Because $\Delta G = \Delta H - T\Delta S$, at this temperature $\Delta G = 0$, indicating that the liquid and vapor phases are in equilibrium. The normal boiling point of water is therefore 373.15 K. Above the normal boiling point, the $T\Delta S$ term is greater than ΔH, making $\Delta G < 0$; hence, liquid water evaporates spontaneously. Below the normal boiling point, the ΔH term is greater than $T\Delta S$, making $\Delta G > 0$. Thus, liquid water does not evaporate spontaneously, but water vapor spontaneously condenses to liquid.

The Relationship Between ΔG and Work

As we've just seen, the value of ΔG allows us to predict the spontaneity of a physical or chemical change. In addition, the magnitude of ΔG for a process provides other important information. The change in free energy, ΔG, is equal to the maximum amount of work that a system can perform on the surroundings while undergoing a spontaneous change (at constant temperature and pressure): $\Delta G = w_{max}$. To see why this is true, let's look again at the relationships among free energy, enthalpy, and entropy expressed in Equation 18.20. We can rearrange this equation to give

$$\Delta H = \Delta G + T\Delta S \tag{18.22}$$

This equation tells us that when energy is released during an exothermic process ($\Delta H < 0$), such as during the combustion of a fuel, for example, some of that energy can be used to do work ($\Delta G < 0$), while some is used to increase the entropy of the universe ($T\Delta S > 0$). Only if the process occurs infinitely slowly in a perfectly reversible manner will the entropy of the universe be unchanged (Section 18.4). Because no real system is perfectly reversible, the entropy of the universe increases during all processes that produce energy. As a result, no process that utilizes stored energy can ever be 100% efficient; that is, ΔH will never equal ΔG because ΔS has a positive value.

One of the major challenges facing engineers is to maximize the efficiency of converting stored energy to useful work or of converting one form of energy to another. As indicated in Table 18.2, the efficiencies of various energy-converting devices vary widely. For example, an internal combustion engine typically uses only 25–30% of the energy stored in the hydrocarbon fuel to perform work; the rest of the stored energy is released in an unusable form as heat. In contrast, gas–electric hybrid engines, now used in several models of automobiles, deliver approximately 50% greater fuel efficiency. A large electrical generator is highly efficient (approximately 99%) in converting mechanical to electrical energy, but a typical incandescent light bulb is one of the least efficient devices known (only approximately 5% of the electrical energy is converted to light). In contrast, a mammalian liver cell is a relatively efficient machine, able to utilize fuels such as glucose with an efficiency of 30–50%.

Standard Free-Energy Change

We have seen that there is no way to measure absolute enthalpies, although we can measure *changes* in enthalpy, ΔH, during a chemical reaction. Because enthalpy is one of the components of Gibbs free energy, we are consequently unable to measure absolute free energies either, only changes in free energy. The **standard free-energy change**, $\Delta G°$, is the change in free energy when one substance or set of substances in their standard states is converted to one or more other substances, also in their standard states. The standard free-energy change can be calculated from the definition of free energy, if the standard enthalpy and entropy changes are known, using

$$\Delta G° = \Delta H° - T\Delta S° \qquad (18.23)$$

TABLE 18.2 Approximate thermodynamic efficiencies of various devices

Device	Energy Conversion	Approximate Efficiency, %
Large electrical generator	Mechanical ⟶ electrical	99
Chemical battery	Chemical ⟶ electrical	90
Home furnace	Chemical ⟶ heat	65
Small electric tool	Electrical ⟶ mechanical	60
Space shuttle engine	Chemical ⟶ mechanical	50
Mammalian liver cell	Chemical ⟶ chemical	30–50
Spinach cell	Light ⟶ chemical	30
Internal combustion engine	Chemical ⟶ mechanical	25–30
Fluorescent light	Electrical ⟶ light	20
Solar cell	Light ⟶ electricity	10
Incandescent light bulb	Electricity ⟶ light	5
Yeast cell	Chemical ⟶ chemical	2–4

Thus, according to Equation 18.23, if $\Delta S°$ and $\Delta H°$ for a reaction have the same sign, then the sign of $\Delta G°$ depends on the relative magnitudes of the $\Delta H°$ and $T\Delta S°$ terms. It is important to recognize that a positive value of $\Delta G°$ for a reaction does *not* mean that no products will form if the reactants in their standard states are mixed, only that at equilibrium the concentrations of the products will be lower than the concentrations of the reactants.

Note the pattern

A positive $\Delta G°$ means that at equilibrium the concentrations of products will be lower than the concentrations of reactants.

EXAMPLE 18.8

Calculate the standard free-energy change, $\Delta G°$, at 25°C for the reaction $H_2(g) + O_2(g) \rightleftharpoons H_2O_2(l)$. At 25°C the standard enthalpy change, $\Delta H°$, is −187.78 kJ/mol, and the absolute entropies of the products and reactants are $S°(H_2O_2)$ = 109.6 J/(mol·K), $S°(O_2)$ = 205.2 J/(mol·K), and $S°(H_2)$ = 130.7 J/(mol·K). Is the reaction spontaneous as written?

Given Balanced chemical equation, $\Delta H°$, $S°$ for reactants and products

Asked for Spontaneity of reaction as written

Strategy

A Calculate $\Delta S°$ from the absolute molar entropy values given.
B Use Equation 18.23, the calculated value of $\Delta S°$, and other data given to calculate $\Delta G°$ for the reaction. Use the value of $\Delta G°$ to determine whether the reaction is spontaneous as written.

Solution

A To calculate $\Delta G°$ for the reaction, we need to know $\Delta H°$, $\Delta S°$, and T. We are given $\Delta H°$, and we know that T = 298.15 K. We can calculate $\Delta S°$ from the absolute molar entropy values provided using the "products minus reactants" rule:

$$\begin{aligned}\Delta S° &= S°(H_2O_2) - [S°(O_2) + S°(H_2)] \\ &= [1\ \cancel{\text{mol}}\ H_2O_2 \times 109.6\ \text{J}/(\cancel{\text{mol}} \cdot \text{K})] - \{[1\ \cancel{\text{mol}}\ H_2 \times 130.7\ \text{J}/(\cancel{\text{mol}} \cdot \text{K})] + [(1\ \cancel{\text{mol}}\ O_2 \times 205.2\ \text{J}/(\cancel{\text{mol}} \cdot \text{K})]\} \\ &= -226.3\ \text{J/K (per mole of } H_2O_2)\end{aligned}$$

Note that $\Delta S°$ is very negative for this reaction, as we might expect for a reaction in which 2 mol of gas is converted to 1 mol of a much more ordered liquid. **B** Substituting the appropriate quantities into Equation 18.23 gives

$$\begin{aligned}\Delta G° = \Delta H° - T\Delta S° &= -187.78\ \text{kJ/mol} - (298.15\ \cancel{\text{K}})[-226.3\ \text{J}/(\text{mol} \cdot \cancel{\text{K}})] \\ &= -187.78\ \text{kJ/mol} + 67.47\ \text{kJ/mol} = -120.31\ \text{kJ/mol}\end{aligned}$$

The negative value of $\Delta G°$ indicates that the reaction is spontaneous as written. Because $\Delta S°$ and $\Delta H°$ for this reaction have the same sign, the sign of $\Delta G°$ depends on the relative magnitudes of the $\Delta H°$ and $T\Delta S°$ terms. Thus, in this particular case, the enthalpy term dominates, indicating that the strength of the bonds formed in the product more than compensates for the unfavorable $\Delta S°$ term and for the energy needed to break bonds in the reactants.

EXERCISE 18.8

Calculate the standard free-energy change, $\Delta G°$, at 25°C for the reaction $2H_2(g) + N_2(g) \rightleftharpoons N_2H_4(l)$. At 25°C the standard enthalpy change, $\Delta H°$, is 50.6 kJ/mol, and the absolute entropies of the products and reactants are $S°(N_2H_4)$ = 121.2 J/(mol·K), $S°(N_2)$ = 191.6 J/(mol·K), and $S°(H_2)$ = 130.7 J/ (mol·K). Is the reaction spontaneous as written?

Answer 149.5 kJ/mol; no

Tabulated values of standard free energies of formation allow chemists to calculate the values of $\Delta G°$ for a wide variety of chemical reactions rather than having to measure them in the laboratory. The **standard free energy of formation (ΔG°_f)** of a compound is the change in free energy that occurs when 1 mol of a substance in its standard state is formed from the elements in their standard states. By definition, the standard free energy of formation of an element in its standard state is zero at 298.15 K. One mole of Cl_2 gas at 298.15 K, for example, has $\Delta G^\circ_f = 0$. The standard free energy of formation of a compound can be calculated from the standard enthalpy of formation, ΔH°_f, and the standard entropy of formation, ΔS°_f, using the definition of free energy:

$$\Delta G^\circ_f = \Delta H^\circ_f - T\Delta S^\circ_f \tag{18.24}$$

Using standard free energies of formation to calculate the standard free energy of a reaction is analogous to calculating standard enthalpy changes from standard enthalpies of formation using the familiar "products minus reactants" rule:

Note the pattern

The $\Delta G°$ of a reaction can be calculated from tabulated ΔG°_f values using the "products minus reactants" rule.

$$\Delta G^\circ_{rxn} = \Sigma m\Delta G^\circ_f(\text{products}) - \Sigma n\Delta G^\circ_f(\text{reactants}) \tag{18.25}$$

where m and n are the stoichiometric coefficients of each product and reactant in the balanced chemical equation. A very large negative $\Delta G°$ indicates a strong tendency for products to form spontaneously from reactants; *it does not, however, necessarily indicate that the reaction will occur rapidly*. To make this determination, we need to evaluate the kinetics of the reaction (see Chapter 14).

EXAMPLE 18.9

Calculate $\Delta G°$ for the reaction of isooctane with oxygen gas to give carbon dioxide and water (described in Example 18.7). Use the following data: ΔG°_f (isooctane) = −353.2 kJ/mol, $\Delta G^\circ_f(CO_2)$ = −394.4 kJ/mol, and $\Delta G^\circ_f(H_2O)$ = −237.1 kJ/mol. Is the reaction spontaneous as written?

Given Balanced chemical equation, values of ΔG°_f for isooctane, CO_2, and H_2O

Asked for Spontaneity of reaction as written

Strategy

Use the "products minus reactants" rule to obtain ΔG°_{rxn}, remembering that ΔG°_f for an element in its standard state is zero. From the calculated value, determine whether the reaction is spontaneous as written.

Solution

From Example 18.7, we know that the balanced chemical equation for the reaction is $C_8H_{18}(l) + \frac{25}{2}O_2(g) \rightarrow 8CO_2(g) + 9H_2O(l)$. We are given ΔG°_f values for all the products and reactants except $O_2(g)$. Because oxygen gas is an element in its standard state, $\Delta G^\circ_f(O_2)$ is zero. Using the "products minus reactants" rule, we get

$$\begin{aligned}\Delta G° &= [8\Delta G^\circ_f(CO_2) + 9\Delta G^\circ_f(H_2O)] - \left[1\Delta G^\circ_f(C_8H_{18}) + \frac{25}{2}\Delta G^\circ_f(O_2)\right] \\ &= [(8\ \cancel{\text{mol}})(-394.4\ \text{kJ}/\cancel{\text{mol}}) + (9\ \cancel{\text{mol}})(-237.1\ \text{kJ}/\cancel{\text{mol}})] \\ &\quad - [(1\ \cancel{\text{mol}})(-353.2\ \text{kJ}/\cancel{\text{mol}}) + \left(\frac{25}{2}\ \cancel{\text{mol}}\right)(0\ \text{kJ}/\cancel{\text{mol}})] \\ &= -4935.9\ \text{kJ (per mol of } C_8H_{18})\end{aligned}$$

Because $\Delta G°$ is a large negative number, there is a strong tendency for the spontaneous formation of products from reactants (though not necessarily at a rapid rate). Also notice that the magnitude of $\Delta G°$ is largely determined by the $\Delta G_f°$ of the stable products, water and carbon dioxide.

EXERCISE 18.9

Calculate $\Delta G°$ for the reaction of benzene with hydrogen gas to give cyclohexane using the data $\Delta G_f°$ (benzene) = 124.5 kJ/mol and $\Delta G_f°$ (cyclohexane) = 217.3 kJ/mol. Is the reaction spontaneous as written?

Answer 92.8 kJ; no

Calculated values of $\Delta G°$ are extremely useful in predicting whether a reaction will occur spontaneously if the reactants and products are mixed under standard conditions. We should note, however, that very few reactions are actually carried out under standard conditions, and calculated values of $\Delta G°$ may not tell us whether a given reaction will occur spontaneously under nonstandard conditions. What determines whether a reaction will occur spontaneously is the free-energy change ΔG under the actual experimental conditions, which are usually different from $\Delta G°$. If the ΔH and $T\Delta S$ terms for a reaction have the same sign, for example, then it may be possible to reverse the sign of ΔG by changing the temperature, thereby converting a reaction that is not thermodynamically spontaneous to one that is, or vice versa. Because ΔH and ΔS usually do not vary greatly with temperature in the absence of a phase change, we can use tabulated values of $\Delta H°$ and $\Delta S°$ to calculate $\Delta G°$ at various temperatures, as long as no phase change occurs over the temperature range being considered.

EXAMPLE 18.10

Calculate **(a)** $\Delta G°$ and **(b)** $\Delta G_{300°C}$ for the reaction $N_2(g) + 3H_2(g) \rightleftharpoons 2NH_3(g)$, assuming that ΔH and ΔS do not change between 25°C and 300°C. Use these data: $S°(N_2)$ = 191.6 J/(mol·K), $S°(H_2)$ = 130.7 J/(mol·K), $S°(NH_3)$ = 192.8 J/(mol·K), and $\Delta H_f°$ (NH_3) = −45.9 kJ/mol.

Given Balanced chemical equation, temperatures, $S°$ values, $\Delta H_f°$ for NH_3

Asked for $\Delta G°$ and ΔG at 300°C

Strategy

Ⓐ Convert each temperature to Kelvins. Then calculate $\Delta S°$ for the reaction. Calculate $\Delta H°$ for the reaction, recalling that $\Delta H_f°$ for any element in its standard state is zero.

Ⓑ Substitute the appropriate values into Equation 18.23 to obtain $\Delta G°$ for the reaction.

Ⓒ Assuming that ΔH and ΔS are independent of temperature, substitute values into Equation 18.20 to obtain ΔG for the reaction at 300°C.

Solution

(a) Ⓐ To calculate $\Delta G°$ for the reaction using Equation 18.23, we must know the temperature as well as the values of $\Delta S°$ and $\Delta H°$. At standard conditions, the temperature is 25°C, or 298.15 K. We can calculate $\Delta S°$ for the reaction from the absolute molar entropy values given for the reactants and products using the "products minus reactants" rule:

$$\begin{aligned}\Delta S^\circ_{rxn} &= 2S^\circ(NH_3) - [S^\circ(N_2) + 3S^\circ(H_2)]\\ &= [2\ \text{mol}\ NH_3 \times 192.8\ \text{J/(mol}\cdot\text{K)}]\\ &\quad - \{[1\ \text{mol}\ N_2 \times 191.6\ \text{J/(mol}\cdot\text{K)}] + [3\ \text{mol}\ H_2 \times 130.7\ \text{J/(mol}\cdot\text{K)}]\}\\ &= -198.1\ \text{J/K (per mole of}\ N_2)\end{aligned}$$

We can also calculate ΔH° for the reaction using the "products minus reactants" rule. The value of $\Delta H^\circ_f(NH_3)$ is given and ΔH°_f is zero for both N_2 and H_2:

$$\begin{aligned}\Delta H^\circ_{rxn} &= 2\Delta H^\circ_f(NH_3) - [\Delta H^\circ_f(N_2) + 3\Delta H^\circ_f(H_2)]\\ &= (2 \times -45.9\ \text{kJ/mol}) - [(1 \times 0\ \text{kJ/mol}) + (3 \times 0\ \text{kJ/mol})]\\ &= -91.8\ \text{kJ (per mole of}\ N_2)\end{aligned}$$

B Inserting the appropriate values into Equation 18.23 gives

$$\Delta G^\circ_{rxn} = \Delta H^\circ - T\Delta S^\circ = (-91.8\ \text{kJ}) - (298.15\ \text{K})(-198.1\ \text{J/K}) = -32.7\ \text{kJ (per mole of}\ N_2)$$

(b) **C** To calculate ΔG for this reaction at 300°C, we assume that ΔH and ΔS are independent of temperature (that is, $\Delta H_{300°C} = \Delta H^\circ$ and $\Delta S_{300°C} = \Delta S^\circ$) and insert the appropriate temperature (573 K) into Equation 18.20:

$$\begin{aligned}\Delta G_{300°C} &= \Delta H_{300°C} - (573\ \text{K})(\Delta S_{300°C}) = \Delta H^\circ - (573\ \text{K})\Delta S^\circ\\ &= (-91.8\ \text{kJ}) - (573\ \text{K})(-198.1\ \text{J/K}) = 21.7\ \text{kJ (per mole of}\ N_2)\end{aligned}$$

In this example, changing the temperature has a major effect on the thermodynamic spontaneity of the reaction. Under standard conditions, the reaction of nitrogen and hydrogen gas to produce ammonia is thermodynamically spontaneous, but in practice, it is too slow to be useful industrially. Increasing the temperature in an attempt to make this reaction occur more rapidly also changes the thermodynamics by causing the $-T\Delta S^\circ$ term to dominate, and the reaction is no longer spontaneous at high temperatures. This is a classic example of the conflict that is encountered in real systems between thermodynamics and kinetics, which is often unavoidable.

EXERCISE 18.10

Calculate **(a)** ΔG° and **(b)** $\Delta G_{750°C}$ for the reaction $2NO(g) + O_2(g) \rightleftharpoons 2NO_2(g)$, which is important in the formation of urban smog. Assume that ΔH and ΔS do not change between 25.0°C and 750°C, and use these data: $S^\circ(NO) = 210.8$ J/(mol · K), $S^\circ(O_2) = 205.2$ J/(mol · K), $S^\circ(NO_2) = 240.1$ J/(mol · K), $\Delta H^\circ_f(NO_2) = 33.2$ kJ/mol, and $\Delta H^\circ_f(NO) = 91.3$ kJ/mol.

Answer **(a)** −70.5 kJ/mol of O_2; **(b)** 35.8 kJ/mol of O_2

The effect of temperature on the spontaneity of a reaction, an important factor in the design of an experiment or an industrial process, depends on the sign and magnitude of both ΔH° and ΔS°. The temperature at which a given reaction is at equilibrium can be calculated by setting $\Delta G^\circ = 0$ in Equation 18.23, as illustrated in the next example.

EXAMPLE 18.11

As you saw in Example 18.10, the reaction of nitrogen and hydrogen gas to produce ammonia is one in which ΔH° and ΔS° are both negative. Such reactions are predicted to be thermodynamically spontaneous at low temperatures, but nonspontaneous at high temperatures. Use the data in Example 18.10 to calculate the temperature at which this reaction changes from spontaneous to nonspontaneous, assuming that ΔH° and ΔS° are independent of temperature.

Given $\Delta H°$, $\Delta S°$

Asked for Temperature at which reaction changes from spontaneous to nonspontaneous

Strategy

Set $\Delta G°$ equal to zero in Equation 18.23 and solve for T, the temperature at which the reaction becomes nonspontaneous.

Solution

In Example 18.10, we calculated that $\Delta H°$ is -91.8 kJ/mol of N_2 and $\Delta S°$ is -198.1 J/K per mole of N_2, corresponding to $\Delta G° = -32.7$ kJ/mol of N_2 at 25°C. Thus, the reaction is indeed spontaneous at low temperatures, as expected based on the signs of $\Delta H°$ and $\Delta S°$. The temperature at which the reaction becomes nonspontaneous is found by setting $\Delta G°$ equal to zero and rearranging Equation 18.23 to solve for T:

$$\Delta G° = \Delta H° - T\Delta S° = 0$$

$$\Delta H° = T\Delta S°$$

$$T = \frac{\Delta H°}{\Delta S°} = \frac{(-91.8\ \cancel{\text{kJ}})(1000\ \cancel{\text{J}}/\cancel{\text{kJ}})}{-198.1\ \cancel{\text{J}}/\text{K}} = 463\ \text{K}$$

This is a case in which a chemical engineer is severely limited by thermodynamics. Any attempt to increase the rate of reaction of nitrogen with hydrogen by increasing the temperature will cause reactants to be favored over products above 463 K.

EXERCISE 18.11

As you found in working Exercise 18.10, $\Delta H°$ and $\Delta S°$ are both negative for the reaction of nitric oxide and oxygen to form nitrogen dioxide. Use the data in Exercise 18.10 to calculate the temperature at which this reaction changes from spontaneous to nonspontaneous.

Answer 792.6 K

18.6 • Spontaneity and Equilibrium

Thus far we have identified three criteria for whether a given reaction will occur spontaneously: $\Delta S_{univ} > 0$, $\Delta G_{sys} < 0$ (both discussed in this chapter), and the relative magnitude of the reaction quotient Q versus the equilibrium constant K (discussed in Chapter 15). Recall that if $Q < K$, then the reaction proceeds spontaneously to the right as written, resulting in the net conversion of reactants to products. Conversely, if $Q > K$, then the reaction proceeds spontaneously to the left as written, resulting in the net conversion of products to reactants. If $Q = K$, then the system is at equilibrium, and no net reaction occurs. Table 18.3 summarizes these criteria and their relative values for spontaneous, nonspontaneous, and equilibrium processes. Because all three

TABLE 18.3 Criteria for the spontaneity of a process

Spontaneous	Equilibrium	Nonspontaneous[a]
$\Delta S_{univ} > 0$	$\Delta S_{univ} = 0$	$\Delta S_{univ} < 0$
$\Delta G_{sys} < 0$	$\Delta G_{sys} = 0$	$\Delta G_{sys} > 0$
$Q < K$	$Q = K$	$Q > K$

[a]Spontaneous in the *reverse* direction.

criteria are assessing the same thing—the spontaneity of the process—it would be most surprising indeed if they were not related. The relationship between ΔS_{univ} and ΔG_{sys} was described in Section 18.5. In this section, we explore the relationship between the standard free energy of reaction $\Delta G°$ and the equilibrium constant K.

Free Energy and the Equilibrium Constant

Because $\Delta H°$ and $\Delta S°$ determine the magnitude of $\Delta G°$ (Equation 18.23), and because the equilibrium constant K is a measure of the ratio of the concentrations of products to the concentrations of reactants, we should be able to express K in terms of $\Delta G°$, and vice versa. As you learned in Section 18.5, ΔG is equal to the maximum amount of work a system can perform on the surroundings while undergoing a spontaneous change. For a reversible process that does not involve external work, we can express the change in free energy in terms of volume, pressure, entropy, and temperature, thereby eliminating ΔH from the equation for ΔG. The general relationship can be shown as

$$\Delta G = V\Delta P - S\Delta T \qquad (18.26)$$

If a reaction is carried out at constant temperature ($\Delta T = 0$), then Equation 18.26 simplifies to

$$\Delta G = V\Delta P \qquad (18.27)$$

Under normal conditions, the pressure dependence of free energy is not important for solids and liquids because of their small molar volumes. For reactions that involve gases, however, the effect of pressure on free energy is important.

Assuming ideal gas behavior, we can replace the V in Equation 18.27 by nRT/P (where n is the number of moles of gas and R is the ideal gas constant) and express ΔG in terms of the initial and final pressures (P_i and P_f, respectively) as in Equation 18.17:

$$\Delta G = \left(\frac{nRT}{P}\right)\Delta P = nRT\frac{\Delta P}{P} = nRT\ln\left(\frac{P_f}{P_i}\right) \qquad (18.28)$$

If the initial state is the standard state with $P_i = 1$ atm, then the change in free energy of a substance upon going from the standard state to any other state with a pressure P can be written as

$$G - G° = nRT\ln P$$

which can be rearranged to

$$G = G° + nRT\ln P \qquad (18.29)$$

As you will soon discover, Equation 18.29 allows us to relate $\Delta G°$ and K_p. Any relationship that is true for K_p must also be true for K because K_p and K are simply different ways of expressing the equilibrium constant using different units.

Let's consider the following hypothetical reaction, in which all the reactants and products are ideal gases and the lowercase letters correspond to the stoichiometric coefficients for the various species:

$$a\mathrm{A} + b\mathrm{B} \rightleftharpoons c\mathrm{C} + d\mathrm{D} \qquad (18.30)$$

Because the free-energy change for a reaction is the difference between the sum of the free energies of the products and the reactants, we can write this expression for ΔG:

$$\Delta G = \Sigma mG_{products} - \Sigma nG_{reactants} = (cG_C + dG_D) - (aG_A + bG_B) \qquad (18.31)$$

Substituting Equation 18.29 for each term into Equation 18.31 gives

$$\Delta G = [(cG_C° + cRT\ln P_C) + (dG_D° + dRT\ln P_D)] - [(aG_A° + aRT\ln P_A) + (bG_B° + bRT\ln P_B)]$$

Combining terms gives the following relationship between ΔG and and the reaction quotient Q:

$$\Delta G = \Delta G^\circ + RT \ln\frac{P_C^c P_D^d}{P_A^a P_B^b} = \Delta G^\circ + RT \ln Q \tag{18.32}$$

where ΔG° indicates that all reactants and products are in their standard states. In Chapter 15, you learned that for gases $Q = K_p$ at equilibrium, and as you've learned in this chapter, $\Delta G = 0$ for a system at equilibrium. Therefore, we can describe the relationship between ΔG° and K_p for gases as

$$0 = \Delta G^\circ + RT \ln K_p \qquad \Delta G^\circ = -RT \ln K_p \tag{18.33}$$

If the products and reactants are in their standard states and $\Delta G^\circ < 0$, then $K_p > 1$ and products are favored over reactants. Conversely, if $\Delta G^\circ > 0$, then $K_p < 1$ and reactants are favored over products. If $\Delta G^\circ = 0$, then $K_p = 1$ and neither reactants nor products are favored: the system is at equilibrium.

Note the pattern

For a spontaneous process under standard conditions, $K > 1$.

EXAMPLE 18.12

In Example 18.10, we calculated that $\Delta G^\circ = -32.7$ kJ/mol of N_2 for the reaction $N_2(g) + 3H_2(g) \rightleftharpoons 2NH_3(g)$. This calculation was for the reaction under standard conditions—that is, with all gases present at a partial pressure of 1 atm and a temperature of 25°C. Calculate ΔG for the same reaction under these nonstandard conditions: $P_{N_2} = 2.00$ atm, $P_{H_2} = 7.00$ atm, $P_{NH_3} = 0.021$ atm, and $T = 100°C$. Does the reaction favor products or reactants?

Given Balanced chemical equation, partial pressure of each species, temperature, ΔG°

Asked for Whether products or reactants are favored

Strategy

A Using the values given and Equation 18.32, calculate Q.

B Substitute the values of ΔG° and Q into Equation 18.32 to obtain ΔG for the reaction under nonstandard conditions.

Solution

A The relationship between ΔG° and ΔG under nonstandard conditions is given in Equation 18.32. Substituting the partial pressures given, we can calculate Q:

$$Q = \frac{P_{NH_3}^2}{P_{N_2} P_{H_2}^3} = \frac{(0.021)^2}{(2.00)(7.00)^3} = 6.4 \times 10^{-7}$$

B Substituting the values of ΔG° and Q into Equation 18.32 gives

$$\Delta G = \Delta G^\circ + RT \ln Q = -32.7 \text{ kJ} + \left[(8.314 \text{ J/K})(373 \text{ K})\left(\frac{1 \text{ kJ}}{1000 \text{ J}}\right)\ln(6.4 \times 10^{-7})\right]$$

$$= -32.7 \text{ kJ} + (-44.2 \text{ kJ}) = -76.9 \text{ kJ/mol of } N_2$$

Because $\Delta G < 0$ and $Q < 1.0$, the reaction is spontaneous to the right as written, so products are favored over reactants.

EXERCISE 18.12

Calculate ΔG for the reaction of nitric oxide with oxygen to give nitrogen dioxide under these conditions: $T = 50°C$, $P_{NO} = 0.0100$ atm, $P_{O_2} = 0.200$ atm, and $P_{NO_2} = 1.00 \times 10^{-4}$ atm. The value of $\Delta G°$ for this reaction is -72.5 kJ/mol of O_2. Are products or reactants favored?

Answer -92.9 kJ/mol of O_2. The reaction is spontaneous to the right as written, so products are favored.

EXAMPLE 18.13

Calculate the equilibrium constant K_p for the reaction of H_2 with N_2 to give NH_3 at 25°C. As calculated in Example 18.10, $\Delta G°$ for this reaction is -32.7 kJ/mol of N_2.

Given Balanced chemical equation from Example 18.10, $\Delta G°$, temperature

Asked for K_p

Strategy

Substitute values for $\Delta G°$ and T (in kelvins) into Equation 18.33 to calculate K_p, the equilibrium constant for the formation of ammonia.

Solution

In Example 18.10, we used tabulated values of $\Delta G_f°$ to calculate $\Delta G°$ for this reaction (-32.7 kJ/mol of N_2). For equilibrium conditions, rearranging Equation 18.33 gives

$$\Delta G° = -RT \ln K_p$$

$$\ln K_p = \frac{-\Delta G°}{RT}$$

Inserting the value of $\Delta G°$ and the temperature (25°C = 298.15 K) into this equation gives

$$\ln K_p = -\frac{(-32.8 \text{ kJ})(1000 \text{ J/kJ})}{(8.314 \text{ J/K})(298.15 \text{ K})} = 13.2$$

$$K_p = 5.6 \times 10^5$$

Thus, the equilibrium constant for the formation of ammonia at room temperature is favorable. As we saw in Chapter 15, however, the *rate* at which the reaction occurs at room temperature is too slow to be useful.

EXERCISE 18.13

Calculate K_p for the reaction of NO with O_2 to give NO_2 at 25°C. As calculated in Exercise 18.10, $\Delta G°$ for this reaction is -72.5 kJ/mol of O_2.

Answer 5.0×10^{12}

Although K_p is defined in terms of the partial pressures of reactants and products, the equilibrium constant K is defined in terms of the concentrations of reactants and products. We described the relationship between the numerical magnitude of K_p and K in Chapter 15 and showed that they are related:

$$K_p = K(RT)^{\Delta n} \tag{18.34}$$

where Δn is the number of moles of gaseous product minus the number of moles of gaseous reactant. For reactions that involve only solutions, liquids, and solids, $\Delta n = 0$ and hence $K_p = K$. Thus, for all reactions that do not involve a change in the number of moles of gas present, the relationship in Equation 18.33 can be written in a more general form:

$$\Delta G° = -RT \ln K \qquad (18.35)$$

Only when a reaction results in a net production or consumption of gases is it necessary to correct Equation 18.35 for the difference between K_p and K.*

Combining Equations 18.23 and 18.35 provides insight into how the components of $\Delta G°$ influence the magnitude of the equilibrium constant:

$$\Delta G° = \Delta H° - T\Delta S° = -RT \ln K \qquad (18.36)$$

Notice that K becomes larger as $\Delta S°$ becomes more positive, indicating that the magnitude of the equilibrium constant is directly influenced by the tendency of the system to move toward maximum disorder. Moreover, K increases as $\Delta H°$ decreases. Thus, the magnitude of the equilibrium constant is also directly influenced by the tendency of the system to seek the lowest energy state possible.

Note the pattern

The magnitude of the equilibrium constant is directly influenced by the tendency of the system to move toward maximum disorder and to seek the lowest energy state possible.

Temperature Dependence of the Equilibrium Constant

The fact that $\Delta G°$ and K are related provides us with another explanation of why equilibrium constants are temperature dependent. This relationship is shown explicitly in Equation 18.36, which can be rearranged to give

$$\ln K = -\frac{\Delta H°}{RT} + \frac{\Delta S°}{R} \qquad (18.37)$$

Thus, assuming $\Delta H°$ and $\Delta S°$ are temperature independent, for an exothermic reaction ($\Delta H° < 0$), the magnitude of K decreases with increasing temperature, whereas for an endothermic reaction ($\Delta H° > 0$), the magnitude of K increases with increasing temperature. Note that the quantitative relationship expressed in Equation 18.37 agrees with the qualitative predictions made by applying Le Châtelier's principle, which we discussed in Chapter 15. Because heat is produced in an exothermic reaction, adding heat (by increasing the temperature) will shift the equilibrium to the left, favoring the reactants and decreasing the magnitude of K. Conversely, since heat is consumed in an endothermic reaction, adding heat will shift the equilibrium to the right, favoring the products and increasing the magnitude of K. Equation 18.37 also shows that the *magnitude* of $\Delta H°$ dictates how rapidly K changes as a function of temperature. In contrast, the magnitude and sign of $\Delta S°$ affect the magnitude of K, but not its temperature dependence.

If we know the value of K at a given temperature and the value of $\Delta H°$ for a reaction, we can estimate the value of K at any other temperature even in the absence of information on $\Delta S°$. Suppose, for example, that K_1 and K_2 are the equilibrium constants for a reaction at temperatures T_1 and T_2, respectively. Applying Equation 18.37 gives the following relationship at each temperature:

$$\ln K_1 = \frac{-\Delta H°}{RT_1} + \frac{\Delta S°}{R}$$

$$\ln K_2 = \frac{-\Delta H°}{RT_2} + \frac{\Delta S°}{R}$$

* Although we typically use concentrations or pressures in our equilibrium calculations, recall that equilibrium constants are generally expressed as unitless numbers because of the use of activities or fugacities in precise thermodynamic work. Systems that contain gases at high pressures or concentrated solutions that deviate substantially from ideal behavior require the use of fugacities or activities, respectively.

Subtracting $\ln K_1$ from $\ln K_2$ gives

$$\ln K_2 - \ln K_1 = \ln\frac{K_2}{K_1} = \frac{\Delta H^\circ}{R}\left(\frac{1}{T_1} - \frac{1}{T_2}\right) \tag{18.38}$$

Thus, calculating ΔH° from tabulated enthalpies of formation and measuring the equilibrium constant at one temperature, K_1, allow us to calculate the value of the equilibrium constant at any other temperature, K_2, assuming that ΔH° and ΔS° are independent of temperature.

EXAMPLE 18.14

The equilibrium constant for the formation of NH_3 from H_2 and N_2 at 25°C was calculated to be $K_p = 5.6 \times 10^5$ in Example 18.13. What is K_p at 500°C? (Use the data from Example 18.10.)

Given Balanced chemical equation, ΔH°, initial and final T, K_p at 25°C

Asked for K_p at 500°C

Strategy

Convert the initial and final temperatures to kelvins. Then substitute appropriate values into Equation 18.38 to obtain K_2, the equilibrium constant at the final temperature.

Solution

The value of ΔH° for the reaction obtained using Hess's law is −91.8 kJ/mol of N_2. If we set T_1 = 25°C = 298.15 K and T_2 = 500°C = 773 K, then from Equation 18.38 we obtain

$$\ln\frac{K_2}{K_1} = \frac{\Delta H^\circ}{R}\left(\frac{1}{T_1} - \frac{1}{T_2}\right) = \frac{(-91.8\ \cancel{\text{kJ}})(1000\ \cancel{\text{J}}/\cancel{\text{kJ}})}{8.314\ \cancel{\text{J}}/\cancel{\text{K}}}\left(\frac{1}{298.15\ \cancel{\text{K}}} - \frac{1}{773\ \cancel{\text{K}}}\right) = -22.8$$

$$\frac{K_2}{K_1} = 1.3 \times 10^{-10}$$

$$K_2 = (5.6 \times 10^5)(1.3 \times 10^{-10}) = 7.3 \times 10^{-5}$$

Thus, at 500°C, the equilibrium strongly favors the reactants over the products.

EXERCISE 18.14

In Exercise 18.13, you calculated $K_p = 5.0 \times 10^{12}$ for the reaction of NO with O_2 to give NO_2 at 25°C. Use the ΔH°_f values in Exercise 18.10 to calculate K_p for this reaction at 1000°C.

Answer 1.3×10^{-3}

Note the pattern

Thermodynamics focuses on the energetics of the products and reactants, whereas kinetics focuses on the pathway from reactants to products.

18.7 • Comparing Thermodynamics and Kinetics

Because thermodynamics deals with state functions, it can be used to describe the overall properties, behavior, and equilibrium composition of a system. It is *not* concerned with the particular pathway by which physical or chemical changes occur, however, so it cannot address the *rate* at which a particular process will occur. Thus, although thermodynamics provides a significant constraint on what *can* occur during a reaction process, it does not describe the detailed steps of what *actually* occurs on an atomic or molecular level.

TABLE 18.4 The relationship between K and $\Delta G°$ at 25°C

$\Delta G°$, kJ/mol	K	Physical Significance
500 100	3×10^{-88} 3×10^{-18}	For all practical purposes, the reaction does not proceed in the forward direction: only reactants are present at equilibrium.
10 0 −10	2×10^{-2} 1 60	Both forward and reverse reactions occur: significant amounts of both products and reactants are present at equilibrium.
−100 −500	3×10^{17} 4×10^{87}	For all practical purposes, the forward reaction proceeds to completion: only products are present at equilibrium.

Table 18.4 gives the numerical values of the equilibrium constant K that correspond to various approximate values of $\Delta G°$. Note that $\Delta G° \geq +10$ kJ/mol or $\Delta G° \leq -10$ kJ/mol ensures that an equilibrium lies essentially all the way to the left or to the right, respectively, under standard conditions, corresponding to a reactant-to-product ratio of approxi-mately 10,000:1 (or vice versa). Only if $\Delta G°$ is quite small (± 10 kJ/mol) are significant amounts of both products and reactants present at equilibrium. Most reactions that we encounter have equilibrium constants substantially greater or less than 1, with the equilibrium strongly favoring either products or reactants. In many cases, we will encounter reactions that are strongly favored by thermodynamics but do not occur at a measurable rate. In contrast, we may encounter reactions that are not thermodynamically favored under standard conditions but nonetheless do occur under certain nonstandard conditions.

A typical challenge in industrial processes is a reaction that has a large negative value of $\Delta G°$ and hence a large value of K, but that is too slow to be practically useful. In such cases, mixing the reactants results in only a physical mixture, not a chemical reaction. An example is the reaction of carbon tetrachloride with water to produce carbon dioxide and hydrochloric acid, for which $\Delta G°$ is -232 kJ/mol:

$$CCl_4(l) + 2H_2O(l) \rightleftharpoons CO_2(g) + 2HCl(g) \qquad (18.39)$$

The value of K for this reaction is 5×10^{40} at 25°C, yet when CCl_4 and water are shaken vigorously at 25°C, nothing happens: the two immiscible liquids form separate layers, with the denser CCl_4 on the bottom. In comparison, the analogous reaction of $SiCl_4$ with water to give SiO_2 and HCl, which has a similarly large equilibrium constant, occurs almost explosively. Although the two reactions have comparable thermodyna-mics, they have very different kinetics!

There are also many reactions for which $\Delta G° \ll 0$ but that do not occur as written because another possible reaction occurs more rapidly. For example, consider the reaction of lead sulfide with hydrogen peroxide. One possible reaction is

$$PbS(s) + 4H_2O_2(l) \rightleftharpoons PbO_2(s) + SO_2(g) + 4H_2O(l) \qquad (18.40)$$

for which $\Delta G°$ is -886 kJ/mol and K is 10^{161}. Yet when lead sulfide is mixed with hydrogen peroxide, the ensuing vigorous reaction does *not* produce PbO_2 and SO_2. Instead, the reaction that actually occurs is

$$PbS(s) + 4H_2O_2(l) \rightleftharpoons PbSO_4(s) + 4H_2O(l) \qquad (18.41)$$

This reaction has a $\Delta G°$ value of -1181 kJ/mol, within the same order of magnitude as the previous reaction, but it occurs much more rapidly.

Now consider reactions with $\Delta G^\circ > 0$. Thermodynamically, such reactions do not occur spontaneously under standard conditions. Nonetheless, these reactions can be made to occur under nonstandard conditions. An example is the reduction of chromium(III) chloride by hydrogen gas:

$$CrCl_3(s) + \frac{1}{2}H_2(g) \rightleftharpoons CrCl_2(s) + HCl(g) \tag{18.42}$$

At 25°C, $\Delta G^\circ = 35$ kJ/mol and $K_p = 7 \times 10^{-7}$. However, at 730°C, $\Delta G^\circ = -52$ kJ/mol and $K_p = 5 \times 10^2$; at this elevated temperature, the reaction is a convenient way of preparing chromium(II) chloride in the laboratory. Moreover, removing HCl gas from the system drives the reaction to completion, as predicted by Le Chatelier's principle. Thus, although the reaction is not thermodynamically spontaneous under standard conditions, it becomes spontaneous under nonstandard conditions.

There are also cases in which a compound whose formation appears to be thermodynamically prohibited can be prepared using a different reaction. The reaction for the preparation of chlorine monoxide from the elements, for example, is

$$\frac{1}{2}O_2(g) + Cl_2(g) \rightleftharpoons Cl_2O(g) \tag{18.43}$$

for which ΔG_f° is 97.9 kJ/mol. The large positive value of ΔG_f° for this reaction indicates that mixtures of chlorine and oxygen do not react to any extent to form Cl_2O. Nonetheless, Cl_2O is easily prepared using the reaction

$$HgO(s) + 2Cl_2(g) \rightleftharpoons Cl_2O(g) + HgCl_2(s) \tag{18.44}$$

which has a ΔG° of -22.2 kJ/mol and a K_p of approximately 1×10^4.

Finally, the ΔG° values for some reactions are so positive that the only way to make them proceed in the desired direction is to supply external energy, often in the form of electricity. Consider, for example, the formation of metallic lithium from molten lithium chloride:

$$LiCl(l) \rightleftharpoons Li(l) + \frac{1}{2}Cl_2(g) \tag{18.45}$$

Even at 1000°C, ΔG is very positive (324 kJ/mol), and there is no obvious way to obtain lithium metal using a different reaction. Hence, in the industrial preparation of metallic lithium, electrical energy is used to drive the reaction to the right, as described in Chapter 19.

Often reactions that are not thermodynamically spontaneous under standard conditions can be made to occur spontaneously if coupled, or connected, in some way to another reaction for which $\Delta G^\circ \ll 0$. Because the overall value of ΔG° for a series of reactions is the sum of the ΔG° values for the individual reactions, virtually any unfavorable reaction can be made to occur by chemically coupling it to a sufficiently favorable reaction or reactions. In the preparation of chlorine monoxide from mercuric oxide and chlorine (Equation 18.44), we have already encountered one example of this phenomenon of *coupled reactions*, although we did not describe it as such at the time. We can see how the chemical coupling works if we write Equation 18.44 as the sum of three separate reactions:

$$\begin{array}{rcll}
\frac{1}{2}\cancel{O_2(g)} + Cl_2(g) & \rightleftharpoons & Cl_2O(g) & \Delta G^\circ = 97.9\ \text{kJ/mol} \quad (1) \\
HgO(s) & \rightleftharpoons & \cancel{Hg(l)} + \frac{1}{2}\cancel{O_2(g)} & \Delta G^\circ = 58.5\ \text{kJ/mol} \quad (2) \\
\cancel{Hg(l)} + Cl_2(g) & \rightleftharpoons & HgCl_2(s) & \Delta G^\circ = -178.6\ \text{kJ/mol} \quad (3) \\
\hline
HgO(s) + 2Cl_2(g) & \rightleftharpoons & Cl_2O(g) + HgCl_2(s) & \Delta G^\circ_{rxn} = -22.2\ \text{kJ/mol}
\end{array}$$

Comparing the $\Delta G°$ values for the three reactions shows that reaction (3) is so energetically favorable that it more than compensates for the other two energetically unfavorable reactions. Hence, the overall reaction is indeed thermodynamically spontaneous as written.

EXAMPLE 18.15

Bronze Age metallurgists were accomplished practical chemists who unknowingly utilized coupled reactions to obtain metals from their ores. Realizing that different ores of the same metal required different treatments, they heated copper oxide ore in the presence of charcoal (carbon) to obtain copper metal, whereas they pumped air into the reaction system if the ore was copper sulfide. Assume that a particular copper ore consists of pure cuprous oxide (Cu_2O). Using the $\Delta G_f°$ values given, calculate **(a)** $\Delta G°$ and K_p for the decomposition of Cu_2O to metallic copper and oxygen gas [$\Delta G_f°(Cu_2O) = -146.0$ kJ/mol] and **(b)** $\Delta G°$ and K_p for the reaction of Cu_2O with carbon to produce metallic copper and carbon monoxide [$\Delta G_f°(CO) = -137.2$ kJ/mol].

Given Reactants and products, $\Delta G_f°$ values for Cu_2O and CO, temperature

Asked for $\Delta G°$ and K_p for the formation of metallic copper from Cu_2O in the absence and presence of carbon

Strategy

A Write the balanced equilibrium equation for each reaction. Using the "products minus reactants" rule, calculate $\Delta G°$ for the reaction.

B Substitute appropriate values into Equation 18.33 to obtain K_p.

Solution

(a) **A** The chemical equation for the decomposition of cuprous oxide is

$$Cu_2O(s) \rightleftharpoons 2Cu(s) + \frac{1}{2}O_2(g)$$

The substances on the right side of this equation are pure elements in their standard states, so their $\Delta G_f°$ values are zero. $\Delta G°$ for the reaction is therefore

$$\Delta G° = \left[2\Delta G_f°(Cu) + \frac{1}{2}\Delta G_f°(O_2)\right] - \Delta G_f°(Cu_2O)$$

$$= \left[(2\ \cancel{mol})(0\ kJ/\cancel{mol}) + \left(\frac{1}{2}\ \cancel{mol}\right)(0\ kJ/\cancel{mol})\right] - [(1\ \cancel{mol})(-146.0\ kJ/\cancel{mol})]$$

$$= 146.0\ kJ$$

B Rearranging and substituting the appropriate values into Equation 18.33, we obtain

$$\ln K_p = -\frac{\Delta G°}{RT} = -\frac{(146.0\ \cancel{kJ})(1000\ \cancel{J/kJ})}{(8.314\ \cancel{J/K})(298.15\ \cancel{K})} = -58.90$$

$$K_p = 2.60 \times 10^{-26}$$

This is a very small number, indicating that Cu_2O does not spontaneously decompose to a significant extent at room temperature.

(b) **A** The O_2 produced in the decomposition of Cu_2O can react with carbon to form CO:

$$\frac{1}{2}O_2(g) + C(s) \rightleftharpoons CO(g)$$

Because $\Delta G°$ for this reaction is equal to $\Delta G_f°$ for CO (−137.2 kJ/mol), it is energetically more feasible to produce metallic copper from cuprous oxide by coupling the two reactions:

$$Cu_2O(s) \rightleftharpoons 2Cu(s) + \cancel{\frac{1}{2}O_2(g)} \qquad \Delta G° = 146.0\ kJ/mol$$

$$\cancel{\frac{1}{2}O_2(g)} + C(s) \rightleftharpoons CO(g) \qquad \Delta G° = -137.2\ kJ/mol$$

$$Cu_2O(s) + C(s) \rightleftharpoons 2Cu(s) + CO(g) \qquad \Delta G° = 8.8\ kJ/mol$$

B✓ We can find the corresponding value of K_p:

$$\ln K_p = -\frac{\Delta G°}{RT} = -\frac{(8.8 \text{ kJ})(1000 \text{ J/kJ})}{(8.314 \text{ J/K})(298.15 \text{ K})} = -3.6$$

$$K_p = 0.03$$

Although this value is still less than 1, indicating that reactants are favored over products at room temperature, it is about 24 orders of magnitude greater than K_p for the production of copper metal in the absence of carbon. Because both $\Delta H°$ and $\Delta S°$ are positive for this reaction, it becomes thermodynamically feasible at slightly elevated temperatures (higher than about 80°C). At temperatures of a few hundred degrees Celsius, the reaction occurs spontaneously, proceeding smoothly and rapidly to the right as written and producing metallic copper and carbon monoxide from cuprous oxide and carbon.

EXERCISE 18.15

Use the ΔG_f° values given to calculate $\Delta G°$ and K_p (where appropriate) for **(a)** the decomposition of cuprous sulfide to copper metal and elemental sulfur ΔG_f° for Cu_2S, −86.2 kJ/mol) and **(b)** the reaction of cuprous sulfide with oxygen gas to produce sulfur dioxide and copper metal [ΔG_f° for $SO_2(g)$, −300.1 kJ/mol].

Answer **(a)** $\Delta G° = 86.2$ kJ/mol; **(b)** $\Delta G° = -213.9$ kJ/mol, $K_p = 3 \times 10^{37}$

18.8 ∘ Thermodynamics and Life

In a thermodynamic sense, a living cell can be viewed as a low-entropy system that is *not* in equilibrium with its surroundings and is capable of replicating itself. A constant input of energy is needed to maintain the cell's highly organized structure, its wide array of precisely folded biomolecules, and its intricate system of thousands of chemical reactions. A cell also needs energy to synthesize complex molecules from simple precursors (for example, to make proteins from amino acids), to create and maintain differences in the concentrations of various substances inside and outside of the cell, and to do mechanical work (such as muscle contraction). In this section, we examine the nature of the energy flow between a cell and its environment as well as some of the chemical strategies cells use to extract energy from their surroundings and to store that energy.

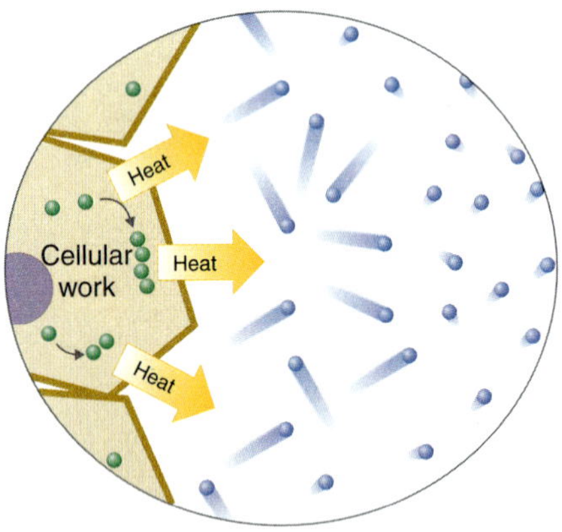

Figure 18.17 Life and entropy. A living cell is in a low-entropy, nonequilibrium state characterized by a high degree of structural organization. To maintain this state, a cell must release some of the energy it obtains from its environment as heat, thereby increasing S_{surr} sufficiently that the second law of thermodynamics is not violated. In this example, the cell combines smaller components into larger, more ordered structures; the accompanying release of heat increases the entropy of the surrounding environment so that $S_{univ} > 0$.

Energy Flow Between the Cell and Its Surroundings

One implication of the first and second laws of thermodynamics is that any closed system must eventually reach equilibrium. With no external input, a clock will run down, a battery will lose its charge, and a mixture of aqueous acid and base will achieve a uniform intermediate pH value. In contrast, a cell is an open system that can exchange matter with its surroundings as well as absorb energy from its environment in the form of heat or light. Cells utilize the energy obtained in these ways to maintain the nonequilibrium state that is essential for life.

Because cells are open systems, they cannot be described using the concepts of classical thermodynamics that we have discussed in this chapter, which have focused on reversible processes occurring in closed chemical systems that can exchange energy, but not matter, with their surroundings. Consequently, a relatively new subdiscipline called *nonequilibrium thermodynamics* has been developed to quantitatively describe open systems such as living cells.

Because a cell cannot violate the second law of thermodynamics, the only way it can maintain a low-entropy, nonequilibrium state characterized by a high degree of structural organization is to increase the entropy of its surroundings. A cell releases some of the energy that it obtains from its environment as heat that is transferred to its surroundings, thereby resulting in an increase in S_{surr} (Figure 18.17). As long as ΔS_{surr} is positive and greater than ΔS_{sys}, the entropy of the universe increases and the second law of thermodynamics is not violated. Releasing heat to the surroundings is necessary but not sufficient for life: the release of energy must be coupled to processes that increase the degree of order within the cell. For example, a wood fire releases heat to its surroundings, but unless energy from the burning wood is also used to do work, there is no increase in order of any portion of the universe.

Note the pattern

Any organism in equilibrium with its environment is dead.

Extracting Energy from the Environment

Although organisms utilize a wide range of specific strategies to obtain the energy they need to live and reproduce, they can generally be divided into two categories. Organisms are either *phototrophs* (from the Greek *photos*, "light," and *trophos*, "feeder"), whose energy source is sunlight, or *chemotrophs*, whose energy source is chemical compounds, usually obtained by consuming or breaking down other organisms. Phototrophs, such as plants, algae, and photosynthetic bacteria, use the radiant energy of the sun directly, converting water and carbon dioxide to energy-rich organic compounds, whereas chemotrophs, such as animals, fungi, and many non-photosynthetic bacteria, obtain energy-rich organic compounds from their environment. Regardless of the nature of their energy and carbon sources, all organisms utilize oxidation–reduction, or redox, reactions to drive the synthesis of complex biomolecules. Organisms that can use only O_2 as the oxidant (a group that includes most animals) are *aerobic organisms* that cannot survive in the absence of O_2. Many organisms that use other oxidants (such as SO_4^{2-}, NO_3^-, or CO_3^{2-}) or oxidized organic compounds can live only in the absence of O_2, which is a deadly poison for them; such species are called *anaerobic organisms*.

The fundamental reaction by which all green plants and algae obtain energy from sunlight is **photosynthesis**, the photochemical reduction of CO_2 to a carbon compound such as glucose. Concurrently, oxygen in water is oxidized to O_2 (recall that $h\nu$ is energy from light):

$$\underset{\text{Photosynthesis}}{6CO_2 + 6H_2O \xrightarrow{h\nu} C_6H_{12}O_6 + 6O_2} \tag{18.46}$$

This reaction is not a spontaneous process as written, so energy from sunlight is used to drive the reaction. Photosynthesis is critical to life on Earth—it produces all the oxygen in our atmosphere.

In many ways, chemotrophs are more diverse than phototrophs because the nature of both the reductant (the nutrient) and the oxidant can vary. The most familiar chemotrophic strategy uses compounds such as glucose as the reductant and molecular oxygen as the oxidant in a process called **respiration** (see Chapter 5). The overall reaction of respiration is the reverse of photosynthesis:

$$\underset{\text{Respiration}}{C_6H_{12}O_6 + 6O_2 \longrightarrow 6CO_2 + 6H_2O} \tag{18.47}$$

An alternative strategy utilizes **fermentation** reactions, in which an organic compound is simultaneously oxidized and reduced. Common examples are alcohol fermentation, used in making wine, beer, and bread, and lactic acid fermentation, used in making yogurt:

$$\underset{\text{Alcohol fermentation}}{C_6H_{12}O_6 \longrightarrow 2CO_2 + 2CH_3CH_2OH} \tag{18.48}$$

$$\underset{\text{Lactic acid fermentation}}{C_6H_{12}O_6 \longrightarrow 2CH_3CH(OH)CO_2H} \tag{18.49}$$

In these reactions, some of the carbon atoms of glucose are oxidized, while others are reduced. Recall that a reaction in which a single species is both oxidized and reduced is called a *disproportionation* reaction.

The Role of NADH and ATP in Metabolism

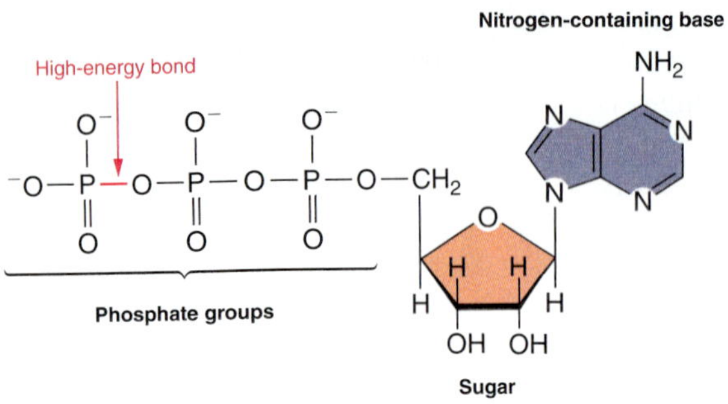

Figure 18.18 Adenosine triphosphate (ATP), the universal energy currency of all cells. The ATP molecule contains a nitrogen-containing base, a sugar, and three phosphate groups, as well as a high-energy phosphoric acid anhydride bond.

Regardless of the identity of the substances from which an organism obtains energy, the energy must be released in very small increments if it is to be useful to the cell. Otherwise, the temperature of the cell would rise to lethal levels. Cells store part of the energy that is released as ATP (adenosine triphosphate), a compound that is the universal energy currency of all living organisms (Figure 18.18).

Most organisms use a number of intermediate species to shuttle electrons between the terminal reductant (such as glucose) and the terminal oxidant (such as O_2). In virtually all cases, an intermediate species oxidizes the energy-rich reduced compound, and the now-reduced intermediate then migrates to another site where it is oxidized. The most important of these electron-carrying intermediates is NAD^+ (Figure 18.19), whose reduced form, formally containing H^-, is NADH. The reduction of NAD^+ to NADH can be written as

$$NAD^+ + H^+ + 2e^- \longrightarrow NADH \qquad (18.50)$$

Most organisms use NAD^+ to oxidize energy-rich nutrients such as glucose to CO_2 and water; NADH is then oxidized to NAD^+ using an oxidant such as O_2. During the oxidation, a fraction of the energy obtained from the oxidation of the nutrient is stored as ATP. The phosphoric acid anhydride bonds in ATP can then be hydrolyzed by water, releasing energy and forming ADP (adenosine diphosphate). It is through this sequence of reactions that energy from the oxidation of nutrients is made available to cells. Thus, ATP has a central role in metabolism: it is synthesized by the oxidation of nutrients, and its energy is then used by the cell to drive synthetic reactions and to perform work (Figure 18.20).

Note the pattern

In biochemical reactions, standard state conditions are defined as pH = 7, 1 M, and 1 atm.

Under standard conditions in aqueous solutions, all reactants are present in concentrations of 1 *M*. For H^+, this corresponds to a pH of zero, and very little biochemistry occurs at pH = 0. For biochemical reactions, chemists have therefore defined a new standard state in which the H^+ concentration is 1×10^{-7} *M* (pH 7) and all other reactants and products are present in their usual standard-state conditions (1 *M* or 1 atm). The free-energy change and corresponding equilibrium constant for a reaction under these new standard conditions are denoted by the addition of a prime sign (′) to the conventional symbol: $\Delta G^{\circ\prime}$ and K'. If protons do not participate in a biological reaction, then $\Delta G^{\circ\prime} = \Delta G^\circ$. Otherwise, the relationship between $\Delta G^{\circ\prime}$ and ΔG° is

$$\Delta G^{\circ'} = \Delta G^\circ + RT \ln(10^{-7})^n \qquad (18.51)$$

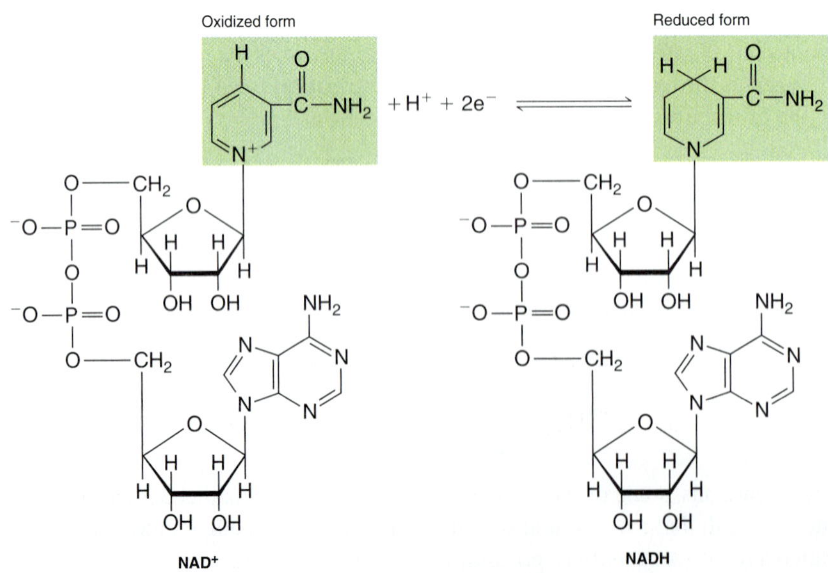

Figure 18.19 Nicotinamide adenine dinucleotide (NAD^+) and its reduced form (NADH). This electron carrier is used by biological systems to transfer electrons from one species to another. The oxidized form, NAD^+, is reduced to NADH by energy-rich nutrients such as glucose, and the reduced form, NADH, is in turn oxidized to NAD^+ by O_2 during respiration.

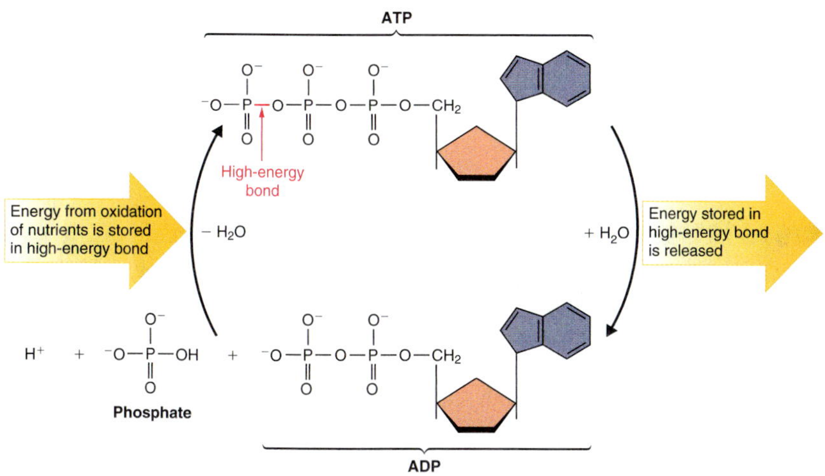

Figure 18.20 The ATP cycle. The high-energy phosphoric acid anhydride bond in ATP stores energy released during the oxidation of nutrients. Hydrolysis of the high-energy bond in ATP releases energy, forming adenosine diphosphate (ADP) and phosphate.

where $\Delta G^{\circ\prime}$ and ΔG° are in kilojoules per mole and n is the number of protons produced in the reaction. At 298 K, this simplifies to

$$\Delta G^{\circ'} = \Delta G^\circ - 39.96\,n \tag{18.52}$$

Thus, any reaction that involves the release of protons is thermodynamically more favorable at pH 7 than at pH 0.

The chemical equation that corresponds to the hydrolysis of ATP to ADP and phosphate is

$$\mathrm{ATP^{4-} + H_2O \rightleftharpoons ADP^{3-} + HPO_4{}^{2-} + H^+} \tag{18.53}$$

This reaction has a $\Delta G^{\circ\prime}$ of -34.54 kJ/mol, but under typical physiological (or biochemical) conditions, the actual value of $\Delta G'$ for the hydrolysis of ATP is about -50 kJ/mol. Organisms use this energy to drive reactions that are energetically uphill, thereby coupling the reactions to the hydrolysis of ATP. One example is found in the biochemical pathway of *glycolysis*, in which the 6-carbon sugar glucose ($C_6H_{12}O_6$) is split into two 3-carbon fragments that are then used as the fuel for the cell. Initially, a phosphate group is added to glucose to form a phosphate ester, glucose-6-phosphate (abbreviated glucose-6-P), in a reaction analogous to that of an alcohol and phosphoric acid:

$$\begin{array}{llll} \mathrm{glucose}(aq) & + \mathrm{HPO_4{}^{2-}}(aq) \rightleftharpoons & \mathrm{glucose\text{-}6\text{-}P^{2-}}(aq) & + \mathrm{H_2O}(l) \\ \mathrm{ROH} & + \mathbf{HO}\mathrm{PO_3{}^{2-}} \rightleftharpoons & \mathrm{ROPO_3{}^{2-}} & + \mathbf{H_2O} \end{array} \tag{18.54}$$

Due to its electrical charge, the phosphate ester is unable to escape from the cell by diffusing back through the membrane surrounding the cell, ensuring that it remains available for further reactions. For the reaction in Equation 18.54, ΔG° is 17.8 kJ/mol and K is 7.6×10^{-4}, indicating that the equilibrium lies far to the left. To force this reaction to occur as written, it is coupled to a thermodynamically favorable reaction, the hydrolysis of ATP to ADP:

$$\begin{array}{lll} \mathrm{glucose} + \cancel{\mathrm{HPO_4{}^{2-}}} \rightleftharpoons \mathrm{glucose\text{-}6\text{-}P^{2-}} + \cancel{\mathrm{H_2O}} & \Delta G^{\circ\prime} = 17.8\ \mathrm{kJ/mol} & K_1 = 7.6 \times 10^{-4} \\ \mathrm{ATP^{4-}} + \cancel{\mathrm{H_2O}} \rightleftharpoons \mathrm{ADP^{3-}} + \cancel{\mathrm{HPO_4{}^{2-}}} + \mathrm{H^+} & \Delta G^{\circ\prime} = -34.54\ \mathrm{kJ/mol} & K_2 = 2.2 \times 10^{5} \\ \hline \mathrm{glucose} + \mathrm{ATP^{4-}} \rightleftharpoons \mathrm{glucose\text{-}6\text{-}P^{2-}} + \mathrm{ADP^{3-}} + \mathrm{H^+} & \Delta G^{\circ\prime} = -16.7\ \mathrm{kJ/mol} & K_3 = 1.7 \times 10^{2} \end{array}$$

Thus, the formation of glucose-6-phosphate is thermodynamically spontaneous if ATP is used as the source of phosphate.

The formation of glucose-6-phosphate is only one of many examples of how cells use ATP to drive an otherwise nonspontaneous biochemical reaction. Under nonstandard physiological conditions, each ATP hydrolyzed actually results in approximately a 10^8 increase in the magnitude of the equilibrium constant, compared with the equilibrium constant of the reaction in the absence of ATP. Thus, a reaction in which two ATPs are converted to ADP increases K by about 10^{16}, three ATPs by 10^{24}, and so forth. Virtually any energetically unfavorable reaction or sequence of reactions can be made to occur spontaneously by coupling it to the hydrolysis of a sufficiently large number of ATP molecules.

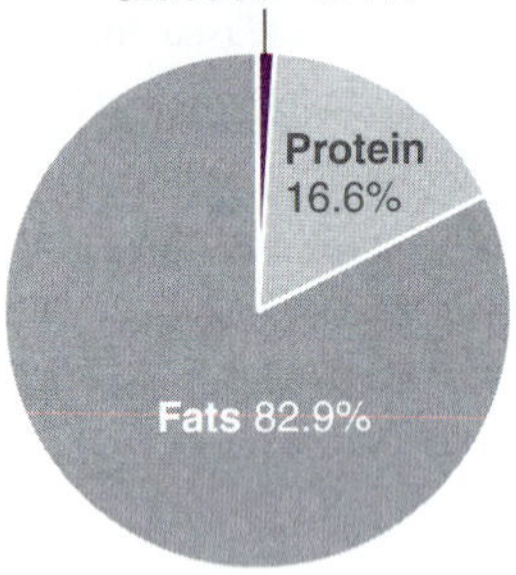

Figure 18.21 Percent distribution of forms of energy storage in adult humans. An average 70-kg adult stores about 6×10^5 kJ of energy in glucose, glycogen, protein, and fats. Fats are by far the most abundant and most efficient form for storing energy.

Energy Storage in Cells

Although all organisms use ATP as the *immediate* free-energy source in biochemical reactions, ATP is not an efficient form in which to store energy on a long-term basis. If the caloric intake of an average resting adult human were stored as ATP, two-thirds of the body weight would have to consist of ATP. Instead, a typical 70-kg adult human has a total of only about 50 g of both ATP and ADP, and, far from being used for long-term storage, each molecule of ATP is turned over about 860 times per day. The entire ATP supply would be exhausted in less than 2 minutes if it were not continuously regenerated.

How does the body store energy for the eventual production of ATP? Three primary means are as sugars, proteins, and fats. Combustion of sugars and proteins yields about 17 kJ of energy per gram, whereas combustion of fats yields more than twice as much energy per gram, about 39 kJ/g. Moreover, sugars and proteins are hydrophilic and contain about 2 g of water per gram of fuel even in very concentrated form. In contrast, fats are hydrophobic and can be stored in essentially anhydrous form. As a result, organisms can store about six times more energy per gram as fats than in any other form. A typical 70-kg adult human has about 170 kJ of energy in the form of glucose circulating in the blood, about 2600 kJ of energy stored in the muscles and liver as glycogen (a polymeric form of glucose), about 100,000 kJ stored in the form of protein (primarily muscle tissue), and almost 500,000 kJ in the form of fats (Figure 18.21). Thus, fats constitute by far the greatest energy reserve, while accounting for only about 12 kg of the 70-kg body mass. To store this amount of energy in the form of sugars would require a total body mass of about 144 kg, more than half of which would be sugar.

EXAMPLE 18.16

Glucose is one form in which the body stores energy. **(a)** Calculate $\Delta G^{\circ\prime}$ for the respiration of glucose to CO_2 and H_2O using these values of ΔG°_f: −910.4 kJ/mol for glucose, −394.4 kJ/mol for $CO_2(g)$, and −237.1 kJ/mol for $H_2O(l)$. **(b)** Assuming 50% efficiency in the conversion of the released energy to ATP, how many molecules of ATP can be synthesized by the combustion of one molecule of glucose? At 298.15 K $\Delta G^{\circ\prime}$ for the hydrolysis of ATP is −34.54 kJ/mol.

Given Balanced chemical equation (Equation 18.47), values of ΔG°_f, conversion efficiency, $\Delta G^{\circ\prime}$ for hydrolysis of ATP

Asked for $\Delta G^{\circ\prime}$ for the combustion reaction, molecules of ATP that can be synthesized

Strategy

Ⓐ Using the "products minus reactants" rule, calculate ΔG°_{rxn} for the respiration reaction.
Ⓑ Multiply the calculated value of ΔG°_{rxn} by the efficiency to obtain the number of kilojoules available for ATP synthesis. Then divide this value by $\Delta G^{\circ\prime}$ for the hydrolysis of ATP to find the maximum number of ATP molecules that can be synthesized.

Solution

(a) Ⓐ Protons are not released or consumed in the reaction, so $\Delta G^{\circ\prime} = \Delta G^\circ$. We begin by using the balanced equation in Equation 18.47:

$$C_6H_{12}O_6 + 6O_2 \longrightarrow 6CO_2 + 6H_2O$$

From the given values of ΔG°_f (remember that ΔG°_f is zero for an element such as O_2 in its standard state), we can calculate ΔG°_{rxn}:

$$\begin{aligned}\Delta G^\circ_{rxn} &= \Sigma m\Delta G^\circ_f(\text{products}) - \Sigma n\Delta G^\circ_f(\text{reactants})\\ &= [6(-394.4) + 6(-237.2)] - [(-1333) + 0]\\ &= -2457 \text{ kJ/mol of glucose}\end{aligned}$$

(b) If we assume that only 50% of the available energy is used, then about 1230 kJ/mol of glucose is available for ATP synthesis. The value of $\Delta G^{\circ\prime}$ for the hydrolysis of ATP under biochemical conditions is −34.54 kJ/mol, so in principle an organism could synthesize

$$\frac{1230 \text{ kJ/mol glucose}}{34.54 \text{ kJ/mol ATP}} = 35.6 \approx 36 \text{ ATP/glucose}$$

Most aerobic organisms actually synthesize about 32 molecules of ATP per molecule of glucose, for an efficiency of about 45%.

EXERCISE 18.16

Some bacteria synthesize methane using this redox reaction:

$$CO_2(g) + 4H_2(g) \longrightarrow CH_4(g) + 2H_2O(g)$$

(a) Calculate $\Delta G^{\circ\prime}$ for this reaction using values of ΔG°_f in Appendix A, and **(b)** calculate how many ATP molecules could be synthesized per mol of CO_2 reduced if the efficiency of the process were 100%.

Answer **(a)** −113.3 kJ/mol CO_2; **(b)** 3 ATP/mol CO_2

SUMMARY AND KEY TERMS

18.1 Thermodynamics and Work (p. 811)

Thermodynamics is the study of the interrelationships among heat, work, and the energy content of a system at equilibrium. The sum of the potential and the kinetic energy of all the components of a system is the **internal energy** (*E*) of the system, which is a **state function**. When the pressure or volume of a gas is changed, any mechanical work done is called *PV work*. Work done by the system on the surroundings is given a negative value, whereas work done on the system by the surroundings has a positive value.

18.2 The First Law of Thermodynamics (p. 814)

The **first law of thermodynamics** states that the energy of the universe is constant. The change in the internal energy of a system is the sum of the heat transferred and the work done. At constant pressure, the relationship between the heat flow q and the internal energy E of a system is its **enthalpy (H)**. The heat flow is equal to the change in the internal energy of the system plus the *PV* work done. When the volume of a system is constant, changes in its internal energy can be calculated by substituting the ideal gas law into the equation for ΔE.

18.3 The Second Law of Thermodynamics (p. 818)

A measure of the disorder of a system is its **entropy** (*S*), a state function whose value increases with an increase in the number of available microstates. A **reversible process** is one for which all intermediate states between extremes are equilibrium states; it can change direction at any time. In contrast, an **irreversible process** occurs in one direction only. The change in entropy of the system or the surroundings is the quantity of heat transferred divided by the temperature. The **second law of thermodynamics** states that in a reversible process, the entropy of the universe is constant, whereas in an irreversible process, such as the transfer of heat from a hot object to a cold object, the entropy of the universe increases.

18.4 Entropy Changes and the Third Law of Thermodynamics (p. 825)

The **third law of thermodynamics** states that the entropy of any perfectly ordered, crystalline substance at absolute zero is zero. Thus, at temperatures higher than absolute zero, entropy has a positive value, which allows us to measure the *absolute entropy* of a substance. Measurements of the heat capacity of a substance and the enthalpies of fusion or vaporization can be used to calculate the changes in entropy that accompany a physical change. The entropy of 1 mol of a substance at a standard temperature of 298 K is its **standard molar entropy** (S°). We can use the "products minus reactants" rule to calculate the standard entropy change ΔS° for a reaction using tabulated values of S° for the reactants and products.

18.5 Free Energy (p. 829)

We can predict whether a reaction will occur spontaneously by combining the entropy, enthalpy, and temperature of a system in a new state function called **Gibbs free energy** (*G*). The change in free energy, ΔG, is the difference between the heat released during a reversible or irreversible process and the heat released for the same process occurring in a reversible manner. If a system is at equilibrium, $\Delta G = 0$. If the process is spontaneous, $\Delta G < 0$. If the process is not spontaneous as written but is spontaneous in the reverse direction, $\Delta G > 0$. At constant temperature and pressure, ΔG is equal to the maximum amount of work a system can perform on the surroundings while undergoing a spontaneous change. The **standard free energy change** (ΔG°) is the change in free energy when one substance or set of substances in their standard states is converted to one or more other substances, also in their standard states. The **standard free energy of formation** (ΔG_f°), is the change in free energy that occurs when 1 mol of a substance in its standard state is formed from the elements in their standard states. Tabulated values of standard free energies of formation are used to calculate ΔG° for a reaction.

18.6 Spontaneity and Equilibrium (p. 837)

For a reversible process that does not involve external work, we can express the change in free energy in terms of volume, pressure, entropy, and temperature. If we assume ideal gas behavior, the ideal gas law allows us to express ΔG in terms of the partial pressures of the reactants and products, which gives us a relationship between ΔG and K_p, the equilibrium constant of a reaction involving gases, or *K*, the equilibrium constant expressed in terms of concentrations. If $\Delta G^\circ < 0$, then *K* or $K_p > 1$ and products are favored over reactants. If $\Delta G^\circ > 0$, then *K* or $K_p < 1$ and reactants are favored over products. If $\Delta G^\circ = 0$, then *K* or $K_p = 1$ and the system is at equilibrium. We can use the measured equilibrium constant *K* at one temperature and ΔH° to estimate the equilibrium constant for a reaction at any other temperature.

18.7 Comparing Thermodynamics and Kinetics (p. 842)

Thermodynamics is used to describe the overall properties, behavior, and equilibrium composition of a system; kinetics describes the rate at which a particular process will occur and the pathway by which it will occur. Whereas thermodynamics tells us what can occur during a reaction process, kinetics tells us what actually occurs on an atomic or molecular level. A reaction that is not thermodynamically spontaneous under standard conditions can often be made to occur spontaneously by varying reaction conditions, by using a different reaction to obtain the same product, by supplying external energy, such as electricity, or by coupling the unfavorable reaction to another reaction for which $\Delta G^\circ << 0$.

18.8 Thermodynamics and Life (p. 846)

A living cell is a system that is not in equilibrium with its surroundings; it requires a constant input of energy to maintain its nonequilibrium state. Cells maintain a low-entropy state by increasing the entropy of their surroundings. *Aerobic organisms* cannot survive in the absence of O_2, whereas *anaerobic organisms* can live only in the absence of O_2. Green plants and algae are *phototrophs*, which extract energy from the environment through a process called **photosynthesis**, the photochemical reduction of CO_2 to a reduced carbon compound. Other species, called *chemotrophs*, extract energy from chemical compounds. One of the main processes chemotrophs use to obtain energy is **respiration**, which is the reverse of photosynthesis. Alternatively, some chemotrophs obtain energy by **fermentation**, in which an organic compound is both the oxidant and the reductant. Intermediates used by organisms to shuttle electrons between the reductant and the oxidant include NAD^+ and NADH. Energy from the oxidation of nutrients is made available to cells through the synthesis of ATP, the energy currency of the cell. Its energy is used by the cell to synthesize substances through coupled reactions and to perform work. The body stores energy as sugars, protein, or fats before using it to produce ATP.

KEY EQUATIONS

Definition of *PV* work	$w = -P_{ext}\Delta V$	(18.4)
Internal energy change	$\Delta E_{sys} = q + w$	(18.6)
Enthalpy change	$\Delta H = \Delta E + \Delta(PV)$	(18.8)
Relationship between ΔH and ΔE for an ideal gas	$\Delta E = \Delta H - RT\Delta n$	(18.10)
Entropy change	$\Delta S = \frac{q_{rev}}{T}$	(18.12)
Temperature dependence of entropy at constant pressure	$\Delta S = nC_p \ln\frac{T_2}{T_1}$	(18.17)
Temperature dependence of entropy at constant volume	$\Delta S = nC_v \ln\frac{T_2}{T_1}$	(18.18)
Free-energy change	$\Delta G = \Delta H - T\Delta S$	(18.20)
Standard free-energy change	$\Delta G^\circ = \Delta H^\circ - T\Delta S^\circ$	(18.23)
Relationship between standard free-energy change and equilibrium constant	$\Delta G^\circ = -RT\ln K$	(18.35)
Temperature dependence of equilibrium constant	$\ln K = \frac{-\Delta H^\circ}{RT} + \frac{\Delta S^\circ}{R}$	(18.37)
Calculation of *K* at second temperature	$\ln\frac{K_2}{K_1} = \frac{\Delta H^\circ}{R}\left(\frac{1}{T_1} - \frac{1}{T_2}\right)$	(18.38)

QUESTIONS AND PROBLEMS

Questions and Problems with colored numbers have answers in the Appendix and complete solutions in the Student Solutions Manual.

CONCEPTUAL

18.1 Thermodynamics and Work

1. Thermodynamics focuses on the energetics of the reactants and products, and provides information about the composition of the reaction system at equilibrium. What information on reaction systems is not provided by thermodynamics?
2. Given a system in which a substance can produce either of two possible products, $A \longrightarrow B$ or $A \longrightarrow C$, which of the following can be predicted using chemical thermodynamics?
   (a) At equilibrium, the concentration of product C is greater than the concentration of product B.
   (b) Product C forms more quickly than product B.
   (c) The reaction $A \longrightarrow C$ is exothermic.
   (d) Low-energy intermediates are formed in the reaction $A \longrightarrow B$.
   (e) The reaction $A \longrightarrow C$ is spontaneous.
3. In what two ways can a closed system exchange energy with its surroundings? Are these two processes path dependent or path independent?
4. A microwave oven operates by providing enough energy to rotate water molecules, which in turn produces heat. Can the change in the internal energy of a cup of water heated in a microwave oven be described as a state function? Can the heat produced be described as a state function?

18.2 The First Law of Thermodynamics

5. Describe how a swinging pendulum that slows with time illustrates the first law of thermodynamics.
6. When air is pumped into a bicycle tire, the air is compressed. Assuming that the volume is constant, express the change in internal energy in terms of q and w.
7. What is the relationship between enthalpy and internal energy for a reaction that occurs at constant pressure?
8. An intrepid scientist placed an unknown salt in a small amount of water. All the salt dissolved in the water, and the temperature of the solution dropped several degrees.
   (a) What is the sign of the enthalpy change for this reaction?
   (b) Assuming the heat capacity of the solution is the same as that of pure water, how would the scientist calculate the molar enthalpy change?
   (c) Propose an explanation for the decrease in temperature.
9. For years, chemists and physicists focused on enthalpy changes as a way to measure the spontaneity of a reaction. What arguments would you use to convince them not to use this method?
10. What is the relationship between enthalpy and internal energy for a reaction that occurs at constant volume?
11. The reaction of methyl chloride with water produces methanol and HCl gas at room temperature, despite the fact that $\Delta H^\circ_{rxn} = 7.3$ kcal/mol. Using thermodynamic arguments, why does methanol form?
12. The *enthalpy of combustion* is defined thermodynamically as the enthalpy change for complete oxidation. The complete oxidation of hydrocarbons is represented by this general equation: hydro-carbon + $O_2(g) \longrightarrow CO_2(g) + H_2O(g)$. Enthalpies of combustion from reactions like this one can be measured experimentally with a high degree of precision. It has been found that the less stable the reactant, the more heat is evolved, so the more negative the value of ΔH_{comb}. In each of the following pairs of hydrocarbons, which member do you expect to have the greater (more negative) heat of combustion? (a) cyclopropane, cyclopentane; (b) butane, 2-methylpropane; (c) hexane, cyclohexane. Justify your answers.
13. Using structural argument, explain why cyclopropane has a positive ΔH°_f (12.7 kJ/mol), whereas cyclopentane has a negative ΔH°_f (−18.4 kJ/mol).
14. Using structural argument, explain why the trans isomer of 2-butene is more stable than the cis isomer. The enthalpies of formation of cis- and trans-2-butene are −7.1 kJ/mol and −11.4 kJ/mol, respectively.

18.3 The Second Law of Thermodynamics

15. A Russian space vehicle developed a leak, which resulted in an internal pressure drop from 1 atm to 0.85 atm. Is this an example of a reversible expansion? Has work been done?
16. Which member of each of pair do you expect to have a higher entropy, and why? (a) solid phenol, liquid phenol; (b) 1-butanol, butane; (c) cyclohexane, cyclohexanol; (d) 1 mol of N_2 mixed with 2 mol of O_2, 2 mol of NO_2; (e) 1 mol of O_2, 1 mol of O_3; (f) 1 mol of propane at 1 atm, 1 mol of propane at 2 atm
17. Which processes are reversible and which are not? (a) ice melting at 0°C; (b) salt crystallizing from a saline solution; (c) evaporation of a liquid in equilibrium with its vapor in a sealed flask; (d) a neutralization reaction
18. Determine whether each process is reversible or irreversible: (a) cooking spaghetti; (b) the reaction between sodium metal and water; (c) oxygen uptake by hemoglobin; (d) evaporation of water at its boiling point.
19. Explain why increasing the temperature of a gas increases its entropy. What effect does this have on the internal energy of the gas?
20. For a series of related compounds, does ΔS_{vap} increase or decrease with an increase in the strength of intermolecular interactions in the liquid state? Why?
21. Is the change in the enthalpy of reaction or the change in entropy of reaction more sensitive to changes in temperature? Explain your reasoning.
22. Solid potassium chloride has a highly ordered lattice structure. Do you expect ΔS_{soln} to be greater or less than zero? Why? What opposing factors must be considered in making your prediction?
23. Aniline ($C_6H_5NH_2$) is an oily liquid at 25°C that darkens on exposure to air and light. It is used in dying fabrics and in staining wood black. One gram of aniline dissolves in 28.6 mL of water, but aniline is completely miscible with ethanol. Do you expect ΔS_{soln} in H_2O to be greater than, less than, or equal to ΔS_{soln} in CH_3CH_2OH? Why?

18.4 Entropy Changes and the Third Law of Thermodynamics

24. Crystalline $MgCl_2$ has $S° = 89.63$ J/(mol·K), whereas aqueous $MgCl_2$ has $S° = -25.1$ J/(mol·K). Is this consistent with the third law of thermodynamics? Explain your answer.

25. Why is it possible to measure absolute entropies but not absolute enthalpies?

26. How many microstates are available to a system at absolute zero? How many are available to a substance in its liquid state?

27. Substance A has a higher heat capacity than substance B. Do you expect the absolute entropy of substance A to be lower than, similar to, or higher than that of substance B? Why? As the two substances are heated, for which substance do you predict the entropy to increase more rapidly?

18.5 Free Energy

28. How does each example illustrate the fact that no process is 100% efficient? (a) burning a log to stay warm; (b) the respiration of glucose to provide energy; (c) burning a candle to provide light

29. Neither the change in enthalpy nor the change in entropy is by itself sufficient to determine whether a reaction will occur spontaneously. Why?

30. If a system is at equilibrium, what must be the relationship between ΔH and ΔS?

31. The equilibrium $2AB \rightleftharpoons A_2B_2$ is exothermic in the forward direction. Which has the higher entropy: the product or the reactant? Why? Which is favored at high temperatures?

32. Is ΔG a state function that describes the system or the surroundings? Do its components, ΔH and ΔS, describe the system or the surroundings?

33. How can you use ΔG to determine the temperature of a phase transition, such as the boiling point of a liquid or the melting point of a solid?

34. Occasionally, an inventor claims to have invented a "perpetual motion" machine, which requires no additional input of energy once the machine has been put into motion. Using your knowledge of thermodynamics, how would you respond to such a claim? Justify your arguments.

35. Must the entropy of the universe increase in a spontaneous process? If not, why is there no process that is 100% efficient?

18.6 Spontaneity and Equilibrium

36. Do you expect products or reactants to dominate at equilibrium in a reaction for which $\Delta G°$ is equal to (a) 1.4 kJ/mol; (b) 105 kJ/mol; (c) −34 kJ/mol?

37. The change in free energy enables us to determine whether a reaction will proceed spontaneously. How is this related to the extent to which a reaction proceeds?

38. What happens to the free energy of the reaction $N_2(g) + 3F_2(g) \longrightarrow 2NF_3(g)$ if the pressure is increased while the temperature remains constant? If the temperature is increased at constant pressure? Why are these effects not so important for reactions that involve liquids and solids?

39. Compare the expressions for the relationship between the free energy of a reaction and its equilibrium constant where the reactants are gases versus liquids. What are the differences between these expressions?

18.7 Comparing Thermodynamics and Kinetics

40. You are in charge of finding conditions to make the reaction $A(l) + B(l) \longrightarrow C(l) + D(g)$ favorable because it is a critical step in the synthesis of your company's key product. You have calculated that $\Delta G°$ for the reaction is negative, yet the ratio of products to reactants is very small. What have you overlooked in your scheme? What can you do to drive the reaction to increase your product yield?

18.8 Thermodynamics and Life

41. The tricarboxylic acid (TCA) cycle in aerobic organisms is one of four pathways responsible for the stepwise oxidation of organic intermediates. The final reaction in the TCA cycle has $\Delta G° = 29.7$ kJ/mol, so it should not occur spontaneously. Suggest an explanation for why this reaction proceeds in the forward direction in living cells.

NUMERICAL

This section includes paired problems (marked by brackets) that require similar problem-solving skills.

18.1 Thermodynamics and Work

42. Calculate the work done in joules in each process:
(a) Compressing 12.8 L of hydrogen gas at an external pressure of 1.00 atm to 8.4 L at a constant temperature.
(b) Expanding 21.9 L of oxygen gas at an external pressure of 0.71 atm to 23.7 L at a constant temperature.

43. How much work in joules per mole is done when oxygen is compressed from a volume of 22.8 L and an external pressure of 1.20 atm to 12.0 L at a constant temperature? Was work done by the system or the surroundings?

44. Champagne is bottled at a CO_2 pressure of about 5 atm. What is the force on the cork if its cross-sectional area is 2.0 cm^2? How much work is done if a 2.0-g cork flies a distance of 8.2 ft straight up when the cork is popped? Was work done by the system or the surroundings?

45. One mole of water is converted to steam at 1.00 atm pressure and 100°C. Assuming ideal behavior, what is the change in volume when the water is converted from a liquid to a gas? If this transformation took place in a cylinder with a piston, how much work could be done by vaporizing the water at 1.00 atm? Is work done by the system or the surroundings?

46. Recall that force can be expressed as mass times acceleration ($F = ma$). Acceleration due to gravity on the earth's surface is 9.8 m/s^2.
(a) What is the gravitational force on a person who weighs 52 kg?
(b) How much work is done if the person leaps from a burning building out of a window 20 m above the ground?
(c) If the person lands on a large rescue cushion fitted with a pressure-release valve that maintains an internal pressure of 1.5 atm, how much air is forced out of the cushion?

47. Acceleration due to gravity on the earth's surface is 9.8 m/s^2. How much work is done by a 175-lb person going over Niagara Falls (approximately 520 ft high) in a barrel that weighs 145 lb?

48. One mole of an ideal gas is allowed to expand reversibly from an initial pressure of 1.0 atm to a final pressure of 0.62 atm at constant temperature against an external pressure of 1.0 atm. How much work has been done?

49. A gas is allowed to expand from a volume of 2.3 L to a volume of 5.8 L at constant temperature. During the process, 138 J of heat is transferred from the surroundings to the gas.

(a) How much work has been done if the gas expands against a vacuum?
(b) How much work has been done if the gas expands against a pressure of 1.3 atm?
(c) What is the change in the internal energy of the system?

18.2 The First Law of Thermodynamics

50. Zinc and HCl react according to the equation

$$Zn(s) + 2HCl(aq) \longrightarrow Zn^{2+}(aq) + 2Cl^-(aq) + H_2(g)$$

When 3.00 g of zinc metal is added to a dilute HCl solution at 1.00 atm and 25°C, and the above reaction is allowed to go to completion at constant pressure, 6.99 kJ of heat must be removed to return the final solution to its original temperature. What are the values of q and w, and what is the change in internal energy?

51. A block of CO_2 weighing 15 g evaporates in a 5.0-L container at 25°C. How much work has been done if the gas is allowed to expand against an external pressure of 0.98 atm under isothermal conditions? The enthalpy of sublimation of CO_2 is 25.1 kJ/mol. What is the change in internal energy (kJ/mol) for the sublimation of CO_2 under these conditions?

52. Acetylene torches, used industrially to cut and weld metals, reach flame temperatures as high as 3000°C. The combustion reaction is

$$2C_2H_2(g) + 5O_2(g) \longrightarrow 4CO_2(g) + 2H_2O(l) \quad \Delta H = -2599 \text{ kJ}$$

Calculate the amount of work done against a pressure of 1.0 atm when 4.0 mol of acetylene is allowed to react with 10 mol of O_2 at 1.0 atm at 20°C. What is the change in internal energy for the reaction?

53. When iron dissolves in 1.00 M aqueous HCl, the products are $FeCl_2(aq)$ and hydrogen gas. Calculate the work done if 30 g of Fe reacts with excess hydrochloric acid in a closed vessel at 20°C. How much work is done if the reaction takes place in an open vessel with an external pressure of 1 atm?

18.3 The Second Law of Thermodynamics

54. Liquid nitrogen, which has a boiling point of −195.79°C, is used as a coolant and as a preservative for biological tissues. Is the entropy of nitrogen higher or lower at −200°C than at −190°C? Explain your answer. Liquid nitrogen freezes to a white solid at −210.00°C, with an enthalpy of fusion of 0.71 kJ/mol. What is its entropy of fusion? Is freezing biological tissue in liquid nitrogen an example of a reversible or irreversible process?

55. Using the second law of thermodynamics, explain why heat flows from a hot body to a cold body but not from a cold body to a hot body.

56. One test of the spontaneity of a reaction is whether the entropy of the universe increases, $\Delta S_{univ} > 0$. Using an entropic argument, show that this reaction is spontaneous at 25°C:

$$4Fe(s) + 3O_2(g) \longrightarrow 2Fe_2O_3(s)$$

Why does the entropy of the universe increase in this reaction even though gaseous molecules, which have a high entropy, are consumed?

57. Calculate the missing data in the table.

Compound	ΔH_{fus}, kJ/mol	ΔS_{fus}, J/(mol · K)	Melting Point, °C
Acetic acid	11.7		16.6
CH_3CN	8.2	35.9	
CH_4	0.94		−182.5
CH_3OH		18.2	−97.7
Formic acid	12.7	45.1	

Based on this table, can you conclude that entropy is related to the nature of functional groups? Explain your reasoning.

58. Calculate the missing data in the table.

Compound	ΔH_{vap}, kJ/mol	ΔS_{vap}, J/(mol · K)	Melting Point, °C
Hexanoic acid	71.1		105.7
Hexane	28.9	85.5	
Formic acid		60.7	100.8
1-Hexanol	44.5		157.5

The text states that the magnitude of ΔS_{vap} tends to be similar for a wide variety of compounds. Based on the values in the table, do you agree?

18.4 Entropy Changes and the Third Law of Thermodynamics

59. Phase transitions must be considered when calculating entropy changes. Why? What is the final temperature of the water when 5.20 g of ice at 0.0°C is added to 250 mL of water in an insulated thermos at 30.0°C? The value of ΔH_{fus} for water is 6.01 kJ/mol, and the heat capacity of liquid water is 75.3 J/(mol·°C). What is the entropy change for this process?

60. Calculate the change in both enthalpy and entropy when a 3.0-g block of ice melts at 0.0°C [$\Delta H_{fus}(H_2O)$ = 6.01 kJ/mol]. For the same block of ice, calculate the entropy change for the system when the ice is warmed from 0.0°C to 25°C. The heat capacity of liquid water over this temperature range is 75.3 J/(mol·°C).

61. Use the data in Table 18.1 and Appendix A to calculate $\Delta S°$ for each reaction:
(a) $H_2(g) + \frac{1}{2}O_2(g) \longrightarrow H_2O(l)$
(b) $CH_3OH(l) + HCl(g) \longrightarrow CH_3Cl(g) + H_2O(l)$
(c) $H_2(g) + Br_2(l) \longrightarrow 2HBr(g)$
(d) $Zn(s) + 2HCl(aq) \longrightarrow ZnCl_2(aq) + H_2(g)$

62. Calculate the entropy change (J/K) when 4.35 g of liquid bromine is heated from 30.0°C to 50.0°C if the molar heat capacity of liquid bromine, C_p, is 75.1 kJ/(mol·K).

63. Calculate the molar heat capacity of titanium, C_p[kJ/(mol·K)], if the change in entropy when a 6.00-g sample of $TiCl_4(l)$ is heated from 25.0°C to 40.0°C is 0.153 J/K.

64. When a 1.00-g sample of lead is heated from 298.2 K to its melting temperature of 600.5 K, the change in entropy is 0.130 J/(mol·K). Under standard conditions, the enthalpy of fusion of lead is 4.98 kJ/mol. Determine the molar heat capacity of lead over this temperature range.

65. Phosphorus oxychloride ($POCl_3$) is a chlorinating agent that is frequently used in organic chemistry to replace oxygen with chlorine. Given the following information, does $POCl_3$ spontaneously convert from a liquid to a gas at 110°C? $\Delta S_{vap} =$ 93.08 J/(mol · K), $\Delta H_{vap} = 35.2$ kJ/mol. Does it spontaneously crystallize at 0.0°C if $\Delta H_{fus} = 34.3$ kJ/mol and $\Delta S_{fus} = 125$ J/K? Using the information provided, calculate the melting point of $POCl_3$.

66. A useful reagent for the fluorination of alcohols, carboxylic acids, and carbonyl compounds is selenium tetrafluoride (SeF_4). One must be careful when using this compound, however, because it is known to attack glass (such as the glass of a reaction vessel).
(a) Is SeF_4 a liquid or a gas at 100°C given that $\Delta H_{vap} =$ 46.9 kJ/mol and $\Delta S_{vap} = 124$ J/K?
(b) Determine the boiling point of SeF_4.
(c) Would you use SeF_4 for a solution reaction at 0°C if $\Delta H_{fus} = 46$ kJ/mol and $\Delta S_{fus} = 178$ J/(mol · K)?

18.5 Free Energy

67. Use the tables in the text to determine whether the following reactions are spontaneous under standard conditions. If the reaction is not spontaneous, write the corresponding spontaneous reaction.
(a) $H_2(g) + \frac{1}{2}O_2(g) \longrightarrow H_2O(l)$
(b) $2H_2(g) + C_2H_2(g) \longrightarrow C_2H_6(g)$
(c) $(CH_3)_2O(g) + H_2O(g) \longrightarrow 2CH_3OH(l)$
(d) $CH_4(g) + H_2O(g) \longrightarrow CO(g) + 3H_2(g)$

68. Use the tables in the text to determine whether the following reactions are spontaneous under standard conditions. If the reaction is not spontaneous, write the corresponding spontaneous reaction.
(a) $K_2O_2(s) \longrightarrow 2K(s) + O_2(g)$
(b) $PbCO_3(s) \longrightarrow PbO(s) + CO_2(g)$
(c) $P_4(s) + 6H_2(g) \longrightarrow 4PH_3(g)$
(d) $2AgCl(s) + H_2S(g) \longrightarrow Ag_2S(s) + 2HCl(g)$

69. Nitrogen fixation is the process by which nitrogen in the atmosphere is reduced to NH_3 for use by organisms. Several reactions are associated with this process; three are listed in the table. Which of these are spontaneous at 25°C? If a reaction is not spontaneous, at what temperature does it become spontaneous?

Reaction	ΔH°_{298}, kcal/mol	ΔS°_{298}, cal/(deg · mol)
$\frac{1}{2}N_2 + O_2 \longrightarrow NO_2$	8.0	−14.4
$\frac{1}{2}N_2 + \frac{1}{2}O_2 \longrightarrow NO$	21.6	2.9
$\frac{1}{2}N_2 + \frac{3}{2}H_2 \longrightarrow NH_3$	−11.0	−23.7

70. A student was asked to propose three reactions for the oxidation of carbon or a carbon compound to CO or CO_2. The reactions are listed in the table. Are any of these reactions spontaneous at 25°C? If a reaction does not occur spontaneously at 25°C, at what temperature does it become spontaneous?

Reaction	ΔH°_{298}, kcal/mol	ΔS°_{298}, cal/(deg · mol)
$C(s) + H_2O(g) \longrightarrow CO(g) + H_2(g)$	42	32
$CO(g) + H_2O(g) \longrightarrow CO_2(g) + H_2(g)$	−9.8	−10.1
$CH_4(g) + H_2O(g) \longrightarrow CO(g) + 3H_2(g)$	49.3	51.3

71. Tungsten trioxide (WO_3) is a dense yellow powder that, because of its bright color, is used as a pigment in oil paints and water colors (although cadmium yellow is more commonly used in artists' paints). Tungsten metal can be isolated by the reaction of WO_3 with H_2 at 1100°C according to the equation $WO_3(s) + 3H_2(g) \longrightarrow W(s) + 3H_2O(g)$. What is the lowest temperature at which the reaction occurs spontaneously? $\Delta H^\circ = 27.4$ kJ/mol and $\Delta S^\circ = 29.8$ J/K.

72. Sulfur trioxide (SO_3) is produced in large quantities in the industrial synthesis of sulfuric acid. Sulfur dioxide is converted to sulfur trioxide by reaction with oxygen gas.
(a) Write a balanced equation for the reaction of SO_2 with $O_2(g)$ and determine its ΔG°.
(b) What is the value of the equilibrium constant at 600°C.
(c) If you had to rely on the equilibrium concentrations alone, would you obtain a higher yield of product at 400°C or at 600°C?

73. Calculate ΔG° for the general reaction $MCO_3(s) \longrightarrow MO(s) + CO_2(g)$ at 25°C, where M is Mg or Ba. At what temperature does each of these reactions become spontaneous?

Compound	ΔH°_f, kJ/mol	S°, J/(mol · K)
MCO_3		
Mg	−1111	65.85
Ba	−1213.0	112.1
MO		
Mg	−601.6	27.0
Ba	−548.0	72.1
CO_2	−393.5	213.8

74. The reaction of aqueous solutions of barium nitrate with sodium iodide is described by the equation

$$Ba(NO_3)_2(aq) + 2NaI(aq) \longrightarrow BaI_2(aq) + 2NaNO_3(aq)$$

You are interested in determining the absolute entropy of BaI_2, but that information is not listed in your tables. However, you have been able to obtain the following information:

	$Ba(NO_3)_2$	NaI	BaI_2	$NaNO_3$
ΔH°_f, kJ/mol	−952.36	−295.31	−605.4	−447.5
S°, J/(mol · K)	302.5	170.3		205.4

You know that ΔG° for the reaction at 25°C is 22.64 kJ/mol. What is ΔH° for this reaction? What is S° for BaI_2?

18.6 Spontaneity and Equilibrium

75. Carbon monoxide, a toxic product from the incomplete combustion of fossil fuels, reacts with water to form CO_2 and H_2, as shown in the equation $CO(g) + H_2O(g) \rightleftharpoons CO_2(g) + H_2(g)$, for which $\Delta H^\circ = -41.0$ KJ/mol and $\Delta S^\circ = -42.3$ J cal/(mol · K) at 25°C and 1 atm.
(a) What is ΔG° for this reaction?
(b) What is ΔG if the gases have the following partial pressures: $P_{CO} = 1.3$ atm, $P_{H_2O} = 0.8$ atm, $P_{CO_2} = 2.0$ atm, and $P_{H_2} = 1.3$ atm?
(c) What is ΔG if the temperature is increased to 150°C assuming no change in pressure?

76. Methane and water react to form CO and H_2 according to the equation $CH_4(g) + H_2O(g) \rightleftharpoons CO(g) + 3H_2(g)$.
(a) What is the standard free energy change for this reaction?
(b) What is K_p for this reaction under standard conditions?
(c) What is the CO pressure if 1.3 atm of methane reacts with 0.8 atm of water, producing 1.8 atm of hydrogen gas?
(d) What is the hydrogen gas pressure if 2.0 atm of methane is allowed to react with 1.1 atm of water?
(e) At what temperature does the reaction become spontaneous?

77. Calculate the equilibrium constant at 25°C for each equilibrium reaction, and comment on the extent of the reaction:
(a) $CCl_4(g) + 6H_2O(l) \rightleftharpoons CO_2(g) + 4HCl(aq)$; $\Delta G° = -377$ kJ/mol
(b) $Xe(g) + 2F_2(g) \rightleftharpoons XeF_4(s)$; $\Delta H° = -66.3$ kJ/mol, $\Delta S° = -102.3$ J/(mol · K)
(c) $PCl_3(g) + S \rightleftharpoons PSCl_3(l)$; $\Delta G_f°$: $PCl_3 = -272.4$ kJ/mol, $\Delta G_f°$ $PSCl_3 = -363.2$ kJ/mol

78. Calculate the equilibrium constant at 25°C for each equilibrium reaction, and comment on the extent of the reaction:
(a) $2KClO_3(s) \rightleftharpoons 2KCl(s) + 3O_2(g)$; $\Delta G° = -225.8$ kJ/mol
(b) $CoCl_2(s) + 6H_2O(g) \rightleftharpoons CoCl_2{\cdot}6H_2O(s)$; $\Delta H°_{rxn} = -352$ kJ/mol, $\Delta S°_{rxn} = -899$ J/(mol · K)
(c) $2PCl_3(g) + O_2(g) \rightleftharpoons 2POCl_3(g)$; $\Delta G_f°$: $PCl_3 = -272.4$ kJ/mol, $\Delta G_f°$ $POCl_3 = -558.5$ kJ/mol

79. The gas-phase decomposition of N_2O_4 to NO_2 is an equilibrium reaction with $K_p = 4.66 \times 10^{-3}$. Calculate the standard free energy change for the equilibrium reaction between N_2O_4 and NO_2.

80. The standard free energy for the dissolution $K_4Fe(CN)_6 \cdot H_2O(s) \rightleftharpoons 4K^+(aq) + Fe(CN)_6^{4-}(aq) + H_2O(l)$ is 26.1 kJ/mol. What is the equilibrium constant for this process at 25°C?

81. Ammonia reacts with water in liquid ammonia solution (*amm*) according to the equation $NH_3(g) + H_2O(amm) \rightleftharpoons NH_4^+(amm) + OH^-(amm)$. The change in enthalpy for this reaction is 21 kJ/mol, and $\Delta S° = -303$ J/(mol · K). What is the equilibrium constant for the reaction at the boiling point of liquid ammonia (−31°C)?

82. At 25°C, a saturated solution of barium carbonate is found to have a concentration of $[Ba^{2+}] = [CO_3^{2-}] = 5.08 \times 10^{-5}$ *M*. Determine $\Delta G°$ for the dissolution of $BaCO_3$.

83. Lead phosphates are believed to play a major role in controlling the overall solubility of lead in acidic soils. One of the dissolution reactions is $Pb_3(PO_4)_2(s) + 4H^+(aq) \rightleftharpoons 3Pb^{2+}(aq) + 2H_2PO_4^-(aq)$, for which $\log K = -1.80$. What is $\Delta G°$ for this reaction?

84. The conversion of butane to 2-methylpropane is an equilibrium process with $\Delta H° = -2.05$ kcal/mol and $\Delta G° = -0.89$ kcal/mol.
(a) What is the change in entropy for this conversion?
(b) Based on structural arguments, are the sign and magnitude of the entropy change what you would expect? Why?
(c) What is the equilibrium constant for this reaction?

85. The reaction of $CaCO_3(s)$ to produce $CaO(s)$ and $CO_2(g)$ has an equilibrium constant at 25°C of 2×10^{-23}. Values of $\Delta H_f°$ are: $CaCO_3$, −1207.6 kJ/mol; CaO, −634.9 kJ/mol; CO_2, −393.5 kJ/mol.
(a) What is $\Delta G°$ for this reaction?
(b) What is the equilibrium constant at 900°C?
(c) What is the partial pressure of $CO_2(g)$ in equilibrium with CaO and $CaCO_3$ at this temperature?
(d) Are reactants or products favored at the lower temperature? At the higher temperature?

86. In acidic soils, dissolved Al^{3+} undergoes a complex formation reaction with SO_4^{2-} to form $[AlSO_4^+]$. The equilibrium constant at 25°C for the reaction $Al^{3+}(aq) + SO_4^{2-}(aq) \rightleftharpoons AlSO_4^+(aq)$ is 1585.
(a) What is $\Delta G°$ for this reaction?
(b) How does this value compare with $\Delta G°$ for the reaction $Al^{3+}(aq) + F^-(aq) \rightleftharpoons AlF^{2+}(aq)$, for which $K = 10^7$ at 25°C?
(c) Which is the better ligand to use to trap Al^{3+} from the soil?

APPLICATIONS

87. Electric utilities have been exploring thermal energy storage as a potentially attractive energy-storage solution for peak use. Thermal energy is extracted as steam and stored in hot rock, oil, or hot water for later conversion to electricity via heat exchangers. Which steps involve heat transfer? Which involve work done?

88. During World War II, German scientists developed the first rocket-powered airplane to be flown in combat, the *Messerschmitt 163 Komet.* The "Comet" was powered by the reaction of liquid hydrogen peroxide (H_2O_2) and hydrazine (N_2H_4) according to the reaction

$$2H_2O_2(l) + N_2H_4(l) \longrightarrow N_2(g) + 4H_2O(g)$$

(a) Determine the standard molar enthalpy of reaction.
(b) What amount of work can be done by reacting 1.00 kg of hydrazine with a stoichiometric amount of hydrogen peroxide at 25°C at a constant pressure of 1.00 atm?
(c) What is the change in internal energy under these conditions?

89. During the 1950s, pentaborane-9 was tested as a potential rocket fuel. However, the idea was abandoned when it was discovered that B_2O_3, the product of the reaction of pentaborane-9 with O_2, was an abrasive that destroyed rocket nozzles. The reaction is represented by the equation

$$2B_5H_9(l) + 12O_2(g) \longrightarrow 5B_2O_3(s) + 9H_2O(g)$$

(a) Determine the standard molar enthalpy of reaction.
(b) How much work is done against a pressure of 1.0 atm when 50 lb of fuel is allowed to react with sufficient oxygen to cause the reaction to go to completion at 25°C?
(c) What is ΔE for the reaction?

90. Polar explorers must be particularly careful to keep their clothes from becoming damp because the resulting heat loss could be fatal. If a polar explorer's clothes absorbed 1.0 kg of water and the clothes dried from the polar wind, what would be the heat loss? $\Delta H_{vap}(H_2O) = 44$ kJ/mol. How much glucose must be consumed to make up for this heat loss to prevent death? $\Delta H_{comb}(\text{glucose}) = -802$ kJ/mol.

91. Propane gas is generally preferred to kerosene as a fuel for stoves in the boating industry because kerosene stoves require more maintenance. Propane, however, is much more flammable than kerosene and imposes an added risk because it is denser than air and can collect in the bottom of a boat and ignite. The complete combustion of propane produces CO_2 gas and H_2O vapor and has a value of $\Delta H = -2220$ kJ/mol at 25°C. What is ΔE?

92. The propane in Problem 93 can be produced from the hydrogenation of propene according to the reaction $C_3H_6(g) + H_2(g) \longrightarrow C_3H_8(g)$; $\Delta H = -124$ kJ/mol. Given that the reaction $H_2(g) + \frac{1}{2}O_2(g) \longrightarrow H_2O(g)$ has $\Delta H = -285.8$ kJ/mol, what is (a) the standard enthalpy of combustion of propene, and (b) the value of ΔE for the combustion of propene?

93. The anaerobic conversion of sucrose, a sweetening agent, to lactic acid, which is associated with sour milk, can be described by the equation

$$\underset{\text{Sucrose}}{C_{12}H_{22}O_{11}} + H_2O \longrightarrow \underset{\text{Lactic acid}}{4CH_3CH(OH)CO_2H}$$

The combustion of sucrose, however, occurs as follows:

$$C_{12}H_{22}O_{11}(s) + 12O_2(g) \longrightarrow 12CO_2(g) + 11H_2O(l)$$

(a) Which reaction is thermodynamically more favorable: anaerobic conversion of sucrose to lactic acid, or aerobic oxidation to CO_2 and H_2O? Values of ΔH_f are: lactic acid, −694.1 kJ/mol; sucrose, −222 kJ/mol; CO_2, −393.5 kJ/mol; and H_2O, −285.8 kJ/mol.

(b) What is ΔE for the combustion of 12.0 g of sucrose at normal body temperature (37°C)?

94. Phosphorus exists as several allotropes, the most common being red, black, and white phosphorus. White phosphorus consists of tetrahedral P_4 molecules and melts at 44.15°C; it is converted to red phosphorus by heating at 400°C for several hours. The chemical differences between red and white phosphorus are considerable: white phosphorus burns in air, whereas red phosphorus is stable; white phosphorus is soluble in organic compounds, whereas red phosphorus is not; white phosphorus melts at 44.15°C, whereas red phosphorus melts at 597°C. If the enthalpy of fusion of white phosphorus is 0.659 kJ/mol. What is its ΔS? Black phosphorus is even less reactive than red. Based on this information, which allotrope would you predict to have the highest entropy? The lowest? Why?

95. Ruby and sapphire have a common mineral name, corundum (Al_2O_3). Although they are crystalline versions of the same compound, the nature of the imperfections determines the identity of the gem. Outline a method for measuring and comparing the entropy of a ruby with the entropy of a sapphire. How would you expect the entropies to compare with the entropy of a perfect corundum crystal?

96. Tin has two crystalline forms, α and β, represented in the equilibrium equation

$$\underset{\text{Gray}}{\alpha\text{-tin}} \underset{}{\overset{18°C}{\rightleftharpoons}} \underset{\text{White}}{\beta\text{-tin}} \overset{232°C}{\rightleftharpoons} Sn(l)$$

The earliest known tin artifacts were discovered in Egyptian tombs of the 18th dynasty (1580–1350 B.C.), although archeologists are surprised that so few tin objects exist from earlier eras. It has been suggested that many early tin objects were either oxidized to a mixture of stannous and stannic oxides or transformed to powdery, gray tin. Sketch a thermodynamic cycle similar to Figure 18.15b to show the conversion of liquid tin to gray tin. Then calculate the change in entropy that accompanies the conversion of $Sn(l)$ to α-Sn using the following data: C_p(white) = 26.99, C_p(gray) = 25.77 J/(mol · K), $\Delta H_{fus} = 7.0$ kJ/mol, $\Delta H_{\beta \longrightarrow \alpha} =$ −2.2 kJ/mol.

97. The reaction of SO_2 with O_2 to produce SO_3 has great industrial significance because SO_3 is converted to H_2SO_4 by reaction with water. Unfortunately, the reaction is also environmentally important because SO_3 from industrial smokestacks is a primary source of acid rain. ΔH for the reaction of SO_2 with O_2 to form SO_3 is −23.49 kJ/mol, and ΔS is −22.66 J/K. Does this reaction occur spontaneously at 25°C? Does it occur spontaneously at 800°C assuming no change in ΔH and ΔS? Why is this reaction usually carried out at elevated temperatures?

98. Pollutants from industrial societies pose health risks to individuals from exposure to metals such as lead, mercury, and cadmium. The biological effects of a toxic metal may be reduced by removing it from the system using a chelating agent, which binds to the metal and forms a complex that is eliminated from the system without causing more damage. In designing a suitable chelating agent, one must be careful, however, because some chelating agents form metal complexes that are more toxic than the metal itself. Both methylamine (CH_3NH_2) and ethylenediamine ($H_2NCH_2CH_2NH_2$, abbreviated en) are commonly used to treat heavy metal poisoning. In the case of cadmium, the reactions are as follows:

$$Cd^{2+} + 2CH_3NH_2 \longrightarrow Cd(CH_3NH_2)_2^{2+}$$
$$\Delta H = -7.03 \text{ kcal/mol}, \Delta S = -1.58 \text{ cal/(mol · K)}$$
$$Cd^{2+} + en \longrightarrow Cd(en)^{2+}$$
$$\Delta H = -7.02 \text{ kcal/mol}, \Delta S = 3.15 \text{ cal/(mol · K)}$$

Based strictly on thermodynamic arguments, which would you choose to administer to a patient suffering from cadmium toxicity, and why? Assume a body temperature of 37°C.

99. Explosive reactions often have a large negative enthalpy change and a large positive entropy change, but the reaction must also be kinetically favorable. For example, the following equation represents the reaction between hydrazine, a rocket propellant, and the oxidizer dinitrogen tetroxide:

$$2N_2H_4(l) + N_2O_4(l) \longrightarrow 4H_2O(g) + 3N_2(g)$$
$$\Delta H° = -249 \text{ kJ/mol}, \Delta S° = 218 \text{ J/(mol · K)}$$

(a) How much free energy is produced from this reaction at 25°C?

(b) Is the reaction thermodynamically favorable?

(c) What is K?

(d) This reaction requires thermal ignition. Why?

100. Cesium, a silvery-white metal used in the manufacture of vacuum tubes, is produced industrially by the reaction of CsCl with CaC_2:

$$2CsCl(l) + CaC_2(s) \longrightarrow CaCl_2(l) + 2C(s) + 2Cs(g)$$

Compare the free energy produced from this reaction at 25°C and at 1227°C, the temperature at which it is normally run, given these values: $\Delta H°_{298\text{ K}} = 32.0$ kJ/mol, $\Delta S°_{298\text{ K}} = 8.0$ J/(mol·K); $\Delta H°_{1500\text{ K}} = -0.6$ kJ/mol, $\Delta S°_{1500\text{ K}} = 3.6$ J/(mol·K).

(a) If you wanted to minimize energy costs in your production facility by reducing the temperature of the reaction, what is the lowest temperature at which products are favored over reactants (assuming the reaction is kinetically favorable at the lower temperature)? Assume ΔH and ΔS do not vary with temperature.

(b) What is the ratio $K_{1500\ K}/K_{298\ K}$?

101. Dessicants (drying agents) can often be regenerated by heating, although it is generally not economically worthwhile to do so. A dessicant that is commonly regenerated is $CaSO_4 \cdot 2H_2O$:

$$CaSO_4 \cdot 2H_2O(s) \longrightarrow CaSO_4(s) + 2H_2O(g)$$
$$\Delta H^\circ_{298\ K} = 25.1 \text{ kJ/mol},\ \Delta S^\circ_{298\ K} = 69.3 \text{ J/(mol} \cdot \text{K)}$$

Regeneration is carried out at 250°C.

(a) What is ΔG° for this reaction?

(b) What is the equilibrium constant at 25°C?

(c) What is the ratio $K_{250°C}/K_{25°C}$?

(d) What is the equilibrium constant at 250°C?

(e) Is regeneration of $CaSO_4 \cdot 2H_2O$ an enthalpy- or entropy-driven process? Explain your answer.

102. Nitrogen triiodide ($NI_3 \cdot NH_3$) is a simple explosive that, when detonated, produces N_2 and I_2. It can be painted on surfaces when wet, but it is shock sensitive when dry (even touching it with a feather can cause an explosion!). Do you expect ΔG for the explosion reaction to be positive or negative? Why doesn't NI_3 explode spontaneously?

103. Adenosine triphosphate (ATP) contains high-energy phosphate bonds that are used in energy metabolism, coupling energy-yielding and energy-requiring processes. Cleaving a phosphate link by hydrolysis (ATP hydrolysis) can be described by the reaction ATP + $H_2O \rightleftharpoons$ ADP + P_i + H^+, where P_i symbolizes phosphate. Glycerol and ATP react to form glycerol-3-phosphate, ADP, and H^+, with an overall $K = 6.61 \times 10^5$ at 37°C. The reaction of glycerol with phosphoric acid to form glycerol-3-phosphate and water has an equilibrium constant of 2.82×10^{-2}. What is the equilibrium constant for ATP hydrolysis? How much free energy is released from the hydrolysis of ATP?

104. Consider the biological reduction of molecular nitrogen, for which the following is the minimal reaction stoichiometry under optimal conditions (P_i = phosphate):

$$8H^+ + 8e^- + N_2 + 16ATP \longrightarrow H_2 + 2NH_3 + 16ADP + 16P_i$$

(a) What is the approximate ratio of K_{eq} for the reaction shown above and for the same reaction in the absence of ATP?

(b) Given the fact that at pH 7 both the reaction of protons and electrons to give H_2 and the reaction of H_2 with N_2 to give ammonia are thermodynamically spontaneous (that is, $K \gg 1$), suggest a reason that nitrogen-fixing bacteria use such a large energy input to drive a reaction that is already spontaneous.

19 Electrochemistry

LEARNING OBJECTIVES

- To distinguish between galvanic and electrolytic cells
- To predict spontaneous reactions using redox potentials
- To balance redox reactions using half-reactions
- To understand the relationship between cell potential and the equilibrium constant
- To be able to measure solution concentrations using cell potentials
- To describe how commercial galvanic cells operate
- To describe the process of corrosion
- To understand electrolysis and to be able to describe it quantitatively

CHAPTER OUTLINE

In oxidation–reduction (redox) reactions, electrons are transferred from one species (the reductant) to another (the oxidant). This transfer of electrons provides a means for converting chemical energy to electrical energy, or vice versa. The study of the relationship between electricity and chemical reactions is called **electrochemistry**, an area of chemistry we introduced in Chapters 4 and 5. In this chapter, we describe electrochemical reactions in more depth and explore some of their applications.

In the first three sections, we review redox reactions; describe how they can be used to generate an electrical potential, or *voltage*; and discuss factors that affect the magnitude of the potential. We then explore the relationships among the electrical potential, the change in free energy, and the equilibrium constant for a redox reaction, which are all measures of the thermodynamic driving force for a reaction. Finally, we examine two kinds of applications of electrochemical

A view from the top of the Statue of Liberty, showing the green patina coating the statue. The patina is formed by corrosion of the copper skin of the statue, which forms a thin layer of an insoluble compound that contains copper(II), sulfate, and hydroxide ions.

principles: those in which a spontaneous reaction is used to provide electricity, and those in which electrical energy is used to drive a thermodynamically nonspontaneous reaction. By the end of this chapter, you will understand why different kinds of batteries are used in cars, flashlights, cameras, and portable computers; how rechargeable batteries operate; and why corrosion occurs and how to slow, if not prevent, it. You will also discover how metal objects can be plated with silver or chromium for protection; how silver polish removes tarnish; and how to calculate the amount of electricity needed to produce aluminum, chlorine, copper, and sodium on an industrial scale.

19.1 Describing Electrochemical Cells

In any electrochemical process, electrons flow from one chemical substance to another, driven by an oxidation–reduction (redox) reaction. As we described in Chapter 3, a redox reaction occurs when electrons are transferred from a substance that is oxidized to one that is being reduced. The **reductant** is the substance that loses electrons and is oxidized in the process; the **oxidant** is the species that gains electrons and is reduced in the process. The associated potential energy is determined by the potential difference between the valence electrons in atoms of different elements (see Section 7.3).

Because it is impossible to have a reduction without an oxidation, and vice versa, a redox reaction can be described as two **half-reactions**, one representing the oxidation process and one the reduction process. For the reaction of zinc with bromine, the half-reactions are

$$\text{Reduction half-reaction:} \quad Br_2(aq) + 2e^- \longrightarrow 2Br^-(aq) \qquad (19.1a)$$

$$\text{Oxidation half-reaction:} \quad Zn(s) \longrightarrow Zn^{2+}(aq) + 2e^- \qquad (19.1b)$$

Notice that each reaction is written to show what is actually occurring in the system. Adding the two half-reactions gives the overall chemical reaction:

$$\text{Overall reaction:} \quad \underset{\text{Reductant}}{Zn(s)} + \underset{\text{Oxidant}}{Br_2(aq)} \longrightarrow Zn^{2+}(aq) + 2Br^-(aq) \qquad (19.1c)$$

A redox reaction is balanced when *the number of electrons lost by the reductant is equal to the number of electrons gained by the oxidant*. Hence, like any balanced chemical equation, the overall process is electrically neutral; that is, the net charge is the same on both sides of the equation.

Note the pattern

In any redox reaction, the number of electrons lost by the reductant is equal to the number of electrons gained by the oxidant.

In most of our discussions of chemical reactions, we have assumed that the reactants are in intimate physical contact with one another. Acid–base reactions, for example, are usually carried out with the acid and the base dispersed in a single phase, such as a liquid solution. With redox reactions, however, it is possible to physically separate the oxidation and reduction half-reactions in space, as long as there is a complete circuit, including an external electrical connection, such as a wire, between the two half-reactions. As the reaction progresses, the electrons flow from reductant to oxidant over this electrical connection, producing an electric current that can be used to do work. An apparatus that is used to generate electricity from a spontaneous redox reaction or, conversely, that uses electricity to drive a nonspontaneous redox reaction is called an **electrochemical cell**.

There are two types of electrochemical cells: galvanic cells and electrolytic cells. A **galvanic cell*** uses the energy released during a spontaneous redox reaction ($\Delta G < 0$) to *generate* electricity. This type of electrochemical cell is often called a **voltaic cell** after its inventor, the Italian physicist Alessandro Volta (1745–1827). In contrast, an **electrolytic cell** *consumes* electrical energy from an external source, using it to cause a

* Named for the Italian physicist and physician Luigi Galvani (1737–1798), who observed that dissected frog leg muscles twitched when a small electric shock was applied, demonstrating the electrical nature of nerve impulses.

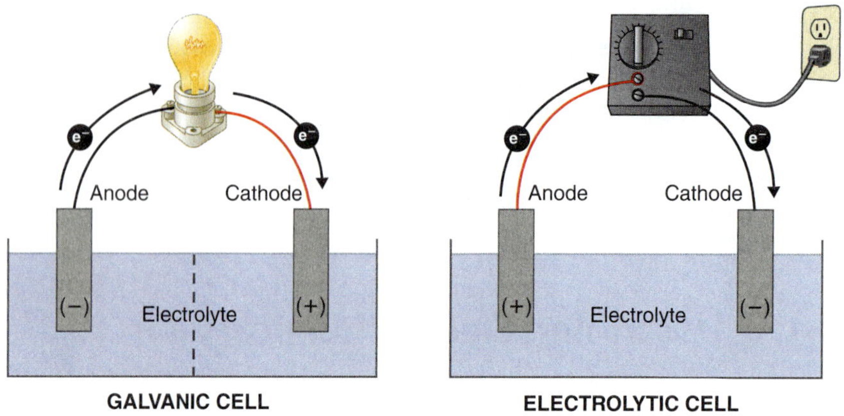

Energy released by spontaneous redox reaction is converted to electrical energy

Oxidation half-reaction:
$Y \rightarrow Y^+ + e^-$

Reduction half-reaction:
$Z + e^- \rightarrow Z^-$

Overall cell reaction:
$Y + Z \rightarrow Y^+ + Z^-$ ($G < 0$)

Electrical energy is used to drive nonspontaneous redox reaction

Oxidation half-reaction:
$Z^- \rightarrow Z + e^-$

Reduction half-reaction:
$Y^+ + e^- \rightarrow Y$

Overall cell reaction:
$Y^+ + Z^- \rightarrow Y + Z$ ($G > 0$)

Figure 19.1 Electrochemical cells. A galvanic cell (*left*) transforms the energy released by a spontaneous redox reaction into electrical energy that can be used to perform work. The oxidative and reductive half-reactions usually occur in separate compartments that are connected by an external electrical circuit; in addition, a second connection that allows ions to flow between the compartments (shown here as a vertical dashed line to represent a porous barrier) is necessary to maintain electrical neutrality. The potential difference between the electrodes (voltage) causes electrons to flow from the reductant to the oxidant through the external circuit, generating an electric current. In an electrolytic cell (*right*), an external source of electrical energy is used to generate a potential difference between the electrodes that forces electrons to flow, driving a nonspontaneous redox reaction; only a single compartment is employed in most applications. In both kinds of electrochemical cell, the anode is the electrode at which the oxidation half-reaction occurs, and the cathode is the electrode at which the reduction half-reaction occurs.

nonspontaneous redox reaction to occur ($\Delta G > 0$). Both types contain two **electrodes**, which are solid metals connected to an external circuit that provides an electrical connection between the two parts of the system (Figure 19.1). The oxidation half-reaction occurs at one electrode, the **anode**, and the reduction half-reaction occurs at the other, the **cathode**. When the circuit is closed, electrons flow from the anode to the cathode. The electrodes are also connected by an *electrolyte*, an ionic substance or solution that allows ions to transfer between the electrode compartments, thereby maintaining the system's electrical neutrality. In this section, we focus on reactions that occur in galvanic cells. We discuss electrolytic cells in Section 19.7.

$Zn(s) + Cu^{2+}(aq) \rightarrow Zn^{2+}(aq) + Cu(s)$

Figure 19.2 Reaction of metallic zinc with aqueous copper(II) ions in a single compartment. When a zinc rod is inserted into a beaker that contains an aqueous solution of copper(II) sulfate, a spontaneous redox reaction occurs: the zinc electrode dissolves to give $Zn^{2+}(aq)$ ions, while $Cu^{2+}(aq)$ ions are simultaneously reduced to metallic copper. The reaction occurs so rapidly that the copper is deposited as very fine particles that appear black, rather than the usual reddish color of copper.

Galvanic (Voltaic) Cells

To illustrate the basic principles of a galvanic cell, let's consider the reaction of metallic zinc with cupric ion (Cu^{2+}) to give copper metal and Zn^{2+} ion. The balanced chemical equation is

$$Zn(s) + Cu^{2+}(aq) \longrightarrow Zn^{2+}(aq) + Cu(s) \qquad (19.2)$$

We can cause this reaction to occur by inserting a zinc rod into an aqueous solution of copper(II) sulfate. As the reaction proceeds, the zinc rod dissolves and a mass of metallic copper forms (Figure 19.2). These changes occur spontaneously, but all the energy released is in the form of heat rather than in a form that can be used to do work.

This same reaction can be carried out using the galvanic cell illustrated in Figure 19.3a. To assemble the cell, a copper strip is inserted into a beaker that contains a 1 *M* solution of Cu^{2+} ions, and a zinc strip is inserted into a different beaker that contains a 1 *M* solution of Zn^{2+} ions. The two metal strips, which serve as electrodes, are connected by a wire, and the compartments are connected by a **salt bridge**, a U-shaped tube inserted into both solutions that contains a concentrated liquid or gelled electrolyte. The ions in the salt bridge are selected so that they do not interfere with the electrochemical reaction by being oxidized or reduced themselves or by forming a precipitate or complex; commonly used cations and anions are Na^+ or K^+ and NO_3^- or SO_4^{2-}, respectively. (Note that the ions in the salt bridge do *not* have to be the same as those in the redox

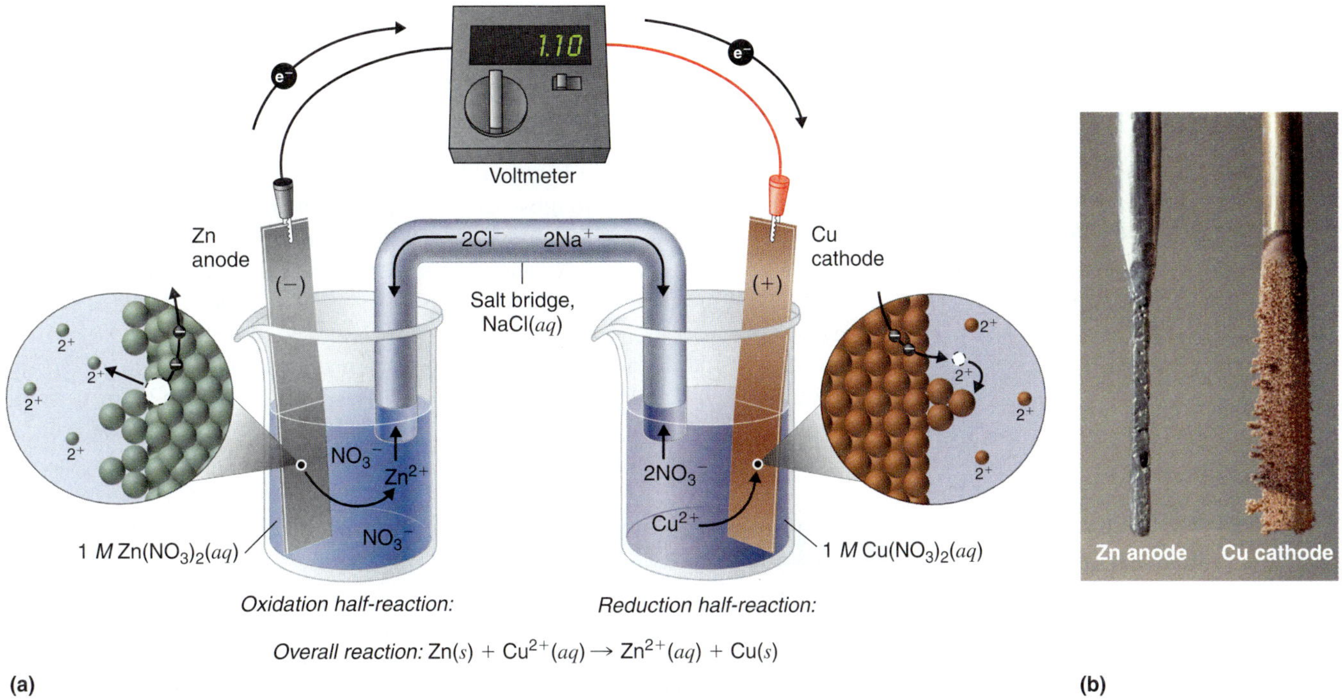

Figure 19.3 Reaction of metallic zinc with aqueous copper(II) ions in a galvanic cell. (a) A galvanic cell can be constructed by inserting a copper strip into a beaker that contains an aqueous 1 *M* solution of Cu^{2+} ions, and a zinc strip into a different beaker that contains an aqueous 1 *M* solution of Zn^{2+} ions. The two metal strips are connected by a wire that allows electricity to flow, and the beakers are connected by a salt bridge that maintains electrical neutrality. When the switch is closed to complete the circuit, the zinc electrode (the anode) is spontaneously oxidized to Zn^{2+} ions in the left compartment, while Cu^{2+} ions are simultaneously reduced to copper metal at the copper electrode (the cathode). (b) As the reaction progresses, the Zn anode loses mass as it dissolves to give $Zn^{2+}(aq)$ ions, while the Cu cathode gains mass as $Cu^{2+}(aq)$ ions are reduced to copper metal that is deposited on the cathode.

couple in either compartment.) When the circuit is closed, a spontaneous reaction occurs: zinc metal is oxidized to Zn^{2+} ions at the zinc electrode (the anode), and Cu^{2+} ions are reduced to Cu metal at the copper electrode (the cathode). As the reaction progresses, the zinc strip dissolves and the concentration of Zn^{2+} ions in the Zn^{2+} solution increases; simultaneously, the copper strip gains mass and the concentration of Cu^{2+} ions in the Cu^{2+} solution decreases (Figure 19.3b). Thus, we have carried out the same reaction as we did using a single beaker, but this time the oxidative and reductive half-reactions are physically separated from each other. The electrons that are released at the anode flow through the wire, producing an electric current. Galvanic cells therefore transform chemical energy into electrical energy that can then be used to do work.

The electrolyte in the salt bridge serves two purposes: to complete the circuit by carrying electrical charge and to maintain electrical neutrality in both solutions. The identity of the salt in a salt bridge is unimportant, as long as the component ions do not react or undergo a redox reaction under the operating conditions of the cell. The salt bridge maintains electrical neutrality in both compartments by allowing ions to migrate between the two solutions. Without such a connection, the total positive charge in the Zn^{2+} solution would increase as the zinc metal dissolves, and the total positive charge in the Cu^{2+} solution would decrease. The salt bridge allows charges to be neutralized by a flow of anions into the Zn^{2+} solution and a flow of cations into the Cu^{2+} solution. In the absence of a salt bridge or some other similar connection, the reaction would rapidly cease because electrical neutrality could not be maintained.

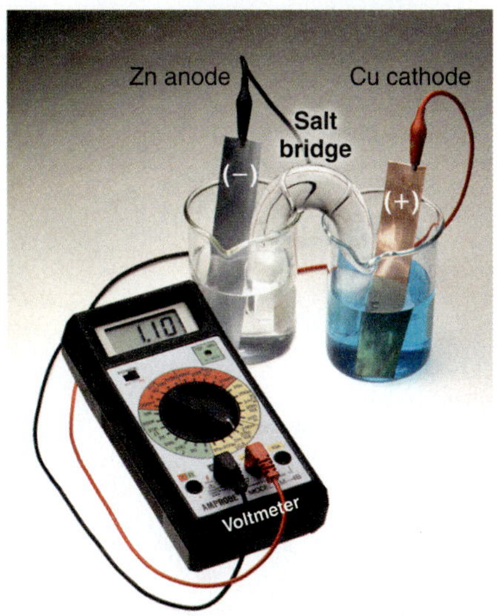

A galvanic cell with a salt bridge.

A *voltmeter* can be used to measure the difference in electrical potential between the two compartments. Opening the switch that connects the wires to the anode and cathode prevents a current from flowing, so no chemical reaction occurs. With the switch closed, however, the external circuit is closed, and an electric current can flow from the anode to the cathode. The **potential** of the cell, measured in volts, is the difference in electrical potential between the two half-reactions; electrical potential is related to the energy needed to move a charged particle in an electric field. In the cell we have described, the voltmeter indicates a potential of 1.10 V (Figure 19.3a). Because electrons from the oxidation half-reaction are released at the anode, the anode in a galvanic cell is negatively charged. The cathode, which attracts electrons, is positively charged.

Note the pattern

A galvanic cell converts the energy released by a spontaneous chemical reaction to electrical energy.

Not all electrodes undergo a chemical tranformation during a redox reaction. To prevent it from reacting during a redox process, the electrode can be made from an inert, highly conducting metal such as platinum, which does not appear in the overall electrochemical reaction. This phenomenon is illustrated in the next example.

EXAMPLE 19.1

A chemist has constructed a galvanic cell consisting of two beakers. One beaker contains a strip of tin immersed in aqueous sulfuric acid, and the other contains a platinum electrode immersed in aqueous nitric acid. The two solutions are connected by a salt bridge, and the electrodes are connected by a wire. Current begins to flow, and bubbles of a gas appear at the platinum electrode. The spontaneous redox reaction that occurs is described by this balanced equation:

$$3Sn(s) + 2NO_3^-(aq) + 8H^+(aq) \longrightarrow 3Sn^{2+}(aq) + 2NO(g) + 4H_2O(l)$$

For this galvanic cell, **(a)** write the half-reaction that occurs at each electrode; **(b)** indicate which electrode is the cathode and which is the anode; and **(c)** indicate which electrode is the positive electrode and which is the negative electrode.

Given Galvanic cell, redox reaction

Asked for Half-reactions, identity of anode and cathode, electrode assignment as positive or negative

Strategy

Ⓐ Identify the oxidation half-reaction and the reduction half-reaction. Then identify the anode and cathode from the half-reaction that occurs at each electrode.

Ⓑ From the direction of electron flow, assign each electrode as either positive or negative.

Solution

(a) Ⓐ In the reduction half-reaction, nitrate is reduced to nitric oxide. (The nitric oxide would then react with oxygen in the air to form NO_2, with its characteristic red-brown color.) In the oxidation half-reaction, metallic tin is oxidized. The half-reactions corresponding to the actual reactions that occur in the system are

Reduction: $NO_3^-(aq) + 4H^+(aq) + 3e^- \longrightarrow NO(g) + 2H_2O(l)$

Oxidation: $Sn(s) \longrightarrow Sn^{2+}(aq) + 2e^-$

Thus, nitrate is reduced to NO, while the tin electrode is oxidized to Sn^{2+}.

(b) Because the reduction reaction occurs at the Pt electrode, it is the cathode. Conversely, the oxidation reaction occurs at the tin electrode, so it is the anode.

(c) Ⓑ Electrons flow from the tin electrode through the wire to the platinum electrode, where they transfer to nitrate. The electric circuit is completed by the salt bridge, which permits the diffusion of cations toward the cathode and of anions toward the anode. Because electrons flow *out* of the tin electrode, it must be electrically negative. In contrast, electrons flow *toward* the Pt electrode, so that electrode must be electrically positive.

EXERCISE 19.1

Consider a simple galvanic cell consisting of two beakers connected by a salt bridge. One beaker contains a solution of MnO_4^- in dilute sulfuric acid and has a Pt electrode. The other beaker contains a solution of Sn^{2+} in dilute sulfuric acid, also with a Pt electrode. When the two electrodes are connected by a wire, current flows and a spontaneous reaction occurs that is described by the balanced equation

$$2MnO_4^-(aq) + Sn^{2+}(aq) + 16H^+(aq) \longrightarrow 2Mn^{2+}(aq) + Sn^{4+}(aq) + 8H_2O(l)$$

For this galvanic cell, **(a)** write the half-reaction that occurs at each electrode; **(b)** indicate which electrode is the cathode and which is the anode; and **(c)** indicate which electrode is positive and which is negative.

Answer **(a) (b)** $MnO_4^-(aq) + 8H^+(aq) + 5e^- \longrightarrow Mn^{2+}(aq) + 4H_2O(l)$ occurs at the cathode; $Sn^{2+}(aq) \longrightarrow Sn^{4+}(aq) + 2e^-$ occurs at the anode; **(c)** the cathode (electrode in beaker that contains the permanganate solution) is positive, and the anode (electrode in beaker that contains the tin solution) is negative.

Constructing a Cell Diagram

Because it is somewhat cumbersome to describe any given galvanic cell in words, a more convenient notation has been developed. In this line notation, called a *cell diagram,* the identity of the electrodes and the chemical contents of the compartments are indicated by their chemical formulas, with the anode written on the far left and the cathode on the far right. Phase boundaries are shown by single vertical lines, and the salt bridge, which has two phase boundaries, by a double vertical line. Thus, the cell diagram for the Zn/Cu cell shown in Figure 19.3a is written as

Phase boundary — Two phase boundaries — Phase boundary

$$Zn(s)\,|\,Zn^{2+}(aq,\,1\,M)\,||\,Cu^{2+}(aq,\,1\,M)\,|\,Cu(s) \qquad (19.3)$$

Anode — **Salt bridge** — **Cathode**

Notice that a cell diagram includes solution concentrations when they are provided.

Galvanic cells can have arrangements other than the examples we have seen so far. For example, the voltage produced by a redox reaction can be measured more accurately using two electrodes immersed in a single beaker containing an electrolyte that completes the circuit. This arrangement reduces errors caused by resistance to the flow of charge at a boundary, called the *junction potential*. One example of this type of galvanic cell is

$$Pt(s)\,|\,H_2(g)\,|\,HCl(aq)\,|\,AgCl(s)\,|\,Ag(s) \qquad (19.4)$$

Notice that this cell diagram does not include a double vertical line representing a salt bridge because there is no salt bridge providing a junction between two dissimilar solutions. Moreover, solution concentrations have not been specified, so they are not included in the cell diagram. The half-reactions and the overall reaction for this cell are

Cathode reaction: $AgCl(s) + e^- \longrightarrow Ag(s) + Cl^-(aq)$ (19.5a)

Anode reaction: $\frac{1}{2}H_2(g) \longrightarrow H^+(aq) + e^-$ (19.5b)

Overall: $AgCl(s) + \frac{1}{2}H_2(g) \longrightarrow Ag(s) + Cl^-(aq) + H^+(aq)$ (19.5c)

A single-compartment galvanic cell will initially exhibit the same voltage as a galvanic cell constructed using separate compartments, but it will discharge rapidly because of the direct reaction of the reactant at the anode with the oxidized member of the cathodic redox couple. Consequently, cells of this type are not particularly useful for producing electricity.

EXAMPLE 19.2

Draw a cell diagram for the galvanic cell described in Example 19.1.

Given Galvanic cell, redox reaction

Asked for Cell diagram

Strategy

Using the symbols described, write the cell diagram beginning with the oxidation half-reaction on the left.

Solution

The anode is the tin strip, and the cathode is the Pt electrode. Beginning on the left with the anode, we indicate the phase boundary between the electrode and the tin solution by a vertical bar. The anode compartment is thus $Sn(s)\,|\,Sn^{2+}(aq)$. We could include $H_2SO_4(aq)$ with the contents of the anode compartment, but the sulfate ion (as HSO_4^-) does not participate in the overall reaction and does not need to be indicated specifically. The cathode compartment contains aqueous nitric acid, which does participate in the overall reaction, together with the product of the reaction, NO, and the Pt electrode. These are written as $HNO_3(aq)\,|\,NO(g)\,|\,Pt(s)$, with single vertical bars indicating the phase boundaries. Combining the two compartments and using a double vertical bar to indicate the salt bridge give

$$Sn(s)\,|\,Sn^{2+}(aq)\,\|\,HNO_3(aq)\,|\,NO(g)\,|\,Pt(s)$$

Notice that solution concentrations were not specified, so they are not included in this cell diagram.

EXERCISE 19.2

Draw a cell diagram for the following reaction, assuming the concentration of Ag^+ and Mg^{2+} are each 1 M:

$$Mg(s) + 2Ag^+(aq) \longrightarrow Mg^{2+}(aq) + 2Ag(s)$$

Answer $Mg(s)\,|\,Mg^{2+}(aq, 1\,M)\,\|\,Ag^+(aq, 1\,M)\,|\,Ag(s)$

19.2 Standard Potentials

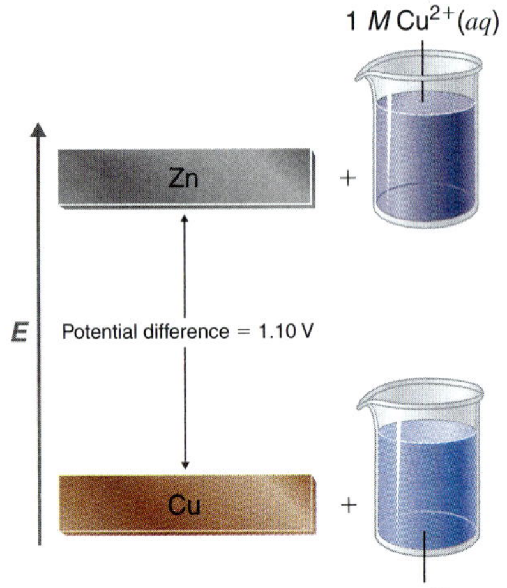

Figure 19.4 Potential energy difference in the Zn/Cu system. The potential energy of a system consisting of metallic zinc and aqueous Cu^{2+} is higher than the potential energy of a system consisting of metallic copper and aqueous Zn^{2+}. Much of this potential energy difference is because the valence electrons of metallic zinc are higher in energy than the valence electrons of metallic copper. Because the $Zn(s) + Cu^{2+}(aq)$ system is higher in energy by 1.10 V than the $Cu(s) + Zn^{2+}(aq)$ system, energy is released when electrons are transferred from zinc to Cu^{2+} to form copper and Zn^{2+}.

In a galvanic cell, current is produced when electrons flow externally through the circuit from the anode to the cathode because of a difference in potential energy between the two electrodes in the electrochemical cell. In the Zn/Cu system, the valence electrons in zinc have a substantially higher potential energy than the valence electrons in copper because of shielding of the *s* electrons of zinc by the electrons in filled *d* orbitals (see Section 6.5). Hence, electrons flow spontaneously from zinc to copper(II) ions, forming zinc(II) ions and metallic copper (Figure 19.4). Just like water flowing spontaneously downhill, which can be made to do work by forcing a waterwheel, the flow of electrons from a higher potential energy to a lower one can also be harnessed to perform work.

Because the potential energy of valence electrons differs greatly from one substance to another, the voltage of a galvanic cell depends partly on the identity of the reacting substances. If we construct a galvanic cell similar to the one in Figure 19.3a but instead of copper use a strip of cobalt metal and 1 M Co^{2+} in the cathode compartment, the measured voltage is not 1.10 V but 0.51 V. Thus, we can conclude that the difference in potential energy between the valence electrons of cobalt and of zinc is less than the difference between the valence electrons of copper and of zinc by 0.59 V.

The measured potential of a cell also depends strongly on the *concentrations* of the reacting species and the *temperature* of the system. Thus, in order to develop a scale of relative potentials that will allow us to predict the direction of an electrochemical reaction and the magnitude of the driving force for the reaction, the potentials for oxidations and reductions of different substances must be measured under comparable conditions. To do this, chemists use the **standard cell potential, E°_{cell}**, defined as the potential of a cell measured under standard conditions—that is, with all species in their standard states (1 *M* for solutions,* 1 atm for gases, pure solids or pure liquids for other substances) and at a fixed temperature, usually 25°C.

Measuring Standard Electrode Potentials

It is physically impossible to measure the potential of a single electrode: only the *difference* between the potentials of two electrodes can be measured. (This is analogous to measuring absolute enthalpies or free energies. Recall from Chapter 18 that only *differences* in enthalpy and free energy can be measured.) We can, however, compare the standard cell potentials for two different galvanic cells that have one kind of electrode in common. This allows us to measure the potential difference between two dissimilar electrodes. For example, the measured standard cell potential, E°, for the Zn/Cu system is 1.10 V, whereas E° for the corresponding Zn/Co system is 0.51 V. This implies that the potential difference between the Co and Cu electrodes is (1.10 V − 0.51 V) = 0.59 V. In fact, that is exactly the potential measured under standard conditions if a cell is constructed with this cell diagram:

$$Co(s) \mid Co^{2+}(aq, 1\ M) \parallel Cu^{2+}(aq, 1\ M) \mid Cu(s) \qquad E^\circ = 0.59\ V \tag{19.6}$$

This cell diagram corresponds to the oxidation of a cobalt anode and the reduction of Cu^{2+} in solution at the copper cathode.

All tabulated values of *standard electrode potentials* by convention are listed for a reaction written as a reduction, not as an oxidation, in order to be able to compare standard potentials for different substances (see Appendix E). The standard cell potential, E°_{cell}, is therefore the *difference* between the tabulated reduction potentials of the two half-reactions, not their sum:

$$E^\circ_{cell} = E^\circ_{cathode} - E^\circ_{anode} \tag{19.7}$$

In contrast, recall that half-reactions are written to show the reduction and oxidation reactions that actually occur in the cell, so the overall cell reaction is written as the *sum* of the two half-reactions. According to Equation 19.7, when we know the standard potential for any single half-reaction, we can obtain the value of the standard potential of many other half-reactions by measuring the standard potential of the corresponding cell.

Although it is impossible to measure the potential of any electrode directly, we can choose a reference electrode whose potential is defined as 0 V under standard conditions. The **standard hydrogen electrode (SHE)** is universally used for this purpose and is assigned a standard potential of 0 V. It consists of a strip of platinum wire in contact with an aqueous solution containing 1 *M* H^+. The $[H^+]$ in solution is in equilibrium with H_2

Note the pattern

The overall cell reaction is the sum of the two half-reactions, but the cell potential is the difference between the reduction potentials: $E^\circ_{cell} = E^\circ_{cathode} - E^\circ_{anode}$.

* Concentrated solutions of salts (about 1 *M*) generally do not exhibit ideal behavior, and the actual standard state corresponds to an *activity* of 1 rather than a concentration of 1 *M*. Corrections for nonideal behavior are important for precise quantitative work, but not for the more qualitative approach that we are taking here.

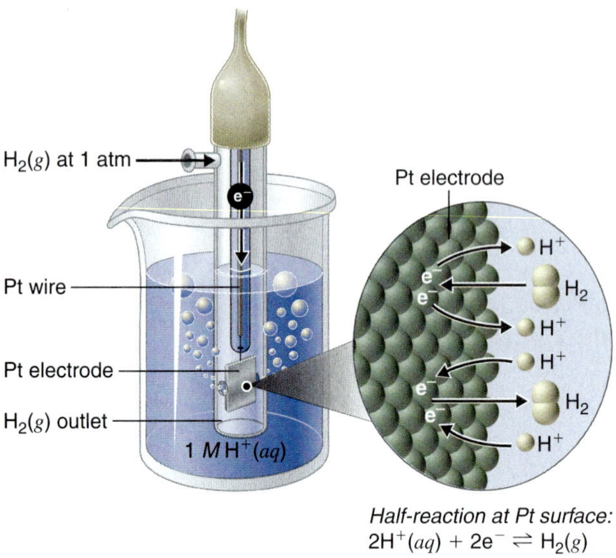

Figure 19.5 The standard hydrogen electrode (SHE). The SHE consists of platinum wire that is connected to a Pt surface in contact with an aqueous solution containing 1 M H^+ in equilibrium with H_2 gas at a pressure of 1 atm. As shown in the molecular view, the Pt surface catalyzes the oxidation of hydrogen molecules to protons or the reduction of protons to hydrogen gas. (Water is omitted for clarity.) The standard potential of the SHE is arbitrarily assigned a value of 0 V.

gas at a pressure of 1 atm at the Pt–solution interface (Figure 19.5). Protons are reduced or hydrogen molecules are oxidized at the Pt surface according to the equation

$$2H^+(aq) + 2e^- \rightleftharpoons H_2(g) \qquad (19.8)$$

One especially attractive feature of the SHE is that the Pt metal electrode is not consumed during the reaction.

Figure 19.6 shows a galvanic cell that consists of a SHE in one beaker and a Zn strip in another beaker containing a solution of Zn^{2+} ions. When the circuit is closed, the voltmeter indicates a potential of 0.76 V. The zinc electrode begins to dissolve to form Zn^{2+}, and H^+ ions are reduced to H_2 in the other compartment. Thus, the hydrogen electrode is the cathode, and the zinc electrode is the anode. The diagram for this galvanic cell is

$$Zn(s) \mid Zn^{2+}(aq) \parallel H^+(aq, 1\ M) \mid H_2(g, 1\ \text{atm}) \mid Pt(s) \qquad (19.9)$$

The half-reactions that actually occur in the cell and their corresponding electrode potentials are

$$\begin{aligned}
&\text{Cathode:} && 2H^+(aq) + 2e^- \longrightarrow H_2(g) && E^\circ_{\text{cathode}} = 0\ \text{V} && (19.10a)\\
&\text{Anode:} && Zn(s) \longrightarrow Zn^{2+}(aq) + 2e^- && E^\circ_{\text{anode}} = -0.76\ \text{V} && (19.10b)\\
&\text{Overall:} && Zn(s) + 2H^+(aq) \longrightarrow Zn^{2+}(aq) + H_2(g) && && (19.10c)\\
& && E^\circ_{\text{cell}} = E^\circ_{\text{cathode}} - E^\circ_{\text{anode}} = 0.76\ \text{V}
\end{aligned}$$

Notice that although the reaction at the anode is an oxidation, which means that the sum of the two half-reactions gives the overall reaction, by convention its tabulated E° value is reported as a reduction potential. We must therefore subtract E°_{anode} from E°_{cathode} to obtain E°_{cell}.

The potential of a half-reaction measured against the SHE under standard conditions is called the **standard electrode potential** for that half-reaction. The standard electrode potential for $Zn^{2+}(aq) + 2e^- \longrightarrow Zn(s)$ is -0.76 V; thus, the standard electrode potential for the *oxidation* of Zn to Zn^{2+}, often called the Zn/Zn^{2+} *redox*

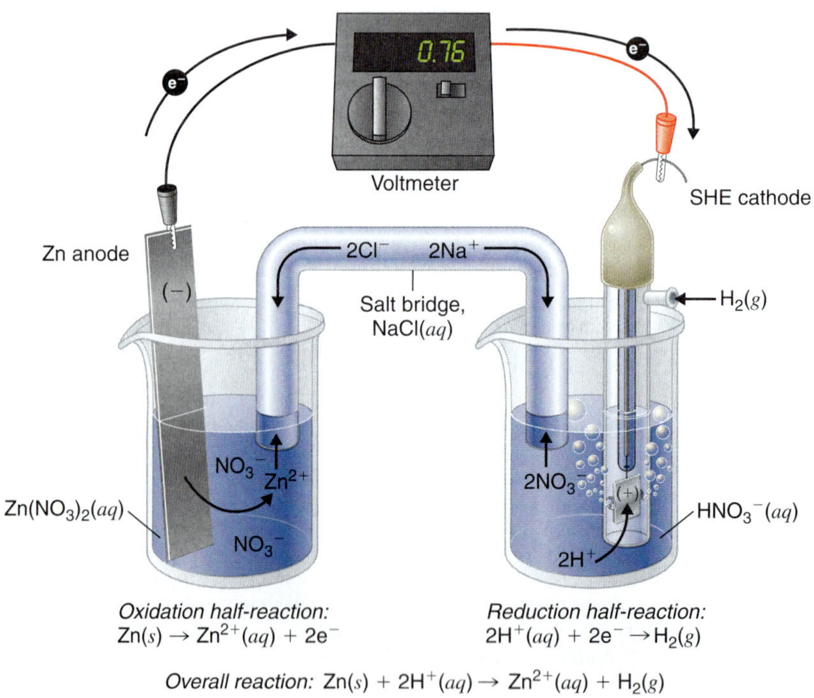

Figure 19.6 Determining a standard electrode potential using a standard hydrogen electrode (SHE). The voltmeter shows that the standard cell potential of a galvanic cell consisting of a SHE and a Zn/Zn^{2+} couple is $E^\circ_{\text{cell}} = 0.76$ V. Because the zinc electrode in this cell dissolves spontaneously to form $Zn^{2+}(aq)$ ions while $H^+(aq)$ ions are reduced to H_2 at the platinum surface, the standard electrode potential of the Zn^{2+}/Zn couple is -0.76 V.

couple, or the Zn/Zn^{2+} *couple*, is $-(-0.76\text{ V}) = 0.76\text{ V}$. Because electrical potential is the energy needed to move a charged particle in an electric field, standard electrode potentials for half-reactions are intensive properties and do not depend on the amount of substance involved. Consequently, *$E°$ values are independent of the stoichiometric coefficients for the half-reaction*, and, most important, the coefficients used to produce a balanced overall reaction do not affect the value of the cell potential.

Note the pattern

$E°$ values do not depend on the stoichiometric coefficients for a half-reaction.

To measure the potential of the Cu/Cu^{2+} couple, we can construct a galvanic cell analogous to the one shown in Figure 19.6, but containing a Cu/Cu^{2+} couple in the sample compartment instead of Zn/Zn^{2+}. When we close the circuit this time, the measured potential for the cell is *negative* (−0.34 V) rather than positive. The negative value of $E°_{cell}$ indicates that the direction of spontaneous electron flow is the *opposite* of that for the Zn/Zn^{2+} couple. Hence, the reactions that occur spontaneously are the reduction of Cu^{2+} to Cu at the copper electrode, which gains mass as the reaction proceeds, and the oxidation of H_2 to H^+ at the platinum electrode. Hence, in this cell the copper strip is the *cathode*, and the hydrogen electrode is the *anode*. The cell diagram therefore is written with the SHE on the left and the Cu^{2+}/Cu couple on the right:

$$Pt(s) \mid H_2(g, 1\text{ atm}) \parallel Cu^{2+}(aq, 1\,M) \mid Cu(s) \tag{19.11}$$

The half-cell reactions and potentials of the spontaneous reaction are

$$\text{Cathode:} \quad Cu^{2+}(aq) + 2e^- \longrightarrow Cu(s) \qquad E°_{cathode} = 0.34\text{ V} \tag{19.12a}$$

$$\text{Anode:} \quad H_2(g) \longrightarrow 2H^+(aq) + 2e^- \qquad E°_{anode} = 0\text{ V} \tag{19.12b}$$

$$\text{Overall:} \quad H_2(g) + Cu^{2+}(aq) \longrightarrow 2H^+(aq) + Cu(s) \tag{19.12c}$$

$$E°_{cell} = E°_{cathode} - E°_{anode} = 0.34\text{ V}$$

Thus, the standard electrode potential for the Cu^{2+}/Cu couple is 0.34 V.

Balancing Redox Reactions Using the Half-Reaction Method

In Chapter 4, we described a method for balancing redox reactions using oxidation numbers. Oxidation numbers were assigned to each atom in a redox reaction to identify any changes in the oxidation states. Here we present an alternative approach to balancing redox reactions, the half-reaction method, in which the overall redox reaction is divided into an oxidation half-reaction and a reduction half-reaction, each balanced for mass and charge. This method more closely reflects the events that take place in an electrochemical cell, where the two half-reactions may be physically separated from each other.

We can illustrate how to balance a redox reaction using half-reactions with the reaction that occurs when the commercial solid drain cleaner, Drano, is poured into a clogged drain. Drano contains a mixture of sodium hydroxide and powdered aluminum, which in solution reacts to produce hydrogen gas:

$$Al(s) + OH^-(aq) \longrightarrow Al(OH)_4^-(aq) + H_2(g) \tag{19.13}$$

In this reaction, $Al(s)$ is oxidized to Al^{3+} and H^+ is reduced to H_2 gas, which bubbles through the solution, agitating it and breaking up the clogs.

The overall redox reaction is composed of a reduction half-reaction and an oxidation half-reaction. From the standard electrode potentials listed in Appendix E, we find the corresponding half-reactions that describe the reduction of H^+ ions in water to H_2 and the oxidation of Al to Al^{3+} in basic solution:

$$\text{Reduction:} \quad 2H_2O(l) + 2e^- \longrightarrow 2OH^-(aq) + H_2(g) \tag{19.14a}$$

$$\text{Oxidation:} \quad Al(s) + 4OH^-(aq) \longrightarrow Al(OH)_4^-(aq) + 3e^- \tag{19.14b}$$

The half-reactions chosen must exactly reflect the reaction conditions, such as the basic conditions shown here. Moreover, the physical states of the reactants and products must be identical to those given in the overall reaction, whether gaseous, liquid, solid, or in solution.

In Equation 19.14a, two H^+ ions gain one electron each in the reduction, and in Equation 19.14b, the aluminum atom loses three electrons in the oxidation. The charges are balanced by multiplying the reduction half-reaction (Equation 19.14a) by 3 and the oxidation half-reaction (Equation 19.14b) by 2 to give the same number of electrons in both half-reactions:

$$\text{Reduction:}\quad 6H_2O(l) + 6e^- \longrightarrow 6OH^-(aq) + 3H_2(g) \qquad (19.15a)$$

$$\text{Oxidation:}\quad 2Al(s) + 8OH^-(aq) \longrightarrow 2Al(OH)_4^-(aq) + 6e^- \qquad (19.15b)$$

Adding the two half-reactions gives

$$6H_2O(l) + 2Al(s) + 8OH^-(aq) \longrightarrow 2Al(OH)_4^-(aq) + 3H_2(g) + 6OH^-(aq) \qquad (19.16)$$

Simplifying by canceling substances that appear on both sides of the equation gives

$$6H_2O(l) + 2Al(s) + 2OH^-(aq) \longrightarrow 2Al(OH)_4^-(aq) + 3H_2(g) \qquad (19.17)$$

We have a -2 charge on the left side of the equation and a -2 charge on the right side. Thus, the charges are balanced, but we must also check that atoms are balanced:

$$2Al + 8O + 14H = 2Al + 8O + 14H \qquad (19.18)$$

The atoms also balance, so Equation 19.17 is a balanced chemical equation for the redox reaction depicted in Equation 19.13.

We can also balance a redox reaction by first balancing the atoms in each half-reaction and then balancing the charges. With this alternative method, we do not need to use the half-reactions listed in Appendix E, but instead focus on the atoms whose oxidation states change, as illustrated in the following steps:

Step 1 **Write the reduction half-reaction and the oxidation half-reaction.** For the reaction shown in Equation 19.13, hydrogen is reduced from H^+ in OH^- to H_2, and aluminum is oxidized from Al^0 to Al^{3+}:

$$\text{Reduction:}\quad OH^-(aq) \longrightarrow H_2(g) \qquad (19.19a)$$

$$\text{Oxidation:}\quad Al(s) \longrightarrow Al(OH)_4^-(aq) \qquad (19.19b)$$

Step 2 **Balance the atoms by balancing elements other than O and H. Then balance O atoms by adding H_2O, and finally balance H atoms by adding H^+.** Elements other than O and H in the last two equations are balanced as written, so we proceed with balancing O atoms. We can do this by adding water to the appropriate side of each half-reaction:

$$\text{Reduction:}\quad OH^-(aq) \longrightarrow H_2(g) + H_2O(l) \qquad (19.20a)$$

$$\text{Oxidation:}\quad Al(s) + 4H_2O(l) \longrightarrow Al(OH)_4^-(aq) \qquad (19.20b)$$

Balancing H atoms by adding H^+, we obtain

$$\text{Reduction:}\quad OH^-(aq) + 3H^+(aq) \longrightarrow H_2(g) + H_2O(l) \qquad (19.21a)$$

$$\text{Oxidation:}\quad Al(s) + 4H_2O(l) \longrightarrow Al(OH)_4^-(aq) + 4H^+(aq) \qquad (19.21b)$$

We have now balanced the atoms in each half-reaction, but the charges are not balanced.

Step 3 **Balance the charges in each half-reaction by adding electrons.** Two electrons are gained in the reduction of H^+ ions to H_2, and three electrons are lost during the oxidation of Al^0 to Al^{3+}:

$$\text{Reduction:}\quad OH^-(aq) + 3H^+(aq) + 2e^- \longrightarrow H_2(g) + H_2O(l) \qquad (19.22a)$$

$$\text{Oxidation:}\quad Al(s) + 4H_2O(l) \longrightarrow Al(OH)_4^-(aq) + 4H^+(aq) + 3e^- \qquad (19.22b)$$

Step 4 **Multiply the reductive and oxidative half-reactions by appropriate integers to obtain the same number of electrons in both half-reactions.** In this case, we multiply Equation 19.22a (the reductive half-reaction) by 3 and Equation 19.22b (the oxidative half-reaction) by 2 to obtain the same number of electrons in both half-reactions:

Reduction: $$3OH^-(aq) + 9H^+(aq) + 6e^- \longrightarrow 3H_2(g) + 3H_2O(l) \quad (19.23a)$$

Oxidation: $$2Al(s) + 8H_2O(l) \longrightarrow 2Al(OH)_4^-(aq) + 8H^+(aq) + 6e^- \quad (19.23b)$$

Step 5 **Add the two half-reactions and cancel substances that appear on both sides of the equation.** Adding and, in this case, canceling $8H^+$, $3H_2O$, and $6e^-$ give

$$2Al(s) + 5H_2O(l) + 3OH^-(aq) + H^+(aq) \longrightarrow 2Al(OH)_4^-(aq) + 3H_2(g) \quad (19.24)$$

Notice that we have three OH^- and one H^+ on the left side. Neutralizing the H^+ gives us a total of $5H_2O + H_2O = 6H_2O$, and leaves $2OH^-$ on the left side:

$$2Al(s) + 6H_2O(l) + 2OH^-(aq) \longrightarrow 2Al(OH)_4^-(aq) + 3H_2(g) \quad (19.25)$$

Step 6 **Check to make sure that all atoms and charges are balanced.** Equation 19.25 is identical to Equation 19.17, obtained using the first method, so the charges and numbers of atoms on each side of the equation balance.

EXAMPLE 19.3

In acidic solution, the redox reaction of dichromate ion ($Cr_2O_7^{2-}$) and iodide (I^-) can be monitored visually. The yellow dichromate solution reacts with the colorless iodide solution to produce a solution that is deep amber due to the presence of a green $Cr^{3+}(aq)$ complex and brown $I_2(aq)$ ions (Figure 19.7):

$$Cr_2O_7^{2-}(aq) + I^-(aq) \longrightarrow Cr^{3+}(aq) + I_2(aq)$$

Balance this equation using half-reactions.

Given Redox reaction, Appendix E

Asked for Balanced equation using half-reactions

Strategy

Follow the steps outlined above to balance the redox reaction using the half-reaction method.

Figure 19.7 The reaction of dichromate with iodide. The reaction of a yellow solution of sodium dichromate with a colorless solution of sodium iodide produces a deep amber solution that contains a green $Cr^{3+}(aq)$ complex and brown $I_2(aq)$ ions.

Solution

From the standard electrode potentials listed in Appendix E, we find the half-reactions corresponding to the overall reaction:

Reduction: $Cr_2O_7^{2-}(aq) + 14H^+(aq) + 6e^- \longrightarrow 2Cr^{3+}(aq) + 7H_2O(l)$

Oxidation: $2I^-(aq) \longrightarrow I_2(aq) + 2e^-$

Balancing the number of electrons by multiplying the oxidation reaction by 3 gives

Oxidation: $6I^-(aq) \longrightarrow 3I_2(aq) + 6e^-$

Adding the two half-reactions and canceling electrons give

$$Cr_2O_7^{2-}(aq) + 14H^+(aq) + 6I^-(aq) \longrightarrow 2Cr^{3+}(aq) + 7H_2O(l) + 3I_2(aq)$$

We must now check to make sure the charges and atoms on each side of the equation balance:

$$(-2) + 14 + (-6) = +6$$
$$+6 = +6$$
$$2Cr + 7O + 14H + 6I = 2Cr + 7O + 14H + 6I$$

The charges and atoms balance, so our equation is balanced.

We can also use the alternative procedure, which does not require the use of the half-reactions listed in Appendix E.

Step 1 Chromium is reduced from Cr^{6+} in $Cr_2O_7^{2-}$ to Cr^{3+}, and I^- ions are oxidized to I_2. Dividing the reaction into two half-reactions gives

Reduction: $Cr_2O_7^{2-}(aq) \longrightarrow Cr^{3+}(aq)$

Oxidation: $I^-(aq) \longrightarrow I_2(aq)$

Step 2 Balancing the atoms other than oxygen and hydrogen gives

Reduction: $Cr_2O_7^{2-}(aq) \longrightarrow 2Cr^{3+}(aq)$

Oxidation: $2I^-(aq) \longrightarrow I_2(aq)$

We now balance the O atoms by adding H_2O—in this case, to the right side of the reduction half-reaction. Because the oxidation half-reaction does not contain oxygen, it can be ignored in this step.

Reduction: $Cr_2O_7^{2-}(aq) \longrightarrow 2Cr^{3+}(aq) + 7H_2O(l)$

Next we balance the H atoms by adding H^+ to the left side of the reduction half-reaction. Again, we can ignore the oxidation half-reaction.

Reduction: $Cr_2O_7^{2-}(aq) + 14H^+(aq) \longrightarrow 2Cr^{3+}(aq) + 7H_2O(l)$

Step 3 We must now add electrons to balance the charges. The reduction half-reaction ($2Cr^{+6}$ to $2Cr^{+3}$) has a $+12$ charge on the left and a $+6$ charge on the right, so six electrons are needed to balance the charge. The oxidation half-reaction ($2I^-$ to I_2) has a -2 charge on the left side and a 0 charge on the right, so it needs two electrons to balance the charge:

Reduction: $Cr_2O_7^{2-}(aq) + 14H^+(aq) + 6e^- \longrightarrow 2Cr^{3+}(aq) + 7H_2O(l)$

Oxidation: $2I^-(aq) \longrightarrow I_2(aq) + 2e^-$

Step 4 To have the same number of electrons in both half-reactions, we must multiply the oxidation half-reaction by 3:

Oxidation: $6I^-(aq) \longrightarrow 3I_2(s) + 6e^-$

Step 5 Adding the two half-reactions and canceling substances that appear in both reactions give

$$Cr_2O_7^{2-}(aq) + 14H^+(aq) + 6I^-(aq) \longrightarrow 2Cr^{3+}(aq) + 7H_2O(l) + 3I_2(aq)$$

Step 6 This is the same equation we obtained using the first method. Thus, the charges and atoms on each side of the equation balance.

EXERCISE 19.3

Copper is commonly found as the mineral *covellite*, CuS. The first step in extracting the copper is to dissolve the mineral in nitric acid (HNO_3), which oxidizes sulfide to sulfate and reduces nitric acid to NO:

$$CuS(s) + HNO_3(aq) \longrightarrow NO(g) + CuSO_4(aq)$$

Balance this equation using the half-reaction method.

Answer $3CuS(s) + 8HNO_3(aq) \longrightarrow 8NO(g) + 3CuSO_4(aq) + 4H_2O(l)$

Calculating Standard Cell Potentials

The standard cell potential for a redox reaction, E°_{cell}, is a measure of the tendency of the reactants in their standard states to form the products in their standard states; consequently, it is a measure of the driving force for the reaction, which earlier we called *voltage*. We can use the two standard electrode potentials we found earlier to calculate the standard potential for the Zn/Cu cell represented by this cell diagram:

$$Zn(s) \mid Zn^{2+}(aq, 1\ M) \parallel Cu^{2+}(aq, 1\ M) \mid Cu(s) \quad (19.26)$$

We know the values of E°_{anode} for the reduction of Zn^{2+} and $E^\circ_{cathode}$ for the reduction of Cu^{2+}, so we can calculate E°_{cell}:

Cathode:	$Cu^{2+}(aq) + 2e^- \longrightarrow Cu(s)$	$E^\circ_{cathode} = 0.34\ V$	(19.27a)
Anode:	$Zn(s) \longrightarrow Zn^{2+}(aq, 1\ M) + 2e^-$	$E^\circ_{anode} = -0.76\ V$	(19.27b)
Overall:	$Zn(s) + Cu^{2+}(aq) \longrightarrow Zn^{2+}(aq) + Cu(s)$	$E^\circ_{cell} = E^\circ_{cathode} - E^\circ_{anode} = 1.10\ V$	(19.27c)

This is the same value that is observed experimentally. If the value of E°_{cell} (the standard cell potential) is positive, the reaction will occur spontaneously as written. If the value of E°_{cell} is negative, then the reaction is not spontaneous, and it will not occur as written under standard conditions; it will, however, proceed spontaneously in the opposite direction.* The next example and exercise illustrate how we can use measured cell potentials to calculate standard potentials for redox couples.

Note the pattern

A positive E°_{cell} means that the reaction will occur spontaneously as written. A negative E°_{cell} means that the reaction will proceed spontaneously in the opposite direction.

EXAMPLE 19.4

A galvanic cell with a measured standard cell potential of 0.27 V is constructed using two beakers connected by a salt bridge. One beaker contains a strip of gallium metal immersed in a 1 *M* solution of $GaCl_3$, and the other contains a piece of nickel immersed in a 1 *M* solution of $NiCl_2$. The half-reactions that occur when the compartments are connected are

Cathode: $Ni^{2+}(aq) + 2e^- \longrightarrow Ni(s)$

Anode: $Ga(s) \longrightarrow Ga^{3+}(aq) + 3e^-$

If the potential for the oxidation of Ga to Ga^{3+} is 0.55 V under standard conditions, what is the potential for the oxidation of Ni to Ni^{2+}?

Given Galvanic cell, half-reactions, standard cell potential, potential for the oxidation half-reaction under standard conditions

Asked for Standard electrode potential of reaction occurring at the cathode

* As we shall see in Section 19.7, this does not mean that the reaction cannot be made to occur at all under standard conditions. With a sufficient input of electrical energy, virtually any reaction can be forced to occur.

Strategy

Ⓐ Write the equation for the half-reaction that occurs at the anode along with the value of the standard electrode potential for the half-reaction.

Ⓑ Use Equation 19.7 to calculate the standard electrode potential for the half-reaction that occurs at the cathode. Then reverse the sign to obtain the potential for the corresponding oxidation half-reaction under standard conditions.

Solution

Ⓐ We have been given the potential for the oxidation of Ga to Ga^{3+} under standard conditions, but to report the standard electrode potential, we must reverse the sign. Thus, for the reduction reaction $Ga^{3+}(aq) + 3e^- \longrightarrow Ga(s)$, $E^\circ_{anode} = -0.55$ V. Ⓑ Using the value given for E°_{cell} and the calculated value of E°_{anode}, we can calculate the standard potential for the reduction of Ni^{2+} to Ni from Equation 19.7:

$$\begin{aligned} E^\circ_{cell} &= E^\circ_{cathode} - E^\circ_{anode} \\ 0.27\text{ V} &= E^\circ_{cathode} - (-0.55\text{ V}) \\ E^\circ_{cathode} &= -0.28\text{ V} \end{aligned}$$

This is the standard electrode potential for the reaction $Ni^{2+}(aq) + 2e^- \longrightarrow Ni(s)$. Because we are asked for the potential for the oxidation of Ni to Ni^{2+} under standard conditions, we must reverse the sign of $E^\circ_{cathode}$. Thus, $E^\circ = -(-0.28\text{ V}) = 0.28$ V for the oxidation. With three electrons consumed in the reduction and two produced in the oxidation, the overall reaction is not balanced. Recall, however, that standard potentials are *independent* of stoichiometry.

EXERCISE 19.4

A galvanic cell is constructed with one compartment that contains a mercury electrode immersed in a 1 *M* solution of mercuric acetate [$Hg(NO_3)_2$] and one compartment that contains a strip of magnesium immersed in a 1 *M* solution of $MgCl_2$. When the compartments are connected, a potential of 3.22 V is measured and these half-reactions occur:

$$\text{Cathode:}\quad Hg^{2+}(aq) + 2e^- \longrightarrow Hg(l)$$

$$\text{Anode:}\quad Mg(s) \longrightarrow Mg^{2+}(aq) + 2e^-$$

If the potential for the oxidation of Mg to Mg^{2+} is 2.37 V under standard conditions, what is the standard electrode potential for the reaction that occurs at the anode?

Answer 0.85 V

Reference Electrodes and Measuring Concentrations

When using a galvanic cell to measure the concentration of a substance, we are generally interested in the potential of only one of the electrodes of the cell, the so-called **indicator electrode**, whose potential is related to the concentration of the substance being measured. To ensure that any change in the measured potential of the cell is due to only the substance being analyzed, the potential of the other electrode, the **reference electrode**, must be constant. You are already familiar with one example of a reference electrode, the SHE. The potential of a reference electrode must be unaffected by the properties of the solution and, if possible, it should be physically isolated from the solution of interest. To measure the potential of a solution, we select a reference electrode and an appropriate indicator electrode. Whether reduction or oxidation of the substance being analyzed occurs depends on the potential of the half-reaction for the substance of interest (the sample) and the potential of the reference electrode.

There are many possible choices of reference electrode other than the SHE. The SHE requires a constant flow of highly flammable hydrogen gas, which makes it inconvenient to use. Consequently, two other electrodes are commonly chosen as reference electrodes. One is the **silver–silver chloride electrode**, which consists of a silver wire coated with a very thin layer of AgCl that is dipped into a chloride ion solution with a fixed concentration. The cell diagram and reduction half-reaction are

$$Cl^-(aq) \mid AgCl(s) \mid Ag(s)$$

$$AgCl(s) + e^- \longrightarrow Ag(s) + Cl^-(aq) \qquad (19.28)$$

If a saturated solution of KCl is used as the chloride solution, the potential of the silver–silver chloride electrode is 0.197 V versus the SHE. That is, 0.197 V must be subtracted from the measured value to obtain the standard electrode potential measured against the SHE.

A second common reference electrode is the **saturated calomel electrode (SCE)**, which has the same general form as the silver–silver chloride electrode. The SCE consists of a platinum wire inserted into a moist paste of liquid mercury, Hg_2Cl_2 (called *calomel* in the old chemical literature), and KCl. This interior cell is surrounded by an aqueous KCl solution, which acts as a salt bridge between the interior cell and the exterior solution (Figure 19.8a). Although it sounds and looks complex, this cell is actually easy to prepare and maintain, and its potential is highly reproducible. The SCE cell diagram and corresponding half-reaction are

$$Pt(s) \mid Hg_2Cl_2(s) \mid KCl(aq, sat) \qquad (19.29)$$

$$Hg_2Cl_2(s) + 2e^- \longrightarrow 2Hg(l) + 2Cl^-(aq) \qquad (19.30)$$

At 25°C, the potential of the SCE is 0.2415 V versus the SHE, which means that 0.2415 V must be subtracted from the potential versus an SCE to obtain the standard electrode potential.

One of the most common uses of electrochemistry is to measure the H^+ ion concentration of a solution. A **glass electrode** is generally used for this purpose, in which an internal Ag/AgCl electrode is immersed in a 0.10 *M* HCl solution that is separated from the solution by a very thin glass membrane (Figure 19.8b). The glass membrane

Figure 19.8 Three common types of electrodes. (a) The saturated calomel electrode (SCE) is a reference electrode that consists of a platinum wire inserted into a moist paste of liquid mercury, calomel (Hg_2Cl_2), and KCl. The interior cell is surrounded by an aqueous KCl solution, which acts as a salt bridge between the interior cell and the exterior solution. (b) In a glass electrode, an internal Ag/AgCl electrode is immersed in a 1 *M* HCl solution that is separated from the sample solution by a very thin glass membrane. The potential of the electrode depends on the H^+ ion concentration of the sample. (c) The potential of an ion-selective electrode depends on the concentration of only a single ionic species in solution.

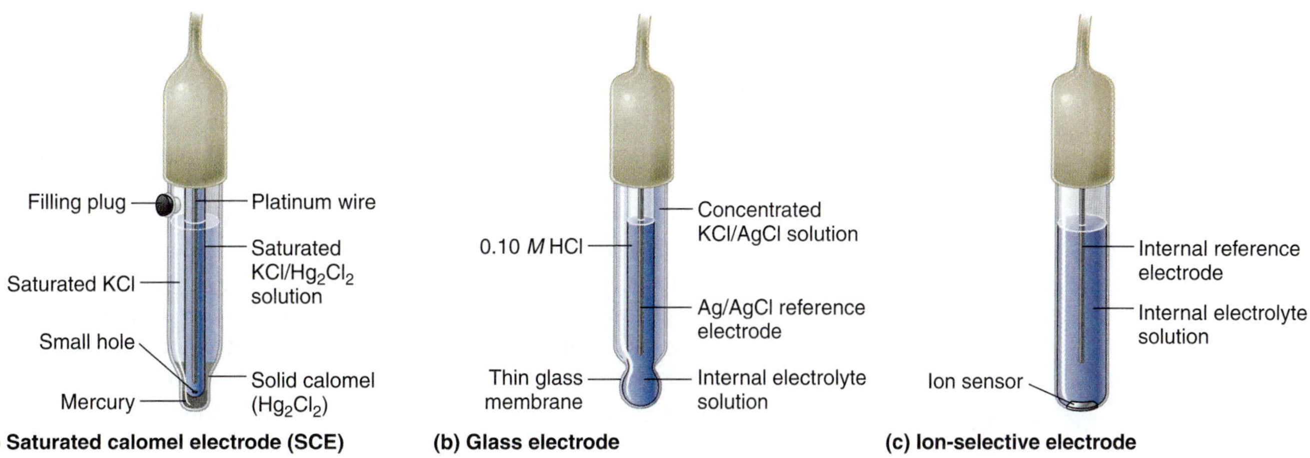

TABLE 19.1 Some species whose aqueous concentrations can be measured using electrochemical methods.

Species	Type of Sample
H^+	Laboratory samples, blood, soil, ground and surface water
NH_3/NH_4^+	Waste water, runoff water
K^+	Blood, wine, soil
CO_2/HCO_3^-	Blood, groundwater
F^-	Groundwater, drinking water, soil
Br^-	Grains, plant extracts
I^-	Milk, pharmaceuticals
NO_3^-	Groundwater, drinking water, soil, fertilizer

absorbs protons, which affects the measured potential. The extent of the adsorption on the inner side is fixed because $[H^+]$ is fixed inside the electrode, but the adsorption of protons on the outer surface depends on the pH of the solution. The potential of the glass electrode depends on $[H^+]$ as follows:

$$E_{glass} = E' + (0.0591\ V \times \log[H^+]) = E' - 0.0591\ V\ (pH) \qquad (19.31)$$

The voltage E' is a constant that depends on the exact construction of the electrode. Although it can be measured, in practice a glass electrode is *calibrated*; that is, it is inserted into a solution of known pH, and the display on the pH meter is adjusted to the known value. Once the electrode is properly calibrated, it can be placed in a solution and used to determine an unknown pH.

Ion-selective electrodes are used to measure the concentration of a particular species in solution; they are designed so that their potential depends on only the concentration of the desired species (Figure 19.8c). These electrodes usually contain an internal reference electrode that is connected by a solution of an electrolyte to a crystalline inorganic material or a membrane, which acts as the sensor. For example, one type of ion-selective electrode uses a single crystal of Eu-doped LaF_3 as the inorganic material. When fluoride ions in solution diffuse to the surface of the solid, the potential of the electrode changes, resulting in a so-called *fluoride electrode*. Similar electrodes are used to measure the concentrations of other species in solution. Some of the species whose concentrations can be determined in aqueous solution using ion-selective electrodes and similar devices are listed in Table 19.1.

19.3 Comparing Strengths of Oxidants and Reductants

We can use the procedure described in Section 19.2 to measure the standard potentials for a wide variety of chemical substances, some of which are listed in Table 19.2. (Appendix E contains a more extensive listing.) These data allow us to compare the oxidative and reductive strengths of a variety of substances. Note that the half-reaction for the standard hydrogen electrode lies more than halfway down the list in Table 19.2. All reactants that lie *above* it in the table are stronger oxidants than H^+, and all those that lie below it are weaker. The strongest oxidant in the table is F_2, with a standard electrode potential of 2.87 V. This high value is consistent with the high electronegativity of fluorine, and tells us that fluorine has a stronger tendency to accept electrons (it is a stronger oxidant) than any other element.

Similarly, all species in Table 19.2 that lie *below* H_2 are stronger reductants than H_2, and those that lie above H_2 are weaker. The strongest reductant in the table is thus metallic lithium, with a standard electrode potential of −3.04 V. This fact might be surprising because cesium is the least electronegative element, not lithium. The apparent anomaly can be explained by the fact that electrode potentials are measured in aqueous solution, where intermolecular interactions are important, whereas ionization potentials and electron affinities are measured in the gas phase. Due to its small size, the Li^+ ion is stabilized in aqueous solution by strong electrostatic interactions with the negative end of the dipoles of water molecules. These interactions result in a significantly greater $\Delta H_{hydration}$ for Li^+ compared with Cs^+. Lithium metal is therefore the strongest reductant (most easily oxidized) of the alkali metals in aqueous solution.

Note the pattern

Species in Table 19.2 that lie below H_2 are stronger reductants (more easily oxidized) than H_2. Species that lie above H_2 are weaker (more easily reduced).

Because the half-reactions shown in Table 19.2 are arranged in order of their $E°$ values, we can use the table to quickly predict the relative strengths of various oxidants and reductants. Any species on the left side of a half-reaction will spontaneously oxidize any species on the right side of another half-reaction that lies *below* it in the table. Conversely, any species on the right side of one half-reaction

TABLE 19.2 Standard potentials for selected reduction half-reactions at 25°C

Half-Reaction	$E°$, V
$F_2(g) + 2e^- \longrightarrow 2F^-(aq)$	2.87
$H_2O_2(aq) + 2H^+(aq) + 2e^- \longrightarrow 2H_2O(l)$	1.78
$Ce^{4+}(aq) + e^- \longrightarrow Ce^{3+}(aq)$	1.72
$PbO_2(s) + HSO_4^-(aq) + 3H^+(aq) + 2e^- \longrightarrow PbSO_4(s) + 2H_2O(l)$	1.69
$Cl_2(g) + 2e^- \longrightarrow 2Cl^-(aq)$	1.36
$Cr_2O_7^-(aq) + 14H^+(aq) + 6e^- \longrightarrow 2Cr^{3+}(aq) + 7H_2O(l)$	1.23
$O_2(g) + 4H^+(aq) + 4e^- \longrightarrow 2H_2O(l)$	1.23
$MnO_2(s) + 4H^+(aq) + 2e^- \longrightarrow Mn^{2+}(aq) + 2H_2O(l)$	1.22
$Br_2(aq) + 2e^- \longrightarrow 2Br^-(aq)$	1.09
$NO_3^-(aq) + 3H^+(aq) + 2e^- \longrightarrow HNO_2(aq) + H_2O(l)$	0.93
$Ag^+(aq) + e^- \longrightarrow Ag(s)$	0.80
$Fe^{3+}(aq) + e^- \longrightarrow Fe^{2+}(aq)$	0.77
$H_2SeO_3(aq) + 4H^+(aq) + 4e^- \longrightarrow Se(s) + 3H_2O(l)$	0.74
$O_2(g) + 2H^+(aq) + 2e^- \longrightarrow H_2O_2(aq)$	0.70
$MnO_4^-(aq) + 2H_2O(l) + 3e^- \longrightarrow MnO_2(s) + 4OH^-(aq)$	0.60
$MnO_4^{2-}(aq) + 2H_2O(l) + 2e^- \longrightarrow MnO_2(s) + 4OH^-(aq)$	0.60
$I_2(s) + 2e^- \longrightarrow 2I^-(aq)$	0.54
$H_2SO_3(aq) + 4H^+(aq) + 4e^- \longrightarrow S(s) + 3H_2O(l)$	0.45
$O_2(g) + 2H_2O(l) + 4e^- \longrightarrow 4OH^-(aq)$	0.40
$Cu^{2+}(aq) + 2e^- \longrightarrow Cu(s)$	0.34
$AgCl(s) + e^- \longrightarrow Ag(s) + Cl^-(aq)$	0.22
$Cu^{2+}(aq) + e^- \longrightarrow Cu^+(aq)$	0.15
$Sn^{4+}(aq) + 2e^- \longrightarrow Sn^{2+}(aq)$	0.15
$2H^+(aq) + 2e^- \longrightarrow H_2(g)$	0.00
$Sn^{2+}(aq) + 2e^- \longrightarrow Sn(s)$	−0.14
$2SO_4^{2-}(aq) + 4H^+(aq) + 2e^- \longrightarrow S_2O_6^{2-}(aq) + H_2O(l)$	−0.22
$Ni^{2+}(aq) + 2e^- \longrightarrow Ni(s)$	−0.26
$PbSO_4(s) + 2e^- \longrightarrow Pb(s) + SO_4^{2-}(aq)$	−0.36
$Cd^{2+}(aq) + 2e^- \longrightarrow Cd(s)$	−0.40
$Fe^{2+}(aq) + 2e^- \longrightarrow Fe(s)$	−0.45
$Ag_2S(s) + 2e^- \longrightarrow 2Ag(s) + S^{2-}(aq)$	−0.69
$Zn^{2+}(aq) + 2e^- \longrightarrow Zn(s)$	−0.76
$Be^{2+}(aq) + 2e^- \longrightarrow Be(s)$	−1.85
$Li^+(aq) + e^- \longrightarrow Li(s)$	−3.04

will spontaneously reduce any species on the left side of another half-reaction that lies *above* it in the table. We can use these generalizations to predict the spontaneity of a wide variety of redox reactions ($E°_{cell} > 0$), as illustrated in the next example.

EXAMPLE 19.5

The black tarnish that forms on silver objects is primarily Ag_2S. The half-reaction for reversing the tarnishing process is

$$Ag_2S(s) + 2e^- \longrightarrow 2Ag(s) + S^{2-}(aq) \qquad E° = -0.69 \text{ V}$$

Referring to Table 19.2, **(a)** predict which of these species can reduce Ag_2S to Ag under standard conditions: $H_2O_2(aq)$, $Zn(s)$, $I^-(aq)$, $Sn^{2+}(aq)$. **(b)** Of these species, identify which is the strongest reducing agent in aqueous solution and thus the best candidate for a commercial product. **(c)** From the data in Table 19.2, suggest an alternative reducing agent that is readily available, inexpensive, and possibly more effective at removing tarnish.

Given Reduction half-reaction and standard electrode potential, list of possible reductants

Asked for Reductants for Ag_2S, strongest reductant, potential reducing agent for removing tarnish

Strategy

A From their positions in Table 19.2, decide which of the species can reduce Ag_2S. Determine which of these is the strongest reductant.

B Use Table 19.2 to identify a reductant for Ag_2S that is a common household product.

Solution

We can solve the problem in one of two ways: by comparing the relative positions of the four possible reductants with that of the Ag_2S/Ag couple in Table 19.2, or by comparing $E°$ for each of the species with $E°$ for the Ag_2S/Ag couple (−0.69 V).

(a) **A** ✓ The species in Table 19.2 are arranged from top to bottom in order of increasing reducing strength. Of the four species given in the problem, $I^-(aq)$, $Sn^{2+}(aq)$, and $H_2O_2(aq)$ lie above Ag_2S, and one, $Zn(s)$, lies below it. We can therefore conclude that $Zn(s)$ can reduce $Ag_2S(s)$ under standard conditions, whereas $I^-(aq)$, $Sn^{2+}(aq)$, and $H_2O_2(aq)$ cannot. Notice that $Sn^{2+}(aq)$ and $H_2O_2(aq)$ appear *twice* in the table: on the left side (oxidant) in one half-reaction and on the right side (reductant) in another.

(b) The strongest reductant is $Zn(s)$, the species on the right side of the half-reaction that lies closer to the bottom of Table 19.2 than the half-reactions involving $I^-(aq)$, $Sn^{2+}(aq)$, and $H_2O_2(aq)$. (Commercial products that use a piece of zinc are often marketed as a "miracle product" for removing tarnish from silver. All that is required is to add warm water and salt for electrical conductivity.)

(c) **B** ✓ Of the reductants that lie *below* $Zn(s)$ in Table 19.2, and therefore are stronger reductants, only one is commonly available in household products: $Al(s)$, which is sold as aluminum foil for wrapping foods.

EXERCISE 19.5

Refer to Table 19.2 to predict **(a)** which of these species can oxidize $MnO_2(s)$ to $MnO_4^-(aq)$ under standard conditions, and **(b)** which is the strongest oxidizing agent in aqueous solution: $Sn^{4+}(aq)$, $Cl^-(aq)$, $Ag^+(aq)$, $Cr^{3+}(aq)$, $H_2O_2(aq)$.

Answer **(a)** $Ag^+(aq)$, $H_2O_2(aq)$; **(b)** $H_2O_2(aq)$

EXAMPLE 19.6

Use the data in Table 19.2 to determine whether each of these reactions is likely to occur spontaneously under standard conditions:

(a) $Sn(s) + Be^{2+}(aq) \longrightarrow Sn^{2+}(aq) + Be(s)$

(b) $MnO_2(s) + H_2O_2(aq) + 2H^+(aq) \longrightarrow O_2(g) + Mn^{2+}(aq) + 2H_2O(l)$

Given Redox reaction, list of standard electrode potentials (Table 19.2)

Asked for Reaction spontaneity

Strategy

Ⓐ Identify the half-reactions in each equation. Using Table 19.2, determine the standard potentials for the half-reactions in the appropriate direction.
Ⓑ Use Equation 19.7 to calculate the standard cell potential for the overall reaction. From this value, determine whether the overall reaction is spontaneous.

Solution

(a) Ⓐ Metallic tin is oxidized to $Sn^{2+}(aq)$, and $Be^{2+}(aq)$ is reduced to elemental beryllium. We can find the standard electrode potentials for the latter (reduction) half-reaction, −1.85 V, and for the former (oxidation) half-reaction, −0.14 V, directly from Table 19.2. Ⓑ Adding the two half-reactions gives the overall reaction:

Cathode:	$Be^{2+}(aq) + 2e^- \longrightarrow Be(s)$	$E^\circ_{cathode} = -1.85$ V
Anode:	$Sn(s) \longrightarrow Sn^{2+}(aq) + 2e^-$	$E^\circ_{anode} = -0.14$ V
Overall:	$Sn(s) + Be^{2+}(aq) \longrightarrow Sn^{2+}(aq) + Be(s)$	$E^\circ_{cell} = E^\circ_{cathode} - E^\circ_{anode} = -1.71$ V

The standard cell potential is quite negative, so the reaction will *not* occur spontaneously as written. That is, metallic tin cannot be used to reduce Be^{2+} to beryllium metal under standard conditions. Instead, the reverse process, the reduction of stannous ions (Sn^{2+}) by metallic beryllium, which has a positive value of E°_{cell}, will occur spontaneously.

(b) Ⓐ MnO_2 is the oxidant (Mn^{4+} is reduced to Mn^{2+}), while H_2O_2 is the reductant (O^{2-} is oxidized to O_2). We can obtain the standard electrode potentials for the reduction and oxidation half-reactions directly from Table 19.2. Ⓑ The two half-reactions and their corresponding potentials are

Cathode:	$MnO_2(s) + 4H^+(aq) + 2e^- \longrightarrow Mn^{2+}(aq) + 2H_2O(l)$	$E^\circ_{cathode} = 1.22$ V
Anode:	$H_2O_2(aq) \longrightarrow O_2(g) + 2H^+(aq) + 2e^-$	$E^\circ_{anode} = 0.70$ V
Overall:	$MnO_2(s) + H_2O_2(aq) + 2H^+(aq) \longrightarrow O_2(g) + Mn^{2+}(aq) + 2H_2O(l)$	$E^\circ_{cell} = E^\circ_{cathode} - E^\circ_{anode} = 0.53$ V

The standard potential for the reaction is positive, indicating that under standard conditions it will occur spontaneously as written. Hydrogen peroxide will reduce MnO_2, and oxygen gas will evolve from the solution.

EXERCISE 19.6

Use the data in Table 19.2 to determine whether each reaction is likely to occur spontaneously under standard conditions:

(a) $2Ce^{4+}(aq) + 2Cl^-(aq) \longrightarrow 2Ce^{3+}(aq) + Cl_2(g)$
(b) $4MnO_2(s) + 3O_2(g) + 4OH^-(aq) \longrightarrow 4MnO_4^-(aq) + 2H_2O$

Answer **(a)** spontaneous ($E^\circ_{cell} = 0.36$ V); **(b)** nonspontaneous ($E^\circ_{cell} = -0.20$ V)

Although the sign of E°_{cell} tells us whether a particular redox reaction will occur spontaneously under standard conditions, it does not tell us to what *extent* the reaction proceeds, nor does it tell us what will happen under nonstandard conditions. To answer these questions requires a more quantitative understanding of the relationship between electrochemical cell potential and chemical thermodynamics, as described in the next section.

Michael Faraday. Faraday (1791–1867) was a British physicist and chemist who was arguably one of the greatest experimental scientists in history. The son of a blacksmith, Faraday was self-educated and became an apprentice bookbinder at age 14 before turning to science. His experiments in electricity and magnetism made electricity a routine tool in science and led to both the electric motor and the electric generator. He discovered the phenomenon of electrolysis and laid the foundations of electrochemistry. In fact, most of the specialized terms introduced in this chapter (*electrode, anode, cathode*, and so on) are due to Faraday. In addition, he discovered benzene and invented the system of oxidation state numbers that we use today. Faraday is probably best known for "The Chemical History of a Candle," a series of public lectures on the chemistry and physics of flames.

19.4 Electrochemical Cells and Thermodynamics

Changes in reaction conditions can have a tremendous effect on the course of a redox reaction. For example, under standard conditions, the reaction of Co(*s*) with Ni^{2+}(*aq*) to form Ni(*s*) and Co^{2+}(*aq*) occurs spontaneously, but if we reduce the concentration of Ni^{2+} by a factor of 100, so that $[Ni^{2+}]$ is 0.01 *M*, then the reverse reaction occurs spontaneously instead. The relationship between voltage and concentration is one of the factors that must be understood in order to predict whether a reaction will be spontaneous.

The Relationship Between Cell Potential and Free Energy

Electrochemical cells convert chemical to electrical energy, and vice versa. The total amount of energy produced by an electrochemical cell, and thus the amount of energy available to do electrical work, depends on both the cell potential and the total number of electrons that are transferred from the reductant to the oxidant during the course of the reaction. The resulting electric current is measured in **coulombs (C)**, an SI unit that measures the number of electrons passing a given point in 1 s. A coulomb relates electrical potential (in volts) to energy (in joules). Electric current is measured in **amperes (A)**; 1 ampere is defined as the flow of 1 coulomb per second past a given point ($1\ C = 1\ A \cdot s$):

$$\frac{1\ J}{1\ V} = 1\ C = A \cdot s \tag{19.32}$$

In chemical reactions, however, we need to relate the coulomb to the charge on a *mole* of electrons. Multiplying the charge on the electron by Avogadro's number gives us the charge on 1 mol of electrons, which is called the **faraday (*F*)**, named after the English physicist and chemist Michael Faraday (1791–1867):

$$\begin{aligned} F &= (1.60218 \times 10^{-19}\ C)\left(\frac{6.02214 \times 10^{23}}{1\ mol\ e^-}\right) \\ &= 9.64855 \times 10^4\ C/mol\ e^- \simeq 96{,}486\ J\ (V \cdot mol\ e^-) \end{aligned} \tag{19.33}$$

The total charge transferred from the reductant to the oxidant is therefore *nF*, where *n* is the number of moles of electrons.

The maximum amount of work that can be produced by an electrochemical cell, w_{max}, is equal to the product of the cell potential, E_{cell}, and the total charge transferred during the reaction, *nF*:

$$w_{max} = -n\,F\,E_{cell} \tag{19.34}$$

Work is expressed as a negative number because work is being done *by* the system (an electrochemical cell with a positive potential) *on* the surroundings.

As you learned in Chapter 18, ΔG is also a measure of the maximum amount of work that can be performed during a chemical process ($\Delta G = w_{max}$). Consequently, there must be a relationship between the potential of an electrochemical cell and the change in free energy, ΔG, the most important thermodynamic quantity discussed in Chapter 18. This relationship is

$$\Delta G = -n\,F\,E_{cell} \tag{19.35}$$

Note the pattern

A spontaneous redox reaction is characterized by a negative value of $\Delta G°$, which corresponds to a positive value of $E°_{cell}$.

A spontaneous redox reaction is therefore characterized by a *negative* value of ΔG and a *positive* value of E_{cell}, consistent with our earlier discussions. When both reactants and products are in their standard states, the relationship between $\Delta G°$ and $E°_{cell}$ is

$$\Delta G° = -n\,F\,E°_{cell} \tag{19.36}$$

EXAMPLE 19.7

Suppose you want to prepare elemental bromine from bromide using the dichromate ion as an oxidant. Using data in Table 19.2, calculate the free-energy change, $\Delta G°$, for this redox reaction under standard conditions. Is the reaction spontaneous?

Given Redox reaction

Asked for $\Delta G°$ for the reaction, spontaneity

Strategy

Ⓐ From the relevant half-reactions and the corresponding values of $E°$, write the overall reaction and calculate $E°_{cell}$ using Equation 19.7.

Ⓑ Determine the number of electrons transferred in the overall reaction. Then use Equation 19.36 to calculate $\Delta G°$. If $\Delta G°$ is negative, then the reaction is spontaneous.

Solution

Ⓐ✓ As always, the first step is to write the relevant half-reactions and use them to obtain the overall reaction and the magnitude of $E°$. From Table 19.2, we can find the reduction and oxidation half-reactions and corresponding $E°$ values:

Cathode:	$Cr_2O_7^{2-}(aq) + 14H^+(aq) + 6e^- \longrightarrow 2Cr^{3+}(aq) + 7H_2O(l)$	$E°_{cathode} = 1.23\ V$
Anode:	$2Br^-(aq) \longrightarrow Br_2(aq) + 2e^-$	$E°_{anode} = 1.09\ V$

To obtain the overall balanced reaction, we must multiply both sides of the oxidation half-reaction by 3 to obtain the same number of electrons as in the reduction half-reaction, remembering that the magnitude of $E°$ is not affected:

Cathode:	$Cr_2O_7^{2-}(aq) + 14H^+(aq) + 6e^- \longrightarrow 2Cr^{3+}(aq) + 7H_2O(l)$	$E°_{cathode} = 1.23\ V$
Anode:	$6Br^-(aq) \longrightarrow 3Br_2(aq) + 6e^-$	$E°_{anode} = 1.09\ V$
Overall:	$Cr_2O_7^{2-}(aq) + 6Br^-(aq) + 14H^+(aq) \longrightarrow 2Cr^{3+}(aq) + 3Br_2(aq) + 7H_2O(l)$	$E°_{cell} = E°_{cathode} - E°_{anode} = 0.14\ V$

Ⓑ✓ We can now calculate $\Delta G°$ using Equation 19.36. Because *six* electrons are transferred in the overall reaction, the value of n is 6:

$$\begin{aligned}\Delta G° = -nFE°_{cell} &= -(6\ \cancel{mole})[96{,}486\ J/(\cancel{V}\cdot\cancel{mol})](0.14\ \cancel{V})\\ &= -8.1\times10^4\ J\\ &= -81\ kJ/mol\ Cr_2O_7^{2-}\end{aligned}$$

Thus, $\Delta G°$ is -81 kJ for the reaction as written, and the reaction is spontaneous.

EXERCISE 19.7

Use the data in Table 19.2 to calculate $\Delta G°$ for the reduction of ferric ion by iodide:

$$2Fe^{3+}(aq) + 2I^-(aq) \longrightarrow 2Fe^{2+}(aq) + I_2(s)$$

Is the reaction spontaneous?

Answer -44 kJ/mol I_2; yes

Potentials for the Sums of Half-Reactions

Only a limited number of half-reactions are given in Table 19.2. When the standard potential for a half-reaction is not available, we can use relationships between standard potentials and free energy to obtain the potential of any other half-reaction that can be written as the sum of two or more half-reactions whose standard potentials are available.

For example, the potential for the reduction of $Fe^{3+}(aq)$ to Fe(s) is not listed in the table, but two related reductions are given:

$$Fe^{3+}(aq) + e^- \longrightarrow Fe^{2+}(aq) \qquad E^\circ = 0.77\text{ V} \tag{19.37a}$$

$$Fe^{2+}(aq) + 2e^- \longrightarrow Fe(s) \qquad E^\circ = -0.45\text{ V} \tag{19.37b}$$

Although the sum of these two half-reactions gives the desired half-reaction, we cannot simply add the potentials of two reductive half-reactions to obtain the potential of a third reductive half-reaction *because E° is not a state function*. However, because ΔG° is a state function, the sum of the ΔG° values for the individual reactions gives us ΔG° for the overall reaction, which is proportional to *both* the potential *and* the number of electrons (n) transferred. To obtain the value of E° for the overall half-reaction, we first must add the values of ΔG° ($= -n\,F\,E^\circ$) for each half-reaction (Equations 19.37a and 19.37b) to obtain ΔG° for the overall half-reaction:

$$\begin{array}{lll} Fe^{3+}(aq) + e^- \longrightarrow \cancel{Fe^{2+}(aq)} & \Delta G^\circ = -(1)\,F(0.77\text{ V}) & \\ \cancel{Fe^{2+}(aq)} + 2e^- \longrightarrow Fe(s) & \Delta G^\circ = -(2)\,F(-0.45\text{ V}) & \\ \hline Fe^{3+}(aq) + 3e^- \longrightarrow Fe(s) & \Delta G^\circ = [-(1)\,F(0.77\text{ V})] + [-(2)\,F(-0.45\text{ V})] & (19.38) \end{array}$$

Solving the last expression for ΔG° for the overall half-reaction gives

$$\Delta G^\circ = F[(-0.77\text{ V}) + (-2)(-0.45\text{ V})] = F(0.13\text{ V}) \tag{19.39}$$

Note the pattern

Values of E° for half-reactions cannot be added to give E° for the sum of the half-reactions; only values of $\Delta G^\circ = -n\,F\,E^\circ$ for half-reactions can be added.

Three electrons ($n = 3$) are transferred in the overall reaction (Equation 19.38), so substituting into Equation 19.36 and solving for E° give

$$\begin{aligned} \Delta G^\circ &= -n\,F\,E^\circ \\ F(0.13\text{ V}) &= -3\,F\,E^\circ \\ E^\circ &= -\frac{0.13\text{ V}}{3} = -0.043\text{ V} \end{aligned}$$

Notice that the value of E° is *very* different from the value that is obtained by simply adding the potentials for the two half-reactions (0.32 V), and even has the opposite sign!

The Relationship Between Cell Potential and the Equilibrium Constant

We can use the relationship between ΔG° and the equilibrium constant K, defined in Chapter 18, to obtain a relationship between E°_{cell} and K. Recall that for a general reaction of the type $a\text{A} + b\text{B} \longrightarrow c\text{C} + d\text{D}$, the standard free-energy change and the equilibrium constant are related by the equation

$$\Delta G^\circ = -RT\ln K \tag{19.40}$$

Given the relationship between the standard free-energy change and the standard cell potential (Equation 19.36), we can write

$$-n\,F\,E^\circ_{cell} = -RT\ln K \tag{19.41}$$

Rearranging this equation gives

$$E^\circ_{cell} = \left(\frac{RT}{n\,F}\right)\ln K \tag{19.42}$$

For $T = 298$ K, Equation 19.42 can be simplified to

$$E^\circ_{cell} = \left(\frac{RT}{n\,F}\right)\ln K = \left[\frac{[8.314\ \cancel{\text{J}}/(\cancel{\text{mol}}\cdot\cancel{\text{K}})](298\ \cancel{\text{K}})}{n[96{,}486\ \cancel{\text{J}}/(\text{V}\cdot\cancel{\text{mol}})]}\right]2.303\log K = \left(\frac{0.0591\text{ V}}{n}\right)\log K \tag{19.43}$$

Thus, the standard cell potential, E°_{cell}, is directly proportional to the logarithm of the equilibrium constant. This means that large equilibrium constants correspond to large positive values of E°_{cell}, and vice versa.

EXAMPLE 19.8

Use the data in Table 19.2 to calculate the equilibrium constant for the reaction of metallic lead with PbO_2 in the presence of sulfate ions to give $PbSO_4$ under standard conditions. (This is the reaction that occurs when a car battery is discharged.) Report your answer to two significant figures.

Given Redox reaction

Asked for Equilibrium constant K

Strategy

A Write the relevant half-reactions and potentials. From these, obtain the overall reaction and E°_{cell}.

B Determine the number of electrons transferred in the overall reaction. Use Equation 19.43 to solve for log K and then K.

Solution

A The relevant half-reactions and potentials from Table 19.2 are

Cathode:	$PbO_2(s) + SO_4^{2-}(aq) + 4H^+(aq) + 2e^- \longrightarrow PbSO_4(s) + 2H_2O(l)$	$E^\circ_{\text{cathode}} = 1.69\text{ V}$
Anode:	$Pb(s) + SO_4^{2-}(aq) \longrightarrow PbSO_4(s) + 2e^-$	$E^\circ_{\text{anode}} = -0.36\text{ V}$
Overall:	$Pb(s) + PbO_2(s) + 2SO_4^{2-}(aq) + 4H^+(aq) \longrightarrow 2PbSO_4(s) + 2H_2O(l)$	$E^\circ_{\text{cell}} = E^\circ_{\text{cathode}} - E^\circ_{\text{anode}} = 2.05\text{ V}$

B Two electrons are transferred in the overall reaction, so $n = 2$. Solving Equation 19.43 for log K and inserting the values of n and E° give

$$\log K = \frac{nE^\circ}{0.0591\text{ V}} = \frac{2(2.05\text{ V})}{0.0591\text{ V}} = 69.37$$

$$K = 2.3 \times 10^{69}$$

Thus, the equilibrium lies far to the right, favoring a discharged battery (as anyone who has ever tried unsuccessfully to start a car after letting it sit for a long time will know).

EXERCISE 19.8

Use data in Table 19.2 to calculate the equilibrium constant for the reaction of $Sn^{2+}(aq)$ with oxygen to produce $Sn^{4+}(aq)$ and water under standard conditions. Report your answer to two significant figures. The reaction is

$$2Sn^{2+}(aq) + O_2(g) + 4H^+(aq) \rightleftharpoons 2Sn^{4+}(aq) + 2H_2O(l)$$

Answer 1.2×10^{73}

Figure 19.9 summarizes the relationships that we have developed based on properties of the *system*—that is, based on the equilibrium constant, standard free-energy change, and standard cell potential—and the criteria for spontaneity ($\Delta G^\circ < 0$). Unfortunately, these criteria apply only to systems in which all reactants and products are present in their standard states, a situation that is seldom encountered in the real world. A more generally useful relationship between cell potential and reactant and product concentrations, as we are about to see, utilizes the relationship between ΔG and the reaction quotient Q developed in Chapter 18.

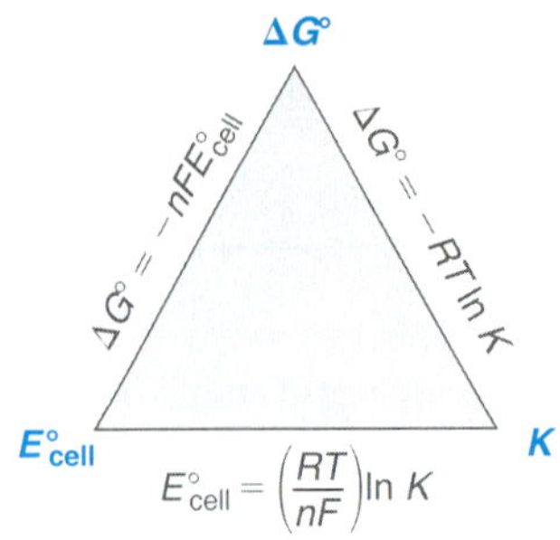

ΔG°	E°_{cell}	K	Direction of Reaction
<0	>0	>1	Spontaneous in forward direction
>0	<0	<1	Spontaneous in reverse direction
0	0	1	No net reaction: system at equilibrium

Figure 19.9 The relationships among criteria for thermodynamic spontaneity. The three properties of a system that can be used to predict the spontaneity of a redox reaction under standard conditions are the equilibrium constant K, the standard free-energy change ΔG°, and the standard cell potential E°_{cell}. If we know the value of one of these quantities, then these relationships enable us to calculate the value of the other two. The signs of ΔG° and E°_{cell} and the magnitude of K determine the direction of spontaneous reaction under standard conditions.

The Effect of Concentration on Cell Potential: The Nernst Equation

Recall from Chapter 18 that the *actual* free-energy change for a reaction under nonstandard conditions, ΔG, is given by

$$\Delta G = \Delta G^\circ + RT \ln Q \tag{19.44}$$

We also know that $\Delta G = -nFE_{cell}$ and $\Delta G^\circ = -nFE^\circ_{cell}$. Substituting these expressions into Equation 19.44, we obtain

$$-nFE_{cell} = -nFE^\circ_{cell} + RT \ln Q \tag{19.45}$$

Dividing both sides of this equation by $-nF$ gives

$$E_{cell} = E^\circ_{cell} - \left(\frac{RT}{nF}\right) \ln Q \tag{19.46}$$

Equation 19.46 is called the **Nernst equation**, after the German physicist and chemist Walter Nernst (1864–1941), who first derived it. The Nernst equation is arguably the most important relationship in electrochemistry. When a redox reaction is at equilibrium ($\Delta G = 0$), Equation 19.46 reduces to Equation 19.42 because $Q = K$ and there is no net transfer of electrons (that is, $E_{cell} = 0$).

Note the pattern

The Nernst equation can be used to determine the value of E_{cell}, and thus the direction of spontaneous reaction, for any redox reaction under any conditions.

Substituting the values of the constants into Equation 19.46 with $T = 298$ K and converting to base-10 logarithms give the relationship of the actual cell potential (E_{cell}), the standard cell potential (E°_{cell}), and the reactant and product concentrations at room temperature (contained in Q):

$$E_{cell} = E^\circ_{cell} - \left(\frac{0.0591\ \text{V}}{n}\right) \log Q \tag{19.47}$$

Equation 19.47 allows us to calculate the potential associated with any electrochemical cell at 298 K for any combination of reactant and product concentrations under any conditions. We can therefore determine the spontaneous direction of any redox reaction under any conditions, as long as we have tabulated values for the relevant standard electrode potentials. Notice in Equation 19.47 that the cell potential changes by $0.0591/n$ V for each 10-fold change in the value of Q because $\log 10 = 1$.

EXAMPLE 19.9

In Exercise 19.6, you determined that the following reaction proceeds spontaneously under standard conditions because $E^\circ_{cell} > 0$ (which you now know means that $\Delta G^\circ < 0$):

$$2Ce^{4+}(aq) + 2Cl^-(aq) \longrightarrow 2Ce^{3+}(aq) + Cl_2(g) \qquad E^\circ_{cell} = 0.25\ \text{V}$$

Calculate E for this reaction under the following nonstandard conditions, and determine whether it will occur spontaneously: $[Ce^{4+}] = 0.013\ M$, $[Ce^{3+}] = 0.60\ M$, $[Cl^-] = 0.0030\ M$, $P_{Cl_2} = 1.0$ atm, and $T = 25°C$.

Given Balanced redox reaction, standard cell potential, nonstandard conditions

Asked for Cell potential

Strategy

Determine the number of electrons transferred during the redox process. Then use the Nernst equation to find the cell potential under the nonstandard conditions.

Solution

We can use the information given and the Nernst equation to calculate E_{cell}. Moreover, because the temperature is 25°C (298 K), we can use Equation 19.47 instead of 19.46. The overall reaction involves the net transfer of two electrons:

$$2Ce^{4+}(aq) + 2e^- \longrightarrow 2Ce^{3+}(aq) \quad \text{and} \quad 2Cl^-(aq) \longrightarrow Cl_2(g) + 2e^-$$

so $n = 2$. Substituting the concentrations given in the problem, the partial pressure of Cl_2, and the value of E°_{cell} into Equation 19.47 gives

$$\begin{aligned} E_{cell} &= E^\circ_{cell} - \left(\frac{0.0591 \text{ V}}{n}\right) \log Q \\ &= 0.25 \text{ V} - \left(\frac{0.0591 \text{ V}}{2}\right) \log \left(\frac{[Ce^{3+}]^2 P_{Cl_2}}{[Ce^{4+}]^2 [Cl^-]^2}\right) \\ &= 0.25 \text{ V} - [(0.0296 \text{ V})(8.37)] = 0.00 \text{ V} \end{aligned}$$

Thus, the reaction will *not* occur spontaneously under these conditions (because $E = 0$ V, $\Delta G = 0$). The composition specified is that of an equilibrium mixture.

EXERCISE 19.9

In Exercise 19.6, you determined that molecular oxygen will not oxidize MnO_2 to permanganate via the reaction

$$4MnO_2(s) + 3O_2(g) + 4OH^-(aq) \longrightarrow 4MnO_4^-(aq) + 2H_2O(l) \qquad E^\circ_{cell} = -0.20 \text{ V}$$

Calculate E_{cell} for the reaction under the following nonstandard conditions, and decide whether the reaction will occur spontaneously: pH 10, 0.20 atm O_2, 1.0×10^{-4} M MnO_4^-, and $T = 25$°C.

Answer $E_{cell} = -0.22$ V; reaction will not occur spontaneously

Applying the Nernst equation to a simple electrochemical cell such as the Zn/Cu cell discussed earlier allows us to see how the cell voltage varies as the reaction progresses and the concentrations of the dissolved ions change. Recall that the overall reaction for this cell is

$$Zn(s) + Cu^{2+}(aq) \longrightarrow Zn^{2+}(aq) + Cu(s) \qquad E^\circ_{cell} = 1.10 \text{ V} \quad (19.48)$$

The reaction quotient is therefore $Q = [Zn^{2+}]/[Cu^{2+}]$. Suppose that the cell initially contains 1.0 M Cu^{2+} and 1.0×10^{-6} M Zn^{2+}. The initial voltage measured when the cell is connected can be calculated from Equation 19.47:

$$\begin{aligned} E_{cell} &= E^\circ_{cell} - \left(\frac{0.0591 \text{ V}}{n}\right) \log \frac{[Zn^{2+}]}{[Cu^{2+}]} \\ &= 1.10 \text{ V} - \left(\frac{0.0591 \text{ V}}{2}\right) \log \left(\frac{1.0 \times 10^{-6}}{1.0}\right) = 1.28 \text{ V} \end{aligned} \quad (19.49)$$

Thus, the initial voltage is *higher* than E° because $Q < 1$. As the reaction proceeds, $[Zn^{2+}]$ in the anode compartment increases as the zinc electrode dissolves, while $[Cu^{2+}]$ in the cathode compartment decreases as metallic copper is deposited on the electrode. During

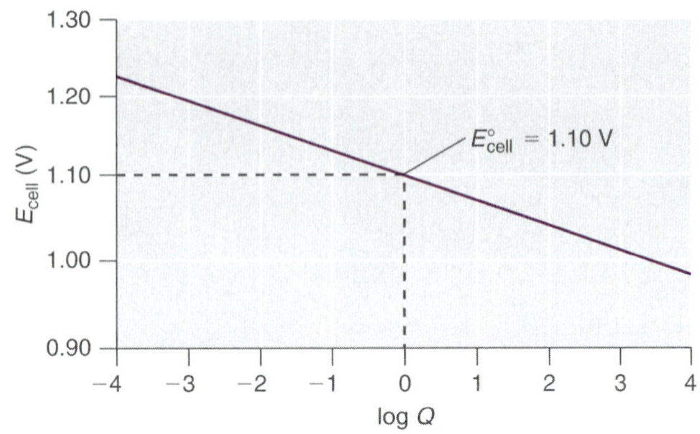

Figure 19.10 The variation of E_{cell} with log Q for a Zn/Cu cell. Initially, log $Q < 0$ and the voltage of the cell is higher than E°_{cell}. As the reaction progresses, log Q increases and E_{cell} decreases. When $[Zn^{2+}] = [Cu^{2+}]$, log $Q = 0$ and $E_{cell} = E^\circ_{cell} = 1.10$ V. As long as the electrical circuit remains intact, the reaction will continue and log Q will increase until $Q = K$ and the cell voltage reaches zero. At this point, the system will have reached equilibrium.

this process, the ratio $Q = [Zn^{2+}]/[Cu^{2+}]$ steadily increases, and the cell voltage therefore steadily decreases. Eventually, $[Zn^{2+}] = [Cu^{2+}]$, so $Q = 1$ and $E_{cell} = E^\circ_{cell}$. Beyond this point, $[Zn^{2+}]$ will continue to increase in the anode compartment and $[Cu^{2+}]$ will continue to decrease in the cathode compartment. Thus, the value of Q will increase further, leading to a further decrease in E_{cell}. When the concentrations in the two compartments are the opposite of the initial concentrations (that is, 1.0 M Zn^{2+} and 1.0×10^{-6} M Cu^{2+}), $Q = 1.0 \times 10^6$ and the cell potential will be reduced to 0.92 V.

The variation of E_{cell} with log Q over this range is linear with a slope of $-0.0591/n$, as illustrated in Figure 19.10. As the reaction proceeds still further, Q continues to increase and E_{cell} continues to decrease. If neither of the electrodes dissolves completely, thereby breaking the electrical circuit, the cell voltage will eventually reach zero. This is the situation that occurs when a battery is "dead." The value of Q when $E_{cell} = 0$ is calculated as follows:

$$E_{cell} = E^\circ_{cell} - \left(\frac{0.0591\text{ V}}{n}\right)\log Q = 0 \tag{19.50}$$

$$E^\circ = \left(\frac{0.0591\text{ V}}{n}\right)\log Q$$

$$\log Q = \frac{E^\circ n}{0.0591\text{ V}} = \frac{(1.10\text{ V})(2)}{0.0591\text{ V}} = 37.23$$

$$Q = 1.7 \times 10^{37}$$

Recall that at equilibrium, $Q = K$. Thus, the equilibrium constant for the reaction of Zn metal with Cu^{2+} to give Cu metal and Zn^{2+} is 1.7×10^{37} at 25°C.

Concentration Cells

A voltage can also be generated by constructing an electrochemical cell in which each compartment contains the same redox active solution but at different concentrations. The voltage is produced as the concentrations equilibrate. Suppose, for example, we have a cell with 0.01 M $AgNO_3$ in one compartment and 1.0 M $AgNO_3$ in the other. The cell diagram and corresponding half-reactions are

$$Ag(s) \mid Ag^+(aq, 0.010\ M) \parallel Ag^+(aq, 1.0\ M) \mid Ag(s) \tag{19.51}$$

$$\text{Cathode:}\quad Ag^+(aq, 1.0\ M) + e^- \longrightarrow Ag(s) \tag{19.52a}$$

$$\text{Anode:}\quad Ag(s) \longrightarrow Ag^+(aq, 0.01\ M) + e^- \tag{19.52b}$$

$$\text{Overall:}\quad Ag^+(aq, 1.0\ M) \longrightarrow Ag^+(aq, 0.010\ M) \tag{19.52c}$$

As the reaction progresses, the concentration of Ag^+ will increase in the left (oxidation) compartment as the silver electrode dissolves, while the Ag^+ concentration in the right (reduction) compartment decreases as the electrode in that compartment gains mass. The *total* mass of Ag(s) in the cell will remain constant, however. We can calculate the potential of the cell using the Nernst equation, inserting 0 for E°_{cell} because $E^\circ_{cathode} = -E^\circ_{anode}$:

$$E_{cell} = E^\circ_{cell} - \left(\frac{0.0591\text{ V}}{n}\right)\log Q = 0 - \left(\frac{0.0591\text{ V}}{1}\right)\log\left(\frac{0.010}{1.0}\right) = 0.12\text{ V}$$

An electrochemical cell of this type, in which the anode and cathode compartments are identical except for the concentration of a reactant, is called a **concentration cell**. As the reaction proceeds, the difference between the concentrations of Ag^+ in the two compartments will decrease, and so will E_{cell}. Finally, when the concentration of Ag^+ is the same in both compartments, equilibrium will have been reached, and the measured potential difference between the two compartments will be zero ($E_{cell} = 0$).

EXAMPLE 19.10

Calculate the voltage in a galvanic cell that contains a manganese electrode immersed in a 2.0 *M* solution of $MnCl_2$ as the cathode, and a manganese electrode immersed in a 5.2×10^{-2} *M* solution of $MnSO_4$ as the anode ($T = 25°C$).

Given Galvanic cell, identities of the electrodes, solution concentrations

Asked for Voltage

Strategy

Ⓐ Write the overall reaction that occurs in the cell.
Ⓑ Determine the number of electrons transferred. Substitute this value into the Nernst equation to calculate the voltage.

Solution

Ⓐ This is a concentration cell, in which the electrode compartments contain the same redox active substance but at different concentrations. The anions (Cl^- and SO_4^{2-}) do not participate in the reaction, so their identity is not important. The overall reaction is

$$Mn^{2+}(aq, 2.0\ M) \longrightarrow Mn^{2+}(aq, 5.2 \times 10^{-2}\ M)$$

Ⓑ For the reduction of $Mn^{2+}(aq)$ to $Mn(s)$, $n = 2$. We substitute this value and the given Mn^{2+} concentrations into Equation 19.47:

$$E_{cell} = E^{\circ}_{cell} - \left(\frac{0.0591\ V}{n}\right)\log Q = 0\ V - \left(\frac{0.0591\ V}{2}\right)\log\left(\frac{5.2 \times 10^{-2}}{2.0}\right) = 0.047\ V$$

Thus, manganese will dissolve from the electrode in the compartment that contains the more dilute solution and will be deposited on the electrode in the compartment that contains the more concentrated solution.

EXERCISE 19.10

Suppose we construct a galvanic cell by placing two identical platinum electrodes in two beakers that are connected by a salt bridge. One beaker contains 1.0 *M* HCl, and the other a 0.010 *M* solution of Na_2SO_4 at pH 7.0. Both cells are in contact with the atmosphere, with $P_{O_2} = 0.20$ atm. If the relevant electrochemical reaction in both compartments is the four-electron reduction of oxygen to water, $O_2(g) + 4H^+(aq) + 4e^- \longrightarrow 2H_2O(l)$, what will be the potential when the circuit is closed?

Answer 0.41 V

Using Cell Potentials to Measure Solubility Products

Because voltages are relatively easy to measure accurately using a voltmeter, electrochemical methods provide a convenient way to determine the concentrations of very dilute solutions and the solubility products, K_{sp}, of sparingly soluble substances. As you learned in Chapter 17, solubility products can be very small, with values of $\leq 10^{-30}$. Equilibrium constants of this magnitude are virtually impossible to measure accurately by direct methods, so we must use alternative methods that are more sensitive, such as electrochemical methods.

To understand how an electrochemical cell is used to measure a solubility product, consider the cell shown in Figure 19.11, which is designed to measure the solubility product of silver chloride, $K_{sp} = [Ag^+][Cl^-]$. In one compartment, the cell contains a silver wire inserted into a 1.0 *M* solution of Ag^+; the other compartment contains a silver wire inserted into a 1.0 *M* Cl^- solution saturated with AgCl. In this system, the Ag^+ ion

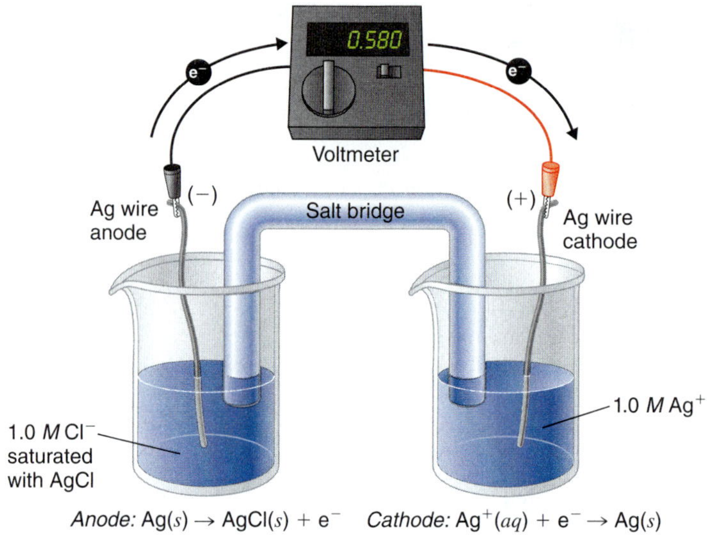

Figure 19.11 A galvanic cell for measuring the solubility product of AgCl. One compartment contains a silver wire inserted into a 1.0 *M* solution of Ag^+, and the other compartment contains a silver wire inserted into a 1.0 *M* Cl^- solution saturated with AgCl. The potential due to the difference in $[Ag^+]$ between the two cells can be used to determine K_{sp}.

concentration in the first compartment equals K_{sp}. We can see this by dividing both sides of the equation for K_{sp} by $[Cl^-]$ and substituting: $[Ag^+] = K_{sp}/[Cl^-] = K_{sp}/1.0 = K_{sp}$. The overall cell reaction is $Ag^+(aq, \text{concentrated}) \longrightarrow Ag^+(aq, \text{dilute})$. Thus, the voltage of the concentration cell due to the difference in $[Ag^+]$ between the two cells is

$$E_{cell} = 0\text{ V} - \left(\frac{0.0591\text{ V}}{1}\right)\log\left(\frac{[Ag^+]_{dilute}}{[Ag^+]_{concentrated}}\right) = -0.0591\text{ V}\log\left(\frac{K_{sp}}{1.0}\right) = -0.0591\text{ V}\log K_{sp} \qquad (19.53)$$

By closing the circuit, we can measure the potential caused by the difference in $[Ag^+]$ in the two cells. In this case, the experimentally measured voltage of the concentration cell at 25°C is 0.580 V. Solving Equation 19.53 for K_{sp} gives

$$\log K_{sp} = \frac{-E_{cell}}{0.0591\text{ V}} = \frac{-0.580\ \cancel{\text{V}}}{0.0591\ \cancel{\text{V}}} = -9.81 \qquad (19.54)$$

$$K_{sp} = 1.5 \times 10^{-10}$$

Thus, a single potential measurement can provide the information we need to determine the value of the solubility product of a sparingly soluble salt.

EXAMPLE 19.11

To measure the solubility product of lead(II) sulfate ($PbSO_4$) at 25°C, you construct a galvanic cell like the one shown in Figure 19.11, which contains a 1.0 *M* solution of a very soluble Pb^{2+} salt [lead(II) acetate trihydrate] in one compartment that is connected by a salt bridge to a 1.0 *M* solution of Na_2SO_4 saturated with $PbSO_4$ in the other. You then insert a Pb electrode into each compartment and close the circuit. Your voltmeter shows a voltage of 230 mV. What is K_{sp} for $PbSO_4$? Report your answer to two significant figures.

Given Galvanic cell, solution concentrations, electrodes, voltage

Asked for K_{sp}

Strategy

Ⓐ From the information given, write the equation for K_{sp}. Express this equation in terms of the concentration of Pb^{2+}.

Ⓑ Determine the number of electrons transferred in the electrochemical reaction. Substitute the appropriate values into Equation 19.53 and solve for K_{sp}.

Solution

Ⓐ You have constructed a concentration cell, with one compartment containing a 1.0 *M* solution of Pb^{2+} and the other containing a dilute solution of Pb^{2+} in 1.0 *M* Na_2SO_4. As for any concentration cell, the voltage between the two compartments can be calculated using the Nernst equation. The first step is to relate the concentration of Pb^{2+} in the dilute solution to K_{sp}:

$$[Pb^{2+}][SO_4^{2-}] = K_{sp}$$

$$[Pb^{2+}] = \frac{K_{sp}}{[SO_4^{2-}]} = \frac{K_{sp}}{1.0\ M} = K_{sp}$$

B The reduction of Pb^{2+} to Pb is a two-electron process, and proceeds according to the reaction Pb^{2+} (*aq*, concentrated) $\longrightarrow$ Pb^{2+} (*aq*, dilute), so

$$E_{cell} = E^{\circ}_{cell} - \left(\frac{0.0591\ V}{n}\right) \log Q$$

$$0.230\ V = 0\ V - \left(\frac{0.0591\ V}{2}\right) \log\left(\frac{[Pb^{2+}]_{dilute}}{[Pb^{2+}]_{concentrated}}\right) = -0.0296\ V \log\left(\frac{K_{sp}}{1.0}\right)$$

$$K_{sp} = 1.7 \times 10^{-8}$$

EXERCISE 19.11

A concentration cell similar to the one described in Example 19.11 contains a 1.0 *M* solution of lanthanum nitrate [$La(NO_3)_3$] in one compartment and a 1.0 *M* solution of sodium fluoride saturated with LaF_3 in the other. A metallic La strip is inserted into each compartment, and the circuit is closed. The measured potential is 0.32 V. What is the value of K_{sp} for LaF_3? Report your answer to two significant figures.

Answer 5.7×10^{-17}

Using Cell Potentials to Measure Concentrations

Another use for the Nernst equation is to calculate the concentration of a species given a measured potential and the concentrations of all the other species. We saw an example of this in Example 19.11, in which the experimental conditions were defined in such a way that the concentration of the metal ion was equal to K_{sp}. Potential measurements can be used to obtain the concentrations of dissolved species under other conditions as well, which explains the widespread use of electrochemical cells in many analytical devices. Perhaps the most common application is in the determination of $[H^+]$ using a pH meter, as illustrated in the next example.

EXAMPLE 19.12

Suppose a galvanic cell is constructed with a standard Zn/Zn^{2+} couple in one compartment and a modified hydrogen electrode in the second compartment (Figure 19.6). The pressure of hydrogen gas is 1.0 atm, but $[H^+]$ in the second compartment is unknown. The cell diagram is

$$Zn(s)\,|\,Zn^{2+}(aq,\ 1.0\ M)\,\|\,H^+(aq,\ ?\ M)\,|\,H_2(g,\ 1.0\ atm)\,|\,Pt(s)$$

What is the pH of the solution in the second compartment if the measured potential in the cell is 0.26 V at 25°C?

Given Galvanic cell, cell diagram, cell potential

Asked for pH of the solution

Strategy

A Write the overall cell reaction.
B Substitute appropriate values into the Nernst equation and solve for $-\log[H^+]$ to obtain the pH.

Solution

A Under standard conditions, the overall reaction that occurs is the reduction of protons by zinc to give H_2 (note that Zn lies below H_2 in Table 19.2):

$$Zn(s) + 2H^+(aq) \longrightarrow Zn^{2+}(aq) + H_2(g) \qquad E^\circ = 0.76\ V$$

B By substituting the given values into the simplified Nernst equation (Equation 19.47), we can calculate $[H^+]$ under nonstandard conditions:

$$E_{cell} = E^\circ_{cell} - \left(\frac{0.0591\ V}{n}\right)\log\left(\frac{[Zn^{2+}]P_{H_2}}{[H^+]^2}\right)$$

$$0.26\ V = 0.76\ V - \left(\frac{0.0591\ V}{2}\right)\log\left(\frac{(1.0)(1.0)}{[H^+]^2}\right)$$

$$16.9 = \log\left(\frac{1}{[H^+]^2}\right) = \log[H^+]^{-2} = (-2)\log[H^+]$$

$$8.46 = -\log[H^+]$$

$$pH = 8.5$$

Thus, the potential of a galvanic cell can be used to measure the pH of a solution.

EXERCISE 19.12

Suppose you work for an environmental laboratory and you want to use an electrochemical method to measure the concentration of Pb^{2+} in groundwater. You construct a galvanic cell using a standard oxygen electrode in one compartment ($E^\circ_{cathode} = 1.23\ V$). The other compartment contains a strip of lead in a sample of groundwater to which you have added sufficient acetic acid, a weak organic acid, to ensure electrical conductivity. The cell diagram is

$$Pb(s)\,\|\,Pb^{2+}(aq,\ ?\ M)\,\|\,H^+(aq),\ 1.0\ M\,|\,O_2(g,\ 1.0\ atm)\,|\,Pt(s)$$

When the circuit is closed, the cell has a measured potential of 1.62 V. Use Table 19.3 and Appendix E to determine the concentration of Pb^{2+} in the groundwater.

Answer $1.18 \times 10^{-9}\ M$

19.5 • Commercial Galvanic Cells

Because galvanic cells can be self-contained and portable, they can be used as batteries and fuel cells. A **battery**, also called a *storage cell*, is a galvanic cell (or a series of galvanic cells) that contains all the reactants needed to produce electricity. In contrast, a **fuel cell** is a galvanic cell that requires a constant external supply of one or more reactants in order to generate electricity. In this section, we describe the chemistry behind some of the more common types of batteries and fuel cells.

Batteries

There are two basic kinds of batteries: *disposable*, or *primary*, batteries, in which the electrode reactions are effectively irreversible and which cannot be recharged; and *rechargeable*, or *secondary*, batteries, which form an insoluble product that adheres to the electrodes. These batteries can be recharged by applying an electrical potential in the reverse direction. The recharging process temporarily converts a rechargeable battery from a galvanic cell to an electrolytic cell.

Batteries are cleverly engineered devices that are based on the same fundamental laws as galvanic cells. The major difference between batteries and the galvanic cells we have described is that commercial batteries use solids or pastes rather than solutions as reactants to maximize the electrical output per unit mass. The use of highly concentrated or solid reactants has another beneficial effect: the concentrations of the reactants and products do not change greatly as the battery is discharged, and consequently the output voltage remains remarkably constant during the discharge process. This behavior is in contrast to that of the Zn/Cu cell, whose output decreases logarithmically as the reaction proceeds (Figure 19.10).

When a battery consists of more than one galvanic cell, the cells are usually connected in series—that is, with the positive (+) terminal of one cell connected to the negative (−) terminal of the next, and so on. The overall voltage of the battery is therefore the sum of the voltages of the individual cells.

Leclanché Dry Cell

The dry cell, by far the most common type of battery, is used in flashlights, Walkmen, and GameBoys. Although the dry cell was patented in 1866 by the French chemist Georges Leclanché and more than 5 *billion* such cells are sold every year, the details of its electrode chemistry are still not completely understood. In spite of its name, the Leclanché dry cell is actually a "wet cell": the electrolyte is an acidic water-based paste containing MnO_2, NH_4Cl, $ZnCl_2$, graphite, and starch (Figure 19.12a). The half-reactions at the anode and cathode can be summarized as follows:

$$\text{Cathode:} \quad 2MnO_2(s) + 2NH_4^+(aq) + 2e^- \longrightarrow Mn_2O_3(s) + 2NH_3(aq) + H_2O(l) \qquad (19.55a)$$

$$\text{Anode:} \quad Zn(s) \longrightarrow Zn^{2+}(aq) + 2e^- \qquad (19.55b)$$

The Zn^{2+} ions formed by the oxidation of Zn(*s*) at the anode react with NH_3 formed at the cathode and Cl^- ions present in solution, so the overall cell reaction is

$$\text{Overall:} \quad 2MnO_2(s) + 2NH_4Cl(aq) + Zn(s) \longrightarrow Mn_2O_3(s) + Zn(NH_3)_2Cl_2(s) + H_2O(l) \qquad (19.55c)$$

Figure 19.12 Three kinds of primary (nonrechargeable) batteries. (a) A Leclanché dry cell is actually a "wet cell," in which the electrolyte is an acidic water-based paste containing MnO_2, NH_4Cl, $ZnCl_2$, graphite, and starch. Though inexpensive to manufacture, the cell is not very efficient in producing electrical energy and has a limited shelf life. (b) In a "button" battery, the anode is a zinc–mercury amalgam, and the cathode can be either HgO (shown here) or Ag_2O as the oxidant. Button batteries are reliable and have a high output-to-mass ratio, which allows them to be used in applications such as calculators and watches, where their small size is crucial. (c) A lithium–iodine battery consists of two cells separated by a metallic nickel mesh that collects charge from the anodes. The anode is lithium metal, and the cathode is a solid complex of I_2. The electrolyte is a layer of solid LiI that allows Li^+ ions to diffuse from the cathode to the anode. Although this type of battery produces only a relatively small current, it is highly reliable and long-lived.

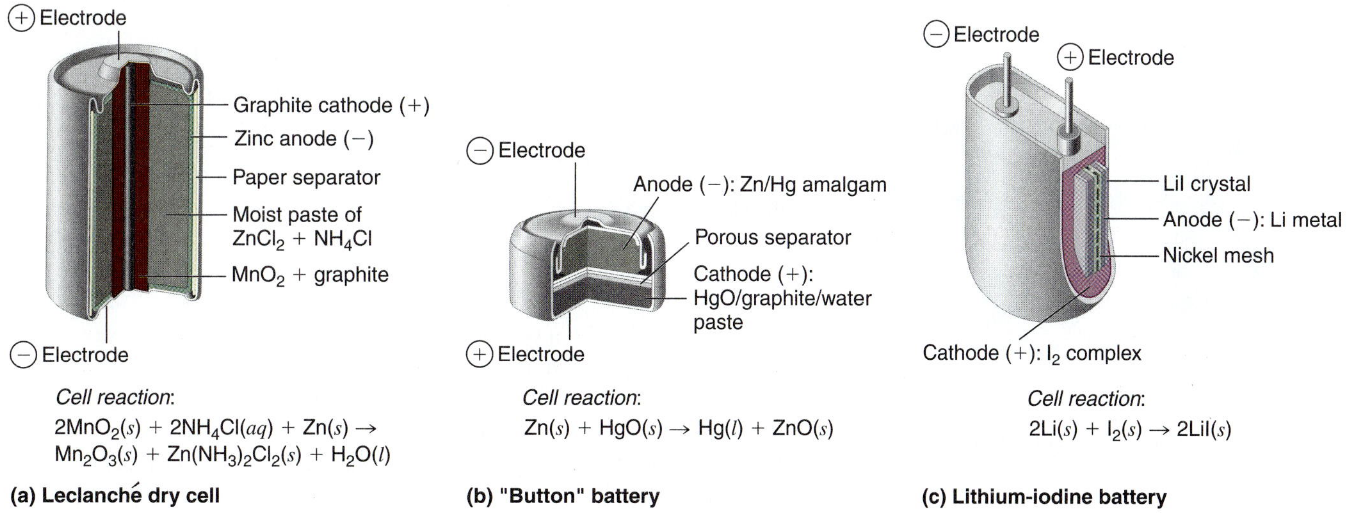

The dry cell produces about 1.55 V and is inexpensive to manufacture. It is not, however, very efficient in producing electrical energy because only the relatively small fraction of the MnO_2 that is near the cathode is actually reduced and only a small fraction of the zinc cathode is actually consumed as the cell discharges. In addition, dry cells have a limited shelf life because the Zn anode reacts spontaneously with NH_4Cl in the electrolyte, causing the case to corrode and allowing the contents to leak out.

The **alkaline battery** is essentially a Leclanché cell adapted to operate under alkaline, or basic, conditions. The half-reactions that occur in an alkaline battery are

$$\text{Cathode:}\quad 2MnO_2(s) + H_2O(l) + 2e^- \longrightarrow Mn_2O_3(s) + 2OH^-(aq) \qquad (19.56a)$$

$$\text{Anode:}\quad Zn(s) + 2OH^-(aq) \longrightarrow ZnO(s) + H_2O(l) + 2e^- \qquad (19.56b)$$

$$\text{Overall:}\quad Zn(s) + 2MnO_2(s) \longrightarrow ZnO(s) + Mn_2O_3(s) \qquad (19.56c)$$

This battery also produces about 1.5 V, but it has a longer shelf life and more constant output voltage as the cell is discharged than the Leclanché dry cell. Although the alkaline battery is more expensive to produce than the Leclanché dry cell, the improved performance makes this battery more cost effective.

"Button" Batteries

Although some of the small *"button" batteries* used to power watches, calculators, and cameras are miniature alkaline cells, most are based on completely different chemistry. In these batteries, the anode is a zinc–mercury amalgam rather than pure zinc, and the cathode uses either HgO or Ag_2O as the oxidant rather than MnO_2 (Figure 19.12b). The cathodic and overall reactions and cell output for these two types of button batteries are

$$\text{Cathode (Hg):}\quad HgO(s) + H_2O(l) + 2e^- \longrightarrow Hg(l) + 2OH^-(aq) \qquad (19.57a)$$

$$\text{Overall (Hg):}\quad Zn(s) + HgO(s) \longrightarrow Hg(l) + ZnO(s) \qquad E_{cell} = 1.35\text{ V} \qquad (19.57b)$$

$$\text{Cathode (Ag):}\quad Ag_2O(s) + H_2O(l) + 2e^- \longrightarrow 2Ag(s) + 2OH^-(aq) \qquad (19.57c)$$

$$\text{Overall (Ag):}\quad Zn(s) + Ag_2O(s) \longrightarrow 2Ag(s) + ZnO(s) \qquad E_{cell} = 1.6\text{ V} \qquad (19.57d)$$

The major advantages of the mercury and silver cells are their reliability and the high output-to-mass ratio. These factors make them ideal for applications where small size is crucial, as in cameras and hearing aids. The disadvantages are the expense and the environmental problems caused by the disposal of heavy metals such as Hg and Ag.

Lithium–Iodine Battery

None of the batteries described above is actually "dry." They all contain small amounts of liquid water, which adds significant mass and causes potential corrosion problems. Consequently, substantial effort has been expended to develop water-free batteries.

One of the few commercially successful water-free batteries is the **lithium–iodine battery**. The anode is lithium metal, and the cathode is a solid complex of I_2. Separating them is a layer of solid LiI, which acts as the electrolyte by allowing the diffusion of Li^+ ions. The electrode reactions are

$$\text{Cathode:}\quad I_2(s) + 2e^- \longrightarrow 2I^-(\text{LiI}) \qquad (19.58a)$$

$$\text{Anode:}\quad Li(s) \longrightarrow Li^+(\text{LiI}) + e^- \qquad (19.58b)$$

$$\text{Overall:}\quad 2Li(s) + I_2(s) \longrightarrow 2LiI(s) \qquad E_{cell} = 3.5\text{ V} \qquad (19.58c)$$

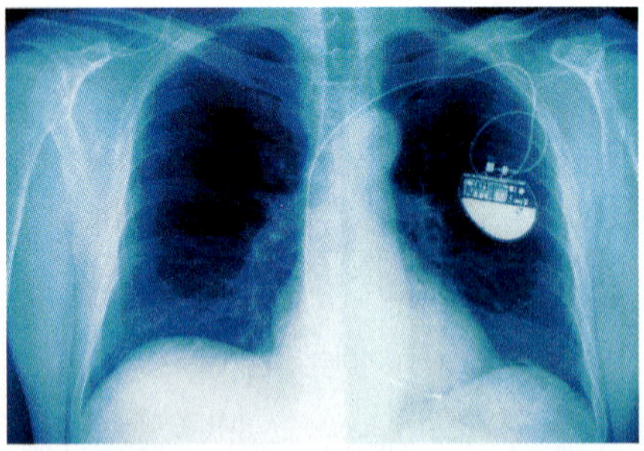

Cardiac pacemaker. An X ray of a patient showing the location and size of a pacemaker powered by a lithium–iodine battery.

As shown in Figure 19.12c, a typical lithium–iodine battery consists of two cells separated by a nickel metal mesh that collects charge from the anode. Because of the high internal resistance caused by the solid electrolyte, only a low current can be drawn. Nonetheless, such batteries have proven to be long-lived (up to 10 years) and reliable. They are therefore used in applications where frequent replacement is difficult or undesirable, such as in cardiac pacemakers and other medical implants and in computers for memory protection. These batteries are also used in security transmitters and smoke alarms. Other batteries based on lithium anodes and solid electrolytes are under development, using TiS_2, for example, for the cathode.

Dry cells, button batteries, and lithium–iodine batteries are all disposable and cannot be recharged once they are discharged. Rechargeable batteries, in contrast, offer significant economic and environmental advantages because they can be recharged and discharged numerous times. As a result, manufacturing and disposal costs drop dramatically for a given number of hours of battery usage. Two common rechargeable batteries are the nickel–cadmium battery and the lead–acid battery, which we describe next.

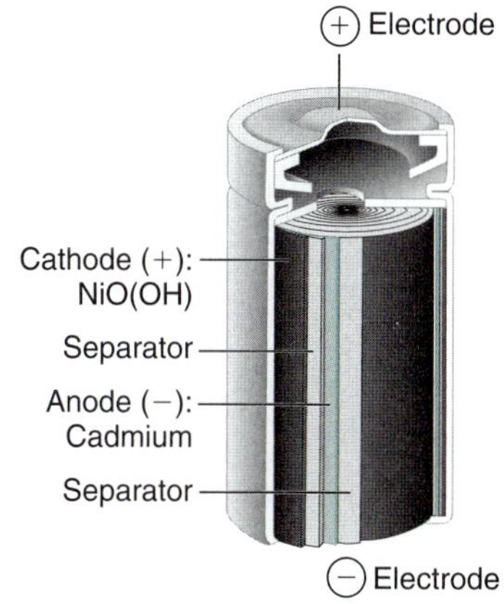

Cell reaction:
$Cd(s) + 2NiO(OH)(s) + 2H_2O(l) \rightarrow Cd(OH)_2(s) + 2Ni(OH_2)(s)$

Figure 19.13 The nickel–cadmium (Nicad) battery, a rechargeable battery. Nicad batteries contain a cadmium anode and a highly oxidized nickel cathode. This design maximizes the surface area of the electrodes and minimizes the distance between them, which gives the battery both a high discharge current and a high capacity.

Nickel–Cadmium (Nicad) Battery

The *nickel–cadmium,* or *Nicad, battery* is used in small electrical appliances and devices like drills, portable vacuum cleaners, and AM and FM digital tuners. It is a water-based cell with a cadmium anode and a highly oxidized nickel cathode that is usually described as the nickel(III) oxo-hydroxide, NiO(OH). As shown in Figure 19.13, the design maximizes the surface area of the electrodes and minimizes the distance between them, which decreases internal resistance and makes a rather high discharge current possible.

The electrode reactions during the discharge of a Nicad battery are

$$\text{Cathode:}\quad 2NiO(OH)(s) + 2H_2O(l) + 2e^- \longrightarrow 2Ni(OH)_2(s) + 2OH^-(aq) \qquad (19.59a)$$

$$\text{Anode:}\quad Cd(s) + 2OH^-(aq) \longrightarrow Cd(OH)_2(s) + 2e^- \qquad (19.59b)$$

$$\text{Overall:}\quad Cd(s) + 2NiO(OH)(s) + 2H_2O(l) \longrightarrow Cd(OH)_2(s) + 2Ni(OH_2)(s) \qquad E_{cell} = 1.4\text{ V} \qquad (19.59c)$$

Because the products of the discharge half-reactions are solids that adhere to the electrodes [$Cd(OH)_2$ and $2Ni(OH_2)$], the overall reaction is readily reversed when the cell is recharged. Although Nicad cells are lightweight, rechargeable, and high capacity, they have certain disadvantages. For example, they tend to lose capacity quickly if not allowed to discharge fully before recharging, they do not store well for long periods when fully charged, and they present significant environmental and disposal problems because of the toxicity of cadmium.

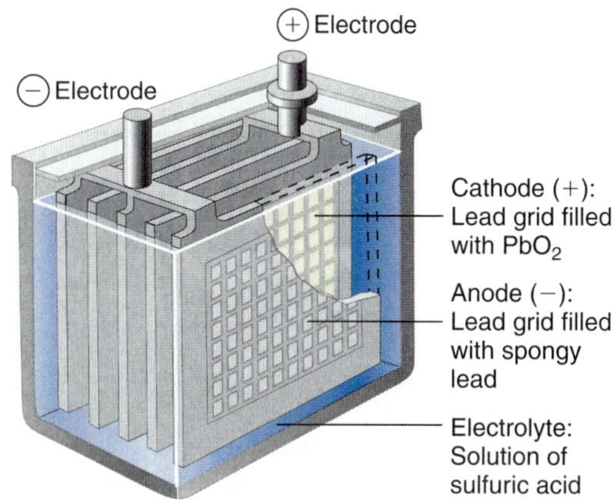

Cell reaction:
$Pb(s) + PbO_2(s) + 2HSO_4^-(aq) + 2H^+(aq) \rightarrow 2PbSO_4(s) + 2H_2O(l)$

Figure 19.14 One cell of a lead–acid battery. The anodes in each cell of a rechargeable battery are plates or grids of lead containing spongy lead metal, while the cathodes are similar grids containing powdered lead dioxide, PbO_2. The electrolyte is an aqueous solution of sulfuric acid. The value of $E°$ for such a cell is about 2 V. Connecting three such cells in series produces a 6-V battery, whereas a typical 12-V car battery contains six. When treated properly, this type of high-capacity battery can be discharged and recharged many times over.

Lead–Acid (Lead Storage) Battery

The lead–acid battery is used to provide the starting power in virtually every automobile and marine engine on the market. Marine and car batteries typically consist of multiple cells connected in series. The total voltage generated by the battery is the potential per cell, $E°_{cell}$, times the number of cells. As shown in Figure 19.14, the anode of each cell in a lead storage battery is a plate or grid of spongy lead metal, and the cathode is a similar grid containing powdered lead dioxide (PbO_2). The electrolyte is usually an approximately 37% solution (by mass) of sulfuric acid in water, with a density of 1.28 g/mL (about 4.5 *M* H_2SO_4). Because the redox active species are solids, there is no need to separate the electrodes. The electrode reactions in each cell during discharge are

$$\text{Cathode:}\quad PbO_2(s) + HSO_4^-(aq) + 3H^+(aq) + 2e^- \longrightarrow PbSO_4(s) + 2H_2O(l) \qquad E°_{cathode} = 1.685\text{ V} \qquad (19.60a)$$

$$\text{Anode:}\quad Pb(s) + HSO_4^-(aq) \longrightarrow PbSO_4(s) + H^+(aq) + 2e^- \qquad E°_{anode} = -0.356\text{ V} \qquad (19.60b)$$

$$\text{Overall:}\quad Pb(s) + PbO_2(s) + 2HSO_4^-(aq) + 2H^+(aq) \longrightarrow 2PbSO_4(s) + 2H_2O(l) \qquad E°_{cell} = 2.041\text{ V} \qquad (19.60c)$$

As the cell is discharged, a powder of $PbSO_4$ forms on the electrodes. Moreover, sulfuric acid is consumed and water is produced, decreasing the density of the electrolyte and providing a convenient way of monitoring the status of a battery by simply measuring the density of the electrolyte.

When an external voltage in excess of 2.04 V per cell is applied to a lead–acid battery, the electrode reactions reverse, and $PbSO_4$ is converted back to metallic lead and PbO_2. If the battery is recharged too vigorously, however, electrolysis of water can occur (Section 19.7), resulting in the evolution of potentially explosive hydrogen gas. The gas bubbles formed in this way can dislodge some of the $PbSO_4$ or PbO_2 particles from the grids, allowing them to fall to the bottom of the cell, where they can build up and cause an internal short circuit. Thus, the recharging process must be carefully monitored to optimize the life of the battery. With proper care, however, a lead–acid battery can be discharged and recharged thousands of times. In automobiles, the alternator supplies the electric current that causes the discharge reaction to reverse.

Fuel Cells

A fuel cell is a galvanic cell that requires a constant external supply of reactants because the products of the reaction are continuously removed. Unlike a battery, it does not store chemical or electrical energy, but instead allows electrical energy to be extracted directly from a chemical reaction. In principle, this should be a more efficient process than, for example, burning the fuel to drive an internal combustion engine that turns a generator, which is typically less than 40% efficient. Unfortunately, significant cost and reliability problems have hindered the wide-scale adoption of fuel cells. In practice, their use has been restricted to applications in which mass may be a significant cost factor, such as U.S. manned space vehicles.

These space vehicles use a hydrogen/oxygen fuel cell that requires a continuous input of $H_2(g)$ and $O_2(g)$, as illustrated in Figure 19.15. The electrode reactions are

$$\text{Cathode:}\quad O_2(g) + 4H^+ + 4e^- \longrightarrow 2H_2O(g) \qquad (19.61a)$$

$$\text{Anode:}\quad 2H_2(g) \longrightarrow 4H^+ + 4e^- \qquad (19.61b)$$

$$\text{Overall:}\quad 2H_2(g) + O_2(g) \longrightarrow 2H_2O(g) \qquad (19.61c)$$

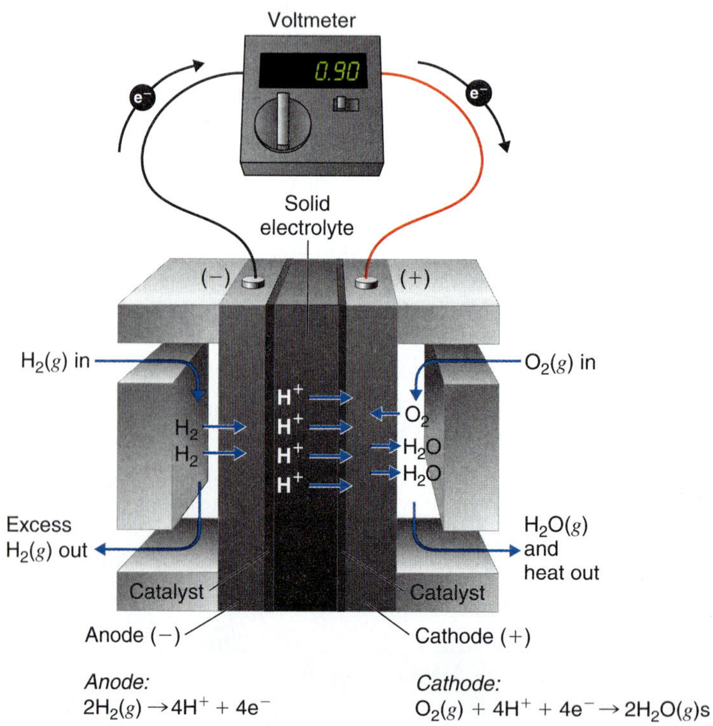

Figure 19.15 A hydrogen fuel cell produces electrical energy directly from a chemical reaction. Hydrogen is oxidized to protons at the anode, and the electrons are transferred through an external circuit to the cathode, where oxygen is reduced and combines with H^+ to form water. A solid electrolyte allows the protons to diffuse from the anode to the cathode. Although fuel cells are an essentially pollution-free means of obtaining electrical energy, their expense and technological complexity have thus far limited their applications.

The overall reaction represents an essentially pollution-free conversion of hydrogen and oxygen to water, which in space vehicles is then collected and used. Although this type of fuel cell should produce 1.23 V under standard conditions, in practice the device achieves only about 0.9 V. One of the major barriers to achieving greater efficiency is the fact that the four-electron reduction of $O_2(g)$ at the cathode is intrinsically rather slow, which limits current that can be achieved. All major automobile manufacturers have major research programs involving fuel cells: one of the most important goals is the development of a better catalyst for the reduction of O_2.

19.6 Corrosion

Note the pattern

Corrosion is a galvanic process.

Corrosion is a galvanic process by which metals deteriorate through oxidation, usually but not always to their oxides. For example, when exposed to air, iron rusts, silver tarnishes, and copper and brass acquire a bluish-green surface called a *patina*. Of the various metals subject to corrosion, iron is by far the most important commercially. An estimated \$100 billion per year is spent in the United States alone to replace iron objects destroyed by corrosion. Consequently, the development of methods for protecting metal surfaces from corrosion constitutes a very active area of industrial research. In this section, we describe some of the chemical and electrochemical processes responsible for corrosion. We also examine the chemical basis for some common methods for preventing corrosion and treating corroded metals.

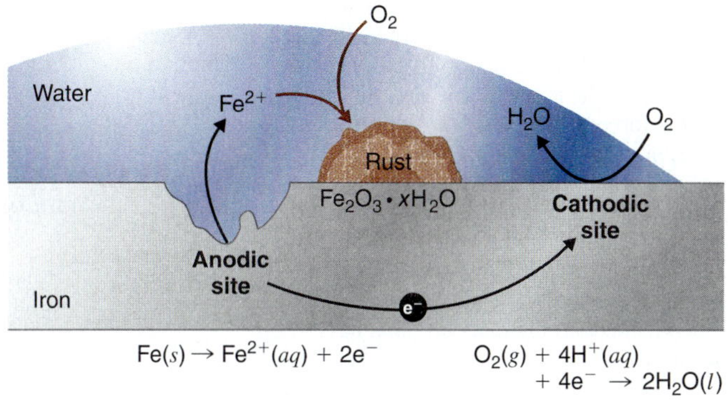

Figure 19.16 Rust, the result of corrosion of metallic iron. Iron is oxidized to $Fe^{2+}(aq)$ at an anodic site on the surface of the iron, which is often an impurity or a lattice defect. Oxygen is reduced to water at a different site on the surface of the iron, which acts as the cathode. Electrons are transferred from the anode to the cathode through the electrically conductive metal. Water provides a salt bridge between the two sites, as well as acting as a solvent for the Fe^{2+} that is produced initially. Rust ($Fe_2O_3{\cdot}xH_2O$) is formed by the subsequent nonelectrochemical oxidation of Fe^{2+} by atmospheric oxygen.

Under ambient conditions, the oxidation of most metals is thermodynamically spontaneous, with the notable exception of gold and platinum. Hence, it is actually somewhat surprising that any metals are useful at all in Earth's moist, oxygen-rich atmosphere. Some metals, however, are resistant to corrosion for kinetic reasons. For example, the aluminum in soft-drink cans and airplanes is protected by a thin coating of metal oxide that forms on the surface of the metal and acts as an impenetrable barrier that prevents further destruction. Aluminium cans also have a thin plastic layer to prevent reaction of the oxide with acid in the soft-drink. Chromium, magnesium, and nickel also form protective oxide films. Stainless steels are remarkably resistant to corrosion because they usually contain a significant proportion of chromium, nickel, or both.

In contrast to these metals, when iron corrodes, it forms a red-brown hydrated metal oxide, $Fe_2O_3{\cdot}xH_2O$, commonly known as *rust*, that does not provide a tight protective film (Figure 19.16). Instead, the rust continually flakes off to expose a fresh metal surface vulnerable to reaction with oxygen and water. Because both oxygen and water are required for rust to form, an iron nail immersed in deoxygenated water will not rust even over a period of weeks. Similarly, a nail immersed in an organic solvent such as kerosene or mineral oil saturated with oxygen will not rust because of the absence of water.

In the corrosion process, iron metal acts as the anode in a galvanic cell and is oxidized to Fe^{2+}; oxygen is reduced to water at the cathode. The relevant reactions are

$$\text{Cathode:}\quad O_2(g) + 4H^+(aq) + 4e^- \longrightarrow 2H_2O(l) \qquad E^\circ = 1.23\ \text{V} \qquad (19.62a)$$

$$\text{Anode:}\quad Fe(s) \longrightarrow Fe^{2+}(aq) + 2e^- \qquad E^\circ = -0.45\ \text{V} \qquad (19.62b)$$

$$\text{Overall:}\quad 2Fe(s) + O_2(g) + 4H^+(aq) \longrightarrow 2Fe^{2+}(aq) + 2H_2O(l) \qquad E^\circ = 1.68\ \text{V} \qquad (19.62c)$$

The Fe^{2+} ions produced in the initial reaction are then oxidized by atmospheric oxygen in a nonelectrochemical process to produce the insoluble hydrated oxide containing Fe^{3+}, as represented in the equation

$$4Fe^{2+}(aq) + O_2(g) + (2 + 4x)H_2O \longrightarrow 2Fe_2O_3 \cdot xH_2O + 4H^+(aq) \qquad (19.63)$$

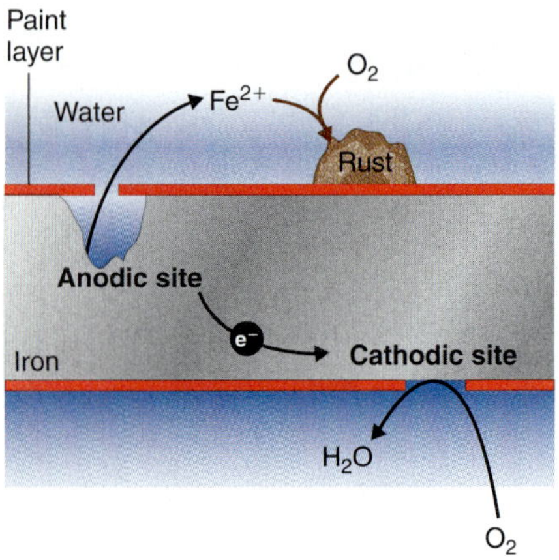

Figure 19.17 Small scratches in a protective paint coating can lead to rapid corrosion of iron. Holes in a protective coating allow oxygen to be reduced at the surface with the greater exposure to air (the cathode), while metallic iron is oxidized to $Fe^{2+}(aq)$ at the less exposed site (the anode). Rust is formed when $Fe^{2+}(aq)$ diffuses to a location where it can react with atmospheric oxygen, which is often remote from the anode. The electrochemical interaction between cathodic and anodic sites can cause a large pit to form *under* a painted surface, eventually resulting in sudden failure with little visible warning that corrosion has occurred.

The sign and magnitude of $E°$ for the corrosion process (Equation 19.62c) indicate that there is a strong driving force for the oxidation of iron by O_2 under standard conditions (1 M H^+). Under neutral conditions, the driving force is somewhat less, but still appreciable (E = 1.25 V at pH 7). Normally, the reaction of atmospheric CO_2 with water to form H^+ and HCO_3^- provides a low enough pH to enhance the rate of the reaction, as does acid rain (see Section 4.7).

Automobile manufacturers spend a great deal of time and money developing paints that adhere tightly to the car's metal surface to prevent oxygenated water, acid, and salt from coming into contact with the underlying metal. Unfortunately, even the best paint is subject to scratching or denting, and the electrochemical nature of the corrosion process means that two scratches relatively remote from each other can operate together as anode and cathode, leading to sudden mechanical failure (Figure 19.17).

One of the most common techniques used to prevent the corrosion of iron is applying a protective coating of another metal that is more difficult to oxidize. Faucets and some external parts of automobiles, for example, are often coated with a thin layer of chromium using an electrolytic process that will be discussed in the next section. With the increased use of polymeric materials in cars, however, the use of chrome-plated steel has diminished in recent years. Similarly, the "tin cans" that hold soups and other foods are actually made of steel coated with a thin layer of tin. Neither chromium nor tin is intrinsically resistant to corrosion, but both form protective oxide coatings.

As with a protective paint, scratching a protective metal coating will allow corrosion to occur. In this case, however, the presence of the second metal can actually increase the rate of corrosion. The values of the standard electrode potentials for Sn^{2+} ($E°$ = −0.14 V) and Fe^{2+} ($E°$ = −0.45 V) in Table 19.2 show that Fe is more easily oxidized than Sn. As a result, the more corrosion-resistant metal (in this case, tin) accelerates the corrosion of iron by acting as the cathode and providing a large surface area for the reduction of oxygen (Figure 19.18). This process is seen in some older homes where copper and iron pipes have been directly connected to each other. The less easily oxidized copper acts as the cathode, causing iron to dissolve rapidly near the connection and occasionally resulting in a catastrophic plumbing failure.

One way to avoid these problems is to use a more easily oxidized metal to protect iron from corrosion. In this approach, called *cathodic protection,* a more reactive metal such as Zn ($E°$ = −0.76 V for $Zn^{2+} + 2e^- \longrightarrow Zn$) becomes the anode and iron becomes the cathode. This prevents oxidation of the iron and protects the iron object from corrosion. The reactions that occur under these conditions are

$$\text{Cathode:} \quad O_2(g) + 4e^- + 4H^+(aq) \longrightarrow 2H_2O(l) \qquad (19.64a)$$

$$\text{Anode:} \quad Zn(s) \longrightarrow Zn^{2+}(aq) + 2e^- \qquad (19.64b)$$

$$\text{Overall:} \quad 2Zn(s) + O_2(g) + 4H^+(aq) \longrightarrow 2Zn^{2+}(aq) + 2H_2O(l) \qquad (19.64c)$$

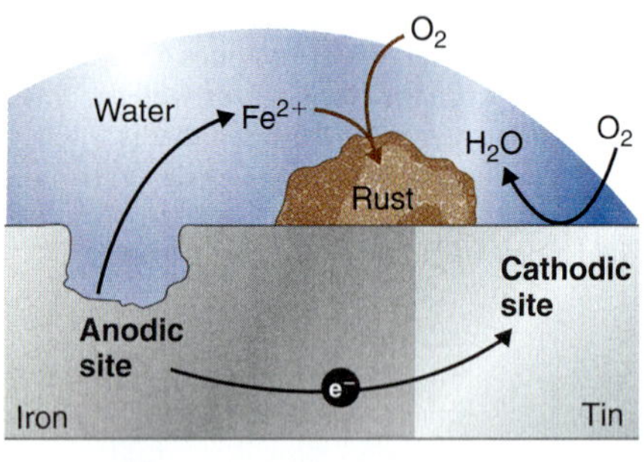

Figure 19.18 Galvanic corrosion. If iron is in contact with a more corrosion-resistant metal such as tin, copper, or lead, the other metal can act as a large cathode that greatly increases the rate of reduction of oxygen. Because reduction of oxygen is coupled to oxidation of iron, this can result in a dramatic increase in the rate at which iron is oxidized at the anode. Galvanic corrosion is likely to occur whenever two dissimilar metals are connected directly, allowing electrons to be transferred from one to the other.

The more reactive metal reacts with oxygen and will eventually dissolve, "sacrificing" itself to protect the iron object. Cathodic protection is the principle underlying *galvanized steel*, which is steel protected by a thin layer of zinc. Galvanized steel is used in objects ranging from nails to garbage cans. In a similar strategy, *sacrificial electrodes* using magnesium, for example, are used to protect underground tanks or pipes (Figure 19.19). Replacing the sacrificial electrodes is more cost effective than replacing the iron objects they are protecting.

EXAMPLE 19.13

Suppose an old wooden sailboat, held together with iron screws, has a bronze propeller (recall that bronze is an alloy of copper containing about 7–10% tin). (a) If the boat is immersed in seawater, what corrosion reaction will occur? What is $E°_{cell}$? (b) How could you prevent this corrosion from occurring?

Given Identity of metals

Asked for Corrosion reaction, E°_{cell}, preventive measures

Strategy

Ⓐ Write the reactions that occur at the anode and the cathode. From these, write the overall cell reaction and calculate E°_{cell}.
Ⓑ Based on the relative redox activity of various substances, suggest possible preventive measures.

Solution

(a) Ⓐ According to Table 19.2, both copper and tin are less active metals than iron (that is, they have higher positive values of E° than iron). Thus, if tin or copper is brought into electrical contact (by seawater) with iron in the presence of oxygen, corrosion will occur. We therefore anticipate that the bronze propeller will act as the cathode at which O_2 is reduced, and the iron screws will act as anodes at which iron dissolves:

Cathode:	$O_2(g) + 4H^+(aq) + 4e^- \longrightarrow 2H_2O(l)$	$E^\circ_{cathode} = 1.23$ V
Anode:	$Fe(s) \longrightarrow Fe^{2+}(aq) + 2e^-$	$E^\circ_{anode} = -0.45$ V
Overall:	$2Fe(s) + O_2(g) + 4H^+(aq) \longrightarrow 2Fe^{2+}(aq) + 2H_2O(l)$	$E^\circ_{cell} = 1.68$ V

Over time, the iron screws will dissolve and the boat will fall apart.

(b) Ⓑ Possible ways to prevent corrosion, in order of decreasing cost and inconvenience, are: disassembling the boat and rebuilding it with bronze screws; removing the boat from the water and storing it in a dry place; or attaching an inexpensive piece of zinc metal to the propeller shaft to act as a sacrificial electrode and replacing it once or twice a year. Because zinc is a more active metal than iron, it will act as the anode in the electrochemical cell and dissolve (Equation 19.64c).

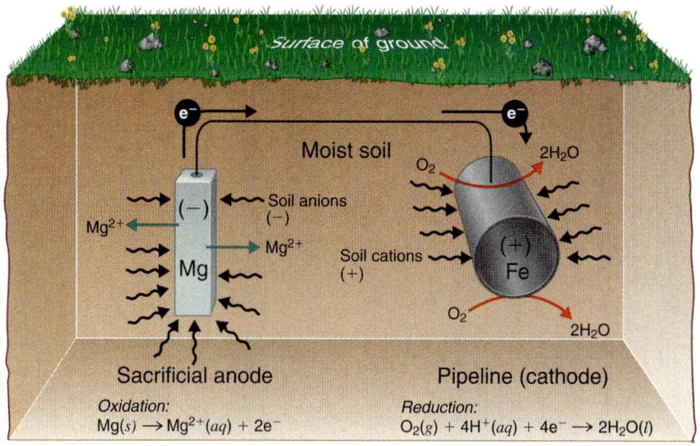

Figure 19.19 The use of a sacrificial electrode to protect against corrosion. Connecting a magnesium rod to an underground steel pipeline protects the pipeline from corrosion. Because magnesium ($E^\circ = -2.37$ V) is much more easily oxidized than iron ($E^\circ = -0.45$ V), the Mg rod acts as the anode in a galvanic cell. The pipeline is therefore forced to act as the cathode at which oxygen is reduced. The soil between the anode and the cathode acts as a salt bridge that completes the electrical circuit and maintains electrical neutrality. As Mg(s) is oxidized to Mg^{2+} at the anode, anions such as nitrate in the soil diffuse toward the anode to neutralize the positive charge. Simultaneously, cations in the soil such as H^+ or NH_4^+ diffuse toward the cathode, where they replenish the protons that are consumed as oxygen is reduced. A similar strategy utilizes many miles of somewhat less reactive zinc wire to protect the Alaska oil pipeline.

EXERCISE 19.13

Suppose the water pipes leading into your house are made of lead, while the rest of the plumbing in the house is iron. To eliminate the possibility of lead poisoning, you call a plumber to replace the lead pipes. He quotes you a very low price if he can use up his existing supply of copper pipe to do the job. **(a)** Do you accept his proposal? **(b)** What else should you have the plumber do while there?

Answer

(a) Not unless you plan to sell the house very soon, because the Cu/Fe pipe joints will lead to rapid corrosion.
(b) Any existing Pb/Fe joints should be examined carefully for corrosion of the iron pipes due to the Pb–Fe junction; the less active Pb will have served as the cathode for the reduction of O_2, promoting oxidation of the more active Fe nearby.

19.7 ○ Electrolysis

So far we have described various galvanic cells in which a spontaneous chemical reaction is used to generate electrical energy. In an electrolytic cell, however, the opposite process, called **electrolysis**, occurs: an external voltage is applied to drive a nonspontaneous reaction (Figure 19.1). In this section, we look at how electrolytic cells are constructed and explore some of their many commercial applications.

Note the pattern

In an electrolytic cell, an external voltage is applied to drive a nonspontaneous reaction.

Electrolytic Cells

If we construct an electrochemical cell in which one electrode is copper metal immersed in a 1 *M* Cu^{2+} solution and the other electrode is cadmium metal immersed in a 1 *M* Cd^{2+}

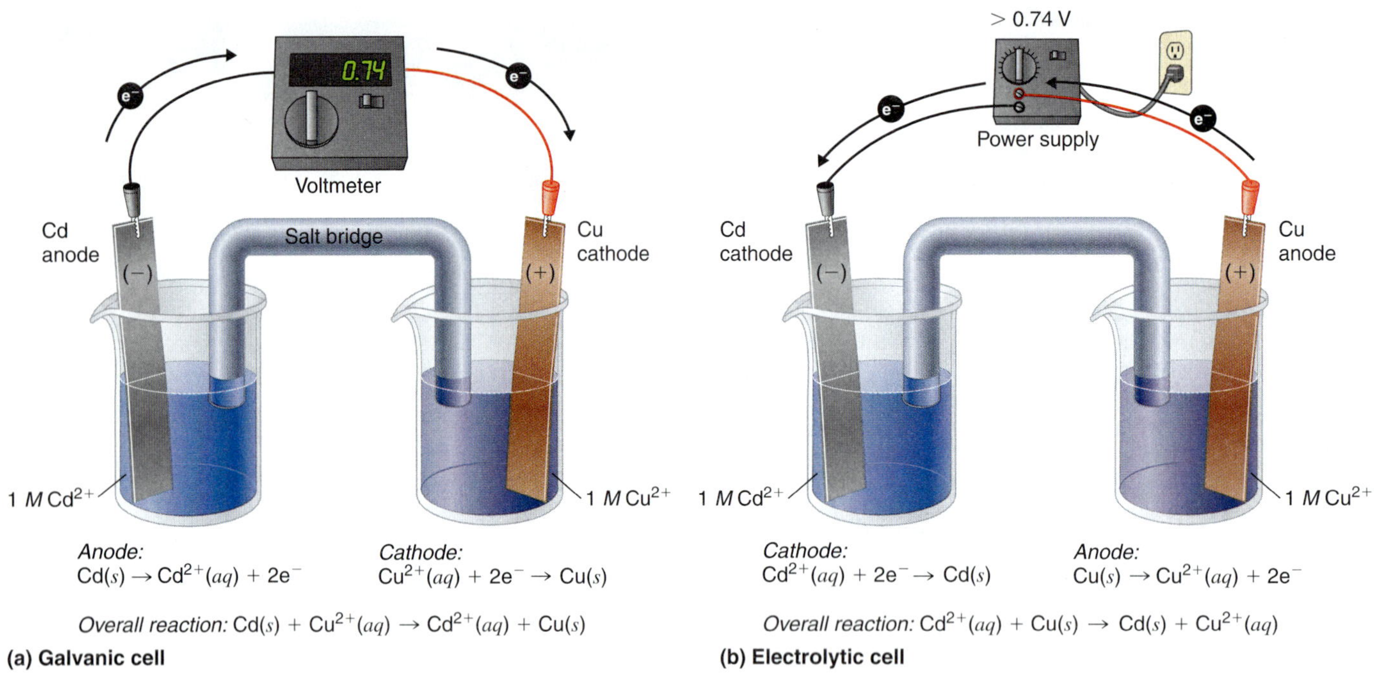

Figure 19.20 An applied voltage can reverse the flow of electrons in a galvanic Cd/Cu cell. (a) When compartments that contain a Cd electrode immersed in 1 *M* $Cd^{2+}(aq)$ and a Cu electrode immersed in 1 *M* $Cu^{2+}(aq)$ are connected to create a galvanic cell, Cd(*s*) is spontaneously oxidized to $Cd^{2+}(aq)$ at the anode, and $Cu^{2+}(aq)$ is spontaneously reduced to Cu(*s*) at the cathode. The potential of the galvanic cell is 0.74 V. (b) Applying an external potential greater than 0.74 V in the reverse direction forces electrons to flow out of the Cu electrode [which is now the anode, at which metallic Cu(*s*) is oxidized to $Cu^{2+}(aq)$] and into the Cd electrode [which is now the cathode, at which $Cd^{2+}(aq)$ is reduced to Cd(*s*)]. Note that the anode in an electrolytic cell is *positive* because electrons are flowing out of it, whereas the cathode is *negative* because electrons are flowing into it.

solution and then close the circuit, the potential difference between the two compartments will be 0.74 V. The cadmium electrode will begin to dissolve (Cd is oxidized to Cd^{2+}) and thus is the anode, while metallic copper will be deposited on the copper electrode (Cu^{2+} is reduced to Cu), which is the cathode (Figure 19.20a). The overall reaction

$$\mathrm{Cd}(s) + \mathrm{Cu}^{2+}(aq) \longrightarrow \mathrm{Cd}^{2+}(aq) + \mathrm{Cu}(s) \qquad E^\circ_{\mathrm{cell}} = 0.74\ \mathrm{V} \qquad (19.65)$$

is thermodynamically spontaneous as written ($\Delta G^\circ < 0$):

$$\Delta G^\circ = -nFE^\circ = -(2\ \cancel{\mathrm{mol}}\ \mathrm{e}^-)[96{,}486\ \mathrm{J}/(\cancel{\mathrm{V}}\cdot\cancel{\mathrm{mol}})](0.74\ \cancel{\mathrm{V}}) = -140\ \mathrm{kJ\ (per\ mole\ Cd)} \qquad (19.66)$$

In this direction, the system is acting as a galvanic cell.

The reverse reaction, reduction of Cd^{2+} by Cu, is thermodynamically nonspontaneous and will occur only with an *input* of 140 kJ. We can force the reaction to proceed in the reverse direction by applying an electrical potential greater than 0.74 V from an external power supply. The applied voltage forces electrons through the circuit in the reverse direction, converting a galvanic cell to an electrolytic cell. Thus, the copper electrode is now the anode (Cu is oxidized), and the cadmium electrode is now the cathode (Cd^{2+} is reduced) (Figure 19.20b). Notice that the signs of the cathode and anode have switched to reflect the flow of electrons in the circuit. The half-reactions that occur at the cathode and the anode are

$$\begin{array}{llll}
\text{Cathode:} & \mathrm{Cd}^{2+}(aq) + 2\mathrm{e}^- \longrightarrow \mathrm{Cd}(s) & E^\circ_{\mathrm{cathode}} = -0.40\ \mathrm{V} & (19.67a) \\
\text{Anode:} & \mathrm{Cu}(s) \longrightarrow \mathrm{Cu}^{2+}(aq) + 2\mathrm{e}^- & E^\circ_{\mathrm{anode}} = 0.34\ \mathrm{V} & (19.67b) \\
\hline
\text{Overall:} & \mathrm{Cd}^{2+}(aq) + \mathrm{Cu}(s) \longrightarrow \mathrm{Cd}(s) + \mathrm{Cu}^{2+}(aq) & E^\circ_{\mathrm{cell}} = -0.74\ \mathrm{V} & (19.67c)
\end{array}$$

Because $E^\circ_{cell} < 0$, the overall reaction—the reduction of Cd^{2+} by Cu—clearly cannot occur spontaneously and proceeds only when sufficient electrical energy is applied. The differences between galvanic and electrolytic cells are summarized in Table 19.3.

TABLE 19.3 Comparison of galvanic and electrolytic cells

Property	Galvanic Cell	Electrolytic Cell
ΔG	<0	>0
E_{cell}	>0	<0
Electrode process:		
Anode	Oxidation	Oxidation
Cathode	Reduction	Reduction
Sign of electrode:		
Anode	−	+
Cathode	+	−

Electrolytic Reactions

At sufficiently high temperatures, ionic solids melt to form liquids that conduct electricity extremely well due to the high concentrations of ions. If two inert electrodes are inserted into molten NaCl, for example, and an electrical potential is applied, Cl^- is oxidized at the anode and Na^+ is reduced at the cathode. The overall reaction is

$$2NaCl(l) \longrightarrow 2Na(l) + Cl_2(g) \qquad (19.68)$$

This is the reverse of the formation of NaCl from its elements. Note that the product of the reduction reaction is *liquid* sodium because the melting point of sodium metal is 97.8°C, well below that of NaCl (801°C). Approximately 20,000 tons of sodium metal are produced commercially in the United States each year by electrolysis of molten NaCl in a *Downs cell* (Figure 19.21). In this specialized cell, $CaCl_2$ (mp 772°C) is first added to the NaCl to lower the melting point of the mixture to about 600°C, thereby lowering the operating costs.

Similarly, in the *Hall–Heroult process* used to produce aluminum commercially, a molten mixture of about 5% aluminum oxide (Al_2O_3, mp 2054°C) and 95% cryolite (Na_3AlF_6, mp 1012°C) is electrolyzed at about 1000°C, producing molten aluminum at the cathode and CO_2 gas at the carbon anode. The overall reaction is

$$2Al_2O_3(l) + 3C(s) \longrightarrow 4Al(l) + 3CO_2(g) \qquad (19.69)$$

Oxide ions react with oxidized carbon at the anode, producing $CO_2(g)$.

There are two important points to make about these two commercial processes and about the electrolysis of molten salts in general. First, the electrode potentials for molten salts are likely to be very different from the standard cell potentials listed in Table 19.2 and Appendix E, which are compiled for the reduction of the hydrated ions in aqueous solutions under standard conditions. Second, using a mixed salt system means there is a possibility of competition between different electrolytic reactions. When a mixture of NaCl and $CaCl_2$ is electrolyzed, Cl^- is oxidized because it is the

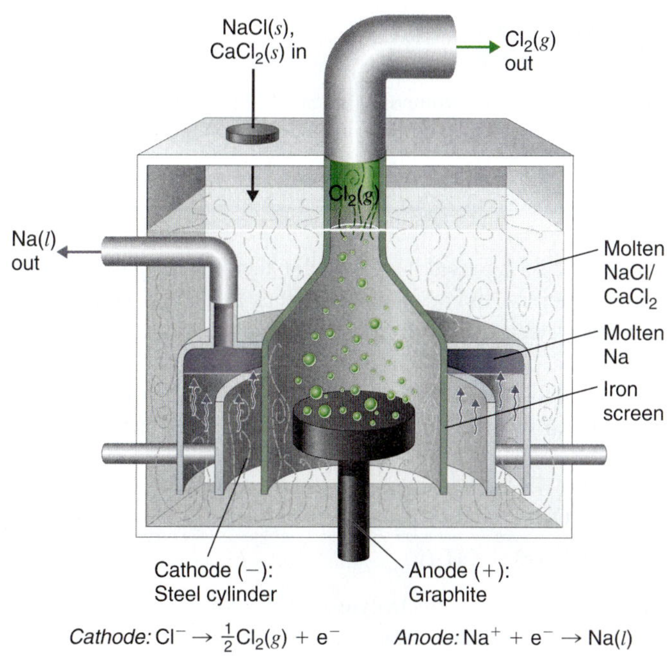

Figure 19.21 A Downs cell for the electrolysis of molten NaCl. Electrolysis of a molten mixture of NaCl and $CaCl_2$ results in the formation of elemental sodium and chlorine gas. Because sodium is a liquid under these conditions and liquid sodium is less dense than molten sodium chloride, the sodium floats to the top of the melt and is collected in concentric capped iron cylinders surrounding the cathode. Gaseous chlorine collects in the inverted cone over the anode. An iron screen separating the cathode and anode compartments ensures that the molten sodium and gaseous chlorine do not come into contact.

only anion present, but either Na^+ or Ca^{2+} can be reduced. Conversely, in the Hall–Heroult process, only one cation is present that can be reduced, Al^{3+}, but there are three species that can be oxidized, C, O^{2-}, and F^-.

In the Hall–Heroult process, C is oxidized instead of O^{2-} or F^- because oxygen and fluorine are more electronegative than carbon, which means that C is a weaker oxidant than either O_2 or F_2. Similarly, in the Downs cell, we might expect electrolysis of a $NaCl/CaCl_2$ mixture to produce calcium rather than sodium because Na is slightly less electronegative than Ca ($\chi = 0.93$ vs. 1.00, respectively), making Na easier to oxidize and, conversely, Na^+ more difficult to reduce. In fact, reduction of Na^+ to Na is the observed reaction. In cases where the electronegativities of two species are similar, other factors, such as the formation of complex ions, become important and may determine the outcome.

EXAMPLE 19.14

If a molten mixture of $MgCl_2$ and KBr is electrolyzed, what products will form at the cathode and the anode, respectively?

Given Identity of salts

Asked for Electrolysis products

Strategy

Ⓐ List all possible reduction and oxidation products. Based on the electronegativity values shown in Figure 7.15, determine which species will be reduced and which species will be oxidized.

Ⓑ Identify the products that will form at each electrode.

Solution

Ⓐ The possible reduction products are Mg and K, and the possible oxidation products are Cl_2 and Br_2. Because Mg is more electronegative than K ($\chi = 1.31$ vs. 0.82), it is likely that Mg will be reduced rather than K. Because Cl is more electronegative than Br (3.16 vs. 2.96), Cl_2 is a stronger oxidant than Br_2.

Ⓑ Electrolysis will therefore produce Br_2 at the anode and Mg at the cathode.

EXERCISE 19.14

Predict the products if a molten mixture of $AlBr_3$ and LiF is electrolyzed.

Answer Br_2 and Al

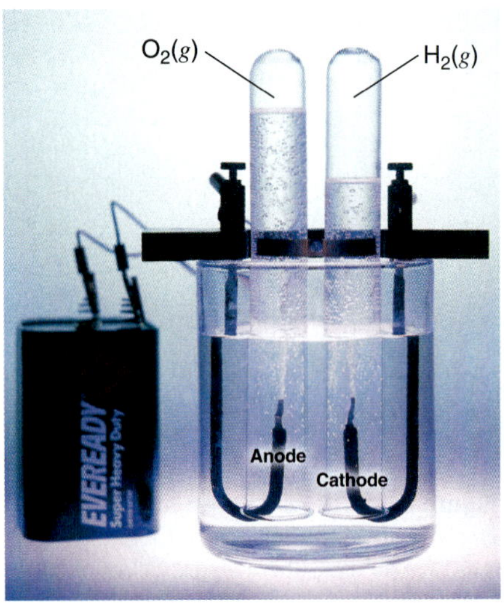

Anode: $2H_2O(l) \rightarrow O_2(g) + 4H^+(aq) + 4e^-$

Cathode: $2H^+(aq) + 2e^- \rightarrow H_2(g)$

Overall: $2H_2O(l) \rightarrow O_2(g) + 2H_2(g)$

Figure 19.22 The electrolysis of water. Applying an external potential of about 1.7–1.9 V to two inert electrodes immersed in an aqueous solution of an electrolyte such as H_2SO_4 or Na_2SO_4 drives the thermodynamically nonspontaneous decomposition of water into H_2 at the cathode and O_2 at the anode.

Electrolysis can also be used to drive the thermodynamically nonspontaneous decomposition of water into its constituent elements, H_2 and O_2. However, because pure water is a very poor electrical conductor, a small amount of an ionic solute (such as H_2SO_4 or Na_2SO_4) must first be added to make it electrically conductive. Inserting inert electrodes into the solution and applying a voltage between them will result in the rapid evolution of bubbles of H_2 and O_2 (Figure 19.22). The reactions that occur are

$$\text{Cathode:}\quad 2H^+(aq) + 2e^- \longrightarrow H_2(g) \qquad E^\circ_{\text{cathode}} = 0\text{ V} \tag{19.70a}$$

$$\text{Anode:}\quad 2H_2O(l) \longrightarrow O_2(g) + 4H^+(aq) + 4e^- \qquad E^\circ_{\text{anode}} = 1.23\text{ V} \tag{19.70b}$$

$$\text{Overall:}\quad 2H_2O(l) \longrightarrow O_2(g) + 2H_2(g) \qquad E^\circ_{\text{cell}} = -1.23\text{ V} \tag{19.70c}$$

For a system that contains an electrolyte such as Na_2SO_4, which has a negligible effect on the ionization equilibrium of liquid water, the pH of the solution will be 7 and $[H^+] = [OH^-] = 1.0 \times 10^{-7}$. Assuming that $P_{O_2} = P_{H_2} = 1$ atm, we can use the standard potentials and Equation 19.47 to calculate E for the overall reaction:

$$E_{cell} = E^\circ_{cell} - \left(\frac{0.0591\ V}{n}\right)\log(P_{O_2}P^2_{H_2})$$

$$E_{cell} = -1.23\ V - \left(\frac{0.0591\ V}{4}\right)\log(1) = -1.23\ V \qquad (19.71)$$

Thus, E_{cell} is -1.23 V, which is the value of E°_{cell} if the reaction is carried out in the presence of 1 M H^+ rather than at pH 7.

In practice, a voltage about 0.4–0.6 V higher than the calculated value is needed to electrolyze water. This added voltage, called an **overvoltage**, represents the additional driving force required to overcome barriers such as the large activation energy for the formation of a gas at a metal surface. Overvoltages are needed in all electrolytic processes, which explains why, for example, approximately 14 V must be applied to recharge the 12-V battery in your car.

In general, any metal that does *not* react readily with water to produce hydrogen can be produced by the electrolytic reduction of an aqueous solution that contains the metal cation. The p-block metals and most of the transition metals are in this category, but metals in high oxidation states, which form oxoanions, cannot be reduced to the metal by simple electrolysis. Active metals, such as aluminum and those of Groups 1 and 2, react so readily with water that they can be prepared only by the electrolysis of molten salts. Similarly, any nonmetallic element that does not readily oxidize water to O_2 can be prepared by the electrolytic oxidation of an aqueous solution that contains an appropriate anion. In practice, among the nonmetals, only F_2 cannot be prepared using this method. Oxoanions of nonmetals in their highest oxidation states, such as NO_3^-, SO_4^{2-}, PO_4^{3-}, are usually difficult to reduce electrochemically and usually behave like spectator ions that remain in solution during electrolysis.

Note the pattern

In general, any metal that does not react readily with water to produce hydrogen can be produced by the electrolytic reduction of an aqueous solution that contains the metal cation.

Electroplating

In a process called **electroplating**, a layer of a second metal is deposited on the metal electrode that acts as the cathode during electrolysis. Electroplating is used to enhance the appearance of metal objects and protect them from corrosion. Examples of electroplating include the chromium layer found on many bathroom fixtures or (in earlier days) on the bumpers and hubcaps of cars, as well as the thin layer of precious metal that coats silver-plated dinnerware or jewelry. In all cases, the basic concept is the same. A schematic view of an apparatus for electroplating silverware and a photograph of a commercial electroplating cell are shown in Figure 19.23.

The half-reactions in electroplating a fork, for example, with silver are

Cathode (fork): $Ag^+(aq) + e^- \longrightarrow Ag(s) \qquad E^\circ_{cathode} = 0.80\ V \qquad (19.72a)$

Anode (silver bar): $Ag(s) \longrightarrow Ag^+(aq) + e^- \qquad E^\circ_{anode} = 0.80\ V \qquad (19.72b)$

The overall reaction is the transfer of silver metal from one electrode (a silver bar acting as the anode) to another (a fork acting as the cathode). Because $E^\circ_{cell} = 0$ V, it takes only a small applied voltage to drive the electroplating process. In practice, various other substances may be added to the plating solution to control its electrical conductivity and regulate the concentration of free metal ions, thus ensuring a smooth, even coating.

Quantitative Considerations

If we know the stoichiometry of an electrolysis reaction, the amount of current passed, and the length of time, we can calculate the amount of material consumed or produced in the reaction. Conversely, we can use the stoichiometry to determine the combination of current and time needed to produce a given amount of material.

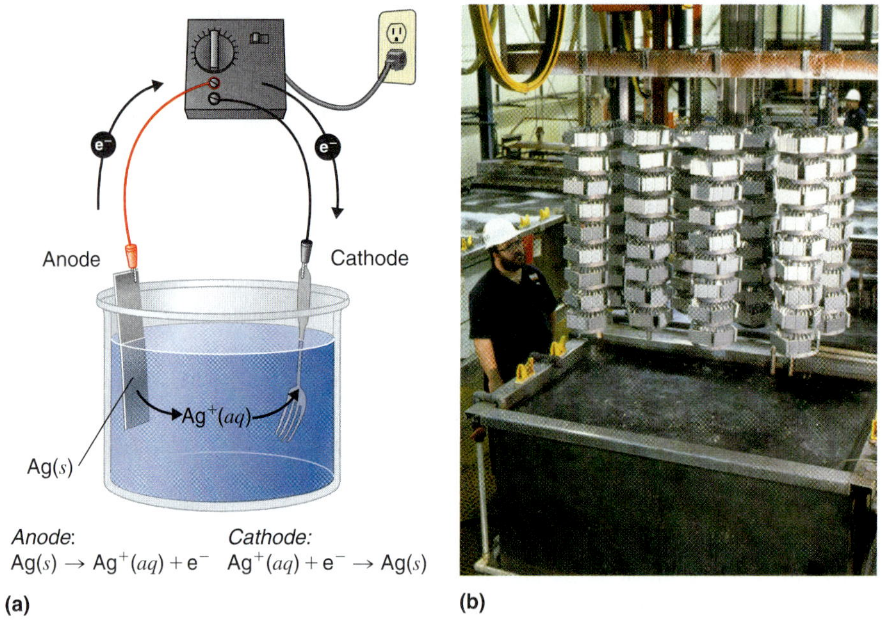

Figure 19.23 Electroplating. (a) Electroplating uses an electrolytic cell in which the object to be plated, such as a fork, is immersed in a solution of the metal to be deposited. The object being plated acts as the cathode, upon which the desired metal is deposited in a thin layer, while the anode usually consists of the metal that is being deposited (in this case silver) that maintains the solution concentration as it dissolves. (b) In this commercial electroplating apparatus, a large number of objects can be plated simultaneously by lowering the rack into the Ag^+ solution and applying the correct potential.

The quantity of material that is oxidized or reduced at an electrode during an electrochemical reaction is determined by the stoichiometry of the reaction and the amount of charge that is transferred. For example, in the reaction $Ag^+(aq) + e^- \longrightarrow Ag(s)$, 1 mol of electrons reduces 1 mol of Ag^+ to Ag metal. In contrast, in the reaction $Cu^{2+}(aq) + 2e^- \longrightarrow Cu(s)$, 1 mol of electrons reduces only 0.5 mol of Cu^{2+} to Cu metal. Recall that the charge on 1 mol of electrons is 1 faraday (1*F*), which is equal to 96,486 C. We can therefore calculate the number of moles of electrons transferred when a known current is passed through a cell for a given period of time. The total charge transferred is the product of the current and the time:

$$\text{charge (C)} = \text{current (A)} \times \text{time (s)} \tag{19.73}$$

The stoichiometry of the reaction and the total charge transferred enable us to calculate the amount of product formed during an electrolysis reaction or the amount of metal deposited in an electroplating process.

For example, if a current of 0.60 A passes through an aqueous solution of $CuSO_4$ for 6.0 min, the total number of coulombs of charge that passes through the cell is

$$\text{charge} = (0.600\ \text{A})(6.00\ \cancel{\text{min}})(60\ \text{s}/\cancel{\text{min}}) = 216\ \text{A}\cdot\text{s} = 216\ \text{C} \tag{19.74}$$

The number of moles of electrons transferred to Cu^{2+} is therefore

$$\text{moles e}^- = \frac{216\ \cancel{\text{C}}}{96{,}486\ \cancel{\text{C}}/\text{mol}} = 2.24 \times 10^{-3}\ \text{mol e}^- \tag{19.75}$$

Because two electrons are required to reduce a single Cu^{2+} ion, the total number of moles of Cu produced is half the number of moles of electrons transferred, or 1.12×10^{-3} mol. This corresponds to 0.711 g of Cu. In commercial electrorefining processes, much higher currents (≥50,000 A) are used, corresponding to approximately 0.5 *F*/s, and reaction times are on the order of 3 to 4 weeks.

EXAMPLE 19.15

A silver-plated spoon typically contains about 2.00 g of Ag. If 12.0 hours are required to achieve the desired thickness of the Ag coating, what is the average current per spoon that must flow during the electroplating process, assuming an efficiency of 100%?

Given Mass of metal, time, efficiency

Asked for Current required

Strategy

Ⓐ Calculate the number of moles of metal corresponding to the given mass transferred.
Ⓑ Write the reaction, and determine the number of moles of electrons required for the electroplating process.
Ⓒ Use the definition of the faraday to calculate the number of coulombs required. Then convert coulombs to current in amperes.

Solution

Ⓐ We must first determine the number of moles of Ag corresponding to 2.00 g of Ag:

$$\text{moles Ag} = \frac{2.00\ \cancel{g}}{107.868\ \cancel{g}/\text{mol}} = 1.85 \times 10^{-2}\ \text{mol Ag}$$

Ⓑ The reduction reaction is $Ag^{+}(aq) + e^{-} \longrightarrow Ag(s)$, so 1 mol of electrons produces 1 mol of silver. Ⓒ Using the definition of the faraday gives

$$\text{coulombs} = (1.85 \times 10^{-2}\ \cancel{\text{mol e}^-})(96{,}486\ \text{C}/\cancel{\text{mol e}^-}) = 1.78 \times 10^{3}\ \text{C}$$

The current in amperes needed to deliver this amount of charge in 12 h is therefore

$$\text{amperes} = \frac{1.78 \times 10^{3}\ \text{C}}{(12.0\ \cancel{h})(60\ \cancel{\text{min}}/\cancel{h})(60\ \text{s}/\cancel{\text{min}})} = 4.12 \times 10^{-2}\ \text{C/s} = 4.12 \times 10^{-2}\ \text{A}$$

Because the electroplating process is usually much less than 100% efficient (typical values are closer to 30%), the actual current necessary is higher than 0.1 A.

EXERCISE 19.15

A typical aluminum soft-drink can weighs about 29 g. How much time is needed to produce this amount of $Al(s)$ in the Hall–Heroult process, using a current of 15 A to reduce a molten Al_2O_3/Na_3AlF_6 mixture?

Answer 5.8 h

SUMMARY AND KEY TERMS

19.1 Describing Electrochemical Cells (p. 861)

Electrochemistry is the study of the relationship between electricity and chemical reactions. The oxidation–reduction reaction that occurs during an electrochemical process consists of two **half-reactions**, one representing the oxidation process and one the reduction process. The sum of the half-reactions gives the overall chemical reaction. The overall redox reaction is balanced when the number of electrons lost by the **reductant** is equal to the number of electrons gained by the **oxidant**. An electric current is produced from the flow of electrons from reductant to oxidant. An **electrochemical cell** can either generate electricity from a spontaneous redox reaction or consume electricity to drive a nonspontaneous reaction. In a **galvanic (voltaic) cell**, the energy from a spontaneous reaction generates electricity, whereas in an **electrolytic cell**, electrical energy is consumed to drive a nonspontaneous redox reaction. Both types of cells use two **electrodes** that provide an electrical connection between systems that are separated in space. The oxidative half-reaction occurs at the **anode**, and the reductive half-reaction occurs at the **cathode**. A **salt bridge** connects the separated solutions, allowing ions to migrate to either solution to ensure the system's electrical neutrality. A *voltmeter* is a device used to measure the flow of electric current between the two half-reactions. The **potential** of the cell, measured in volts, is the energy needed to move a charged particle in an electric field. An electrochemical cell can be described using line notation called a *cell diagram*, in which vertical lines indicate phase boundaries and the location of the salt bridge. Resistance to the flow of charge at a boundary is called the *junction potential.*

19.2 Standard Potentials (p. 866)

The flow of electrons in an electrochemical cell depends on the identity of the reacting substances, the difference in the potential energy of their valence electrons, and their concentrations. The potential of the cell under standard conditions (1 *M* for solutions, 1 atm for gases, pure solids or liquids for other substances) and at a fixed temperature (25°C) is called the **standard cell potential, E°_{cell}**. Only the *difference* between the potentials of two electrodes can be measured. By convention, all tabulated values of *standard electrode potentials* are listed as standard reduction potentials. The overall cell potential is the reduction potential of the reductive half-reaction minus the reduction potential of the oxidative half-reaction ($E^\circ_{cell} = E^\circ_{cathode} - E^\circ_{anode}$). The potential of the **standard hydrogen electrode (SHE)** is defined as 0 V under standard conditions. The potential of a half-reaction measured against the SHE under standard conditions is called its **standard electrode potential.** The standard cell potential is a measure of the driving force for a given redox reaction. All E° values are independent of the stoichiometric coefficients for the half-reaction. Redox reactions can be balanced using the half-reaction method, in which the overall redox reaction is divided into an oxidation half-reaction and a reduction half-reaction, each one balanced for mass and charge. The half-reactions selected from tabulated lists must exactly reflect reaction conditions. In an alternative method, the atoms in each half-reaction are balanced and then the charges are balanced. Whenever a half-reaction is reversed, the sign of E° corresponding to that reaction must also be reversed. If E°_{cell} is positive, the reaction will occur spontaneously under standard conditions. If E°_{cell} is negative, then the reaction is not spontaneous under standard conditions, although it will proceed spontaneously in the opposite direction. The potential of an **indicator electrode** is related to the concentration of the substance being measured, whereas the potential of the **reference electrode** is held constant. Whether reduction or oxidation occurs depends on the potential of the sample versus the potential of the reference electrode. In addition to the SHE, other reference electrodes are the **silver–silver chloride electrode**, the **saturated calomel electrode (SCE)**, the **glass electrode**, which is commonly used to measure pH, and **ion-selective electrodes**, which depend on the concentration of a single ionic species in solution. Differences in potential between the SHE and other reference electrodes must be included when calculating values for E°.

19.3 Comparing Strengths of Oxidants and Reductants (p. 876)

The oxidative and reductive strengths of a variety of substances can be compared using standard electrode potentials. Apparent anomalies can be explained by the fact that electrode potentials are measured in aqueous solution, which allows for strong intermolecular electrostatic interactions, and are not measured in the gas phase.

19.4 Electrochemical Cells and Thermodynamics (p. 880)

A **coulomb (C)** relates electrical potential, expressed in volts, and energy, expressed in joules. The current generated from a redox reaction is measured in **amperes (A)**, where 1 A is defined as the flow of 1 C/s past a given point. The **faraday (*F*)** is Avogadro's number multiplied by the charge on an electron, and corresponds to the charge on 1 mol of electrons. The product of the cell potential and the total charge is the maximum amount of energy available to do work, which is related to the change in free energy that occurs during the chemical process. Summing ΔG values for half-reactions gives ΔG for the overall reaction, which is proportional to *both* the potential *and* the number of electrons (n) transferred. Spontaneous redox reactions have a negative ΔG and therefore a positive E_{cell}. Because the equilibrium constant K is related to ΔG, E°_{cell} and K are also related. Large equilibrium constants correspond to large positive values of E°. The **Nernst equation** allows us to determine the spontaneous direction of any redox reaction under any reaction conditions from values of the relevant standard electrode potentials. **Concentration cells** consist of anode and cathode compartments that are identical except for the concentrations of the reactant. Because $\Delta G = 0$ at equilibrium, the measured potential of a concentration cell is zero at equilibrium (the concentrations are equal). A galvanic cell can also be used to measure the solubility product of a sparingly soluble substance, and to calculate the concentration of a species given a measured potential and the concentrations of all the other species.

19.5 Commercial Galvanic Cells (p. 890)

A **battery** is a contained unit that produces electricity, whereas a **fuel cell** is a galvanic cell that requires a constant external supply of one or more reactants to generate electricity. One type of battery is the *Leclanché dry cell*, which contains an electrolyte in an acidic water-based paste. This battery is called an **alkaline battery** when adapted to operate under alkaline conditions. *"Button" batteries* have a high output-to-mass ratio; **lithium–iodine batteries** consist of a solid electrolyte; the *nickel–cadmium (Nicad) battery* is rechargeable; and the *lead–acid battery*, which is also rechargeable, does not require the electrodes to be in separate compartments. A fuel cell requires an external supply of reactants as the products of the reaction are continuously removed. In a fuel cell, energy is not stored, but instead electrical energy is provided by a chemical reaction.

19.6 Corrosion (p. 895)

The deterioration of metals through oxidation is a galvanic process called **corrosion**. Protective coatings consist of a second metal that is more difficult to oxidize than the metal being protected. Alternatively, a more easily oxidized metal can be applied to a metal surface, thus providing *cathodic protection* of the surface. *Galvanized steel* is protected by a thin layer of zinc. *Sacrificial electrodes* can also be attached to an object to protect it.

19.7 Electrolysis (p. 897)

In **electrolysis**, an external voltage is applied to drive a nonspontaneous reaction. A *Downs cell* is used to produce sodium metal from a mixture of salts, and the *Hall–Heroult process* is used to produce aluminum commercially. Electrolysis can also be used to produce H_2 and O_2 from water. In practice, an additional voltage, called an **overvoltage**, must be applied to overcome factors such as a large activation energy and a junction potential. **Electroplating** is the process by which a second metal is deposited on a metal surface, thereby enhancing an object's appearance or providing protection from corrosion. The amount of material consumed or produced in a reaction can be calculated from the stoichiometry of an electrolysis reaction, the amount of current passed, and the duration of the electrolytic reaction.

KEY EQUATIONS

Standard cell potential	$E^\circ_{cell} = E^\circ_{cathode} - E^\circ_{anode}$	(19.7)
Charge on a mole of electrons (faraday)	$F \approx 96{,}486\ \text{J/(V·mol)}$	(19.33)
Maximum work from an electrochemical cell	$w_{max} = -nFE_{cell}$	(19.34)
Relationship between $\Delta G°$ and $E°$	$\Delta G° = -nFE^\circ_{cell}$	(19.36)
Relationship between $\Delta G°$ and K for a redox reaction	$\Delta G° = -RT \ln K$	(19.40)
Relationship between $E°$ and K for a redox reaction at 25°C	$E^\circ_{cell} = \left(\frac{RT}{nF}\right) \ln K$	(19.42)
	$E^\circ_{cell} = \left(\frac{0.0591\ \text{V}}{n}\right) \log K$	(19.43)
Relationship between ΔG and Q	$\Delta G = \Delta G° + RT \ln Q$	(19.44)
Relationship between E_{cell} and Q at 25°C	$E_{cell} = E^\circ_{cell} - \left(\frac{0.0591\ \text{V}}{n}\right) \log Q$	(19.47)
Relationship of charge, current, and time	charge (C) = current (A) × time (s)	(19.73)

QUESTIONS AND PROBLEMS

MGC For instructor-assigned homework, go to **www.masteringgeneralchemistry.com**

Questions and Problems with colored numbers have answers in the Appendix and complete solutions in the Student Solutions Manual.

CONCEPTUAL

19.1 Describing Electrochemical Cells

1. Is $2NaOH(aq) + H_2SO_4(aq) \longrightarrow Na_2SO_4(aq) + 2H_2O(l)$ an oxidation–reduction reaction? Why or why not?

2. If the two half-reactions are physically separated, how is it possible for a redox reaction to occur? What is the name of the apparatus in which the two half-reactions are carried out simultaneously?

3. What is the difference between a galvanic cell and an electrolytic cell? Which would you use to generate electricity?

4. What is the purpose of a salt bridge in a galvanic cell? Is it always necessary to use a salt bridge in a galvanic cell?

5. One criterion for a good salt bridge is that it contains ions that have similar rates of diffusion in aqueous solution as K^+ and Cl^- ions do. What would happen if the diffusion rates of the anions and cations differed significantly?

6. It is often more accurate to measure the potential of a redox reaction by immersing two electrodes in a single beaker rather than in two beakers. Why?

19.2 Standard Potentials

7. Is a hydrogen electrode chemically inert? What is the major disadvantage to using a hydrogen electrode?

8. List two factors that affect the measured potential of an electrochemical cell, and explain their impact on the measurements.

9. What is the relationship between electron flow and the potential energy of valence electrons? If the valence electrons of substance A have a higher potential energy than those of substance B, what is the direction of electron flow between them in a galvanic cell?

10. If the components of a galvanic cell include aluminum and bromine, what is the predicted direction of electron flow? Why?

11. Write a cell diagram representing a cell that contains the Ni/Ni^{2+} couple in one compartment and the SHE in the other compartment. What are the values of $E^\circ_{cathode}$, E°_{anode}, and E°_{cell}?

12. Explain why $E°$ values are independent of the stoichiometric coefficients in the corresponding half-reaction.

13. Identify the oxidants and reductants in each redox reaction:

(a) $Cr(s) + Ni^{2+}(aq) \longrightarrow Cr^{2+}(aq) + Ni(s)$

(b) $Cl_2(g) + Sn^{2+}(aq) \longrightarrow 2Cl^-(aq) + Sn^{4+}(aq)$

(c) $H_3AsO_4(aq) + 8H^+(aq) + 4Zn(s) \longrightarrow AsH_3(g) + 4H_2O(l) + 4Zn^{2+}(aq)$

(d) $2NO_2(g) + 2OH^-(aq) \longrightarrow NO_2^-(aq) + NO_3^-(aq) + H_2O(l)$

14. Identify the oxidants and reductants in each redox reaction:

(a) $Br_2(l) + 2I^-(aq) \longrightarrow 2Br^-(aq) + I_2(s)$

(b) $Cu^{2+}(aq) + 2Ag(s) \longrightarrow Cu(s) + 2Ag^+(aq)$

(c) $H^+(aq) + 2MnO_4^-(aq) + 5H_2SO_3(aq) \longrightarrow 2Mn^{2+}(aq) + 3H_2O(l) + 5HSO_4^-(aq)$

(d) $IO_3^-(aq) + 5I^-(aq) + 6H^+(aq) \longrightarrow 3I_2(s) + 3H_2O(l)$

15. All reference electrodes must conform to certain requirements. List the requirements and explain their significance.

16. For each application, describe the reference electrode you would use and explain why. In each case, how would the measured potential compare with the corresponding $E°$?

(a) measuring the potential of a Cl^-/Cl_2 couple

(b) measuring the pH of a solution

(c) measuring the potential of a MnO_4^-/Mn^{2+} couple

19.3 Comparing Strengths of Oxidants and Reductants

17. The order of electrode potentials cannot always be predicted by ionization potentials and electron affinities. Why? Do you expect sodium metal to have a higher or lower electrode potential than predicted from its ionization potential? What is its approximate electrode potential?

18. Without referring to tabulated data, which of the following couples would you expect to have the *least* negative electrode potential and which the *most* negative: Br^-/Br_2, Ca^{2+}/Ca, O_2/OH^-, Al^{3+}/Al? Why?

19. Because of the sulfur-containing amino acids present in egg whites, eating eggs with a silver fork will tarnish the fork. As a chemist, you have all kinds of interesting cleaning products in your cabinet, including a 1 *M* solution of oxalic acid ($H_2C_2O_4$). Would you choose this solution to clean the fork that you have tarnished from eating scrambled eggs?

20. The electrode potential for the reaction $Cu^{2+}(aq) + 2e^- \longrightarrow Cu(s)$ is 0.34 V under standard conditions. Is the potential for the oxidation of 0.5 mol of Cu equal to −0.34/2 V? Explain your answer.

19.4 Electrochemical Cells and Thermodynamics

21. State whether you agree or disagree with this reasoning and explain your answer: Standard electrode potentials arise from the number of electrons transferred. The greater the number of electrons transferred, the greater the measured potential difference. If 1 mol of a substance produces 0.76 V when 2 mol of electrons are transferred—as in $Zn(s) \longrightarrow Zn^{2+}(aq) + 2e^-$—then 0.5 mol of the substance will produce 0.76/2 V because only 1 mol of electrons is transferred.

22. What is the relationship between the *measured cell potential* and the *total charge* that passes through the cell? Which of these is dependent on concentration? Which is dependent on the identity of the oxidant or reductant? Which is dependent on the number of electrons transferred?

23. In the equation $w_{max} = -nFE_{cell}$, which quantities are extensive properties and which are intensive properties?

24. For any spontaneous redox reaction, $E°$ is positive. Use thermodynamic arguments to explain why this is true.

25. State whether you agree or disagree with this statement and explain your answer: Electrochemical methods are especially useful in determining the reversibility or irreversibility of reactions that take place in the cell.

26. Although the sum of two half-reactions gives another half-reaction, the sum of the potentials of the two half-reactions cannot be used to obtain the potential of the net half-reaction. Why? When does the sum of two half-reactions correspond to the overall reaction, and why?

27. Occasionally, you will find high-quality electronic equipment that has its electronic components plated in gold. What is the advantage of this?

28. Blood analyzers, which measure pH, P_{CO_2}, and P_{O_2}, are frequently used in clinical emergencies. For example, blood P_{CO_2} is measured with a pH electrode covered with a plastic membrane that is permeable to CO_2. Based on your knowledge of how electrodes function, explain how such an electrode might work. *Hint*: $CO_2(g) + H_2O(l) \longrightarrow HCO_3^-(aq) + H^+(aq)$.

29. Concentration cells contain the same species in solution in two different compartments. Explain what produces a voltage in a concentration cell. When does V = 0 in such a cell?

30. Describe how an electrochemical cell can be used to measure the solubility of a sparingly soluble salt.

19.5 Commercial Galvanic Cells

31. What advantage is there to using an alkaline battery rather than a Leclanché dry cell?

32. Why does the density of the fluid in lead–acid batteries drop when the battery is discharged?

33. What type of battery would you use for each application and why?
(a) powering an electric motor scooter
(b) a backup battery for a personal digital assistant (PDA)
(c) powering an iPod

34. Why are galvanic cells used as batteries and fuel cells? What is the difference between a battery and a fuel cell? What is the advantage to using highly concentrated or solid reactants in a battery?

19.6 Corrosion

35. Do you expect a bent nail to corrode more or less rapidly than a straight nail? Why?

36. What does it mean when a metal is described as being coated with a sacrificial layer? Is this different from galvanic protection?

37. Why is it important for automobile manufacturers to apply paint to the metal surface of a car? Why is this process particularly important for vehicles in northern climates, where salt is used on icy roads?

19.7 Electrolysis

38. Why might an electrochemical reaction that is thermodynamically favored require an overvoltage in order to occur?

39. How could you use an electrolytic cell to make quantitative comparisons of the strengths of various oxidants and reductants?

40. Why are mixtures of molten salts generally used during electrolysis rather than a pure salt?

41. Two solutions, one containing $Fe(NO_3)_2 \cdot 6H_2O$ and the other the same molar concentration of $Fe(NO_3)_3 \cdot 6H_2O$, were electrolyzed under identical conditions. Which solution produced the most metal, and justify your answer?

NUMERICAL

This section includes paired problems (marked by brackets) that require similar problem-solving skills.

19.1 Describing Electrochemical Cells

42. Copper sulfate forms a bright blue solution in water. If a piece of zinc metal is placed in a beaker of aqueous $CuSO_4$ solution, the blue color fades with time, the zinc strip begins to erode, and a black solid forms around the zinc strip. What is happening? Write half-reactions to show the chemical changes that are occurring. What will happen if a piece of copper metal is placed in a colorless aqueous solution of $ZnCl_2$?

43. Consider this spontaneous redox reaction: $NO_3^-(aq) + SO_3^{2-}(aq) + H^+(aq) \longrightarrow SO_4^{2-}(aq) + HNO_2(aq)$.
(a) Write the two half-reactions for this overall reaction?
(b) If the reaction is carried out in a galvanic cell using an inert electrode in each compartment, which electrode corresponds to which half-reaction?
(c) Which electrode is negatively charged, and which is positively charged?

44. The reaction $Pb(s) + 2VO^{2+}(aq) + 4H^+(aq) \longrightarrow Pb^{2+}(aq) + 2V^{3+}(aq) + 2H_2O(l)$ occurs spontaneously.
(a) Write the two half-reactions for this redox reaction.
(b) If the reaction is carried out in a galvanic cell using an inert electrode in each compartment, which reaction occurs at the cathode and which occurs at the anode?
(c) Which electrode is positively charged, and which is negatively charged?

45. Sulfate is reduced to HS^- in the presence of glucose, which is oxidized to bicarbonate. Write the two half-reactions corresponding to this process. What is the equation for the overall reaction?

46. Phenolphthalein is an indicator that turns pink under basic conditions. When an iron nail is placed in a gel that contains $[Fe(CN)_6]^{3-}$, the gel around the nail begins to turn pink. What is occurring? Write the half-reactions, and then write the overall redox reaction.

47. For each galvanic cell represented by a cell diagram, write the two half-reactions and the overall reaction. Indicate which reaction occurs at the anode and which occurs at the cathode.
(a) $Zn(s)|Zn^{2+}(aq)||H^+(aq)|H_2(g), Pt(s)$
(b) $Ag(s)|AgCl(s)|Cl^-(aq)||H^+(aq)|H_2(g)|Pt(s)$
(c) $Pt(s)|H_2(g)|H^+(aq)||Fe^{2+}(aq), Fe^{3+}(aq)|Pt(s)$

48. Write the half-reactions and the overall reaction for each cell diagram. State which half-reaction occurs at the anode and which occurs at the cathode.
(a) $Pb(s)|PbSO_4(s)|SO_4^{2-}(aq)||Cu^{2+}(aq)|Cu(s)$
(b) $Hg(s)|Hg_2Cl_2(s)|Cl^-(aq)||Cd^{2+}(aq)|Cd(s)$

49. For each redox reaction, write the half-reactions and draw the cell diagram for a galvanic cell in which the overall reaction occurs as written. Identify each electrode as either positive or negative.
(a) $Ag(s) + Fe^{3+}(aq) \longrightarrow Ag^+(aq) + Fe^{2+}(aq)$
(b) $Fe^{3+}(aq) + \frac{1}{2}H_2(g) \longrightarrow Fe^{2+}(aq) + H^+(aq)$

50. Write the half-reactions for each overall reaction, decide whether the reaction will occur spontaneously, and construct a cell diagram for a galvanic cell in which a spontaneous reaction will occur:
(a) $2Cl^-(aq) + Br_2(l) \longrightarrow Cl_2(g) + 2Br^-(aq)$
(b) $2NO_2(g) + 2OH^-(aq) \longrightarrow NO_2^-(aq) + NO_3^-(aq) + H_2O(l)$
(c) $2H_2O(l) + 2Cl^-(aq) \longrightarrow H_2(g) + Cl_2(g) + 2OH^-(aq)$
(d) $C_3H_8(g) + 5O_2(g) \longrightarrow 3CO_2(g) + 4H_2O(g)$

51. Write the half-reactions for each overall reaction, decide whether the reaction will occur spontaneously, and construct a cell diagram for a galvanic cell in which a spontaneous reaction will occur:
(a) $Co(s) + Fe^{2+}(aq) \longrightarrow Co^{2+}(aq) + Fe(s)$
(b) $O_2(g) + 4H^+(aq) + 4Fe^{2+}(aq) \longrightarrow 2H_2O(l) + 4Fe^{3+}(aq)$
(c) $6Hg^{2+}(aq) + 2NO_3^-(aq) + 8H^+ \longrightarrow 3Hg_2^{2+}(aq) + 2NO(g) + 4H_2O(l)$
(d) $CH_4(g) + 2O_2(g) \longrightarrow CO_2(g) + 2H_2O(g)$

19.2 Standard Potentials

52. Write the cell diagram for a galvanic cell with a SHE and a zinc electrode that carries out this overall reaction: $Zn(s) + 2H^+(aq) \longrightarrow Zn^{2+}(aq) + H_2(g)$.

53. Draw the cell diagram for a galvanic cell utilizing a SHE and a copper electrode that carries out the reaction $H_2(g) + Cu^{2+}(aq) \longrightarrow 2H^+(aq) + Cu(s)$.

54. Balance each reaction and calculate the standard reduction potential for each:
(a) $Cu^+(aq) + Ag^+(aq) \longrightarrow Cu^{2+}(aq) + Ag(s)$
(b) $Sn(s) + Fe^{3+}(aq) \longrightarrow Sn^{2+}(aq) + Fe^{2+}(aq)$
(c) $Mg(s) + Br_2(l) \longrightarrow 2Br^-(aq) + Mg^{2+}(aq)$

55. Balance each reaction and calculate the standard electrode potential for each. Be sure to include the physical state of each product and reactant.
(a) $Cl_2(g) + H_2(g) \longrightarrow 2Cl^- + 2H^+$
(b) $Br_2 + Fe^{2+} \longrightarrow 2Br^- + Fe^{3+}$
(c) $Fe^{3+} + Cd \longrightarrow Fe^{2+} + Cd^{2+}$

56. Write a balanced equation for each redox reaction:
(a) $H_2PO_2^-(aq) + SbO_2^-(aq) \longrightarrow HPO_3^{2-}(aq) + Sb(s)$ in basic solution
(b) $HNO_2(aq) + I^-(aq) \longrightarrow NO(g) + I_2(s)$ in acidic solution
(c) $N_2O(g) + ClO^-(aq) \longrightarrow Cl^-(aq) + NO_2^-(aq)$ in basic solution
(d) $Br_2(l) \longrightarrow Br^-(aq) + BrO_3^-(aq)$ in basic solution
(e) $Cl(CH_2)_2OH(aq) + K_2Cr_2O_7(aq) \longrightarrow ClCH_2CO_2H(aq) + Cr^{3+}(aq)$ in acidic solution

57. Balance each redox reaction:
(a) $I^-(aq) + HClO_2(aq) \longrightarrow IO_3^-(aq) + Cl_2(g)$ in acidic solution
(b) $Cr^{2+}(aq) + O_2(g) \longrightarrow Cr^{3+}(aq) + H_2O(l)$ in acidic solution
(c) $CrO_2^-(aq) + ClO^-(aq) \longrightarrow CrO_4^{2-}(aq) + Cl^-(aq)$ in basic solution
(d) $S(s) + HNO_2(aq) \longrightarrow H_2SO_3(aq) + N_2O(g)$ in acidic solution
(e) $F(CH_2)_2OH(aq) + K_2Cr_2O_7(aq) \longrightarrow FCH_2CO_2H(aq) + Cr^{3+}(aq)$ in acidic solution

58. The standard cell potential for the oxidation of Pb to Pb^{2+} with the concomitant reduction of Cu^+ to Cu is 0.39 V. You know that $E°$ for the Pb^{2+}/Pb couple is −0.13 V. What is $E°$ for the Cu^+/Cu couple?

59. You have built a galvanic cell similar to the one in Figure 19.6 using an iron nail, a solution of $FeCl_2$, and a standard hydrogen electrode. When the cell is connected, you notice that the iron nail begins to corrode. What else do you observe? Under standard conditions, what is E_{cell}?

60. Carbon is used to reduce iron ore to metallic iron. The overall reaction is

$$2Fe_2O_3 \cdot xH_2O(s) + 3C(s) \longrightarrow 4Fe(s) + 3CO_2(g) + 2xH_2O(g)$$

Write the two half-reactions for this overall reaction.

61. Will each reaction occur spontaneously under standard conditions?
(a) $Cu(s) + 2H^+(aq) \longrightarrow Cu^{2+}(aq) + H_2(g)$
(b) $Zn^{2+}(aq) + Pb(s) \longrightarrow Zn(s) + Pb^{2+}(aq)$

62. Each reaction takes place in acidic solution. Balance each reaction, and then determine whether it occurs spontaneously as written under standard conditions.
(a) $Se(s) + Br_2(l) \longrightarrow H_2SeO_3(aq) + Br^-(aq)$
(b) $NO_3^-(aq) + S(s) \longrightarrow HNO_2(aq) + H_2SO_3(aq)$
(c) $Fe^{3+}(aq) + Cr^{3+}(aq) \longrightarrow Fe^{2+}(aq) + Cr_2O_7^{2-}(aq)$

63. Calculate $E°_{cell}$ and $\Delta G°$ for the redox reaction represented by the cell diagram $Pt(s)|Cl_2(g, 1\ atm)|ZnCl_2(aq, 1\ M)|Zn(s)$. Will this reaction occur spontaneously?

64. If you place Zn-coated (galvanized) tacks in a glass and add an aqueous solution of iodine, the brown color of the iodine solution fades to a pale yellow. What has happened? Write the two half-reactions and the overall balanced equation for this reaction. What is E°_{cell}?

65. Your lab partner wants to recover solid silver from silver chloride by using a 1.0 *M* solution of HCl and 1 atm H_2 under standard conditions. Will this plan work?

19.4 Electrochemical Cells and Thermodynamics

66. The chemical equation for the combustion of butane is

$$C_4H_{10}(g) + \frac{13}{2}O_2(g) \longrightarrow 4CO_2(g) + 5H_2O(g)$$

This reaction has $\Delta H^\circ = -2877$ kJ/mol. Calculate E°_{cell} and then determine ΔG°. Is this a spontaneous process? What is the change in entropy that accompanies this process at 298 K?

67. For the cell represented as $Al(s)\,|\,Al^{3+}(aq)\,\|\,Sn^{2+}(aq), Sn^{4+}(aq)\,|\,Pt(s)$, how many electrons are transferred in the redox reaction? What is the standard cell potential? Is this a spontaneous process? What is ΔG°?

68. How many electrons are transferred during the reaction $Pb(s) + Hg_2Cl_2(aq) \longrightarrow PbCl_2(aq) + 2Hg(l)$? What is the standard cell potential? Is the oxidation of Pb by Hg_2Cl_2 spontaneous? Calculate ΔG° for this reaction.

69. Explain why the sum of the potentials for the half-reactions $Sn^{2+}(aq) + 2e^- \longrightarrow Sn(s)$ and $Sn^{4+}(aq) + 2e^- \longrightarrow Sn^{2+}(aq)$ does not equal the potential for the reaction $Sn^{4+}(aq) + 4e^- \longrightarrow Sn(s)$. What is the net cell potential? Compare the values of ΔG° for the sum of the potentials and the actual net cell potential.

70. Based on tables in this text, do you agree that the potential for the half-reaction $Cu^{2+} + 2e^- \longrightarrow Cu(s)$ equals 0.68 V? Why or why not?

71. For each reaction, calculate E°_{cell} and then determine ΔG°. Indicate whether each reaction is spontaneous.
(a) $2Na(s) + 2H_2O(l) \longrightarrow 2NaOH(aq) + H_2(g)$
(b) $K_2S_2O_6(aq) + 2KI(aq) \longrightarrow I_2(s) + 2K_2SO_4(aq)$
(c) $Sn(s) + CuSO_4(aq) \longrightarrow Cu(s) + SnSO_4(aq)$

72. What is the change in free energy for the reaction between Ca^{2+} and sodium metal to give calcium metal and Na^+? Do the sign and magnitude of ΔG agree with what you would expect based on the positions of these elements in the periodic table? Why or why not?

73. In acidic solution, permanganate (MnO_4^-) oxidizes Cl^- to chlorine gas, and MnO_4^- is reduced to $Mn^{2+}(aq)$. (a) Write the balanced equation for this reaction; (b) determine E°_{cell}; and (c) calculate the equilibrium constant.

74. Potentiometric titrations are an efficient method for determining the endpoint of a redox titration. In such a titration, the potential of the solution is monitored as measured volumes of an oxidant or a reductant are added. Data for a typical titration, the potentiometric titration of Fe(II) with a 0.1 *M* solution of Ce(IV), are given in the table. The starting potential has been arbitrarily set equal to zero because it is the *change* in potential with the addition of the oxidant that is important.

Titrant, mL	E, mV
2.00	50
6.00	100
9.00	255
10.00	960
11.00	1325
12.00	1625
14.00	1875

(a) Write the balanced equation for the oxidation of Fe^{2+} by Ce^{4+}.
(b) Plot the data, and then locate the endpoint.
(c) How many millimoles of Fe^{2+} did the solution being titrated originally contain?

75. The standard electrode potential, E°, for the half-reaction $Ni^{2+}(aq) + 2e^- \longrightarrow Ni(s)$ is -0.25 V. What pH is needed for this reaction to take place in the presence of 1.00 atm $H_2(g)$ as the reductant if $[Ni^{2+}]$ is 1.00 *M*?

76. The reduction of Mn(VII) to Mn(*s*) by $H_2(g)$ proceeds in five steps that can be readily followed by changes in the color of the solution. Here is the redox chemistry:
(1) $MnO_4^-(aq) + e^- \longrightarrow MnO_4^{2-}(aq)$ $E^\circ = +0.56$ V (purple ⟶ dark green)
(2) $MnO_4^{2-}(aq) + 2e^- + 4H^+(aq) \longrightarrow MnO_2(s)$ $E^\circ = +2.26$ V (dark green ⟶ dark brown solid)
(3) $MnO_2(s) + e^- + 4H^+(aq) \longrightarrow Mn^{3+}(aq)$ $E^\circ = +0.95$ V (dark brown solid ⟶ red-violet)
(4) $Mn^{3+}(aq) + e^- \longrightarrow Mn^{2+}(aq)$ $E^\circ = +1.51$ V (red-violet ⟶ pale pink)
(5) $Mn^{2+}(aq) + 2e^- \longrightarrow Mn(s)$ $E^\circ = -1.18$ V (pale pink ⟶ colorless)
(a) Is the reduction of $MnO_4^-(aq)$ to $Mn^{3+}(aq)$ by $H_2(g)$ spontaneous under standard conditions? What is E°_{cell}?
(b) Is the reduction of $Mn^{3+}(aq)$ to Mn(*s*) by $H_2(g)$ spontaneous under standard conditions? What is E°_{cell}?

77. Mn(III) can disproportionate (both oxidize and reduce itself) by means of these half-reactions:

$$Mn^{3+}(aq) + e^- \longrightarrow Mn^{2+}(aq) \qquad E^\circ = 1.51 \text{ V}$$

$$Mn^{3+}(aq) + 2H_2O(l) \longrightarrow MnO_2(s) + 4H^+(aq) + e^- \qquad E^\circ = 0.95 \text{ V}$$

(a) What is E° for the disproportionation reaction?
(b) Is disproportionation more or less thermodynamically favored at low pH than at pH 7? Explain your answer.
(c) How could you prevent the disproportionation reaction from occurring?

78. For the reduction of oxygen to water, $E^\circ = 1.23$ V. What is the potential for this half-reaction at pH 7.00? What is the potential in a 0.85 *M* solution of NaOH?

79. Given the following biologically relevant half-reactions, will FAD, a molecule used to transfer electrons, be an effective oxidant for the conversion of acetaldehyde to acetate at pH 4.00?

$$\text{acetate} + 2H^+ + 2e^- \longrightarrow \text{acetaldehyde} + H_2O \qquad E = -0.58 \text{ V}$$

$$\text{FAD} + 2H^+ + 2e^- \longrightarrow \text{FADH}_2 \qquad E = -0.18 \text{ V}$$

80. The biological molecule abbreviated as NADH can be oxidized to NAD^+ via the half-reaction $NADH \longrightarrow NAD^+ + H^+ + 2e^-$,

$E = +0.32$ V. Would NADH be able to reduce (a) acetate to pyruvate or (b) pyruvate to lactate? (c) What potential is needed to convert acetate to lactate?

$$\text{acetate} + CO_2 + 2H^+ + 2e^- \longrightarrow \text{pyruvate} + H_2O \quad E = -0.70\text{ V}$$

$$\text{pyruvate} + 2H^+ + 2e^- \longrightarrow \text{lactate} \quad E = -0.185\text{ V}$$

81. Under acidic conditions, ideally any half-reaction with $E° > 1.23$ V will oxidize water via the reaction $O_2(g) + 4H^+(aq) + 4e^- \longrightarrow 2H_2O(l)$.
(a) Will aqueous acidic $KMnO_4$ evolve oxygen with the formation of MnO_2?
(b) At pH 14.00, what is $E°$ for the oxidation of water by aqueous $KMnO_4$ (1 *M*) with the formation of MnO_2?
(c) At pH 14.00, will water be oxidized if you are trying to form MnO_2 from MnO_4^{2-} via the reaction $2MnO_4^{2-}(aq) + 2H_2O(l) \longrightarrow 2MnO_2(s) + O_2(g) + 4OH^-(aq)$?

82. Ideally, any half-reaction with $E° > 1.23$ V will oxidize water as a result of the half-reaction $O_2(g) + 4H^+(aq) + 4e^- \longrightarrow 2H_2O(l)$.
(a) Will FeO_4^{2-} oxidize water if the half-reaction for the reduction of Fe(VI) $\longrightarrow$ Fe(III) is $FeO_4^{2-}(aq) + 8H^+(aq) + 3e^- \longrightarrow Fe^{3+}(aq) + 4H_2O$, $E° = 1.9$ V?
(b) What pH is needed to cause this reaction to proceed spontaneously if $[Fe^{3+}] = [FeO_4^{2-}] = 1.0$ *M* and $P_{O_2} = 1.0$ atm?

83. Complexing agents can bind to metals and result in net stabilization of the complexed species. What is the net thermodynamic stabilization energy that results from using CN^- as a complexing agent for Mn^{3+}/Mn^{2+}?

$$Mn^{3+}(aq) + e^- \longrightarrow Mn^{2+}(aq) \quad E° = 1.51\text{ V}$$

$$Mn(CN)_6^{3-}(aq) + e^- \longrightarrow Mn(CN)_6^{4-}(aq) \quad E° = -0.24\text{ V}$$

84. You have constructed a cell with zinc and lead amalgam electrodes described by the cell diagram $Zn(Hg)(s)\,|\,ZnSO_4(aq), PbSO_4(aq)\,|\,Pb(Hg)(s)$. If you vary the concentration of $ZnSO_4$ and measure the potential at different concentrations, you obtain these data:

$[ZnSO_4]$, mol/L	E_{cell}, V
0.0005	0.6114
0.002	0.5832
0.01	0.5535

(a) Write the half-reactions that occur in this cell.
(b) What is the overall redox reaction?
(c) $E°$ of the cell is 0.411 V. What is ΔG?
(d) What is the equilibrium constant for this redox reaction?

85. Hydrogen gas reduces Ni^{2+} according to the reaction $Ni^{2+}(aq) + H_2(g) \longrightarrow Ni(s) + 2\,H^+(aq)$; $E°_{cell} = -0.25$ V, $\Delta H° = 54$ kJ/mol.
(a) What is *K* for this redox reaction?
(b) Is this reaction likely to occur?
(c) What conditions can be changed to increase the likelihood that the reaction will occur as written?
(d) Is the reaction more likely to occur at higher or lower pH?

86. The silver–silver bromide electrode has a standard potential of 0.073 V. What is K_{sp} of AgBr?

19.5 Commercial Galvanic Cells

87. This reaction is characteristic of a lead storage battery:

$$Pb(s) + PbO_2(s) + 2H_2SO_4(aq) \longrightarrow 2PbSO_4(s) + 2H_2O(l)$$

If you have a battery with an electrolyte that has a density of 1.15 g/cm^3 and contains 30.0% sulfuric acid by mass, is the potential greater or less than that of the standard cell?

19.7 Electrolysis

88. The electrolysis of molten salts is frequently used in industry to obtain pure metals. How many grams of metal are deposited from these salts with each mole of electrons? (a) $AlCl_3$; (b) $MgCl_2$; (c) $FeCl_3$

89. Electrolysis is the most direct way of recovering a metal from its ores. However, the $Na^+(aq)/Na(s)$, $Mg^{2+}(aq)/Mg(s)$, and $Al^{3+}(aq)/Al(s)$ couples all have standard electrode potentials, $E°$, less than zero, indicating that these metals can never by obtained by electrolysis of aqueous solutions of their salts. Why? What reaction would occur instead?

90. Calculate the mass of copper metal that is deposited if a 5.12-A current is passed through a copper nitrate solution for 1.5 h.

91. What volume of chlorine gas at STP is evolved when a solution of $MgCl_2$ is electrolyzed using a current of 12.4 A for 1.0 h?

92. What mass of PbO_2 is reduced when a current of 5.0 A is withdrawn over a period of 2 h from a lead storage battery?

93. Electrolysis of $Cr^{3+}(aq)$ produces $Cr^{2+}(aq)$. If you had 500 mL of a 0.15 *M* solution of $Cr^{3+}(aq)$, how long would it take to reduce the Cr^{3+} to Cr^{2+} using a 0.158-A current?

94. Predict the products when aqueous solutions of the following are electrolyzed: (a) $MgBr_2$; (b) $Hg(CH_3CO_2)_2$; (c) $Al_2(SO_4)_3$.

95. Predict the products obtained at each electrode when aqueous solutions of the following are electrolyzed: (a) $AgNO_3$; (b) RbI.

APPLICATIONS

96. The percent efficiency of a fuel cell is defined as $\Delta G^0/\Delta H^0 \times 100$. If hydrogen gas were distributed for domestic and industrial use from a central electrolysis facility, the gas could be piped to consumers much as methane is piped today. Conventional nuclear power stations have an efficiency of 25–30%. Use tabulated data to calculate the efficiency of a fuel cell in which the reaction $H_2(g) + \frac{1}{2}O_2(g) \longrightarrow H_2O(g)$ occurs under standard conditions.

97. You are about to run an organic reaction and you need a strong oxidant. Although you have BrO_3^- at your disposal, you prefer to use MnO_4^-. You notice you also have MnO_2 in the lab.
(a) Predict whether you will be able to synthesize MnO_4^- using the materials available to you.
(b) Write the overall reaction for the synthesis of MnO_4^-.
(c) What is $\Delta G°$ for this reaction?
(d) What is the equilibrium constant?

98. It is possible to construct a galvanic cell using amalgams as electrodes, each containing different concentrations of the same metal. One example is the $Pb(Hg)(a_1)\,|\,PbSO_4\,(soln)\,|\,Pb(Hg)(a_2)$ cell, in which a_1 and a_2 represent the concentrations of lead in the amalgams. Notice that no chemical change occurs; rather, the reaction transfers lead from one amalgam to the other, thus altering the Pb concentration in both amalgams. Write an equation for *E* for such a cell.

99. The biological electron transport chain provides for an orderly, stepwise transfer of electrons. Both NADH and $FADH_2$ are energy-rich molecules that liberate a large amount of energy upon oxidation. Free energy released during the transfer of electrons from either of these molecules to oxygen drives the synthesis of ATP formed during respiratory metabolism. The reactions are

$$NADH + H^+ + \tfrac{1}{2}O_2 \longrightarrow NAD^+ + H_2O \qquad \Delta G^{\circ\prime} = -52.6 \text{ kcal/mol}$$

$$FADH_2 + \tfrac{1}{2}O_2 \longrightarrow FAD + H_2O \qquad \Delta G^{\circ\prime} = -43.4 \text{ kcal/mol}$$

The standard potential ($E^{\circ\prime}$) for a biological process is defined at pH = 7.
(a) What is $E^{\circ\prime}_{cell}$ for each reaction?
(b) What is $E^{\circ\prime}_{anode}$ for the reaction that uses NADH?
(c) How does $E^{\circ\prime}_{anode}$ of $FADH_2$ compare with that of NADH?

100. The oldest known metallurgical artifacts are beads made from alloys of copper, produced in Egypt, Mesopotamia, and the Indus Valley in 3000 B.C. To determine the copper content of alloys such as brass, a brass sample is dissolved in nitric acid to obtain $Cu^{2+}(aq)$, and then the pH is adjusted to a value of 4–5. Excess KI is used to reduce the Cu^{2+} to Cu^+ with concomitant oxidation of I^- to I_2. The iodine that is produced is then titrated with thiosulfate solution to determine the amount of Cu^{2+} in the original solution. These reactions are involved in the procedure:
(1) $Cu^{2+}(aq) + I^-(aq) + e^- \longrightarrow CuI(s)$ $E^\circ = 0.86$ V
(2) $S_4O_6^{2-}(aq) + 2e^- \longrightarrow 2S_2O_3^{2-}(aq)$ $E^\circ = 0.08$ V
(3) $NO_3^-(aq) + 2H^+(aq) + 2e^- \longrightarrow NO_2^-(aq) + H_2O(l)$ $E^\circ = 0.94$ V
(4) $I_2(s) + 2e^- \longrightarrow 2I^-(aq)$ $E^\circ = 0.54$ V
(a) Write a balanced equation for the reaction between nitric acid and Cu(*s*).
(b) What is E°_{cell} for this reaction?
(c) What is E°_{cell} for the reaction between thiosulfate and iodine?
(d) When the pH of reaction solution 3 is adjusted to 7.00, what is *E* for this half-reaction?
(e) Why is it necessary to adjust the pH of reaction solution 3 before proceeding to step 4?
(f) Use tabulated data to explain why rust (Fe_2O_3) is considered a contaminant that renders this method useless in determining Cu concentrations?

101. While working at a nuclear reactor site, you have been put in charge of reprocessing spent nuclear fuel elements. Your specific task is to reduce U(VI) to elemental U without reducing Pu(VI) to elemental Pu. You have the following information at your disposal:

$$E^\circ \quad UO_2^{2+} \xrightarrow{0.05} UO_2^{+} \xrightarrow{0.62} U^{4+} \xrightarrow{-0.61} U^{3+} \xrightarrow{-1.85} U$$

$$E^\circ \quad PuO_2^{2+} \xrightarrow{0.91} PuO_2^{+} \xrightarrow{1.17} Pu^{4+} \xrightarrow{0.98} Pu^{3+} \xrightarrow{-2.03} Pu$$

Use tabulated data to decide what reagent will accomplish your task in a 1 *M* acidic solution.

102. Stainless steels typically contain 11% Cr and are resistant to corrosion because of the formation of an oxide layer that can be approximately described as $FeCr_2O_4$, where the iron is Fe(II). The protective layer forms when Cr(II) is oxidized to Cr(III) and Fe is oxidized to Fe(II). Explain how this film prevents the corrosion of Fe to rust, which has the formula Fe_2O_3.

103. Ion-selective electrodes are a powerful tool for measuring specific concentrations of ions in solution. For example, they are used to measure iodide in milk, copper-ion levels in drinking water, fluoride concentrations in toothpastes, and the silver-ion concentration in photographic emulsions and spent fixing solutions. Describe how ion-selective electrodes work, and then propose a design for an ion-selective electrode that can be used for measuring water hardness (Ca^{2+}, Mg^{2+}) in water-conditioning systems.

104. Enzymes are proteins that catalyze a specific reaction with a high degree of specificity. An example is the hydrolysis of urea by urease:

$$NH_2CONH_2 + 2H_2O + H^+ \xrightarrow{\text{Urease}} 2NH_4^+ + HCO_3^-$$

An enzyme electrode for measuring urea concentrations can be made by coating the surface of a glass electrode with a gel that contains urease.
(a) Explain what occurs when the electrode is placed in contact with a solution that contains urea.
(b) What species diffuses through the gel?
(c) What would be an effective reference electrode?

105. Gas-sensing electrodes can be constructed using a combination electrode that is surrounded by a gas-permeable membrane. For example, to measure CO_2, a pH electrode and a reference electrode are placed in solution on the "inner" side of a CO_2-permeable membrane, and the sample solution is placed on the "external" side. As CO_2 diffuses through the membrane, the pH of the internal solution changes due to the reaction $CO_2(g) + H_2O(l) \longrightarrow HCO_3^-(aq) + H^+(aq)$. Thus, the pH of the internal solution varies directly with the CO_2 vapor pressure in the external sample. Ammonia electrodes operate in the same manner. Describe an electrode that would test for ammonia levels in seawater.

106. U.S. submarines that are not nuclear-powered use a combination of batteries and diesel engines for their power. When submerged, they are battery driven; when on the surface, they are diesel driven. Why aren't batteries used when the submarines are on the surface?

107. List some practical considerations in designing a battery to power an electric car.

108. It is possible to run a digital clock using the power supplied by two potatoes. The clock is connected to two wires, one attached to a copper plate and the other to a zinc plate. Each plate is pushed into a different potato, and when the two potatoes are connected by a wire, the clock begins to run as if it were connected to a battery. (a) Draw a cell diagram of the potato clock. (b) Explain why the clock runs.

109. The silver–zinc battery has the highest energy density of any rechargeable battery available today. Its use is presently limited to military applications, primarily in portable communications, aerospace, and torpedo-propulsion systems. The disadvantages of these cells are their limited life (they typically last no more than about 2 years) and their high cost, which restricts their use to situations in which cost is only a

minor factor. The generally accepted equations representing this type of battery are

$$2AgO(s) + Zn(s) + H_2O(l) \longrightarrow Ag_2O(s) + Zn(OH)_2(aq) \quad E° = 1.85\ V$$

$$Ag_2O(s) + Zn(s) + H_2O(l) \longrightarrow 2Ag(s) + Zn(OH)_2(aq) \quad E° = 1.59\ V$$

(a) Write the overall cell reaction, and calculate $E°_{cell}$.
(b) If the cell is 75% efficient, what is the maximum amount of work that can be generated from this type of battery?
(c) Use tabulated data to calculate the maximum work that can be generated by a lead storage cell. If a silver–zinc battery is operating at 100% efficiency, how do the two batteries compare?

110. All metals used in boats and ships are subject to corrosion, particularly when the vessels are operated in salt water, which is a good electrolyte. Based on the data in the table, where potentials are measured using a glass electrode, explain why (a) iron or steel should not be used in bolts in a lead ballast keel; (b) ordinary brass should not be used as a structural fastening, particularly below the waterline; (c) an aluminum hull should not be painted with a copper-based antifouling paint; (d) magnesium sacrificial anodes are preferred over zinc when a vessel is kept in fresh water, and (e) Monel (an alloy that contains mostly nickel and copper) is preferred over stainless steel for freshwater tanks.

Metal	***E* vs. Ag/AgCl, V**
Titanium	0.02
Monel [Ni(Cu)]	−0.06
Ni(Al) bronze	−0.16
Lead	−0.20
Manganese bronze	−0.29
Brass	−0.30
Copper	−0.31
Tin	−0.31
Stainless steel	−0.49
Aluminum	−0.87
Zinc	−1.00
Magnesium	−1.60

111. Parents often coat a baby's first shoes with copper to preserve them for posterity. A conducting powder, such as graphite, is rubbed on the shoe, and then copper is electroplated on the shoe. How much copper is deposited on a shoe if the electrolytic process is run for 60 min at 1.2 A from a 1.0 *M* solution of $CuSO_4$?

112. Before 1886, metallic aluminum was so rare that a bar of it was displayed next to the Crown Jewels at the Paris Exposition of 1855. Today it is obtained commercially from aluminum oxide by the Hall–Heroult process, an electrolytic process that uses molten Al_2O_3 and cryolite (Na_3AlF_6). As the operation proceeds, molten Al sinks to the bottom of the cell. The overall reaction is $2Al_2O_3(l) + 3C(s) \longrightarrow 4Al(l) + 3CO_2(g)$; however, the process is only approximately 90% efficient. (a) Why is cryolite added to the Al_2O_3? (b) How much aluminum is deposited if electrodeposition occurs for 24 h at 1.8 A?

113. Compact audio discs (CDs) are manufactured by electroplating. Information is stored on a CD master in a patter of "pits" (depressions, which correspond to an audio track) and "lands" (the raised areas between depressions). A laser beam cuts the pits into a plastic or glass material. The material is cleaned, sprayed with $[Ag(NH_3)_2]^+$, and then washed with a formaldehyde solution that reduces the complex and leaves a thin silver coating. Nickel is electrodeposited on the disk and then peeled away to produce a master disk, which is used to stamp copies.
(a) Write the half-reactions that correspond to the electrodeposition reaction.
(b) If a CD has a radius of 12 cm and an interior hole with a diameter of 2.5 cm, how long does it take to deposit a 50-μm layer of nickel on one side of the CD using a 1.0 *M* solution of $NiSO_4$ and a current of 0.8 A?

114. One of the most important electrolytic processes used in industry is the electrolytic reduction of acrylonitrile ($CH_2{=}CHCN$) to adiponitrile [$NC(CH_2)_4CN$]. The product is then hydrogenated to hexamethylenediamine [$H_2N(CH_2)_6NH_2$], a key component of one form of nylon. Monsanto produces about 200,000 metric tons of adiponitrile annually using this process. The cathode reaction in the electrochemical cell is

$$2CH_2{=}CHCN + 2H^+ + 2e^- \longrightarrow NC(CH_2)_4CN$$

The cost of the electricity consumed is a major portion of the overall cost of the product. Calculate the total number of kilowatt-hours of electricity used by Monsanto each year in this process, assuming a continuous applied potential of 5.0 V and an electrochemical efficiency of 50%.

115. Calculate the total amount of energy consumed in the electrolysis reaction used to make the 16×10^6 metric tons of aluminum produced annually worldwide, assuming a continuous applied potential of 5.0 V and an efficiency of 50%. Express your answer in kilojoules and in kilowatt-hours.

20 Nuclear Chemistry

LEARNING OBJECTIVES

- To understand the factors that affect nuclear stability
- To know the different kinds of radioactive decay
- To be able to balance a nuclear reaction
- To be able to interpret a radioactive decay series
- To know the differences between ionizing and nonionizing radiation and their effects on matter
- To be able to identify natural and artificial sources of radiation
- To be able to calculate a mass–energy balance and a nuclear binding energy
- To understand the differences between nuclear fission and fusion
- To understand how nuclear reactors operate
- To understand how nuclear transmutation reactions led to the formation of the elements in stars and how they can be used to synthesize transuranium elements

CHAPTER OUTLINE

Until now, you have been studying chemical processes in which atoms share or transfer electrons to form new compounds, leaving the atomic nuclei largely unaffected. In this chapter, we examine some properties of the atomic nucleus and the changes that can occur in atomic nuclei.

Nuclear reactions differ from other chemical processes in one critical way: in a nuclear reaction, the identities of the elements change. In addition, nuclear reactions are often accompanied by the release of enormous amounts of energy, as much as a *billion* times more than the energy released by chemical reactions. Moreover, the yields and rates of a nuclear reaction are generally unaffected by changes in temperature, pressure, or the presence of a catalyst.

We begin our discussion by examining the structure of the atomic nucleus and the factors that determine whether a particular nucleus is stable or decays spontaneously to another element. We then discuss the major kinds

The glow caused by intense radiation. The high-energy particles ejected into the surrounding water or air by an intense radioactive source such as this nuclear reactor core produce a ghostly bluish glow.

of nuclear decay reactions as well as the properties and uses of the radiation that is emitted when nuclei decay. In these discussions, you will learn how radioactive emissions can be used to study the mechanisms of chemical reactions and biological processes, and how to calculate the amount of energy released during a nuclear reaction. You will also discover why houses are tested for radon gas, how radiation is used to probe organs such as the brain, and how the energy from nuclear reactions can be harnessed to produce electricity. Last, we explore the nuclear chemistry that takes place in stars, and we describe the role that stars play in producing most of the elements in the universe.

20.1 The Components of the Nucleus

Although most of the known elements have at least one isotope whose atomic nucleus is stable indefinitely, all elements have isotopes that are unstable and disintegrate, or *decay*, at measurable rates by emitting radiation. Some elements have no stable isotopes and eventually decay to other elements. In contrast to the chemical reactions that were the main focus of earlier chapters and are due to changes in the arrangements of the valence electrons of atoms, the process of nuclear decay is a nuclear reaction that results in changes inside the atomic nucleus. We begin our discussion of nuclear reactions by reviewing the conventions used to describe the components of the nucleus.

The Atomic Nucleus

As you learned in Chapter 1, each element can be represented by the notation ${}^{A}_{Z}X$, where A, the mass number, is the sum of the numbers of protons and neutrons, and Z, the atomic number, is the number of protons. The protons and neutrons that make up the nucleus of an atom are called **nucleons**, and an atom with a particular number of protons and neutrons is called a **nuclide**. Nuclides that have the same number of protons but different numbers of neutrons are called *isotopes*. Isotopes can also be represented by an alternative notation that uses the name of the element followed by the mass number, as we see in *carbon-12*. The stable isotopes of oxygen, for example, can be represented in any of these ways:

${}^{A}_{Z}X$:	${}^{16}_{8}O$	${}^{17}_{8}O$	${}^{18}_{8}O$
${}^{A}X$:	${}^{16}O$	${}^{17}O$	${}^{18}O$
Element-A:	Oxygen-16	Oxygen-17	Oxygen-18

Because the number of neutrons is equal to $A - Z$, we see that the first isotope above has eight neutrons, the second nine, and the third ten. Isotopes of all naturally occurring elements on Earth are present in nearly fixed proportions with each proportion constituting an isotope's *natural abundance*. For example, in a typical terrestrial sample of oxygen, 99.76% of the O atoms are oxygen-16, 0.20% are oxygen-18, and 0.04% are oxygen-17.

Any nucleus that is unstable and decays spontaneously is said to be **radioactive**, emitting subatomic particles and electromagnetic radiation. The emissions are collectively called *radioactivity* and can be measured. Isotopes that emit radiation are called **radioisotopes**. As you learned in Chapter 14, the rate at which radioactive decay occurs is characteristic of the isotope and is generally reported as a *half-life* ($t_{1/2}$), the amount of time required for half the initial number of nuclei present to decay in a first-order reaction (see Section 14.5). An isotope's half-life can range from fractions of a second to billions of years and among other applications, can be used to measure the age of ancient objects. The next example and exercise review the calculations involving radioactive decay rates and half-lives.

EXAMPLE 20.1

Fort Rock Cave in Oregon is the site where archaeologists discovered several Indian sandals, the oldest ever found in Oregon. Analysis of the ^{14}C content of the sagebrush used to make the sandals gave an average decay rate of 5.1 disintegrations per minute (dpm) per gram carbon. The current $^{14}C/^{12}C$ ratio in living organisms is 1.3×10^{-12}, with a decay rate of 15 dpm/g C. How long ago was the sagebrush in the sandals cut? The half-life of ^{14}C is 5730 years.

Given Radioisotope, current $^{14}C/^{12}C$ ratio, initial decay rate, final decay rate, half-life

Asked for Age

Strategy

A Use Equation 14.34 to calculate N_0/N, the ratio of the number of atoms of ^{14}C originally present in the sample to the number of atoms now present.

B Substitute the value for the half-life of ^{14}C into Equation 14.32 to obtain the rate constant for the reaction.

C Substitute the calculated values for N_0/N and the rate constant into Equation 14.36 to obtain the elapsed time t.

Solution

We can use the integrated rate law for a first-order nuclear reaction (Equation 14.36) to calculate the amount of time that has passed since the sagebrush was cut to make the sandals:

$$\ln\frac{N}{N_0} = -kt$$

A From Equation 14.34, we know that $A = kN$. We can therefore use the initial and final activities ($A_0 = 15$ and $A = 5.1$) to calculate N_0/N:

$$\frac{A_0}{A} = \frac{kN_0}{kN} = \frac{N_0}{N} = \frac{15}{5.1}$$

B Now we can calculate the rate constant k from the half-life of the reaction (5730 yr) using Equation 14.32:

$$t_{1/2} = \frac{0.693}{k}$$

Rearranging this equation to solve for k gives us

$$k = \frac{0.693}{t_{1/2}} = \frac{0.693}{5730 \text{ yr}} = 1.21 \times 10^{-4} \text{ yr}^{-1}$$

C Substituting the calculated values into the equation for t gives

$$t = \frac{\ln(N_0/N)}{k} = \frac{\ln(15/5.1)}{1.21 \times 10^{-4} \text{ yr}^{-1}} = 8910 \text{ yr}$$

Thus, the sagebrush in the sandals is about 8900 years old.

EXERCISE 20.1

While trying to find a suitable way to protect his own burial chamber, the ancient Egyptian pharoah Sneferu developed the pyramid, a burial structure that protected desert graves from thieves and exposure to wind. Analysis of the ^{14}C content of several items in pyramids built during his reign gave an average decay rate of 8.6 dpm/g C. When were the objects in the chamber created?

Answer About 4600 years ago, 2600 B.C.

Nuclear Stability

As discussed in Chapter 1, the nucleus of an atom occupies a tiny fraction of the volume of the atom and contains the number of protons and neutrons that is characteristic of a given isotope. Electrostatic repulsions would normally cause the positively charged protons to repel each other, but the nucleus does not fly apart because of the **strong nuclear force**, an extremely powerful but very short-range attractive force between nucleons (Figure 20.1). All stable nuclei except the hydrogen-1 nucleus (^{1}H) contain at least one neutron in order to overcome the electrostatic repulsion between protons. As the number of protons in the nucleus increases, the number of neutrons needed for a stable nucleus increases even more rapidly. Too many protons (or too few neutrons) in the nucleus result in an imbalance between forces, which leads to nuclear instability.

The relationship between the numbers of protons and neutrons in stable nuclei, arbitrarily defined as having a half-life longer than ten times the age of the Earth, is shown graphically in Figure 20.2. Note that the stable isotopes form a "peninsula of stability" in a "sea of instability." Only two stable isotopes, ^{1}H and ^{3}He, have a neutron-to-proton ratio less than 1. Several stable isotopes of light atoms have a neutron-to-proton ratio equal to 1 (for example, ${}^{4}_{2}He$, ${}^{10}_{5}B$, and ${}^{40}_{20}Ca$). All other stable nuclei have a higher neutron-to-proton ratio, which increases steadily to about 1.5 for the heaviest nuclei. Regardless of the number of neutrons, however, all elements with $Z > 83$ are unstable and radioactive.

As shown in Figure 20.3, more than half of the stable nuclei (166 out of 279) have *even* numbers of both neutrons and protons; only six of the 279 stable nuclei do not have odd numbers of both. Moreover, certain numbers of neutrons or protons result in especially stable nuclei; these are the so-called *magic numbers* 2, 8, 20, 50, 82, and 126. For example, tin ($Z = 50$) has ten stable isotopes, but the elements on either side of tin in the periodic table, indium ($Z = 49$) and antimony ($Z = 51$), have only two stable isotopes each. Nuclei with magic numbers of *both* protons *and* neutrons are said to be

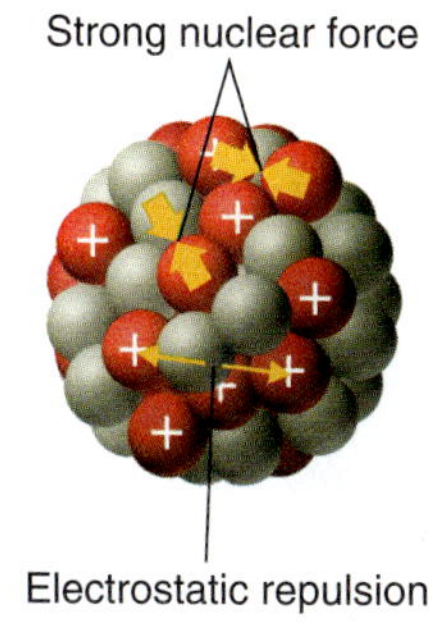

Figure 20.1 Competing interactions within the atomic nucleus. Electrostatic repulsions between positively charged protons would normally cause the nuclei of atoms (except H) to fly apart. In stable atomic nuclei, these repulsions are overcome by the strong nuclear force, a short-range but powerful attractive interaction between nucleons. If the attractive interactions due to the strong nuclear force are weaker than the electrostatic repulsions between protons, the nucleus is unstable, and it will eventually decay.

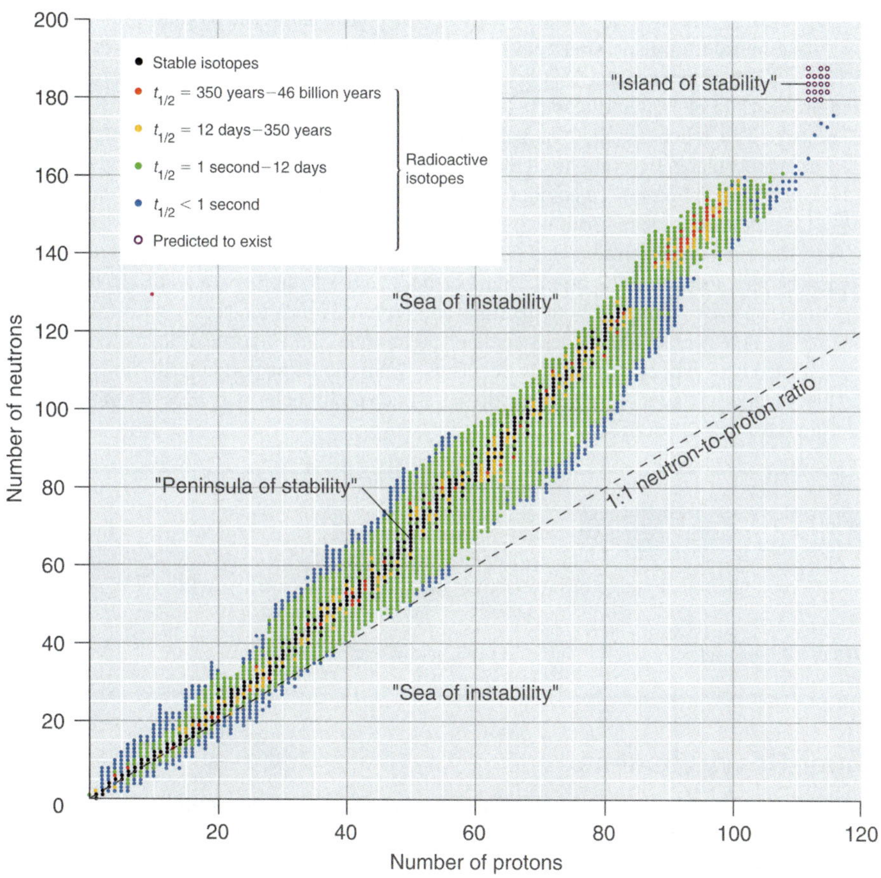

Figure 20.2 The relationship between nuclear stability and the neutron-to-proton ratio. In this plot of the number of neutrons versus the number of protons, each black point corresponds to a stable nucleus. In this classification, a stable nucleus is arbitrarily defined as one with a half-life longer than 46 billion years (ten times the age of Earth). Note that as the number of protons (the atomic number) increases, the number of neutrons required for a stable nucleus increases even more rapidly. Isotopes shown in red, yellow, green, and blue are progressively less stable and more radioactive; the farther an isotope is from the diagonal band of stable isotopes, the shorter its half-life. The purple dots indicate super-heavy nuclei that are predicted to be relatively stable, meaning that they are expected to be radioactive but to have relatively long half-lives. In most cases, these elements have not yet been observed or synthesized. [Data source: National Nuclear Data Center, Brookhaven National Laboratory. Evaluated Nuclear Structure Data File (ENSDF), March 2005.]

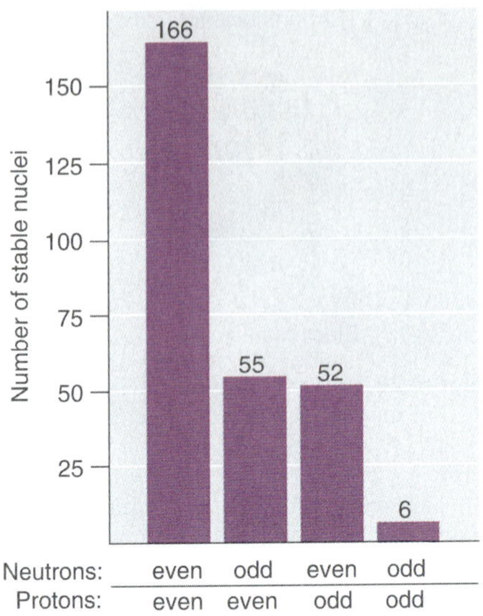

Figure 20.3 **The relationship between the numbers of protons and neutrons and nuclear stability.** Note that most stable nuclei contain even numbers of *both* neutrons and protons.

"doubly magic" and are even more stable. Examples of elements with doubly magic nuclei are $^{4}_{2}He$, with two protons and two neutrons, and $^{208}_{82}Pb$, with 82 protons and 126 neutrons, which is the heaviest known stable isotope of any element.

The pattern of stability suggested by the magic numbers of nucleons is reminiscent of the stability associated with the closed-shell electron configurations of the noble gases in Group 18, and has led to the hypothesis that the nucleus contains shells of nucleons that are in some ways analogous to the shells occupied by electrons in an atom. As shown in Figure 20.2, the "peninsula" of stable isotopes is surrounded by a "reef" of radioactive isotopes, which are stable enough to exist for varying lengths of time before they eventually decay to produce other nuclei.

EXAMPLE 20.2

Which of these nuclides do you predict should be stable, and which should be radioactive? **(a)** $^{30}_{15}P$; **(b)** $^{98}_{43}Tc$; **(c)** tin-118; **(d)** $^{239}_{94}Pu$

Given Mass number and atomic number

Asked for Predicted nuclear stability

Strategy

Use the number of protons, the neutron-to-proton ratio, and the presence of even or odd numbers of neutrons and protons to predict the stability or radioactivity of each nuclide.

Solution

(a) This isotope of phosphorus, ^{30}P, contains 15 neutrons and 15 protons, giving a neutron-to-proton ratio of 1.0. Although the atomic number, 15, is much less than the value of 83 above which all nuclides are unstable, the neutron-to-proton ratio is lower than expected for stability for an element with this mass. As shown in Figure 20.2, its neutron-to-proton ratio should be greater than 1. Moreover, this isotope has an odd number of both neutrons and protons, which also tends to make a nuclide unstable. Consequently, $^{30}_{15}P$ is predicted to be radioactive, and it is.

(b) This isotope of technetium, ^{98}Tc, contains 55 neutrons and 43 protons, giving a neutron-to-proton ratio of 1.28 and placing $^{98}_{43}Tc$ near the edge of the band of stability. The atomic number, 55, is much less than the value of 83 above which all isotopes are unstable. These facts suggest that $^{98}_{43}Tc$ might be stable. Note, however, that $^{98}_{43}Tc$ contains an odd number of both neutrons and protons, a combination that seldom gives a stable nucleus. Consequently, $^{98}_{43}Tc$ is predicted to be radioactive (which is indeed the case).

(c) Tin-118 contains 68 neutrons and 50 protons, for a neutron-to-proton ratio of 1.36. As in part (b), this value and the atomic number both suggest stability. In addition, the isotope contains an *even* number of both neutrons and protons, which tends to increase nuclear stability. Most important, the nucleus contains 50 protons, and 50 is one of the magic numbers associated with especially stable nuclei. Thus, $^{118}_{50}Sn$ should be particularly stable.

(d) This nuclide has an atomic number of 94. Because all nuclei with $Z > 83$ are unstable, $^{239}_{94}Pu$ must be radioactive.

EXERCISE 20.2

Classify each nuclide as stable or radioactive: **(a)** $^{232}_{90}Th$; **(b)** $^{40}_{20}Ca$; **(c)** $^{15}_{8}O$; **(d)** ^{139}La.

Answer **(a)** radioactive; **(b)** stable; **(c)** radioactive; **(d)** stable

Super-Heavy Elements

In addition to the "peninsula of stability," Figure 20.2 shows a small "island of stability" that is predicted to exist in the upper right corner. This island corresponds to the **super-heavy elements**, with atomic numbers near the magic number 126. Because the next magic number for neutrons should be 184, it has been suggested that an element that contains 114 protons and 184 neutrons might even be stable enough to exist in nature. Although these claims were met with skepticism for many years, since 1999 a few atoms of isotopes with $Z = 114$ and $Z = 116$ have been prepared and found to be surprisingly stable. Thus, a number of relatively long-lived nuclei may well be accessible among the super-heavy elements.

20.2 ∘ Nuclear Reactions

The two general kinds of nuclear reactions are nuclear decay reactions and nuclear transmutation reactions. In a **nuclear decay reaction**, also called *radioactive decay*, an unstable nucleus emits radiation and is transformed into the nucleus of one or more other elements. The resulting *daughter nuclei* have a lower mass and are lower in energy (more stable) than the *parent nucleus* that decayed. In contrast, in a **nuclear transmutation reaction**, a nucleus reacts with a subatomic particle or another nucleus to form a product nucleus that is *more massive* than the starting material. As we shall see, nuclear decay reactions occur spontaneously under all conditions, but nuclear transmutation reactions occur only under very special conditions, such as the collision of a beam of energetic particles with a target nucleus or in the interior of stars. We begin this section by considering the different classes of radioactive nuclei, along with their characteristic nuclear decay reactions and the radiation they emit.

Classes of Radioactive Nuclei

Each of the three general classes of radioactive nuclei is characterized by a different decay process or set of processes:

1. ***Neutron-rich nuclei.*** The nuclei on the upper left side of the band of stable nuclei in Figure 20.2 have a neutron-to-proton ratio that is too high to give a stable nucleus. These nuclei decay by a process that *converts a neutron to a proton*, thereby *decreasing* the neutron-to-proton ratio.
2. ***Neutron-poor nuclei.*** Nuclei on the lower right side of the band of stable nuclei have a neutron-to-proton ratio that is too low to give a stable nucleus. These nuclei decay by processes that have the net effect of *converting a proton to a neutron*, thereby *increasing* the neutron-to-proton ratio.
3. ***Heavy nuclei.*** With very few exceptions, heavy nuclei (those with $A \geq 200$) are intrinsically unstable regardless of the neutron-to-proton ratio, and all nuclei with $Z > 83$ are unstable. This is presumably due to the cumulative effects of electrostatic repulsions between the large number of positively charged protons, which cannot be totally overcome by the strong nuclear force, regardless of the number of neutrons present. Such nuclei tend to decay by emitting an **α particle** (a helium nucleus, $^{4}_{2}He$), which decreases the number of protons and neutrons in the original nucleus by 2. Since the neutron-to-proton ratio in an α particle is 1, the net result of alpha emission is an increase in the neutron-to-proton ratio.

Note the pattern

Nuclear decay reactions always produce daughter nuclei that have a more favorable neutron-to-proton ratio and hence are more stable than the parent nucleus.

Nuclear Decay Reactions

Just as we use the number and type of *atoms* present to balance a chemical equation, we can use the number and type of *nucleons* present to write a balanced equation for a nuclear decay reaction. This procedure also allows us to predict the identity of either the parent or the daughter nucleus if the identity of only one is known. Regardless of the mode of decay, *the total number of nucleons is conserved* in all nuclear reactions.

TABLE 20.1 Nuclear decay emissions and their symbols

Identity	Symbol	Charge	Mass, amu
Helium nucleus	${}^{4}_{2}\alpha$	+2	4.001506
Electron	${}^{0}_{-1}\beta$ or β^-	−1	0.000549
Photon	${}^{0}_{0}\gamma$	n/a	n/a
Neutron	${}^{1}_{0}\mathrm{n}$	0	1.008665
Proton	${}^{1}_{1}\mathrm{p}$	+1	1.007276
Positron	${}^{0}_{+1}\beta$ or β^+	+1	0.000549

To describe nuclear decay reactions, chemists have extended the ${}^{A}_{Z}\mathrm{X}$ notation for nuclides to include radioactive emissions. Table 20.1 lists the name and symbol for each type of emitted radiation. We introduced the most common of these, α and β particles and γ rays, in Chapters 1 and 14. The most notable addition is the **positron**, a particle that has the same mass as an electron, but a *positive* charge rather than a negative charge.

Like the notation used to indicate isotopes, the left superscript in the symbol for a particle gives the mass number, which is the total number of protons and neutrons. For a proton or a neutron, $A = 1$. Because neither an electron nor a positron contains protons or neutrons, its mass number is 0. The numbers should not be taken literally, however, as meaning that these particles have zero mass; ejection of a beta particle (an electron) simply has a negligible effect on the mass of a nucleus.

Similarly, the left subscript gives the charge of the particle. Because protons carry a positive charge, $Z = +1$ for a proton. In contrast, a neutron contains no protons and is electrically neutral, so $Z = 0$. In the case of an electron, $Z = -1$, and for a positron, $Z = +1$. Because γ rays are high-energy photons, both A and Z are 0.

In some cases, two different symbols are used for particles that are identical but produced in different ways. For example, the symbol ${}^{0}_{-1}\mathrm{e}$, which is usually simplified to e^-, represents a free electron or an electron associated with an *atom*, whereas the symbol ${}^{0}_{-1}\beta$, often simplified to β^-, denotes an electron that originates from within the *nucleus*, which is a β particle. Similarly, ${}^{4}_{2}\mathrm{He}^{2+}$ refers to the nucleus of a helium atom, and ${}^{4}_{2}\alpha$ denotes an identical particle that has been ejected from a heavier nucleus.

There are six fundamentally different kinds of nuclear decay reactions, each of which releases a different kind of particle or energy. The essential features of each are shown in Table 20.2. The most common are *alpha* and *beta decay* and *gamma emission*, but the others are essential to an understanding of nuclear decay reactions.

Alpha Decay

Many nuclei with mass numbers greater than 200 undergo **alpha decay**, which results in the emission of a helium-4 nucleus as an α particle, ${}^{4}_{2}\alpha$. The general reaction is

$$\underset{\text{Parent}}{{}^{A}_{Z}\mathrm{X}} \longrightarrow \underset{\text{Daughter}}{{}^{A-4}_{Z-2}\mathrm{X}'} + \underset{\text{Alpha particle}}{{}^{4}_{2}\alpha} \qquad (20.1)$$

The daughter nuclide contains two fewer protons and two fewer neutrons than the parent. Thus, α-particle emission produces a daughter nucleus with a mass number $A - 4$ and a nuclear charge $Z - 2$ compared to the parent nucleus. Radium-226, for example, undergoes alpha decay to form radon-222:

$${}^{226}_{88}\mathrm{Ra} \longrightarrow {}^{222}_{86}\mathrm{Rn} + {}^{4}_{2}\alpha \qquad (20.2)$$

TABLE 20.2 Common modes of nuclear decay

Decay Type	Radiation Emitted	Generic Equation	Model
Alpha decay	${}^{4}_{2}\alpha$	${}^{A}_{Z}X \longrightarrow {}^{A-4}_{Z-2}X' + {}^{4}_{2}\alpha$	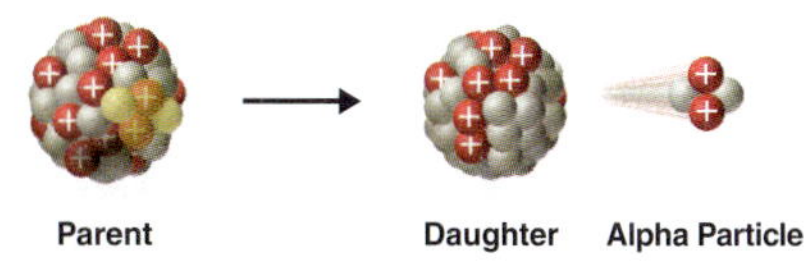
Beta decay	${}^{\;\,0}_{-1}\beta$	${}^{A}_{Z}X \longrightarrow {}^{A}_{Z+1}X' + {}^{\;\,0}_{-1}\beta$	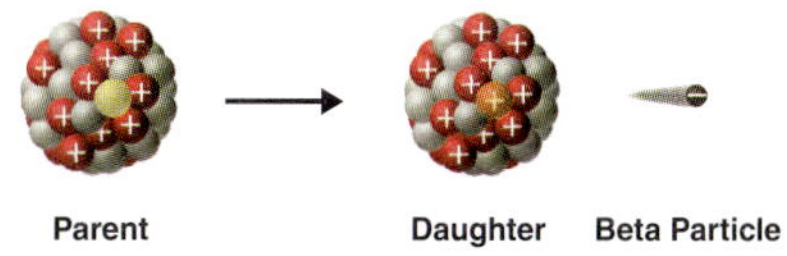
Positron emission	${}^{\;\,0}_{+1}\beta$	${}^{A}_{Z}X \longrightarrow {}^{A}_{Z-1}X' + {}^{\;\,0}_{+1}\beta$	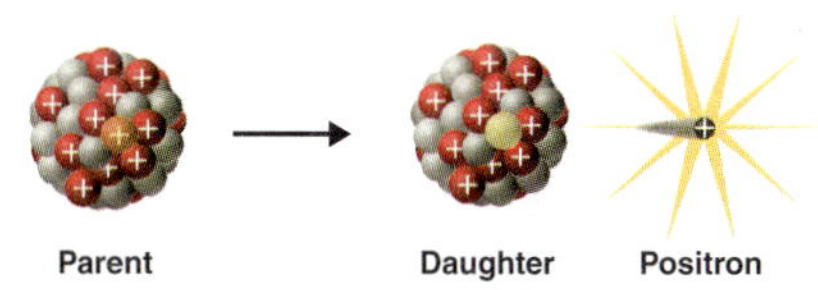
Electron capture	X rays	${}^{A}_{Z}X + {}^{\;\,0}_{-1}e \longrightarrow {}^{A}_{Z-1}X' + \text{X ray}$	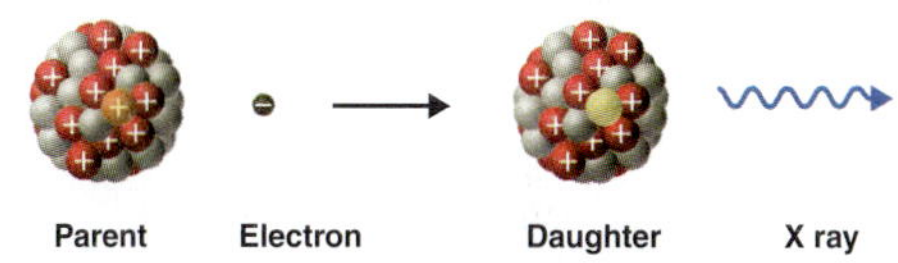
Gamma emission	${}^{0}_{0}\gamma$	${}^{A}_{Z}X^{*} \xrightarrow{\text{Relaxation}} {}^{A}_{Z}X + {}^{0}_{0}\gamma$	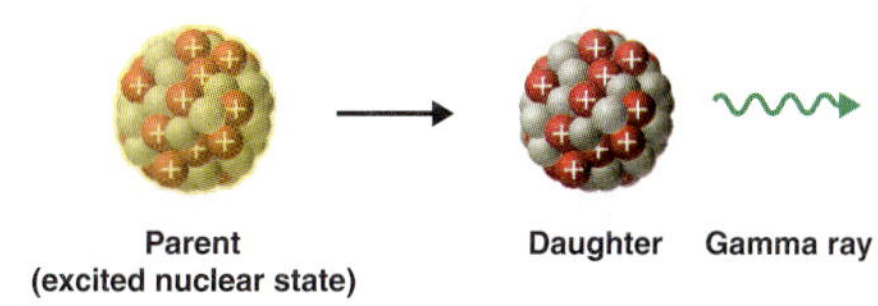
Spontaneous fission	Neutrons	${}^{A+B+C}_{Z+Y}X \longrightarrow {}^{A}_{Z}X' + {}^{B}_{Y}X' + C\,{}^{1}_{0}n$	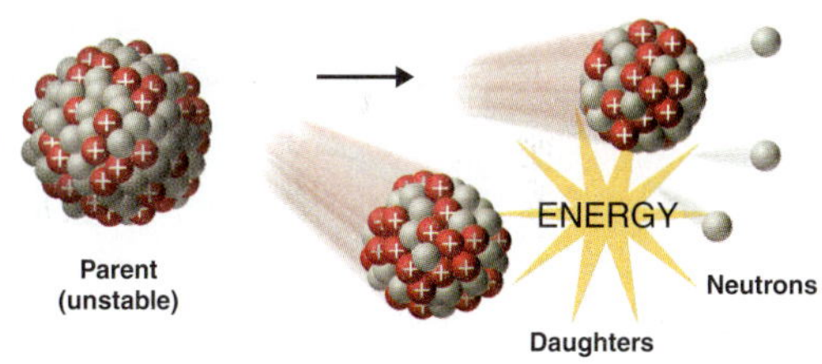

Because nucleons are conserved in this and all other nuclear reactions, the sum of the mass numbers of the products, $222 + 4 = 226$, equals the mass number of the parent. Similarly, the sum of the atomic numbers of the products, $86 + 2 = 88$, equals the atomic number of the parent. Thus, the nuclear equation is balanced.

Beta Decay

Nuclei that contain too many neutrons often undergo **beta decay**, in which a neutron is converted to a proton and a high-energy electron that is ejected from the nucleus as a β particle:

$$\underset{\substack{\text{Unstable}\\\text{neutron in}\\\text{nucleus}}}{{}^{1}_{0}n} \longrightarrow \underset{\substack{\text{Proton}\\\text{retained}\\\text{by nucleus}}}{{}^{1}_{1}p} + \underset{\substack{\text{Beta particle}\\\text{emitted by}\\\text{nucleus}}}{{}^{\;\,0}_{-1}\beta} \qquad (20.3)$$

The general reaction for beta decay is therefore

$$\underset{\text{Parent}}{{}^{A}_{Z}X} \longrightarrow \underset{\text{Daughter}}{{}^{\;\;\;A}_{Z+1}X'} + \underset{\text{Beta particle}}{{}^{\;\,0}_{-1}\beta} \qquad (20.4)$$

Although beta decay does *not* change the mass number of the nucleus, it does result in an increase of +1 in the atomic number because of the addition of a proton in the daughter nucleus. Thus, beta decay decreases the neutron-to-proton ratio, moving the nucleus toward the band of stable nuclei. For example, carbon-14 undergoes beta decay to form nitrogen-14:

$$ {}^{14}_{6}\mathrm{C} \longrightarrow {}^{14}_{7}\mathrm{N} + {}^{0}_{-1}\beta \tag{20.5} $$

Once again, the number of nucleons is conserved and the charges are balanced. The parent and the daughter nucleus have the same mass number, 14, and the sum of the atomic numbers of the products is 6, the same as the atomic number of the carbon-14 parent.

Positron Emission

Because a positron has the same mass as an electron but opposite charge, **positron emission** is the opposite of beta decay. Thus, positron emission is characteristic of neutron-poor nuclei, which decay by the transformation of a proton to a neutron and a high-energy positron that is emitted:

$$ {}^{1}_{1}\mathrm{p} \longrightarrow {}^{1}_{0}\mathrm{n} + {}^{0}_{+1}\beta \tag{20.6} $$

The general reaction for positron emission is therefore

$$ \underset{\text{Parent}}{{}^{A}_{Z}\mathrm{X}} \longrightarrow \underset{\text{Daughter}}{{}^{A}_{Z-1}\mathrm{X}'} + \underset{\text{Positron}}{{}^{0}_{+1}\beta} \tag{20.7} $$

Like beta decay, positron emission does not change the mass number of the nucleus. In this case, however, the atomic number of the daughter nucleus is *lower* by 1 than that of the parent. Thus, the neutron-to-proton ratio has increased, again moving the nucleus closer to the band of stable nuclei. For example, carbon-11 undergoes positron emission to form boron-11:

$$ {}^{11}_{6}\mathrm{C} \longrightarrow {}^{11}_{5}\mathrm{B} + {}^{0}_{+1}\beta \tag{20.8} $$

Nucleons are conserved, and the charges balance. The mass number, 11, does not change, and the sum of the atomic numbers of the products is 6, the same as the atomic number of the parent carbon-11 nuclide.

Electron Capture

A neutron-poor nucleus can decay by either positron emission or **electron capture** (EC), in which an electron in an inner shell reacts with a proton to produce a neutron:

$$ {}^{1}_{1}\mathrm{p} + {}^{0}_{-1}\mathrm{e} \longrightarrow {}^{1}_{0}\mathrm{n} \tag{20.9} $$

When a second electron moves from an outer shell to take the place of the lower-energy electron that was absorbed by the nucleus, an X ray is emitted. The overall reaction for electron capture is thus

$$ \underset{\text{Parent}}{{}^{A}_{Z}\mathrm{X}} + \underset{\text{Electron}}{{}^{0}_{-1}\mathrm{e}} \longrightarrow \underset{\text{Daughter}}{{}^{A}_{Z-1}\mathrm{X}'} + \text{X ray} \tag{20.10} $$

Electron capture does not change the mass number of the nucleus because both the proton that is lost and the neutron that is formed have a mass number of 1. As with positron emission, however, the atomic number of the daughter nucleus is *lower* by 1 than that of the parent. Once again, the neutron-to-proton ratio has increased, moving the nucleus toward the band of stable nuclei. For example, iron-55 decays by electron capture to form manganese-55, which is often written as

$$ {}^{55}_{26}\mathrm{Fe} \xrightarrow{\text{EC}} {}^{55}_{25}\mathrm{Mn} + \text{X ray} \tag{20.11} $$

Notice that the atomic numbers of the parent and daughter nuclides differ in Equation 20.11, although the mass numbers are the same. To write a balanced reaction for this reaction, we must explicitly include the captured electron in the equation:

$$^{55}_{26}\text{Fe} + {}^{0}_{-1}\text{e} \longrightarrow {}^{55}_{25}\text{Mn} + \text{X ray} \tag{20.12}$$

Both positron emission and electron capture are usually observed for nuclides with low neutron-to-proton ratios, but the decay rates for the two processes can be very different.

Gamma Emission

Many nuclear decay reactions produce daughter nuclei that are in a nuclear excited state, which is similar to an atom in which an electron has been excited to a higher-energy orbital to give an electronic excited state. Just as an electron in an electronic excited state emits energy in the form of a photon when it returns to the ground state (see Chapter 6), a nucleus in an excited state releases energy in the form of a photon when it returns to the ground state. These high-energy photons are γ rays. **Gamma emission** can occur virtually instantaneously, as it does in the alpha decay of uranium-238 to thorium-234:

$$^{238}_{92}\text{U} \longrightarrow \underbrace{^{234}_{90}\text{Th}^*}_{\text{Excited nuclear state}} + \underbrace{^{4}_{2}\alpha}_{\text{Emitted alpha particle}} \xrightarrow{\text{Relaxation}} {}^{234}_{90}\text{Th} + {}^{0}_{0}\gamma \tag{20.13}$$

or, if we disregard the decay event that created the excited nucleus:

$$^{234}_{88}\text{Th}^* \longrightarrow {}^{234}_{88}\text{Th} + {}^{0}_{0}\gamma \tag{20.14a}$$

or more generally

$$^{A}_{Z}\text{X}^* \longrightarrow {}^{A}_{Z}\text{X} + {}^{0}_{0}\gamma \tag{20.14b}$$

Gamma emission can also occur after a significant delay. For example, technetium-99*m* (the *m* is for metastable, as explained in Chapter 14) has a half-life of about 6 hours before emitting a γ ray to form technetium-99.

Because γ rays are energy, their emission does not affect either the mass number or the atomic number of the daughter nuclide. Gamma-ray emission is therefore the only kind of radiation that does not necessarily involve the conversion of one element to another, although it is almost always observed in conjunction with some other nuclear decay reaction.

Spontaneous Fission

Only very massive nuclei with high neutron-to-proton ratios can undergo **spontaneous fission**, in which the nucleus breaks into two pieces that have different atomic numbers and atomic masses. This process is most important for the trans-actinide elements, with $Z \geq 104$. Spontaneous fission is invariably accompanied by the release of large amounts of energy, and it is usually accompanied by the emission of several neutrons as well. An example is the spontaneous fission of $^{254}_{98}\text{Cf}$, which gives a distribution of fission products; one possible set of products is shown in the equation

$$^{254}_{98}\text{Cf} \longrightarrow {}^{118}_{46}\text{Pd} + {}^{132}_{52}\text{Te} + 4{}^{1}_{0}\text{n} \tag{20.15}$$

Once again, the number of nucleons is conserved. Thus, the sum of the mass numbers of the products ($118 + 132 + 4 = 254$) equals the mass number of the reactant. Similarly, the sum of the atomic numbers of the products [$46 + 52 + (4 \times 0) = 98$] is the same as the atomic number of the parent nuclide.

Note the pattern

The total number of nucleons is conserved in all nuclear reactions.

EXAMPLE 20.3

Write a balanced equation to describe each nuclear reaction: **(a)** the beta decay of $^{35}_{16}S$; **(b)** the decay of $^{201}_{80}Hg$ by electron capture; **(c)** the decay of $^{30}_{15}P$ by positron emission.

Given Radioactive nuclide and mode of decay

Asked for Balanced equation

Strategy

A Identify the reactants and products from the information given.
B Use the values of A and Z to identify any missing components needed to balance the equation.

Solution

(a) A We know the identities of the reactant and one of the products (a β particle). We can therefore begin by writing an equation that shows the reactant and one of the products and indicates the unknown product as $^{A}_{Z}X$:

$$^{35}_{16}S \longrightarrow {}^{A}_{Z}X + {}^{0}_{-1}\beta$$

B Because both protons and neutrons must be conserved in a nuclear reaction, the unknown product must have a mass number of $A = 35 - 0 = 35$ and an atomic number of $Z = 16 - (-1) = 17$. The element with $Z = 17$ is chlorine, so the balanced equation for the nuclear reaction is

$$^{35}_{16}S \longrightarrow {}^{35}_{17}Cl + {}^{0}_{-1}\beta$$

(b) A We know the identities of both reactants: $^{201}_{80}Hg$ and an inner electron, $^{0}_{-1}e$. The reaction is

$$^{201}_{80}Hg + {}^{0}_{-1}e \longrightarrow {}^{A}_{Z}X$$

B Both protons and neutrons are conserved, so the mass number of the product must be $A = 201 + 0 = 201$, and the atomic number of the product must be $Z = 80 + (-1) = 79$, which corresponds to the element gold. The balanced equation for the reaction is thus

$$^{201}_{80}Hg + {}^{0}_{-1}e \longrightarrow {}^{201}_{79}Au$$

(c) A As in part (a), we are given the identities of the reactant and one of the products—in this case, a positron. The unbalanced equation is therefore

$$^{30}_{15}P \longrightarrow {}^{A}_{Z}X + {}^{0}_{+1}\beta$$

B The mass number of the second product is $A = 30 - 0 = 30$, and its atomic number is $Z = 15 - 1 = 14$, which corresponds to silicon. The balanced equation for the reaction is

$$^{30}_{15}P \longrightarrow {}^{30}_{14}Si + {}^{0}_{+1}\beta$$

EXERCISE 20.3

Write a balanced equation for each nuclear reaction: **(a)** the decay of $^{11}_{6}C$ by positron emission; **(b)** the beta decay of molybdenum-99; **(c)** the emission of an α particle followed by gamma emission from $^{185}_{74}W$.

Answer **(a)** $^{11}_{6}C \longrightarrow {}^{11}_{5}B + {}^{0}_{+1}\beta$; **(b)** $^{99}_{42}Mo \longrightarrow {}^{99m}_{43}Tc + {}^{0}_{-1}\beta$; **(c)** $^{185}_{74}W \longrightarrow {}^{181}_{72}Hf + {}^{4}_{2}\alpha + {}^{0}_{0}\gamma$

EXAMPLE 20.4

Predict the kind of nuclear change each unstable nuclide undergoes when it decays: **(a)** $^{45}_{22}Ti$; **(b)** $^{242}_{94}Pu$; **(c)** $^{12}_{5}B$; **(d)** $^{256}_{100}Fm$.

Given Nuclide

Asked for Type of nuclear decay

Strategy

Based on the neutron-to-proton ratio and the value of *Z*, predict the type of nuclear decay reaction that will produce a more stable nuclide.

Solution

(a) This nuclide has a neutron-to-proton ratio of only 1.05, much lower than the requirement for stability for an element with an atomic number in this range. Nuclei that have low neutron-to-proton ratios decay by converting a proton to a neutron. The two possibilities are positron emission, which converts a proton to a neutron and a positron, and electron capture, which converts a proton and a core electron to a neutron. In this case, both are observed, with positron emission occurring about 86% of the time and electron capture about 14% of the time.

(b) Nuclei with $Z > 83$ are too heavy to be stable and usually undergo alpha decay, which decreases both the mass number and the atomic number. Thus, $^{242}_{94}Pu$ is expected to decay by alpha emission.

(c) This nuclide has a neutron-to-proton ratio of 1.4, which is very high for a light element. Nuclei with high neutron-to-proton ratios decay by converting a neutron to a proton and an electron. The electron is emitted as a β particle, and the proton remains in the nucleus, causing an increase in the atomic number with no change in the mass number. We therefore predict that $^{12}_{5}B$ will undergo beta decay.

(d) This is a massive nuclide, with an atomic number of 100 and a mass number much higher than 200. Nuclides with $A \geq 200$ tend to decay by alpha emission, and even heavier nuclei tend to undergo spontaneous fission. We therefore predict that $^{256}_{100}Fm$ will decay by either or both of these two processes. In fact, it decays by both spontaneous fission and alpha emission, in a 97:3 ratio.

EXERCISE 20.4

Predict the most probable nuclear decay reaction each nuclide undergoes: **(a)** $^{32}_{14}Si$; **(b)** $^{43}_{21}Sc$; **(c)** $^{231}_{91}Pa$.

Answer **(a)** beta decay; **(b)** positron emission or electron capture; **(c)** alpha decay

Radioactive Decay Series

The nuclei of all elements with atomic numbers greater than 83 are unstable. Thus, all isotopes of all elements beyond bismuth in the periodic table are radioactive. Because alpha decay decreases *Z* by only 2, and positron emission or electron capture decreases *Z* by only 1, it is impossible for any nuclide with $Z > 85$ to decay to a stable daughter nuclide in a single step, except via nuclear fission. Consequently, radioactive isotopes with $Z > 85$ usually decay to a daughter nucleus that is radiaoctive, which in turn decays to a second radioactive daughter nucleus, and so forth, until a stable nucleus finally results. This series of sequential alpha- and beta-decay reactions is called a **radioactive decay series**. The most common is the uranium-238 decay series, which produces lead-206 in a series of 14 sequential alpha- and beta-decay reactions (Figure 20.4). Although a radioactive decay series can be written for almost any isotope with $Z > 85$, only two others occur naturally: the decay of uranium-235 to lead-207

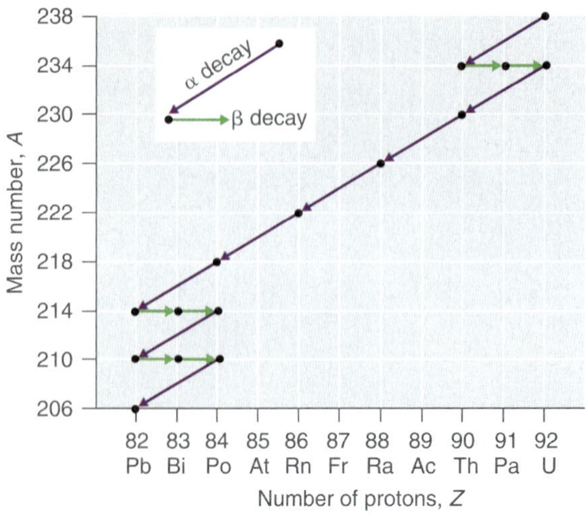

Figure 20.4 A radioactive decay series. The uranium-238 decay series consists of 14 sequential alpha- and beta-decay reactions that eventually produce lead-206. Two other naturally occurring radioactive decay series are known: the decay of uranium-235 to lead-207, which requires 11 steps, and the decay of thorium-232 to lead-208, which takes 10 steps.

(in 11 steps) and thorium-232 to lead-208 (in 10 steps). A fourth series, the decay of neptunium-237 to bismuth-209 in 11 steps, is known to have occurred on the primitive Earth. With a half-life of "only" 2.14 million years, all the neptunium-237 present when the Earth was formed decayed long ago, and today all the neptunium on Earth is synthetic.

Due to these radioactive decay series, small amounts of very unstable isotopes are found in ores that contain uranium or thorium. These rare, unstable isotopes should have decayed long ago to stable nuclei with lower atomic number, and they would no longer be found on Earth. Because they are generated continuously by the decay of uranium or thorium, however, their amounts have reached a steady state, in which their rate of formation is equal to their rate of decay. In some cases, the abundance of the daughter isotopes can be used to date a material or identify its origin, as described in Chapter 14.

Induced Nuclear Reactions

The discovery of radioactivity in the late 19th century showed that some nuclei spontaneously transform into nuclei with a different number of protons, thereby producing a different element. When scientists realized that these naturally occurring radioactive isotopes decayed by emitting subatomic particles, they realized that in principle it should be possible to carry out the reverse reaction, converting a stable nucleus to another more massive nucleus by bombarding it with subatomic particles in a nuclear transmutation reaction.

The first successful nuclear transmutation reaction was carried out in 1919 by Ernest Rutherford, who showed that α particles emitted by radium could react with nitrogen nuclei to form oxygen nuclei. As shown in this equation, a proton is emitted in the process:

$$ {}^{4}_{2}\alpha + {}^{14}_{7}\text{N} \longrightarrow {}^{17}_{8}\text{O} + {}^{1}_{1}\text{p} \qquad (20.16)$$

Rutherford's nuclear transmutation experiments led to the discovery of the neutron. He found that bombarding the nucleus of a light target element with an α particle usually converted the target nucleus to a product that had an atomic number higher by 1 and a mass number higher by 3 than the target nucleus. Such behavior is consistent with the emission of a proton after reaction with the α particle. Very light targets such as Li, Be, and B reacted differently, however, emitting a new kind of highly penetrating radiation rather than a proton. Because these high-energy particles could not be deflected by a magnetic or electrical field, Rutherford concluded that they were electrically neutral (see Chapter 1). Other observations suggested that the mass of the neutral particle was similar to the mass of the proton. In 1932, James Chadwick (Nobel Prize, 1935), who was a student of Rutherford's at the time, named these neutral particles *neutrons* and proposed that they were fundamental building blocks of the atom. The reaction used initially by Chadwick to explain the production of neutrons was

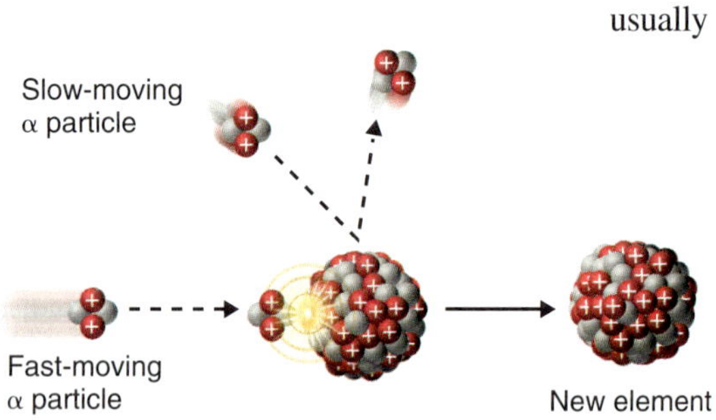

Figure 20.5 A nuclear transmutation reaction. Bombarding a target of one element with high-energy nuclei or subatomic particles can create new elements. Electrostatic repulsions normally prevent a positively charged particle from colliding and reacting with a positively charged nucleus. If the positively charged particle is moving at a very high speed, however, its kinetic energy may be great enough to overcome the electrostatic repulsions and it may collide with the target nucleus. Such collisions can result in a nuclear transmutation reaction.

$$ {}^{4}_{2}\alpha + {}^{9}_{4}\text{Be} \longrightarrow {}^{12}_{6}\text{C} + {}^{1}_{0}\text{n} \qquad (20.17)$$

Because α particles and atomic nuclei are both positively charged, electrostatic forces cause them to repel each other. Only α particles with very high kinetic energy can overcome this repulsion and collide with a nucleus (Figure 20.5). Because neutrons have no electrical charge, however, they are not repelled by the nucleus. Hence, bombardment with neutrons is a much easier way to prepare new isotopes of the lighter elements. In fact, carbon-14 is formed naturally in the atmosphere by the bombardment of nitrogen-14 with neutrons generated by cosmic rays:

$$ {}^{1}_{0}\text{n} + {}^{14}_{7}\text{N} \longrightarrow {}^{14}_{6}\text{C} + {}^{1}_{1}\text{p} \qquad (20.18)$$

EXAMPLE 20.5

In 1933, Frédéric Joliot and Iréne Joliot-Curie (daughter of Marie and Pierre Curie) prepared the first artificial radioactive isotope by bombarding aluminum-27 with α particles. For each ^{27}Al that reacted one neutron was released. Identify the product nuclide, and write a balanced equation for this nuclear transmutation reaction.

Given Reactants in nuclear transmutation reaction

Asked for Product nuclide, balanced equation

Strategy

A Based on the reactants and one product, identify the other product of the reaction. Use conservation of mass and charge to determine the values of Z and A of the product nuclide and thus its identity.

B Write the balanced equation for the nuclear reaction.

Solution

A Bombarding an element with α particles usually produces an element with an atomic number that is 2 greater than the atomic number of the target nucleus. Thus, we expect that aluminum (Z = 13) will be converted to phosphorus (Z = 15). With one neutron released, conservation of mass requires that the mass number of the other product be 3 greater than the mass number of the target. In this case, the mass number of the target is 27, so the mass number of the product will be 30. The second product is therefore phosphorus-30, $^{30}_{15}$P. **B** The balanced equation for the nuclear reaction is

$$^{27}_{13}\text{Al} + {}^{4}_{2}\alpha \longrightarrow {}^{30}_{15}\text{P} + {}^{1}_{0}\text{n}$$

EXERCISE 20.5

Because all isotopes of technetium are radioactive and have short half-lives, it does not exist in nature. Technetium can, however, be prepared by nuclear transmutation reactions. For example, bombarding a molybdenum-96 target with deuterium nuclei ($^{2}_{1}$H) produces technetium-97. Identify the other product of the reaction, and write a balanced equation for this transmutation reaction.

Answer neutron, $^{1}_{0}$n; $^{96}_{42}\text{Mo} + {}^{2}_{1}\text{H} \longrightarrow {}^{97}_{43}\text{Tc} + {}^{1}_{0}\text{n}$

We noted in Section 20.2 that very heavy nuclides, corresponding to $Z \geq 104$, tend to decay by spontaneous fission. Nuclides with slightly lower values of Z, such as the isotopes of uranium ($Z = 92$) and plutonium ($Z = 94$), do not undergo spontaneous fission at any significant rate. Some isotopes of these elements, however, such as $^{235}_{92}$U and $^{239}_{94}$Pu undergo *induced nuclear fission* when they are bombarded with relatively low-energy neutrons, as shown in this equation for uranium-235 and in Figure 20.6:

$$^{235}_{92}\text{U} + {}^{1}_{0}\text{n} \longrightarrow {}^{236}_{92}\text{U} \longrightarrow {}^{141}_{56}\text{Ba} + {}^{92}_{36}\text{Kr} + 3\,{}^{1}_{0}\text{n} \qquad (20.19)$$

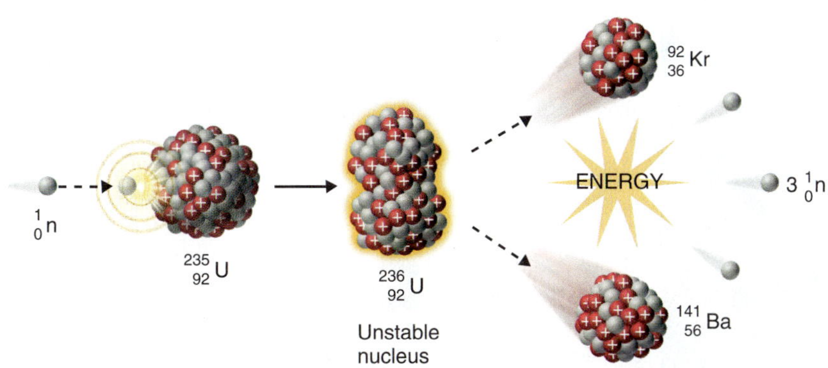

Figure 20.6 Neutron-induced nuclear fission. Collision of a relatively slow-moving neutron with a fissile nucleus can split it into two smaller nuclei with the same or different masses, as shown here. Neutrons are also released in the process, along with a great deal of energy.

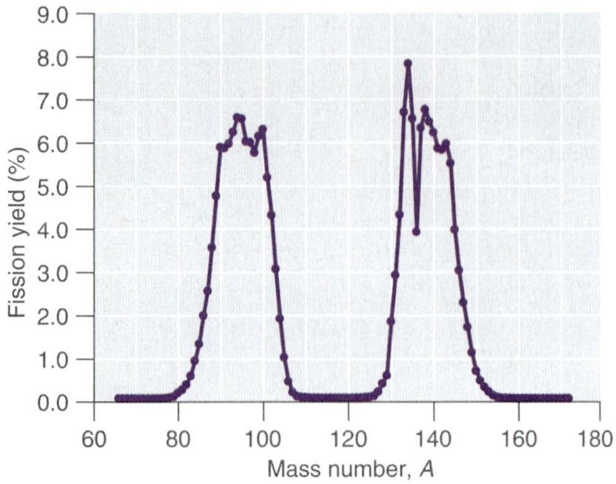

Figure 20.7 Mass distribution of nuclear fission products of ^{235}U. Nuclear fission usually produces a range of products with different masses and yields, although the mass ratio of each pair of fission products from a fission event is approximately 3:2. As shown in this plot, more than 50 different fission products are known for ^{235}U. [Data source: T. R. England and B. F. Rider, Los Alamos National Laboratory, ENDF-349 (1994).]

Any isotope that can undergo a nuclear fission reaction when bombarded with neutrons is called a *fissile isotope*.

During nuclear fission, the nucleus usually divides asymmetrically rather than into two equal parts, as shown in Figure 20.6. Moreover, every fission event of a given nuclide does not give the same products; more than 50 different fission modes have been identified for uranium-235, for example. Consequently, nuclear fission of a fissile nuclide can never be described by a single equation. Instead, as shown in Figure 20.7, a distribution of many pairs of fission products with different yields is obtained, but the mass ratio of each pair of fission products produced by a single fission event is always roughly 3:2.

TABLE 20.3 Some reactions used to synthesize transuranium elements

Reaction
$^{239}_{94}\text{Pu} + ^{4}_{2}\alpha \longrightarrow ^{242}_{96}\text{Cm} + ^{1}_{0}\text{n}$
$^{239}_{94}\text{Pu} + ^{4}_{2}\alpha \longrightarrow ^{241}_{95}\text{Am} + ^{1}_{1}\text{p} + ^{1}_{0}\text{n}$
$^{242}_{96}\text{Cm} + ^{4}_{2}\alpha \longrightarrow ^{243}_{97}\text{Bk} + ^{1}_{1}\text{p} + 2^{1}_{0}\text{n}$
$^{253}_{99}\text{Es} + ^{4}_{2}\alpha \longrightarrow ^{256}_{101}\text{Md} + ^{1}_{0}\text{n}$
$^{238}_{92}\text{U} + ^{12}_{6}\text{C} \longrightarrow ^{246}_{98}\text{Cf} + 4^{1}_{0}\text{n}$
$^{252}_{98}\text{Cf} + ^{10}_{5}\text{B} \longrightarrow ^{256}_{103}\text{Lr} + 6^{1}_{0}\text{n}$

Synthesis of Transuranium Elements

Uranium ($Z = 92$) is the heaviest naturally occurring element. Consequently, all the elements with $Z > 92$, the **transuranium elements**, are artificial and have been prepared by bombardment of suitable target nuclei with smaller particles. The first of the transuranium elements to be prepared was neptunium ($Z = 93$), which was synthesized in 1940 by bombarding a ^{238}U target with neutrons. As shown in Equation 20.20, this reaction occurs in two steps. Initially, a neutron combines with a ^{238}U nucleus to form ^{239}U, which is unstable and undergoes beta decay to produce ^{239}Np:

$$^{238}_{92}\text{U} + ^{1}_{0}\text{n} \longrightarrow ^{239}_{92}\text{U} \longrightarrow ^{239}_{93}\text{Np} + ^{0}_{-1}\beta \tag{20.20}$$

Subsequent beta decay of ^{239}Np produces the second transuranium element, plutonium ($Z = 94$):

$$^{239}_{93}\text{Np} \longrightarrow ^{239}_{94}\text{Pu} + ^{0}_{-1}\beta \tag{20.21}$$

Bombarding the target with more massive nuclei creates elements that have atomic numbers significantly greater than that of the target nucleus (Table 20.3). Such techniques have resulted in creation of the super-heavy elements 114 and 116, both of which lie in or near the "island of stability" discussed in Section 20.1.

Accelerating positively charged particles to the speeds needed to overcome the electrostatic repulsions between them and the target nuclei requires a device called a *particle accelerator*, which uses electrical and magnetic fields to accelerate the particles. Operationally, the simplest particle accelerator is the linear accelerator (Figure 20.8), in which

Figure 20.8 A linear particle accelerator. (a) An aerial view of the Stanford Linear Accelerator (SLAC), the longest linear particle accelerator in the world; the overall length of the tunnel is 2 miles. (b) Rapidly reversing the polarity of the electrodes in the tube causes the charged particles to be alternately attracted as they enter one section of the tube and repelled as they leave that section. As a result, the particles are continuously accelerated along the length of the tube.

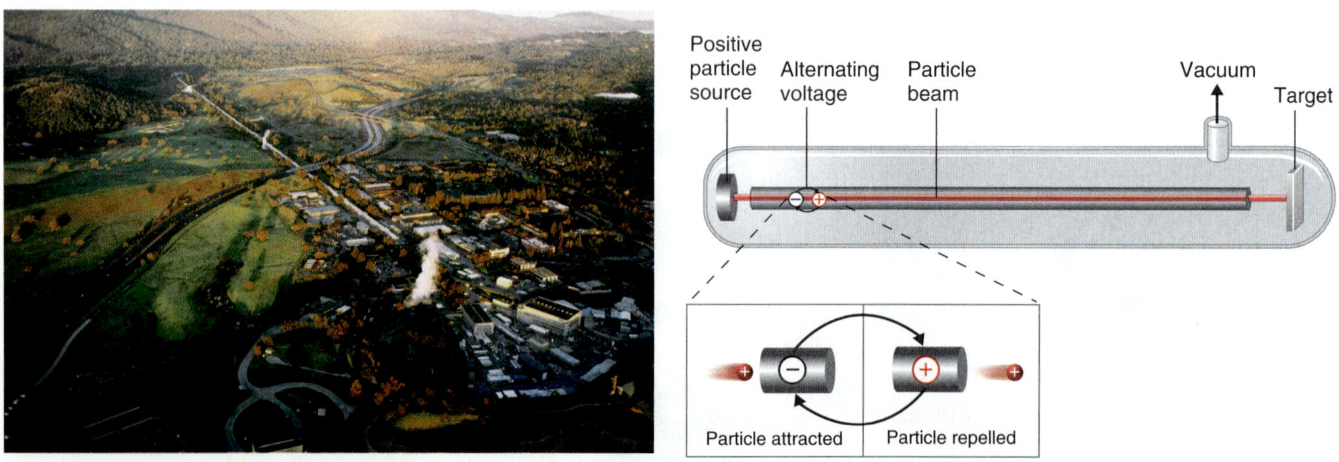

(a) (b)

a beam of particles is injected at one end of a long evacuated tube. Rapid alternation of the polarity of the electrodes along the tube causes the particles to be alternately accelerated toward a region of opposite charge and repelled by a region with the same charge, resulting in a tremendous acceleration as the particle travels down the tube. A modern linear accelerator such as SLAC at Stanford University may be 2 miles long.

To achieve the same outcome in less space, a particle accelerator called a *cyclotron* forces the charged particles to travel in a circular path rather than a linear one. The particles are injected into the center of a ring and accelerated by rapidly alternating the polarity of two large D-shaped electrodes above and below the ring, which accelerates the particles outward along a spiral path toward the target.

The length of a linear accelerator and the size of the D-shaped electrodes in a cyclotron severely limit the kinetic energy that particles can attain in these device. These limitations can be overcome by using a *synchrotron*, a hybrid of the two designs. A synchrotron contains an evacuated tube similar to that of a linear accelerator, but the tube is circular and can be more than a mile in diameter (Figure 20.9). Charged particles are accelerated around the circle by a series of magnets whose polarities rapidly alternate.

Figure 20.9 A synchrotron. An aerial photograph of what is currently the world's most powerful particle accelerator, the Tevatron at the Fermi National Accelerator Laboratory in Illinois. The large tube characteristic of a synchrotron is 4 miles in circumference, contains 1000 superconducting magnets cooled by liquid helium, and can accelerate a beam of protons to almost the speed of light, giving them an energy greater than 1 TeV (teraelectronvolt = 10^3 GeV = 10^{12} eV) for collisions with other particles.

20.3 • The Interaction of Nuclear Radiation with Matter

Because nuclear reactions do not typically affect the valence electrons of the atom (although electron capture draws an electron from an orbital of the lowest energy level), they do not cause chemical changes directly. Nonetheless, the particles and photons emitted during nuclear decay are very energetic, and they can indirectly produce chemical changes in the matter surrounding the nucleus that has decayed. For instance, an α particle is an ionized helium nucleus (He^{2+}), which can act as a powerful oxidant. In this section, we describe how radiation interacts with matter and the some of the chemical and biological effects of radiation.

Ionizing Versus Nonionizing Radiation

The effects of radiation on matter are determined primarily by the energy of the radiation, which depends on the nuclear decay reaction that produced it. **Nonionizing radiation** is relatively low in energy, and when it collides with an atom in a molecule or an ion, most or all of its energy can be absorbed without causing a structural or chemical change. Instead, the kinetic energy of the radiation is transferred to the atom or molecule with which it collides, causing it to rotate, vibrate, or move more rapidly. Because this energy can be transferred to adjacent molecules or ions in the form of heat, many radioactive substances are warm to the touch. Highly radioactive elements such as polonium, for example, have been used as heat sources in the U.S. space program. As long as the intensity of the nonionizing radiation is not great enough to cause overheating, it is relatively harmless, and its effects can be neutralized by cooling.

Figure 20.10 Radiation damage. When high-energy particles emitted by radioactive decay interact with matter, they can break bonds or ionize molecules, resulting in changes in physical properties such as ductility or color. The glass electrical insulator on the left has not been exposed to radiation, but the insulator on the right has received intense radiation doses over a long period of time. Radiation damage changed the chemical structure of the glass, causing it to become bright blue.

In contrast, **ionizing radiation** is higher in energy, and some of its energy can be transferred to one or more atoms with which it collides as it passes through matter. If enough energy is transferred, electrons can be excited to very high energy levels, resulting in the formation of positively charged ions:

$$\text{atom} + \text{ionizing radiation} \longrightarrow \text{ion}^+ + e^- \qquad (20.22)$$

Molecules that have been ionized in this way are often highly reactive, and they can decompose or undergo other chemical changes that create a cascade of reactive molecules that can damage biological tissues and other materials (Figure 20.10). Because the energy of ionizing radiation is very high, we often report its energy in units such as *megaelectronvolts* (*MeV*) per particle: 1 MeV/particle = 96 *billion* J/mol!

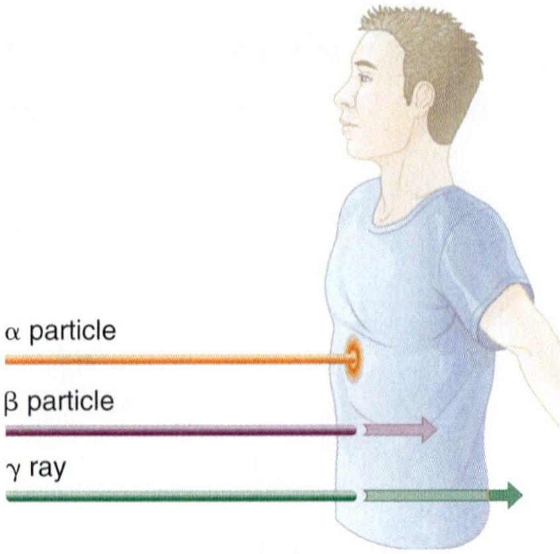

Figure 20.11 Depth of penetration of ionizing radiation. The depth of penetration of alpha, beta, and gamma radiation varies with the particle. Because α particles interact strongly with matter, they do not penetrate deeply into the human body. In contrast, β particles do not interact as strongly with matter and penetrate more deeply. Gamma rays, which have no charge, are stopped by only very dense materials and can pass right through the human body without being absorbed.

The Effects of Ionizing Radiation on Matter

The effects of ionizing radiation depend on four factors:

1. The type of radiation, which dictates how far it can penetrate into matter
2. The energy of the individual particles or photons
3. The number of particles or photons that strike a given area per unit time
4. The chemical nature of the substance exposed to the radiation

The relative abilities of the various forms of ionizing radiation to penetrate biological tissues are illustrated in Figure 20.11. Because of its high charge and mass, α radiation interacts strongly with matter. Consequently, it does not penetrate deeply into an object, and it can be stopped by a piece of paper, clothing, or skin. In contrast, γ rays, with no charge and essentially no mass, do not interact strongly with matter and penetrate deeply into most objects, including the human body. Several inches of lead or more than 12 inches of special concrete is needed to completely stop γ rays. Because β particles are intermediate in mass and charge between α particles and γ rays, their interaction with matter is also intermediate. Beta particles readily penetrate paper or skin, but they can be stopped by a piece of wood or a relatively thin sheet of metal.

Because of their great penetrating ability, γ rays are by far the most dangerous type of radiation when they come from a source *outside* the body. Alpha particles, however, are the most damaging if their source is *inside* the body because all of their energy is absorbed by internal tissues. Recognizing that the danger of radiation depends strongly on its nature and the kind of exposure allows scientists to safely handle many radioactive materials, as long as they take precautions to avoid, for example, inhaling fine particulate dust that contains alpha emitters. Some properties of ionizing radiation are summarized in Table 20.4.

There are many different ways to measure radiation exposure, or the *dose*. The **roentgen (R)**, which is used to measure the amount of energy absorbed by dry air, can be used to describe exposure quantitatively.* Damage to biological tissues, however, is proportional to the amount of energy absorbed by *tissues*, not air. The most common unit used to measure the effects of radiation on biological tissue is the **rad** (radiation absorbed dose); the SI equivalent is the gray (Gy). The *rad* is defined as the amount of radiation that causes 0.01 J of energy to be absorbed by 1 kg of matter, and the *gray* is defined as the amount of radiation that causes 1 J of energy to be absorbed per kilogram:

$$1 \text{ rad} = 0.010 \text{ J/kg} \qquad 1 \text{ Gy} = 1 \text{ J/kg} \tag{20.23}$$

William Röentgen. Born in the Lower Rhine Province of Germany, Röentgen was the only child of a cloth manufacturer and merchant. His family moved to the Netherlands where he showed no particular aptitude in school, but where he was fond of roaming the countryside. Röentgen was expelled from technical school in Utrecht after being unjustly accused of drawing a caricature of one of the teachers. He began studying mechanical engineering in Zurich, which he could enter without having the credentials of a regular student, and received a Ph.D. at the University of Zurich in 1869. In 1876 he became Professor of Physics at Strasborg University.

TABLE 20.4 Some properties of ionizing radiation

Type	Energy Range, MeV	Penetration Distance in Water[a]	Penetration Distance in Air[a]
α particles	3–9	<0.05 mm	<10 cm
β particles	≤ 3	<4 mm	1 m
X rays	$<10^{-2}$	<1 cm	<3 m
γ rays	10^{-2}–10^{1}	<20 cm	>3 m

[a]Distance at which half of the radiation has been absorbed.

* Named after the German physicist Wilhelm Röentgen (1845–1923; Nobel Prize, 1901), who discovered X rays. The roentgen is actually defined as the amount of radiation needed to produce an electrical charge of 2.58×10^{-4} C in 1 kg of dry air.

Thus, a 70-kg human who receives a dose of 1.0 rad over his or her entire body absorbs 0.010 J/70 kg = 1.4×10^{-4} J, or 0.14 mJ. To put this in perspective, 0.14 mJ is the amount of energy transferred to your skin by a 3.8×10^{-5} g droplet of boiling water. Because the energy of the droplet of water is transferred to a relatively large area of tissue, it is harmless. A radioactive particle, however, transfers its energy to a single molecule which makes it the atomic equivalent of a bullet fired from a high-powered rifle.

Because α particles have a much higher mass and charge than β particles or γ rays, the difference in mass between α and β particles is analogous to being hit by a bowling ball instead of a Ping-Pong ball traveling at the same speed. Thus, the amount of tissue damage caused by 1 rad of α particles is much greater than the damage caused by 1 rad of β particles or γ rays. Thus, a unit called the **rem** (roentgen equivalent in man) was devised to describe the *actual* amount of tissue damage caused by a given amount of radiation. The number of rems of radiation is equal to the number of rads multiplied by the *RBE* (relative biological effectiveness) factor, which is 1 for β particles, γ rays, and X rays, and about 20 for α particles. Because actual radiation doses tend to be very small, most measurements are reported in *millirems* (1 mrem = 10^{-3} rem).

Note the pattern

An alpha emitter is the least dangerous type of radiation, unless it is ingested, inhaled, or injected.

Natural Sources of Radiation

We are continuously exposed to measurable background radiation from a variety of natural sources, which on average is equal to about 150–600 mrem/yr (Figure 20.12). One component of background radiation is *cosmic rays*, high-energy particles and γ rays emitted by the sun and other stars that bombard the Earth continuously. Because cosmic rays are partially absorbed by the atmosphere before they reach Earth's surface, the exposure of people living at sea level (about 30 mrem/yr) is significantly lower than the exposure of people living at higher altitudes (about 50 mrem/yr in Denver, Colorado). Every 4 hours spent in an airplane at higher than 30,000 ft adds about 1 mrem to a person's annual radiation exposure.

A second component of background radiation is *cosmogenic radiation*, produced by the interaction of cosmic rays with gases in the upper atmosphere. When high-energy cosmic rays collide with oxygen and nitrogen atoms, neutrons and protons are released. These, in turn, react with other atoms to produce radioactive isotopes such as ^{14}C:

$$^{14}_{7}\text{N} + ^{1}_{0}\text{n} \longrightarrow ^{14}_{6}\text{C} + ^{1}_{1}\text{p} \qquad (20.24)$$

The carbon atoms react with oxygen atoms to form CO_2, which is eventually washed to Earth's surface in rain and taken up by plants. About 1 atom in 1×10^{12} of the carbon atoms in our bodies is radioactive ^{14}C, which decays by beta emission. About 10,000 ^{14}C nuclei disintegrated in your body during the 30 seconds or so that it took you to read this paragraph. Tritium (^{3}H) is also produced in the upper atmosphere and falls to Earth in precipitation. The total radiation dose attributable to ^{14}C is estimated to be 1 mrem/yr, while that due to ^{3}H is about a thousand times less.

The third major component of background radiation is *terrestrial radiation*, which is due to the remnants of radioactive elements that were present on the primordial Earth and their decay products. For example, many of the rocks and minerals in the soil contain small amounts of radioactive isotopes such as ^{232}Th and ^{238}U, as well as radioactive daughter isotopes such as ^{226}Ra. The amount of background radiation from these sources is about the same as that from cosmic rays (~30 mrem/yr). These isotopes are also found in small amounts in building materials derived from rocks and minerals, which significantly increases the radiation exposure for people who live in brick or concrete-block houses (60–160 mrem/yr) instead of houses made of wood (10–20 mrem/yr).

Our tissues also absorb radiation (about 40 mrem/yr) from naturally occurring radioactive elements that are present in our bodies. For example, the average adult contains about 140 g of potassium as the K^+ ion. Naturally occurring potassium

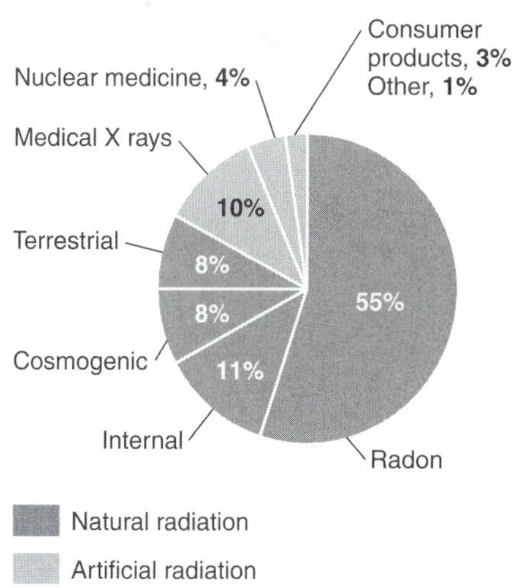

Figure 20.12 The radiation exposure of a typical adult in the United States. The average radiation dose from natural sources for an adult in the United States is about 150–600 mrem/yr. Radon accounts for more than half of an adult's total radiation exposure, whereas background radiation (terrestrial and cosmogenic) and exposure from medical sources account for about 15% each. [Data source: Office of Civilian Radioactive Waste Management.]

contains 0.0117% ^{40}K, which decays by emitting both a β particle and a γ ray. In the last 20 seconds, about the time it took you to read this paragraph, approximately 40,000 ^{40}K nuclei disintegrated in your body.

By far the most important source of background radiation is *radon*, the heaviest of the noble gases (Group 18). Radon-222 is produced during the decay of ^{238}U, and other isotopes of radon are produced by the decay of other heavy elements. Even though radon is chemically inert, all its isotopes are radioactive. For example, ^{222}Rn undergoes two successive alpha-decay events to give ^{214}Pb:

$$^{222}_{86}\text{Rn} \longrightarrow {}^{4}_{2}\alpha + {}^{218}_{84}\text{Po} \longrightarrow {}^{4}_{2}\alpha + {}^{214}_{82}\text{Pb} \qquad (20.25)$$

Because radon is a dense gas, it tends to accumulate in enclosed spaces such as basements, especially in locations where the soil contains greater-than-average amounts of naturally occurring uranium minerals. Under most conditions, radioactive decay of radon poses no problems because of the very short range of the emitted α particle. If an atom of radon happens to be in your lungs when it decays, however, the chemically reactive daughter isotope polonium-218 can become irreversibly bound to molecules in the lung tissue. Subsequent decay of ^{218}Po releases an α particle directly into one of the cells lining the lung, and the resulting damage can eventually cause lung cancer. The ^{218}Po isotope is also readily absorbed by particles in cigarette smoke, which adhere to the surface of the lungs and can hold the radioactive isotope in place. Recent estimates suggest that radon exposure is a contributing factor in about 15% of deaths due to lung cancer. Because of the potential health problem radon poses, many states require houses to be tested for radon before they can be sold. By current estimates, radon accounts for more than half of the radiation exposure of a typical adult in the United States.

Artificial Sources of Radiation

In addition to naturally occurring background radiation, humans are exposed to small amounts of radiation from a variety of artificial sources. The most important of these are the X rays used for diagnostic purposes in medicine and dentistry, which are photons with much lower energy than γ rays. A single chest X ray provides a radiation dose of about 0.07 mrem, and a dental X ray about 0.5 mrem. Other minor sources include television screens and computer monitors with cathode-ray tubes, which also produce X rays. Luminescent paints for watch dials originally used radium, a highly toxic alpha emitter if ingested by those painting the dials. Radium was replaced by tritium (^{3}H) and promethium (^{147}Pr), which emit low-energy β particles that are absorbed by the watch crystal or the glass covering the instrument. Radiation exposure from television screens, monitors, and luminescent dials totals about 2 mrem/yr. Residual fallout from previous atmospheric nuclear-weapons testing is estimated to account for about twice this amount, and the nuclear power industry accounts for less than 1 mrem/yr (about the same as a single 4-hour jet flight).

EXAMPLE 20.6

Calculate the annual radiation dose in rads a typical 70-kg chemistry student receives from the naturally occurring ^{40}K in his or her body, which contains about 140 g of potassium (as the K^{+} ion). The natural abundance of ^{40}K is 0.0117%. Each 1.00 mol of ^{40}K undergoes 1.05×10^{7} decays per second, and each decay event is accompanied by the emission of a 1.32-MeV β particle.

Given Mass of student, mass of isotope, natural abundance, rate of decay, energy of particle

Asked for Annual radiation dose in rads

Strategy

A Calculate the number of moles of ^{40}K present using its mass, molar mass, and natural abundance.
B Determine the number of decays per year for this amount of ^{40}K.
C Multiply the number of decays per year by the energy associated with each decay event. To obtain the annual radiation dose, use the mass of the student to convert this value to rads.

Solution

A✓ The number of moles of ^{40}K present in the body is the total number of potassium atoms times the natural abundance of potassium atoms present as ^{40}K divided by the atomic mass of ^{40}K:

$$\text{moles } ^{40}\text{K} = 140\ \cancel{\text{g K}} \times \frac{0.0117\ \text{mol } ^{40}\text{K}}{100\ \cancel{\text{mol K}}} \times \frac{1\ \cancel{\text{mol K}}}{40.0\ \cancel{\text{g K}}} = 4.10 \times 10^{-4}\ \text{mol } ^{40}\text{K}$$

B✓ We are given the number of atoms of ^{40}K that decay per second in 1.00 mol of ^{40}K, so the number of decays per year is

$$\frac{\text{decays}}{\text{year}} = 4.10 \times 10^{-4}\ \cancel{\text{mol } ^{40}\text{K}} \times \frac{1.05 \times 10^{7}\ \text{decays}/\cancel{\text{s}}}{1.00\ \cancel{\text{mol } ^{40}\text{K}}} \times \frac{60\ \cancel{\text{s}}}{1\ \cancel{\text{min}}} \times \frac{60\ \cancel{\text{min}}}{1\ \cancel{\text{h}}} \times \frac{24\ \cancel{\text{h}}}{1\ \cancel{\text{day}}} \times \frac{365\ \cancel{\text{days}}}{1\ \text{yr}}$$
$$= 1.36 \times 10^{11}\ \text{decays/yr}$$

C✓ The total energy the body receives per year from the decay of ^{40}K is equal to the total number of decays per year multiplied by the energy associated with each decay event:

$$\text{total energy per year} = \frac{1.36 \times 10^{11}\ \cancel{\text{decays}}}{\text{yr}} \times \frac{1.32\ \cancel{\text{MeV}}}{\cancel{\text{decays}}} \times \frac{10^{6}\ \cancel{\text{eV}}}{\cancel{\text{MeV}}} \times \frac{1.602 \times 10^{-19}\ \text{J}}{\cancel{\text{eV}}} = 2.87 \times 10^{-2}\ \text{J/yr}$$

We use the definition of the rad (1 rad = 10^{-2} J/kg of tissue) to convert this figure to a radiation dose in rads. If we assume the dose is equally distributed throughout the body, then the radiation dose per year is

$$\text{radiation dose per year} = \frac{2.87 \times 10^{-2}\ \cancel{\text{J}}/\text{yr}}{70.0\ \cancel{\text{kg}}} \times \frac{1\ \text{rad}}{1 \times 10^{-2}\ \cancel{\text{J/kg}}} = 4.10 \times 10^{-2}\ \text{rad/yr} = 41\ \text{mrad/yr}$$

This corresponds to almost half of the normal background radiation most people experience.

TABLE 20.5 The effects of a single radiation dose on a 70-kg human

Dose, rem	Symptoms/Effects
<5	No observable effect
5–20	Possible chromosomal damage
20–100	Temporary reduction in white bloodcell count
50–100	Temporary sterility in men (up to a year)
100–200	Mild radiation sickness, vomiting, diarrhea, fatigue. Immune system suppressed and bone growth in children retarded
>300	Permanent sterility in women
>500	Fatal to 50% within 30 days. Destruction of bone marrow and intestine
>3000	Fatal within hours

EXERCISE 20.6

Because strontium is chemically similar to calcium, small amounts of the Sr^{2+} ion are taken up by the body and deposited in calcium-rich tissues such as bone, using the same mechanism that is responsible for the absorption of Ca^{2+}. Consequently, the radioactive strontium (^{90}Sr) found in fission waste and released by atmospheric nuclear-weapons testing is a major health concern. A normal 70-kg human contains about 280 mg of strontium, and each mole of ^{90}Sr undergoes 4.55×10^{14} decays per second by the emission of a 0.546-MeV β particle. What would be the annual radiation dose in rads for a 70-kg person if 0.10% of the strontium ingested were ^{90}Sr?

Answer 5.7×10^{3} rad/yr (10 times the fatal dose!)

Assessing the Impact of Radiation Exposure

One of the more controversial public policy issues debated today is whether the radiation exposure from artificial sources, when combined with the exposure from natural sources, poses a significant risk to human health. The effects of single radiation doses of different magnitudes on humans are listed in Table 20.5. Because of the many factors involved in radiation exposure (length of exposure, intensity of the source, energy and type of particle), it is difficult to quantify the specific dangers of one radioisotope versus another. Nonetheless, some general conclusions regarding the effects of radiation exposure are generally accepted as valid.

Radiation doses of 600 rem and higher are invariably fatal, while a dose of 500 rem kills half the exposed subjects within 30 days. Smaller doses (≤50 rem) appear to cause only limited health effects, even though they correspond to tens of years of natural radiation. This does not, however, mean that such doses have no ill effects; they may cause long-term health problems such as cancer or genetic changes that affect offspring. The possible detrimental effects of the much smaller doses attributable to artificial sources (<100 mrem/yr) are more difficult to assess.

The tissues most affected by large, whole-body exposures are bone marrow, intestinal tissue, hair follicles, and reproductive organs, all of which contain rapidly dividing cells. The susceptibility of rapidly dividing cells to radiation exposure explains why cancers are often treated by radiation: because cancer cells divide faster than normal cells, they are destroyed preferentially by radiation. Long-term radiation-exposure studies on fruit flies show a linear relationship between the number of genetic defects and both the magnitude of the dose and the exposure time. In contrast, similar studies on mice show a much lower number of defects when a given dose of radiation is spread out over a long period of time rather than received all at once. Both patterns are plotted in Figure 20.13, but which of the two is applicable to humans? According to one hypothesis, mice have very low risk from low doses because their bodies have ways of dealing with the damage caused by natural radiation. At much higher doses, however, their natural repair mechanisms are overwhelmed, leading to irreversible damage. Because mice are biochemically much more similar to humans than are fruit flies, many scientists believe that this model also applies to humans. In contrast, the linear model assumes that *all* exposure to radiation is intrinsically damaging, and suggests that stringent regulation of low-level radiation exposure is necessary. Which of these views is more accurate is still in question, and the answer has extremely important consequences for regulating radiation exposure.

Note the pattern

A large dose of radiation spread out over time is less harmful than the same total amount of radiation administered over a short time.

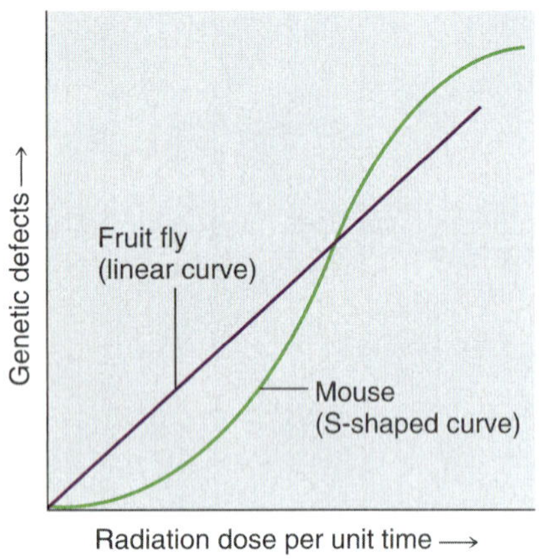

Figure 20.13 Two possible relationships between the number of genetic defects and radiation exposure. Studies on fruit flies show a linear relationship between the number of genetic defects and the magnitude of the radiation dose and exposure time, which is consistent with a cumulative effect of radiation. In contrast, studies on mice show an S-shaped curve, which suggests that the number of defects is lower when radiation exposure occurs over a longer time. Which of these relationships is more applicable to humans is a matter of considerable debate.

20.4 • Thermodynamic Stability of the Atomic Nucleus

Nuclear reactions, like chemical reactions, are accompanied by changes in energy. The energy changes in nuclear reactions, however, are enormous compared with those of even the most energetic chemical reactions. In fact, the energy changes in a typical nuclear reaction are so large that they result in a measurable change of mass. In this section, we describe the relationship between mass and energy in nuclear reactions, and show how the seemingly small changes in mass that accompany nuclear reactions result in the release of enormous amounts of energy.

Mass–Energy Balance

The relationship between mass m and energy E was introduced in Chapter 6 and is expressed in the equation

$$E = mc^2 \tag{20.26}$$

where c is the speed of light (2.998×10^8 m/s), and energy and mass are expressed in units of joules and kilograms, respectively. This relationship was first derived by Albert Einstein in 1905 as part of his special theory of relativity; it states that the mass of a particle is directly proportional to its energy. Thus, according to Equation 20.26, every mass has an associated energy, and similarly, any reaction that involves a change in energy must be accompanied by a change in mass. This implies that all exothermic reactions should be accompanied by a decrease in mass, and all endothermic reactions should be accompanied by an increase in mass. Given the law of conservation of mass (see Chapter 3), how can this be true? The solution to this apparent contradiction is that *chemical reactions are indeed accompanied by changes in mass*, but these changes are simply too small to be detected.*

For example, the chemical equation for the combustion of graphite to produce carbon dioxide is

$$\text{C (graphite)} + \frac{1}{2}\text{O}_2(g) \longrightarrow \text{CO}_2(g) \qquad \Delta H^\circ = -393.5 \text{ kJ/mol} \quad (20.27)$$

Combustion reactions are typically carried out at constant pressure, and under these conditions, the heat released or absorbed is equal to ΔH. As you learned in Chapter 18, when a reaction is carried out at constant volume, the heat released or absorbed is equal to ΔE. For most chemical reactions, however, $\Delta E \approx \Delta H$. If we rewrite Einstein's equation as

$$\Delta E = (\Delta m)c^2 \quad (20.28)$$

we can rearrange the equation to obtain the following relationship between the change in mass and the change in energy:

$$\Delta m = \frac{\Delta E}{c^2} \quad (20.29)$$

Because 1 J = 1 $(\text{kg} \cdot \text{m}^2)/\text{s}^2$, the change in mass is

$$\Delta m = \frac{-393.5 \text{ kJ/mol}}{(2.998 \times 10^8 \text{ m/s})^2} = \frac{-3.935 \times 10^5 (\text{kg} \cdot \cancel{\text{m}^2})/(\cancel{\text{s}^2} \cdot \text{mol})}{(2.998 \times 10^8 \cancel{\text{m/s}})^2} = -4.38 \times 10^{-12} \text{ kg/mol} \quad (20.30)$$

This is a mass change of about 3.6×10^{-10} g/g carbon that is burned, or about one -hundred-millionths of the mass of an electron per atom of carbon. In practice, this mass change is much too small to be measured experimentally, and it is utterly negligible.

In contrast, a typical nuclear reaction, the radioactive decay of ^{14}C to ^{14}N and an electron (a β particle), produces a much larger change in mass:

$$^{14}\text{C} \longrightarrow {}^{14}\text{N} + {}^{0}_{-1}\beta \quad (20.31)$$

We can use the experimentally measured masses of subatomic particles and common isotopes given in Tables 20.1 and 20.6 to calculate the change in mass directly. The reaction involves the conversion of a neutral ^{14}C *atom* to a positively charged ^{14}N *ion* (with six, not seven, electrons) and a negatively charged β particle (an electron), so the mass of the products is identical to the mass of a neutral ^{14}N atom. The total change in mass during the reaction is therefore the difference between the mass of a neutral ^{14}N atom (14.003074 amu) and the mass of a ^{14}C atom (14.003242 amu):

$$\begin{aligned}\Delta m &= \text{mass}_{\text{products}} - \text{mass}_{\text{reactants}} \\ &= 14.003074 \text{ amu} - 14.003242 \text{ amu} = -0.000168 \text{ amu}\end{aligned} \quad (20.32)$$

The difference in mass, which has been released as energy, corresponds to almost one-third of an electron. The change in mass for the decay of 1 mol of ^{14}C is $-0.000168 \text{ g} = -1.68 \times 10^{-4} \text{ g} = -1.68 \times 10^{-7}$ kg. Although a mass change of this magnitude may seem small, it is about a thousand times larger than the mass change for the combustion of graphite. The energy change is

$$\begin{aligned}\Delta E &= (\Delta m)c^2 = (-1.68 \times 10^{-7} \text{ kg})(2.998 \times 10^8 \text{ m/s})^2 \\ &= -1.51 \times 10^{10} (\text{kg} \cdot \text{m}^2)/\text{s}^2 = -1.51 \times 10^{10} \text{ J} = -1.51 \times 10^7 \text{ kJ}\end{aligned} \quad (20.33)$$

* This situation is similar to the wave–particle duality discussed in Chapter 6. As you may recall, all particles exhibit wavelike behavior, but the wavelength is inversely proportional to the mass of the particle (actually, to its momentum, the product of its mass and velocity). Consequently, wavelike behavior is detectable only for particles with very small masses, such as electrons.

TABLE 20.6 Experimentally measured masses of selected isotopes

Isotope	Mass, amu	Isotope	Mass, amu	Isotope	Mass, amu
^{1}H	1.007825	^{14}N	14.003074	^{208}Po	207.981246
^{2}H	2.014102	^{16}O	15.994915	^{210}Po	209.982874
^{3}H	3.016049	^{52}Cr	51.940508	^{222}Rn	222.017578
^{3}He	3.016029	^{56}Fe	55.934938	^{226}Ra	226.025410
^{4}He	4.002603	^{59}Co	58.933195	^{230}Th	230.033134
^{6}Li	6.015123	^{58}Ni	57.935343	^{234}Th	234.043601
^{7}Li	7.016005	^{60}Ni	59.930786	^{234}Pa	234.043308
^{9}Be	9.012182	^{90}Rb	89.914802	^{233}U	233.039635
^{10}B	10.012937	^{144}Cs	143.932077	^{234}U	234.040952
^{11}B	11.009305	^{206}Pb	205.974465	^{235}U	235.043930
^{12}C	12	^{207}Pb	206.975897	^{238}U	238.050788
^{13}C	13.003355	^{208}Pb	207.976652	^{239}Pu	239.052163
^{14}C	14.003242				

Data source: G. Audi, A. H. Wapstra, and C. Thibault, *The AME2003 atomic mass evaluation*.

The energy released in this nuclear reaction is more than 100,000 times greater than that of a typical chemical reaction, even though the decay of ^{14}C is a relatively low-energy nuclear reaction.

Because the energy changes in nuclear reactions are so large, they are often expressed in kiloelectronvolts (1 keV = 10^3 eV), megaelectronvolts (1 MeV = 10^6 eV), and even gigaelectronvolts (1 GeV = 10^9 eV) *per atom or particle*. The change in energy that accompanies a nuclear reaction can be calculated from the change in mass using the relationship 1 amu = 931 MeV. The energy released by the decay of one atom of ^{14}C is thus

$$(-1.68 \times 10^{-4}\ \cancel{\text{amu}})(931\ \text{MeV}/\cancel{\text{amu}}) = -0.156\ \text{MeV} = -156\ \text{keV} \qquad (20.34)$$

EXAMPLE 20.7

Calculate the changes in mass (in amu) and energy (in J/mol and eV/atom) that accompany the radioactive decay of ^{238}U to ^{234}Th and an α particle. The α particle absorbs two electrons from the surrounding matter to form a helium atom.

Given Nuclear decay reaction

Asked for Changes in mass and energy

Strategy

Ⓐ Use the mass values in Tables 20.1 and 20.6 to calculate the change in mass for the decay reaction in amu.
Ⓑ Use Equation 20.29 to calculate the change in energy in J/mol.
Ⓒ Use the relationship between amu and MeV to calculate the change in energy in eV/atom.

Solution

Ⓐ✓ Using particle and isotope masses from Tables 20.1 and 20.6, we can calculate the change in mass as follows:

$$\Delta m = \text{mass}_{\text{products}} - \text{mass}_{\text{reactants}} = (\text{mass }^{234}\text{Th} + \text{mass }^{4}_{2}\text{He}) - \text{mass }^{238}\text{U}$$
$$= (234.043601\ \text{amu} + 4.002603\ \text{amu}) - 238.050788\ \text{amu} = -0.004584\ \text{amu}$$

B Thus, the change in mass for 1 mol of ^{238}U is -0.004584 g or -4.584×10^{-6} kg. The change in energy in J/mol is:

$$\Delta E = (\Delta M)c^2 = (-4.584 \times 10^{-6}\ \text{kg})(2.998 \times 10^8\ \text{m/s})^2 = -4.120 \times 10^{11}\ \text{J/mol}$$

C The change in energy in eV/atom is:

$$\Delta E = -4.584 \times 10^{-3}\ \cancel{\text{amu}} \times \frac{931\ \cancel{\text{MeV}}}{\cancel{\text{amu}}} \times \frac{1 \times 10^6\ \text{eV}}{1\ \cancel{\text{MeV}}} = -4.27 \times 10^6\ \text{eV/atom}$$

EXERCISE 20.7

Calculate the changes in mass (in amu) and energy (in kJ/mol and keV/atom) that accompany the radioactive decay of tritium (^3H) to ^3He and a β particle.

Answer $\Delta m = -2.0 \times 10^{-5}$ amu; $\Delta E = -1.9 \times 10^6$ kJ/mol $= -19$ keV/atom

Nuclear Binding Energies

We have seen that energy changes in both chemical and nuclear reactions are accompanied by changes in mass. Einstein's equation, which allows us to interconvert mass and energy, has another interesting consequence: The mass of an atom is always less than the sum of the masses of its component particles. The only exception to this rule is hydrogen-1 (^1H), whose measured mass of 1.007825 amu is identical to the sum of the masses of a proton and an electron. In contrast, the experimentally measured mass of an atom of deuterium (^2H) is 2.014102 amu, although its calculated mass is 2.016490 amu:

$$\begin{aligned} m_{^2H} &= m_{\text{neutron}} + m_{\text{proton}} + m_{\text{electron}} \\ &= 1.008665\ \text{amu} + 1.007276\ \text{amu} + 0.000549\ \text{amu} = 2.016490\ \text{amu} \end{aligned} \tag{20.35}$$

The difference between the sum of the masses of the components and the measured atomic mass is called the **mass defect** of the nucleus. Just as a molecule is more stable than its isolated atoms, a nucleus is more stable (lower in energy) than its isolated components. Consequently, when isolated nucleons assemble into a stable nucleus, energy is released. According to Equation 20.29, this release of energy must be accompanied by a decrease in the mass of the nucleus.

The amount of energy released when a nucleus forms from its component nucleons is the **nuclear binding energy** (Figure 20.14). In the case of deuterium, the mass defect is 0.002388 amu, which corresponds to a nuclear binding energy of 2.22 MeV for the deuterium nucleus. Because the magnitude of the mass defect is proportional to the nuclear binding energy, both values indicate the stability of the nucleus.

Not all nuclei are equally stable. Chemists describe the *relative* stability of different nuclei by comparing the binding energy *per nucleon*, which is obtained by dividing the nuclear binding energy by the mass number A of the nucleus. As shown in Figure 20.15, the binding energy per nucleon increases rapidly with increasing atomic number until about $Z = 26$, where it levels off to about 8–9 MeV per nucleon and then decreases slowly. The initial increase in binding energy is not a smooth curve, but exhibits sharp peaks corresponding to the light nuclei that have equal numbers of protons and neutrons (for example, ^4He, ^{12}C, ^{16}O). As mentioned earlier, these are particularly stable combinations.

Because the maximum binding energy per nucleon is reached at ^{56}Fe, all other nuclei are thermodynamically unstable with regard to the formation of ^{56}Fe. Consequently, heavier nuclei (toward the right in Figure 20.15) should spontaneously

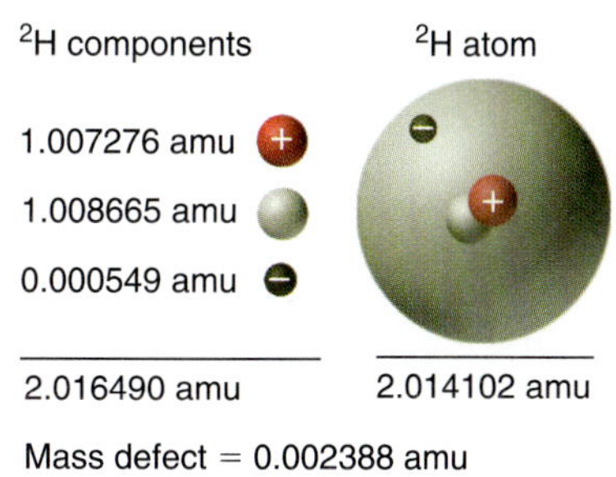

Figure 20.14 Nuclear binding energy in deuterium. The mass of a ^2H atom is less than the sum of the masses of a proton, a neutron, and an electron by 0.002388 amu; the difference in mass corresponds to the nuclear binding energy. The larger the value of the mass defect, the greater the nuclear binding energy and the more stable the nucleus.

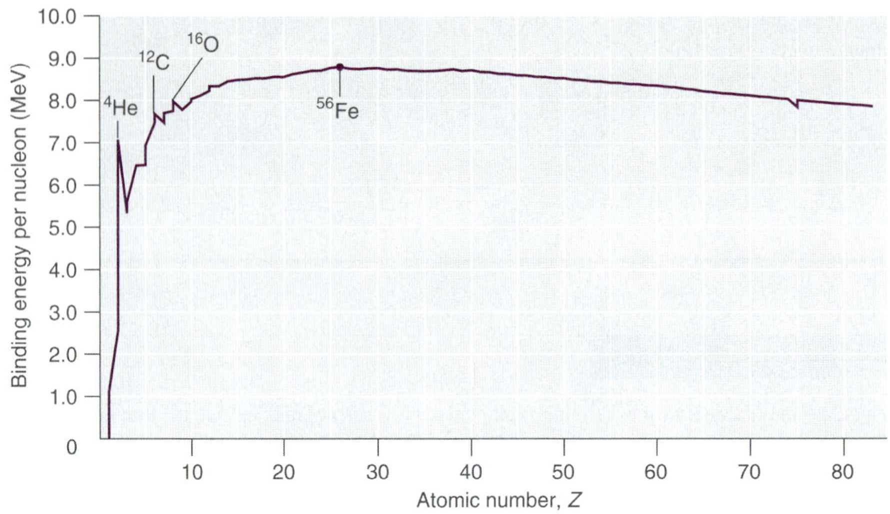

Figure 20.15 The curve of nuclear binding energy. This plot of the average binding energy per nucleon as a function of atomic number shows that the binding energy per nucleon increases with increasing atomic number until about $Z = 26$, levels off, and then decreases. The sharp peaks correspond to light nuclei that have equal numbers of protons and neutrons.

undergo reactions such as alpha decay, which result in a decrease in atomic number. Conversely, lighter elements (on the left in Figure 20.15) should spontaneously undergo reactions that result in an increase in atomic number. This is indeed the observed pattern.

Note the pattern

Heavier nuclei spontaneously undergo nuclear reactions that decrease their atomic number. Lighter nuclei spontaneously undergo nuclear reactions that increase their atomic number.

EXAMPLE 20.8

Calculate the total nuclear binding energy (in MeV) and the binding energy per nucleon for ^{56}Fe. The experimental mass of the nuclide is given in Table 20.6.

Given Nuclide and mass

Asked for Nuclear binding energy, binding energy per nucleon

Strategy

Ⓐ Sum the masses of the protons, electrons, and neutrons or, alternatively, use the mass of the appropriate number of ^{1}H atoms (since its mass is the same as the mass of one electron and one proton).

Ⓑ Calculate the mass defect by subtracting the experimental mass from the calculated mass.

Ⓒ Determine the nuclear binding energy by multiplying the mass defect by the change in energy in electron volts per atom. Divide this value by the number of nucleons to obtain the binding energy per nucleon.

Solution

Ⓐ An iron-56 atom contains 26 protons, 26 electrons, and 30 neutrons. We could add the masses of these three sets of particles, but noting that 26 protons and 26 electrons are equivalent to 26 ^{1}H atoms, we can calculate the sum of the masses more quickly as follows:

$$\text{calculated mass} = 26(\text{mass } {}^{1}_{1}\text{H}) + 30(\text{mass } {}^{1}_{0}\text{n})$$
$$= 26(1.007825)\text{ amu} + 30(1.008665)\text{ amu} = 56.463400\text{ amu}$$
$$\text{experimental mass} = 55.934938\text{ amu}$$

Ⓑ We subtract to find the mass defect:

$$\text{mass defect} = \text{calculated mass} - \text{experimental mass}$$
$$= 56.463400\text{ amu} - 55.934938\text{ amu} = 0.528462\text{ amu}$$

Ⓒ The nuclear binding energy is thus 0.528462 amu × 931 MeV/amu = 492 MeV. The binding energy per nucleon is 492 MeV/56 nucleons = 8.79 MeV/nucleon.

EXERCISE 20.8

Calculate the total nuclear binding energy (in MeV) and the binding energy per nucleon for ^{238}U.

Answer 1800 MeV/^{238}U; 7.57 MeV/nucleon

Nuclear Fission and Fusion

As discussed in Section 20.2, **nuclear fission** is the splitting of a heavy nucleus into two lighter ones. Fission was discovered in 1938 by the German scientists Otto Hahn, Lise Meitner, and Fritz Strassmann, who bombarded a sample of uranium with neutrons in an attempt to produce new elements with $Z > 92$. They observed that lighter elements such as barium ($Z = 56$) were formed during the reaction, and they realized that such products had to originate from the neutron-induced fission of uranium-235:

$$^{235}_{92}U + ^{1}_{0}n \longrightarrow ^{141}_{56}Ba + ^{92}_{36}Kr + 3^{1}_{0}n \tag{20.36}$$

This hypothesis was confirmed by detection of the krypton-92 fission product. As discussed in Section 20.2, the nucleus usually divides asymmetrically rather than into two equal parts, and the fission of a given nuclide does not give the same products every time.

In a typical nuclear fission reaction, more than one neutron is released by each dividing nucleus. When these neutrons collide with and induce fission in other neighboring nuclei, a self-sustaining series of nuclear fission reactions known as a **nuclear chain reaction** can result (Figure 20.16). For example, the fission of ^{235}U releases two to three neutrons per fission event. If absorbed by other ^{235}U nuclei, those neutrons induce additional fission events, and the rate of the fission reaction increases geometrically. Each series of events is called a *generation*. Experimentally, it is found that some minimum mass of a fissile isotope is required to sustain a nuclear chain reaction; if the mass is too low, too many neutrons are able to escape without being captured and inducing a fission reaction. The minimum mass capable of supporting sustained fission is called the **critical mass**. This amount depends on the purity of the material and the shape of the mass,

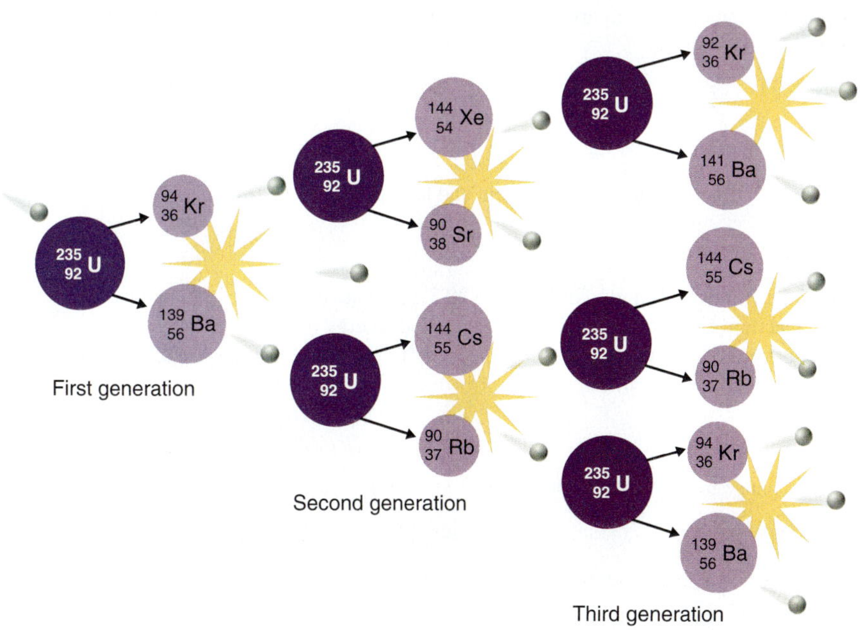

Figure 20.16 A nuclear chain reaction. The process is initiated by the collision of a single neutron with a ^{235}U nucleus, which undergoes fission as shown in Figure 20.6. Because each neutron released can cause the fission of another ^{235}U nucleus, the rate of the fission reaction accelerates geometrically. Each series of events is a generation.

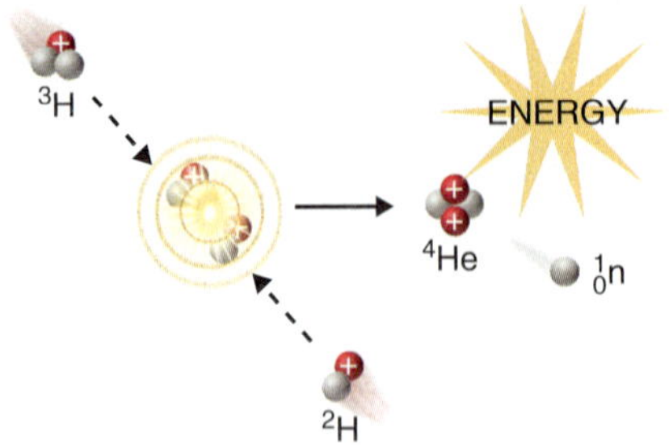

Figure 20.17 Nuclear fusion. In a nuclear fusion reaction, lighter nuclei combine to produce a heavier nucleus. As shown, fusion of ^{3}H and ^{2}H to give ^{4}He and a neutron releases an enormous amount of energy. In principle, nuclear fusion can produce much more energy than fission, but very high temperatures are required to overcome electrostatic repulsions between the positively charged nuclei and initiate the fusion reaction.

which corresponds to the amount of surface area available from which neutrons can escape, and on the identity of the isotope. If the mass of the fissile isotope is greater than the critical mass, then under the right conditions the resulting *supercritical mass* can release energy explosively. The enormous energy released from nuclear chain reactions is responsible for the massive destruction caused by the detonation of nuclear weapons such as fission bombs, but it also forms the basis of the nuclear power industry.

Nuclear fusion, in which two light nuclei combine to produce a heavier, more stable nucleus, is the opposite of a nuclear fission reaction. As in the nuclear transmutation reactions discussed in Section 20.2, the positive charge on both nuclei results in a large electrostatic energy barrier to fusion. This barrier can be overcome if one or both particles have sufficient kinetic energy to overcome the electrostatic repulsions, allowing the two nuclei to approach close enough for a fusion reaction to occur. The principle is similar to adding heat to increase the rate of a chemical reaction (see Chapter 14).

As shown in the plot of nuclear binding energy per nucleon versus atomic number in Figure 20.15, fusion reactions are most exothermic for the lightest element. For example, in a typical fusion reaction, two deuterium atoms combine to produce helium-3:

$$2\,^2_1\text{H} \longrightarrow \,^3_2\text{He} + \,^1_0\text{n} \tag{20.37}$$

In another reaction, a deuterium atom and a tritium atom fuse to produce helium-4 (Figure 20.17):

$$^2_1\text{H} + \,^3_1\text{H} \longrightarrow \,^4_2\text{He} + \,^1_0\text{n} \tag{20.38}$$

Initiating these reactions, however, requires a temperature comparable to that in the interior of the sun (approximately 1.5×10^7 K). Currently, the only method available on Earth to achieve such a temperature is the detonation of a fission bomb. Thus, the so-called hydrogen bomb (or H-bomb) is actually a deuterium-tritium bomb (a D-T bomb), which uses a nuclear fission reaction to create the very high temperatures needed to initiate fusion of solid lithium deuteride (^{6}LiD), which releases neutrons that then react with ^{6}Li, producing tritium. The deuterium-tritium reaction releases energy explosively. The next example and exercise demonstrate the enormous amounts of energy produced by nuclear fission and fusion reactions. In fact, fusion reactions are the power sources for all stars, including our sun.

EXAMPLE 20.9

Calculate the amount of energy (in eV/atom and kJ/mol) released when the neutron-induced fission of ^{235}U produces ^{144}Cs, ^{90}Rb, and two neutrons:

$$^{235}_{92}\text{U} + \,^1_0\text{n} \longrightarrow \,^{144}_{55}\text{Cs} + \,^{90}_{37}\text{Rb} + 2\,^1_0\text{n}$$

Given Balanced nuclear reaction

Asked for Energy released in eV/atom and kJ/mol

Strategy

Ⓐ Following the method used in Example 20.7, calculate the change in mass that accompanies the reaction. Convert this value to the change in energy in eV/atom.

Ⓑ Calculate the change in mass per mole of ^{235}U. Then use Equation 20.28 to calculate the change in energy in kJ/mol.

Solution

Ⓐ The change in mass that accompanies the reaction is

$$\Delta m = \text{mass}_{\text{products}} - \text{mass}_{\text{reactants}} = \text{mass}\left(^{144}_{55}\text{Cs} + \,^{90}_{37}\text{Rb} + \,^1_0\text{n}\right) - \text{mass}\,^{235}_{92}\text{U}$$

$$= (143.932077 \text{ amu} + 89.914802 \text{ amu} + 1.008665 \text{ amu}) - 235.043930 \text{ amu}$$

$$= -0.188386 \text{ amu}$$

The change in energy in eV/atom is

$$\Delta E = (-0.188386 \text{ amu})(931 \text{ MeV/amu}) = -175 \text{ MeV}$$

B ✓ The change in mass per mole of $^{235}_{92}\text{U}$ is -0.188386 g $= -1.88386 \times 10^{-4}$ kg, so the change in energy in kJ/mol is

$$\Delta E = (\Delta m)c^2 = (-1.88386 \times 10^{-4} \text{ kg})(2.998 \times 10^8 \text{ m/s})^2$$
$$= -1.693 \times 10^{13} \text{ J/mol} = -1.693 \times 10^{10} \text{ kJ/mol}$$

EXERCISE 20.9

Calculate the amount of energy (in eV/atom and kJ/mol) released when deuterium and tritium fuse to give helium-4 and a neutron:

$$^2_1\text{H} + ^3_1\text{H} \longrightarrow ^4_2\text{He} + ^1_0\text{n}$$

Answer $\Delta E = -17.6$ MeV/atom $= -1.697 \times 10^9$ kJ/mol

20.5 • Applied Nuclear Chemistry

The ever-increasing energy needs of modern societies have led scientists and engineers to develop ways of harnessing the energy released by nuclear reactions. To date, all practical applications of nuclear power have been based on nuclear fission reactions. Although in principle nuclear fusion offers many advantages, technical difficulties in achieving the high energies required to initiate nuclear fusion reactions have thus far precluded the use of fusion for the controlled release of energy. In this section, we describe the various types of nuclear power plants that are currently used to generate electricity from nuclear reactions, along with some possible ways to harness fusion energy in the future. In addition, we discuss some of the applications of nuclear radiation and radioisotopes, which have innumerable uses in medicine, biology, chemistry, and industry.

Pitchblende, the major uranium ore, consisting mainly of uranium oxide.

Nuclear Reactors

When a critical mass of a fissile isotope has been achieved, the resulting flux of neutrons can lead to a self-sustaining reaction. A variety of techniques can be used to control the flow of neutrons from such a reaction, which allows nuclear fission reactions to be maintained at safe levels. Many levels of control are required, along with a fail-safe design, because otherwise the chain reaction can accelerate so rapidly that it releases enough heat to melt or vaporize the fuel and the container, a situation that can release enough radiation to contaminate the surrounding area. Uncontrolled nuclear fission reactions are relatively rare, but they have occurred at least 18 times in the past.

Figure 20.18 A "fossil nuclear reactor" in a uranium mine near Oklo in Gabon, West Africa. More than a billion years ago, a number of uranium-rich deposits in West Africa apparently "went critical," initiating uncontrolled nuclear fission reactions that may have continued intermittently for more than 100,000 years, until the concentration of uranium-235 became too low to support a chain reaction. This photo shows a geologist standing in a mine dug to extract the concentrated uranium ore. Commercial interest waned rapidly after it was recognized that the uranium ore was severely depleted in U-235, the isotope of interest.

There is compelling evidence that uncontrolled nuclear chain reactions occurred naturally in the early history of our planet, about 1.7 billion years ago in uranium deposits near Oklo in Gabon, West Africa (Figure 20.18). The natural abundance of ^{235}U 2 billion years ago was about 3% compared with 0.72% today; in contrast, the "fossil nuclear reactor" deposits in Gabon now contain only 0.44% ^{235}U. An unusual combination of geologic phenomena in this region apparently resulted in the formation of deposits of essentially pure uranium oxide containing 3% ^{235}U, which coincidentally happens to be identical to the fuel used in many modern nuclear plants. When rainwater or groundwater saturated one of these deposits, the water acted as a natural *moderator* that decreased the kinetic energy of the neutrons emitted by radioactive decay of ^{235}U, allowing the neutrons to initiate a chain reaction. As a result, the entire deposit "went critical" and became an uncontrolled nuclear chain reaction, which is estimated to have produced about 100 kW of power. It is thought that these natural nuclear reactors operated only intermittently, however, because the heat released would have vaporized the water. Removing the water would have shut down the reactor until the rocks cooled enough to allow water to reenter the deposit, at which point the chain reaction would begin again. This on–off cycle was repeated for more than 100,000 years, until so much ^{235}U was consumed that the deposits could no longer support a chain reaction.

Figure 20.19 The Chernobyl nuclear power plant. In 1986, mechanical and operational failures during testing at the Chernobyl power plant in the Ukraine caused an uncontrolled nuclear chain reaction. The resulting fire destroyed much of the facility and severely damaged the core of the reactor, resulting in the release of large amounts of radiation that was spread over the surrounding area by the prevailing winds. The effects were devastating to the health of the population in the region and to the Soviet economy.

The most recent instance of an uncontrolled nuclear chain reaction occurred on April 25–26, 1986, at the Chernobyl nuclear power plant in the former USSR (now Ukraine), resulting in the worst nuclear power accident in history (Figure 20.19). During testing of the reactor's turbine generator, a series of mechanical and operational failures caused a chain reaction that quickly went out of control, destroying the reactor core and igniting a fire that destroyed much of the facility and released a large amount of radioactivity. Thirty people were killed immediately, and the high levels of radiation in a 20-mile radius forced nearly 350,000 people to be resettled or evacuated. In addition, the accident caused a disruption to the Soviet economy that is estimated to have cost almost $13 billion. It is somewhat surprising, however, that the long-term health effects on the 600,000 people affected by the accident appear to be much less severe than originally anticipated. Initially, it was predicted that the accident would result in tens of thousands of premature deaths, but an exhaustive study almost 20 years after the event suggests that "only" 4,000 people will die prematurely from radiation exposure due to the accident. Although that may seem like a lot, in fact it represents only about a 3% increase in the cancer rate among the 600,000 people most affected, of whom about a quarter would be expected to eventually die of cancer even if the accident had not occurred.

If, on the other hand, the neutron flow in a reactor is carefully regulated so that only enough heat is released to boil water, then the resulting steam can be used to produce electricity. Thus, a nuclear reactor is similar in many respects to the conventional power plants discussed in Chapter 5, which burn coal or natural gas to generate electricity—the only difference is the source of the heat that is used to convert water to steam.

Light-Water Reactors

We begin our description of nuclear power plants with light-water reactors, which are used extensively to produce electricity in countries such as Japan, Israel, South Korea, Taiwan, and France that lack large reserves of fossil fuels. The essential components of a light-water reactor are depicted in Figure 20.20. All existing nuclear power plants have similar components, although different designs utilize different fuels and operating conditions. *Fuel rods* containing a fissile isotope in a structurally stabilized form (such as uranium oxide pellets encased in a corrosion-resistant zirconium alloy) are suspended in a cooling bath that transfers the heat generated by the fission reaction to a secondary cooling system. The heat is used to generate steam for the production of electricity. In addition, *control rods* are utilized to absorb neutrons and thereby control the rate of the nuclear chain reaction. Control rods are made of a substance that efficiently absorbs neutrons, such as boron, cadmium, or, in nuclear submarines, hafnium. Pulling the control rods out increases the neutron flux, allowing the reactor to generate more heat, whereas inserting the rods completely stops the reaction, a process called "scramming the reactor."

Despite this apparent simplicity, many technical hurdles must be overcome for nuclear power to be an efficient and safe source of energy. Uranium contains only 0.72% uranium-235, which is the only naturally occurring fissile isotope of uranium. Because this abundance is not enough to support a chain reaction, the uranium fuel must be at least partially enriched in ^{235}U, to a concentration of about 3%, for it to be able to sustain a chain reaction. At this level of enrichment, a nuclear explosion is

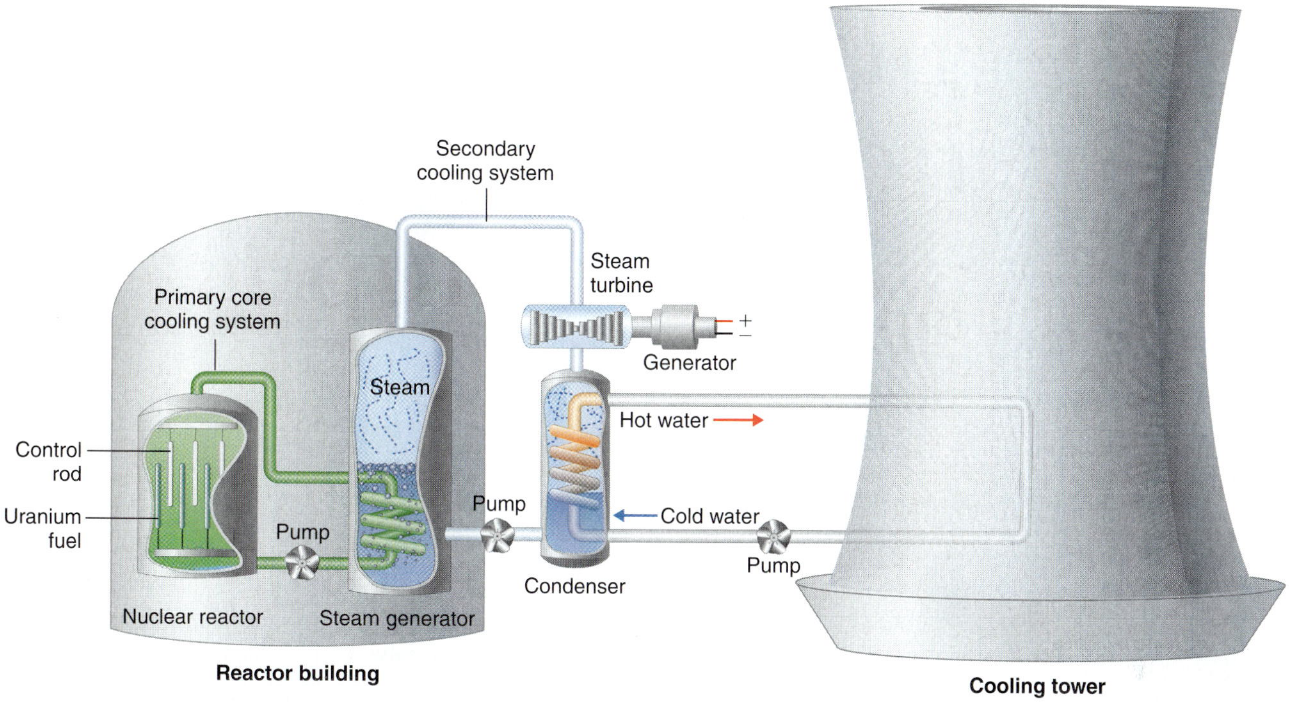

Figure 20.20 A light-water nuclear fission reactor for the production of electric power. The fuel rods are made of a corrosion-resistant alloy that encases the partially enriched uranium fuel; controlled fission of ^{235}U in the fuel produces heat. Water surrounds the fuel rods and moderates the kinetic energy of the neutrons, slowing them to increase the probability that they will induce fission. Control rods that contain elements such as boron, cadmium, or hafnium, which are very effective at absorbing neutrons, are used to control the rate of the fission reaction. A heat exchanger is used to boil water in a secondary cooling system, creating steam to drive the turbine and produce electricity. The large hyperbolic cooling tower, which is the most visible portion of the facility, is used to condense the steam in the secondary cooling circuit; it is often located at some distance from the actual reactor.

impossible; far higher levels of enrichment ($\geq$90%) are required for military applications such as nuclear weapons or the nuclear reactors in submarines. Enrichment is accomplished by converting uranium oxide to UF_6, which is volatile and contains discrete UF_6 molecules. Because $^{235}UF_6$ and $^{238}UF_6$ have different masses, they have different rates of effusion and diffusion, and they can be separated using a gas diffusion process, as described in Chapter 12. Another difficulty is that neutrons produced by nuclear fission are too energetic to be absorbed by neighboring nuclei, and they escape from the material without inducing fission in nearby ^{235}U nuclei. Consequently, a moderator must be used to slow the neutrons enough to allow them to be captured by other ^{235}U nuclei. High-speed neutrons are scattered by substances such as water or graphite, which decreases their kinetic energy and increases the probability that they will react with another ^{235}U nucleus. The moderator in a light-water reactor is the water that is used as the primary coolant. The system is highly pressurized to about 100 atm to keep the water from boiling normal.

All nuclear reactors also require a powerful cooling system to absorb the heat generated in the reactor core and create steam that is used to drive a turbine that generates electricity. In 1979, an accident occurred when the main water pumps used for cooling at the nuclear power plant at Three Mile Island in Pennsylvania stopped running, which prevented the steam generators from removing heat. Eventually, the zirconium casing of the fuel rods ruptured, resulting in a meltdown of about half of the reactor core. Although there was no loss of life and only a small release of radioactivity, the accident produced sweeping changes in nuclear power plant operations. The U.S. Nuclear Regulatory Commission tightened its oversight to improve safety.

The main disadvantage of nuclear fission reactors is that the spent fuel, which contains too little of the fissile isotope for power generation, is much more radioactive than the unused fuel due to the presence of many daughter nuclei with shorter half-lives than ^{235}U. The decay of these daughter isotopes generates so much heat that the spent fuel rods must be stored in water for as long as 5 years before they can be handled. Even then, the radiation levels are so high that the rods must be stored for many, many more years to allow the daughter isotopes to decay to nonhazardous levels. How to store these spent fuel rods for hundreds of years is a pressing issue that has not yet been successfully resolved. As a result, some people are convinced that nuclear power is not a viable option for providing our future energy needs, although a number of other countries continue to rely on nuclear reactors for a large fraction of their energy.

Heavy-Water Reactors

Deuterium (^2H) absorbs neutrons much less effectively than does hydrogen (^1H), but it is about twice as effective at slowing neutrons. Consequently, a nuclear reactor that uses D_2O instead of H_2O as the moderator is so efficient that it can use *unenriched* uranium as fuel. Using a lower grade of uranium reduces the operating costs and eliminates the need for plants that produce enriched uranium. Because of the expense of D_2O, however, only countries like Canada, which has abundant supplies of hydroelectric power for generating D_2O by electrolysis (see Chapter 19), have made a major investment in heavy-water reactors.

Breeder Reactors

A *breeder reactor* is a nuclear fission reactor that produces more fissionable fuel than it consumes. This does not violate the first law of thermodynamics because the fuel produced is not the same as the fuel consumed. Under heavy neutron bombardment, the nonfissile ^{238}U isotope is converted to ^{239}Pu, which can undergo fission:

$$^{238}_{92}U + ^{1}_{0}n \longrightarrow ^{239}_{92}U \longrightarrow ^{239}_{93}Np + ^{0}_{-1}\beta \qquad (20.39)$$

$$^{239}_{93}Np \longrightarrow ^{239}_{94}Pu + ^{0}_{-1}\beta$$

The overall reaction is thus the conversion of nonfissile ^{238}U to fissile ^{239}Pu, which can be isolated chemically and used to fuel a new reactor. An analogous series of reactions can be used to convert nonfissile ^{232}Th to ^{233}U, which can also be used as a fuel for a nuclear reactor. Typically, about 8 to 10 years are required for a breeder reactor to produce twice as much fissile material as it consumes, which is enough to fuel a replacement for the original reactor plus a new reactor. The products of fission of ^{239}Pu, however, have substantially longer half-lives than the fission products formed in light-water reactors.

Nuclear Fusion Reactors

Although nuclear fusion reactions such as Equations 20.37 and 20.38 are thermodynamically spontaneous, the positive charge on both nuclei results in a large electrostatic energy barrier to the reaction. (As you learned in Chapter 18, thermodynamic spontaneity has no relationship to reaction rate.) Extraordinarily high temperatures (about 1.0×10^8°C!) are required to overcome the electrostatic repulsions and initiate a fusion reaction. Even the most feasible such reaction, so-called D-T fusion (Equation 20.38), requires a temperature of about 4.0×10^7°C. Achieving these temperatures and controlling the materials to be fused are extraordinarily difficult problems, as is extracting the energy released by the fusion reaction, because a commercial fusion reactor would require such high temperatures to be maintained for long periods of time. Several different technologies are currently being explored, including the use of intense magnetic fields to contain ions in the form of a dense, high-energy plasma at a temperature high enough to sustain fusion (Figure 20.21a).

(a)

(b)

Figure 20.21 Two possible designs for a nuclear fusion reactor. The extraordinarily high temperatures needed to initiate a nuclear fusion reaction would immediately destroy a container made of any known material. (a) One way to avoid contact with the container walls is to use a high-energy plasma as the fuel. Because a plasma is essentially a gas composed of ionized particles, it can be confined using a strong magnetic field shaped like a torus (a hollow donut). (b) Another approach to nuclear fusion is inertial confinement, which utilizes an icosahedral array of powerful lasers to heat and compress a tiny fuel pellet (a mixture of solid LiD and LiT) to induce fusion.

Another concept utilizes focused laser beams to heat and compress fuel pellets in controlled miniature fusion explosions (Figure 20.21b).

Nuclear reactions such as these are called **thermonuclear reactions** because a great deal of thermal energy must be invested to initiate the reaction. The amount of energy released by the reaction, however, is several orders of magnitude greater than the energy needed to initiate it. Thus, in principle, a nuclear fusion reaction should result in a significant net production of energy. In addition, the Earth's oceans contain an essentially inexhaustible supply of both deuterium and tritium, which suggests that nuclear fusion could provide a limitless supply of energy. Unfortunately, however, the technical requirements for a successful nuclear fusion reaction are so great that net power generation by controlled fusion has yet to be achieved.

Uses of Radioisotopes

Nuclear radiation can damage biological molecules, thereby disrupting normal functions such as cell division. Because radiation is particularly destructive to rapidly dividing cells such as tumor cells and bacteria, it has been used medically to treat cancer since 1904, when radium-226 was first used to treat a cancerous tumor. Many radioisotopes are now available for medical use, and each has specific advantages for certain applications.

In modern *radiation therapy*, radiation is often delivered by a source planted inside the body. For example, tiny capsules containing an isotope such as ^{192}Ir, coated with a thin layer of chemically inert platinum, are inserted into the middle of a tumor that cannot be removed surgically. The capsules are removed when the treatment is over. In some cases, physicians take advantage of the body's own chemistry to deliver a radioisotope to the desired location. For example, the thyroid glands in the lower front of the neck are the only organs in the body that utilize iodine. Consequently, radioactive iodine is taken up almost exclusively by the thyroid (Figure 20.22a). Thus, when radioactive isotopes of iodine (^{125}I or ^{131}I) are injected into the blood of a patient suffering from thyroid cancer, the thyroid glands filter the radioisotope out of the blood and concentrate it in the tissue to be destroyed. In cases where a tumor is surgically inaccessible (for example, when it is located deep in the brain), an external radiation source such as a ^{60}Co "gun" is

Figure 20.22 Medical imaging and treatment with radioisotopes. (a) Radioactive iodine is used both to obtain images of the thyroid and to treat thyroid cancer. Injected iodine-123 or iodine-131 is selectively taken up by the thyroid gland, where it is incorporated into the thyroid hormone, thyroxine. Because iodine-131 emits low-energy β particles that are absorbed by the surrounding tissue, it can be used to destroy malignant tissue in the thyroid. In contrast, iodine-123 emits higher-energy γ rays that penetrate tissues readily, enabling it to image the thyroid gland, as shown in the photo. (b) Some technetium compounds are selectively absorbed by cancerous cells within bones. The yellow spots in this figure show that a primary cancer has metastasized (spread) to the patient's spine (*lower center*) and ribs (*right center*).

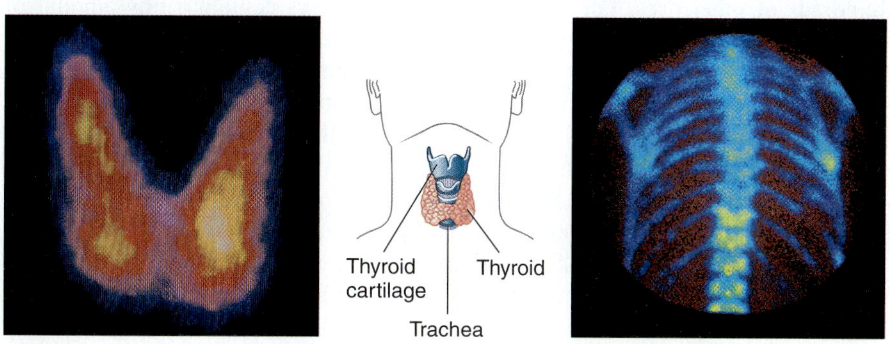

(a) (b)

TABLE 20.7 Radioisotopes used in medical imaging and treatment

Isotope	Half-life	Tissue
^{18}F	110 min	Brain
^{24}Na	15 h	Circulatory system
^{32}P	14 days	Eyes, liver, tumors
^{59}Fe	46 days	Blood, spleen
^{60}Co	10.5 mo	External radiotherapy
^{99m}Tc	6 h	Heart, thyroid, liver, kidney, lungs, skeleton
^{125}I	60 days	Thyroid, prostate, brain
^{131}I	8 days	Thyroid
^{201}Tl	3 days	Heart
^{133}Xe	5 days	Lungs

used to aim a tightly focused beam of γ rays at it. Unfortunately, radiation therapy damages healthy tissue in addition to the target tumor and results in severe side effects, such as nausea, hair loss, and a weakened immune system. Although radiation therapy is generally not a pleasant experience, in many cases it is the only choice.

A second major medical use of radioisotopes is *medical imaging*, in which a radioisotope is temporarily localized in a particular tissue or organ, where its emissions provide a "map" of the tissue or organ. Medical imaging uses radioisotopes that cause little or no tissue damage, but are easily detected. One of the most important radioisotopes for medical imaging is ^{99m}Tc. Depending on the particular chemical form in which it is administered, technetium tends to localize in bones and in soft tissues, such as the heart or kidneys, which are almost invisible in conventional X rays (Figure 20.22b). Some properties of other radioisotopes used for medical imaging are listed in Table 20.7.

Because γ rays produced by isotopes such as ^{131}I and ^{99m}Tc are emitted randomly in all directions, it is impossible to achieve high levels of resolution in images that use such isotopes. However, remarkably detailed three-dimensional images can be obtained using an imaging technique called *positron emission tomography (PET)*. The technique uses radioisotopes that decay by positron emission (Section 20.2), and the resulting positron is annihilated when it collides with an electron in the surrounding matter. In the annihilation process, both particles are converted to energy in the form of two γ rays that are emitted simultaneously and at 180° to each other:

$$ {}^{0}_{+1}\beta + {}^{0}_{-1}e \longrightarrow 2\,{}^{0}_{0}\gamma \quad (20.40)$$

With PET, biological molecules that have been "tagged" with a positron-emitting isotope such as ^{18}F or ^{11}C can be used to probe the functions of organs such as the brain.

Another major health-related use of ionizing radiation is the irradiation of food, an effective way to kill bacteria such as *Salmonella* in chicken and eggs and potentially lethal strains of *E. coli* in beef. Collectively, such organisms cause almost 3 million cases of food poisoning annually in the United States, resulting in hundreds of deaths. Figure 20.23 shows how irradiation dramatically extends the storage life of foods such as strawberries. Although U.S. health authorities have given only limited approval of this technique, the growing number of illnesses caused by antibiotic-resistant bacteria is increasing the pressure to expand the scope of food irradiation.

Figure 20.23 Preservation of strawberries using ionizing radiation. Fruits such as strawberries can be irradiated by high-energy γ rays to kill bacteria and prolong their storage life. The nonirradiated strawberries on the left are completely spoiled after 15 days in storage, but the irradiated strawberries on the right show no visible signs of spoilage under the same conditions.

One of the more unusual effects of radioisotopes is in dentistry. Because dental enamels contain a mineral called feldspar, $KAlSi_3O_8$ (which is also found in granite rocks), teeth contain a small amount of naturally occurring radioactive ^{40}K. The radiation caused by the decay of ^{40}K results in the emission of light (*fluorescence*), which gives the highly desired "pearly white" appearance associated with healthy teeth.

In addition to the medical uses of radioisotopes, radioisotopes have literally hundreds of other uses. For example, smoke detectors contain a tiny amount of ^{241}Am, which ionizes the air in the detector so an electric current can flow through it. Smoke particles reduce the number of ionized particles and decrease the electric current, which triggers an alarm. Another application is the "go-devil" used to detect leaks in long pipelines. It is a packaged radiation detector that is inserted into a pipeline and propelled through the pipe by the flowing liquid. Sources of ^{60}Co are attached to the pipe at regular intervals; as the detector travels along the pipeline, it sends a radio signal each time it passes a source. When a massive leak causes the go-devil to stop, the repair crews know immediately which section of the pipeline is damaged. Finally, radioactive substances are used in gauges that measure and control the thickness of sheets and films. As shown in Figure 20.24, thickness gauges rely on the absorption of either β particles (by paper, plastic, and very thin metal foils) or γ rays (for thicker metal sheets); the amount of radiation absorbed can be measured accurately and is directly proportional to the thickness of the material.

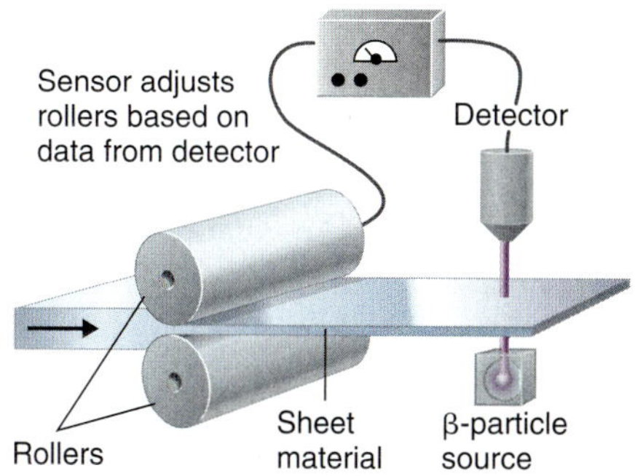

Figure 20.24 Using radiation to control the thickness of a material. Because the amount of radiation absorbed by a material is proportional to its thickness, radiation can be used to control the thickness of plastic film, tin foil, or paper. As shown, a beta emitter is placed on one side of the material being produced and a detector on the other. An increase in the amount of radiation that reaches the detector indicates a decrease in the thickness of the material, and vice versa. The output of the detector can thus be used to control the thickness of the material.

20.6 The Origin of the Elements

The relative abundances of the elements in the universe vary by more than 12 orders of magnitude. For the most part, these differences in abundance cannot be explained by differences in nuclear stability. Although the ^{56}Fe nucleus is the most stable nucleus known, the most abundant element in the universe is not iron, but hydrogen (^1H), which accounts for about 90% of all atoms. In fact, ^1H is the raw material from which all other elements are formed.

In this section, we explain why ^1H and ^2He together account for at least 99% of all the atoms in the universe. We also describe the nuclear reactions that take place in stars, which transform one nucleus into another and create all the naturally occurring elements.

Relative Abundances of the Elements on Earth and in the Universe

The relative abundances of the elements in the universe and on Earth relative to silicon are shown in Figure 20.25. The data are estimates based on the characteristic emission spectra of the elements in stars, the absorption spectra of matter in clouds of interstellar dust, and the approximate composition of the Earth as measured by geologists. The data in Figure 20.25 illustrate two important points. First, except for hydrogen, the most abundant elements have *even* atomic numbers. Not only is this consistent with the trends in nuclear stability discussed in Section 20.1, but it also suggests that heavier elements are formed by combining helium nuclei (Z = 2). Second, the relative

Figure 20.25 The relative abundances of the elements in the universe and on Earth. In this logarithmic plot, the relative abundances of the elements relative to that of silicon (arbitrarily set equal to 1) in the universe (green bars) and on Earth (purple bars) are shown as a function of atomic number. Note that elements with even atomic numbers are generally more abundant in the universe than elements with odd atomic numbers. Note also that the relative abundances of many elements in the universe are *very* different from their relative abundances on Earth.

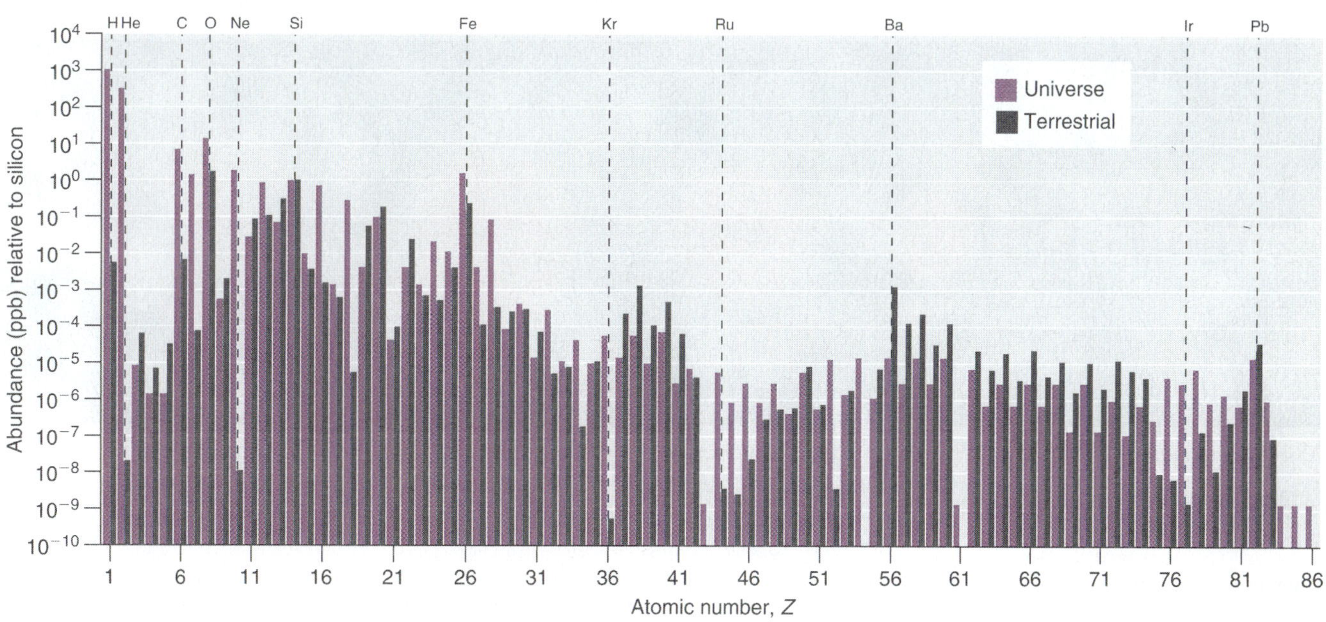

TABLE 20.8 Relative abundances of elements on Earth and in the Universe

Element	Terrestrial/Universal Abundance Ratio
H	0.0020
He	2.4×10^{-8}
C	0.36
N	0.02
O	46
Ne	1.9×10^{-6}
Na	1200
Mg	48
Al	1600
Si	390
S	0.84
K	5000
Ca	710
Ti	2200
Fe	57

abundances of the elements in the universe and on Earth are often very different, as indicated by the data in Table 20.8 for some common elements. Some of these differences are easily explained. For example, nonmetals such as H, He, C, N, O, Ne, and Kr are much less abundant relative to silicon on Earth than they are in the rest of the universe. These elements are either noble gases (He, Ne, Kr) or elements that form volatile hydrides, such as NH_3, CH_4, and H_2O. Because Earth's gravity is not strong enough to hold such light substances in the atmosphere, these elements have been slowly diffusing into outer space ever since our planet was formed. Argon (Ar) is an exception; it is relatively abundant on Earth compared with the other noble gases because it is continuously produced in rocks by the radioactive decay of isotopes such as ^{40}K. In contrast, many metals, such as Al, Na, Fe, Ca, Mg, K, and Ti, are relatively abundant on Earth because they form nonvolatile compounds such as oxides that cannot escape into space. Other metals, however, are much *less* abundant on Earth than in the universe; examples are Ru and Ir. You may recall from Chapter 1 that the anomalously high iridium content of a 66-million-year-old rock layer was a key finding in the development of the asteroid-impact theory for the extinction of the dinosaurs. This section explains some of the reasons for the great differences in abundances of the metallic elements.

All the elements originally present on Earth (and on other planets) were synthesized from hydrogen and helium nuclei in the interiors of stars that have long since exploded and disappeared. Six of the most abundant elements in the universe (carbon, oxygen, neon, magnesium, silicon, and iron) have nuclei that are integral multiples of the helium-4 nucleus, which suggests that helium-4 is the primary building block for heavier nuclei.

Synthesis of the Elements in Stars

Elements are synthesized in discrete stages during the lifetime of a star, and some steps occur only in the most massive stars known (Figure 20.26). Initially, all stars are formed by the aggregation of interstellar "dust," which is mostly hydrogen. As the cloud of dust slowly contracts due to gravitational attraction, its density eventually

Figure 20.26 Nuclear reactions during the life cycle of a massive star. At each stage in the lifetime of a star, a different fuel is used for nuclear fusion, resulting in the formation of different elements. Fusion of hydrogen to give helium is the primary fusion reaction in young stars. As the star ages, helium accumulates and begins to "burn," undergoing fusion to form heavier elements such as carbon and oxygen. As the adolescent star matures, significant amounts of iron and nickel are formed by fusion of the heavier elements formed previously. The heaviest elements are formed only during the final death throes of the star, formation of a nova or supernova.

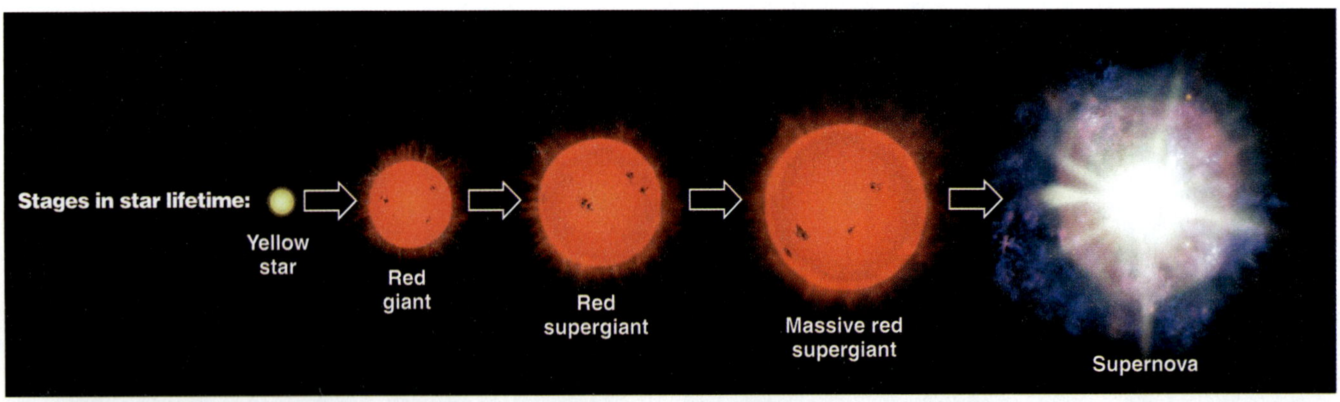

	Yellow star	Red giant	Red supergiant	Massive red supergiant	Supernova
Core Temperature:	1.5×10^7 K	2×10^8 K	7×10^8 K	3×10^9 K	1×10^{11} K
Primary Nuclear Reaction:	^1H fusion	^4He fusion	$^4He + {}^{12}C$ $^{12}C + {}^{12}C$ $^{12}C + {}^{16}O$	Proton–neutron exchange reactions	Multiple neutron captures
Elements Formed:	He	C, O, Ne, Mg	Na, Si, S, Ar, Ca	Fe, Ni	Elements with $Z > 28$

reaches about 100 g/cm^3 and the temperature increases to about 1.5×10^7 K, forming a dense plasma of ionized hydrogen nuclei. At this point, self-sustaining nuclear reactions begin, and the star "ignites," creating a *yellow star* like our sun.

In the first stage of its life, the star is powered by a series of nuclear fusion reactions that convert hydrogen to helium:

$$\begin{aligned} {}^1_1\text{H} + {}^1_1\text{H} &\longrightarrow {}^2_1\text{H} + {}^{\ 0}_{+1}\beta \\ {}^2_1\text{H} + {}^1_1\text{H} &\longrightarrow {}^3_2\text{He} + {}^0_0\gamma \\ {}^3_2\text{He} + {}^3_2\text{He} &\longrightarrow {}^4_2\text{He} + 2\,{}^1_1\text{H} \end{aligned} \tag{20.41}$$

The overall reaction is the conversion of four hydrogen nuclei to a helium-4 nucleus, which is accompanied by the release of two positrons, two γ rays, and a great deal of energy:

$$4\,{}^1_1\text{H} \longrightarrow {}^4_2\text{He} + 2\,{}^{\ 0}_{+1}\beta + 2\,{}^0_0\gamma \tag{20.42}$$

These reactions are responsible for most of the enormous amount of energy that is released as sunlight and solar heat. It takes several billion years, depending on the size of the star, to convert about 10% of the hydrogen to helium.

Once large amounts of helium-4 have been formed, they become concentrated in the core of the star, which slowly becomes denser and hotter. At a temperature of about 2×10^8 K, the helium-4 nuclei begin to fuse, producing beryllium-8:

$$2\,{}^4_2\text{He} \longrightarrow {}^8_4\text{Be} \tag{20.43}$$

Although beryllium-8 has both an even mass number and an even atomic number, the low neutron-to-proton ratio makes it *very* unstable, decomposing in only about 10^{-16} s. Nonetheless, this is long enough for it to react with a third helium-4 nucleus to form carbon-12, which is very stable. Sequential reactions of carbon-12 with helium-4 produce the elements with even numbers of protons and neutrons up to magnesium-24:

$${}^8_4\text{Be} \xrightarrow{{}^4_2\text{He}} {}^{12}_6\text{C} \xrightarrow{{}^4_2\text{He}} {}^{16}_8\text{O} \xrightarrow{{}^4_2\text{He}} {}^{20}_{10}\text{Ne} \xrightarrow{{}^4_2\text{He}} {}^{24}_{12}\text{Mg} \tag{20.44}$$

So much energy is released by these reactions that it causes the surrounding mass of hydrogen to expand, producing a *red giant* that is about 100 times larger than the original yellow star.

As the star expands, the heavier nuclei accumulate in its core, which contracts further to a density of about 50,000 g/cm^3 and becomes yet hotter. At a temperature of about 7×10^8 K, carbon and oxygen nuclei undergo nuclear fusion reactions to produce sodium and silicon nuclei:

$${}^{12}_6\text{C} + {}^{12}_6\text{C} \longrightarrow {}^{23}_{11}\text{Na} + {}^1_1\text{H} \tag{20.45}$$

$${}^{12}_6\text{C} + {}^{16}_8\text{O} \longrightarrow {}^{28}_{14}\text{Si} + {}^0_0\gamma \tag{20.46}$$

At these temperatures, carbon-12 reacts with helium-4 to initiate a series of reactions that produce more oxygen-16, neon-20, magnesium-24, and silicon-28, as well as heavier nuclides such as sulfur-32, argon-36, and calcium-40:

$${}^{12}_6\text{C} \xrightarrow{{}^4_2\text{He}} {}^{16}_8\text{O} \xrightarrow{{}^4_2\text{He}} {}^{20}_{10}\text{Ne} \xrightarrow{{}^4_2\text{He}} {}^{24}_{12}\text{Mg} \xrightarrow{{}^4_2\text{He}} {}^{28}_{14}\text{Si} \xrightarrow{{}^4_2\text{He}} {}^{32}_{16}\text{S} \xrightarrow{{}^4_2\text{He}} {}^{36}_{18}\text{Ar} \xrightarrow{{}^4_2\text{He}} {}^{40}_{20}\text{Ca} \tag{20.47}$$

The energy released by these reactions causes a further expansion of the star to form a *red supergiant*, and the core temperature increases steadily. At a temperature of about 3×10^9 K, the nuclei that have been formed exchange protons and neutrons freely. This equilibration process forms heavier elements up to iron-56 and nickel-58, which have the most stable nuclei known.

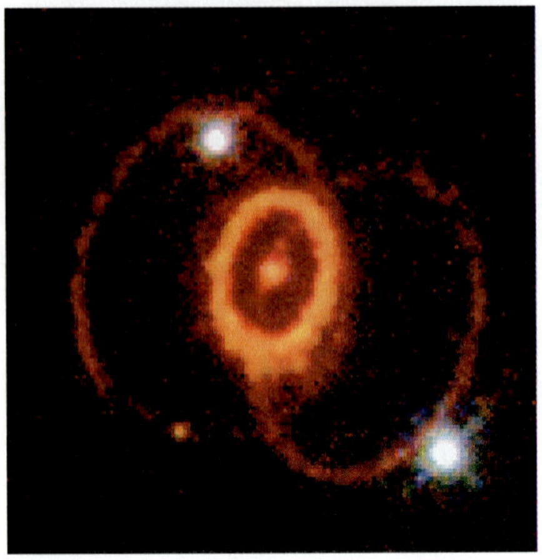

Figure 20.27 **A supernova.** A view of the remains of Supernova 1987A, showing the circular halo of expanding debris produced by the explosion. Multiple neutron-capture events occur during a supernova explosion, forming both the heaviest elements and many of the less stable nuclides.

Formation of Heavier Elements in Supernovas

None of the processes described so far produces nuclei with $Z > 28$. All naturally occurring elements heavier than nickel are formed in the rare but spectacular cataclysmic explosions called *supernovas* (Figure 20.26). When the fuel in the core of a very massive star has been consumed, its gravity causes it to collapse in about 1 s. As the core is compressed, the iron and nickel nuclei within it disintegrate to protons and neutrons, and many of the protons capture electrons to form neutrons. The resulting *neutron star* is a dark object that is so dense that atoms no longer exist. Simultaneously, the energy released by the collapse of the core causes the supernova to explode in what is arguably the single most violent event in the universe. The force of the explosion blows most of the star's matter into space, creating a gigantic and rapidly expanding dust cloud, a *nebula* (Figure 20.27). During the extraordinarily short duration of this event, the concentration of neutrons is so great that multiple neutron-capture events occur, leading to the production of the heaviest elements and many of the less stable nuclides. Under these conditions, for example, an iron-56 nucleus can absorb as many as 64 neutrons, briefly forming an extraordinarily unstable iron isotope that can then undergo multiple rapid β-decay processes to produce tin-120:

$$^{56}_{26}\text{Fe} + 64\,^{1}_{0}\text{n} \longrightarrow \left[^{120}_{26}\text{Fe}\right] \longrightarrow {}^{120}_{50}\text{Sn} + 24\,^{0}_{-1}\beta \qquad (20.48)$$

Although a supernova occurs only every few hundred years in a galaxy such as the Milky Way, these rare explosions provide the only conditions under which elements heavier than nickel can be formed. The force of the explosions distributes these elements throughout the galaxy surrounding the supernova, and eventually they are captured in the dust that condenses to form new stars. Based on its elemental composition, our sun is thought to be a second- or third-generation star, which contains a considerable amount of cosmic debris from the explosion of supernovas in the remote past.

EXAMPLE 20.10

The reaction of two carbon-12 nuclei in a carbon-burning star can produce elements other than sodium. Write balanced equations for the formation of **(a)** magnesium-24 and **(b)** neon-20 from two carbon-12 nuclei.

Given Reactant and product nuclides

Asked for Balanced equation

Strategy

Use conservation of mass and charge to determine the type of nuclear reaction that will convert the reactant to the indicated product. Write the balanced equation for the reaction.

Solution

(a) A magnesium-24 nucleus ($Z = 12$, $A = 24$) has the same nucleons as two carbon-12 nuclei ($Z = 6$, $A = 12$). The reaction is therefore a fusion of two carbon-12 nuclei, and no other particles are produced: $^{12}_{6}\text{C} + {}^{12}_{6}\text{C} \longrightarrow {}^{24}_{12}\text{Mg}$.

(b) The neon-20 product has $Z = 10$ and $A = 20$. Conservation of mass requires that the other product have $A = (2 \times 12) - 20 = 4$; because of conservation of charge, it must have $Z = (2 \times 6) - 10 = 2$. These are the characteristics of an α particle. The reaction is therefore $^{12}_{6}\text{C} + {}^{12}_{6}\text{C} \longrightarrow {}^{20}_{10}\text{Ne} + {}^{4}_{2}\alpha$.

EXERCISE 20.10

How many neutrons must an iron-56 nucleus absorb during a supernova explosion to produce an arsenic-75 nucleus? Write a balanced equation for the reaction.

Answer 19 neutrons; $^{56}_{26}\text{Fe} + 19\ ^{1}_{0}\text{n} \longrightarrow \left[^{75}_{26}\text{Fe}\right] \longrightarrow\ ^{75}_{33}\text{As} + 7\ ^{0}_{-1}\beta$

SUMMARY AND KEY TERMS

20.1 The Components of the Nucleus (p. 913)

Subatomic particles of the nucleus (protons and neutrons) are called **nucleons**. A **nuclide** is an atom with a particular number of protons and neutrons. An unstable nucleus that decays spontaneously is **radioactive** whose emissions are collectively called *radioactivity*. Isotopes that emit radiation are called **radioisotopes**. Each nucleon is attracted to other nucleons by the **strong nuclear force**. Stable nuclei generally have even numbers of both protons and neutrons and a neutron-to-proton ratio of at least 1. Nuclei that contain *magic numbers* of protons and neutrons are often especially stable. **Superheavy elements**, with atomic numbers near 126, may even be stable enough to exist in nature.

20.2 Nuclear Reactions (p. 917)

In **nuclear decay reactions**, the *parent nucleus* is converted to a more stable *daughter nucleus*. Nuclei with too many neutrons decay by converting a neutron to a proton, whereas nuclei with too few neutrons decay by converting a proton to a neutron. Very heavy nuclei (with $A \geq 200$ and $Z > 83$) are unstable and tend to decay by emitting an **α particle**. When an unstable nuclide undergoes radioactive decay, the total number of nucleons is conserved, as is the total positive charge. Six different kinds of nuclear decay reactions are known. **Alpha decay** results in the emission of an α particle, $^{4}_{2}\alpha$, and produces a daughter nucleus with a mass number that is lower by 4 and an atomic number that is lower by 2 than the parent nucleus. **Beta decay** converts a neutron to a proton and emits a high-energy electron, producing a daughter nucleus with the same mass number as the parent and an atomic number that is higher by 1. **Positron emission** is the opposite of beta decay, and converts a proton to a neutron plus a positron. Positron emission does not change the mass number of the nucleus, but the atomic number of the daughter nucleus is lower by 1 than the parent. In **electron capture**, an electron in an inner shell reacts with a proton to produce a neutron, with emission of an X ray. The mass number does not change, but the atomic number of the daughter is lower by 1 than the parent. In **gamma emission**, a daughter nucleus in a *nuclear excited state* undergoes a transition to a lower-energy state by emitting a γ ray. Very heavy nuclei with high neutron-to-proton ratios can undergo **spontaneous fission**, in which the nucleus breaks into two pieces that can have different atomic numbers and atomic masses with the release of neutrons. Many very heavy nuclei decay via a **radioactive decay series**—a succession of some combination of alpha- and beta-decay reactions. In **nuclear transmutation reactions**, a target nucleus is bombarded with energetic subatomic particles to give a product nucleus more massive than the original. All the **transuranium elements**—elements with $Z > 92$—are artificial and must be prepared by nuclear transmutation reactions. These reactions are carried out in *particle accelerators* such as *linear accelerators*, *cyclotrons*, and *synchrotrons*.

20.3 The Interaction of Nuclear Radiation with Matter (p. 927)

The effects of radiation on matter depend on the energy of the radiation. **Nonionizing radiation** is relatively low in energy, and the energy is transferred to matter in the form of heat. **Ionizing radiation** is relatively high in energy, and when it collides with an atom, it can completely remove an electron to form a positively charged ion that can damage biological tissues. Alpha particles do not penetrate very far into matter, whereas γ rays penetrate more deeply. Common units of radiation exposure or dose are the **roentgen (R)**, the amount of energy absorbed by dry air, and the **rad**, the amount of radiation that produces 0.01 J of energy in 1 kg of matter. The **rem** measures the actual amount of tissue damage caused by a given amount of radiation. Natural sources of radiation include *cosmic radiation*, consisting of high-energy particles and γ rays emitted by the sun and other stars; *cosmogenic radiation*, which is produced by the interaction of cosmic rays with gases in the upper atmosphere; and *terrestrial radiation*, from radioactive elements present on the primordial Earth and their decay products. The risks of ionizing radiation depend on the intensity of the radiation, the mode of exposure, and the duration of the exposure.

20.4 Thermodynamic Stability of the Atomic Nucleus (p. 932)

Nuclear reactions are accompanied by large changes in energy, which result in detectable changes in mass. The change in mass is related to the change in energy using Einstein's equation, $\Delta E = (\Delta m)c^2$. Large changes in energy are usually reported in units of keV or MeV (thousands or millions of electronvolts). With the exception of ^{1}H, the experimentally determined mass of an atom is always *less* than the sum of the masses of the component particles (protons, neutrons, and electrons) by an amount called the **mass defect** of the nucleus. The energy corresponding to the mass defect is the **nuclear binding energy**, the amount of energy released when a nucleus forms from its component particles. In **nuclear fission**, nuclei split into lighter nuclei with an accompanying release of multiple neutrons and large amounts of energy. The **critical mass** is the minimum mass required to support a self-sustaining **nuclear chain reaction**. **Nuclear fusion** is a process in which two light nuclei combine to produce a heavier nucleus plus a great deal of energy.

20.5 Applied Nuclear Chemistry (p. 939)

In nuclear power plants, nuclear reactions are used to generate electricity. *Light-water reactors* use enriched uranium as a fuel. They include fuel rods, a moderator, control rods, and a powerful cooling system to absorb the heat generated in the reactor core. *Heavy-water reactors* are able to use unenriched uranium as a fuel because they use D_2O as the moderator, which scatters and slows neutrons much more effectively than H_2O. A *breeder reactor* produces more fissionable fuel than it consumes. A *nuclear fusion reactor* requires very high temperatures. Fusion reactions are **thermonuclear reactions** because they require high temperatures for initiation. Radioisotopes are used in both radiation therapy and *medical imaging*.

20.6 The Origin of the Elements (p. 945)

By far the most abundant element in the universe is hydrogen. The fusion of hydrogen nuclei to form helium nuclei is the major process that fuels young stars such as the sun. Elements heavier than helium are formed from hydrogen and helium in the interiors of stars. Successive fusion reactions of helium nuclei at higher temperatures create elements with even numbers of protons and neutrons up to magnesium, then up to calcium. Eventually, the elements up to iron-56 and nickel-58 are formed by exchange processes at even higher temperatures. Heavier elements can only be made by a process that involves multiple neutron-capture events, which can occur only during the explosion of a supernova.

KEY EQUATIONS

Alpha decay	${}^{A}_{Z}X \longrightarrow {}^{(A-4)}_{(Z-2)}X' + {}^{4}_{2}\alpha$	(20.1)
Beta decay	${}^{A}_{Z}X \longrightarrow {}_{(Z+1)}^{\;\;\;A}X' + {}^{\;\;0}_{-1}\beta$	(20.4)
Positron emission	${}^{A}_{Z}X \longrightarrow {}_{(Z-1)}^{\;\;\;A}X' + {}^{\;\;0}_{+1}\beta$	(20.7)
Electron capture	${}^{A}_{Z}X + {}^{\;\;0}_{-1}e \longrightarrow {}_{(Z-1)}^{\;\;\;A}X' + \text{X ray}$	(20.10)
Gamma emission	${}^{A}_{Z}X^* \longrightarrow {}^{A}_{Z}X + {}^{0}_{0}\gamma$	(20.14b)
Rad	1 rad = 0.01 J/kg	(20.23)

QUESTIONS AND PROBLEMS

MGC For instructor-assigned homework, go to **www.masteringgeneralchemistry.com**

Questions and Problems with colored number have answers in the Appendix and complete solutions in the Student Solutions Manual.

CONCEPTUAL

20.1 The Components of the Nucleus

1. What distinguishes a nuclear reaction from a chemical reaction? Use an example of each to illustrate the differences.
2. What do chemists mean when they say a substance is *radioactive*?
3. What characterizes an isotope? How is the mass of an isotope of an element related to the atomic mass of the element shown in the periodic table?
4. In a typical nucleus, why doesn't electrostatic repulsion between protons destabilize the nucleus? How does the neutron-to-proton ratio affect the stability of an isotope? Why are all isotopes with $Z > 82$ unstable?
5. What is the significance of a *magic number* of protons or neutrons? What is the relationship between the number of stable isotopes of an element and whether the element has a magic number of protons?
6. Do you expect Bi to have a large number of stable isotopes? Does Ca? Explain your answers.
7. Potassium has three common isotopes, ^{39}K, ^{40}K, and ^{41}K, but only potassium-40 is radioactive (a beta emitter). Suggest two reasons for the instability of ^{40}K.
8. Samarium has 11 relatively stable isotopes, but only four are nonradioactive. One of these four is ^{144}Sm, which has a lower neutron-to-proton ratio than lighter, radioactive isotopes of samarium. Why is ^{144}Sm more stable?

20.2 Nuclear Reactions

9. Describe the six classifications of nuclear reactions. What is the most common mode of decay for elements that have heavy nuclei? Why?
10. Complete the table for these five nuclear reactions.

	Alpha Decay	Beta Decay	Gamma Emission	Positron Emission	Electron Capture
Identity of Particle or Radiation	Helium-4 nucleus	_____	_____	_____	_____
Mass Number of Parent − Mass Number of Daughter	4	_____	_____	_____	_____
Atomic Number of Parent − Atomic Number of Daughter	_____	−1	_____	_____	_____
Effect on Neutron-to-Proton Ratio	_____	Decreases	_____	_____	_____

11. What is the most common decay process for elements in row 5 of the periodic table that contain too few neutrons for the number of protons present? Why?
12. Explain the difference between the symbols e^- and β^-. What is the difference in meaning between the symbols ^4_2He and $^4_2\alpha$?
13. What is a mass number? Which particles have a mass number of zero?
14. What are the key differences between the equations written for chemical reactions and for nuclear reactions? How are they similar?
15. Can all the kinds of nuclear decay reactions discussed be characterized by the general equation: parent $\longrightarrow$ daughter + particle? Explain your answer.
16. Which types of nuclear decay reactions conserve both mass number and atomic number? Which conserve mass number but not atomic number? Which do not convert one element to another?
17. Describe a radioactive decay series. How many series occur naturally? Of these, which one no longer occurs in nature? Why?
18. Only nine naturally occurring elements have an atomic number greater than 83, and all of them are radioactive. Except for some isotopes of uranium that have a very long half-life, the half-lives of the heavy elements are so short that these elements should have been completely converted to lighter, more stable elements long ago. Why are these elements still present in nature?
19. Why are neutrons preferred to protons in the preparation of new isotopes of the lighter elements?
20. Why are particle accelerators and cyclotrons needed to create the transuranium elements?

20.3 The Interaction of Nuclear Radiation with Matter

21. Why are many radioactive substances warm to the touch? Why do many radioactive substances glow?
22. Describe the differences between nonionizing and ionizing radiation in terms of the intensity of energy emitted and the effect each has on an atom or molecule after collision. Which nuclear decay reactions are more likely to produce ionizing radiation? Which are more likely to produce nonionizing radiation?
23. Would you expect nonionizing or ionizing radiation to be more effective at treating cancer? Why?
24. Historically, concrete shelters have been used to protect people from nuclear blasts. Comment on the effectiveness of these shelters.
25. Gamma rays are a very high-energy radiation, yet α particles inflict more damage on biological tissue. Why?
26. List the three primary sources of naturally occurring radiation. Explain the factors that influence the dose that one receives throughout the year. Which is the largest contributor to overall exposure? Which is the most hazardous?
27. Because radon is a noble gas, it is inert and generally unreactive. Despite this, exposure to even low concentrations of radon in air is quite dangerous. Describe the physical consequences of exposure to radon gas. Why are people who smoke more susceptible to these effects?
28. Most medical imaging uses isotopes that have extremely short half-lives. These isotopes usually undergo only one kind of nuclear decay reactions. Which kind of decay reaction is usually utilized, and why? Why would a short half-life be preferred in these cases?
29. Which would you prefer: one exposure of 100 rem, or ten exposures of 10 rem each? Explain your rationale.

20.4 Thermodynamic Stability of the Atomic Nucleus

30. How do chemical reactions compare with nuclear reactions with respect to mass changes? Does either type of reaction violate the law of conservation of mass? Explain your answers.
31. Why is the amount of energy released by a nuclear reaction so much greater than the amount of energy released by a chemical reaction?
32. Explain why the mass of an atom is less than the sum of the masses of its component particles.
33. The stability of a nucleus can be described using two values. What are they, and how do they differ from each other?
34. In the days before true chemistry, ancient scholars (*alchemists*) attempted to find the *philosopher's stone*, a material that would enable them to turn lead into gold. Is the conversion of Pb $\longrightarrow$ Au energetically favorable? Explain why or why not.
35. Describe the energy barrier to nuclear fusion reactions, and explain how it can be overcome.
36. Imagine that the universe is dying, the stars have burned out, and all the elements have undergone fusion or radioactive decay. What would be the most abundant element in this future universe, and why?
37. Numerous elements can undergo fission, but only a few can be utilized as fuels in a reactor. What aspect of nuclear fission allows a nuclear chain reaction to occur?
38. How are transmutation reactions and fusion reactions related? Describe the main impediment to fusion reactions, and suggest one or two ways to surmount this difficulty.

20.5 Applied Nuclear Chemistry

39. In nuclear reactors, two different but interrelated factors must be controlled to prevent a mishap that could cause the release of unwanted radiation. How are these factors controlled?
40. What are the three principal components of a nuclear reactor? What is the function of each component?
41. What is meant by the term *enrichment* with regard to uranium for fission reactors? Why does the fuel in a conventional nuclear reactor have to be "enriched"?
42. The plot in a recent spy/action movie involved the threat of introducing stolen "weapons-grade" uranium, which consists of 93.3% ^{235}U, into a fission reactor that normally uses a fuel containing about 3% ^{235}U. Explain why this could be catastrophic.
43. Compare a heavy-water reactor with a light-water reactor. Why are heavy-water reactors less widely used? How do these two reactor designs compare with a breeder reactor?
44. Conventional light-water fission reactors require enriched fuel. An alternative reactor is the so-called heavy-water reactor. The components of the two different reactors are the same except that instead of using water, H_2O, the moderator in a heavy-water reactor is D_2O, "heavy water." Because D_2O is more efficient than H_2O at slowing neutrons, heavy-water reactors do not require that the fuel be enriched to support fission. Why is D_2O more effective at slowing neutrons, and why does this allow unenriched fuels to be used?
45. Isotopes emit γ rays in random directions. What difficulties does this present for medical imaging, and how are these difficulties overcome?
46. If you needed to measure the thickness of 1.0-mm plastic sheets, what type of radiation would you use? Would the radiation source be the same if you were measuring steel of a similar thickness? What is your rationale? Would you want an isotope with a long or short half-life for this device?

20.6 The Origin of the Elements

47. Why do scientists believe that hydrogen is the building block of all the other elements? Why do scientists believe that helium-4 is the building block of the heavier elements?

48. How does a star produce such enormous amounts of heat and light? How are elements heavier than Ni formed?

49. Propose an explanation for the observation that elements with even atomic numbers are more abundant than elements with odd atomic numbers.

50. During the lifetime of a star, different reactions that form different elements are utilized to power the fusion furnace that keeps the star "lit." Explain the different reactions that dominate in the different stages of a star's life cycle and their effect on the temperature of the star.

51. A line in a popular song from the 1960's by Joni Mitchell stated: "We are stardust. . .". Does this statement have any merit, or is it just poetic? Justify your answer.

52. If the laws of physics were different and the primary element in the universe were boron-11 ($Z = 5$), what would be the next four most abundant elements? Propose nuclear reactions for their formation.

NUMERICAL

This section includes paired problems (marked by brackets) that require similar problem-solving skills.

20.1 The Components of the Nucleus

53. Use ^A_ZX notation to write a nuclear symbol for each isotope: (a) chlorine-39; (b) lithium-8; (c) osmium-183; (d) zinc-71.

54. Write the nuclear symbol for each isotope using ^A_ZX notation: (a) lead-212; (b) helium-5; (c) oxygen-19; (d) plutonium-242.

55. Give the number of protons, the number of neutrons, and the neutron-to-proton ratio for each isotope: (a) iron-57; (b) ^{185}W; (c) potassium-39; (d) ^{131}Xe.

56. Give the number of protons, the number of neutrons, and the neutron-to-proton ratio for each isotope: (a) technetium-99*m*; (b) ^{140}La; (c) radium-227; (d) ^{208}Bi.

57. Which of these nuclides do you expect to be radioactive? Explain your reasoning. (a) ^{20}Ne; (b) tungsten-184; (c) ^{106}Ti

58. Which of these nuclides do you expect to be radioactive? Explain your reasoning. (a) ^{107}Ag; (b) ^{50}V; (c) lutetium-176

20.2 Nuclear Reactions

59. What type of particle is emitted in each nuclear reaction? (a) $^{218}\text{Po} \longrightarrow {}^{214}\text{Po}$; (b) $^{32}\text{Si} \longrightarrow {}^{32}\text{P}$; (c) $^{18}\text{F} \longrightarrow {}^{18}\text{O}$; (d) $^{206}\text{Tl} \longrightarrow {}^{206}\text{Pb}$

60. Identify the type of particle emitted in each nuclear reaction: (a) $^{230}\text{Th} \longrightarrow {}^{226}\text{Ra}$; (b) $^{218}\text{Po} \longrightarrow {}^{218}\text{At}$; (c) $^{210}\text{Bi} \longrightarrow {}^{206}\text{Tl}$; (d) $^{36}\text{Cl} \longrightarrow {}^{36}\text{S}$.

61. Predict the mode of decay and write a balanced nuclear reaction for each of the following: (a) ^{235}U; (b) ^{254}Es; (c) ^{36}S; (d) ^{99}Mo.

62. Predict the mode of decay and write a balanced nuclear reaction for each of the following: (a) ^{13}N; (b) ^{231}Pa; (c) ^{7}Be; (d) ^{77}Ge.

63. Balance each nuclear reaction:
(a) $^{208}\text{Po} \longrightarrow \alpha + \text{Pb}$
(b) $^{226}\text{Ra} \longrightarrow \alpha + \text{Rn}$
(c) $^{228}\text{Th} \longrightarrow \text{Ra} + \alpha + \gamma$
(d) $^{231}\text{Pa} \longrightarrow \text{Ac} + \alpha + \gamma$
(e) $\text{Ho} \longrightarrow {}^{166}\text{Er} + \beta^- + \gamma$
(f) $\text{Ac} \longrightarrow {}^{223}\text{Th} + 3\beta^- + \gamma + \alpha$

64. Complete the nuclear reactions:
(a) $^{210}_{84}\text{Po} \longrightarrow {}^{206}\text{Pb}$
(b) $^{217}_{85}\text{At} \longrightarrow \text{Bi} + \alpha$
(c) $\text{Ra} \longrightarrow {}^{220}_{86}\text{Rn} + \alpha$
(d) $^{208}\text{Tl} \longrightarrow {}_{82}\text{Pb} + \beta^-$
(e) $\text{Np} \longrightarrow {}^{239}\text{Pu} + \beta^-$
(f) $\text{Fe} \longrightarrow {}^{52}\text{Mn} + \beta^+ + \gamma$

65. Write a balanced equation for each nuclear reaction: (a) β^- decay of ^{87}Rb; (b) β^+ decay of ^{20}Mg; (c) α decay of ^{268}Mt.

66. Write a balanced equation for each nuclear reaction: (a) β^- decay of ^{45}K; (b) β^+ decay of ^{41}Sc; (c) α decay of ^{146}Sm.

67. The decay products of several isotopes are listed here. Identify the type of radiation emitted, and write a balanced nuclear equation for each. (a) $^{218}\text{Po} \longrightarrow {}^{214}\text{Pb}$; (b) $^{32}\text{Si} \longrightarrow {}^{32}\text{P}$; (c) an excited state of an iron-57 nucleus decaying to its ground state; (d) conversion of thallium-204 to lead-204.

68. The decay products of several isotopes are listed here. Identify the type of radiation emitted, and write a balanced nuclear equation for each: (a) $^{218}\text{Po} \longrightarrow {}^{218}\text{At}$; (b) $^{230}\text{Th} \longrightarrow {}^{226}\text{Ra}$; (c) bismuth-211 converted to thallium-207; (d) americium-242 converted to rhodium-107 with the emission of four neutrons.

69. Predict the most likely mode of decay and write a balanced nuclear reaction for each isotope: (a) ^{238}U; (b) ^{208}Po; (c) ^{40}S; (d) molybdenum-93*m*.

70. Predict the most likely mode of decay and write a balanced nuclear reaction for each isotope: (a) ^{16}N; (b) ^{231}Pa; (c) ^{118}In; (d) ^{77}Ge.

71. For each nuclear reaction, identify the type(s) of decay, and write a balanced nuclear equation:
(a) $^{216}\text{Po} \longrightarrow ? + \text{At}$
(b) $? \longrightarrow \alpha + {}^{231}\text{Pa}$
(c) $^{228}\text{Th} \longrightarrow ? + \alpha + \gamma$
(d) $^{231}\text{Pa} \longrightarrow ? + \beta^- + \gamma$

72. For each nuclear reaction, identify the type(s) of decay, and write a balanced nuclear reaction sequence for each decay:
(a) $^{210}\text{Po} \longrightarrow {}^{206}\text{Pb} + ?$
(b) $^{192}\text{Ir} \longrightarrow \text{Pt} + ?$
(c) $^{241}\text{Am} \longrightarrow {}^{57}\text{Fe} + {}^{184}? + ?$
(d) $\text{Ge} \longrightarrow {}^{77}\text{Ge} + ?$

73. Identify the parent isotope, and write a balanced nuclear reaction for each of the following: (a) lead-205 is formed via an alpha emission; (b) titanium-46 is formed via beta and gamma emission; (c) krypton-36 is formed via a beta decay and a gamma emission.

74. Identify the parent isotope, and write a balanced nuclear reaction for each of the following: (a) iodine-130 formed by ejection of an electron and a gamma ray from a nucleus; (b) uranium-240 formed by alpha decay; (c) curium-247 formed with the release of a helium dication and a gamma ray.

75. Write a balanced nuclear equation for each:
(a) Bromine undergoes a decay and produces a gas with an atomic mass of 81 amu.

(b) An element decays into two metals while emitting two neutrons, each of which can be extracted and converted to chlorides with the formula MCl_2. The masses of the two salts are 162.9 and 210.9 g/mol, respectively.

76. Write a balanced nuclear equation for each case:
(a) An unknown element emits γ rays plus particles that are readily blocked by paper. The sample also contains a substantial quantity of indium-115.
(b) An unstable element undergoes two different decay reactions: beta decay to produce a material with a mass of 222 amu, and alpha decay to astatine.

77. Bombarding ^{249}Cf with ^{12}C produced a transuranium element with a mass of 257 amu, plus several neutral subatomic particles. Identify the element, and write a nuclear reaction for this transmutation.

78. One transuranium element, ^{253}Es, is prepared by bombarding ^{238}U with 15 neutrons. What is the other product of this reaction? Write a balanced transmutation reaction for this conversion.

79. Complete this radioactive decay series:

$$^{223}_{88}\text{Ra} \xrightarrow{\alpha} \text{Rn} \xrightarrow{\alpha} \text{Po} \xrightarrow{\alpha} \text{Pb} \xrightarrow{\beta^-} \text{Bi} \xrightarrow{\alpha} \text{Tl} \xrightarrow{\beta^-} \text{Pb}$$

80. Complete each nuclear fission reaction:
(a) $^{235}_{92}\text{U} + ^{1}_{0}\text{n} \longrightarrow ^{90}_{36}\text{Kr} + \_\_ + 2^{1}_{0}\text{n}$
(b) $\_\_ + ^{1}_{0}\text{n} \longrightarrow ^{140}\text{Cs} + ^{96}\text{Y} + 4^{1}_{0}\text{n}$

81. Complete the nuclear fission reactions:
(a) $^{235}_{92}\text{U} + ^{1}_{0}\text{n} \longrightarrow ^{145}_{57}\text{La} + \_\_ + 3^{1}_{0}\text{n}$
(b) $\_\_ + ^{1}_{0}\text{n} \longrightarrow ^{95}_{42}\text{Mo} + ^{139}_{57}\text{La} + 2^{1}_{0}\text{n} + 7^{\ 0}_{-1}\text{e}$

82. A stable nuclide absorbs a neutron, emits an electron, and then splits into two α particles. Identify the nuclide.

83. Using ^{18}O, how would you synthesize an element with atomic number 106 from ^{249}Cf? Write a balanced equation for the reaction.

84. How would you synthesize an element with atomic number 103 using ^{10}B and ^{252}Cf? Write a balanced equation for the reaction.

20.3 The Interaction of Nuclear Radiation with Matter

85. A 2.14-kg sample of rock contains 0.0985 g of uranium. How much energy is emitted over 25 years if 99.27% of the uranium is ^{238}U, which has a half-life of 4.46×10^9 years, if each decay event is accompanied by the release of 4.039 MeV? If a 180-lb individual absorbes all of the emitted radiation, how much radiation has been absorbed in rads?

86. There is a story about a "radioactive boy scout" who attempted to convert thorium-232, which he isolated from about a thousand gas lantern mantles, to uranium-233 by bombarding the thorium with neutrons. The neutrons were generated via bombardment of an aluminum target with α particles from the decay of americium-241, which was isolated from a hundred smoke detectors. Write balanced nuclear reactions for these processes. The "radioactive boy scout" spent approximately 2 hours a day with his experiment for 2 years. Assuming that the alpha emission of americium has an energy of 5.24 MeV/particle and that the americium-241 was undergoing 3.5×10^6 decays per second, what was the exposure of the 60.0-kg scout in rad? The intrepid scientist has shown no ill effects from this exposure. Why?

20.4 Thermodynamic Stability of the Atomic Nucleus

This table is provided as a reference.

Isotope	Mass (amu)	Isotope	Mass (amu)
^{8}B	8.024607	^{209}Fr	208.99592
^{40}K	39.963998	^{210}Po	209.982874
^{52}Cr	51.940508	^{212}At	211.990745
^{58}Ni	57.935343	^{214}Pb	213.999797
^{59}Co	58.933195	^{214}Bi	213.998712
^{60}Co	59.933817	^{216}Fr	216.003198
^{60}Ni	59.930786	^{199}Pb	198.972917
^{90}Sr	89.907738	^{222}Rn	222.017578
^{92}Kr	91.926156	^{226}Ra	226.025410
^{141}Ba	140.914411	^{227}Ra	227.029178
^{143}Xe	142.935110	^{228}Ac	228.031021
^{167}Os	166.971550	^{230}Th	230.033134
^{171}Pt	170.981240	^{233}U	233.039635
^{194}Hg	193.965439	^{234}Th	234.043601
^{194}Tl	193.971200	^{234}Pa	234.043308
^{199}Pb	198.972917	^{233}U	233.039635
^{199}Bi	198.977672	^{234}U	234.040952
^{206}Pb	205.974465	^{235}U	235.043930
^{207}Pb	206.975897	^{238}Pa	238.054500
^{208}Pb	207.976652	^{238}U	238.050788
^{208}Bi	207.979742	^{239}Pu	239.052163
^{208}Po	207.981246	^{245}Pu	245.067747

87. Using information provided in Table 20.6 and the table above, complete each reaction and calculate the amount of energy released from each in kJ:
(a) $^{238}\text{Pa} \longrightarrow \_\_ + \beta^-$
(b) $^{209}\text{Fr} \longrightarrow \_\_ + \alpha$
(c) $^{199}\text{Bi} \longrightarrow \_\_ + \beta^+$

88. Using information provided in Table 20.6 and the table above, complete each reaction and calculate the amount of energy released from each in kJ:
(a) $^{194}\text{Tl} \longrightarrow \_\_ + \beta^+$
(b) $^{171}\text{Pt} \longrightarrow \_\_ + \alpha$
(c) $^{214}\text{Pb} \longrightarrow \_\_ + \beta^-$

89. Using information provided in Table 20.6 and the table above, complete each reaction and calculate the amount of energy released from each in kJ/mol:
(a) $^{234}_{91}\text{Pa} \longrightarrow ? + ^{\ 0}_{-1}\beta$
(b) $^{226}_{88}\text{Ra} \longrightarrow ? + ^{4}_{2}\alpha$

90. Using information provided in Table 20.6 and the table above, complete the reactions and then calculate the amount of energy released from each in kJ/mol:
(a) $^{60}_{27}\text{Co} \longrightarrow ? + ^{\ 0}_{-1}\beta$ (The mass of cobalt-60 is 59.933817 amu.)
(b) technicium-94 (mass = 93.909657) undergoing spontaneous fission to produce chromium-52 and potassium-40

91. Using information provided in Table 20.6 and the table above, predict whether each reaction is favorable and the amount of energy released or required in MeV and kJ/mol:
(a) the beta decay of bismuth-208 (mass = 207.979742 amu)
(b) the formation of lead-206 by alpha decay

92. Using information provided, predict whether each reaction is favorable and the amount of energy released or required in MeV and kJ/mol:
(a) the alpha decay of oxygen-16
(b) alpha decay to produce chromium-52

93. Calculate the total nuclear binding energy (in MeV) and the binding energy per nucleon for ^{87}Sr if the measured mass of ^{87}Sr is 86.908877 amu.

94. Calculate the total nuclear binding energy (in MeV) and the binding energy per nucleon for ^{60}Ni and ^{87}Sr.

95. The experimentally determined mass of ^{53}Mn is 52.941290 amu. What is (a) the calculated mass; (b) the mass defect; (c) the nuclear binding energy; and (d) the nuclear binding energy per nucleon?

96. The experimental mass of ^{29}S is 28.996610 amu. What is (a) the calculated mass; (b) the mass defect; (c) the nuclear binding energy; and (d) the nuclear binding energy per nucleon?

97. Calculate the amount of energy that is released by the neutron-induced fission of ^{235}U to give ^{141}Ba, ^{92}Kr (mass = 91.926156 amu), and three neutrons. Report your answer in eV/atom and kJ/mol.

98. Calculate the amount of energy that is released by the neutron-induced fission of ^{235}U to give ^{90}Sr, ^{143}Xe, and three neutrons. Report your answer in eV/atom and kJ/mol.

99. Calculate the amount of energy that is released or required by the fusion of helium-4 to produce the unstable beryllium-8 (mass = 8.00530510 amu). Report your answer in kJ/mol. Do you expect this to be a spontaneous reaction?

100. Calculate the amount of energy released by the fusion of ^{6}Li and deuterium to give two helium-4 isotopes. Express your answer in eV/atom and kJ/mol.

101. How much energy is released by the fusion of two deuterium nuclei to give one tritium nucleus and one proton? How does this amount compare with the energy released by the fusion of a deuterium nucleus and a tritium nucleus, which is accompanied by ejection of a neutron? Express your answer in MeV and kJ/mol. Pound for pound, which is a better choice for a fusion reactor fuel mixture?

20.5 Applied Nuclear Chemistry

102. Palladium-103, which decays via electron capture and emits X rays with an energy of 3.97×10^{-2} MeV, is often used to treat prostate cancer. Small pellets of the radioactive metal are embedded *in* the prostate gland. This provides a localized source of radiation to a very small area, even though only about 1% of the X rays are absorbed by the tissue. Due to its short half-life, all of the palladium will decay to a stable isotope in less than a year. If a doctor embeds 50 pellets containing 2.50 mg of ^{103}Pd in the prostate gland of a 73.9-kg patient, what is his radiation exposure over the course of a year?

103. A number of medical treatments utilize cobalt-60*m*, which is formed by bombardment of cobalt with neutrons to produce a highly radioactive *gamma* emitter that undergoes 4.23×10^{16} emissions/s per kg of pure cobalt-60. The energy of the gamma emission is 5.86×10^{-2} MeV. Write the balanced nuclear equation for the formation of this isotope. Is this a transmutation reaction? If a 55.3-kg patient received a 0.50-s exposure to a 0.30-kg cobalt-60 source, what would the exposure be in rads? Predict potential side effects of this dose.

20.6 The Origin of the Elements

104. Write a balanced nuclear reaction for the formation of ^{26}Al from two ^{12}C nuclei and of ^{9}Be from two ^{4}He nuclei.

105. At the end of a star's life cycle, it can collapse, resulting in a supernova explosion that leads to the formation of heavy elements by multiple neutron-capture events. Write reactions for the formation of (a) ^{106}Pd from nickel-58 and (b) selenium-79 from iron-56 during such an explosion.

106. When a star reaches middle age, helium-4 is converted to short-lived beryllium-8, which reacts with another helium-4 to produce carbon. How much energy is released in each of these two reactions (in MeV)? How many atoms of helium must be "burned" in this process to produce the same amount of energy obtained from the fusion of 1 mol of hydrogen atoms to give deuterium?

APPLICATIONS

107. Until the 1940s, uranium glazes were popular on ceramic dishware. One brand, Fiestaware, had bright orange glazes that could contain up to 20% uranium by mass. Although this practice is less common today due to the negative association of radiation, it is still possible to buy Depression-era glassware that is quite radioactive. Aqueous solutions in contact with this "hot" glassware can reach uranium concentrations up to 10 ppm by mass. If 1 g of uranium undergoes 1.11×10^{7} decays/s, each to an α particle with an energy of 4.03 MeV, what would be your exposure in rem and rad if you drank 1 L of water that had been sitting for an extended time in a Fiestaware pitcher? Assume that the water and contaminants are excreted only after 18 h and that you weigh 70.0 kg.

108. Neutrography is a technique used to take the picture of an object using a beam of neutrons. How does the penetrating power of a neutron compare with *alpha*, *beta*, and *gamma* radiation? Do you expect similar penetration for protons? How would the biological damage of each particle compare with the other types of radiation? (Recall that a neutron's mass is approximately 2000 times the mass of an electron.)

109. Spent fuel elements in a nuclear reactor contain radioactive fission products in addition to heavy metals. The conversion of nuclear fuel in a reactor is shown in the diagram:

3.2% ^{235}U → 0.76% ^{235}U; → 0.44% ^{236}U; → 2.0% fission products

96.8% ^{238}U → 1.5% fission products; → 94.3% ^{238}U; → 0.9% Pu; → 0.1% Np, Am, Cm

Neglecting the fission products, write balanced nuclear reactions for the conversion of the original fuel to each product.

110. The first atomic bomb used ^{235}U as a fissile material, but there were immense difficulties in obtaining sufficient quantities of pure ^{235}U. A second fissile element, plutonium, was discovered in 1940, and it rapidly became important as a nuclear fuel. This element was produced by irradiating ^{238}U with neutrons in a nuclear reactor. Complete the series that produced plutonium, all isotopes of which are fissile:

$$^{238}_{92}U + ^{1}_{0}n \longrightarrow U \longrightarrow Np \longrightarrow Pu$$

111. Boron neutron capture therapy is a potential treatment for many diseases. As the name implies, when boron-10, one of the naturally occurring isotopes of boron, is bombarded with neutrons, it absorbs a neutron and emits an α particle. Write a balanced nuclear reaction for this reaction. One advantage of this process is that neutrons cause little damage on their own, but when they are absorbed by boron-10, they can cause localized emission of alpha radiation. Comment on the utility of this treatment and its potential difficulties.

112. An airline pilot typically flies approximately 80 h per month. If 75% of that time is spent at an altitude of about 30,000 ft, how much radiation is that pilot receiving in one month? Over a 30-year career? Is the pilot receiving toxic doses of radiation?

113. At a breeder reactor plant, a 72-kg employee accidentally inhaled 2.8 mg of ^{239}Pu dust. The isotope decays by alpha decay and has a half-life of 24,100 yr. If the energy of the emitted α particles is 5.2 MeV and the dust stays in the employee's body for 18 h, determine (a) the number of plutonium atoms inhaled; (b) the energy absorbed by the body; (c) the physical dose in rads; and (d) the dose in rems. Will the occurrence be fatal?

114. For many years, the standard source for radiation therapy in the treatment of cancer was radioactive ^{60}Co, which undergoes beta decay to ^{60}Ni and emits two γ rays, each with an energy of 1.2 MeV. Show the sequence of reactions. If the half-life of beta decay is 5.27 yr, how many ^{60}Co nuclei are present in a typical source emitting 6,000 disintegrations per second that is used by hospitals? The mass of ^{60}C is 59.93 amu.

115. It is possible to utilize radioactive materials as heat sources to produce electricity. These radioisotope thermoelectric generators (RTGs) have been used in spacecraft and many other applications. Certain Cold War–era Russian-made RTGs used a 5.0-kg strontium-90 source. One mole of strontium-90 releases β particles with an energy of 0.545 MeV and undergoes 2.7×10^{13} decays/s. How many watts of power are available from this RTG? (1 watt = 1 J/s)

116. Potassium consists of three isotopes (potassium-39, potassium-40, and potassium-41). Potassium-40 is the least abundant, and it is radioactive, decaying to argon-40, a stable, nonradioactive isotope, by the emission of a β particle with a half-life of precisely 1.25×10^9 yr. Thus, the ratio of potassium-40 to argon-40 in any potassium-40–containing material can be used to date the sample. In 1952, fragments of an early hominid, *Meganthropus*, were discovered near Modjokerto in Java. The bone fragments were lying on volcanic rock that was believed to be the same age as the bones. Potassium–argon dating on samples of the volcanic material showed that the argon-40-to-potassium-40 molar ratio was 0.00281:1. How old were the rock fragments? Could the bones also be the same age? Could radiocarbon dating have been used to date the fragments?

APPENDIX A

Standard Thermodynamic Quantities for Chemical Substances at 25°C

Substance	ΔH_f° (kJ/mol)	ΔG_f° (kJ/mol)	S° (J/mol·K)
Aluminum:			
Al(*s*)	0.0	0.0	28.3
Al(*g*)	330.0	289.4	164.6
$AlCl_3(s)$	−704.2	−628.8	109.3
$Al_2O_3(s)$	−1675.7	−1582.3	50.9
Barium:			
Ba(*s*)	0.0	0.0	62.5
Ba(*g*)	180.0	146.0	170.2
BaO(*s*)	−548.0	−520.3	72.1
$BaCO_3(s)$	−1213.0	−1134.4	112.1
$BaSO_4(s)$	−1473.2	−1362.2	132.2
Beryllium:			
Be(*s*)	0.0	0.0	9.5
Be(*g*)	324.0	286.6	136.3
$Be(OH)_2(s)$	−902.5	−815.0	45.5
BeO(*s*)	−609.4	−580.1	13.8
Bismuth:			
Bi(*s*)	0.0	0.0	56.7
Bi(*g*)	207.1	168.2	187.0
Bromine:			
Br(*g*)	111.9	82.4	175.0
$Br_2(l)$	0.0	0.0	152.2
$Br^-(aq)$	−121.6	−104.0	82.4
$Br_2(g)$	30.9	3.1	245.5
HBr(*g*)	−36.3	−53.4	198.7
HBr(*aq*)	−121.6	−104.0	82.4
Cadmium:			
Cd(*s*)	0.0	0.0	51.8
Cd(*g*)	111.8	—	167.7
$CdCl_2(s)$	−391.5	−343.9	115.3
CdS(*s*)	−161.9	−156.5	64.9
Calcium:			
Ca(*s*)	0.0	0.0	41.6
Ca(*g*)	177.8	144.0	154.9
$CaCl_2(s)$	−795.4	−748.8	108.4
$CaF_2(s)$	−1228.0	−1175.6	68.5
$Ca(OH)_2(s)$	−985.2	−897.5	83.4
CaO(*s*)	−634.9	−603.3	38.1
$CaSO_4(s)$	−1434.5	−1322.0	106.5
$CaCO_3$(*s*, calcite)	−1207.6	−1129.1	91.7
$CaCO_3$(*s*, aragonite)	−1207.8	−1128.2	88.0

Substance	ΔH°_f (kJ/mol)	ΔG°_f (kJ/mol)	S° (J/mol·K)
Carbon:			
C(*s*, graphite)	0.0	0.0	5.7
C(*s*, diamond)	1.9	2.9	2.4
C(*s*, fullerene—C_{60})	2327.0	2302.0	426.0
C(*s*, fullerene—C_{70})	2555.0	2537.0	464.0
C(*g*)	716.7	671.3	158.1
C(*g*, fullerene—C_{60})	2502.0	2442.0	544.0
C(*g*, fullerene—C_{70})	2755.0	2692.0	614.0
$CBr_4(s)$	29.4	47.7	212.5
$CBr_4(g)$	83.9	67.0	358.1
$CCl_2F_2(g)$	−477.4	−439.4	300.8
$CCl_2O(g)$	−219.1	−204.9	283.5
$CCl_4(l)$	−128.2	−62.6	216.2
$CCl_4(g)$	−95.7	−53.6	309.9
$CF_4(g)$	−933.6	−888.3	261.6
$CHCl_3(l)$	−134.1	−73.7	201.7
$CHCl_3(g)$	−102.7	6.0	295.7
$CH_2Cl_2(l)$	−124.2	—	177.8
$CH_2Cl_2(g)$	−95.4	−68.9	270.2
$CH_3Cl(g)$	−81.9	−58.5	234.6
$CH_4(g)$	−74.6	−50.5	186.3
$CH_3COOH(l)$	−484.3	−389.9	159.8
$CH_3OH(l)$	−239.2	−166.6	126.8
$CH_3OH(g)$	−201.0	−162.3	239.9
$CH_3NH_2(l)$	−47.3	35.7	150.2
$CH_3NH_2(g)$	−22.5	32.7	242.9
$CH_3CN(l)$	40.6	86.5	149.6
$CH_3CN(g)$	74.0	91.9	243.4
$CO(g)$	−110.5	−137.2	197.7
$CO_2(g)$	−393.5	−394.4	213.8
$CS_2(l)$	89.0	64.6	151.3
$CS_2(g)$	116.7	67.1	237.8
$C_2H_2(g)$	227.4	209.9	200.9
$C_2H_4(g)$	52.4	68.4	219.3
$C_2H_6(g)$	−84.0	−32.0	229.2
$C_3H_8(g)$	−103.8	−23.4	270.3
$C_3H_6O_3(l)$ (lactic acid)	−674.5	−518.2	192.1
$C_6H_6(l)$	49.1	124.5	173.4
$C_6H_6(g)$	82.9	129.7	269.2
$C_6H_{12}O_6(s)$ (glucose)	−1273.3	−910.4	212.1
$C_2H_5OH(l)$	−277.6	−174.8	160.7
$C_2H_5OH(g)$	−234.8	−167.9	281.6
$(CH_3)_2O(l)$	−203.3	—	—
$(CH_3)_2O(g)$	−184.1	−112.6	266.4
$CH_3CO_2^-(aq)$	−486.0	−369.3	86.6
n-$C_{12}H_{26}(l)$ (dodecane)	−350.9	28.1	490.6
Cesium:			
Cs(*s*)	0.0	0.0	85.2
Cs(*g*)	76.5	49.6	175.6
CsCl(*s*)	−443.0	−414.5	101.2
Chlorine:			
Cl(*g*)	121.3	105.3	165.2
$Cl_2(g)$	0.0	0.0	223.1
$Cl^-(aq)$	−167.2	−131.2	56.5
HCl(*g*)	−92.3	−95.3	186.9
HCl(*aq*)	−167.2	−131.2	56.5
$ClF_3(g)$	−163.2	−123.0	281.6

Substance	ΔH_f° (kJ/mol)	ΔG_f° (kJ/mol)	S° (J/mol·K)
Chromium:			
$Cr(s)$	0.0	0.0	23.8
$Cr(g)$	396.6	351.8	174.5
$CrCl_3(s)$	−556.5	−486.1	123.0
$CrO_3(g)$	−292.9	—	266.2
$Cr_2O_3(s)$	−1139.7	−1058.1	81.2
Cobalt:			
$Co(s)$	0.0	0.0	30.0
$Co(g)$	424.7	380.3	179.5
$CoCl_2(s)$	−312.5	−269.8	109.2
Copper:			
$Cu(s)$	0.0	0.0	33.2
$Cu(g)$	337.4	297.7	166.4
$CuCl(s)$	−137.2	−119.9	86.2
$CuCl_2(s)$	−220.1	−175.7	108.1
$CuO(s)$	−157.3	−129.7	42.6
$Cu_2O(s)$	−168.6	−146.0	93.1
$CuS(s)$	−53.1	−53.6	66.5
$Cu_2S(s)$	−79.5	−86.2	120.9
$CuCN(s)$	96.2	111.3	84.5
Fluorine:			
$F(g)$	79.4	62.3	158.8
$F^-(aq)$	−332.6	−278.8	−13.8
$F_2(g)$	0.0	0.0	202.8
$HF(g)$	−273.3	−275.4	173.8
$HF(aq)$	−332.6	−278.8	−13.8
Hydrogen:			
$H(g)$	218.0	203.3	114.7
$H_2(g)$	0.0	0.0	130.7
$H^+(aq)$	0.0	0.0	0.0
Iodine:			
$I(g)$	106.8	70.2	180.8
$I^-(aq)$	−55.2	−51.6	111.3
$I_2(s)$	0.0	0.0	116.1
$I_2(g)$	62.4	19.3	260.7
$HI(g)$	26.5	1.7	206.6
$HI(aq)$	−55.2	−51.6	111.3
Iron:			
$Fe(s)$	0.0	0.0	27.3
$Fe(g)$	416.3	370.7	180.5
$Fe^{2+}(aq)$	−89.1	−78.9	−137.7
$Fe^{3+}(aq)$	−48.5	−4.7	−315.9
$FeCl_2(s)$	−341.8	−302.3	118.0
$FeCl_3(s)$	−399.5	−334.0	142.3
$FeO(s)$	−272.0	−251.4	60.7
$Fe_2O_3(s)$	−824.2	−742.2	87.4
$Fe_3O_4(s)$	−1118.4	−1015.4	146.4
$FeS_2(s)$	−178.2	−166.9	52.9
$FeCO_3(s)$	−740.6	−666.7	92.9
Lead:			
$Pb(s)$	0.0	0.0	64.8
$Pb(g)$	195.2	162.2	175.4

Substance	ΔH_f° (kJ/mol)	ΔG_f° (kJ/mol)	S° (J/mol·K)
Lead (continued):			
PbO(*s*, red or litharge)	−219.0	−188.9	66.5
PbO(*s*, yellow or massicot)	−217.3	−187.9	68.7
$PbO_2(s)$	−277.4	−217.3	68.6
$PbCl_2(s)$	−359.4	−314.1	136.0
PbS(*s*)	−100.4	−98.7	91.2
$PbSO_4(s)$	−920.0	−813.0	148.5
$PbCO_3(s)$	−699.1	−625.5	131.0
$Pb(NO_3)_2(s)$	−451.9	—	—
$Pb(NO_3)_2(aq)$	−416.3	−246.9	303.3
Lithium:			
Li(*s*)	0.0	0.0	29.1
Li(*g*)	159.3	126.6	138.8
$Li^+(aq)$	−278.5	−293.3	13.4
LiCl(*s*)	−408.6	−384.4	59.3
$Li_2O(s)$	−597.9	−561.2	37.6
Magnesium:			
Mg(*s*)	0.0	0.0	32.7
Mg(*g*)	147.1	112.5	148.6
$MgCl_2(s)$	−641.3	−591.8	89.6
MgO(*s*)	−601.6	−569.3	27.0
$Mg(OH)_2(s)$	−924.5	−833.5	63.2
$MgSO_4(s)$	−1284.9	−1170.6	91.6
MgS(*s*)	−346.0	−341.8	50.3
Manganese:			
Mn(*s*)	0.0	0.0	32.0
Mn(*g*)	280.7	238.5	173.7
$MnCl_2(s)$	−481.3	−440.5	118.2
MnO(*s*)	−385.2	−362.9	59.7
$MnO_2(s)$	−520.0	−465.1	53.1
$KMnO_4(s)$	−837.2	−737.6	171.7
$MnO_4^-(aq)$	−541.4	−447.2	191.2
Mercury:			
Hg(*l*)	0.0	0.0	75.9
Hg(*g*)	61.4	31.8	175.0
$HgCl_2(s)$	−224.3	−178.6	146.0
$Hg_2Cl_2(s)$	−265.4	−210.7	191.6
HgO(*s*)	−90.8	−58.5	70.3
HgS(*s*, red)	−58.2	−50.6	82.4
$Hg_2(g)$	108.8	68.2	288.1
Molybdenum:			
Mo(*s*)	0.0	0.0	28.7
Mo(*g*)	658.1	612.5	182.0
$MoO_2(s)$	−588.9	−533.0	46.3
$MoO_3(s)$	−745.1	−668.0	77.7
Nickel:			
Ni(*s*)	0.0	0.0	29.9
Ni(*g*)	429.7	384.5	182.2
$NiCl_2(s)$	−305.3	−259.0	97.7
$Ni(OH)_2(s)$	−529.7	−447.2	88.0
Nitrogen:			
N(*g*)	472.7	455.5	153.3
$N_2(g)$	0.0	0.0	191.6

Substance	ΔH_f° (kJ/mol)	ΔG_f° (kJ/mol)	S° (J/mol·K)
Nitrogen (continued):			
$NH_3(g)$	−45.9	−16.4	192.8
$NH_4^+(aq)$	−132.5	−79.3	113.4
$N_2H_4(l)$	50.6	149.3	121.2
$N_2H_4(g)$	95.4	159.4	238.5
$NH_4Cl(s)$	−314.4	−202.9	94.6
$NH_4OH(l)$	−361.2	−254.0	165.6
$NH_4NO_3(s)$	−365.6	−183.9	151.1
$(NH_4)_2SO_4(s)$	−1180.9	−901.7	220.1
$NO(g)$	91.3	87.6	210.8
$NO_2(g)$	33.2	51.3	240.1
$N_2O(g)$	81.6	103.7	220.0
$N_2O_4(l)$	−19.5	97.5	209.2
$N_2O_4(g)$	11.1	99.8	304.4
$N_2O_5(g)$	13.3	117.1	355.7
$HNO_2(g)$	−79.5	−46.0	254.1
$HNO_3(l)$	−174.1	−80.7	155.6
$HNO_3(g)$	−133.9	−73.5	266.9
$HNO_3(aq)$	−207.4	−111.3	146.4
$NF_3(g)$	−132.1	−90.6	260.8
$HCN(l)$	108.9	125.0	112.8
$HCN(g)$	135.1	124.7	201.8
Osmium:			
$Os(s)$	0.0	0.0	32.6
$Os(g)$	791.0	745.0	192.6
$OsO_4(s)$	−394.1	−304.9	143.9
$OsO_4(g)$	−337.2	−292.8	293.8
Oxygen:			
$O(g)$	249.2	231.7	161.1
$O_2(g)$	0.0	0.0	205.2
$O_3(g)$	142.7	163.2	238.9
$OH^-(aq)$	−230.0	−157.2	−10.8
$H_2O(l)$	−285.8	−237.1	70.0
$H_2O(g)$	−241.8	−228.6	188.8
$H_2O_2(l)$	−187.8	−120.4	109.6
$H_2O_2(g)$	−136.3	−105.6	232.7
Phosphorus:			
P(*s*, white)	0.0	0.0	41.1
P(*s*, red)	−17.6	−12.5	22.8
P(*s*, black)	−39.3	—	—
P(*g*, white)	316.5	280.1	163.2
$P_2(g)$	144.0	103.5	218.1
$P_4(g)$	58.9	24.4	280.0
$PCl_3(l)$	−319.7	−272.3	217.1
$PCl_3(g)$	−287.0	−267.8	311.8
$POCl_3(l)$	−597.1	−520.8	222.5
$POCl_3(g)$	−558.5	−512.9	325.5
$PCl_5(g)$	−374.9	−305.0	364.6
$PH_3(g)$	5.4	13.5	210.2
$H_3PO_4(s)$	−1284.4	−1124.3	110.5
$H_3PO_4(l)$	−1271.7	−1123.6	150.8
$H_3PO_4(aq)$	−1277.4	−1018.8	222.0
Potassium:			
$K(s)$	0.0	0.0	64.7
$K(g)$	89.0	60.5	160.3
$KBr(s)$	−393.8	−380.7	95.9
$KCl(s)$	−436.5	−408.5	82.6
$KClO_3(s)$	−397.7	−296.3	143.1

Substance	ΔH_f° (kJ/mol)	ΔG_f° (kJ/mol)	S° (J/mol·K)
Potassium (continued):			
$K_2O(s)$	−361.5	−322.1	94.1
$K_2O_2(s)$	−494.1	−425.1	102.1
$KNO_2(s)$	−369.8	−306.6	152.1
$KNO_3(s)$	−494.6	−394.9	133.1
$KSCN(s)$	−200.2	−178.3	124.3
$K_2CO_3(s)$	−1151.0	−1063.5	155.5
$K_2SO_4(s)$	−1437.8	−1321.4	175.6
Rubidium:			
$Rb(s)$	0.0	0.0	76.8
$Rb(g)$	80.9	53.1	170.1
$RbCl(s)$	−435.4	−407.8	95.9
Selenium:			
$Se(s$, gray)	0.0	0.0	42.4
$Se(g$, gray)	227.1	187.0	176.7
$H_2Se(g)$	29.7	15.9	219.0
Silicon:			
$Si(s)$	0.0	0.0	18.8
$Si(g)$	450.0	405.5	168.0
$SiCl_4(l)$	−687.0	−619.8	239.7
$SiCl_4(g)$	−657.0	−617.0	330.7
$SiH_4(g)$	34.3	56.9	204.6
$SiC(s$, cubic)	−65.3	−62.8	16.6
$SiC(s$, hexagonal)	−62.8	−60.2	16.5
Silver:			
$Ag(s)$	0.0	0.0	42.6
$Ag(g)$	284.9	246.0	173.0
$Ag^+(aq)$	105.6	77.1	72.7
$AgBr(s)$	−100.4	−96.9	107.1
$AgCl(s)$	−127.0	−109.8	96.3
$AgNO_3(s)$	−124.4	−33.4	140.9
$Ag_2O(s)$	−31.1	−11.2	121.3
$Ag_2S(s)$	−32.6	−40.7	144.0
Sodium:			
$Na(s)$	0.0	0.0	51.3
$Na(g)$	107.5	77.0	153.7
$Na^+(aq)$	−240.1	−261.9	59.0
$NaF(s)$	−576.6	−546.3	51.1
$NaF(aq)$	−572.8	−540.7	45.2
$NaCl(s)$	−411.2	−384.1	72.1
$NaCl(aq)$	−407.3	−393.1	115.5
$NaBr(s)$	−361.1	−349.0	86.8
$NaBr(g)$	−143.1	−177.1	241.2
$NaBr(aq)$	−361.7	−365.8	141.4
$NaO_2\ (s)$	−260.2	−218.4	115.9
$Na_2O(s)$	−414.2	−375.5	75.1
$Na_2O_2(s)$	−510.9	−447.7	95.0
$NaCN(s)$	−87.5	−76.4	115.6
$NaNO_3(aq)$	−447.5	−373.2	205.4
$NaNO_3(s)$	−467.9	−367.0	116.5
$NaN_3(s)$	21.7	93.8	96.9
$Na_2CO_3(s)$	−1130.7	−1044.4	135.0
$Na_2SO_3(s)$	−1100.8	−1012.5	145.9
$Na_2SO_4(s)$	−1387.1	−1270.2	149.6

Substance	ΔH_f° (kJ/mol)	ΔG_f° (kJ/mol)	S° (J/mol·K)
Sulfur:			
S(*s*, rhombic)	0.0	0.0	32.1
S(*g*, rhombic)	277.2	236.7	167.8
$SO_2(g)$	−296.8	−300.1	248.2
$SO_3(g)$	−395.7	−371.1	256.8
$SO_4^{2-}(aq)$	−909.3	−744.5	20.1
$SOCl_2(g)$	−212.5	−198.3	309.8
$H_2S(g)$	−20.6	−33.4	205.8
$H_2SO_4(aq)$	−909.3	−744.5	20.1
Tin:			
Sn(*s*, white)	0.0	0.0	51.2
Sn(*s*, gray)	−2.1	0.1	44.1
Sn(*g*, white)	301.2	266.2	168.5
$SnCl_4(l)$	−511.3	−440.1	258.6
$SnCl_4(g)$	−471.5	−432.2	365.8
$SnO_2(s)$	−577.6	−515.8	49.0
Titanium:			
Ti(*s*)	0.0	0.0	30.7
Ti(*g*)	473.0	428.4	180.3
$TiCl_2(s)$	−513.8	−464.4	87.4
$TiCl_3(s)$	−720.9	−653.5	139.7
$TiCl_4(l)$	−804.2	−737.2	252.3
$TiCl_4(g)$	−763.2	−726.3	353.2
$TiO_2(s)$	−944.0	−888.8	50.6
Uranium:			
U(*s*)	0.0	0.0	50.2
U(*g*)	533.0	488.4	199.8
$UO_2(s)$	−1085.0	−1031.8	77.0
$UO_2(g)$	−465.7	−471.5	274.6
$UF_4(s)$	−1914.2	−1823.3	151.7
$UF_4(g)$	−1598.7	−1572.7	368.0
$UF_6(s)$	−2197.0	−2068.5	227.6
$UF_6(g)$	−2147.4	−2063.7	377.9
Vanadium:			
V(*s*)	0.0	0.0	28.9
V(*g*)	514.2	754.4	182.3
$VCl_3(s)$	−580.7	−511.2	131.0
$VCl_4(l)$	−569.4	−503.7	255.0
$VCl_4(g)$	−525.5	−492.0	362.4
$V_2O_5(s)$	−1550.6	−1419.5	131.0
Zinc:			
Zn(*s*)	0.0	0.0	41.6
Zn(*g*)	130.4	94.8	161.0
$ZnCl_2(s)$	−415.1	−369.4	111.5
$Zn(NO_3)_2(s)$	−483.7	—	—
ZnS(*s*, sphalerite)	−206.0	−201.3	57.7
$ZnSO_4(s)$	−982.8	−871.5	110.5
Zirconium:			
Zr(*s*)	0.0	0.0	39.0
Zr(*g*)	608.8	566.5	181.4
$ZrCl_2(s)$	−502.0	−386	110
$ZrCl_4(s)$	−980.5	−889.9	181.6

Source of data: *CRC Handbook of Chemistry and Physics,* 84th Edition (2004).

APPENDIX B

Solubility-Product Constants (K_{sp}) for Compounds at 25°C

Compound Name	Compound Formula	K_{sp}
Aluminum phosphate	$AlPO_4$	9.84×10^{-21}
Barium bromate	$Ba(BrO_3)_2$	2.43×10^{-4}
Barium carbonate	$BaCO_3$	2.58×10^{-9}
Barium chromate	$BaCrO_4$	1.17×10^{-10}
Barium fluoride	BaF_2	1.84×10^{-7}
Barium iodate	$Ba(IO_3)_2$	4.01×10^{-9}
Barium nitrate	$Ba(NO_3)_2$	4.64×10^{-3}
Barium sulfate	$BaSO_4$	1.08×10^{-10}
Barium sulfite	$BaSO_3$	5.0×10^{-10}
Beryllium hydroxide	$Be(OH)_2$	6.92×10^{-22}
Bismuth arsenate	$BiAsO_4$	4.43×10^{-10}
Bismuth iodide	BiI_3	7.71×10^{-19}
Cadmium carbonate	$CdCO_3$	1.0×10^{-12}
Cadmium fluoride	CdF_2	6.44×10^{-3}
Cadmium hydroxide	$Cd(OH)_2$	7.2×10^{-15}
Cadmium iodate	$Cd(IO_3)_2$	2.5×10^{-8}
Cadmium phosphate	$Cd_3(PO_4)_2$	2.53×10^{-33}
Cadmium sulfide	CdS	8.0×10^{-27}
Calcium carbonate	$CaCO_3$	3.36×10^{-9}
Calcium fluoride	CaF_2	3.45×10^{-11}
Calcium hydroxide	$Ca(OH)_2$	5.02×10^{-6}
Calcium iodate	$Ca(IO_3)_2$	6.47×10^{-6}
Calcium phosphate	$Ca_3(PO_4)_2$	2.07×10^{-33}
Calcium sulfate	$CaSO_4$	4.93×10^{-5}
Cesium perchlorate	$CsClO_4$	3.95×10^{-3}
Cesium periodate	$CsIO_4$	5.16×10^{-6}
Cobalt(II) arsenate	$Co_3(AsO_4)_2$	6.80×10^{-29}
Cobalt(II) hydroxide	$Co(OH)_2$	5.92×10^{-15}
Cobalt(II) phosphate	$Co_3(PO_4)_2$	2.05×10^{-35}
Copper(I) bromide	$CuBr$	6.27×10^{-9}
Copper(I) chloride	$CuCl$	1.72×10^{-7}
Copper(I) cyanide	$CuCN$	3.47×10^{-20}
Copper(I) iodide	CuI	1.27×10^{-12}
Copper(I) thiocyanate	$CuSCN$	1.77×10^{-13}
Copper(II) arsenate	$Cu_3(AsO_4)_2$	7.95×10^{-36}
Copper(II) oxalate	CuC_2O_4	4.43×10^{-10}
Copper(II) phosphate	$Cu_3(PO_4)_2$	1.40×10^{-37}
Copper(II) sulfide	CuS	6.3×10^{-36}
Europium(III) hydroxide	$Eu(OH)_3$	9.38×10^{-27}
Gallium(III) hydroxide	$Ga(OH)_3$	7.28×10^{-36}
Iron(II) carbonate	$FeCO_3$	3.13×10^{-11}
Iron(II) fluoride	FeF_2	2.36×10^{-6}
Iron(II) hydroxide	$Fe(OH)_2$	4.87×10^{-17}
Iron(III) hydroxide	$Fe(OH)_3$	2.79×10^{-39}
Iron(III) sulfide	FeS	6.3×10^{-18}
Lanthanum iodate	$La(IO_3)_3$	7.50×10^{-12}
Lead(II) bromide	$PbBr_2$	6.60×10^{-6}
Lead(II) carbonate	$PbCO_3$	7.40×10^{-14}
Lead(II) chloride	$PbCl_2$	1.70×10^{-5}
Lead(II) fluoride	PbF_2	3.3×10^{-8}
Lead(II) hydroxide	$Pb(OH)_2$	1.43×10^{-20}
Lead(II) iodate	$Pb(IO_3)_2$	3.69×10^{-13}
Lead(II) iodide	PbI_2	9.8×10^{-9}

Compound Name	Compound Formula	K_{sp}
Lead(II) selenate	$PbSeO_4$	1.37×10^{-7}
Lead(II) sulfate	$PbSO_4$	2.53×10^{-8}
Lead(II) sulfide	PbS	8.0×10^{-28}
Lithium carbonate	Li_2CO_3	8.15×10^{-4}
Lithium fluoride	LiF	1.84×10^{-3}
Lithium phosphate	Li_3PO_4	2.37×10^{-11}
Magnesium carbonate	$MgCO_3$	6.82×10^{-6}
Magnesium fluoride	MgF_2	5.16×10^{-11}
Magnesium hydroxide	$Mg(OH)_2$	5.61×10^{-12}
Magnesium phosphate	$Mg_3(PO_4)_2$	1.04×10^{-24}
Manganese(II) carbonate	$MnCO_3$	2.24×10^{-11}
Manganese(II) iodate	$Mn(IO_3)_2$	4.37×10^{-7}
Mercury(I) bromide	Hg_2Br_2	6.40×10^{-23}
Mercury(I) carbonate	Hg_2CO_3	3.6×10^{-17}
Mercury(I) chloride	Hg_2Cl_2	1.43×10^{-18}
Mercury(I) fluoride	Hg_2F_2	3.10×10^{-6}
Mercury(I) iodide	Hg_2I_2	5.2×10^{-29}
Mercury(I) oxalate	$Hg_2C_2O_4$	1.75×10^{-13}
Mercury(I) sulfate	Hg_2SO_4	6.5×10^{-7}
Mercury(I) thiocyanate	$Hg_2(SCN)_2$	3.2×10^{-20}
Mercury(II) bromide	$HgBr_2$	6.2×10^{-20}
Mercury (II) iodide	HgI_2	2.9×10^{-29}
Mercury(II) sulfide (red)	HgS	4×10^{-53}
Mercury(II) sulfide (black)	HgS	1.6×10^{-52}
Neodymium carbonate	$Nd_2(CO_3)_3$	1.08×10^{-33}
Nickel(II) carbonate	$NiCO_3$	1.42×10^{-7}
Nickel(II) hydroxide	$Ni(OH)_2$	5.48×10^{-16}
Nickel(II) iodate	$Ni(IO_3)_2$	4.71×10^{-5}
Nickel(II) phosphate	$Ni_3(PO_4)_2$	4.74×10^{-32}
Palladium(II) thiocyanate	$Pd(SCN)_2$	4.39×10^{-23}
Potassium hexachloroplatinate	K_2PtCl_6	7.48×10^{-6}
Potassium perchlorate	$KClO_4$	1.05×10^{-2}
Potassium periodate	KIO_4	3.71×10^{-4}
Praseodymium hydroxide	$Pr(OH)_3$	3.39×10^{-24}
Rubidium perchlorate	$RbClO_4$	3.00×10^{-3}
Scandium fluoride	ScF_3	5.81×10^{-24}
Scandium hydroxide	$Sc(OH)_3$	2.22×10^{-31}
Silver(I) acetate	$AgCH_3CO_2$	1.94×10^{-3}
Silver(I) arsenate	Ag_3AsO_4	1.03×10^{-22}
Silver(I) bromate	$AgBrO_3$	5.38×10^{-5}
Silver(I) bromide	$AgBr$	5.35×10^{-13}
Silver(I) carbonate	Ag_2CO_3	8.46×10^{-12}
Silver(I) chloride	$AgCl$	1.77×10^{-10}
Silver(I) chromate	Ag_2CrO_4	1.12×10^{-12}
Silver(I) cyanide	$AgCN$	5.97×10^{-17}
Silver(I) iodate	$AgIO_3$	3.17×10^{-8}
Silver(I) iodide	AgI	8.52×10^{-17}
Silver(I) oxalate	$Ag_2C_2O_4$	5.40×10^{-12}
Silver(I) phosphate	Ag_3PO_4	8.89×10^{-17}
Silver(I) sulfate	Ag_2SO_4	1.20×10^{-5}
Silver(I) sulfide	Ag_2S	6.3×10^{-50}
Silver(I) sulfite	Ag_2SO_3	1.50×10^{-14}
Silver(I) thiocyanate	$AgSCN$	1.03×10^{-12}
Strontium arsenate	$Sr_3(AsO_4)_2$	4.29×10^{-19}
Strontium carbonate	$SrCO_3$	5.60×10^{-10}
Strontium fluoride	SrF_2	4.33×10^{-9}
Strontium iodate	$Sr(IO_3)_2$	1.14×10^{-7}
Strontium sulfate	$SrSO_4$	3.44×10^{-7}
Thallium(I) bromate	$TlBrO_3$	1.10×10^{-4}
Thallium(I) bromide	$TlBr$	3.71×10^{-6}
Thallium(I) chloride	$TlCl$	1.86×10^{-4}
Thallium(I) chromate	Tl_2CrO_4	8.67×10^{-13}
Thallium(I) iodate	$TlIO_3$	3.12×10^{-6}
Thallium(I) iodide	TlI	5.54×10^{-8}

Compound Name	Compound Formula	K_{sp}
Thallium(I) thiocyanate	$TlSCN$	1.57×10^{-4}
Thallium(III) hydroxide	$Tl(OH)_3$	1.68×10^{-44}
Tin(II) hydroxide	$Sn(OH)_2$	5.45×10^{-27}
Tin(II) sulfide	SnS	1.0×10^{-25}
Yttrium carbonate	$Y_2(CO_3)_3$	1.03×10^{-31}
Yttrium fluoride	YF_3	8.62×10^{-21}
Yttrium hydroxide	$Y(OH)_3$	1.00×10^{-22}
Yttrium iodate	$Y(IO_3)_3$	1.12×10^{-10}
Zinc arsenate	$Zn_3(AsO_4)_2$	2.8×10^{-28}
Zinc carbonate	$ZnCO_3$	1.46×10^{-10}
Zinc fluoride	ZnF_2	3.04×10^{-2}
Zinc hydroxide	$Zn(OH)_2$	3×10^{-17}
Zinc selenide	$ZnSe$	3.6×10^{-26}
Zinc sulfide (wurtzite)	ZnS	1.6×10^{-24}
Zinc sulfide (sphalerite)	ZnS	2.5×10^{-22}

Source of data: *CRC Handbook of Chemistry and Physics,* 84th Edition (2004); sulfide data from *Lange's Handbook of Chemistry,* 15th Edition (1999).

APPENDIX C

Ionization Constants and pK_a Values for Acids at 25°C

Name	Formula	K_{a1}	pK_{a1}	K_{a2}	pK_{a2}	K_{a3}	pK_{a3}	K_{a4}	pK_{a4}
Acetic acid	CH_3CO_2H	1.75×10^{-5}	4.756						
Arsenic acid	H_3AsO_4	5.5×10^{-3}	2.26	1.7×10^{-7}	6.76	5.1×10^{-12}	11.29		
Benzoic acid	$C_6H_5CO_2H$	6.25×10^{-5}	4.204						
Boric acid	H_3BO_3	5.4×10^{-10}*	9.27*	$>1 \times 10^{-14}$*	>14*				
Bromoacetic acid	CH_2BrCO_2H	1.3×10^{-3}	2.90						
Carbonic acid	H_2CO_3	4.5×10^{-7}	6.35	4.7×10^{-11}	10.33				
Chloric acid	$HClO_3$	1.00×10^{2}	−2.00						
Chloroacetic acid	CH_2ClCO_2H	1.3×10^{-3}	2.87						
Chlorous acid	$HClO_2$	1.15×10^{-2}	1.94						
Chromic acid	H_2CrO_4	1.8×10^{-1}	0.74	3.2×10^{-7}	6.49				
Citric acid	$C_6H_8O_7$	7.4×10^{-4}	3.13	1.7×10^{-5}	4.76	4.0×10^{-7}	6.40		
Cyanic acid	$HCNO$	3.5×10^{-4}	3.46						
Dichloroacetic acid	$CHCl_2CO_2H$	4.5×10^{-2}	1.35						
Fluoroacetic acid	CH_2FCO_2H	2.6×10^{-3}	2.59						
Formic acid	CH_2O_2	1.8×10^{-4}	3.75						
Hydrazoic acid	HN_3	2.5×10^{-5}	4.6						
Hydrocyanic acid	HCN	6.2×10^{-10}	9.21						
Hydrofluoric acid	HF	6.3×10^{-4}	3.20						
Hydrogen selenide	H_2Se	1.3×10^{-4}	3.89	1.0×10^{-11}	11.0				
Hydrogen sulfide	H_2S	1.1×10^{-7}	6.97	1.3×10^{-13}	12.90				
Hydrogen telluride	H_2Te	2.5×10^{-3}‡	2.6‡	1×10^{-11}	11				
Hypobromous acid	$HBrO$	2.8×10^{-9}	8.55						
Hypochlorous acid	$HClO$	2.88×10^{-8}	7.54						
Hypoiodous acid	HIO	3.2×10^{-11}	10.5						
Iodic acid	HIO_3	1.7×10^{-1}	0.78						
Iodoacetic acid	CH_2ICO_2H	6.6×10^{-4}	3.18						
Nitrous acid	HNO_2	5.6×10^{-4}	3.25						
Oxalic acid	$C_2H_2O_4$	5.6×10^{-2}	1.25	1.5×10^{-4}	3.81				
Perchloric acid	$HClO_4$	2.00×10^{7}	−7.3						
Periodic acid	HIO_4	2.3×10^{-2}	1.64						
Phenol	C_6H_5OH	1.0×10^{-10}	9.99						
Phosphoric acid	H_3PO_4	6.9×10^{-3}	2.16	6.2×10^{-8}	7.21	4.8×10^{-13}	12.32		
Phosphorous acid	H_3PO_3	5.0×10^{-2}*	1.3*	2.0×10^{-7}*	6.70*				
Pyrophosphoric acid	$H_4P_2O_7$	1.2×10^{-1}	0.91	7.9×10^{-3}	2.10	2.0×10^{-7}	6.70	4.8×10^{-10}	9.32
Resorcinol	$C_6H_4(OH)_2$	4.8×10^{-10}	9.32	7.9×10^{-12}	11.1				
Selenic acid	H_2SeO_4	Strong	Strong	2.0×10^{-2}	1.7				
Selenious acid	H_2SeO_3	2.4×10^{-3}	2.62	4.8×10^{-9}	8.32				
Sulfuric acid	H_2SO_4	Strong	Strong	1.0×10^{-2}	1.99				
Sulfurous acid	H_2SO_3	1.4×10^{-2}	1.85	6.3×10^{-8}	7.2				
meso-Tartaric acid	$C_4H_6O_6$	6.8×10^{-4}	3.17	1.2×10^{-5}	4.91				
Telluric acid	H_2TeO_4	2.1×10^{-8}‡	7.68‡	1.0×10^{-11}‡	11.0‡				
Tellurous acid	H_2TeO_3	5.4×10^{-7}	6.27	3.7×10^{-9}	8.43				
Trichloroacetic acid	CCl_3CO_2H	2.2×10^{-1}	0.66						
Trifluoroacetic acid	CF_3CO_2H	3.0×10^{-1}	0.52						

Source of data: *CRC Handbook of Chemistry and Physics,* 84th Edition (2004), except those for chloric acid, chlorous acid, hypochlorous acid, and perchloric acid, which came from J.R. Bowser, *Inorganic Chemistry*, 1993; sulfide data came from *Lange's Handbook of Chemistry*, 15th Edition (1999).

* Measured at 20°C, not 25°C.

‡ Measured at 18°C, not 25°C.

APPENDIX D

Ionization Constants and pK_b Values for Bases at 25°C

Name	Formula	K_b	pK_b
Ammonia	NH_3	1.8×10^{-5}	4.75
Aniline	$C_6H_5NH_2$	7.4×10^{-10}	9.13
n-Butylamine	$C_4H_9NH_2$	4.0×10^{-4}	3.40
sec-Butylamine	$(CH_3)_2CHCH_2NH_2$	3.6×10^{-4}	3.44
tert-Butylamine	$(CH_3)_3CNH_2$	4.8×10^{-4}	3.32
Dimethylamine	$(CH_3)_2NH$	5.4×10^{-4}	3.27
Ethylamine	$C_2H_5NH_2$	4.5×10^{-4}	3.35
Hydrazine	N_2H_4	1.3×10^{-6}	5.9
Hydroxylamine	NH_2OH	8.7×10^{-9}	8.06
Methylamine	CH_3NH_2	4.6×10^{-4}	3.34
Propylamine	$C_3H_7NH_2$	3.5×10^{-4}	3.46
Pyridine	C_5H_5N	1.7×10^{-9}	8.77
Trimethylamine	$(CH_3)_3N$	6.3×10^{-5}	4.20

Source of data: *CRC Handbook of Chemistry and Physics,* 84th Edition (2004).

APPENDIX E

Standard Reduction Potentials at 25°C

Half-Reaction	$E°$ (V)
$Ac^{3+} + 3e^- \longrightarrow Ac$	−2.20
$Ag^+ + e^- \longrightarrow Ag$	0.7996
$AgBr + e^- \longrightarrow Ag + Br^-$	0.07133
$AgCl + e^- \longrightarrow Ag + Cl^-$	0.22233
$Ag_2CrO_4 + 2e^- \longrightarrow 2Ag + CrO_4{}^{2-}$	0.4470
$AgI + e^- \longrightarrow Ag + I^-$	−0.15224
$Ag_2S + 2e^- \longrightarrow 2Ag + S^{2-}$	−0.691
$Ag_2S + 2H^+ + 2e^- \longrightarrow 2Ag + H_2S$	−0.0366
$AgSCN + e^- \longrightarrow Ag + SCN^-$	0.08951
$Al^{3+} + 3e^- \longrightarrow Al$	−1.662
$Al(OH)_4{}^- + 3e^- \longrightarrow Al + 4OH^-$	−2.328
$Am^{3+} + 3e^- \longrightarrow Am$	−2.048
$As + 3H^+ + 3e^- \longrightarrow AsH_3$	−0.608
$H_3AsO_4 + 2H^+ + 2e^- \longrightarrow HAsO_2 + 2H_2O$	0.560
$Au^+ + e^- \longrightarrow Au$	1.692
$Au^{3+} + 3e^- \longrightarrow Au$	1.498
$H_3BO_3 + 3H^+ + 3e^- \longrightarrow B + 3H_2O$	−0.8698
$Ba^{2+} + 2e^- \longrightarrow Ba$	−2.912
$Be^{2+} + 2e^- \longrightarrow Be$	−1.847
$Bi^{3+} + 3e^- \longrightarrow Bi$	0.308
$BiO^+ + 2H^+ + 3e^- \longrightarrow Bi + H_2O$	0.320
$Br_2(aq) + 2e^- \longrightarrow 2Br^-$	1.0873
$Br_2(l) + 2e^- \longrightarrow 2Br^-$	1.066
$BrO_3{}^- + 6H^+ + 5e^- \longrightarrow \frac{1}{2}Br_2 + 3H_2O$	1.482
$BrO_3{}^- + 6H^+ + 6e^- \longrightarrow Br^- + 3H_2O$	1.423
$CO_2 + 2H^+ + 2e^- \longrightarrow HCO_2H$	−0.199
$Ca^{2+} + 2e^- \longrightarrow Ca$	−2.868
$Ca(OH)_2 + 2e^- \longrightarrow Ca + 2OH^-$	−3.02
$Cd^{2+} + 2e^- \longrightarrow Cd$	−0.4030
$CdSO_4 + 2e^- \longrightarrow Cd + SO_4{}^{2-}$	−0.246
$Cd(OH)_4{}^{2-} + 2e^- \longrightarrow Cd + 4OH^-$	−0.658
$Ce^{3+} + 3e^- \longrightarrow Ce$	−2.336
$Ce^{4+} + e^- \longrightarrow Ce^{3+}$	1.72
$Cl_2(g) + 2e^- \longrightarrow 2Cl^-$	1.35827
$HClO + H^+ + e^- \longrightarrow \frac{1}{2}Cl_2 + H_2O$	1.611
$HClO + H^+ + 2e^- \longrightarrow Cl^- + H_2O$	1.482
$ClO^- + H_2O + 2e^- \longrightarrow Cl^- + 2OH^-$	0.81
$ClO_3{}^- + 6H^+ + 5e^- \longrightarrow \frac{1}{2}Cl_2 + 3H_2O$	1.47
$ClO_3{}^- + 6H^+ + 6e^- \longrightarrow Cl^- + 3H_2O$	1.451
$ClO_4{}^- + 8H^+ + 7e^- \longrightarrow \frac{1}{2}Cl_2 + 4H_2O$	1.39
$ClO_4{}^- + 8H^+ + 8e^- \longrightarrow Cl^- + 4H_2O$	1.389
$Co^{2+} + 2e^- \longrightarrow Co$	−0.28
$Co^{3+} + e^- \longrightarrow Co^{2+}$	1.92
$Cr^{2+} + 2e^- \longrightarrow Cr$	−0.913
$Cr^{3+} + e^- \longrightarrow Cr^{2+}$	−0.407
$Cr^{3+} + 3e^- \longrightarrow Cr$	−0.744
$Cr_2O_7{}^- + 14H^+ + 6e^- \longrightarrow 2Cr^{3+} + 7H_2O$	1.232
$CrO_4{}^{2-} + 4H_2O + 3e^- \longrightarrow Cr(OH)_3 + 5OH^-$	−0.13
$Cs^+ + e^- \longrightarrow Cs$	−3.026
$Cu^+ + e^- \longrightarrow Cu$	0.521
$Cu^{2+} + e^- \longrightarrow Cu^+$	0.153
$Cu^{2+} + 2e^- \longrightarrow Cu$	0.3419
$CuI_2{}^- + e^- \longrightarrow Cu + 2I^-$	0.00

Half-Reaction	$E°$ (V)
$Cu_2O + H_2O + 2e^- \longrightarrow 2Cu + 2OH^-$	−0.360
$Dy^{3+} + 3e^- \longrightarrow Dy$	−2.295
$Er^{3+} + 3e^- \longrightarrow Er$	−2.331
$Es^{3+} + 3e^- \longrightarrow Es$	−1.91
$Eu^{2+} + 2e^- \longrightarrow Eu$	−2.812
$Eu^{3+} + 3e^- \longrightarrow Eu$	−1.991
$F_2 + 2e^- \longrightarrow 2F^-$	2.866
$Fe^{2+} + 2e^- \longrightarrow Fe$	−0.447
$Fe^{3+} + 3e^- \longrightarrow Fe$	−0.037
$Fe^{3+} + e^- \longrightarrow Fe^{2+}$	0.771
$[Fe(CN)_6]^{3-} + e^- \longrightarrow [Fe(CN)_6]^{4-}$	0.358
$Fe(OH)_3 + e^- \longrightarrow Fe(OH)_2 + OH^-$	−0.56
$Fm^{3+} + 3e^- \longrightarrow Fm$	−1.89
$Fm^{2+} + 2e^- \longrightarrow Fm$	−2.30
$Ga^{3+} + 3e^- \longrightarrow Ga$	−0.549
$Gd^{3+} + 3e^- \longrightarrow Gd$	−2.279
$Ge^{2+} + 2e^- \longrightarrow Ge$	0.24
$Ge^{4+} + 4e^- \longrightarrow Ge$	0.124
$2H^+ + 2e^- \longrightarrow H_2$	0.00000
$H_2 + 2e^- \longrightarrow 2H^-$	−2.23
$2H_2O + 2e^- \longrightarrow H_2 + 2OH^-$	−0.8277
$H_2O_2 + 2H^+ + 2e^- \longrightarrow 2H_2O$	1.776
$Hf^{4+} + 4e^- \longrightarrow Hf$	−1.55
$Hg^{2+} + 2e^- \longrightarrow Hg$	0.851
$2Hg^{2+} + 2e^- \longrightarrow {Hg_2}^{2+}$	0.920
$Hg_2Cl_2 + 2e^- \longrightarrow 2Hg + 2Cl^-$	0.26808
$Ho^{2+} + 2e^- \longrightarrow Ho$	−2.1
$Ho^{3+} + 3e^- \longrightarrow Ho$	−2.33
$H_2SeO_3 + 4H^+ + 4e^- \longrightarrow Se + 3H_2O$	0.74
$H_2SO_3 + 4H^+ + 4e^- \longrightarrow S + 3H_2O$	0.449
$I_2 + 2e^- \longrightarrow 2I^-$	0.5355
${I_3}^- + 2e^- \longrightarrow 3I^-$	0.536
$2{IO_3}^- + 12H^+ + 10e^- \longrightarrow I_2 + 6H_2O$	1.195
${IO_3}^- + 6H^+ + 6e^- \longrightarrow I^- + 3H_2O$	1.085
$In^+ + e^- \longrightarrow In$	−0.14
$In^{3+} + 2e^- \longrightarrow In^+$	−0.443
$In^{3+} + 3e^- \longrightarrow In$	−0.3382
$Ir^{3+} + 3e^- \longrightarrow Ir$	1.156
$K^+ + e^- \longrightarrow K$	−2.931
$La^{3+} + 3e^- \longrightarrow La$	−2.379
$Li^+ + e^- \longrightarrow Li$	−3.0401
$Lr^{3+} + 3e^- \longrightarrow Lr$	−1.96
$Lu^{3+} + 3e^- \longrightarrow Lu$	−2.28
$Md^{3+} + 3e^- \longrightarrow Md$	−1.65
$Md^{2+} + 2e^- \longrightarrow Md$	−2.40
$Mg^{2+} + 2e^- \longrightarrow Mg$	−2.372
$Mn^{2+} + 2e^- \longrightarrow Mn$	−1.185
$MnO_2 + 4H^+ + 2e^- \longrightarrow Mn^{2+} + 2H_2O$	1.224
${MnO_4}^- + 8H^+ + 5e^- \longrightarrow Mn^{2+} + 4H_2O$	1.507
${MnO_4}^- + 2H_2O + 3e^- \longrightarrow MnO_2 + 4OH^-$	0.595
$Mo^{3+} + 3e^- \longrightarrow Mo$	−0.200
$N_2 + 2H_2O + 6H^+ + 6e^- \longrightarrow 2NH_4OH$	0.092
$HNO_2 + H^+ + e^- \longrightarrow NO + H_2O$	0.983
${NO_3}^- + 3H^+ + 2e^- \longrightarrow HNO_2 + H_2O$	0.934
${NO_3}^- + 4H^+ + 3e^- \longrightarrow NO + 2H_2O$	0.957
$Na^+ + e^- \longrightarrow Na$	−2.71
$Nb^{3+} + 3e^- \longrightarrow Nb$	−1.099
$Nd^{3+} + 3e^- \longrightarrow Nd$	−2.323
$Ni^{2+} + 2e^- \longrightarrow Ni$	−0.257
$No^{3+} + 3e^- \longrightarrow No$	−1.20
$No^{2+} + 2e^- \longrightarrow No$	−2.50
$Np^{3+} + 3e^- \longrightarrow Np$	−1.856
$O_2 + 2H^+ + 2e^- \longrightarrow H_2O_2$	0.695
$O_2 + 4H^+ + 4e^- \longrightarrow 2H_2O$	1.229

Half-Reaction	$E°$ (V)
$O_2 + 2H_2O + 2e^- \longrightarrow H_2O_2 + 2OH^-$	−0.146
$O_3 + 2H^+ + 2e^- \longrightarrow O_2 + H_2O$	2.076
$OsO_4 + 8H^+ + 8e^- \longrightarrow Os + 4H_2O$	0.838
$P + 3H_2O + 3e^- \longrightarrow PH_3(g) + 3OH^-$	−0.87
$PO_4^{3-} + 2H_2O + 2e^- \longrightarrow HPO_3^{2-} + 3OH^-$	−1.05
$Pa^{3+} + 3e^- \longrightarrow Pa$	−1.34
$Pa^{4+} + 4e^- \longrightarrow Pa$	−1.49
$Pb^{2+} + 2e^- \longrightarrow Pb$	−0.1262
$PbO + H_2O + 2e^- \longrightarrow Pb + 2OH^-$	−0.580
$PbO_2 + SO_4^{2-} + 4H^+ + 2e^- \longrightarrow PbSO_4 + 2H_2O$	1.6913
$PbSO_4 + 2e^- \longrightarrow Pb + SO_4^{2-}$	−0.3588
$Pd^{2+} + 2e^- \longrightarrow Pd$	0.951
$Pm^{3+} + 3e^- \longrightarrow Pm$	−2.30
$Po^{4+} + 4e^- \longrightarrow Po$	0.76
$Pr^{3+} + 3e^- \longrightarrow Pr$	−2.353
$Pt^{2+} + 2e^- \longrightarrow Pt$	1.18
$[PtCl_4]^{2-} + 2e^- \longrightarrow Pt + 4Cl^-$	0.755
$Pu^{3+} + 3e^- \longrightarrow Pu$	−2.031
$Ra^{2+} + 2e^- \longrightarrow Ra$	−2.8
$Rb^+ + e^- \longrightarrow Rb$	−2.98
$Re^{3+} + 3e^- \longrightarrow Re$	0.300
$Rh^{3+} + 3e^- \longrightarrow Rh$	0.758
$Ru^{3+} + e^- \longrightarrow Ru^{2+}$	0.2487
$S + 2e^- \longrightarrow S^{2-}$	−0.47627
$S + 2H^+ + 2e^- \longrightarrow H_2S(aq)$	0.142
$2S + 2e^- \longrightarrow S_2^{2-}$	−0.42836
$H_2SO_3 + 4H^+ + 4e^- \longrightarrow S + 3H_2O$	0.449
$SO_4^{2-} + H_2O + 2e^- \longrightarrow SO_3^{2-} + 2OH^-$	−0.93
$Sb + 3H^+ + 3e^- \longrightarrow SbH_3$	−0.510
$Sc^{3+} + 3e^- \longrightarrow Sc$	−2.077
$Se + 2e^- \longrightarrow Se^{2-}$	−0.924
$Se + 2H^+ + 2e^- \longrightarrow H_2Se$	−0.082
$SiF_6^{2-} + 4e^- \longrightarrow Si + 6F^-$	−1.24
$Sm^{3+} + 3e^- \longrightarrow Sm$	−2.304
$Sn^{2+} + 2e^- \longrightarrow Sn$	−0.1375
$Sn^{4+} + 2e^- \longrightarrow Sn^{2+}$	−0.151
$2SO_4^{2-} + 4H^+ + 2e^- \longrightarrow S_2O_6^{2-} + H_2O$	−0.22
$Sr^{2+} + 2e^- \longrightarrow Sr$	−2.899
$Ta^{3+} + 3e^- \longrightarrow Ta$	−0.6
$TcO_4^- + 4H^+ + 3e^- \longrightarrow TcO_2 + 2H_2O$	0.782
$TcO_4^- + 8H^+ + 7e^- \longrightarrow Tc + 4H_2O$	0.472
$Tb^{3+} + 3e^- \longrightarrow Tb$	−2.28
$Te + 2e^- \longrightarrow Te^{2-}$	−1.143
$Te^{4+} + 4e^- \longrightarrow Te$	0.568
$Th^{4+} + 4e^- \longrightarrow Th$	−1.899
$Ti^{2+} + 2e^- \longrightarrow Ti$	−1.630
$Tl^+ + e^- \longrightarrow Tl$	−0.336
$Tl^{3+} + 2e^- \longrightarrow Tl^+$	1.252
$Tl^{3+} + 3e^- \longrightarrow Tl$	0.741
$Tm^{3+} + 3e^- \longrightarrow Tm$	−2.319
$U^{3+} + 3e^- \longrightarrow U$	−1.798
$VO_2^+ + 2H^+ + e^- \longrightarrow VO^{2+} + H_2O$	0.991
$V_2O_5 + 6H^+ + 2e^- \longrightarrow 2VO^{2+} + 3H_2O$	0.957
$W_2O_5 + 2H^+ + 2e^- \longrightarrow 2WO_2 + H_2O$	−0.031
$XeO_3 + 6H^+ + 6e^- \longrightarrow Xe + 3H_2O$	2.10
$Y^{3+} + 3e^- \longrightarrow Y$	−2.372
$Yb^{3+} + 3e^- \longrightarrow Yb$	−2.19
$Zn^{2+} + 2e^- \longrightarrow Zn$	−0.7618
$Zn(OH)_4^{2-} + 2e^- \longrightarrow Zn + 4OH^-$	−1.199
$Zn(OH)_2 + 2e^- \longrightarrow Zn + 2OH^-$	−1.249
$ZrO_2 + 4H^+ + 4e^- \longrightarrow Zr + 2H_2O$	−1.553
$Zr^{4+} + 4e^- \longrightarrow Zr$	−1.45

Source of data: *CRC Handbook of Chemistry and Physics,* 84th Edition (2004).

APPENDIX F

Properties of Water

Density:	0.99984 g/cm^3 at 0°C 0.99970 g/cm^3 at 10°C 0.99821 g/cm^3 at 20°C 0.98803 g/cm^3 at 50°C 0.95840 g/cm^3 at 100°C
Enthalpy (heat) of vaporization:	45.054 kJ/mol at 0°C 43.990 kJ/mol at 25°C 42.482 kJ/mol at 60°C 40.657 kJ/mol at 100°C
Surface tension:	74.23 J/m^2 at 10°C 71.99 J/m^2 at 25°C 67.94 J/m^2 at 50°C 58.91 J/m^2 at 100°C
Viscosity:	1.793 mPa·s at 0°C 0.890 mPa·s at 25°C 0.547 mPa·s at 50°C 0.282 mPa·s at 100°C
Ion-product constant, K_w:	1.15×10^{-15} at 0°C 1.01×10^{-14} at 25°C 5.18×10^{-14} at 50°C 4.99×10^{-13} at 100°C
Specific heat (C_s):	4.2176 J/(g-°C) at 0°C 4.1818 J/(g-°C) at 20°C 4.1806 J/(g-°C) at 50°C 4.2159 J/(g-°C) at 100°C

Vapor pressure of water (kPa)

T(°C)	*P*(kPa)	*T*(°C)	*P*(kPa)	*T*(°C)	*P*(kPa)	*T*(°C)	*P*(kPa)
0	0.61129	30	4.2455	60	19.932	90	70.117
5	0.87260	35	5.6267	65	25.022	95	84.529
10	1.2281	40	7.3814	70	31.176	100	101.32
15	1.7056	45	9.5898	75	38.563	105	120.79
20	2.3388	50	12.344	80	47.373	110	143.24
25	3.1690	55	15.752	85	57.815	115	169.02

Vapor pressure of water (mmHg)

T(°C)	*P*(mmHg)	*T*(°C)	*P*(mmHg)	*T*(°C)	*P*(mmHg)	*T*(°C)	*P*(mmHg)
0	4.585	30	31.844	60	149.50	90	525.91
5	6.545	35	42.203	65	187.68	95	634.01
10	9.211	40	55.364	70	233.84	100	759.95
15	12.793	45	71.929	75	289.24	105	905.99
20	17.542	50	92.59	80	355.32	110	1074.38
25	23.769	55	118.15	85	433.64	115	1267.74

Source of data: *CRC Handbook of Chemistry and Physics,* 84th Edition (2004).

APPENDIX G

Glossary

A

A: See "Activity (*A*)" (*Section 14.5*), "Frequency factor (*A*)" (*Section 14.7*), or "Mass number (*A*)" (*Section 1.6*).

Absolute zero (0 K): The lowest possible temperature that could theoretically be achieved. It corresponds to $-273.15°C$. (*Section 10.3*)

Absorption spectrum: A spectrum produced by the absorption of light by ground-state atoms (c.f. emission spectrum). Each element has a characteristic absorption spectrum, so scientists can use these spectra to analyze the composition of matter. (*Section 6.3*)

Accuracy: The degree to which a measured value is the same as the true value of the quantity. (*Essential Skills 1*)

Accurate: When a measured value is the same as the true value of the quantity. (*Essential Skills 1*)

Acid (Arrhenius definition): A substance with at least one hydrogen atom that can dissociate to form an anion and an H^+ ion (a proton) in aqueous solution, thereby forming an acidic solution (c.f. the Brønsted–Lowry and Lewis definitions of an acid). (*Sections 2.5 and 4.6*)

Acid (Brønsted–Lowry definition): Any substance that can donate a proton (c.f. the Arrhenius and Lewis definitions of an acid). This is essentially the same as the Arrhenius definition, but it is more general because it is not restricted to aqueous solutions. (*Section 4.6*)

Acid (Lewis definition): Any species that can *accept* a pair of electrons (c.f. the Arrhenius and Brønsted–Lowry definitions of an acid). The Lewis definition expands the Brønsted–Lowry definition to include substances other than the H^+ ion. (*Section 8.7*)

Acid–base indicator: A compound added in small amounts to an acid–base titration to signal the equivalence point by changing color. Indicators are weak acids or weak bases that exhibit intense colors that vary with pH. The conjugate acid and conjugate base of a good indicator have different colors so that they can be distinguished easily. The point in the titration at which the indicator changes color is called the endpoint. (*Sections 4.9 and 16.5*)

Acid–base reaction: A reaction of the general form acid + base $\longrightarrow$ salt (see Table 3.1 for an example). (*Section 3.5*)

Acidic oxides: Oxides, such as SO_3, that react with water to produce an acidic solution or are soluble in aqueous base (c.f. basic oxides and amphoteric oxides). Oxides of nonmetallic elements are generally acidic oxides. (*Section 17.4*)

Acid ionization constant (K_a): An equilibrium constant for the ionization (dissociation) of a weak acid (HA) in water, $HA(aq) + H_2O(l) \rightleftharpoons H_3O^+(aq) + A^-(aq)$, in which the concentration of water is treated as a constant. Thus, $K_a = [H_3O^+][A^-]/[HA]$. (*Section 16.2*)

Acid rain: Rain and snow that are dramatically more acidic because of human activities during the last 150 years. Pollutants expelled from industrial plants, such as nitrogen and sulfur oxides, combine with water in the atmosphere to form acids such as HNO_3 and H_2SO_4 that get washed back into surface water supplies via precipitation. (*Section 4.7*)

Actinide element: Any of the 14 elements between $Z = 90$ (thorium) and $Z = 103$ (lawrencium). The 5*f* orbitals are filling for the actinides and their chemistry is dominated by M^{3+} ions. The actinides and lanthanides are usually grouped together in two rows beneath the main body of the periodic table. (*Sections 1.7 and 7.4*)

Activated complex: Also called the transition state of the reaction, the activated complex is the arrangement of atoms that first forms when molecules are able to overcome the activation energy and react. The activated complex is not a reaction intermediate; it does not last long enough to be detected readily. (*Section 14.7*)

Activation energy (E_a): The energy barrier or threshold that corresponds to the amount of energy the particles in a reaction (atoms, molecules, or ions) must have in order to react when they collide. The activation energy is the minimum amount of energy needed for a reaction to occur. Reacting molecules must have enough energy to overcome electrostatic repulsion, and a minimum amount of energy is required to break chemical bonds so that new ones may be formed. (*Section 14.7*)

Active metals: The metals at the top of the activity series (the alkali metals, alkaline earths, and Al), which have the greatest tendency to lose electrons (to be oxidized) (c.f. inert metals). (*Section 4.8*)

Activity (*A*): Also called the rate of decay, the activity of a radioactive isotope is the decrease in the number of the radioisotope's nuclei per unit time (Equation 14.33): $A = -\Delta N/\Delta t$, where N is the number of atoms of the radioactive isotope. Activity is usually measured in units of disintegrations per second (dps) or disintegrations per minute (dpm). (*Section 14.5*)

Activity series: A list of metals and hydrogen in order of their relative tendency to be oxidized (see Table 4.4). The metals at the top of the series (the alkali metals, alkaline earths, and Al) have the greatest tendency to lose electrons (to be oxidized), whereas those at the bottom of the series (Pt, Au, Ag, Cu, and Hg) have the least tendency to be oxidized. (*Section 4.8*)

Actual yield: The measured mass of products obtained from a reaction. It is almost always less than the theoretical yield (often much less). (*Section 3.4*)

Adduct: The product of a reaction between a Lewis acid and a Lewis base. The resulting acid–base adduct contains a coordinate covalent bond because both electrons are provided by only one of the atoms. (*Section 8.7*)

Adhesive forces: The attractive intermolecular forces between a liquid and the substance comprising the surface of a capillary (see capillary action; c.f. cohesive forces). (*Section 11.3*)

Adsorption: A physical process in which a reactant interacts with the solid surface of a heterogeneous catalyst (c.f. desorption). Adsorption causes a chemical bond in the reactant to become weak and then break. (*Section 14.8*)

Aerobic organisms: Organisms, such as humans, that cannot survive in the absence of O_2 (c.f. anaerobic organisms). (*Section 18.8*)

Aerosol: A dispersion of solid or liquid particles in a gas. An aerosol is one of three kinds of colloids (the other two are sols and emulsions). (*Section 13.7*)

Alcohol: A class of organic compounds obtained by replacing one or more of the hydrogen atoms of a hydrocarbon with an $-OH$ group. The simplest alcohol is methanol (CH_3OH). (*Section 2.4*)

Aldehyde: A class of organic compounds that has the general form RCHO, in which the carbon atom of the carbonyl group is bonded to a hydrogen atom and an R group. The R group may be either another hydrogen atom or an alkyl group (c.f. ketone). (*Section 4.1*)

Aliphatic hydrocarbons: Alkanes, alkenes, alkynes, and cyclic hydrocarbons (hydrocarbons that are not aromatic). (*Section 2.4*)

Alkali metal: Any of the elements in Group 1 of the periodic table (Li, Na, K, Rb, Cs, and Fr). All of the Group 1 elements react readily with nonmetals to give ions with a +1 charge, such as Li^+ and Na^+. (*Section 1.7*)

Alkaline battery: A battery that consists of a Leclanché cell adapted to operate under alkaline (basic) conditions. (*Section 19.5*)

Alkaline earths: The elements in Group 2 of the periodic table (Be, Mg, Ca, Sr, Ba, and Ra). All are metals that react readily with non-metals to give ions with a +2 charge, such as Mg^{2+} and Ca^{2+}. (*Section 1.7*)

Alkanes: One of the four major classes of hydrocarbons, alkanes contain only carbon–hydrogen and carbon–carbon single bonds. Alkanes are saturated hydrocarbons. The other three classes of hydrocarbons are the alkenes, alkynes, and aromatics (all of which are unsaturated hydrocarbons). (*Section 2.4*)

Alkenes: One of the four major classes of hydrocarbons, alkenes contain at least one carbon–carbon double bond. Alkenes are unsaturated hydrocarbons. The other three classes of hydrocarbons are the alkanes, alkynes, and aromatics. (*Section 2.4*)

Alkynes: One of the four major classes of hydrocarbons, alkynes contain at least one carbon–carbon triple bond. Alkynes are unsaturated hydrocarbons. The other three classes of hydrocarbons are the alkanes, alkenes, and aromatic compounds. (*Section 2.4*)

Alloy: A solid solution of two or more metals whose properties differ from those of the constituent elements. (*Sections 1.3 and 12.5*)

Alpha (α) decay: A nuclear decay reaction that results in the emission of a helium-4 nucleus as an α particle. The daughter nuclide contains two fewer protons and two fewer neutrons than the parent nuclide. Thus, α decay produces a daughter nucleus with a mass number (A) that is lower by 4 and a nuclear charge (Z) that is lower by 2 than the parent nucleus. (*Section 20.2*)

Alpha (α) particle: A helium nucleus, ${}^4_2He$. (*Section 20.2*)

Amalgams: Solutions (usually solid solutions) of metals in liquid mercury. (*Section 13.3*)

Amide (peptide) bonds: The covalent bond that links one amino acid to another in peptides and proteins. The carbonyl carbon atom of one amino acid residue bonds to the amino nitrogen atom of the other residue, eliminating a molecule of water in the process (see Equation 12.4). Thus, peptide bonds form in a condensation reaction between a carboxylic acid and an amine. (*Sections 3.5 and 12.8*)

Amine: An organic compound that has the general formula RNH_2, where R is an alkyl group. Amines can be thought of as being obtained by replacing one (or more) of the hydrogen atoms of NH_3 (ammonia) with an alkyl group. Amines, like ammonia, are bases. (*Section 2.5*)

Amorphous solid: The type of solid that forms when the atoms, molecules, or ions of a compound aggregate with no particular order (c.f. crystalline solid). Amorphous solids have irregular or curved surfaces, do not give well-resolved X-ray diffraction patterns, and melt over a wide range of temperatures. (*Section 12.1*)

Ampere (A): The fundamental SI unit of electric current, 1 ampere is defined as the flow of 1 coulomb (C) per second past a given point (i.e., 1 A = 1 C/s). (*Section 19.4*)

Amphiprotic: Substances that can behave as either an acid or a base in a chemical reaction, depending on the nature of the other reactant(s). Water is amphiprotic. (*Section 16.1*)

Amphoteric oxides: Oxides that can dissolve in acid to produce water and dissolve in base to produce a soluble complex (c.f. acidic oxides and basic oxides). Most elements whose oxides exhibit amphoteric behavior are located along the diagonal line that separates the metals from the nonmetals in the periodic table. (*Section 17.4*)

Amplification mechanism: A process by which elements that are present in trace amounts can exert large effects on the health of an organism. For example, a molecule containing a trace element may be an essential part of a larger molecule that acts in turn to regulate the concentrations of other molecules. The amplification mechanism enables small variations in the concentration of the trace element to have large biological effects. (*Section 7.5*)

Amplitude: The vertical height of a wave. The amplitude is defined as half the peak-to-trough height. As the energy of a wave with a given frequency increases, so does its amplitude. (*Section 6.1*)

Anaerobic organisms: Organisms that can live only in the absence of O_2 (c.f. aerobic organisms). (*Section 18.8*)

Anion: An ion that contains more electrons than protons, resulting in a net negative charge. Anions may be monatomic (e.g., Cl^-) or polyatomic (e.g., SO_4^{2-}). (*Section 2.1*)

Anisotropic: An arrangement of molecules in which their properties depend on the direction they are measured (c.f. isotropic). Liquid crystals are anisotropic because they are not as disordered as a liquid (their molecules have some degree of alignment). (*Section 11.8*)

Anode: One of two electrodes in an electrochemical cell (galvanic/voltaic or electrolytic). The oxidation half-reaction occurs at the anode (c.f. cathode, the other electrode). (*Section 19.1*)

Antibonding molecular orbital: A molecular orbital that forms when atomic orbitals or orbital lobes of opposite sign interact to give decreased electron probability between the nuclei due to destructive reinforcement of the wave functions. Antibonding molecular orbitals are always higher in energy than the parent atomic orbitals. They are one of three types of molecular orbitals that can form when atomic orbitals interact (the other two are bonding and nonbonding molecular orbitals). (*Section 9.3*)

Aqueous solution: A solution in which water is the solvent (c.f. nonaqueous solution). (*Section 3.3 and Chapter 4 introduction*)

Arenes: See aromatic hydrocarbons. (*Section 2.4*)

Aromatic hydrocarbons: One of the four major classes of hydrocarbons, aromatics usually contain rings of six carbon atoms that can be drawn with alternating single and double bonds. Aromatics are unsaturated hydrocarbons and are sometimes called arenes. The other three classes of hydrocarbons are the alkanes, alkenes, and alkynes. (*Section 2.4*)

Arrhenius equation: An equation (Equation 14.43), $k = Ae^{-E_a/RT}$, that summarizes the collision model of chemical kinetics, where K is the rate constant, A is the frequency factor, E_a is the activation energy, R is the ideal gas constant [8.314 J/(K · mol)], and T is the absolute temperature. (*Section 14.7*)

Atmosphere (atm): Also referred to as standard atmospheric pressure, it is the atmospheric pressure required to support a column of mercury exactly 760 mm tall. The atmosphere is related to other pressure units as follows: 1 atm = 760 torr = 760 millimeters of mercury (mmHg) = 101325 pascals (Pa) = 101.325 kPa. (*Section 10.2*)

Atom: The fundamental, indivisible particles of which matter is composed. (*Section 1.4*)

Atomic mass unit (amu): One-twelfth of the mass of one atom of ^{12}C; 1 amu = 1.66×10^{-24} g. (*Section 1.6*)

Atomic number (Z): The number of protons in the nucleus of an atom of an element. The atomic number is different for each element. (*Section 1.6*)

Atomic orbital: A wave function with an allowed combination of n, l, and m_l quantum numbers; a particular spatial distribution for an electron. Thus, for a given set of quantum numbers, each principal shell contains a fixed number of subshells, and each subshell contains a fixed number of orbitals. (*Section 6.5*)

Aufbau principle: The process used to build up the periodic table by adding protons one by one to the nucleus and by adding the corresponding electrons to the lowest-energy orbital available without violating the Pauli exclusion principle. (*Section 6.6*)

Average reaction rate: The reaction rate calculated for a given time interval from the concentrations of either the reactant or one of the products at the beginning of the interval (time = t_0) and at the end of the interval (t_1) (c.f. instantaneous reaction rate). (*Section 14.2*)

Avogadro's hypothesis: Equal volumes of different gases contain equal numbers of gas particles when measured at the same temperature and pressure (c.f. Avogadro's law). (*Sections 1.4 and 10.3*)

Avogadro's law: A corollary to Avogadro's hypothesis that describes the relationship between the volume and amount of a gas. According to Avogadro's law, the volume (V) of a sample of gas at constant temperature (T) and pressure (P) is directly proportional to the number of moles of gas in the sample (n) (see Equation 10.8): $V \propto n$ (at constant T and P). (*Section 10.3*)

Avogadro's number: The number of units (e.g., atoms, molecules, or formula units) in one mole: 6.022×10^{23}. (*Section 3.1*)

Azimuthal quantum number (*l*): One of three quantum numbers (the other two are n and m_l) used to specify any wave function, the azimuthal quantum number describes the shape of the region of space occupied by the electron. The allowed values of l depend on the value of n and can range from 0 to $n - 1$. (*Section 6.5*)

B

Band gap: In band theory, the difference in energy between the highest level of one energy band and the lowest level of the band above it. The band gap represents a set of forbidden energies that do not correspond to any allowed combinations of atomic orbitals. (*Section 12.6*)

Band theory: A theory used to describe the bonding in metals and semiconductors. Band theory assumes that the valence orbitals of the atoms in the solid interact with one another to generate a set of molecular orbitals that extend throughout the solid. (*Section 12.6*)

Bandwidth: The difference in energy between the highest and lowest energy levels in the energy band (see band theory). The energy band is proportional to the strength of the interaction between orbitals on adjacent atoms: the stronger the interaction, the larger the band width. (*Section 12.6*)

Barometer: A device used to measure atmospheric pressure (c.f. manometer). A barometer may be constructed from a long glass tube that is closed at one end. It is then filled with mercury and placed upside down in a dish of mercury without allowing any air to enter the tube (see Figure 10.4). The height of the mercury column is proportional to the atmospheric pressure. (*Section 10.2*)

Base (Arrhenius definition): A substance that produces one or more hydroxide ions (OH^-) and a cation when dissolved in aqueous solution, thereby forming a basic solution (c.f. the Brønsted–Lowry and Lewis definitions of a base). (*Sections 2.5 and 4.6*)

Base (Brønsted–Lowry definition): Any substance that can accept a proton (c.f. the Arrhenius and Lewis definitions of a base). This definition is far more general than the Arrhenius definition because the hydroxide ion (OH^-) is just one of many substances that can accept a proton. (*Section 4.6*)

Base (Lewis definition): Any species that can *donate* a pair of electrons (c.f. the Arrhenius and Brønsted–Lowry definitions of a base). All Brønsted–Lowry bases (proton acceptors) are also electron-pair donors, so the Lewis definition of a base does not contradict the Brønsted–Lowry definition. (*Section 8.7*)

Base ionization constant (K_b): An equilibrium constant for the reaction of a weak base (B) with water, $B(aq) + H_2O(l) \rightleftharpoons BH^+(aq) + OH^-(aq)$, in which the concentration of water is treated as a constant. Thus, $K_b = [BH^+][OH^-]/[B]$. (*Section 16.2*)

Basic oxides: Oxides, such as Cs_2O, that react with water to produce a basic solution or dissolve readily in aqueous acid (c.f. acidic oxides and amphoteric oxides). Oxides of metallic elements are generally basic oxides. (*Section 17.4*)

Battery: Also called a storage cell, a battery is a galvanic cell (or series of galvanic cells) that contains all the reactants needed to produce electricity (c.f. fuel cell). Batteries can be disposable (primary) or rechargeable (secondary). (*Section 19.5*)

Bent: One of two possible molecular geometries for an AB_2 molecular species (the other is linear; see Table 9.1). In the bent geometry, the B–A–B bond angle is less than 180°. (*Section 9.1*)

Beta (β) decay: A nuclear decay reaction in which a neutron is converted to a proton and a high-energy electron that is ejected from the nucleus as a β particle (c.f. positron emission). Beta decay is common in nuclei that contain too many neutrons. Although β decay does not change the mass number of the nucleus, it does increase the atomic number by 1. Thus, β decay decreases the neutron-to-proton ratio, moving the nucleus toward the band of stable nuclei. (*Section 20.2*)

Beta (β) particle: A high-energy electron that is ejected from a nucleus, ${}_{-1}^{0}\beta$. (*Section 20.2*)

Bilayer: A two-dimensional sheet consisting of a double layer of phospholipid molecules arranged tail to tail. As a result, the hydrophobic tails of the phospholipids are located in the center of the bilayer, where they are *not* in contact with water, and their hydrophilic heads are on the two surfaces, in contact with the surrounding aqueous solution. (*Section 13.7*)

Blackbody radiation: The energy emitted by an object when it is heated. Blackbody radiation is electromagnetic radiation whose wavelength (and color) depends on the temperature of the object (see Figure 6.5). (*Section 6.2*)

Body-centered cubic (bcc) unit cell: A cubic unit cell that contains eight component atoms, molecules, or ions located at the corners of a cube, as well as an identical component in the center of the cube (see Figure 12.5b). (*Section 12.2*)

Boiling-point elevation (ΔT_b): The difference between the boiling point of a solution and the boiling point of the pure solvent (see Equation 13.20): $\Delta T_b = T_b - T_b^\circ$, where T_b is the boiling point of the solution and T_b° is the boiling point of the pure solvent. ΔT_b is proportional to the molality of the solution (see Equation 13.21). (*Section 13.6*)

Boltzmann distribution: A curve that shows the distribution of molecular speeds at a given temperature (see Figure 10.14). The actual values of speed and kinetic energy are not the same for all particles of a gas but are given by a Boltzmann distribution, in which some molecules have higher or lower speeds (and kinetic energies) than average. (*Section 10.7*)

Bomb calorimeter: A device used to measure enthalpy changes in chemical processes at constant volume (calorimetry). A bomb calorimeter is one kind of constant-volume calorimeter. The heat released by a reaction carried out at constant volume is identical to the change in internal energy (ΔE) rather than the enthalpy change (ΔH), although the difference is usually quite small (on the order of a few percent). (*Section 5.3*)

Bond angles: The angles between bonds. (*Section 9.1*)

Bond distance (r_0): The optimal internuclear distance between two bonded atoms in a molecule. (*Section 8.1*)

Bond energy: The enthalpy change that occurs when a given bond in a gaseous molecule is broken (c.f. lattice energy). (*Section 8.1*)

Bonding molecular orbital: A molecular orbital that forms when atomic orbitals or orbital lobes with the same sign interact to give increased electron probability between the nuclei due to constructive reinforcement of the wave functions. Bonding molecular orbitals are always lower in energy than the parent atomic orbitals. They are one of three types of molecular orbitals that can form when atomic orbitals interact (the other two are antibonding and nonbonding molecular orbitals). (*Section 9.3*)

Bonding pair: A pair of electrons in a Lewis structure that is shared by two atoms, thus forming a covalent bond (c.f. lone pair). (*Section 8.5*)

Bond order (Lewis bonding model): The number of electron pairs that hold two atoms together. For a single bond, the bond order is 1; for a double bond, the bond order is 2; and for a triple bond, the bond order is 3. (*Section 8.8*)

Bond order (molecular orbital theory): One-half the net number of bonding electrons in a molecule. When calculating the bond order, electrons in antibonding molecular orbitals cancel electrons in bonding molecular orbitals, while electrons in non-bonding molecular orbitals have no effect and are not counted. (*Section 9.3*)

Born–Haber cycle: A thermochemical cycle developed by Max Born and Fritz Haber in 1919, the Born–Haber cycle describes a process in which an ionic solid is conceptually formed from its component elements in a stepwise manner (see Figure 8.4). The Born–Haber cycle, combined with Hess's law and experimentally determined enthalpy changes for other chemical processes, can be used to determine enthalpy changes for processes that otherwise could *not* be measured experimentally. (*Section 8.3*)

Bragg equation: The equation that describes the relationship between two X-ray beams diffracted from different planes of atoms (see Equation 12.1): $2d \sin \theta = n\lambda$, where d is the distance separating the two planes of atoms, θ is the angle of incidence of the two X-ray beams (i.e., the angle between the X-ray beams and the planes of the crystal), n is an integer (to distinguish one layer from another), and λ is the wavelength of the X rays. Two X rays that are in phase reinforce each other, whereas two X rays that are out of phase interfere destructively, effectively canceling each other. (*Section 12.3*)

Breeder reactor: A nuclear fission reactor that produces more fissionable fuel than it consumes. This process does not violate the first law of thermodynamics because the fuel produced is different from the fuel consumed. (*Section 20.5*)

BSC theory: A theory, first formulated by J. Bardeen, L. Cooper, and J. R. Schrieffer, that attempts to explain the phenomenon of superconductivity. According to BSC theory, electrons can travel through a superconducting material without resistance because they couple to one another to form pairs of electrons (called Cooper pairs). (*Section 12.7*)

Buffers: Solutions that maintain a relatively constant pH when an acid or a base is added. Buffers protect other molecules in solution from the effects of the added acid or base. Buffers contain either a weak acid (HA) and its conjugate base (A^-) or a weak base (B) and its conjugate acid (BH^+). (*Section 16.6*)

Buffer capacity: The amount of strong acid or strong base that a buffer solution can absorb before the pH changes dramatically. The buffer capacity depends solely on the concentrations of the species in the buffered solution. The more concentrated the buffer solution, the greater its buffer capacity. (*Section 16.6*)

C

c: See "Speed of light (c)" (*Section 6.1*).

Calorie (cal): A non-SI unit of energy, 1 calorie = 4.184 joules exactly. As a result, 1 J = 0.2390 cal exactly. (*Section 5.1*)

Calorie (Cal): The nutritional Calorie used to indicate the caloric content of food. It is equal to 1 kilocalorie (kcal). (*Section 5.4*)

Calorimetry: The set of techniques (experimental procedures) used to measure enthalpy changes in chemical processes with devices called calorimeters. (*Section 5.3*)

Capillary action: The tendency of a polar liquid to rise against gravity into a small-diameter glass tube (a capillary). Capillary action is the net result of two opposing forces: cohesive forces (which hinder capillary action) and adhesive forces (which encourage capillary action). (*Section 11.3*)

Carbon cycle: The distribution and flow of carbon throughout the planet (see Figure 5.21). (*Section 5.5*)

Carbonyl group: A carbon atom double-bonded to an oxygen atom. It is a characteristic feature of many organic compounds, including aldehydes, ketones, and carboxylic acids. (*Section 2.5*)

Carboxylic acid: An organic compound that contains an $-OH$ group covalently bonded to the carbon atom of a carbonyl group. The general formula of carboxylic acids is RCO_2H. Carboxylic acids are covalent compounds, but they dissociate to produce H^+ and RCO_2^- ions when they dissolve in water. (*Section 2.5*)

Catalysis: The acceleration of a chemical reaction by a catalyst. (*Section 3.5*)

Catalyst: A substance that participates in a reaction and causes it to occur more rapidly but that can be recovered unchanged at the end of the reaction and reused. Catalysts, which may be homogeneous or heterogeneous, may also control which products are formed in a reaction. Catalysts are not involved in the overall stoichiometry of the reaction, but they must appear in at least one of the elementary steps in the mechanism for the catalyzed reaction. The catalyzed pathway has a lower activation energy (E_a), but the net change in energy that results from the reaction (i.e., the difference between the energy of the reactants and the energy of the products) is unaffected by the presence of a catalyst. (*Sections 3.5 and 14.8*)

Cathode: One of two electrodes in an electrochemical cell (galvanic/voltaic or electrolytic). The reduction half-reaction occurs at the cathode (c.f. anode, the other electrode). (*Section 19.1*)

Cation: An ion that contains fewer electrons than protons, resulting in a net positive charge. Cations may be monatomic (e.g., Na^+) or polyatomic (e.g., NH_4^+). (*Section 2.1*)

Cell: A collection of molecules, capable of reproducing itself, that is surrounded by a phospholipid bilayer. (*Section 13.7*)

Cell membrane: A mixture of phospholipids that form a phospholipid bilayer around the cell. (*Section 13.7*)

Cell potential: See "Potential (E_{cell})" (*Section 19.1*).

Ceramic: Any nonmetallic inorganic solid that is strong enough to be used in structural applications. Ceramics are typically strong and have high melting points, but they are brittle. Ceramics can be classified as ceramic oxides or nonoxide ceramics. (*Section 12.9*)

Ceramic-matrix composites: A composite consisting of reinforcing fibers embedded in a ceramic matrix. The fibers are generally ceramic, too. (*Section 12.9*)

Cesium chloride structure: The unit cell for many ionic compounds that contain relatively large cations and a 1:1 cation:anion ratio (including CsCl). The cesium chloride structure consists of a simple cubic lattice of Cl^- anions with a Cs^+ cation in the center of the cubic cell (i.e., in the cubic hole) (see Figure 12.9). (*Section 12.3*)

Chain reactions: Reaction mechanisms in which one or more elementary reactions that contain a highly reactive species repeat again and again during the reaction process. Chain reactions have three stages: initiation, propagation, and termination. (*Section 14.6*)

Chalcogens: The elements in Group 16 of the periodic table (O, S, Se, Te, and Po). (*Section 7.4*)

Change in enthalpy (ΔH): At constant pressure, the change in enthalpy of a system is identical to the heat transferred from the surroundings to the system (or vice versa): $\Delta H = q_p$. Because enthalpy is a state function, the magnitude of ΔH depends only on the initial and final states of the system, not on the path taken. (*Section 5.2*)

Chemical bond: The attractive interactions between atoms that hold them together in compounds. Chemical bonds are generally divided into two fundamentally different kinds: ionic and covalent. (*Section 2.1*)

Chemical energy: One of the five forms of energy, chemical energy is stored within a chemical compound because of a particular arrangement of atoms. The other four forms of energy are radiant, thermal, nuclear, and electrical. (*Section 5.1*)

Chemical equation: An expression that gives the identities and quantities of the substances in a chemical reaction. Chemical formulas and other symbols are used to indicate the reactants (on the left) and the products (on the right). An arrow points from the reactants to the products. (*Section 3.3*)

Chemical equilibrium: The point at which the rates of the forward and reverse reactions become the same, so that the net composition of the system no longer changes with time. (*Chapter 15 introduction*)

Chemical kinetics: The study of reaction rates. Factors that affect reaction rates include reactant concentrations, temperature, physical states and surface areas of reactants, as well as solvent and catalyst properties if either is present. By studying the kinetics of a reaction, chemists gain insights into how to control reaction conditions to achieve a desired outcome. (*Chapter 14 introduction*)

Chemical nomenclature: The systematic language of chemistry, which makes it possible to recognize and name the most common kinds of compounds. Nomenclature can be used to name a compound from its structure or draw its structure from its name. (*Chapter 2 introduction*)

Chemical property: The characteristic ability of a substance to react to form new substances (e.g., flammability and susceptibility to corrosion). (*Section 1.3*)

Chemical reaction: A process in which a substance is converted to one or more other substances that have different compositions and properties. A chemical reaction changes only the distribution of atoms, *not* the number of atoms. In most chemical reactions, bonds are broken in the reactants and new bonds are formed to create the products. (*Chapter 3 introduction*)

Chemistry: The study of matter and the changes that material substances undergo. (*Section 1.1*)

Chemotrophs: Organisms, such as animals and fungi, whose source of energy is chemical compounds from their environment, usually obtained by consuming or breaking down other organisms (c.f. phototrophs). (*Section 18.8*)

Cholesteric phase: One of three different ways that most liquid crystals can orient themselves (the other two are the nematic and smectic phases). In the cholesteric phase, the molecules are arranged in planes (similar to the smectic phase), but each layer is rotated by a certain amount with respect to those above and below it, giving it a helical structure. (*Section 11.8*)

Clausius–Clapeyron equation: A linear relationship that expresses the nonlinear relationship between the vapor pressure of a liquid and temperature (see Equation 11.1): $\ln P = -\Delta H_{vap}/RT + C$, where P is pressure, ΔH_{vap} is the heat of vaporization, R is the universal gas constant, T is the absolute temperature, and C is a constant. The Clausius–Clapeyron equation can be used to calculate the heat of vaporization of a liquid from its measured vapor pressure at two or more temperatures. (*Section 11.4*)

Cleavage reaction: A chemical reaction that has the general form $AB \longrightarrow A + B$ (see Table 3.1 for examples). Cleavage reactions are the reverse of condensation reactions. (*Section 3.5*)

Closed system: One of three kinds of system, a closed system can exchange energy but not matter with its surroundings. The other kinds of system are open and isolated. (*Sections 5.2 and 18.1*)

Coal: A complex solid material derived primarily from plants that died and were buried hundreds of millions of years ago and were subsequently subjected to high temperatures and pressures. It is used as a fuel. (*Section 5.5*)

Coefficient: The number greater than 1 that precedes a formula in a balanced chemical equation. When no coefficient is written in front of a species, the coefficient is assumed to be 1. The coefficient indicates the number of atoms, molecules, or formula units of a reactant or product in a balanced chemical equation. (*Section 3.3*)

Cohesive forces: The intermolecular forces that hold a liquid together (c.f. adhesive forces). Cohesive forces hinder capillary action. (*Section 11.3*)

Colligative properties: Properties of solutions (e.g., changes in freezing point or changes in boiling point) that depend primarily on the *number* of solute particles rather than the *kind* of solute particles. (*Section 13.6*)

Colloid: A heterogeneous mixture of particles with diameters of about 2–1000 nm that are distributed throughout a second phase. The dispersed particles do not separate from the dispersing phase upon standing (c.f. suspension). Colloids can be classified as sols, aerosols, or emulsions. (*Section 13.7*)

Combustion: The burning of a material in an oxygen atmosphere. (*Section 1.4*)

Combustion reaction: An oxidation–reduction reaction in which the oxidant is O_2. (*Section 3.5*)

Common ion effect: The shift in equilibrium that results when a strong electrolyte is added that contains one ion in common with a reaction system that is at equilibrium. The equilibrium shifts in the direction that reduces the concentration of the common ion (in accordance with Le Châtelier's principle). (*Section 16.6*)

Complete ionic equation: A chemical equation that shows which ions and molecules are hydrated and which are present in other forms and phases. The complete ionic equation shows what is actually going on in solution (c.f. overall equation and net ionic equation). (*Section 4.4*)

Composite materials: Materials that consist of at least two distinct phases, the matrix (which constitutes the bulk of the material) and fibers or granules that are embedded within the matrix and that limit the growth of cracks by pinning defects in the bulk material. Composite materials are stronger, tougher, stiffer, and more resistant to corrosion than either component alone. (*Section 12.9*)

Compound: A pure substance that contains two or more elements and has chemical and physical properties that are usually different from those of the elements of which it is composed. With only a few exceptions, a particular compound has the same elemental composition (the same elements in the same proportions) regardless of its source or history. (*Section 1.3*)

Complex ion: An ionic species formed between a central metal ion and one or more surrounding ligands. A complex ion forms from a metal ion and a ligand because of a Lewis acid–base interaction. The positively charged metal ion acts as the Lewis acid and the ligand acts as the Lewis base. Small, highly charged metal ions have the greatest tendency to act as Lewis acids, so they have the greatest tendency to form complex ions, too. (*Section 17.3*)

Concentration: The quantity of solute that is dissolved in a particular quantity of solvent or solution. (*Section 4.2*)

Concentration cell: An electrochemical cell in which the anode and cathode compartments are identical except for the concentration of a reactant. As the electrochemical reaction proceeds, the difference between the concentrations of the reactant in the two cells decreases, and so does E_{cell}. Equilibrium is reached when the concentration of the reactant is the same in both cells, in which case the measured potential difference between the two compartments is zero ($E_{cell} = 0$). (*Section 19.4*)

Condensation: The physical process by which atoms or molecules in the vapor phase enter the liquid phase. To condense, vapor molecules must first collide with the surface of the liquid. Condensation is the opposite of evaporation or vaporization. (*Section 11.4*)

Condensation reaction: A chemical reaction that has the general form $A + B \longrightarrow AB$ (see Table 3.1 for examples). Condensation reactions are the reverse of cleavage reactions. Some, but not all, condensation reactions are also oxidation–reduction reactions. (*Section 3.5*)

Conduction band: The band of empty molecular orbitals in a semiconductor. Exciting electrons from the filled valence band to the empty conduction band increases the electrical conductivity of a material. (*Section 12.6*)

Conjugate acid: The substance formed when a Brønsted–Lowry base accepts a proton. (*Section 4.6*)

Conjugate acid–base pair: An acid and a base that differ by only one hydrogen ion. All acid–base reactions involve two conjugate acid–base pairs, the Brønsted–Lowry acid and the base it forms after donating its proton, and the Brønsted–Lowry base and the acid it forms after accepting a proton. (*Sections 4.6 and 16.2*)

Conjugate base: The substance formed when a Brønsted–Lowry acid donates a proton. (*Section 4.6*)

Constant-pressure calorimeter: A device used to measure enthalpy changes in chemical processes at constant pressure (calorimetry). Because ΔH is defined as the heat flow at constant pressure, measurements made using a constant-pressure calorimeter give ΔH values directly. (*Section 5.3*)

Cooling curve: A plot of the temperature of a substance versus the heat removed or versus the cooling time at a constant rate of cooling. Cooling curves relate temperature changes to phase transitions (c.f. heating curve). A cooling curve is not necessarily the opposite of a heating curve because supercooled liquids may form. (*Section 11.5*)

Cooper pairs: Pairs of electrons that migrate through a superconducting material as a unit. Cooper pairs may be the reason electrons are able to travel through a superconducting solid without resistance. (*Section 12.7*)

Coordinate covalent bond: A covalent bond in which both electrons come from the same atom (see adduct). Coordinate covalent bonds are particularly common in compounds that contain atoms with fewer than an octet of electrons. Once formed, a coordinate covalent bond behaves like any other covalent single bond. (*Section 8.5*)

Coordination number: The number of nearest neighbors in a solid structure such as the different cubic and close-packed structures. The hexagonal close-packed and cubic close-packed arrangements result in a coordination number of 12 for each atom in the lattice, whereas the simple cubic and body-centered cubic lattices have coordination numbers of 6 and 8, respectively. (*Section 12.2*)

Corrosion: A galvanic process by which metals deteriorate through oxidation, usually but not always to their oxides. (*Section 19.6*)

Coulomb (C): The SI unit of measure for the number of electrons that pass a given point in 1 second, the coulomb is defined as $6.25 \times 10^{18}\ e^-/s$. The coulomb relates electrical potential (in volts) to energy (in joules): $1\ J/1\ V = 1\ \text{coulomb} = 6.25 \times 10^{18}\ e^-/s$ (Equation 19.32). (*Section 19.4*)

Covalent atomic radius: Half the distance between the nuclei of two like atoms joined by a covalent bond in the same molecule (i.e., Cl_2 or N_2; see Figure 7.5a). (*Section 7.2*)

Covalent bond: The electrostatic attraction between the positively charged nuclei of the bonded atoms and the negatively charged electrons they share. (*Section 2.1*) A chemical bond in which the electrons are shared equally between the two bonding atoms in a molecule or polyatomic ion (c.f. polar covalent bond). (*Section 8.1*)

Covalent compound: A compound that consists of discrete molecules (c.f. ionic compound). (*Section 2.1*)

Covalent solid: A solid that consists of two- or three-dimensional networks of atoms held together by covalent bonds (c.f. ionic solid, molecular solid, and metallic solid). Covalent solids tend to be very hard and have high melting points. (*Section 12.5*)

C_p: See "Molar heat capacity (C_p)" (*Section 5.3*).

Cracking: A process in petroleum refining in which the larger and heavier hydrocarbons in the kerosene and higher-boiling-point fractions are heated to temperatures as high as 900°C to convert them to lighter molecules similar to those in the gasoline fraction. The high temperature causes C—C bonds to break ("crack"), thus converting less volatile, lower-value fractions to more volatile, higher-value mixtures that have carefully controlled formulas. (*Section 2.6*)

Critical mass: The minimum mass of a fissile isotope capable of supporting sustained fission (a nuclear chain reaction). The critical mass depends on the purity of the fissile material, the shape of the mass, and the identity of the isotope. (*Section 20.4*)

Critical point: The combination of the critical temperature and the critical pressure of a substance. (*Section 11.6*)

Critical pressure: The minimum pressure needed to liquefy a substance at its critical temperature. (*Section 11.6*)

Critical temperature (T_c): The highest temperature at which a substance can exist as a liquid, regardless of the applied pressure. Above the critical temperature, molecules have too much kinetic energy for their intermolecular attractive forces to be able to hold them together in a separate liquid phase. Instead, the substance forms a single phase that completely occupies the volume of the container. (*Section 11.6*)

Crown ether: A cyclic polyether $[(OCH_2CH_2)_n]$ that is large enough to accommodate a metal ion in its center. Crown ethers are hydrophobic on the outside and hydrophilic in the center, so they are used to dissolve ionic substances in nonpolar solvents (see cryptand). (*Section 13.3*)

Cryogenic liquids: The ultracold liquids formed from the liquefaction of gases. Cryogenic liquids have applications as refrigerants in both industry and biology. (*Section 10.8*)

Cryptand: A compound that can completely surround a cation with lone pairs of electrons on oxygen and nitrogen atoms. Like crown ethers, cryptands are used to dissolve ionic compounds in nonpolar solvents. (*Section 13.3*)

Crystal lattice: The regular, repeating three-dimensional structure that a crystalline solid forms. A crystal lattice is one of two general ways in which the atoms, molecules, or ions of a solid can be arranged (the other is as an amorphous solid). (*Section 12.1*)

Crystalline solid: The type of solid that forms when the atoms, molecules, or ions of a compound produce a crystal lattice (c.f. amorphous solid). Crystalline solids have well-defined edges and faces, diffract X rays, and tend to have sharp melting points. (*Section 12.1*)

Crystallization: A physical process used to separate homogeneous mixtures (solutions) into their component substances. Crystallization separates mixtures based on differences in their solubilities. (*Section 1.3*)

C_s: See "Specific heat (C_s)" (*Section 5.3*).

Cubic close-packed (ccp) structure: One of two variants of the close-packed arrangement, the most efficient way to pack spheres in a lattice [the other is the hexagonal close-packed (hcp) arrangement]. The cubic close-packed structure results from arranging the atoms of the solid so that their positions alternate from layer to layer in the pattern ABCABC . . . (see Figure 12.7b). (*Section 12.2*)

Cubic hole: The hole located at the center of the simple cubic lattice. The hole is equidistant from all eight atoms or ions at the corners of the unit cell. An atom or ion in a cubic hole has a coordination number of 8. (*Section 12.3*)

Cyclic hydrocarbon: A hydrocarbon in which the ends of the carbon chain are connected to form a ring of covalently bonded carbon atoms. The simplest cyclic hydrocarbon is cyclopropane (C_3H_6). (*Section 2.4*)

D

***d*:** See "Density (*d*)" (*Section 1.3*).

Dalton's law of partial pressures: The total pressure exerted by a mixture of gases is the sum of the partial pressures of the component gases. (*Section 10.5*)

***d* block:** The 10 columns of the periodic table that fall between the *s* block and the *p* block. The elements in the *d* block are filling their $(n - 1)d$ orbitals (see Figure 6.34; c.f. *s* block, *p* block, and *f* block). (*Section 6.6*)

***d*-block elements:** The elements of the 10 columns of the periodic table that fall between the *s* block and the *p* block. The elements in the *d* block are filling their $(n - 1)d$ orbitals (see Figure 6.34; c.f. *s*-block elements, *p*-block elements, and *f*-block elements). (*Section 6.6*)

Defects: Errors in an idealized crystal lattice. Real crystals contain large numbers of defects (typically 10^4 per milligram), ranging from variable amounts of impurities to places where atoms or ions are missing or misplaced. Defects can affect a single point in the lattice (point defects), a row of lattice points (line defects), or a plane of points (plane defects). (*Section 12.4*)

Degenerate: Having the same energy. All orbitals with the same value of *n* (e.g., the three 2*p* orbitals) are degenerate because they all have the same energy. (*Section 6.5*)

Density (*d*): An intensive property of matter, density is the mass per unit volume (and is usually expressed in grams per cubic centimeter, g/cm^3). At a given temperature and pressure, the density of a pure substance is a constant. (*Section 1.3*)

Desorption: The process by which products of heterogeneous catalysis leave the surface of the solid catalyst. Desorption is the opposite of adsorption. (*Section 14.8*)

Dialysis: A process that uses a semipermeable membrane with pores large enough to allow small solute molecules and solvent molecules to pass through, but not large solute molecules (e.g., proteins). (*Section 13.6*)

Dielectric constant (ε): A constant that expresses the ability of a bulk substance (e.g., a particular solvent) to decrease the electrostatic forces between two charged particles. The dielectric constant indicates the tendency of a solvent to dissolve ionic compounds. A solvent that has a high dielectric constant causes the charged particles of an ionic substance to behave as if they have been moved farther apart, so their electrostatic attraction is reduced. (*Section 13.3*)

Differential rate law: A rate law that expresses the rate of a reaction in terms of changes in the concentration of one or more reactants, Δ[R], over a specific time interval, Δt (c.f. integrated rate law). (*Section 14.2*)

Diffusion: The gradual mixing of gases due to the motion of their component particles even in the absence of mechanical agitation such as stirring (c.f. effusion). The result is a gas mixture with a uniform composition. The rate of diffusion of a gaseous substance is inversely proportional to the square root of its molar mass: rate $\propto 1/(M)^{1/2}$ (Graham's law). (*Section 10.7*)

Dipole–dipole interactions: A kind of intermolecular interaction (force) that results between molecules that have net dipole moments. These molecules tend to align themselves so that the positive end of one dipole is near the negative end of another, and vice versa (see Figure 11.3a and b). These attractive interactions are more stable than arrangements in which the positive and negative ends of two dipoles are adjacent. Dipole–dipole interactions and London dispersion forces are often called van der Waals forces. (*Section 11.2*)

Dipole moment (μ): Produced by the asymmetrical charge distribution in a polar substance, the dipole moment is the product of the partial charge *Q* on the bonded atoms and the distance *r* between the partial charges: $\mu = Qr$, where *Q* is measured in coulombs (C) and *r* in meters (m). The unit for dipole moments is the debye (D): $1\ D = 3.3356 \times 10^{-30}$ C.m. Any diatomic molecule that contains a polar covalent bond has a dipole moment, but in polyatomic molecules the presence or absence of a net dipole moment depends on the molecular geometry. For some symmetrical AB_n structures, the individual bond dipole moments cancel one another, giving a net dipole moment of zero. (*Sections 8.9 and 9.1*)

Diprotic acid: A compound that can donate two hydrogen ions per molecule in separate steps (e.g., H_2SO_4). Diprotic acids are one kind of polyprotic acids (see also triprotic acid). (*Section 4.6*)

Disposable batteries: Also called primary batteries, the electrode reactions in disposable batteries are effectively irreversible and cannot be recharged (c.f. rechargeable batteries). (*Section 19.5*)

Distillation: A physical process used to separate homogeneous mixtures (solutions) into their component substances. Distillation makes use of differences in the volatilities of the component substances. (*Section 1.3*)

Doping: The process of deliberately introducing small amounts of impurities into commercial semiconductors in order to tune their electrical properties for specific applications (see *n*-type semiconductor and *p*-type semiconductor). (*Section 12.6*)

Double bond: A chemical bond formed when two atoms share two pairs of electrons. (*Section 2.1*)

Ductile: The ability to be pulled into wires. Metals are ductile, whereas nonmetals are usually brittle instead. (*Section 1.7*)

Dynamic equilibrium: A state in which two opposing processes (e.g., evaporation and condensation) occur at the same rate, thus producing no *net* change in the system. (*Section 11.4*)

E

***E*:** See "Internal energy (*E*)" (*Section 5.2*).

***EA*:** See "Electron affinity (*EA*)" (*Section 7.3*).

E_a: See "Activation energy (E_a)" (*Section 14.7*).

E_{cell}: See "Potential (E_{cell})" (*Section 19.1*).

E°_{cell}: See "Standard cell potential (E°_{cell})" (*Section 19.2*).

Edge dislocation: A crystal defect that results from the insertion of an extra plane of atoms into part of the crystal lattice. The edge dislocation causes the planes of atoms in the lattice to become kinked where the extra plane of atoms begins (see Figure 12.16). (*Section 12.4*)

Effective nuclear charge (Z_{eff}): The nuclear charge an electron actually experiences due to shielding from other electrons closer to the nucleus. Z_{eff} is less than the actual nuclear charge (Z) because the intervening electrons neutralize a portion of the positive charge of the nucleus and thereby decrease the attractive interaction between it and the electron farther away. (*Section 6.5*)

Effusion: The escape of a gas through a small (usually microscopic) opening into an evacuated space (c.f. diffusion). The rate of effusion of a gaseous substance is inversely proportional to the square root of its molar mass: rate $\propto 1/(M)^{1/2}$ (Graham's law). Heavy molecules effuse through a porous material more slowly than light molecules (c.f. Figure 10.17). (*Section 10.7*)

Electrical energy: One of the five forms of energy, electrical energy results from the flow of electrically charged particles. The other four forms of energy are radiant, thermal, chemical, and nuclear. (*Section 5.1*)

Electrical insulators: Materials that conduct electricity poorly because their valence bands are full. The energy gap between the highest filled levels and the lowest empty levels is so large in an electrical insulator that the empty levels are inaccessible. Thermal energy cannot excite an electron from a filled level to an empty one. (*Section 12.6*)

Electrochemical cell: An apparatus that is used to generate electricity from a spontaneous oxidation–reduction (redox) reaction or, conversely, that uses electricity to drive a nonspontaneous redox reaction. (*Section 19.1*)

Electrochemistry: The study of the relationship between electricity and chemical reactions. (*Chapter 19 introduction*)

Electrodes: Solid metals connected to an external circuit that provides an electrical connection between systems in an electrochemical cell (galvanic or electrolytic). The oxidation half-reaction occurs at one electrode (the anode) and the reduction half-reaction occurs at the other (the cathode). The electrodes in an electrochemical cell are also connected by an electrolyte. (*Section 19.1*)

Electrolysis: An electrochemical process in which an external voltage is applied in an electrolytic cell to drive a nonspontaneous reaction. Electrolysis is the opposite process to that which occurs in galvanic cells. (*Section 19.7*)

Electrolyte: An ionic substance or solution that allows ions to transfer between the two electrodes of an electrochemical cell (galvanic/voltaic or electrolytic), thereby maintaining electrical neutrality (see salt bridge). (*Sections 4.1 and 19.1*)

Electrolytic cell: One of two types of electrochemical cells, an electrolytic cell consumes electrical energy from an external source, using it to cause a nonspontaneous oxidation–reduction (redox) reaction ($\Delta G > 0$) to occur (c.f. galvanic cell or voltaic cell, the other type of electrochemical cell). (*Section 19.1*)

Electromagnetic radiation: Energy that is transmitted, or radiated, through space in the form of periodic oscillations of an electric and a magnetic field (see Figure 6.3). All forms of electromagnetic radiation (e.g., microwaves, visible light, and gamma rays) consist of perpendicular oscillating electric and magnetic fields. (*Section 6.1*)

Electron: A subatomic particle with a mass of 9.109×10^{-28} g (0.0005486 amu) and an electrical charge of -1.602×10^{-19} C. The charge on the electron is equal in magnitude but opposite in sign to the charge on a proton. (*Section 1.5*)

Electron affinity (*EA*): The energy change that occurs when an electron is added to a gaseous atom (c.f. ionization energy): $E(g) + e^- \longrightarrow E^-(g)$. Electron affinities can be negative (energy is released when an electron is added), positive (energy must be added to produce an anion), or zero (the process is energetically neutral). (*Section 7.3*)

Electron capture: A nuclear decay reaction in which an electron in an inner shell reacts with a proton to produce a neutron. When a second electron moves from an outer shell to take the place of the lower-energy electron that was absorbed by the nucleus, an X ray is emitted. Like positron emission, electron capture occurs in neutron-poor nuclei. (*Section 20.2*)

Electron configuration: The arrangement of an element's electrons in its atomic orbitals. (*Section 6.6*)

Electron-deficient molecules: Compounds that contain fewer than an octet of electrons around one atom. Electron-deficient molecules (e.g., BCl_3) have a strong tendency to gain an additional pair of electrons by reacting with substances that possess a lone pair of electrons. (*Section 8.7*)

Electronegativity (χ): The relative ability of an atom to attract electrons to itself in a chemical compound. Elements with high electronegativities tend to acquire electrons and are found in the upper right corner of the periodic table, whereas elements with low electronegativities tend to lose electrons and are found in the lower left corner of the periodic table. (*Section 7.3*)

Electron-pair geometry: The three-dimensional arrangement of electron pairs around the central atom of a molecule or polyatomic ion (c.f. molecular geometry). (*Section 9.1*)

Electron sea: Valence electrons that are delocalized throughout a metallic solid. A simple model used to describe the bonding in metals is to view the solid as consisting of positively charged nuclei embedded in an electron sea. (*Section 12.5*)

Electron shielding: The effect by which electrons closer to the nucleus neutralize a portion of the positive charge of the nucleus and thereby decrease the attractive interaction between the nucleus and an electron farther away. Because of electron shielding, the electron experiences an effective nuclear charge (Z_{eff}) that is less than the actual nuclear charge (Z). (*Section 6.5*)

Electron spin: The magnetic moment that results when an electron (an electrically charged particle) spins. In an external magnetic field, the electron has two possible orientations, which are described by a fourth quantum number, m_s [the two possible values of m_s are $+\frac{1}{2}$ (up) and $-\frac{1}{2}$ (down)]. (*Section 6.6*)

Electroplating: A process in which a layer of a second metal is deposited on the metal electrode that acts as the cathode during electrolysis. Electroplating is used to enhance the appearance of metal objects and to protect them from corrosion. (*Section 19.7*)

Electrostatic attraction: An electrostatic interaction between oppositely charged species (positive and negative) that results in a force that causes them to move toward each other. (*Section 2.1*)

Electrostatic interaction: An interaction between electrically charged particles such as protons and electrons. (*Section 2.1*)

Electrostatic repulsion: An electrostatic interaction between two species that have the same charge (both positive or both negative) that results in a force that causes them to repel each other. (*Section 2.1*)

Element: A pure substance that cannot be broken down into simpler ones by chemical changes. (*Section 1.3*)

Elementary reaction: Each of the complex series of reactions that take place in a stepwise fashion to convert reactants to products. Each elementary reaction involves one, two, or (rarely) three atoms, molecules, or ions. The overall sequence of elementary reactions is the mechanism of the reaction. The sum of the elementary reactions in a mechanism must give the balanced chemical equation for the overall reaction. (*Section 14.6*)

Emission spectrum: A spectrum produced by the emission of light by atoms in excited states (c.f. absorption spectrum). Each element has a characteristic emission spectrum, so scientists can use these spectra to analyze the composition of matter. (*Section 6.3*)

Empirical formula: A formula for a compound that consists of the atomic symbol for each component element accompanied by a subscript indicating the *relative* number of atoms of that element in the compound, reduced to the smallest whole numbers. An empirical formula is based on experimental measurements of the numbers of atoms in a sample of a compound, so it indicates only the ratios of the numbers of the elements present (c.f. molecular formula). (*Section 2.2*)

Empirical formula mass: Another name for formula mass. (*Section 3.1*)

Emulsion: A dispersion of one liquid phase in another liquid with which it is immiscible. An emulsion is one of three kinds of colloids (the other two are sols and aerosols). (*Section 13.7*)

Endothermic: A process in which heat (q) is transferred to the system from the surroundings (c.f. exothermic). By convention, $q > 0$ for an endothermic reaction. (*Section 5.2*)

Endpoint: The point in a titration at which an indicator changes color. Ideally the endpoint matches the equivalence point of the titration. (*Section 4.9*)

Energy: The capacity to do work. The five forms of energy are radiant, thermal, chemical, nuclear, and electrical. (*Section 5.1*)

Energy band: The continuous set of allowed energy levels generated in band theory when the valence orbitals of the atoms in the solid interact with one another, thus creating a set of molecular orbitals that extend throughout the solid. (*Section 12.6*)

Energy-level diagram: A schematic drawing that compares the energies of the molecular orbitals (bonding, antibonding, and nonbonding) with the energies of the parent atomic orbitals. The molecular orbitals are arranged in order of energy, with the lowest-energy orbitals at the bottom and the highest-energy orbitals at the top. The bonding in a molecule can be described by inserting the total number of valence electrons into the energy-level diagram, filling the orbitals according to the Pauli principle and Hund's rule (i.e., each molecular orbital can accommodate a maximum of two electrons of opposite spins and the orbitals are filled in order of increasing energy). (*Section 9.3*)

Enthalpy (H): The sum of a system's internal energy E and the product of its pressure P and volume V: $H = E + PV$. Because internal energy, pressure, and volume are all state functions, enthalpy is a state function, too. Enthalpy is also an extensive property (like mass), so the magnitude of the enthalpy change for a reaction is proportional to the amounts of the substances that react. (*Sections 5.2 and 18.2*)

Enthalpy of combustion (ΔH_{comb}): The change in enthalpy that occurs during a combustion reaction. (*Section 5.2*)

Enthalpy of formation (ΔH_f): The enthalpy change for the formation of 1 mol of a compound from its component elements, such as the formation of CO_2 from C and O_2. (*Section 5.2*)

Enthalpy of fusion (ΔH_{fus}): The enthalpy change that accompanies the melting (fusion) of 1 mol of a substance. (*Section 5.2*)

Enthalpy of reaction (ΔH_{rxn}): The change in enthalpy that occurs during a chemical reaction. If heat flows from the system to the surroundings, the enthalpy of the system decreases and ΔH_{rxn} is negative. If heat flows from the surroundings to the system, the enthalpy of the system increases and ΔH_{rxn} is positive. Thus, $\Delta H_{rxn} < 0$ for an exothermic reaction, and $\Delta H_{rxn} > 0$ for an endothermic reaction. (*Section 5.2*)

Enthalpy of solution (ΔH_{soln}: The change in enthalpy that occurs when a specified amount of solute dissolves in a given quantity of solvent. (*Section 5.2*)

Enthalpy of sublimation (ΔH_{sub}): The enthalpy change that accompanies the conversion of a solid directly to a gas. ΔH_{sub} is always positive because energy is always required to evaporate a solid. The enthalpy of sublimation of a substance equals the enthalpy of fusion plus the enthalpy of vaporization of the same substance. (*Sections 8.3 and 11.5*)

Enthalpy of vaporization (ΔH_{vap}): The enthalpy change that accompanies the vaporization of 1 mol of a substance. (*Section 5.2*)

Entropy (S): The degree of disorder in a thermodynamic system. The greater the number of possible microstates for a system, the greater the disorder and the higher the entropy. (*Sections 13.2 and 18.3*)

Enzyme: Catalysts that occur naturally in living organisms and that catalyze biological reactions. Most biological reactions do not occur without a biological catalyst (i.e., an enzyme). Almost all enzymes are protein molecules with molecular masses of 20,000–100,000 amu. Some are homogeneous catalysts, while others are heterogeneous catalysts. (*Sections 3.5, 12.8, and 14.8*)

Enzyme inhibitors: Substances that decrease the rate of an enzyme-catalyzed reaction by binding to a specific portion of the enzyme, thus slowing or preventing a reaction from occurring. Irreversible inhibitors are the equivalent of poisons in heterogeneous catalysis. (*Section 14.8*)

Equilibrium constant expression: The right side of the equilibrium equation (Equation 15.5)—namely, $[C]^c[D]^d/[A]^a[B]^b$, where A and B are reactants, C and D are products, and a, b, c, and d are their stoichiometric coefficients in the balanced chemical equation for the reaction. (*Section 15.2*)

Equilibrium constant (K): The ratio of the rate constants for the forward reaction (k_f) and the reverse reaction (k_r); that is, $K = k_f/k_r$. It is also the equilibrium constant calculated from solution concentrations instead of partial pressures [c.f. equilibrium constant (K_p)]. $K = (C)^c(D)^d/(A)^a(B)^b$ for the general reaction $aA + bB \rightleftharpoons cC + dD$, in which all of the components are dissolved in solution. The higher the value of the equilibrium constant, the more stable the products of the reaction. (*Section 15.2*)

Equilibrium constant (K_p): The equilibrium constant calculated from partial pressures instead of solution concentrations [c.f. equilibrium constant (K)]. $K_p = (P_C)^c(P_D)^d/(P_A)^a(P_B)^b$ for the general reaction $aA + bB \rightleftharpoons cC + dD$, in which all of the components are gases. (*Section 15.2*)

Equilibrium equation: The equation that expresses the equilibrium constant as the ratio of the product of the equilibrium concentrations of the products (raised to their stoichiometric coefficients in the balanced chemical equation) to the product of the equilibrium concentrations of the reactants (also raised to their stoichiometric coefficients in the balanced chemical equation). The equilibrium equation is $K = [C]^c[D]^d/[A]^a[B]^b$ (Equation 15.5), where K is the equilibrium constant for the reaction, A and B are reactants, C and D are products, and a, b, c, and d are their stoichiometric coefficients in the balanced chemical equation for the reaction. (*Section 15.2*)

Equilibrium vapor pressure: The pressure exerted by a vapor in dynamic equilibrium with its liquid. (*Section 11.4*)

Equivalence point: The point in a titration where a stoichiometric amount of the titrant has been added (i.e., the amount required to react completely with the unknown, the substance whose concentration is to be determined). In an acid–base titration, the equivalence point is reached when the number of moles of base (or acid) added equals the number of moles of acid (or base) originally present in the solution. (*Sections 4.9 and 16.5*)

Essential element: Any of the 19 elements (H, C, N, O, Na, Mg, P, S, Cl, K, Ca, Mn, Fe, Co, Cu, Zn, Se, Mo, and I) that are absolutely required in the diets of humans in order for them to survive (see Figure 1.26). Seven more elements (F, Si, V, Cr, Ni, As, and Sn) are thought to be essential for humans. (*Section 1.8*) Approximately 28 elements are known to be essential for the growth of at least one biological species. (*Section 7.5*)

Evaporation: The physical process by which atoms or molecules in the liquid phase enter the gas or vapor phase. Evaporation, also called vaporization, occurs only for those atoms or molecules that have enough kinetic energy to overcome the intermolecular attractive forces holding the liquid together. To escape the liquid, though, the atoms or molecules must be at the surface of the liquid. The atoms or molecules that undergo evaporation create the vapor pressure of the liquid. (*Section 11.4*)

Exact number: An integer obtained either by counting objects or from definitions (e.g., 1 in. = 2.54 cm). Exact numbers have infinitely many significant figures. (*Essential Skills 1*)

Excess reactant: The reactant that remains after a reaction has gone to completion (c.f. limiting reactant). (*Section 3.4*)

Exchange reaction: A chemical reaction that has the general form $AB + C \longrightarrow AC + B$ or $AB + CD \longrightarrow AD + CB$ (see Table 3.1 for examples). (*Section 3.5*)

Excited state: Any arrangement of electrons that is higher in energy than the ground state. When an atom in an excited state undergoes a transition to the ground state, it loses energy by emitting a photon whose energy corresponds to the difference in energy between the two states (c.f. ground state). (*Section 6.3*)

Exothermic: A process in which heat (q) is transferred from the system to the surroundings (c.f. endothermic). By convention, $q < 0$ for an exothermic reaction. (*Section 5.2*)

Expanded-valence molecules: Compounds that have more than an octet of electrons around an atom (c.f. octet rule). The expanded valence shell can be achieved only by elements in Period 3 or higher because they have empty d orbitals that can accommodate the additional electrons. (*Section 8.6*)

Experiments: Systematic observations or measurements, preferably made under controlled conditions—that is, conditions in which the variable of interest is clearly distinguished from any others. (*Section 1.2*)

Extensive property: A physical property that varies with the amount of the substance (e.g., mass, weight, and volume; c.f. intensive property). (*Section 1.3*)

F

Face-centered cubic (fcc) unit cell: A cubic unit cell that contains eight component atoms, molecules, or ions located at the corners of a cube, as well as an identical component in the center of each face of the cube (see Figure 12.5c). (*Section 12.2*)

Faraday (F): The charge on 1 mole of electrons, the faraday is obtained by multiplying the charge on the electron by Avogadro's number (see Equation 19.33). (*Section 19.4*)

***f* block:** The 14 columns usually placed below the main body of the periodic table, the f block consists of elements in which the $(n-2)f$ orbitals are being filled (see Figure 6.34; c.f. s block, p block, and d block). (*Section 6.6*)

***f*-block elements:** The elements in the 14 columns usually placed below the main body of the periodic table. The elements in the f block are filling their $(n-2)f$ orbitals (see Figure 6.34; c.f. s-block elements, p-block elements, and d-block elements). (*Section 6.6*)

Fermentation: A process used by some chemotrophs to obtain energy from their environment, fermentation is a chemical reaction in which both the oxidant and the reductant are organic compounds. Common examples include alcohol fermentation (Equation 18.48) and lactic acid fermentation (Equation 18.49). (*Section 18.8*)

Fiber: A particle of a synthetic polymer that is more than 100 times longer than it is wide. Fibers can be formed by pyrolysis. (*Section 12.8*)

First law of thermodynamics: The energy of the universe is constant. Mathematically, $\Delta E_{\text{universe}} = \Delta E_{\text{system}} + \Delta E_{\text{surroundings}} = 0$ (Equation 18.5), so $\Delta E_{\text{system}} = -\Delta E_{\text{surroundings}}$. The first law of thermodynamics relates the energy change of the system to that of the surroundings. (*Section 18.2*)

First-order reaction: A reaction whose rate is directly proportional to the concentration of one of the reactants (c.f. zeroth-order reaction and second-order reaction; see Table 14.6). First-order reactions have the general form $A \longrightarrow$ products. The differential rate law for a first-order reaction is rate $= k[A]$, where k is the rate constant for the reaction. (*Section 14.3*)

Fissile isotope: Any isotope that can undergo a nuclear fission reaction upon bombardment with neutrons. (*Section 20.4*)

Formal charge: The difference between the number of valence electrons in a free atom and the number of electrons assigned to it in a particular Lewis electron structure (see Equation 8.15). The formal charge is a way of computing the charge distribution within a Lewis structure; the sum of the formal charges on the atoms within a molecule or ion must equal the overall charge on the molecule or ion. A formal charge does not represent a true charge on an atom but is simply used to predict the most likely structure when a compound has more than one Lewis structure. (*Section 8.5*)

Formation constant (K_f): The equilibrium constant for the formation of a complex ion from a hydrated metal ion. The equilibrium constant expression for K_f has the same general form as any other equilibrium constant expression; that is, $K_f = [C]^c[D]^d/[A]^a[B]^b$, where A and B are reactants, C and D are products, and *a*, *b*, *c*, and *d* are their stoichiometric coefficients in the balanced chemical equation for the reaction. Water, a pure liquid, does not appear explicitly in the equilibrium constant expression. (*Section 17.3*)

Formula mass: Also called the empirical formula mass, the formula mass is the sum of the atomic masses of all the elements in the empirical formula, each multiplied by its subscript (written or implied). The formula mass is particularly convenient for ionic compounds, which do not have a readily identifiable molecular unit. Thus, the formula mass is directly analogous to the molecular mass of a covalent compound and is expressed in units of amu. (*Section 3.1*)

Formula unit: The absolute grouping of atoms or ions represented by the empirical formula of a compound, either ionic or covalent. Butane, for example, has the empirical formula C_2H_5, but it contains two C_2H_5 formula units, giving it a molecular formula of C_4H_{10}. (*Section 2.2*)

Fractional crystallization: The separation of compounds based on their relative solubilities in a given solvent. (*Section 13.5*)

Free energy (*G*): See "Gibbs free energy (*G*)" (*Section 18.5*).

Freezing-point depression (ΔT_f): The difference between the freezing point of a pure solvent and the freezing point of the solution (see Equation 13.22): $\Delta T_f = T_f^\circ - T_f$, where T_f° is the melting point of the pure solvent and T_f is the melting point of the solution. ΔT_f is proportional to the molality of the solution (see Equation 13.23). (*Section 13.6*)

Frenkel defect: A defect in an ionic lattice that occurs when one of the ions is in the wrong position. A cation, for example, may occupy a tetrahedral hole instead of an octahedral hole in the lattice. To preserve electrical neutrality, one of the normal cation sites (usually octahedral) must be vacant. (*Section 12.4*)

Frequency (*V*): The number of oscillations (i.e., of a wave) that pass a particular point in a given period of time. The usual units of frequency are oscillations per second or $1/s = s^{-1}$, which in the SI system is called the hertz (Hz). (*Section 6.1*)

Frequency factor (*A*): A constant in the Arrhenius equation (Equation 14.43), the frequency factor converts concentrations to collisions per second. The frequency factor represents the product of the collision frequency and the steric factor (the fraction of collisions with the proper orientation for reaction). (*Section 14.7*)

Fuel cell: A galvanic cell that requires a constant external supply of one or more reactants in order to generate electricity (c.f. battery). (*Section 19.5*)

Fullerenes: One of at least four allotropes of carbon (the other three are graphite, diamond, and nanotubes). The fullerenes (e.g., buckminsterfullerene, C_{60}) are a group of related caged structures (see Figure 7.18). (*Section 7.4*)

Fundamental vibration: The lowest-energy standing wave. (*Section 6.4*)

Fusion: The conversion of a solid to a liquid. Fusion is also called melting. (*Section 11.5*)

G

***G*:** See "Gibbs free energy (*G*)" (*Section 18.5*).

Galvanic cell: One of two types of electrochemical cells, a galvanic cell (or voltaic cell) uses the energy released during a spontaneous oxidation–reduction (redox) reaction ($\Delta G < 0$) to generate electricity (c.f. electrolytic cell, the other type of electrochemical cell). (*Section 19.1*)

Gamma (γ) emission: A nuclear decay reaction that results when a nucleus in an excited state releases energy in the form of a high-energy photon (a γ ray) when it returns to the ground state. Gamma rays are energy, so emitting them leaves the mass number and atomic number of the daughter nuclide unaffected. (*Section 20.2*)

Gas: One of three distinct states of matter under normal conditions, gases have neither fixed shapes nor fixed volumes and expand to fill their containers completely. The volume of a gas depends on its temperature and pressure. The other two states of matter are solid and liquid. (*Section 1.3*)

Gas constant (*R*): A proportionality constant in the ideal gas law (among other equations); R = 0.08206 (L.atm)/(K.mol) = 8.3144 J/(K.mol) = 1.9872 cal/(K.mol). (*Section 10.4*)

Gel: A semisolid sol in which all of the liquid phase has been absorbed by the solid particles. (*Section 13.7*)

Gibbs free energy (*G*): A state function that is defined in terms of three other state functions—namely, enthalpy (*H*), entropy (*S*), and temperature (*T*): $G = H - TS$ (Equation 18.19). (*Section 18.5*)

Glass: An amorphous, translucent solid. A glass is a solid that has been cooled too quickly to form ordered crystals. (*Section 12.1*)

Glass electrode: An electrode used to measure the H^+ ion concentration of a solution, the glass electrode consists of an internal Ag/AgCl electrode immersed in a 1 *M* HCl solution that is separated from the solution by a very thin glass membrane. The glass membrane absorbs protons, which affects the measured potential. The extent of adsorption on the inner side is fixed because $[H^+]$ is fixed inside the electrode, but the adsorption of protons on the outer surface depends on the pH of the solution. (*Section 19.2*)

Grain boundary: The place where two grains in a solid intersect. Most materials consist of many microscopic grains that are randomly oriented with respect to one another, and each grain boundary can be viewed as a two-dimensional dislocation. Defect motion tends to stop at grain boundaries, so controlling the size of the grains in a material controls its mechanical properties. (*Section 12.4*)

Gray (Gy): The SI unit used to measure the effects of radiation on biological tissue, the gray is the amount of radiation that causes 1 joule of energy to be absorbed per kilogram of matter (c.f. rad and roentgen). (*Section 20.3*)

Greenhouse effect: The phenomenon in which substances (e.g., CO_2, water vapor, methane, and chlorofluorocarbons) absorb thermal energy radiated by the earth, thus trapping thermal energy in the earth's atmosphere (analogous to the glass in a greenhouse). (*Section 5.5*)

Greenhouse gases: The substances (e.g., CO_2, water vapor, methane, and chlorofluorocarbons) that absorb thermal energy radiated by the earth, thus trapping thermal energy in the earth's atmosphere (analogous to the glass in a greenhouse). (*Section 5.5*)

Ground state: The most stable arrangement of electrons for an element or compound (c.f. excited state). (*Section 6.3*)

Group: Any of the vertical columns of elements in the periodic table. There are 18 groups in the periodic table, numbered from 1 to 18. The rows of elements in the periodic table are arranged such that elements with similar chemical properties reside in the same group (column). (*Section 1.7*)

Group transfer reaction: A reaction in which a recognizable functional group, such as a phosphoryl unit ($-PO_3^-$), is transferred from one molecule to another. (*Section 7.5*)

H

***H*:** See "Enthalpy (*H*)" (*Section 5.2*).

Half-life ($t_{1/2}$): The period of time it takes for the concentration of a reactant to decrease to one-half its initial value. Thus, the half-life of a reaction is the time required for the reactant concentration to decrease from $[A]_0$ to $[A]_0/2$. If two reactions have the same order, the faster reaction will have a shorter half-life and the slower reaction will have a longer half-life. Unlike zeroth-order or second-order reactions, the half-life of a first-order reaction under a given set of reaction conditions is a constant because it is independent of the concentrations of the reactants. (*Section 14.5*)

Half-reactions: Reactions that represent either the oxidation half or the reduction half of an oxidation–reduction (redox) reaction. Redox reactions can be described as two half-reactions (one for oxidation and one for reduction) because it is impossible to have an oxidation without a reduction, and vice versa. (*Section 19.1*)

Halogen: Any of the elements in Group 17 of the periodic table (F, Cl, Br, I, and At). All of the halogens react readily with metals to give ions with a -1 charge, such as Cl^- and Br^-. (*Section 1.7*)

Hardness (of ionic materials): The resistance of ionic materials to scratching or abrasion. Hardness depends on the lattice energy of the compound because hardness is directly related to how tightly the ions are held together electrostatically. (*Section 8.3*)

Heat (q): Thermal energy that can be transformed from an object at one temperature to an object at another temperature. The net transfer of thermal energy stops when the two objects reach the same temperature. (*Sections 5.1 and 18.1*)

Heat capacity (C): The amount of energy needed to raise the temperature of an object 1°C. The units of heat capacity are joules per degree Celsius (J/°C). (*Section 5.3*)

Heating curve: A plot of the temperature of a substance versus the heat added or versus the heating time at a constant rate of heating. Heating curves relate temperature changes to phase transitions (c.f. cooling curve). (*Section 11.5*)

Heisenberg uncertainty principle: A principle stating that the uncertainty in the position of a particle (Δx) multiplied by the uncertainty in its momentum [$\Delta(mv)$] is greater than or equal to Planck's constant (h) divided by 4π: $(\Delta x)[\Delta(mv)] \geq h/4\pi$. In short, the more accurately we know the exact position of a particle (as $\Delta x \longrightarrow 0$), the less accurately we know the speed and hence the kinetic energy of the particle ($\frac{1}{2}mv^2$) because $\Delta(mv) \longrightarrow \infty$, and vice versa. Because Planck's constant is a very small number, the Heisenberg uncertainty principle is important for only very small particles such as electrons. (*Section 6.4*)

Henderson–Hasselbalch equation: A rearranged version of the equilibrium constant expression that provides a direct way to calculate the pH of a buffer solution. The equation is $pH = pK_a + \log([A^-]/[HA])$ or, more generally, $pH = pK_a + \log([base]/[acid])$. (*Section 16.6*)

Henry's law: An equation that quantifies the relationship between the pressure and the solubility of a gas: $C = kP$, where C is the concentration of dissolved gas at equilibrium, P is the partial pressure of the gas, and k is the Henry's law constant, which must be determined experimentally for each combination of gas, solvent, and temperature. (*Section 13.5*)

Hess's law: The enthalpy change (ΔH) for an overall reaction is the sum of the ΔH values for the individual reactions. Hess's law makes it possible to calculate ΔH values for reactions that are difficult to carry out directly by summing known ΔH values for individual steps that give the overall reaction when added, even though the overall reaction may not actually occur via those steps. (*Section 5.2*)

Heterogeneous catalysis: Catalysis in which the catalyst is in a different phase from the reactants (c.f. homogeneous catalysis). At least one of the reactants in heterogeneous catalysis interacts with the solid surface of the catalyst via adsorption, thus leading first to a weakening of a chemical bond and then to a breaking of the bond. (*Section 14.8*)

Heterogeneous catalyst: A catalyst that is in a different physical state than the reactants (c.f. homogeneous catalyst). (*Section 3.5*)

Heterogeneous equilibrium: An equilibrium in which the reactants of an equilibrium reaction, the products, or both are in more than one phase (c.f. homogeneous equilibrium). The reaction of a gas with a solid or liquid would form a heterogeneous equilibrium. (*Section 15.2*)

Heterogeneous mixture: A mixture in which a material is not completely uniform (e.g., chocolate chip cookie dough, blue cheese, and dirt). (*Section 1.3*)

Heteronuclear diatomic molecule: A molecule that consists of two atoms of different elements (e.g., CO or NO). (*Section 9.3*)

Hexagonal close-packed (hcp) structure: One of two variants of the close-packed arrangement, the most efficient way to pack spheres in a lattice [the other is the cubic close-packed (ccp) arrangement]. The hexagonal close-packed structure results from arranging the atoms of the solid so that their positions alternate from layer to layer in the pattern ABABAB . . . (see Figure 12.7a). (*Section 12.2*)

High-temperature superconductors: Materials that become superconductors at temperatures higher than 30 K (i.e., superconductors with superconducting transition temperatures, T_c, higher than 30 K). (*Section 12.7*)

Homogeneous catalysis: Catalysis in which the catalyst is in the same phase as the reactant(s) (c.f. heterogeneous catalysis). The number of collisions between reactants and catalyst is maximized because the catalyst is uniformly dispersed throughout the reaction mixture. (*Section 14.8*)

Homogeneous catalyst: A catalyst that is uniformly dispersed throughout the reactant mixture to form a solution (c.f. heterogeneous catalyst). (*Section 3.5*)

Homogeneous equilibrium: An equilibrium in which the reactants and products of an equilibrium reaction form a single phase, whether gas or liquid (c.f. heterogeneous equilibrium). When the system is a homogeneous equilibrium, the concentrations of the reactants and products can vary over a wide range. (*Section 15.2*)

Homogeneous mixture: A mixture in which all portions of the material are in the same state, have no visible boundaries, and are uniform throughout (e.g., air and tap water). Homogeneous mixtures are also called solutions. (*Section 1.3*)

Homonuclear diatomic molecule: A molecule that consists of two atoms of the same element (e.g., H_2 or Cl_2). (*Section 9.3*)

Hund's rule: Named for F. H. Hund, the rule states that the lowest-energy electron configuration for an atom is the one that has the maximum number of electrons with parallel spins in degenerate orbitals. (*Section 6.6*)

Hybrid atomic orbitals: The new atomic orbitals formed from the process of hybridization. Hybrid atomic orbitals are equivalent in energy and oriented properly for forming bonds. (*Section 9.2*)

Hybridization: A process in which two or more atomic orbitals that are similar in energy but not equivalent are combined mathematically to produce sets of equivalent orbitals that are properly oriented to form bonds. These new combinations are called hybrid atomic orbitals because they are produced by combining (hybridizing) two or more atomic orbitals from the same atom. (*Section 9.2*)

Hydrate: A compound that contains specific ratios of loosely bound water molecules, called waters of hydration. These loosely bound water molecules can often be removed by simply heating the compound. (*Section 2.2*)

Hydrated ions: Individual cations and anions that are each surrounded by their own shell of water molecules. These shells form when an ionic solid dissolves in water because water molecules are polar. The partially negatively charged oxygen atoms of the H_2O molecules surround the cations (e.g., Na^+ in the case of NaCl), whereas the partially positively charged hydrogen atoms surround the anions (e.g., Cl^-). (*Section 4.1*)

Hydration: The process of surrounding solute particles (atoms, molecules, or ions) with water molecules. Solvation when the solvent is water. (*Section 13.2*)

Hydrocarbons: The simplest class of organic molecules, hydrocarbons consist entirely of carbon and hydrogen. The four major classes of hydrocarbons are the alkanes, alkenes, alkynes, and aromatics. Hydrocarbons may be saturated (alkanes) or unsaturated (alkenes, alkynes, and aromatics). Structurally they may be chains (linear or branched) or cyclic. They may also be aliphatic (alkanes, alkenes, alkynes, and cyclic hydrocarbons) or aromatic. (*Section 2.4*)

Hydrogenation: A chemical reaction in which hydrogen atoms are added to the double bond of an alkene, such as ethylene, to give a product that contains C—C single bonds, in this case ethane. (*Section 14.8*)

Hydrogen bonds: Unusually strong dipole–dipole interactions (intermolecular forces) that result when hydrogen is bonded to very electronegative elements such as O, N, and F. A hydrogen bond consists of a hydrogen bond donor (the hydrogen attached to O, N, or F) and a hydrogen bond acceptor (an atom, usually O, N, or F, that has a lone pair of electrons). (*Section 11.2*)

Hydrolysis reactions: Chemical reactions in which a salt reacts with water to yield an acidic or basic solution. Hydrolysis reactions are just acid–base reactions in which the acid is a cation or the base is an anion; they obey the same principles and rules as all other acid–base reactions. (*Section 16.2*)

Hydronium ion: The H_3O^+ ion. H_3O^+ is a more accurate representation of $H^+(aq)$. (*Sections 4.6 and 16.1*)

Hydrophilic: Substances are classified as hydrophilic if they are attracted to water (c.f. hydrophobic). Hydrophilic substances are polar and often contain O—H or N—H groups that can form hydrogen bonds to water. (*Section 13.3*)

Hydrophobic: Substances are classified as hydrophobic if they are repelled by water (c.f. hydrophilic). Hydrophobic substances are non-polar and usually contain C—H bonds that do not interact favorably with water. (*Section 13.3*)

Hypothesis: A tentative explanation for scientific observations. A hypothesis may not be correct, but it puts the scientist's understanding of the system being studied into a form that can be tested. (*Section 1.2*)

I

***I*:** See "Ionization energy (*I*)" (*Section 7.3*).

Ideal gas: A hypothetical gaseous substance whose behavior is independent of attractive and repulsive forces and can be completely described by the ideal gas law. In reality, there is no such thing as an ideal gas, but it is a useful conceptual model that makes it easier to understand how real gases respond to changing conditions. (*Section 10.4*)

Ideal gas law: $PV = nRT$, where P = pressure, V = volume, n = amount of a gas, R = the gas constant, and T = temperature. The ideal gas law can be obtained by combining the empirical relationships among the volume, temperature, pressure, and amount of a gas (i.e., Boyle's law, Charles's law, and Avogadro's law). (*Section 10.4*).

Ideal solution: A solution that obeys Raoult's law (see Equation 13.13). Like an ideal gas, an ideal solution is a hypothetical system whose properties can be described in terms of a simple model. (*Section 13.6*)

Indicator electrode: The electrode of a galvanic (voltaic) cell whose potential is related to the concentration of the substance being measured. To ensure that any change in the measured potential of the cell is due to only the substance being analyzed, the potential of the other electrode (the reference electrode) must be constant. (*Section 19.2*)

Indicators: Intensely colored organic molecules whose colors change dramatically depending on the pH of the solution. They can be used to determine the approximate pH of a solution. (*Section 4.6*)

Induced dipole: The short-lived dipole moment that is created in atoms and nonpolar molecules adjacent to atoms or molecules that have an instantaneous dipole moment (resulting from the shortlived and ever-fluctuating asymmetrical distribution of its electrons). These instantaneous and induced dipole moments form the basis of London dispersion forces. (*Section 11.2*)

Inert metals: The metals at the bottom of the activity series (Pt, Au, Ag, Cu, and Hg), which have the least tendency to be oxidized (c.f. active metals). (*Section 4.8*)

Initiation: The first stage in a chain reaction, initiation produces one or more reactive intermediates (see radicals). The other stages in a chain reaction are propagation and termination. (*Section 14.6*)

Inorganic compound: An ionic or covalent compound that consists primarily of elements other than carbon and hydrogen (c.f. organic compound). (*Section 2.1*)

Instantaneous dipole moment: The short-lived dipole moment in atoms and nonpolar molecules caused by the constant motion of their electrons, which results in an asymmetrical distribution of charge at any given instant. The instantaneous dipole moment on one atom can interact with the electrons of an adjacent atom, inducing a temporary dipole moment in the second atom. These instantaneous and induced dipole moments form the basis of London dispersion forces. (*Section 11.2*)

Instantaneous reaction rate: The rate of a chemical reaction at any given point in time (c.f. average reaction rate). As the period of time used to calculate an average rate of a reaction becomes shorter and shorter, the average rate approaches the instantaneous rate. (*Section 14.2*)

Integrated rate law: A rate law that expresses the rate of a reaction in terms of the initial concentration, $[R]_0$, and the measured concentration of one or more reactants, [R], after a given amount of time, t (c.f. differential rate law). The integrated rate law can be found by using calculus to integrate the differential rate law. (*Section 14.2*)

Intensive property: A physical property that does not depend on the amount of the substance (e.g., color, melting point, boiling point, electrical conductivity, and physical state at a given temperature; c.f. extensive property). (*Section 1.3*)

Intermediate: A species in a reaction mechanism that does not appear in the balanced chemical equation for the overall reaction. Intermediates are formed as products in one step (elementary reaction) of the mechanism and consumed again as reactants in a later step in the mechanism. (*Section 14.6*)

Intermetallic compound: An alloy that consists of certain metals that combine in only specific proportions. The structures and physical properties of intermetallic compounds are frequently quite different from those of their constituent elements. (*Section 12.5*)

Internal energy (*E*): The sum of the kinetic and potential energies of all of a system's components. Additionally, $\Delta E = q + w$, where q is the heat produced by the system and w is the work performed by the system. Internal energy is a state function. (*Sections 5.2 and 18.1*)

Interstitial alloy: An alloy formed by inserting smaller atoms into holes in the metal lattice (c.f. substitutional alloy and intermetallic compound). (*Section 12.5*)

Interstitial impurity: A point defect that results when an impurity atom occupies an octahedral or tetrahedral hole in the lattice between atoms (c.f. vacancy and substitutional impurity). (*Section 12.4*)

Ion: A charged particle produced when one or more electrons is removed from or added to an atom or molecule. Ions may be cations (positively charged) or anions (negatively charged). (*Section 1.6*)

Ionic compound: A compound consisting of positively charged ions (cations) and negatively charged ions (anions) held together by strong electrostatic forces (c.f. covalent compound). The ratio of cations to anions in an ionic compound is such that there is no net electrical charge. In an ionic compound, the cations and anions are arranged in space to form an extended three-dimensional array that maximizes the number of attractive electrostatic interactions and minimizes the number of repulsive electrostatic interactions. (*Section 2.1*)

Ionic liquids: Ionic substances that are liquids at room temperature. Ionic liquids consist of small, symmetrical anions (e.g., PF_6^- and BF_4^-), combined with larger, symmetrical organic cations that prevent the formation of a highly organized structure, resulting in a low melting point. (*Section 11.6*)

Ionic radius: The radius of a cation or anion. The internuclear distance between a cation and an adjacent anion in an ionic compound is measured experimentally and then divided proportionally between the smaller cation and larger anion (see Figure 7.8). The different methods used to proportionally divide the experimentally measured internuclear distance give slightly different values for the ionic radii, but they do give sets of ionic radii that are internally consistent from one ionic compound to another. (*Section 7.2*)

Ionic solid: A solid that consists of positively and negatively charged ions held together by electrostatic forces. Ionic solids tend to have high melting points and to be rather hard and brittle (c.f. molecular solid, covalent solid, and metallic solid). (*Section 12.5*)

Ionization energy (*I*): The minimum amount of energy needed to remove an electron from the gaseous atom E in its ground state. I is therefore the energy change for the reaction, $E(g) + I \longrightarrow E^+(g) + e^-$. Because an input of energy is needed, the ionization energy is always positive ($I > 0$) for the reaction as written here (c.f. electron affinity). Higher values of I mean that the electron is more tightly bound to the atom and harder to remove. (*Section 7.3*)

Ionizing radiation: Higher in energy than nonionizing radiation, some of the energy of ionizing radiation can be transferred to one or more atoms with which it collides as it passes through matter. If enough energy is transferred, electrons can be excited to very high energy levels, resulting in the formation of positively charged ions. (*Section 20.3*)

Ion pair: A cation and anion that are in intimate contact in solution rather than separated by solvent (see Figure 17.2). The ions in an ion pair are held together by the same attractive forces as in ionic solids. The ions in an ion pair migrate in solution as a single unit (whose net charge is the sum of the charges on the ions). An ion pair can be viewed as a species intermediate between the ionic solid (in which each ion participates in many of the cation–anion interactions that hold the ions in a rigid array) and the completely dissociated ions in solution (where each is fully surrounded by solvent molecules and free to migrate independently). (*Sections 13.6 and 17.2*)

Ion-product constant of liquid water (K_w): An equilibrium constant for the autoionization of water, $2H_2O(l) \rightleftharpoons H_3O^+(aq) + OH^-(aq)$, in which the concentration of water is treated as a constant. Thus, $K_w = [H_3O^+][OH^-] = 1.006 \times 10^{-14}$ at 25°C. Like any other equilibrium constant, the value of K_w varies with temperature, ranging from 1.15×10^{-15} at 0°C to 4.99×10^{-13} at 100°C. (*Section 16.1*)

Ion product (*Q*): A quantity that has precisely the same form as the solubility product (K_{sp}) for the dissolution of a sparingly soluble salt, except that the concentrations used for Q are not necessarily equilibrium concentrations. If $Q < K_{sp}$, the solution is unsaturated and more of the ionic solid will dissolve. If $Q = K_{sp}$, the solution is saturated and at equilibrium. If $Q > K_{sp}$, the solution is supersaturated and some of the ionic solid will precipitate. (*Section 17.1*)

Ion pump: A mechanism that selectively transports ions (e.g., Na^+, Ca^{2+}, Cl^-, K^+, Mg^{2+}, and phosphate) across cell membranes. The selectivity of ion pumps is based on differences in ionic radii and ionic charges. (*Section 7.5*)

Ion-selective electrodes: Electrodes used to measure the concentration of a particular species in solution, ion-selective electrodes are designed so that their potential depends on only the concentration of the desired species. (*Section 19.2*)

Irreversible process: A process in which the intermediate states between the extremes are not equilibrium states and change occurs spontaneously in only one direction (c.f. reversible process). A reversible process can change direction at any time, whereas an irreversible process cannot. (*Section 18.3*)

Isoelectronic series: A group of ions or atoms and ions that have the same number of electrons (and thus the same ground-state electron configurations). For example, the isoelectronic series of ions with the neon closed-shell electron configuration ($1s^22s^22p^6$) is N^{3-}, O^{2-}, F^-, Na^+, Mg^{2+}, and Al^{3+}. (*Section 7.2*).

Isolated system: One of three kinds of system, an isolated system can exchange neither energy nor matter with its surroundings. A truly isolated system does not exist, however, because energy is always exchanged between a system and its surroundings. The other kinds of systems are open and closed. (*Section 5.2*)

Isotopes: Atoms that have the same number of protons, and hence the same atomic number (*Z*), but different numbers of neutrons. All isotopes of an element have the same number of protons and electrons, so they exhibit the same chemistry. (*Section 1.6*)

Isotropic: The arrangement of molecules that is equally disordered in all directions (c.f. anisotropic). Normal liquids are isotropic because their molecules possess enough thermal energy to overcome their intermolecular attractive forces and tumble freely. (*Section 11.8*)

J

Joule (J): The SI unit of energy; 1 joule = 1 (kilogram · meter^2)/second^2. (*Section 5.1*)

K

***k*:** See "Rate constant (k)" (*Section 14.2*).

***K*:** See "Equilibrium constant (K)" (*Section 15.2*).

K_a: See "Acid ionization constant (K_a)" (*Section 16.2*).

K_b: See "Base ionization constant (K_b)" (*Section 16.2*).

***KE*:** See "Kinetic energy (KE)" (*Section 5.1*).

Ketone: A class of organic compounds with the general form RC(O)R′, in which the carbon atom of the carbonyl group is bonded to two alkyl groups (c.f. aldehyde). The alkyl groups may be the same or different. (*Section 4.1*)

K_f: See "Formation constant (K_f)" (*Section 17.3*).

Kinetic control: The altering of reaction conditions to control reaction rates, thereby obtaining a single desired product or set of products (c.f. thermodynamic control). (*Section 15.6*)

Kinetic energy (*KE*): Energy due to the motion of an object. $KE = \frac{1}{2}mv^2$, where m is the mass of the object and v is its velocity. (*Section 5.1*)

Kinetic molecular theory of gases: A theory that describes (on the molecular level) why ideal gases behave the way they do. The basic postulates of kinetic molecular theory are that gas particles are in constant random motion, they are so far apart that their individual volumes are negligible, their intermolecular interactions are negligible, all of their collisions are elastic, and the average kinetic energy of the molecules of any gas depends only on the temperature, and at a given temperature, the molecules of all gases have the same average kinetic energy. (*Section 10.7*)

K_p: See "Equilibrium constant (K_p)" (*Section 15.2*).

K_{sp}: See "Solubility product (K_{sp})" (*Section 17.1*).

K_w: See "Ion-product constant of liquid water (K_w)" (*Section 16.1*).

L

***l*:** See "Azimuthal quantum number (l)" (*Section 6.5*).

Lanthanide element: Any of the 14 elements between $Z = 58$ (cerium) and $Z = 71$ (lutetium). The $4f$ orbitals are filling for the lanthanides and their chemistry is dominated by M^{3+} ions. The lanthanides and actinides are usually grouped together in two rows beneath the main body of the periodic table. (*Sections 1.7 and 7.4*)

Lattice energy: The enthalpy change that occurs when a solid ionic compound (whose ions form a three-dimensional array called a lattice) is transformed into gaseous ions (c.f. bond energy). Lattice energies are highest for substances that contain small, highly charged ions. (*Section 8.1*)

Law: A verbal or mathematical description of a phenomenon that allows for general predictions. A law says *what* happens, not *why* it happens. (*Section 1.2*)

Law of conservation of energy: The total amount of energy in the universe remains constant. Energy cannot be created or destroyed. Energy can be converted from one form to another, however. (*Section 5.1*)

Law of conservation of mass: In any chemical reaction, the mass of the substances that react equals the mass of the products that are formed. (*Section 1.4*)

Law of definite proportions: Formulated by the French scientist Joseph Proust (1754–1826), the law states that a chemical substance always contains the same proportions of elements by mass. (*Section 1.2*)

Law of mass action: The ratio of the product of the equilibrium concentrations of the products (raised to their stoichiometric coefficients in the balanced chemical equation) to the product of the equilibrium concentrations of the reactants (also raised to their stoichiometric coefficients in the balanced chemical equation). For any reversible reaction of the general form $aA + bB \rightleftharpoons cC + dD$, where A and B are reactants, C and D are products, and a, b, c, and d are the stoichiometric coefficients in the balanced chemical equation for the reaction, the law of mass action can be stated as $K = [C]^c[D]^d/[A]^a[B]^b$, where K is the equilibrium constant for the reaction. (*Section 15.2*)

Law of multiple proportions: When two elements form a series of compounds, the ratios of the masses of the second element that are present per gram of the first element can almost always be expressed as the ratios of integers. (The same law holds for the mass ratios of compounds forming a series that contains more than two elements.) (*Section 1.4*)

LCAOs: See "Linear combination of atomic orbitals (LCAOs)" (*Section 9.3*).

Le Châtelier's principle: Formulated by the French chemist Henri Louis Le Châtelier (1850–1936), the principle states that if a stress is applied to a system at equilibrium, the composition of the system will change to relieve the applied stress. The types of stresses that can change the composition of an equilibrium mixture are (1) a change in the concentrations (or partial pressures) of the components by the addition or removal of reactants or products, (2) a change in the total pressure or volume, and (3) a change in the temperature of the system. (*Section 15.5*)

Leclanché dry cell: Actually a "wet cell" battery, the electrolyte in the Leclanché cell is an acidic water-based paste that contains MnO_2, NH_4Cl, $ZnCl_2$, graphite, and starch (see Figure 19.12a). (*Section 19.5*)

Leveling effect: The phenomenon that makes H_3O^+ the strongest acid that can exist in water. As a result, any species (e.g., HI or HNO_3) that is a stronger acid than the conjugate acid of water (namely, H_3O^+) is leveled to the strength of H_3O^+ in aqueous solution. The leveling effect makes it impossible to distinguish in aqueous solution between the strengths of acids that are stronger than H_3O^+. (*Section 16.2*)

Lewis acid: Any species that can *accept* a pair of electrons (c.f. the Arrhenius and Brønsted–Lowry definitions of an acid). The Lewis definition expands the Brønsted–Lowry definition to include substances other than the H^+ ion. (*Section 8.7*)

Lewis base: Any species that can *donate* a pair of electrons (c.f. the Arrhenius and Brønsted–Lowry definitions of a base). All Brønsted–Lowry bases (proton acceptors) are also electron-pair donors, so the Lewis definition of a base does not contradict the Brønsted–Lowry definition. (*Section 8.7*)

Lewis dot symbols: A system of symbols developed by G. N. Lewis in the early 20th century that can be used to predict the number of bonds formed by most elements in their compounds. Each Lewis dot symbol consists of the chemical symbol for an element surrounded by dots that represent its valence electrons. (*Section 8.4*)

Ligand: A molecule or ion that contains at least one lone pair of electrons. Ligands form complex ions with metal ions. The ligand acts as a Lewis base and the metal ion acts as a Lewis acid in a Lewis acid–base interaction. (*Section 17.3*)

Limiting reactant: The reactant that restricts the amount of product obtained in a chemical reaction (c.f. excess reactant). (*Section 3.4*)

Linear: The lowest-energy arrangement for compounds that have two electron pairs around the central atom. In the linear arrangement, the electron pairs are on opposite sides of the central atom with a 180° angle between them (see Table 9.1). Linear is also one of two possible molecular geometries for an AB_2 molecular species (the other is bent). (*Section 9.1*)

Linear combination of atomic orbitals (LCAOs): Molecular orbitals created from the sum and the difference of two wave functions (atomic orbitals). According to the LCAO method, a molecule has as many molecular orbitals as there are atomic orbitals. (*Section 9.3*)

Line defect: A defect in a crystal that affects a row of points in the lattice (c.f. point defect and plane defect). (*Section 12.4*)

Line spectrum: A spectrum in which light of only certain wavelengths is emitted or absorbed, rather than a continuous range of wavelengths (colors). The line spectrum of each element is characteristic of that element. (*Section 6.3*)

Liquefaction: The condensation of gases into a liquid form. Gases invariably condense to form liquids because they no longer possess enough kinetic energy to overcome the intermolecular attractive forces. (*Section 10.8*)

Liquid: One of three distinct states of matter under normal conditions, liquids have fixed volumes but flow to assume the shape of their containers. The volume of a liquid is virtually independent of temperature and pressure. The other two states of matter are solid and gas. (*Section 1.3*)

Liquid crystals: Substances that exhibit phases that have properties intermediate between those of a crystalline solid and those of a normal liquid. Liquid crystals possess long-range molecular order (like a solid) but still flow (like a liquid). Liquid crystals typically consist of long, rigid molecules that can interact strongly with one another. (*Section 11.8*)

Lithium–iodine battery: A battery that consists of an anode of lithium metal and a cathode containing a solid complex of I_2. Separating the anode and the cathode is a layer of solid LiI, which acts as the electrolyte by allowing the diffusion of Li^+ ions. A typical lithium–iodine battery is shown in Figure 19.12c. (*Section 19.5*)

London dispersion forces: A kind of intermolecular interaction (force) that results from temporary fluctuations in the electron distribution within atoms and nonpolar molecules. These fluctuations produce instantaneous dipole moments that induce similar dipole moments in adjacent molecules, resulting in attractive forces between otherwise nonpolar substances. These attractive interactions are weak and fall off rapidly with increasing distance (as $1/r^6$, where r is the distance between dipoles). London dispersion forces and dipole–dipole interactions are often called van der Waals forces. (*Section 11.2*)

Lone pair: A pair of electrons in a Lewis structure that is not involved in covalent bonding (c.f. bonding pair). Lone pairs are valence electrons, but they are not used to bond with other atoms. (*Section 8.5*)

Luster: The ability to reflect light. Metals, for instance, have a shiny surface that reflects light (metals are lustrous), whereas nonmetals do not. (*Section 12.5*)

Lustrous: Having a shiny appearance. Metals are lustrous, whereas nonmetals are not. (*Section 1.7*)

M

Macromineral: Any of the six essential elements—sodium, magnesium, potassium, calcium, chlorine, and phosphorus—that provide essential ions in body fluids and form the major structural components of the body. They are found in large amounts in biological tissues and are present as inorganic compounds, either dissolved or precipitated. (*Sections 1.8 and 7.5*)

Magnetic quantum number (m_l): One of three quantum numbers (the other two are n and l) used to specify any wave function, the magnetic quantum number describes the orientation of the region of space occupied by the electron with respect to an applied magnetic field. The allowed values of m_l depend on the value of l and can range from $-l$ to l in integral steps. (*Section 6.5*)

Main group element: Any of the elements in Groups 1, 2, and 13–18 in the periodic table. These groups contain metals, nonmetals, and semimetals. (*Section 1.7*)

Malleable: The ability to be hammered or pressed into thin sheets or foils. Metals are malleable, whereas nonmetals are usually brittle instead. (*Section 1.7*)

Manometer: A device used to measure the pressures of samples of gases contained in an apparatus (c.f. barometer). The key feature of a manometer is a U-shaped tube that contains mercury (or some other nonvolatile liquid). A closed-end manometer is shown schematically in Figure 10.5a. (*Section 10.2*)

Mass: The quantity of matter an object contains. Mass is a fundamental property of an object that does not depend on its location (c.f. weight). (*Section 1.3*)

Mass defect: The difference between the sum of the masses of the components of an atom (protons, neutrons, and electrons) and the measured atomic mass. The mass defect occurs because a nucleus is more stable (lower in energy) than its isolated components. When isolated nucleons assemble into a stable nucleus, energy is released, which must be accompanied by a decrease in mass (because $E = mc^2$, where E is energy, m is mass, and c is the speed of light). (*Section 20.4*)

Mass number (A): The number of protons and neutrons in the nucleus of an atom of an element. The isotopes of an element differ only in their atomic mass, which is given by the mass number (A). (*Section 1.6*)

Mass percentage: A common unit of concentration, the mass percentage of a solution is the ratio of the mass of the solute to the total mass of the solution (see Equation 13.8). (*Section 13.4*)

Matter: Anything that occupies space and possesses mass. (*Section 1.3*)

Mechanical work: The energy required to move an object a distance d when opposed by a force F, such as gravity: work (w) = force (F) $\times$ distance (d). (*Section 5.1*)

Meissner effect: The phenomenon, first described by W. Meissner in 1933, in which a superconductor completely expels a magnetic field from its interior. (*Section 12.7*)

Melting point: The temperature at which the individual ions in a lattice or the individual molecules in a covalent compound have enough kinetic energy to overcome the attractive forces that hold them in place in the solid. At the melting point, the ions or molecules can move freely and the substance becomes a liquid. Thus, the solid and the liquid coexist in equilibrium at the melting point. (*Section 8.3*)

Meniscus: The upper surface of a liquid in a tube. The shape of the meniscus depends on the relative strengths of the cohesive and adhesive forces. Liquids (e.g., water) for which the adhesive forces are stronger than the cohesive forces have a concave meniscus, whereas liquids (e.g., mercury) for which the cohesive forces are stronger than the adhesive forces have a convex meniscus. (*Section 11.3*)

Metal: Any of the elements to the left of the zigzag line in the periodic table that runs from boron (B) down to astatine (At). Metals are good conductors of electricity and heat, they can be pulled into wires (ductility), they can be hammered into thin sheets or foils (malleability), and most are shiny (lustrous). In chemical reactions, metals tend to lose electrons to form positively charged ions (cations). All metals except mercury (Hg) are solids at room temperature and pressure. (*Section 1.7*)

Metallic atomic radius: Half the distance between the nuclei of two adjacent metal atoms (see Figure 7.5b). (*Section 7.2*)

Metallic solid: A solid that consists of metal atoms held together by metallic bonds. Metallic solids have high thermal and electrical conductivities, they are malleable and ductile, and they have luster (c.f. ionic solid, molecular solid, and covalent solid). (*Section 12.5*)

Metal-matrix composites: A composite that consists of reinforcing fibers embedded in a metal or metal alloy matrix. The fibers tend to consist of boron, graphite, or ceramic. (*Section 12.9*)

Micelle: A spherical or cylindrical aggregate of detergents or soaps in water that minimizes contact between the hydrophobic tails of the detergents or soaps and water. Only the hydrophilic heads of the detergents or soaps are in direct contact with water, while the hydrophobic tails are located in the interior of the micelle aggregate. Micelles form spontaneously above a certain concentration. (*Section 13.7*)

Midpoint: The point in an acid–base titration at which exactly enough acid (or base) has been added to neutralize one-half of the base (or acid) originally present. The midpoint occurs halfway to the equivalence point. By definition, $[HA] = [A^-]$ at the midpoint of the titration of an acid (HA). (*Section 16.5*)

Millimeters of mercury (mmHg): A unit of pressure, often called the torr (after Evangelista Torricelli, the inventor of the mercury barometer); 760 mmHg = 760 torr = 1 atmosphere (atm) = 101325 pascals (Pa) = 101.325 kPa. (*Section 10.2*)

Miscible: Capable of forming a single homogeneous phase, regardless of the proportions with which the substances are mixed. Ethanol and water are miscible because they form a solution when mixed in all proportions. (*Section 13.1*)

Mixture: A combination of two or more pure substances in variable proportions in which the individual substances retain their identities. (*Section 1.3*)

m_l: See "Magnetic quantum number (m_l)" (*Section 6.5*).

MO: See "Molecular orbital (MO)" (*Section 9.3*).

Molality (*m*): A common unit of concentration, the molality of a solution is the number of moles of solute present in exactly one kilogram of solvent (c.f. molarity). (*Section 13.4*)

Molar heat capacity (C_p): The amount of energy needed to increase the temperature of 1 mol of a substance by 1° C. The units of C_p are J/(mol·°C). (*Section 5.3*)

Molarity (*M*): A common unit of concentration, the molarity of a solution is the number of moles of solute present in exactly 1 L of solution (c.f. molality). The molarity is also the number of millimoles of solute present in exactly 1 mL of solution. The units of molarity are moles per liter of solution (mol/L), abbreviated as *M*. (*Section 4.2*)

Molar mass: The mass in grams of one mole of a substance. The molar mass of any substance is numerically equivalent to its atomic mass, molecular mass, or formula mass in grams per mole. (*Section 3.1*)

Molar volume: The molar mass (g/mol) of an element divided by its density (g/cm^3). Molar volume, therefore, has units of cm^3/mol. (*Section 7.1*)

Mole (mol): The quantity of a substance that contains the same number of units (e.g., atoms or molecules) as the number of carbon atoms in exactly 12 g of isotopically pure carbon-12. According to the most recent experimental measurements, 12 g of carbon-12 contains 6.0221367×10^{23} atoms. (*Section 3.1*)

Molecular formula: A representation of a covalent compound that consists of the atomic symbol for each component element (in a prescribed order) accompanied by a subscript indicating the number of atoms of that element in the molecule. The subscript is written only if the number is greater than 1. Molecular formulas give only the elemental composition of molecules (see Figure 2.4a; c.f. structural formula and empirical formula). (*Section 2.1*)

Molecular geometry: The arrangement of the bonded atoms in a molecule or polyatomic ion in space (c.f. electron-pair geometry). Knowing the molecular geometry of a compound is crucial to understanding its chemistry. (*Section 9.1*)

Molecular mass: The sum of the average masses of the atoms in one molecule of the substance. The molecular mass is calculated by summing the atomic masses of the elements in the substance, each multiplied by its subscript (written or implied) in the molecular formula, and is expressed in units of amu. It is analogous to the formula mass of an ionic compound. (*Section 3.1*)

Molecularity: The number of molecules that collide during any step (elementary reaction) in a reaction mechanism. An elementary reaction is unimolecular if only a single reactant molecule is involved; it is bimolecular if two reactant molecules are involved; and it is termolecular (a relatively rare situation) if three reactant molecules are involved (see Table 14.8). The reaction order of an elementary reaction is the same as its molecularity. (*Section 14.6*)

Molecular orbital (MO): A particular spatial distribution of electrons in a molecule that is associated with a particular orbital energy. Unlike an atomic orbital, which is localized on a single atom, a molecular orbital extends (is delocalized) over all of the atoms in a molecule or ion. (*Section 9.3*)

Molecular orbital theory: A delocalized bonding model (c.f. valence bond theory). In molecular orbital theory, the linear combination of atomic orbitals creates molecular orbitals that can be used to explain the bonding in molecules and polyatomic ions as well as their concomitant molecular geometries. (*Section 9.3*)

Molecular solid: A solid that consists of molecules held together by relatively weak forces, such as dipole–dipole interactions, hydrogen bonds, and London dispersion forces (c.f. ionic solid, covalent solid, and metallic solid). (*Section 12.5*)

Molecule: A group of atoms in which one or more pairs of electrons are shared between bonded atoms. Covalent compounds consist of molecules. (*Section 2.1*)

Mole fraction (*X*): The ratio of the number of moles of any component of a mixture to the total number of moles of all species present in the mixture (n_t): mole fraction of component A = X_A = moles of A/total moles = n_A/n_t (see Equation 10.29). The mole fraction, a dimensionless quantity between 0 and 1, can be used to describe the composition of a gas mixture. In a mixture of gases, the partial pressure of each of the component gases is the product of the total pressure and the mole fraction of that gas. (*Section 10.5*)

Mole ratio: The ratio of the number of moles of one substance to the number of moles of another, as depicted by a balanced chemical equation. (*Section 3.3*)

Molten salt: A salt that has been heated to its melting point. Molten salts conduct electricity, have high heat capacities, attain very high temperatures as liquids, and are useful as solvents because of their relatively low toxicity. (*Section 11.6*)

Monatomic ion: An ion that contains only a single atom (e.g., Na^+, Al^{3+}, Cl^-, or S^{2-}). (*Section 2.1*)

Monomers: The basic structural units of polymers. Many monomer molecules are connected in polymers, forming chains or networks via covalent bonds. (*Section 12.8*)

Monoprotic acid: A compound that is capable of donating a single proton per molecule (e.g., HF and HNO_3; c.f. polyprotic acid). (*Section 4.6*)

N

***n*:** See "Principal quantum number (n)" (*Section 6.5*).

Nanotubes: One of at least four allotropes of carbon (the other three are graphite, diamond, and the fullerenes). Nanotubes are cylinders of carbon atoms (see Figure 7.18) and are intermediate in structure between graphite and the fullerenes. Carbon nanotubes can be described as sheets of graphite that have been rolled up into a cylinder, or as fullerene cages that have been stretched in one direction. (*Section 7.4*)

Natural logarithm: The power x to which e (an irrational number whose value is approximately 2.7183) must be raised to obtain a particular number. (*Essential Skills 6*)

Nematic phase: One of three different ways that most liquid crystals can orient themselves (the other two are the smectic and cholesteric phases). In the nematic phase, only the long axes of the molecules are aligned, so they are free to rotate or to slide past one another. (*Section 11.8*)

Nernst equation: An equation for calculating cell potentials (E_{cell}) under nonstandard-state conditions, the Nernst equation can be used to determine the direction of spontaneous reaction for any redox reaction under any conditions. $E_{cell} = E^{\circ}_{cell} - (RT/nF) \ln Q$, where E_{cell} is the cell potential, E°_{cell} is the standard cell potential, R is the ideal gas constant, T is the absolute temperature, n is the number of moles of electrons, F is the faraday constant, and Q is the reaction quotient (Equation 19.46). (*Section 19.4*)

Net ionic equation: A chemical equation that shows only those species that participate in the chemical reaction. Canceling the spectator ions from a complete ionic equation gives the net ionic equation. (*Section 4.4*)

Neutralization reaction: A chemical reaction in which an acid and a base react in stoichiometric amounts to produce water and a salt. (*Section 4.6*)

Neutral solution: A solution in which the total positive charge from all of the cations is matched by an identical total negative charge from all of the anions. Pure water is a neutral solution in which $[H^+] = [OH^-] = 1.0 \times 10^{-7}$ M at 25°C, and which has a pH of 7.0. (*Section 4.6*)

Neutron: A subatomic particle that resides in the nucleus of almost all atoms. A neutron has a mass of 1.675×10^{-24} g (1.008665 amu), which is only slightly greater than the mass of a proton, but no charge (neutrons are electrically neutral). Neutrons and protons constitute by far the bulk of the mass of atoms. (*Section 1.5*)

Nickel–cadmium (nicad) battery: A water-based electrochemical cell, the nicad battery has a cadmium anode and a highly oxidized nickel cathode that is usually described as the nickel(III) oxo-hydroxide, NiO(OH). (*Section 19.5*)

Noble gas: Any of the elements in Group 18 of the periodic table (He, Ne, Ar, Kr, Xe, and Rn). All are unreactive monatomic gases at room temperature and pressure. (*Section 1.7*)

Nodes: Points where the amplitude of a wave is zero. For a standing wave, such as a plucked guitar string, the string does not move at the nodes. All overtones of the fundamental standing wave have one or more nodes. (*Section 6.4*)

Nonaqueous solution: A solution in which any substance other than water is the solvent (c.f. aqueous solution). (*Chapter 4 introduction*)

Nonbonding molecular orbital: A molecular orbital that forms when atomic orbitals or orbital lobes interact only very weakly, creating essentially no change in the electron probability density between the nuclei. Nonbonding molecular orbitals have approximately the same energy as the parent atomic orbitals. They are one of three types of molecular orbitals that can form when atomic orbitals interact (the other two are bonding and antibonding molecular orbitals). (*Section 9.3*)

Nonelectrolyte: A substance (e.g., ethanol or glucose) that dissolves in water to form neutral molecules. Nonelectrolytes have essentially no effect on the conductivity of water (c.f. electrolytes). (*Section 4.1*)

Nonionizing radiation: Radiation that is relatively low in energy, so that when it collides with an atom in a molecule or ion, most or all of its energy can be absorbed without causing a structural or chemical change (c.f. ionizing radiation). Instead, the kinetic energy of the radiation is transferred to the atom or molecule with which it collides, causing the atom or molecule to rotate, vibrate, or move more rapidly. (*Section 20.3*)

Nonmetal: Any of the elements to the right of the zigzag line in the periodic table that runs from boron (B) down to astatine (At). Nonmetals are generally poor conductors of heat and electricity, they are not shiny (lustrous), and solid nonmetals tend to be brittle (so they cannot be pulled into wires or hammered into sheets or foils). Nonmetals tend to gain electrons in reactions with metals to form negatively charged ions (anions) or to share electrons in reactions with other nonmetals. Nonmetals may be solids, liquids, or gases at room temperature and pressure. (*Section 1.7*)

Nonstoichiometric compounds: Solids that have intrinsically variable stoichiometries. Nonstoichiometric compounds contain large numbers of defects, usually vacancies, which give rise to stoichiometries that can depart significantly from simple integral ratios without affecting the fundamental structure of the crystal. (*Section 12.4*)

Nonvolatile liquids: Liquids that have relatively low vapor pressures. Nonvolatile liquids tend to evaporate more slowly from an open container than volatile liquids do. (*Section 11.4*)

Normal boiling point: The temperature at which a substance boils when the external pressure is one atmosphere. For water, the normal boiling point is exactly 100°C. (*Section 11.4*)

***n*-Type semiconductor:** A semiconductor that has been doped with an impurity that contains more valence electrons than the atoms of the host lattice (c.f. p-type semiconductor). The increased number of valence electrons helps populate the conduction band, thus increasing the electrical conductivity of the semiconductor. (*Section 12.6*)

Nuclear binding energy: The amount of energy released when a nucleus forms from its component nucleons. The nuclear binding energy is the amount of energy that corresponds to the mass defect of the nucleus because energy and mass are related by $E = mc^2$, where E is energy, m is mass, and c is the speed of light. (*Section 20.4*)

Nuclear chain reaction: A self-sustaining series of nuclear fission reactions. A minimum mass of a fissile isotope is required to sustain a nuclear chain reaction (see critical mass). (*Section 20.4*)

Nuclear decay reaction: Also called radioactive decay, a nuclear decay reaction occurs when an unstable nucleus emits radiation and is transformed into the nucleus of one or more other elements. The resulting daughter nuclei have a lower mass and are lower in energy (more stable) than the parent nucleus that decayed (c.f. nuclear transmutation reaction). (*Section 20.2*)

Nuclear energy: One of the five forms of energy, nuclear energy is stored in the nucleus of an atom. The other four forms of energy are radiant, thermal, chemical, and electrical. (*Section 5.1*)

Nuclear fission: The splitting of a heavy nucleus into two lighter ones (c.f. nuclear fusion). (*Section 20.4*)

Nuclear fusion: The combining of two light nuclei to produce a heavier, more stable nucleus (c.f. nuclear fission). (*Section 20.4*)

Nuclear transmutation reaction: A nuclear reaction in which a nucleus reacts with a subatomic particle or another nucleus to give a product nucleus that is more massive than the starting material (c.f. nuclear decay reaction). An artificially induced reaction that converts an otherwise stable nuclide to another, more massive nuclide. (*Section 20.2*)

Nucleons: The protons and neutrons that make up the nucleus of an atom. (*Section 20.1*)

Nucleus: The central core of an atom, where the protons and any neutrons reside. The nucleus comprises most of the mass of an atom but very little of the volume. (*Section 1.5*)

Nuclide: An atom with a particular number of nucleons (protons and neutrons). Nuclides that have the same number of protons, but different numbers of neutrons, are called isotopes. (*Section 20.1*)

O

Octahedral: The lowest-energy arrangement for compounds that have six electron pairs around the central atom. In the octahedral arrangement, each electron pair is positioned at 90° angles to four adjacent electron pairs and at a 180° angle to the fifth electron pair (that is, they are positioned at the vertices of an octahedron) (see Table 9.1). Octahedral is also the most common molecular geometry for an AB_6 molecular species. (*Section 9.1*)

Octahedral hole: One of two kinds of holes in a face-centered cubic array of atoms or ions (the other is a tetrahedral hole). One octahedral hole is located in the center of the face-centered cubic unit cell, and there is a shared one in the middle of each edge. An atom or ion in an octahedral hole has a coordination number of 6. (*Section 12.3*)

Octane rating: A measure of a fuel's ability to burn in a combustion engine without knocking or pinging (indications of premature combustion). The higher the octane rating, the higher quality the fuel. *n*-Heptane, for example, which causes a great deal of knocking upon combustion, has an octane rating of 0, whereas isooctane, a very smooth-burning fuel, has an octane rating of 100. (*Section 2.6*)

Octaves: Groups of seven elements, corresponding to the horizontal rows in the main group elements (not counting the noble gases, which were unknown at the time). John Newlands arranged the elements that were known in the mid-19th century in order of increasing atomic mass and discovered that every seventh element had similar properties. He thus proposed the "law of octaves" to explain this pattern, but it turned out that the law did not seem to work for elements heavier than calcium. (*Section 7.1*)

Octet rule: The tendency for atoms to lose, gain, or share electrons to reach a total of eight valence electrons (c.f. expanded-valence molecules). The octet rule explains the stoichiometry of most compounds in the *s* and *p* blocks of the periodic table. (*Section 8.4*)

Open system: One of three kinds of systems, an open system can exchange both matter and energy with its surroundings. The other kinds of system are closed and isolated. (*Sections 5.2 and 18.1*)

Organic compound: A covalent compound that contains predominantly carbon and hydrogen (c.f. inorganic compound). (*Section 2.1*)

Osmosis: The net flow of solvent through a semipermeable membrane. The direction of solvent flow is always from the side with the lower concentration of solute to the side with the higher concentration. (*Section 13.6*)

Osmotic pressure (Π): The pressure difference between the two sides of a semipermeable membrane that separates a pure solvent from a solution prepared from the same solvent. The osmostic pressure develops because the flow of solvent through the membrane in opposing directions is unequal. Osmotic pressure is a colligative property of a solution, so it depends on the concentration of dissolved solute particles (see Equation 13.24): $\Pi = MRT$, where M is the molarity of the solution, R is the ideal gas constant, and T is the absolute temperature. (*Section 13.6*)

Overall equation: A chemical equation that shows all of the reactants and products as undissociated, electrically neutral compounds. Although an overall equation gives the identity of the reactants and products, it does not show the identities of the actual species in solution, especially if ionic substances that are strong electrolytes are involved (c.f. complete ionic equation and net ionic equation). (*Section 4.4*)

Overall reaction order: For an experimentally determined rate law of the form rate $= k[A]^m[B]^n$, the overall reaction order is the sum of the exponents of the rate law, $m + n$ (c.f. reaction order). (*Section 14.2*)

Overlapping bands: Molecular orbitals derived from two or more different kinds of valence electrons that have similar energies. Overlapping bands result when the width of adjacent bands is relatively large compared with the energy gap between them. (*Section 12.6*)

Overtones: Vibrations of a standing wave that are higher in energy than the fundamental vibration (the lowest-energy vibration). All overtones have one or more nodes. (*Section 6.4*)

Overvoltage: The voltage that must be applied in electrolysis in addition to the calculated (theoretical) value to overcome factors such as a high activation energy and the formation of bubbles on a surface. Overvoltages are needed in all electrolytic processes. (*Section 19.7*)

Oxidant: A compound that is capable of accepting electrons (c.f. reductant). Oxidants (also called oxidizing agents) can oxidize other compounds. In the process of accepting electrons, an oxidant is reduced. (*Sections 3.5 and 19.1*)

Oxidation: The loss of one or more electrons in a chemical reaction (c.f. reduction). The substance that loses electrons is said to be oxidized. (*Section 3.5*)

Oxidation–reduction (redox) reaction: A chemical reaction in which there is a net transfer of one or more electrons from one reactant to another. This transfer of electrons provides a means for converting chemical energy to electrical energy, or vice versa. The total number of electrons lost by some reactants must equal the total number of electrons gained by other reactants. Thus, oxidation must be accompanied by reduction, and vice versa. (*Section 3.5 and Chapter 19 introduction*)

Oxidation state: The charge that each atom in a compound would have if all of its bonding electrons were transferred to the atom with the greater attraction for electrons. (*Section 3.5*)

Oxidation state method: A procedure for balancing oxidation–reduction (redox) reactions in which the overall reaction is conceptually separated into two parts: an oxidation and a reduction (see Table 4.3 for a step-by-step summary of this method). (*Section 4.8*)

Oxidizing agent: Another name for an oxidant (c.f. reducing agent). (*Section 3.5*)

Oxoacid: An acid in which the dissociable H^+ ion is attached to an oxygen atom of a polyatomic anion (e.g., H_2SO_4 or HNO_3). Oxoacids are occasionally called oxyacids. (*Section 2.5*)

Oxoanion: A polyatomic anion that contains a single metal or nonmetal atom plus one or more oxygen atoms. Sometimes oxoanions are called oxyanions. (*Section 2.3*)

Ozone: An unstable form of oxygen that consists of three oxygen atoms bonded together (O_3). A layer of ozone in the stratosphere helps protect the plants and animals on earth from harmful ultraviolet radiation. Ozone is responsible for the pungent smell we associate with lightning discharges and electric motors. It is also toxic. (*Section 3.6*)

Ozone layer: A concentration of ozone in the stratosphere (about 10^{15} ozone molecules per liter) that acts as a protective screen, absorbing ultraviolet light that would otherwise reach the surface of the earth, where it would harm plants and animals. (*Section 3.6*)

P

***p*:** See "Steric factor (*p*)" (*Section 14.7*).

***P*:** See "Pressure (*P*)" (*Sections 1.3 and 10.2*).

Partial pressure: The pressure a gas in a mixture would exert if it were the only one present (at the same temperature and volume). The pressure exerted by each gas in a mixture (its partial pressure) is independent of the pressure exerted by all other gases present. As a result, the total pressure exerted by a mixture of gases is the sum of the partial pressures of the components (Dalton's law of partial pressures). Additionally, the partial pressure of each of the component gases in a mixture is the product of the total pressure and the mole fraction of that gas. (*Section 10.5*)

Parts per billion (ppb): A common unit of concentration, parts per billion is micrograms of solute per kilogram of solvent (see Equation 13.10; c.f. parts per million). (*Section 13.4*)

Parts per million (ppm): A common unit of concentration, parts per million is milligrams of solute per kilogram of solvent (see Equation 13.9; c.f. parts per billion). (*Section 13.4*)

Pascal (Pa): The SI unit for pressure. Derived from the SI units for force (newtons) and area (square meters), the pascal is newtons per square meter, N/m^2: 1 pascal (Pa) = 1 newton/meter^2 (N/m^2). The pascal is named after the French mathematician Blaise Pascal (1623–1662). (*Section 10.2*)

Pauli exclusion principle: Developed by Wolfgang Pauli, this principle states that no two electrons in an atom can have the same value of all four quantum numbers (n, l, m_l, and m_s). This principle arises from Pauli's determination that each atomic orbital can contain no more than two electrons. (*Section 6.6*)

***p* block:** The six columns on the right side of the periodic table, consisting of the elements in which the *np* orbitals are being filled (see Figure 6.34; c.f. *s* block, *d* block, and *f* block). (*Section 6.6*)

***p*-block elements:** The elements in the six columns on the right side of the periodic table. The elements of the *p* block are filling their *np* orbitals (see Figure 6.34; c.f. *s*-block elements, *d*-block elements, and *f*-block elements). (*Section 6.6*)

***PE*:** See "Potential energy (*PE*)" (*Section 5.1*).

Peptide (amide) bonds: The covalent bond that links one amino acid to another in peptides and proteins. The carbonyl carbon atom of one amino acid residue bonds to the amino nitrogen atom of the other residue, eliminating a molecule of water in the process (see Equation 12.4). Thus, peptide bonds form in a condensation reaction between a carboxylic acid and an amine. (*Sections 3.5 and 12.8*)

Peptides: Biological polymers that contain fewer than about 50 amino acid residues (c.f. proteins). (*Section 12.8*)

Percent composition: The percentage of each element present in a pure substance. With few exceptions, the percent composition of a chemical compound is constant (see law of definite proportions). (*Section 3.2*)

Percent yield: The ratio of the actual yield of a reaction to the theoretical yield, multiplied by 100% to give a percentage. Percent yields can range from 0% to 100%. (*Section 3.4*)

Period: The rows of elements in the periodic table. At present the periodic table consists of seven periods, numbered from 1 to 7. (*Section 1.7*)

Periodic: Phenomena, such as waves, that repeat regularly in both space and time. (*Section 6.1*)

Periodic table: A chart of the chemical elements arranged in rows of increasing atomic number (Z), so that the elements in each column (group) have similar chemical properties. Each element in the periodic table is assigned a unique one-, two-, or three-letter symbol. (*Section 1.7*)

Perovskite structure: A structure that consists of a body-centered array of two metal ions, with one set (M) located at the corners of the cube, and the other set (M′) in the centers of the cube. The anions are in the centers of the square faces (see Figure 12.12). (*Section 12.3*)

pH: The negative base-10 logarithm of the hydrogen ion concentration: $pH = -\log[H^+]$. Because it is a negative logarithm, pH decreases with increasing $[H^+]$. (*Section 4.6*)

Phase changes: The changes of state that occur when any of the three forms of matter (solids, liquids, and gases) is converted to either of the other two (see Figure 11.17). All phase changes are accompanied by changes in the energy of the system. Changes from a more ordered to a less ordered state (e.g., from liquid to gas) are endothermic, whereas changes from a less ordered to a more ordered state (e.g., from liquid to solid) are always exothermic. Phase changes are also called phase transitions. (*Section 11.5*)

Phase diagram: A graphic summary of the physical state of a substance as a function of temperature and pressure in a closed system (see Figure 11.22). A typical phase diagram consists of regions that represent the different phases possible for the substance (solid, liquid, and gas). The solid and liquid regions are separated by the melting curve of the substance, while the liquid and gas phases are separated by the vapor pressure curve (which ends at the critical point). Only a single phase is stable within a given region, but two phases are in equilibrium at the given temperatures and pressures along the lines that separate the regions. (*Section 11.7*)

Phase transitions: Another name for phase changes. (*Section 11.5*)

Phospholipids: A large class of biological, detergent-like molecules that contain a hydrophilic head and two hydrophobic tails (detergents and soaps have just one tail). The additional tail results in a cylindrical shape that prevents phospholipids from forming spherical micelles. Instead, phospholipids form bilayers. (*Section 13.7*)

Photoelectric effect: A phenomenon in which electrons are ejected from the surface of a metal that has been exposed to light. Each metal has a characteristic threshold frequency of light (ν_0). Below the threshold frequency, no electrons are emitted regardless of the light's intensity. Above the threshold frequency, the number of electrons emitted is proportional to the intensity of the light, and the kinetic energy of the ejected electrons is proportional to the frequency of the light. (*Section 6.2*)

Photons: "Particles" (quantums) of light (radiant energy), each of which possesses a particular energy E given by $E = h\nu$. (*Section 6.2*)

Photosynthesis: The fundamental reaction by which all green plants and algae obtain energy from sunlight, photosynthesis is the photochemical reduction of CO_2 to a carbon compound such as glucose (c.f. respiration). Oxygen in water is concurrently oxidized to O_2: $6CO_2 + 6H_2O \xrightarrow{h\nu} C_6H_{12}O_6 + 6O_2$ (Equation 18.46). (*Section 18.8*)

Phototrophs: Organisms, such as green plants and algae, that extract energy from the environment through photosynthesis (c.f. chemotrophs). Sunlight, therefore, is the ultimate energy source for phototrophs. (*Section 18.8*)

pH scale: A logarithmic scale used to express the hydrogen ion (H^+) concentration of a solution, making it possible to describe acidity or basicity quantitatively. The pH scale is convenient because very large and very small concentrations can be expressed as relatively simple numbers. (*Section 4.6*)

Physical property: A characteristic that scientists can measure without changing the composition of the sample under study (e.g., mass, color, and volume). Physical properties can be extensive or intensive. (*Section 1.3*)

Pinning: A process that introduces multiple defects into a material so that the presence of one defect prevents the motion of another. By preventing the motion of the defects, pinning increases the mechanical strength of the material. (*Section 12.4*)

Pi (π) orbital: A bonding molecular orbital formed from the sum of the side-to-side interactions of two or more parallel *np* atomic orbitals. The sum of the interactions increases electron probability in the region above and below a line connecting the nuclei [c.f. pi star (π^*) orbital]. A π molecular orbital, like a π^* molecular orbital, possesses a nodal plane that contains the nuclei. (*Section 9.3*)

Pi star (π^*) orbital: An antibonding molecular orbital formed from the difference of the side-to-side interactions of two or more parallel *np* atomic orbitals. The difference of the interactions decreases electron probability in the region above and below a line connecting the nuclei [c.f. pi (π) orbital]. A π^* molecular orbital, like a π molecular orbital, possesses a nodal plane that contains the nuclei. (*Section 9.3*)

Plane defect: A defect in a crystal that affects a plane of points in the lattice (c.f. point defect and line defect). (*Section 12.4*)

Plastic: The property of a material that allows it to be molded into almost any shape. Although many plastics are polymers, not all polymers are plastics. (*Section 12.8*)

Pnicogens: The elements in Group 15 of the periodic table (N, P, As, Sb, and Bi). (*Section 7.4*)

Point defect: A defect in a crystal that affects a single point in the lattice c.f. line defect and plane defect). Point defects can consist of an atom missing from a site in the crystal (a vacancy) or an impurity atom that occupies either a normal lattice site (a substitutional impurity) or a hole in the lattice between atoms (an interstitial impurity). (*Section 12.4*)

Poisons: Substances that bind irreversibly to catalysts, thus preventing reactants from adsorbing to the catalyst, which reduces or destroys the catalyst's efficiency (see enzyme inhibitors, too). (*Section 14.8*)

Polar bond: A chemical bond in which there is an unequal distribution of charge between the bonding atoms. One atom is electron rich (so it has a partial negative charge, δ^-), whereas the other atom is electron poor (so it has a partial positive charge, δ^+). (*Section 4.1*)

Polar covalent bond: A covalent bond in which the electrons are shared unequally between the bonded atoms (c.f. covalent bond). (*Section 8.9*)

Polyatomic ions: Groups of two or more atoms that have a net electrical charge, although the atoms that make up a polyatomic ion are held together by the same covalent bonds that hold atoms together in molecules. Just as there are many more kinds of molecules than simple elements, there are many more kinds of polyatomic ions than monatomic ions (see Table 2.4). (*Section 2.1*).

Polyatomic molecules: Molecules that contain more than two atoms. (*Section 2.1*)

Polymer-matrix composites: Composites that consist of reinforcing fibers embedded in a polymer matrix. Fiberglass, for instance, is a polymer-fiber matrix consisting of glass fibers embedded in a polymer matrix. (*Section 12.9*)

Polymers: Giant molecules that consist of many basic structural units (monomers) connected in chains or networks by covalent bonds. (*Section 12.8*)

Polyprotic acid: A compound that can donate more than one hydrogen ion per molecule (e.g., H_2SO_4 or H_3PO_4; c.f. monoprotic acid). (*Section 4.6*)

Positron: A particle that has the same mass as an electron, but has a positive charge instead of a negative charge. (*Section 20.2*)

Positron emission: The opposite of β decay, positron emission is a nuclear decay reaction in which a proton is transformed into a neutron and a high-energy positron is emitted. Positron emission is characteristic of neutron-poor nuclei. (*Section 20.2*)

Potential (E_{cell}): Measured in volts, the potential of an electrochemical cell is the difference in electrical potential between the two half-reactions (oxidation at the anode and reduction at the cathode). Electrical potential is related to the energy needed to move a charged particle in an electric field. (*Section 19.1*)

Potential energy (*PE*): Energy stored in an object because of the relative positions or orientations of its components. Electrical, nuclear, and chemical energy are different forms of potential energy. (*Section 5.1*)

Precipitate: The insoluble product that forms in a precipitation reaction. (*Section 4.5*)

Precipitation reaction: A chemical reaction that yields an insoluble product (a precipitate) when two solutions are mixed. Precipitation reactions are a subclass of exchange reactions that occur between ionic compounds when one of the products is insoluble. Because both components of each compound change partners, such reactions are sometimes called double-displacement reactions. (*Section 4.5*)

Precise: When multiple measurements give nearly identical values for a quantity. (*Essential Skills 1*)

Precision: The degree to which multiple measurements give nearly identical values for a quantity. (*Essential Skills 1*)

Pressure (*P*): The amount of force (F) exerted on a given area (A) of surface: $P = \frac{F}{A}$ (*Sections 1.3 and 10.2*)

Principal quantum number (*n*): One of three quantum numbers (the other two are l and m_l) used to specify any wave function, the principal quantum number tells the average relative distance of the electron from the nucleus. The allowed values of n are positive integers ($n = 1, 2, 3, 4, \ldots$). As n increases for a given atom, so does the average distance of the electron from the nucleus. (*Section 6.5*)

Principal shell: All wave functions that have the same value of n constitute a principal shell because those electrons have similar average distances from the nucleus. (*Section 6.5*)

Product: The final compound or compounds produced in a chemical reaction. The products are usually written to the right of the arrow in the chemical equation. (*Section 3.3*)

Promotion: The excitation of an electron from a filled ns^2 atomic orbital to an empty np or $(n-1)d$ valence orbital. The formation of hybrid atomic orbitals can be viewed as occurring via promotion followed by hybridization. (*Section 9.2*)

Propagation: The second stage in a chain reaction, propagation takes the reactive intermediates formed in initiation and continuously consumes and regenerates them in reactions that also form products. The other stages of a chain reaction are initiation and termination. (*Section 14.6*)

Proteins: Biological polymers that contain more than 50 amino acid residues (c.f. peptides). (*Section 12.8*)

Proton: A subatomic particle that resides in the nucleus of all atoms. A proton has a mass of 1.673×10^{-24} g (1.007276 amu) and an electrical charge of -1.602×10^{-19} C. The charge on the proton is equal in magnitude but opposite in sign to the charge on an electron. Protons and neutrons constitute by far the bulk of the mass of atoms. (*Section 1.5*)

Pseudo inert gas configuration: The $(n-1)d^{10}$ and similar electron configurations that are particularly stable and are often encountered in the heavier p-block elements. The electron configuration of Ga^{3+}, for example, is [Ar] $3d^{10}$, not [Ar] $4s^2\,3d^8$. (*Section 7.3*)

***p*-Type semiconductor:** A semiconductor that has been doped with an impurity that contains fewer valence electrons than the atoms of the host lattice (c.f. n-type semiconductor). The reduced number of valence electrons creates holes in the valence band, thus increasing the electrical conductivity of the semiconductor. (*Section 12.6*)

Pyrolysis: A high-temperature decomposition reaction that can be used to form fibers of synthetic polymers. (*Section 12.8*)

Q

q: See "Heat (q)" (*Section 5.1*).

Q: See "Reaction quotient (Q)" (*Section 15.4*) or "Ion product (Q)" (*Section 17.1*).

Q_p: See "Reaction quotient (Q_p)" (*Section 15.4*).

Qualitative analysis: A procedure for determining the identity of metal ions present in a mixture (rather than quantitative information about their amounts). (*Section 17.5*)

Quantitative analysis: A methodology that combines chemical reactions and stoichiometric calculations to determine the amounts or concentrations of substances present in a sample. (*Section 4.9*)

Quantum: The smallest possible unit of energy. The energy of electromagnetic waves is quantized rather than continuous, so energy can be gained or lost only in integral multiples of some smallest unit of energy, a quantum. (*Section 6.2*)

Quantum mechanics: A theory developed by Erwin Schrödinger to describe the energies and spatial distributions of electrons in atoms and molecules. (*Section 6.5*)

Quantum numbers: A unique set of numbers that specifies a wave function (a solution to the Schrödinger equation). Quantum numbers provide important information about the energy and spatial distribution of an electron. (*Section 6.5*)

R

R: The abbreviation used for alkyl groups and aryl groups in general formulas and structures. (*Section 2.4*)

R: See "Gas constant (R)" (*Section 10.4*).

r_0: See "Bond distance (r_0)" (*Section 8.1*).

Rad (radiation absorbed dose): A unit used to measure the effects of radiation on biological tissues, the rad is the amount of radiation that causes 0.01 joule of energy to be absorbed by 1 kilogram of matter (c.f. gray and roentgen). (*Section 20.3*)

Radiant energy: One of the five forms of energy, radiant energy is carried by light, microwaves, and radio waves (the other forms of energy are thermal, chemical, nuclear, and electrical). Objects left in bright sunshine or exposed to microwaves become warm because much of the radiant energy they absorb is converted to thermal energy. (*Section 5.1*)

Radiation dose: The amount of radiation an object has been exposed to (see roentgen, rad, gray, and rem). (*Section 20.3*)

Radicals: Species that have one or more unpaired valence electrons. Radicals are often formed as reaction intermediates in chain reactions. (*Section 14.6*)

Radioactive: Any nucleus that is unstable and decays spontaneously, emitting subatomic particles and electromagnetic radiation. (*Section 20.1*)

Radioactive decay series: A series of sequential alpha- and beta-decay reactions that occur until a stable nucleus is finally obtained. The uranium-238 decay series, for example, produces lead-206 in a series of 14 sequential α- and β-decay reactions (Figure 20.4). (*Section 20.2*)

Radioactivity: The spontaneous emission of energy rays (radiation) by matter. (*Section 1.5*)

Radioisotopes: Isotopes that emit radiation. (*Section 20.1*)

Raoult's law: The equation that quantifies the relationship between solution composition and vapor pressure (see Equation 13.13): $P_A = X_A P_A^\circ$, where P_A is the vapor pressure of component A of the solution, X_A is the mole fraction of component A present in solution, and P_A° is the vapor pressure of pure A. (*Section 13.6*)

Rate constant (k): A proportionality constant, the value of which is characteristic of the reaction and reaction conditions. A given reaction has a particular value of the rate constant under a given set of conditions, such as temperature, pressure, and solvent. Varying the temperature or the solvent usually changes the value of the rate constant. The numerical value of k, however, does not change as the reaction progresses under a given set of conditions. (*Section 14.2*)

Rate-determining step: The slowest step (elementary reaction) in a reaction mechanism. The rate law for an overall reaction is the same as the rate law for the rate-determining step of the reaction. (*Section 14.6*)

Rate laws: Mathematical expressions that describe the relationships between reactant rates and reactant concentrations in a chemical reaction (see differential rate law and integrated rate law). Rate laws are mathematical descriptions of experimentally verifiable data. (*Section 14.2*)

Rate of decay: Also called the activity (A), the rate of decay of a radioactive isotope is the decrease in the number of the radioisotope's nuclei per unit time (Equation 14.33): $A = -\Delta N/\Delta t$, where N is the number of atoms of the radioactive isotope. Rate of decay is usually measured in units of disintegrations per second (dps) or disintegrations per minute (dpm). (*Section 14.5*)

Reactant: The starting material or materials in a chemical reaction. The reactants are usually written to the left of the arrow in the chemical equation. (*Section 3.3*)

Reaction mechanism: The sequence of events (the stepwise changes) that occur at the molecular level during a reaction, the reaction mechanism describes how individual atoms, ions, or molecules interact to form particular products. (*Section 14.6*)

Reaction order: Numbers that indicate the degree to which the rate of the reaction depends on the concentration of each reactant. For an experimentally determined rate law of the form rate = $k[A]^m[B]^n$, the values of m and n are derived from experimental measurements of the changes in reactant concentrations over time and indicate the reaction order (c.f. overall reaction order). The values of m and n need not be integers, and they are unrelated to the stoichiometric coefficients of A and B in the balanced chemical equation. (*Section 14.2*)

Reaction quotient (Q): A quantity that has precisely the same form as the equilibrium constant expression, except that the reaction quotient may be derived from a set of values measured at *any* time during the reaction of *any* mixture of the reactants and products, regardless of whether the system is at equilibrium. Thus, for the general reaction $aA + bB \rightleftharpoons cC + dD$, where A and B are reactants, C and D are products, and a, b, c, and d are the stoichiometric coefficients in the balanced chemical equation for the reaction, $Q = [C]^c[D]^d/[A]^a[B]^b$ (Equation 15.25). (*Section 15.4*)

Reaction quotient (Q_p): A quantity that has precisely the same form as K_p, the equilibrium constant calculated from partial pressures instead of solution concentrations, except that Q_p may be derived from a set of values measured at *any* time during the reaction of *any* mixture of the reactants and products, regardless of whether the system is at equilibrium. $Q_p = (P_C)^c(P_D)^d/(P_A)^a(P_B)^b$ for the general reaction $aA + bB \rightleftharpoons cC + dD$, in which all the components are gases. (*Section 15.4*)

Reaction rates: The changes in the concentrations of reactants and products with time. The factors that affect reaction rates include reactant concentrations, temperature, physical states and surface areas of reactants, as well as solvent and catalyst properties if either is present. (*Chapter 14 introduction*)

Rechargeable batteries: Also called secondary batteries, rechargeable batteries can be recharged by applying an electrical potential in the reverse direction (c.f. disposable batteries). The recharging process temporarily converts a rechargeable battery from a galvanic cell to an electrolytic cell. (*Section 19.5*)

Redox reaction: Another name for an oxidation–reduction reaction. (*Section 3.5*)

Reducing agent: Another name for a reductant (c.f. oxidizing agent). (*Section 3.5*)

Reductant: A compound that is capable of donating electrons (c.f. oxidant). Reductants (also called reducing agents) can reduce other compounds. In the process of donating electrons, a reductant is oxidized. (*Sections 3.5 and 19.1*)

Reduction: The gain of one or more electrons in a chemical reaction (c.f. oxidation). The substance that gains electrons is said to be reduced. (*Section 3.5*)

Reference electrode: An electrode in a galvanic (voltaic) cell whose potential is unaffected by the properties of the solution (c.f. indicator electrode). Ideally, the reference electrode should be physically isolated from the solution whose concentration is being measured. (*Section 19.2*)

Reforming: The chemical conversion of straight-chain alkanes to either branched-chain alkanes or mixtures of aromatic hydrocarbons. Reforming is the second process used in petroleum refining to increase the amounts of more volatile, higher-value products (cracking is the first process). The necessary chemical reactions in reforming are brought about by the use of metal catalysts such as platinum. (*Section 2.6*)

Rem (roentgen equivalent in man): A unit that describes the actual amount of tissue damage caused by a given amount of radiation, the number of rems of radiation equals the number of rads multiplied by a relative biological effectiveness (RBE) factor, which is 1 for β particles, γ rays, and X rays, and about 20 for α particles. (*Section 20.3*)

Resonance structures: Equivalent Lewis dot structures. The positions of the atoms are the same in the various resonance structures of a compound, but the positions of the electrons are different. The different resonance structures of a compound are linked by double-headed arrows. (*Section 8.5*)

Respiration: A process by which chemotrophs obtain energy from their environment, the overall reaction of respiration is the reverse of photosynthesis: $C_6H_{12}O_6 + 6O_2 \longrightarrow 6CO_2 + 6H_2O$ (Equation 18.47). (*Section 18.8*)

Reverse osmosis: A process that uses the application of an external pressure greater than the osmotic pressure of a solution to reverse the flow of solvent through the semipermeable membrane. Reverse osmosis can be used to produce pure water from seawater. (*Section 13.6*)

Reversible process: A process in which every intermediate state between the extremes is an equilibrium state, regardless of the direction of the change (c.f. irreversible process). A reversible process can change direction at any time, whereas an irreversible process cannot. (*Section 18.3*)

Roentgen (R): A unit that describes the amount of energy absorbed by dry air, the roentgen can be used to measure radiation exposure or dose (c.f. rad and gray). (*Section 20.3*)

Root mean square (rms) speed (v_{rms}): The square root of $\overline{v}^2$, where $\overline{v}^2$ is the average of the squares of the speeds of the particles of a gas at a given temperature (see Equation 10.33). The root mean square speed is the speed of a gas particle that has average kinetic energy. As a result, the root mean square speed and the average speed are different (although the difference is typically less than 10%). (*Section 10.7*)

S

S: See "Entropy (S)" (*Sections 13.2 and 18.3*).

Salt: The general term for any ionic substance that does not have OH^- as the anion or H^+ as the cation. (*Section 4.6*)

Salt bridge: A U-shaped tube inserted into both solutions of a galvanic (voltaic) cell that contains a concentrated liquid or gelled electrolyte. The ions in the salt bridge are selected so they do not interfere with the electrochemical reaction by being oxidized or reduced themselves. The salt bridge completes the circuit between the anode and cathode. (*Section 19.1*)

Saturated calomel electrode (SCE): A reference electrode that consists of a platinum wire inserted into a moist paste of liquid mercury, Hg_2Cl_2, and KCl. These substances, which make up the interior cell, are surrounded by an aqueous KCl solution, which acts as a salt bridge between the interior cell and the exterior solution. The half-reaction for the saturated calomel electrode is $Hg_2Cl_2(s) + 2e^- \longrightarrow 2Hg(l) + 2Cl^-(aq)$ (Equation 19.30). (*Section 19.2*)

Saturated hydrocarbons: Alkanes are also called saturated hydrocarbons because they contain only carbon–carbon and carbon–hydrogen single bonds. Each carbon atom has four single bonds, the maximum number possible (c.f. unsaturated hydrocarbons). (*Section 2.4*)

Saturated solution: A solution that contains the maximum possible amount of a solute under a given set of conditions (e.g., temperature and pressure). (*Section 13.3*)

***s* block:** The two columns on the left side of the periodic table, consisting of the elements in which the *ns* orbitals are being filled (see Figure 6.34; c.f. *p* block, *d* block, and *f* block). (*Section 6.6*)

***s*-block elements:** The elements of the two columns on the left side of the periodic table. The elements of the *s* block are filling their *ns* orbitals (see Figure 6.34; c.f. *p*-block elements, *d*-block elements, and *f*-block elements). (*Section 6.6*)

SCE: See "Saturated calomel electrode (SCE)" (*Section 19.2*).

Schottky defects: A coupled pair of vacancies, one cation and one anion, that maintains the electrical neutrality of an ionic solid. (*Section 12.4*)

Scientific method: The procedure scientists use to search for answers to questions and solutions to problems. The procedure consists of making observations, formulating hypotheses, and designing experiments, which lead in turn to additional observations, hypotheses, and experiments in repeated cycles. (*Section 1.2*)

Scientific notation: A system that expresses numbers in the form $N \times 10^n$, where N is greater than or equal to 1 and less than 10 ($1 \leq N \leq 10$) and n is an integer that can be either positive or negative ($10^0 = 1$). The purpose of scientific notation is to simplify the manipulation of numbers with large or small magnitudes. (*Essential Skills 1*)

Second law of thermodynamics: The entropy of the universe remains constant in a reversible process, whereas the entropy of the universe increases in an irreversible (spontaneous) process. (*Section 18.3*)

Second-order reaction: A reaction whose rate is proportional to the square of the concentration of the reactant (for a reaction of the general form $2A \longrightarrow$ products) or is proportional to the product of the concentrations of two reactants (for a reaction of the general form $A + B \longrightarrow$ products) (c.f. zeroth-order reaction and first-order reaction; see Table 14.6). The differential rate law for the second-order reaction $2A \longrightarrow$ products is rate $= k[A]^2$, where k is the rate constant of the reaction. (*Section 14.3*)

Seed crystal: A solid sample of a substance that can be added to a supercooled liquid or a supersaturated solution to help induce crystallization. (*Sections 11.5 and 13.3*)

Semiconductor: Substances such as Si and Ge that have conductivities between those of metals and insulators. (*Section 12.6*)

Semimetal: Any of the seven elements (B, Si, Ge, As, Sb, Te, and At) that lie adjacent to the zigzag line in the periodic table that runs from boron (B) down to astatine (At). Semimetals (also called metalloids in some texts) exhibit properties intermediate between those of metals and nonmetals. (*Section 1.7*)

Semipermeable membrane: A barrier with pores small enough to allow solvent molecules to pass through, but not solute molecules or ions. (*Section 13.6*)

SHE: See "Standard hydrogen electrode (SHE)" (*Section 19.2*).

Sigma (σ) orbital: A bonding molecular orbital in which the electron density along the internuclear axis and between the nuclei has cylindrical symmetry [c.f. sigma star (σ^*) orbital]. All cross-sections perpendicular to the internuclear axis are circles. (*Section 9.3*)

Sigma star (σ^*) orbital: An antibonding molecular orbital in which the electron density along the internuclear axis and between the nuclei has cylindrical symmetry [c.f. sigma (σ) orbital]. All cross-sections perpendicular to the internuclear axis are circles. (*Section 9.3*)

Significant figures: The numbers that describe a value without exaggerating the degree to which it is known to be accurate. An additional figure is often reported to indicate the degree of uncertainty. (*Essential Skills 1*)

Silver–silver chloride electrode: A reference electrode that consists of a silver wire coated with a very thin layer of AgCl that is dipped into a chloride ion solution with a fixed concentration. The half-reaction for the silver–silver chloride electrode is $AgCl(s) + e^- \longrightarrow Ag(s) + Cl^-(aq)$ (Equation 19.28). (*Section 19.2*)

Simple cubic unit cell: A cubic unit cell that consists of eight component atoms, molecules, or ions located at the corners of a cube (see Figure 12.5a). (*Section 12.2*)

Single bond: A chemical bond formed when two atoms share a single pair of electrons. (*Section 2.1*)

Single-displacement reaction: A chemical reaction in which an ion in solution is displaced through oxidation of a metal. Single-displacement reactions include the oxidation of certain metals by aqueous acid and the oxidation of certain metals by aqueous solutions of various metal salts. (*Section 4.8*)

Sintering: A process that fuses the grains of a ceramic into a dense, strong material. Sintering is used to produce high-strength ceramics. (*Section 12.9*)

Slip plane: The plane along which the motion of a deformation of a solid occurs. To shape a solid without shattering it, planes of closepacked atoms must move past one another to a new position that is energetically equivalent to the old one. (*Section 12.4*)

Smectic phase: One of three different ways that most liquid crystals can orient themselves (the other two are the nematic and cholesteric phases). In the smectic phase, the long axes of the molecules are aligned (similar to the nematic phase), but the molecules are arranged in planes, too. (*Section 11.8*)

Sodium chloride structure: The solid structure that results when the octahedral holes of a face-centered cubic lattice of anions are filled with cations. The sodium chloride structure has a 1:1 cation:anion ratio, and each ion has a coordination number of 6. (*Section 12.3*)

Sol: A dispersion of solid particles in a liquid or solid. A sol is one of three kinds of colloids (the other two are aerosols and emulsions). (*Section 13.7*)

Sol–gel process: A process used to manufacture ceramics. The sol–gel process produces the fine powders of ceramic oxides with uniformly sized particles that are necessary to manufacture high-quality ceramics. (*Section 12.9*)

Solid: One of three distinct states of matter under normal conditions, solids are relatively rigid and have fixed shapes and volumes. The volume of a solid is virtually independent of temperature and pressure. The other two states of matter are liquid and gas. (*Section 1.3*)

Solid electrolytes: Solid materials with very high electrical conductivities. Cations in compounds with Frenkel defects are often able to move rapidly from one site in the crystal to another, resulting in the high electrical conductivities. (*Section 12.4*)

Solubility: A measure of how much of a solid substance remains dissolved in a given amount of a specified liquid at a specified temperature and pressure. Most substances are more soluble at higher temperatures and pressures. (*Sections 1.3 and 13.3*)

Solubility product (K_{sp}): The equilibrium constant expression for the dissolution of a sparingly soluble salt. The concentration of a pure solid is a constant, so it does not appear explicitly in the equilibrium constant expression. (*Section 17.1*)

Solute: The substance or substances present in lesser amounts in a solution (c.f. solvent). (*Chapter 4 introduction*)

Solution: A homogeneous mixture of two or more substances in which the substances present in lesser amounts (called solutes) are dispersed uniformly throughout the substance present in greater amount (the solvent). (*Chapters 4 and 13 introductions*)

Solvation: The process of surrounding each solute particle (atom, molecule, or ion) with particles of solvent (see hydration). (*Section 13.2*)

Solvent: The substance present in the greater amount in a solution (c.f. solute). (*Chapter 4 introduction*)

***sp* hybrid orbitals:** The two equivalent hybrid orbitals that result when one *ns* orbital and one *np* orbital are combined (hybridized). The two *sp* hybrid orbitals are oriented at 180° from each other. They are equivalent in energy, and their energy is between the energy values associated with pure *s* and pure *p* orbitals. (*Section 9.2*)

sp^2 hybrid orbitals: The three equivalent hybrid orbitals that result when one *ns* orbital and two *np* orbitals are combined (hybridized). The three sp^2 hybrid orbitals are oriented in a plane at 120° from each other. They are equivalent in energy, and their energy is between the energy values associated with pure *s* and pure *p* orbitals. (*Section 9.2*)

sp^3 hybrid orbitals: The four equivalent hybrid orbitals that result when one *ns* orbital and three *np* orbitals are combined (hybridized). The four sp^3 hybrid orbitals point at the vertices of a tetrahedron, so they are oriented at 109.5° from each other. They are equivalent in energy, and their energy is between the energy values associated with pure *s* and pure *p* orbitals. (*Section 9.2*)

sp^3d hybrid orbitals: The five hybrid orbitals that result when one *ns*, three *np*, and one $(n-1)d$ orbitals are combined (hybridized). The five sp^3d hybrid orbitals point at the vertices of a trigonal bipyramid. As a result, the five hybrid orbitals are not all equivalent: three form a triangular array oriented at 120° angles, while the other two are oriented at 90° to the first three and at 180° to each other. These kinds of hybrid orbitals, along with sp^3d^2 hybrid orbitals, are invoked to explain the bonding in molecules that contain more than an octet of electrons around the central atom. (*Section 9.2*)

sp^3d^2 hybrid orbitals: The six equivalent hybrid orbitals that result when one *ns*, three *np*, and two $(n-1)d$ orbitals are combined (hybridized). The six sp^3d^2 hybrid orbitals point at the vertices of an octahedron, so each is oriented at 90° from the four adjacent orbitals and at 180° from the fifth orbital. These kinds of hybrid orbitals, along with sp^3d hybrid orbitals, are invoked to explain the bonding in molecules that contain more than an octet of electrons around the central atom. (*Section 9.2*)

Specific heat (C_s): The amount of energy needed to increase the temperature of 1 g of a substance by 1°C. The units of C_s are J/(g·°C). (*Section 5.3*)

Spectator ions: Ions that do not participate in the actual reaction. Spectator ions appear on both sides of a complete ionic equation and their coefficients are the same on both sides. Canceling the spectator ions from a complete ionic equation gives a net ionic equation. (*Section 4.4*)

Speed (v) (of a wave): The distance traveled by a wave per unit time. The speed of a wave equals the product of its wavelength and frequency ($v = \lambda\nu$) and is typically measured in meters per second (m/s). (*Section 6.1*)

Speed of light (c): The speed with which *all* forms of electromagnetic radiation (e.g., microwaves, visible light, and gamma rays) travel in a vacuum. The speed of light is a fundamental constant with a value of 2.99792458×10^8 m/s. Because the various kinds of electromagnetic radiation all have the same speed (c), they differ only in wavelength (λ) and frequency (ν); $c = \lambda\nu$. (*Section 6.1*)

Spontaneous fission: A nuclear decay reaction in which the nucleus breaks into two pieces with different atomic numbers and atomic masses. Spontaneous fission is most important for the transactinide elements ($Z \geq 104$)—that is, very massive nuclei with high neutron-to-proton ratios. It is accompanied by the release of large amounts of energy and usually the emission of several neutrons. (*Section 20.2*)

Square planar: One of two possible molecular geometries for an AB_4 molecular species (the other is tetrahedral; see Table 9.1). In the square planar geometry, all five atoms lie in the same plane and all adjacent B—A—B bond angles are 90°. (*Section 9.1*)

Standard atmospheric pressure: The atmospheric pressure required to support a column of mercury exactly 760 mm tall. This pressure is also referred to as 1 atmosphere (atm) and is related to other pressure units as follows: 1 atm = 760 torr = 760 millimeters of mercury (mmHg) = 101325 pascals (Pa) = 101.325 kPa. (*Section 10.2*)

Standard cell potential (E°_{cell}): The potential of an electrochemical cell measured under standard conditions—that is, with all species in their standard states (1 *M* for solutions, 1 atm for gases, and pure solids or pure liquids for other substances) and at a fixed temperature (usually 298 K). (*Section 19.2*)

Standard conditions: The conditions under which most thermochemical data are tabulated. The standard conditions are a pressure of 1 atmosphere (atm) for all gases and a concentration of 1.0 *M* for all species in solution. In addition, each pure substance must be in its standard state. (*Section 5.2*)

Standard electrode potential: The potential of a half-reaction measured against the standard hydrogen electrode (SHE) under standard conditions. (*Section 19.2*)

Standard enthalpy of formation (ΔH°_f): The enthalpy change for the formation of 1 mol of a compound from it component elements, when the component elements are each in their standard states (most stable forms). The standard enthalpy of formation of any element in its most stable form is zero by definition. (*Section 5.2*)

Standard enthalpy of reaction (ΔH°_{rxn}): The enthalpy change that occurs when a reaction is carried out with all reactants and products in their standard states. The magnitude of ΔH°_{rxn} is the sum of the standard enthalpies of formation of the products, each multiplied by its appropriate stoichiometric coefficient, minus the sum of the standard enthalpies of formation of the reactants, also multiplied by their coefficients. (*Section 5.2*)

Standard free-energy change (ΔG°): The change in free energy when one substance or set of substances in their standard states is converted to one or more other substances, also in their standard states. The standard free-energy change can be calculated from the definition of free energy, if the standard enthalpy and entropy changes are known, using Equation 18.23: $\Delta G^\circ = \Delta H^\circ - T\Delta S^\circ$. (*Section 18.5*)

Standard free energy of formation (ΔG°_f): The change in free energy that occurs when 1 mole of a substance in its standard state is formed from the elements in their standard states. By definition, the standard free energy of formation of an element in its standard state is taken to be zero at 298 K. $\Delta G^\circ_f = \Delta H^\circ_f - T\Delta S^\circ_f$ (Equation 18.24). (*Section 18.5*)

Standard hydrogen electrode (SHE): The electrode chosen as a reference for all others, the standard hydrogen electrode has been assigned a standard potential of 0 V. It consists of a strip of platinum wire in contact with an aqueous solution that contains 1 *M* H^+. The $[H^+]$ in solution is in equilibrium with H_2 gas at a pressure of 1 atm at the Pt–solution interface. Protons are reduced or hydrogen molecules are oxidized at the Pt surface according to the equation $2H^+(aq) + 2e^- \rightleftharpoons H_2(g)$ (Equation 19.8). (*Section 19.2*)

Standard molar entropy ($S°$): The entropy of 1 mole of a substance at a standard temperature of 298 K. The units of $S°$ are J/(mol•K). Unlike enthalpy or internal energy, it is possible to obtain absolute entropy values by measuring the entropy change that occurs between the reference point of 0 K, corresponding to $S = 0$, and 298 K. (*Section 18.4*)

Standard molar volume: The volume of 1 mol of an ideal gas at STP (0°C and 1 atm pressure). The standard molar volume corresponds to 22.41 L (see Table 10.3 for the molar volumes of several real gases). (*Section 10.4*)

Standard reduction potentials: The standard cell potentials ($E°_{cell}$) for reactions written as reductions, not oxidations. All tabulated values of standard cell potentials by convention are listed as standard reduction potentials in order to be able to compare standard potentials for different substances. (*Section 19.2*)

Standard solution: A solution whose concentration is known precisely. Only pure crystalline compounds that do not react with water or CO_2 (such as potassium hydrogen phthalate, KHP) are suitable for use in preparing a standard solution. (*Section 4.9*)

Standard state: The most stable form of a pure substance at a pressure of 1 atm at a specified temperature [e.g., 25°C (298 K)]. (*Section 5.2*)

Standard temperature and pressure (STP): A particular set of reference conditions—namely, 0°C (273.15 K) and 1 atm pressure. (*Section 10.4*)

Standing wave: A wave that does not travel in space. An example of a standing wave is the motion of a string of a violin or guitar. When the string is plucked, it vibrates at certain fixed frequencies because it is fastened at both ends. (*Section 6.4*)

State (of a system): A complete description of the system at a given time, including its temperature and pressure, the amount of matter it contains, its chemical composition, and the physical state of the matter. (*Section 5.2*)

State function: A property of a system whose magnitude depends on only the present state of the system and not its previous history. Temperature, pressure, volume, and potential energy are all state functions, whereas heat and work are not (they are path dependent). (*Sections 5.2 and 18.1*)

Steric factor (p): The fraction of orientations of particles (atoms, molecules, or ions) that result in a chemical reaction. In general, the value of the steric factor can range from 0 (no orientations of molecules result in reaction) to 1 (all orientations result in reaction). (*Section 14.7*)

Stock solution: A commercially prepared solution of known concentration. A solution of desired concentration is often prepared by diluting a small volume of a more concentrated stock solution with additional solvent. (*Section 4.2*)

Stoichiometric quantity: The amount of product or reactant specified by the coefficients in a balanced chemical equation. (*Section 3.4*)

Stoichiometry: A collective term for the quantitative relationships among the masses, numbers of moles, and numbers of particles (atoms, molecules, and ions) of the reactants and products in a balanced reaction. (*Section 3.4*)

STP: See "Standard temperature and pressure (STP)" (*Section 10.4*).

Stratosphere: The layer of the atmosphere above the troposphere, the stratosphere extends from an altitude of 13 km (8 miles) to about 44 km (27 miles). Of all of the layers in earth's atmosphere, the stratosphere contains the highest concentration of ozone. (*Section 3.6*)

Strong acid: An acid that reacts essentially completely with water to give H^+ and the corresponding anion (c.f. weak acid). Strong acids are strong electrolytes. (*Section 4.6*)

Strong base: A base that dissociates essentially completely in water to give OH^- and the corresponding cation (c.f. weak base). Strong bases are strong electrolytes, (*Section 4.6*)

Strong electrolyte: An electrolyte that dissociates completely into ions when dissolved in water, thus producing an aqueous solution that conducts electricity very well (e.g., $BaCl_2$ or NaOH). (*Section 4.1*)

Strong nuclear force: An extremely powerful but very short-range attractive force between nucleons that keeps the nucleus of an atom from flying apart (due to electrostatic repulsions between the positively charged protons). (*Section 20.1*)

Structural formula: A representation of a molecule that shows which atoms are bonded to one another and, in some cases, the approximate arrangement of the atoms in space (see Figure 2.4b; c.f. molecular formula). (*Section 2.1*)

Sublimation: The conversion of a solid directly to a gas (without an intervening liquid phase). (*Sections 8.3 and 11.5*)

Subshell: A group of wave functions that have the same values of both n (the principal quantum number) and l (the azimuthal quantum number). The regions of space occupied by electrons in the same subshell usually have the same shape, but they are oriented differently in space. (*Section 6.5*)

Substitutional alloy: An alloy formed by the substitution of one metal atom for another of similar size in the lattice (c.f. interstitial alloy and intermetallic compound). (*Section 12.5*)

Substitutional impurity: A point defect that results when an impurity atom occupies a normal lattice site (c.f. vacancy and interstitial impurity). (*Section 12.4.*)

Substrate: The reactant in an enzyme-catalyzed reaction. (*Section 14.8*)

Superalloys: High-strength alloys, often of complex composition, that are used in applications (e.g., aerospace) that require mechanical strength, high surface stability (minimal flaking or pitting), and resistance to high temperatures. Superalloys are new metal phases based on cobalt, nickel, and iron. (*Section 12.9*)

Superconducting transition temperature (T_c): The temperature at which a material becomes superconducting (i.e., the temperature at which the electrical resistance of a substance drops to zero). (*Section 12.7*)

Superconductivity: The phenomenon in which a solid at low temperatures exhibits zero resistance to the flow of electrical current. (*Section 12.7*)

Superconductors: Solids that at low temperatures exhibit zero resistance to the flow of electrical current. (*Section 12.7*)

Supercooled liquid: A metastable liquid phase that exists below the normal melting point of the substance (c.f. superheated liquid). Supercooled liquids usually crystallize upon standing or when a seed crystal is added. (*Section 11.5*)

Supercritical fluid: The single, dense fluid phase that exists above the critical temperature of a substance. A supercritical fluid resembles a gas (because it completely fills its container), but it has a density comparable to a liquid. (*Section 11.6*)

Supercritical mass: A mass of a fissile isotope that exceeds the critical mass for that isotope. Under the right conditions, a supercritical mass can release energy explosively. (*Section 20.4*)

Superheated liquid: A metastable liquid phase that exists at a temperature and pressure at which the substance should be a gas (i.e., above its normal boiling point; c.f. supercooled liquid). Superheated liquids eventually boil, sometimes violently. (*Section 11.5*)

Super-heavy elements: Elements with atomic numbers near the magic number of 126. The super-heavy elements form a small "island of stability" in the upper right corner of the plot of number of neutrons versus number of protons (Figure 20.2). (*Section 20.1*)

Supersaturated solution: An unstable solution that contains more dissolved solute than it would normally contain under the given set of conditions (e.g., temperature and pressure). (*Section 13.3*)

Surface tension: The energy required to increase the surface area of a liquid by a certain amount. Surface tension is measured in units of energy per area (e.g., J/m^2). The stronger the intermolecular interactions between molecules of the liquid, the higher the surface tension. (*Section 11.3*)

Surfactants: Substances (surface-active agents), such as soaps and detergents, that disrupt the attractive intermolecular interactions between molecules of a polar liquid (e.g., water), thereby reducing the surface tension of the liquid. (*Section 11.3*)

Surroundings: All of the universe that is not the system; that is, system + surroundings = universe (c.f. system). (*Sections 5.2 and 18.1*)

Suspension: A heterogeneous mixture of particles with diameters of about 1 μm (1000 nm) that are distributed throughout a second phase. The dispersed particles separate from the dispersing phase upon standing (c.f. colloid). (*Section 13.7*)

System: The small, well-defined part of the universe in which we are interested (e.g., a chemical reaction; c.f. surroundings). A system can be open, closed, or isolated. (*Sections 5.2 and 18.1*)

Système Internationale d'Unités (SI): The International System of Units is based on metric units and requires measurements to be expressed in decimal form. There are seven base units in the SI system; all other SI units of measurement are derived from these seven. (*Essential Skills 1*)

T

T_c: See "Critical temperature (T_c)" (*Section 11.6*).

$t_{1/2}$: See "Half-life ($t_{1/2}$)" (*Section 14.5*).

Temperature: A measure of an object's thermal energy content. (*Section 5.1*)

Termination: The third stage in a chain reaction, termination consumes the reactive intermediates (see radicals) produced in the initiation and propagation stages, usually by forming stable products. (*Section 14.6*)

Tetrahedral: The lowest-energy arrangement for compounds that have four electron pairs around the central atom. In the tetrahedral arrangement, the electron pairs point at the vertices of a tetrahedron with 109.5° angles between adjacent electron pairs (see Table 9.1). Tetrahedral is also one of two possible molecular geometries for an AB_4 molecular species (the other is square planar). (*Section 9.1*)

Tetrahedral hole: One of two kinds of holes in a face-centered cubic array of atoms or ions (the other is an octahedral hole). Tetrahedral holes are located between an atom at a corner and the three atoms at the centers of the adjacent faces of the face-centered cubic unit cell. An atom or ion in a tetrahedral hole has a coordination number of 4. (*Section 12.3*)

Theoretical yield: The maximum amount of product that can be formed from the reactants in a chemical reaction. The theoretical yield is the amount of product that would be obtained if the reaction occurred perfectly and the method of purifying the product were 100% efficient (c.f. actual yield). (*Section 3.4*)

Theory: A statement that attempts to explain *why* nature behaves the way it does. Theories tend to be incomplete and imperfect, evolving with time to explain new facts as they are discovered. (*Section 1.2*)

Thermal energy: One of five forms of energy, thermal energy results from atomic and molecular motion; the faster the motion, the higher the thermal energy. The other four kinds of energy are radiant, chemical, nuclear, and electrical. (*Section 5.1*)

Thermochemistry: The branch of chemistry that describes the energy changes that occur during chemical reactions. (*Chapter 5 introduction*)

Thermodynamic control: The altering of reaction conditions so that a single desired product or set of products is present in significant quantities at equilibrium (c.f. kinetic control). (*Section 15.6*)

Thermodynamics: The study of the interrelationships among heat, work, and the energy content of a system at equilibrium. Thermodynamics can be used to determine whether a particular reaction is energetically possible in the direction in which it is written and the composition of the reaction system at equilibrium. It cannot be used to determine whether an energetically feasible reaction will actually occur as written, and it reveals nothing about the rate of the reaction or pathway by which it will occur. (*Chapter 18 introduction*)

Thermonuclear reactions: Nuclear reactions, such as fusion, for which a great deal of thermal energy must be invested in order to initiate the reaction. The amount of energy released by the reaction, however, is several orders of magnitude greater than the energy needed to initiate it. (*Section 20.5*)

Third law of thermodynamics: The entropy of any perfectly ordered, crystalline substance at absolute zero is zero. In practice, absolute zero is an ideal temperature that is unobtainable, and a perfect single crystal is also an ideal that cannot be achieved. (*Section 18.4*)

Titrant: The solution of known concentration that is reacted with a compound in a solution of unknown concentration in a titration. For a successful titration, the chemical reaction must be fast, complete, and specific (i.e., only the compound of interest must react with the titrant). (*Section 4.9*)

Titration: An experimental procedure used to determine the concentration of a compound of interest. In a titration, a carefully measured volume of a solution of known concentration (called the titrant) is added to a measured volume of a solution containing a compound whose concentration is to be determined (the unknown). The reaction used in a titration can be an acid–base, precipitation, or oxidation–reduction reaction, as long as it is fast, complete, and specific (i.e., only the compound of interest reacts with the titrant). (*Section 4.9*)

Titration curve: A plot of the pH of the solution being titrated versus the amount of acid or base (of known concentration) added. The shape of the curve is indicative of the species involved in the acid–base reaction (see Figure 16.9 for the titration curves for strong acid–strong base titrations and Figure 16.10 for weak acid–strong base and weak base–strong acid titrations). (*Section 16.5*)

Torr: A unit of pressure named for Evangelista Torricelli, the inventor of the mercury barometer. One torr is the same as one millimeter of mercury (mmHg); 760 torr = 760 mmHg = 1 atmosphere (atm) = 101325 pascals (Pa) = 101.325 kPa. (*Section 10.2*)

Transition element: Any of the elements in Groups 3–12 in the periodic table. All of the transition elements are metals. (*Section 1.7*)

Transition state: Also called the activated complex, the transition state of the reaction is the arrangement of atoms that first forms when molecules are able to overcome the activation energy and react. The transition state is not a reaction intermediate; it does not last long enough to be detected readily. (*Section 14.7*)

Transmutation: The process of converting one element to another. (*Section 1.4*)

Transuranium elements: All the elements with $Z > 92$, the transuranium elements are artificial and have been prepared by bombarding suitable target nuclei with smaller particles. (*Section 20.2*)

Triads: Sets of three elements that have similar properties. Two examples are chlorine, bromine, and iodine, and copper, silver, and gold. (*Section 7.1*)

Trigonal bipyramidal: The lowest-energy arrangement for compounds that have five electron pairs around the central atom. In the trigonal bipyramidal arrangement, three of the electron pairs are in the same plane with 120° angles between them and the other two are above and below the plane, positioned at 90° to the plane and 180° to each other (see Table 9.1). Trigonal bipyramidal is also the most common molecular geometry for an AB_5 molecular species. (*Section 9.1*)

Trigonal planar: The lowest-energy arrangement for compounds that have three electron pairs around the central atom. In the trigonal planar arrangement, the electron pairs are in the same plane with 120° angles between them (see Table 9.1). Trigonal planar is also one of three possible molecular geometries for an AB_3 molecular species (the other two are trigonal pyramidal and T-shaped). (*Section 9.1*)

Trigonal pyramidal: One of three possible molecular geometries for an AB_3 molecular species (the other two are trigonal planar and T-shaped). In the trigonal pyramidal geometry, the central atom lies above the plane of the three atoms bonded to it. As a result, the B—A—B bond angles are less than 120°. (*Section 9.1*)

Triple bond: A chemical bond formed when two atoms share three pairs of electrons. (*Section 2.1*)

Triple point: The point in a phase diagram where the solid/liquid, liquid/gas, and solid/gas lines intersect. The triple point is the only combination of temperature and pressure at which all three phases (solid, liquid, and gas) are in equilibrium and can therefore exist simultaneously. (*Section 11.7*)

Triprotic acid: A compound that can donate three hydrogen ions per molecule in separate steps (e.g., H_3PO_4). Triprotic acids are one kind of polyprotic acids (see also diprotic acid). (*Section 4.6*)

Troposphere: The lowest layer of the atmosphere, the troposphere extends from earth's surface to an altitude of about 11–13 km (7–8 miles). The temperature of the troposphere decreases steadily with increasing altitude. (*Section 3.6*)

T-shaped: One of three possible molecular geometries for an AB_3 molecular species (the other two are trigonal planar and trigonal pyramidal). The T-shaped molecular geometry is achieved when the central atom forms three bonds and has two lone pairs of electrons. The T-shaped geometry is based on the trigonal bipyramidal geometry, with the two lone pairs occupying equatorial positions (thus, the three bonds occupy the two axial positions and the third equatorial position). (*Section 9.1*)

Tyndall effect: The phenomenon of scattering a beam of visible light. The particles of a colloid exhibit the Tyndall effect, but the particles of a solution do not. (*Section 13.7*)

U

Ultraviolet light: Higher-energy radiation than visible light, ultraviolet (uv) light cannot be detected by the human eye but can cause a wide variety of chemical reactions that are harmful to organisms (e.g., sunburn). (*Section 3.6*)

Uncertainty: The estimated degree of error in a measurement. The degree of uncertainty in a measurement can be indicated by reporting all significant figures plus one. (*Essential Skills 1*)

Unit cell: The smallest repeating unit of a crystal lattice. (*Section 12.2*)

Unsaturated hydrocarbons: Hydrocarbons that contain at least one carbon–carbon multiple bond—that is, alkenes, alkynes, and aromatics (c.f. saturated hydrocarbons). (*Section 2.4*)

V

Vacancy: A point defect that consists of a single atom missing from a site in the crystal (c.f. substitutional impurity and interstitial impurity). (*Section 12.4*)

Valence bond theory: A localized bonding model (c.f. molecular orbital theory) that assumes that the strength of a covalent bond is proportional to the amount of overlap between atomic orbitals and that an atom can use different combinations of atomic orbitals (hybrids) to maximize the overlap between bonded atoms. (*Section 9.2*)

Valence electrons: Electrons in the outermost shell of an atom. The chemistry of an atom depends mostly on its valence electrons. (*Section 6.6*)

Valence-shell electron-pair repulsion (VSEPR) model: A model used to predict the shapes of many molecules and polyatomic ions, based on the idea that the lowest-energy arrangement for a compound is the one in which its electron pairs (bonding and non-bonding) are as far apart as possible. The VSEPR model provides no information about bond lengths or the presence of multiple bonds, nor does it attempt to explain any observations about molecular structure. (*Section 9.1*)

van der Waals atomic radius: Half the internuclear distance between two nonbonded atoms in the solid (see Figure 7.5c). The van der Waals atomic radius is particularly useful for elements such as the noble gases, most of which form no stable compounds. (*Section 7.2*)

van der Waals equation: A modification of the ideal gas law designed to describe the behavior of real gases by explicitly including the effects of molecular size and intermolecular forces: $(P + an^2/V^2)(V - nb) = nRT$, where P = pressure, V = volume, n = amount of a gas, R = the gas constant, T = temperature, and a and b are empirical constants that are different for each gas. The pressure term corrects for intermolecular attractive forces that tend to reduce the pressure from that predicted by the ideal gas law. The volume term corrects for the volume occupied by the gaseous molecules. (*Section 10.8*)

van der Waals forces: The intermolecular forces known as dipole–dipole interactions and London dispersion forces. (*Section 11.2*)

van't Hoff factor (i): The ratio of the apparent number of particles in solution to the number predicted by the stoichiometry of the salt (see Equation 13.25): i = (moles of particles in solution)/(moles of solute dissolved). (*Section 13.6*)

Vaporization: The physical process by which atoms or molecules in the liquid phase enter the gas or vapor phase. Vaporization, also called evaporation, occurs only for those atoms or molecules that have enough kinetic energy to overcome the intermolecular attractive forces holding the liquid together. To escape the liquid, though, the atoms or molecules must be at the surface of the liquid. The atoms or molecules that undergo vaporization create the vapor pressure of the liquid. (*Section 11.4*)

Vapor pressure: The pressure created over a liquid by the molecules of the liquid substance that have enough kinetic energy to escape to the vapor phase. (*Section 11.4*)

Viscosity (η): The resistance of a liquid to flow. Liquids that flow readily (e.g., water or gasoline) have a low viscosity, whereas liquids that flow very slowly (e.g., motor oil or molasses) have a high viscosity. Viscosity is expressed in units of the poise (mPa·s). (*Section 11.3*)

Visible light: Radiation that the human eye can detect. (*Section 3.6*)

Volatile liquids: Liquids with relatively high vapor pressures. Volatile liquids tend to evaporate readily from an open container (c.f. nonvolatile liquids). (*Section 11.4*)

Volume: The amount of space occupied by a sample of matter. (*Section 1.3*)

Voltaic cell: One of two types of electrochemical cells, a voltaic cell (or galvanic cell) uses the energy released during a spontaneous oxidation–reduction (redox) reaction ($\Delta G < 0$) to generate electricity (c.f. electrolytic cell, the other type of electrochemical cell). (*Section 19.1*)

W

Waters of hydration: The loosely bound water molecules in hydrate compounds. These waters of hydration can often be removed by simply heating the compound. (*Section 2.2*)

Wave: A periodic oscillation that transmits energy through space. (*Section 6.1*)

Wave function: A mathematical function that relates the location of an electron at a given point in space (identified by x, y, z coordinates) to the amplitude of its wave, which corresponds to its energy. As a result, each wave function Ψ is associated with a particular energy E. That is, wave functions are mathematical equations that describe atomic orbitals. (*Sections 6.5 and 9.3*)

Wavelength (λ): The distance between two corresponding points in a wave—between the midpoints of two peaks, for example, or two troughs. Wavelengths are described by any appropriate unit of distance, such as meters. (*Section 6.1*)

Wave–particle duality: A principle that matter and energy (e.g., light) have properties typical of both waves and particles. (*Section 6.4*)

Weak acid: An acid in which only a fraction of the molecules react with water to produce H^+ and the corresponding anion (c.f. strong acid). Weak acids are weak electrolytes. (*Section 4.6*)

Weak base: A base in which only a fraction of the molecules react with water to produce OH^- and the corresponding cation (c.f. strong base). Weak bases are weak electrolytes. (*Section 4.6*)

Weak electrolyte: A compound (e.g., CH_3CO_2H) that produces relatively few ions when dissolved in water, thus producing an aqueous solution that conducts electricity, but not as well as solutions of strong electrolytes (c.f. strong electrolyte). (*Section 4.1*)

Weight: A force caused by the gravitational attraction that operates on an object. The weight of an object depends on its location (c.f. mass). (*Section 1.3*)

Work (w): The product of a force F acting through a distance d: $w = Fd$ (Equation 18.2). Because work occurs only when an object or a substance moves against an opposing force, work requires that the system and its surroundings be physically connected. (*Section 18.1*)

Work hardening: The practice of introducing a dense network of dislocations throughout a solid, making it very tough and hard. (*Section 12.4*)

X

X: See "Mole fraction (X)" (*Section 10.5*).

X-ray diffraction: An experimental technique used to obtain information about the structures of crystalline substances. A beam of X rays (whose wavelengths are approximately the same magnitude as the distances between atoms in molecules or ions) is aimed at a sample of a crystalline material, and the X rays are diffracted by the layers of atoms in the crystalline lattice. When the beams strike photographic film, an X-ray diffraction pattern is produced, which consists of dark spots on a light background. Interatomic distances in crystals can be obtained mathematically from these diffraction data. (*Section 12.3*)

Z

Z: See "Atomic number (Z)" (*Section 1.6*).

Z_{eff}: See "Effective nuclear charge (Z_{eff})" (*Section 6.5*).

Zeroth-order reaction: A reaction whose rate is independent of concentration (c.f. first-order reaction and second-order reaction; see Table 14.6). The differential rate law of a zeroth-order reaction is rate $= k$, where k is the rate constant of the reaction. (*Section 14.3*)

Zinc blende structure: The solid structure that results when half of the tetrahedral holes in a face-centered cubic lattice of anions are filled with cations. The zinc blende structure has a 1:1 cation:anion ratio and each ion has a coordination number of 4. (*Section 12.3*)

APPENDIX H

Answers to Selected Problems

Chapter 14

1. Kinetics gives information on the rate and mechanism of a reaction; the balanced chemical equation tells only the stoichiometry of the reaction.

3. Reaction rates generally increase with increasing reactant concentration, temperature, and addition of a catalyst. Physical properties such as high solubility also increase reaction rates. Solvent polarity can either increase or decrease the rate of a reaction, but increasing solvent viscosity generally decreases reaction rates.

5. Increasing the temperature increases the average kinetic energy of molecules and ions, causing them to collide more frequently and with greater energy, which increases the rate of reaction. First dissolve sugar in hot tea, and then add ice.

9. Reactant concentrations are highest at the beginning of a reaction. The plot of [C] vs. t is a curve with a slope that becomes steadily less positive.

11. Faster. The H_3O^+/OH^- reaction is faster due to the decreased relative size of reactants and the higher electrostatic attraction between reactants.

15. Only reactions (b) and (d). Zeroth-order reactions are usually those for which the rate is limited by the number of available reaction sites or the amount of available surface.

17. (a) The reaction is first order in that reactant; (b) the overall order is ≥ 1.

19. (a) For a given reaction under particular conditions, the magnitude of the first-order rate constant does not depend on whether a differential or integrated rate law is used. (b) The differential rate law requires multiple experiments to determine reactant order, the integrated rate law needs only one experiment. (c) Using the differential rate law, a graph of concentration versus time is a curve with a slope that becomes less negative with time, whereas for the integrated rate law, a graph of ln[reactant] versus time gives a straight line with slope $= -k$. The integrated rate law allows you to calculate the concentration of a reactant at any time during the reaction; the differential rate law does not.

21. The reaction rate increases as the rate constant increases. We cannot directly compare reaction rates and rate constants for reactions of different orders because they are not mathematically equivalent.

23. The substance with more disintegrations per second has the shorter half-life.

25. The order of a reaction depends upon the molecularity of the rate-determining step; there is no relationship between the order of a reaction and its stoichiometry.

31. (a) An increase in electrostatic repulsions between reactants will increase the activation energy; (b) bond formation in the activated complex tends to lower the activation energy; (c) an increase in the energy of the activated complex increases the activation energy of the reaction.

35. The activation energy is not related to either the temperature or the frequency factor A, but the frequency factor is usually proportional to the square root of the temperature. An increase in the frequency factor increases the rate.

37. A catalyst lowers the activation energy of a reaction. Some catalysts can also orient the reactants and thereby increase the frequency factor. Catalysts have no effect on the change in potential energy for a reaction.

39. In the process of adsorption, a reactant binds tightly to a surface. Because intermolecular interactions between the surface and the reactant weaken or break bonds in the reactant, its reactivity is increased, and the activation energy for a reaction is often decreased.

41. (a) Heterogeneous catalysts are easier to recover; (b) collision frequency is greater for homogeneous catalysts; (c) homogeneous catalysts are often more sensitive to temperature; and (d) homogeneous catalysts are often more expensive.

43. The Mn^{2+} ion donates an electron to Ce^{4+} and then accepts an electron from Tl^+. It functions as an electron conduit.

47. Average rate: 41 mi/h; instantaneous rate was 0 at t = 13 min and 38 mi/h at 28 min.

49. (a) rate will double; (b) rate will double; (c) rate will quadruple; (d) no change in rate.

51. 298 s, 1270 s

53. First order in Fe^{3+}; second order in I^-; third order overall; rate $= k[Fe^{3+}][I^-]^2$.

55. 1.29×10^{-4} M/s; 3.22×10^{-5} M/s

59. No. The reaction is second order: the half-life decreases with increasing reactant concentration according to $t_{1/2} = 1/k[A_0]$.

61. 1.92×10^3 s or 1920 s

63. The k_2 step is likely to be rate limiting; the rate cannot proceed any faster than the second step.

65. $\text{rate} = k_2 \dfrac{k_1}{k_{-1}} \dfrac{[O_2NNH_2]}{[H^+]} = k\dfrac{[O_2NNH_2]}{[H^+]}$

67. Rate will approximately double: 20°C to 30°C, rate increases by about $2^1 = 2$; 20°C to 70°C, rate increases by about $2^5 = 32$-fold. Plot of rate versus temperature will give an exponential increase: rate $\propto 2^{\Delta T/10}$.

69. (a) 1.0×10^{-5} M/s; (b) 6.6×10^{-5} M/s; (c) 3.5×10^{-4} M/s

71. 96 kJ/mol

73. $\dfrac{\Delta[ES]}{\Delta t} = -(k_2 + k_{-1})[ES] + k_1[E][S] + k_{-2}[E][P] \approx 0$; first order in substrate.

75. In both cases, pathway A's product. All of the Z produced in the catalyzed reversible pathway B ($X \rightleftharpoons Z$) will eventually be converted to X as X is converted irreversibly to Y by pathway A:

$$Z \underset{B}{\rightleftharpoons} X \xrightarrow{A} Y.$$

77. rate $= k_f\,[SO_2][CO] - k_r[SO][CO_2]$
rate $= k_f\,[SO][CO] - k_r[S][CO_2]$
rate $= k_f\,[SO][SO_2] - k_r[S][SO_3]$
79. Reaction is second order, first order in O and first order in O_3. Yes, ozone is being produced faster than it is being destroyed. If ozone concentrations are not increasing, then either some other reaction must be consuming some of the ozone produced in this reaction or the ozone-producing reaction does not operate at this rate continuously.
81. Yes, the object is about 2300 years old.
83. (a) $k = 0.0565\ \text{yr}^{-1}$; (b) 0.482 g of ^3H
87. (a) Second order, first order in O and first order in O_3; (b) 17 kJ/mol; (c) 0.44 kJ/mol;
(d)

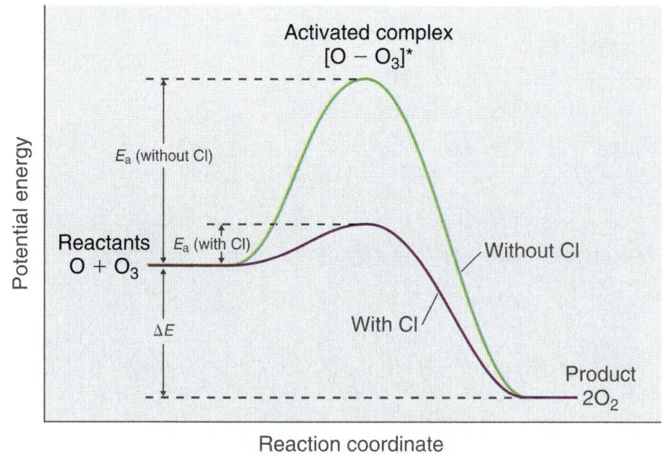

(e) Cl is a potent catalyst for ozone destruction.

89.

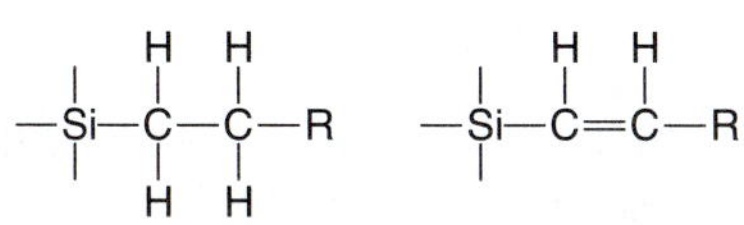

91. $Ni_{1-x}O$ is a nonstoichiometric oxide that contains a fraction of Ni(I) sites. These can react with oxygen to form a Ni(III)-oxide site, which is reduced by CO to give Ni(I) and CO_2.
93. 0.35 gram of ^{32}P.
95. 4.1×10^{15}.

Chapter 15

3. Both forward and reverse reactions occur, but at the same rate. Na^+ and Cl^- ions continuously leave the surface of an NaCl crystal to enter solution, while Na^+ and Cl^- ions in solution precipitate on the surface of the crystal.
5. The equilibrium constant for the reaction written in reverse (K') is $K' = 1/K$.
7. All are heterogeneous.
9. Rapid cooling "quenches" the reaction mixture and prevents the system from reverting to the low-temperature equilibrium composition that favors reactants.

11. $K = k_f/k_r$; $K = \dfrac{[C]^c[D]^d}{[A]^a[B]^b}$

13. (a) $K = \dfrac{[NO_2]^2}{[NO]^2[O_2]}$; $K_p = \dfrac{(P_{NO_2})^2}{(P_{NO})^2\,(P_{O_2})}$

(b) $K = \dfrac{[HI]}{[H_2]^{1/2}[I_2]^{1/2}}$; $K_p = \dfrac{P_{HI}}{(P_{H_2})^{1/2}\,(P_{I_2})^{1/2}}$

(c) $K = \dfrac{[trans\text{-stilbene}]}{[cis\text{-stilbene}]}$; $K_p = \dfrac{P_{trans\text{-stilbene}}}{P_{cis\text{-stilbene}}}$

15. (a) $K = \dfrac{[SO_3]^2}{[O_2]^3}$; $K_p = \dfrac{(P_{SO_3})^2}{(P_{O_2})^3}$

(b) $K = \dfrac{[CO]^2}{[CO_2]}$; $K_p = \dfrac{(P_{CO})^2}{P_{CO_2}}$

(c) $K = \dfrac{[SO_2]^2}{[O_2]^3}$; $K_p = \dfrac{(P_{SO_2})^2}{(P_{O_2})^3}$

17. At equilibrium, $[A] = \sqrt{[B]}$; $\Delta n = -1$, so $K_p = K(RT)^{\Delta n} = \dfrac{K}{RT}$; difference increases as T increases.

23. (a) $K = \dfrac{[CH_4][H_2S]^2}{[CS_2][H_2]^4}$; doubling $[CS_2]$ would require multiplying $[H_2]$ by $(1/2)^4 = 1/16$.

(b) $K = \dfrac{[PCl_3][Cl_2]}{[PCl_5]}$; if $[Cl_2]$ is halved, $[PCl_5]$ must also be halved.

(c) $K = \dfrac{[NO]^4[H_2O]^6}{[NH_3]^4[O_2]^5}$; if [NO] is doubled, $[H_2O]$ is multiplied by $\sqrt[6]{1/16} = 1/2^{2/3} = 1/0.587$

25. $K = \dfrac{[NO]^4[H_2O]^6}{[NH_3]^4[O_2]^5}$; increasing P will favor reactants. Increasing P favors the side with fewer moles of gas.
27. The ratio k_f/k_r will decrease. Increasing P favors products, while increasing T favors reactants, so the effect of increasing both T and P depends on the magnitude of ΔH.
29. (a) No effect; (b) P_{H_2} will decrease; (c) P_{Cl_5} will decrease
33. If $Q < K$, reactants will be converted to products; if $Q > K$, products will be converted to reactants.
35. $K = 0.892$: concentrations of products and reactants are approximately equal at equilibrium;
$K = 3.25 \times 10^8$: concentration of products to reactants at equilibrium is very large;
$K = 5.26 \times 10^{-11}$: concentration of products to reactants at equilibrium is very small.

37. (a) $K' = \dfrac{[NH_3]}{[N_2]^{1/2}[H_2]^{3/2}}$; (b) $K'' = \dfrac{[NH_3]^{2/3}}{[N_2]^{1/3}[H_2]}$; $K = \dfrac{[NH_3]^2}{[N_2][H_2]^3}$ for $N_2(g) + 3H_2(g) \rightleftharpoons 2NH_3(g)$; $K' = \sqrt{K}$, $K'' = \sqrt[3]{K}$

39. (a) $K = \dfrac{[NO_2]^2}{[NO]^2[O_2]}$; (b) $K = \dfrac{[HI]}{[H_2]^{1/2}[I_2]^{1/2}}$;

(c) $K = \dfrac{[Ca^{2+}][OCl\check{S}]^2\,P_{CO_2}}{[HOCl]^2}$

41. (a) $K = 1.25 \times 10^{-5}$, $K_p = 2.40 \times 10^{-4}$; (b) $K = 9.43$, $K_p = 5.60 \times 10^{-3}$.

43. (a) $K = \dfrac{[Cl_2][NO]^2}{[NOCl]^2} = 4.66 \times 10^{-4}$, $K_p = 1.91 \times 10^{-2}$;
(b) $K = \dfrac{[PCl_5]}{[PCl_3][Cl_2]} = 28.2$, $K_p = 0.658$.

45. $\dfrac{NH_3(g) + 5}{4O_2(g)} \rightleftharpoons \dfrac{NO(g) + 3}{2H_2O(g)}$, also written as:
$4NH_3(g) + 5O_2(g) \rightleftharpoons 4NO(g) + 6H_2O(g)$

47. (a) 3.29×10^3; (b) 1.17×10^{14}; (c) 3.04×10^{-4}

49. No; since $\Delta H > 0$, K_p (and P_{O_2}) should increase with increasing T.

51. $K = 54 = K_p$

53. 0.135

55. (a) $K_{375} = 7.8 \times 10^{-2}$, $K_{303} = 1.2 \times 10^{-3}$; (b) 0.26 M; (c) 11 atm

59. $K = 2.50$; 0.71 mol of isobutane, 0.29 mol of n-butane

61. $P_{COCl_2} = 0.0934$ atm, $P_{Cl_2} = P_{CO} = 4.5 \times 10^{-6}$ atm; because K_p is very small, assume that the decrease in P_{COCl_2} is negligible compared with its initial pressure.

63. [IBr] = 0.049 M; $P_{IBr} = 1.5$ atm, $P_{I_2} = 0.12$ atm, $P_{Br_2} = 0.43$ atm

65. Not at equilibrium: in both cases, the sum of the equilibrium partial pressures is *less* than the total pressure, so the reaction will proceed to the right to decrease the pressure.

67. (a) $K = \dfrac{[H_2O]}{[H_2]}$, $K = \dfrac{[CO_2]}{[CO]}$; (b) $P_{H_2O} = 21.1$atm, $P_{H_2} = 0.27$ atm, $P_{CO_2} = 21.3$ atm, $P_{CO} = 0.05$ atm; (c) $K_p = 0.14$; (d) the amount of CoO has no effect on the shape of a graph of products versus reactants *as long as some solid CoO is present.*

69. (a) 1: Q and K do not change; 2: Q does not change, K decreases; 3: Q and K do not change; 4: Q decreases, K does not change.
(b) 1: Q decreases, K does not change; 2: Q does not change, K increases; 3: Q increases, K does not change; 4: Q decreases, K does not change.
(c) 1: Q and K do not change; 2: Q does not change, K increases; 3: Q and K do not change; 4: Q decreases, K does not change.

71. None of the changes would affect K. (a) Q doubles; (b) Q is halved; Q decreases.

73. K would not change; it does not depend on volume.

75. Endothermic; K increases with increasing T.

77. [CO] = [H_2O] = 0.630 M, [CO_2] = [H_2] = 0.703 M; no effect on K

79. (a) At 800°C, [CO] = 0.678 M, [CO_2] = 0.231 M; at 1000°C, [CO] = 0.649 M, [CO_2] = 0.260 M; (b) at 800°C, $P_{CO} = 59.7$ atm, $P_{CO_2} = 20.3$ atm; at 1000°C, $P_{CO} = 67.8$ atm, $P_{CO_2} = 27.2$ atm; (c) removing CO would cause the reaction to shift to the right, causing P_{CO_2} to decrease.

81.

	CH_4	O_2	CO_2	H_2O	Q	K
Initial (moles)	0.45	0.90	0	0	**0**	1.29
At equilibrium	0.215	0.43	0.235	0.47	**1.29**	1.29
Add 0.50 mol of methane	0.715	0.43	0.235	0.47	**0.39**	1.29
New equilibrium	0.665	0.33	0.285	0.57	1.29	**1.29**
Remove water	0.665	0.33	0.285	0	**0**	1.29
New equilibrium	0.57	0.14	0.38	0.19	**1.29**	1.29

83. Use low temperature and high pressure (small volume).

87. (a) reactants; (b) [Cl_2] = [PCl_3] = 0.082 M, [PCl_5] = 0.168 M; (c) increasing pressure favors reactant (PCl_5); (d) 1.60×10^3 kg, 52.6%.

89. Products are favored; $K = \dfrac{[\text{chloral hydrate}]}{[Cl_3CHO][H_2O]}$; high concentrations of water will favor chloral hydrate formation.

93. $K = \dfrac{[CO_2]^4[S_2]}{[CO]^4[SO_2]^2}$; (a) $K' = \dfrac{[CO_2]^2[S_2]^{1/2}}{[CO]^2[SO_2]} = K^{1/2}$;
(b) $K'' = \dfrac{[CO_2][S_2]^{1/4}}{[CO][SO_2]^{1/2}} = K^{1/4}$;
(c) $K_p = 1.6$, $K = 93$, $[S_2] = 5.1 \times 10^{-10}$ M.

95. $P_{CO} = P_{Cl_2} = 0.420$ atm, $P_{COCl_2} = 0.261$ atm; reactants are slightly favored.

99. Both reactions are favored by increasing [HCl] and decreasing volume.

101. Isocitrate is immediately consumed by the next reaction in the cycle, which favors formation of products and drives the conversion of citrate to isocitrate to completion. The reaction is initially driven toward the formation of isocitrate, and then a new equilibrium state is reached in which the isocitrate concentration has increased proportionally.

Chapter 16

1. $K_{auto} = [H_3O^+]\,[OH^-]/[H_2O]^2$
$K_w = [H_3O^+]\,[OH^-] = K_{auto}\,[H_2O]^2$

3. $H_2O(l) + HNO_3(g) \rightarrow H_3O^+(aq) + NO_3^-(aq)$; water is the base
$H_2O(l) + NH_3(g) \rightarrow OH^-(aq) + NH_4^-(aq)$; water is the acid

7. (a) $\underset{\text{acid(1)}}{HSO_4^-(aq)} + \underset{\text{base(2)}}{H_2O(l)} \rightleftharpoons \underset{\text{base(1)}}{SO_4^{2-}(aq)} + \underset{\text{acid(2)}}{H_3O^+(aq)}$

(b) $\underset{\text{base(2)}}{C_3H_7NO_2(aq)} + \underset{\text{acid(1)}}{H_3O^+(aq)} \rightleftharpoons \underset{\text{acid(2)}}{C_3H_8NO_2^+(aq)} + \underset{\text{base(1)}}{H_2O(l)}$

(c) $\underset{\text{acid(1)}}{HOAc(aq)} + \underset{\text{base(2)}}{NH_3(aq)} \rightleftharpoons \underset{\text{base(1)}}{CH_3CO_2^-(aq)} + \underset{\text{acid(2)}}{NH_4^+(aq)}$

(d) $\underset{\text{acid(1)}}{SbF_5(aq)} + \underset{\text{base(2)}}{2HF(aq)} \rightleftharpoons \underset{\text{acid(2)}}{H_2F^+(aq)} + \underset{\text{base(1)}}{SbF_6^-(aq)}$

9. (a) $K_a = \dfrac{[CO_3^{2-}][H^+]}{[HCO_3^-]}$

(b) $K_a = \dfrac{[HCO_2^-][H^+]}{[HCO_2H]}$

(c) $K_a = \dfrac{[H_2PO_4^-][H^+]}{[H_3PO_4]}$

11. (a) right; (b) right; (c) left; (d) right

13. No, only the first dissociation reaction goes to completion; $2.0\ M < [H^+] < 4.0\ M$

17. (a) increase; (b) stay approximately the same; (c) increase; (d) decrease

19. $Sn(H_2O)_6^{4+}$; its higher charge to radius ratio will cause greater polarization of bound water molecules.

25. $CF_3S^- < CH_3S^- < OH^-$ (strongest base)

29. NH_3; Cl atoms withdraw electron density from N in Cl_2NH.

31. $C_{H_3PO_4} = [H_3PO_4] + [H_2PO_4^-] + [HPO_4^{2-}] + [PO_4^{3-}]$

33. Percent ionization increases upon dilution.
37. (a) neutral; (b) acidic; (c) basic
39. At the equivalence point of the titration of a weak acid, the solution contains the conjugate base, which reacts with water to form hydroxide ions, making the solution basic. At the equivalence point of the titration of a weak base, the solution contains the conjugate acid, which reacts with water to form hydronium ions, making the solution acidic.
45. Adding NH_4^+ shifts the ionization equilibrium of ammonia ($H_2O(l) + NH_3(aq) \rightleftharpoons OH^-(aq) + NH_4^+(aq)$) to the left, decreasing $[OH^-]$ and decreasing the pH.
47. (a) Shifts to left, pH decreases; (b) shifts to right, pH increases; (c) shifts to left, pH increases
51. the midpoint
55. (a) Not a buffer; the HCl completely neutralizes the sodium acetate to give acetic acid and NaCl(*aq*).

(b) Buffer; the HCl neutralizes only half of the sodium acetate to give a solution containing equal amounts of acetic acid and sodium acetate.

(c) Not a buffer; the NaOH completely neutralizes the acetic acid to give sodium acetate.

(d) Buffer; the NaOH neutralizes only half of the acetic acid to give a solution containing equal amounts of acetic acid and sodium acetate.

(e) Buffer; the solution will contain a 2:1 ratio of sodium acetate and acetic acid.

59. The CO_2/HCO_3^- buffer is an open system, whose pH and buffering capacity depend on the CO_2 pressure as well as $[HCO_3^-]$. Increasing the CO_2 pressure decreases the pH of blood; increasing the O_2 pressure has no direct effect on the pH of blood.
61. $K_{H_2SO_4} = [H_3SO_4^+][HSO_4^-] = K[H_2SO_4]^2$; $[H_3SO_4^+] = 0.3\ M$; fraction ionized is 0.02.
63. pOH = 8.37, $[OH^-] = 4.3 \times 10^{-9}\ M$; acidic
65. (a) pH = 0.82, pOH = 13.18; (b) pH = 12.5, pOH = 1.5; (c) pH = 2.64, pOH = 11.36; (d) pH = 12.99, pOH = 1.01; (e) pH = 3.77, pOH = 10.23; (f) pH = −0.76, pOH = 14.76
69. 2.88 mg of HCl
71. acid B < acid C < acid A (strongest)
73. (a) $K_a = 6.3 \times 10^{-11}$, $pK_a = 10.2$; (b) $K_a = 7.9 \times 10^{-7}$, $pK_a = 6.1$; (c) $K_a = 0.50$, $pK_a = 0.30$; (d) $K_a = 2.5 \times 10^{-13}$, $pK_a = 12.6$; (e) $K_a = 3.2 \times 10^{-17}$, $pK_a = 16.5$
75. $K_a = 6.3 \times 10^{-11}$; $pK_a = 9.2$.
81. 0.94% dissociated; butanoic acid will have the higher pK_a value; the inductive effect due to the chlorine substituent will make 3-chlorobutanoic acid a somewhat stronger acid.
83. $pK_b = 9.43$; $(CH_3)_2CHNH_2$ will be a stronger base and have a lower pK_b; aniline is a weaker base because the lone pair on the nitrogen atom can be delocalized onto the aromatic ring.
85. 1.79
87. 3.8×10^{-5}
91. 8.12
95. 0.86 mL
99. (a) dilute 8.33 mL of 12.0 *M* HCl to 500.0 mL; (b) about 72 mL; (c) 77.0 mL.
103. pH at equivalence point = 8.46

Volume of Base Added, mL	0	5	10	15	20	25
pH	2.21	3.23	3.67	4.11	4.87	12.36

107. 1.82
111. (a) 1.35×10^{-3}; (b) 4.05; (c) 3.88; (d) 4.30
113. (a) $pK_w = 13.53$, pH = pOH = 6.77; (b) lower
115. 68.4 mL; 1.59; 7.82

Chapter 17

1. (a) $K_{sp} = [Ag^+][I^-]$
(b) $K_{sp} = [Ca^{2+}][F^-]^2$
(c) $K_{sp} = [Pb^{2+}][Cl^-]^2$
(d) $K_{sp} = [Ag^+]^2[CrO_4^{2-}]$
5. For a 1:1 salt, the molar solubility is simply $\sqrt{K_{sp}}$, for a 2:1 salt, the molar solubility is $\sqrt[3]{K_{sp}/4}$. Consequently, the magnitudes of K_{sp} can be correlated with molar solubility *only* if the salts have the same stoichiometry.
7. Because of the common ion effect. Adding a soluble Mg^{2+} salt increases $[Mg^{2+}]$ in solution, and le Châtelier's principle predicts that this will shift the solubility equilibrium of $MgCO_3$ to the left, decreasing its solubility.
9. Because Mg^{2+} is smaller than Ca^{2+}, ion-pair formation will be more important for $Mg_3(PO_4)_2$. Consequently, the difference for $Ca_3(PO_4)_2$ will be less.
13. Mg^{2+} has a higher charge-to-radius ratio than Ba^{2+}. Consequently, Mg^{2+} will be a significantly stronger Lewis acid and will have a greater tendency to form complex ions.
15. Co^{2+} is the Lewis acid, and pyridine, with a lone pair of electrons on the nitrogen atom, is the Lewis base.
17. The S^{2-} ion is a stronger base than the acetate ion. Consequently, SnS will dissolve more readily in dilute HCl.
21. Elements that form amphoteric oxides generally lie on or near the diagonal line dividing metals (which tend to form basic oxides) from nonmetals (which form acidic oxides). Thus, Be, B, and Al are most likely to form amphoteric oxides.
23. (a) $2.05 \times 10^{-4}\ M$; (b) $7.73 \times 10^{-9}\ M$; (c) $3.1 \times 10^{-10}\ M$
25. 3.38 g
27. (a) $4.15 \times 10^{-4}\ M$; (b) $9.3 \times 10^{-6}\ M$; (c) $6.3 \times 10^{-6}\ M$; (d) $6.5 \times 10^{-5}\ M$; (e) $1.03 \times 10^{-3}\ M$
29. 22.3 mg. A secondary reaction occurs, where OH^- from the dissociation of the salt reacts with H^+ from the dissociation of water. This reaction causes further dissociation of the salt (le Châtelier's principle).
31. 1.2×10^{-10}
33. 1.70×10^{-5}
37. (a) 3.86×10^{-71}; (b) 6.7×10^{-9}; (c) 9.0×10^{-8}; (d) 2.15×10^{-9}
39. $7.4 \times 10^{-6}\ M$, 2.1 mg
41. Precipitation will occur in all cases.
43. 290 mL
45. 1.59×10^{-11}. The common ion effect decreases the solubility of the salt.

47.
$$[Fe(H_2O)_6]^{2+}(aq) + OH^-(aq) \rightleftharpoons [Fe(H_2O)_5(OH)]^+(aq) \qquad \log K_1 = 5.56$$
$$[Fe(H_2O)_5(OH)]^+(aq) + OH^-(aq) \rightleftharpoons [Fe(H_2O)_4(OH)_2](aq) \qquad \log K_2 = 9.77$$
$$[Fe(H_2O)_4(OH)_2](aq) + OH^-(aq) \rightleftharpoons [Fe(H_2O)_3(OH)_3]^-(aq) \qquad \log K_3 = 9.67$$
$$[Fe(H_2O)_3(OH)_3]^-(aq) + OH^-(aq) \rightleftharpoons [Fe(OH)_4]^{2-}(aq) + 2H_2O(l) \qquad \log K_4 = 8.58$$
$$[Fe(H_2O)_6]^{2+}(aq) + 4OH^-(aq) \rightleftharpoons [Fe(OH)_4]^{2-}(aq) + 6H_2O(l) \qquad \log K_f = \log K_1 + \log K_2 + \log K_2 + \log K_4 = 33.58$$

$[Fe(OH)_4]^{2-}$ has a large value of K_f, so it should be stable in the presence of excess OH^-.

49. 2.94
51. No, both metal ions will precipitate; AgBr will precipitate as Br^- is added, and CuBr will begin to precipitate at $[Br^-] = 8.6 \times 10^{-6}$ *M*.
53. (a) 4.3 g; (b) 4.9×10^{-3} *M*
55. (a) 4.5×10^{-7} *M*; (b) 0.84 *M*
59. No; these cations would precipitate as sulfides.

Chapter 18

1. Thermodynamics tells us nothing about the rate at which reactants are converted to products.
3. heat and work; path dependent
7. At constant pressure, $\Delta H = \Delta E + P\Delta V$.
11. For a spontaneous process, $\Delta G = \Delta H - T\Delta S < 0$. Since ΔH is positive, $T\Delta S$ must also be positive and greater in magnitude than ΔH.
13. With bond angles of 60°, cyclopropane is highly strained, causing it to be less stable than an unstrained cyclic hydrocarbon like cyclopentane.
15. Irreversible expansion, no work is done because the external pressure is effectively zero.
17. reversible: (a) and (c); irreversible: (b) and (d)
23. ΔS_{soln} in ethanol must be more positive than ΔS_{soln} in water to make ΔG_{soln} more negative for ethanol.
25. The third law of thermodynamics gives an absolute reference point for entropies, but no such absolute reference point is available for enthalpy (or any other function involving energy).
27. At any temperature, A will have a higher value of *S* than B, and the entropy of A will increase faster with increasing temperature.
29. Only a combination of the change in enthalpy and the change in entropy of the system can be used to predict how changes in the system will affect the entropy of the universe.
31. Because there are more molecules of reactants than products, ΔS is negative for the reaction as written. The $-T\Delta S$ term is therefore positive for the reaction and increases in magnitude as *T* increases, making ΔG more positive and favoring reactants.
35. Yes. All real processes involve some loss of heat (due to friction, etc.), which is used to increase the entropy of the universe.
37. Because $\Delta G° = -RT \ln K$, the equilibrium constant becomes larger and the extent of the reaction increases as $\Delta G°$ becomes more negative.
39. $\Delta G° = -RT \ln K$ in both cases. For gases, the equilibrium constant is K_p. *K* and K_p are related by: $K_p = K(RT)^{\Delta n}$, where Δn is the number of moles of gaseous product minus the number of moles of gaseous reactant.
41. It is coupled to another reaction that is spontaneous, which drives this reaction forward (le Châtelier's principle).
43. 13.1 kJ work done on the surroundings
45. $\Delta V = 30.6$ L, $w = -3.1$ kJ, work done on the surroundings
47. 220 kJ
49. (a) 0; (b) −455 J; (c) −317 J
51. −350 J; 8.2 kJ
55. $\Delta S_{univ} = \Delta S_{cold} + \Delta S_{hot} = \frac{q}{T_{cold}} + (-\frac{q}{T_{hot}})$; only if $T_{cold} < T_{hot}$ can ΔS_{univ} be > 0.
59. Phase transitions have large ΔS values; 13.1°C; 4.3 J.
61. (a) −164.3 J/K; (b) −9.1 J/K; (c) 114.5 J/K; (d) −173.2 J/K
63. 25 J/ (mol · K)
65. yes; 274 K
67. (a) −237.1 kJ/mol, spontaneous as written; (b) −100.4 kJ/mol, spontaneous as written; (c) 16.6 kJ/mol, spontaneous in reverse direction; (d) 141.9 kJ/mol, spontaneous in reverse direction.
69. (a) $\frac{1}{2}N_2(g) + O_2(g) \rightarrow NO_2(g)$; not spontaneous at any *T*;
(b) $\frac{1}{2}N_2(g) + \frac{1}{2}O_2(g) \rightarrow NO(g)$; not spontaneous at 25°C, spontaneous above 7500 K;
(c) $\frac{1}{2}N_2(g) + \frac{3}{2}H_2 \rightarrow NH_3(g)$; spontaneous at 25°C
71. 919 K, 646°C
73. $MgCO_3$: $\Delta G° = 63$ kJ/mol, spontaneous above 658 K; $BaCO_3$: $\Delta G° = 220$ kJ/mol, spontaneous above 1560 K
75. (a) $\Delta G° = -28.5$ kJ/mol; (b) −26.4 kJ/mol; (c) −23.0 kJ/mol.
77. (a) 1.21×10^{66}; (b) 1.89×10^{6}; (c) 8.25×10^{15}
79. −13.3 kJ/mol
81. 5.1×10^{-21}
83. 10.3 kJ/mol
85. (a) 129.5 kJ/mol; (b) 6.0; (c) 6.0 atm; (d) products are favored at high *T*, reactants are favored at low *T*.
89. (a) −4520 kJ/mol B_5H_9; (b) 1300 kJ; (c) -1.63×10^6 kJ/50 lb of B_5H_9
91. −2220 kJ/mol
93. (a) aerobic conversion; (b) −268 kJ
97. Yes, reaction is spontaneous at 25°C; not spontaneous at 800°C ($\Delta G = 0.82$ kJ/mol); reaction rate is much faster at 800°C
99. (a) −314 kJ/mol; (b) yes; (c) 1.1×10^{55}; (d) ignition is required to overcome high activation energy to reaction.
101. (a) 4.4 kJ/mol; (b) 0.17; (c) 78; (d) 13; (e) entropy driven
103. 2.3×10^7; −43.7 kJ/mol

Chapter 19

5. A large difference in cation/anion diffusion rates would increase resistance in the salt bridge and limit electron flow through the circuit.

11. $Ni(s) \mid Ni^{2+}(aq) \parallel H^+(aq) \mid H_2(g) \mid Pt(s)$

$E°_{anode}$ $Ni^{2+} + 2e^- \rightarrow Ni$; − 0.257 V

$E°_{cathode}$ $2H^+ + 2e^- \rightarrow H_2$; 0.000 V

$E°_{cell}$ $2H^+ + Ni \rightarrow H_2 + Ni^{2+}$; 0.257 V

13. (a) oxidant: $Ni^{2+}(aq)$; reductant: $Cr(s)$
(b) oxidant: $Cl_2(g)$; reductant: $Sn^{2+}(aq)$
(c) oxidant: $H_3AsO_4(aq)$; reductant: $Zn(s)$
(d) oxidant: $NO_2(g)$; reductant: $NO_2(g)$

19. No. $E° = -0.691$ V for $Ag_2S(s) + 2e^- \rightarrow Ag(s) + S^{2-}(aq)$, which is too negative for Ag_2S to be spontaneously reduced by oxalic acid [$E° = 0.49$ V for $2CO_2(g) + 2H^+(aq) + 2e^- \rightarrow H_2C_2O_4(aq)$].
23. extensive: w_{max} and *n*; intensive: E_{cell}
27. Gold is highly resistant to corrosion because of its very positive reduction potential.
33. (a) lead storage battery; (b) lithium–iodine battery; Nicad or lithium ion battery (rechargeable)
37. Paint keeps oxygen and water from coming into direct contact with the metal, which prevents corrosion. Paint is more necessary because salt is an electrolyte that increases the conductivity of water and facilitates the flow of electric current between anodic and cathodic sites.
41. For the same electrolysis time and current, the $Fe(NO_3)_2 \cdot 6H_2O$ solution will produce a greater mass of metallic iron; reduction of 1 mol of Fe^{2+} ions requires only 2 mol of electrons versus 3 mol of electrons for 1 mol of Fe^{3+} ions.
43. (a) Reductive half-reaction: $NO_3^-(aq) + 3H^+(aq) + 2e^- \rightarrow HNO_2(g) + H_2O(l)$

Oxidative half-reaction: $SO_3^{2-}(g) + 2OH^-(aq) \rightarrow SO_4^{2-}(aq) + H_2O(l) + 2e^-$

(b) Reduction of nitrate occurs at the cathode; oxidation of sulfur dioxide occurs at the cathode.
(c) Cathode is positive, anode is negative.

45.
Reduction: $SO_4^{2-}(aq) + 9H^+(aq) + 8e^- \rightarrow HS^-(aq) + 4H_2O(l)$
Oxidation: $C_6H_{12}O_6(aq) + 12H_2O(l) \rightarrow 6HCO_3^-(g) + 30H^+(aq) + 24e^-$
Overall: $C_6H_{12}O_6(aq) + 3SO_4^{2-}(aq) \rightarrow 6HCO_3^-(g) + 3H^+(aq) + 3HS^-(aq)$

47. (a) Reduction: $2H^+(aq) + 2e^- \rightarrow H_2(g)$; cathode
Oxidation: $Zn(s) \rightarrow Zn^{2+}(aq) + 2e^-$; anode
Overall: $Zn(s) + 2H^+(aq) \rightarrow Zn^{2+}(aq) + H_2(g)$
(b) Reduction: $AgCl(s) + e^- \rightarrow Ag(s) + Cl^-(aq)$; cathode
Oxidation: $H_2(g) \rightarrow 2H^+(aq) + 2e^-$; anode
Overall: $2AgCl(s) + H_2(g) \rightarrow 2H^+(aq) + Ag(s) + Cl^-(aq)$
(c) Reduction: $Fe^{3+}(aq) + e^- \rightarrow Fe^{2+}(aq)$; cathode
Oxidation: $H_2(g) \rightarrow 2H^+(aq) + 2e^-$; anode
Overall: $2Fe^{3+}(aq) + H_2(g) \rightarrow 2H^+(aq) + 2Fe^{2+}(aq)$

53. $Pt(s) \mid H_2(g) \mid H^+(aq) \parallel Cu^{2+}(aq) \mid Cu(s)$

55. (a) $Cl_2(g) + H_2(g) \rightarrow 2Cl^-(aq) + 2H^+(aq)$; $E° = 1.358$ V
(b) $Br_2(l) + 2Fe^{2+}(aq) \rightarrow 2Br^-(aq) + 2Fe^{3+}(aq)$; $E° = 0.316$ V
(c) $2Fe^{3+}(aq) + Cd(s) \rightarrow 2Fe^{2+}(aq) + Cd^{2+}(aq)$; $E° = 1.174$ V

59. (a) $I^-(aq) + 2HClO_2(aq) \rightarrow IO_3^-(aq) + Cl_2(g) + H_2O(l)$
(b) $4Cr^{2+}(aq) + O_2(g) + 4H^+(aq) \rightarrow 4Cr^{3+}(aq) + 2H_2O(l)$
(c) $2CrO_2^-(aq) + 3ClO^-(aq) + 2OH^-(aq) \rightarrow 2CrO_4^{2-}(aq) + 3Cl^-(aq) + H_2O(l)$
(d) $S(s) + 2HNO_2(aq) \rightarrow H_2SO_3(aq) + N_2O(g)$
(e) $3F(CH_2)_2OH(aq) + 2K_2Cr_2O_7(aq) + 16H^+(aq) \rightarrow 4K^+(aq) + 3FCH_2CO_2H(aq) + 4Cr^{3+}(aq) + 11H_2O(l)$

61. (a)no; (b)no

65. Yes, $H_2(g)$ will reduce $AgCl(s)$ under standard conditions.

67. $6e^-$; $E°_{cell} = 1.813$ V; reaction is spontaneous; $\Delta G° = -525$ kJ/mol of Al.

73. (a) $10Cl^-(aq) + 2MnO_4^-(aq) + 16H^+(aq) \rightarrow 5Cl_2(g) + 2Mn^{2+}(aq) + 8H_2O(l)$
(b) $E°_{cell} = 0.149$ V
(c) $K = 1.6 \times 10^{25}$

79. yes, $E° = 0.40$ V

81. (a) yes, $E° = 0.47$ V; (b) 0.194 V; (c) yes, $E° = 0.20$ V

87. $[H_2SO_4] = 3.52\ M$; $E > E°$

91. 5.2 L

95. (a) cathode: $Ag(s)$; anode: $O_2(g)$; (b) cathode: $H_2(g)$; anode: $I_2(s)$

99. (a) $E°'_{cell} = 1.14$ V for NADH, 0.94 V for $FADH_2$; (b) −0.09 V; (c)−0.29 V

107. high power density, light weight, able to be recharged many times, able to hold charge for a long time without damage, long lifetime before replacement, operational safety, lack of toxic components for environmentally friendly disposal of discarded batteries

109. (a) $AgO(s) + Zn(s) + H_2O(l) \rightarrow Ag(s) + Zn(OH)_2(aq)$, $E° = 3.44$ V; (b) 498 kJ/mol of Zn; (c) 396 kJ/mol Pb (lead storage battery) versus 665 kJ/mol of Zn (silver–zinc battery)

111. 1.42 g of Cu

113. (a) $Ni^{2+}(aq) + 2e^- \rightarrow Ni(s)$; $Ni(s) \rightarrow Ni^{2+}(aq) + 2e^-$

Chapter 20

5. Isotopes with "magic numbers" of protons and/or neutrons tend to be especially stable. Elements with magic numbers of protons tend to have more stable isotopes than elements that do not.

7. Potassium-40: 19 protons and 21 neutrons. Nuclei with odd numbers of both protons and neutrons tend to be unstable. In addition, the neutron-to-proton ratio is very low for an element with this mass, which decreases nuclear stability.

11. Both positron decay and electron capture increase the neutron-to-proton ratio; electron capture is more common for heavier elements such those of row 5.

13. The mass number is the sum of the numbers of protons and neutrons present. Particles with a mass number of zero include β particles (electrons) and positrons; gamma rays and X rays also have a mass number of zero.

19. Unlike protons, neutrons have no charge, which minimizes the electrostatic barrier to colliding and reacting with a positively charged nucleus.

23. Ionizing radiation causes greater tissue damage, so it is more likely to destroy cancerous cells.

29. Ten exposures of 10 rem are less likely to cause major damage.

33. The mass defect is proportional to the total nuclear binding energy of the nucleus, but the binding energy *per nucleon* is a better measure of the relative stability of a nucleus.

39. The neutron flow is regulated by the use of control rods that absorb neutrons, whereas the speed of the neutrons produced by fission is controlled by the use of a moderator that slows the neutrons enough to allow them to react with nearby fissile nuclei.

45. It is difficult to pinpoint the exact location of the nucleus that decayed. In contrast, the collision of a positron with an electron causes both particles to be annihilated, and in the process two gamma rays are emitted in opposite directions, which makes it possible to locate precisely where a positron emitter is located and to create detailed images of tissues.

49. The raw material for all elements with $Z > 2$ is helium ($Z = 2$), and fusion of helium nuclei will always produce nuclei with an even number of protons.

53. (a) $^{39}_{17}Cl$; (b) $^{8}_{3}Li$; (c) $^{183}_{76}Os$; (d) $^{71}_{30}Zn$

55. (a) 26 protons, 31 neutrons, 1.19; (b) 74 protons, 111 neutrons, 1.50; (c) 19 protons, 20 neutrons, 1.05; (d) 54 protons, 77 neutrons, 1.43

61. (a) α decay; $^{235}_{92}U \rightarrow {}^{4}_{2}\alpha + {}^{231}_{90}Th$
(b) α decay; $^{254}_{99}Es \rightarrow {}^{4}_{2}\alpha + {}^{250}_{97}Bk$
(c) β decay; $^{36}_{16}S \rightarrow {}^{0}_{-1}\beta + {}^{36}_{17}Cl$
(d) β decay; $^{99}_{42}Mo \rightarrow {}^{0}_{-1}\beta + {}^{99m}_{43}Tc$

63. (a) $^{208}_{84}Po \rightarrow {}^{4}_{2}\alpha + {}^{204}_{82}Pb$
(b) $^{226}_{88}Ra \rightarrow {}^{4}_{2}\alpha + {}^{222}_{86}Rn$
(c) $^{228}_{90}Th \rightarrow {}^{224}_{88}Ra + {}^{4}_{2}\alpha + \gamma$
(d) $^{231}_{91}Pa \rightarrow {}^{227}_{89}Ac + {}^{4}_{2}\alpha + \gamma$
(e) $^{166}_{67}Ho \rightarrow {}^{166}_{68}Er + {}^{0}_{-1}\beta + \gamma$
(f) $^{223}_{89}Ac \rightarrow {}^{223}_{90}Th + {}^{0}_{-1}\beta + \gamma$

65. (a) $^{87}_{37}Rb \rightarrow {}^{87}_{38}Sr + {}^{0}_{-1}\beta$
(b) $^{20}_{12}Mg \rightarrow {}^{20}_{11}Na + {}^{0}_{+1}\beta$
(c) $^{268}_{109}Mt \rightarrow {}^{4}_{2}\alpha + {}^{264}_{107}Bh$

67. (a) α particle; $^{218}_{84}Po \rightarrow {}^{4}_{2}\alpha + {}^{214}_{82}Pb$
(b) β particle; $^{32}_{14}Si \rightarrow {}^{0}_{-1}\beta + {}^{32}_{15}P$
(c) γ ray; $^{57m}_{26}Fe \rightarrow {}^{57}_{26}Fe + \gamma$
(d) β particle; $^{204}_{81}Tl \rightarrow {}^{0}_{-1}\beta + {}^{204}_{82}Pb$

69. (a) α emission; $^{238}_{92}U \rightarrow {}^{4}_{2}\alpha + {}^{234}_{90}Th$
(b) EC; $^{208}_{84}Po \rightarrow {}^{208}_{83}Bi + \gamma$

(c) β emission; $^{40}_{16}S \rightarrow {}^{0}_{-1}\beta + {}^{40}_{17}Cl$

(d) γ emission; $^{93m}_{42}Mo \rightarrow {}^{93}_{42}Mo + \gamma$

71. (a) β decay; $^{216}_{84}Po \rightarrow {}^{0}_{-1}\beta + {}^{216}_{85}At$

(b) α decay; $^{235}_{93}Np \rightarrow {}^{4}_{2}\alpha + {}^{231}_{91}Pa$

(c) α decay; γ emission; $^{228}_{90}Th \rightarrow {}^{224}_{88}Ra + {}^{4}_{2}\alpha + \gamma$

(d) β decay, γ emission; $^{231}_{91}Pa \rightarrow {}^{231}_{92}U + {}^{0}_{-1}\beta + \gamma$

75. (a) $^{81}_{35}Br \rightarrow {}^{81}_{36}Kr + {}^{0}_{-1}\beta$

(b) $^{234}_{94}Pu \rightarrow {}^{140}_{56}Ba + {}^{92}_{38}Sr + 2{}^{1}_{0}n$

77. $^{249}_{98}Cf + {}^{12}_{6}C \rightarrow {}^{257}_{104}Rf + 4{}^{1}_{0}n$

81. (a) $^{235}_{92}U + {}^{1}_{0}n \rightarrow {}^{145}_{57}La + {}^{88}_{35}Br + 3{}^{1}_{0}n$

(b) $^{235}_{92}U + {}^{1}_{0}n \rightarrow {}^{95}_{42}Mo + {}^{139}_{57}La + 2{}^{1}_{0}n + 7{}^{0}_{-1}\beta$

87. (a) $^{238}_{91}Pa \rightarrow {}^{238}_{92}U + {}^{0}_{-1}\beta$; 4.721×10^{-16} kJ

(b) $^{216}_{87}Fr \rightarrow {}^{212}_{85}At + {}^{4}_{2}\alpha$; 1.470×10^{-15} kJ

(c) $^{199}_{83}Bi \rightarrow {}^{199}_{82}Pb + {}^{0}_{+1}\beta$; 6.277×10^{-16} kJ

89. (a) $^{234}_{91}Pa \rightarrow {}^{234}_{92}U + {}^{0}_{-1}\beta$; 1.624×10^{8} kJ/mol

(b) $^{226}_{88}Ra \rightarrow {}^{222}_{86}Rn + {}^{4}_{2}\alpha$; 4.700×10^{8} kJ/mol

91. (a) Beta decay of bismuth-208 to polonium is endothermic (ΔE = 1.415 MeV/atom).

(b) Formation of lead-208 by alpha decay of polonium-210 is exothermic ($\Delta E = -5.408$ MeV/atom).

93. 757 MeV/atom, 8.70 MeV/nucleon

95. (a) calculated mass = 53.438245 amu; (b) mass defect = 0.241235 amu; (c) 463 MeV/atom; (d) 8.73 MeV/nucleon

97. 173 MeV/atom, 1.67×10^{10} kJ/mol

99. $\Delta E = +9.0 \times 10^{6}$ kJ/mol beryllium-8

101. D–D fusion: $\Delta E = -4.03$ MeV/tritium nucleus formed = -3.89×10^{8} kJ/mol tritium

D–T fusion: $\Delta E = -17.6$ MeV/tritium nucleus = -1.70×10^{9} kJ/mol

D–T fusion releases about 3 1/2 times as much energy per unit mass as D–D fusion.

103. $^{59}_{27}Co + {}^{1}_{0}n \rightarrow {}^{60}_{27}Co + \gamma$; not a transmutation reaction; 110 rad; nausea, diarrhea, hair loss, decreased white blood cell count, fatigue

105. (a) $^{58}_{28}Ni + 48{}^{1}_{0}n \rightarrow {}^{106}_{28}Ni \rightarrow {}^{106}_{46}Pd + 18{}^{0}_{-1}\beta$

(b) $^{56}_{26}Fe + 23{}^{1}_{0}n \rightarrow {}^{79}_{26}Fe \rightarrow {}^{79}_{34}Se + 8{}^{0}_{-1}\beta$

107. 6.6×10^{-3} rad, 0.13 rem

109. $^{235}_{92}U + {}^{1}_{0}n \rightarrow {}^{236}_{92}U + \gamma$

$^{238}_{92}U + {}^{1}_{0}n \rightarrow {}^{239}_{92}U + \gamma$

$^{239}_{92}U \rightarrow {}^{239}_{93}Np + {}^{0}_{-1}\beta$

$^{239}_{93}Np \rightarrow {}^{239}_{94}Pu + {}^{0}_{-1}\beta$

$^{239}_{94}Pu \rightarrow {}^{239}_{95}Am + {}^{0}_{-1}\beta$

$^{239}_{95}Am \rightarrow {}^{239}_{96}Cm + {}^{0}_{-1}\beta$

113. (a) 7.1×10^{18} atoms of Pu; (b) 0.35 J; (c) 0.48 rad; (d) 9.6 rem—this dose is unlikely to be fatal.

115. 130 W

Art and Photo Credits

Molecular Models

We wish to thank the Cambridge Crystallographic Data Centre (CCDC) and the Fachinformationszentrum Karlsruhe (FIZ Karlsruhe) for allowing Imagineering Media Services (IMS) to access their databases of atomic coordinates for experimentally determined three-dimensional structures. CCDC's **Cambridge Structural Database (CSD)** is the world repository of small molecule crystal structures (distributed as part of the CSD System), and in FIZ Karlsruhe's **Inorganic Crystal Structure Database (ICSD)** is the world's largest inorganic crystal structure database. The coordinates of organic and organometallic compounds in CSD and inorganic and intermetallic compounds in ICSD were invaluable in ensuring the accuracy of the molecular models produced by IMS for this textbook. The authors, the publisher, and IMS gratefully acknowledge the assistance of both organizations. Any errors in the molecular models in this text are entirely the responsibility of the authors, the publisher, and IMS.

The CSD System: The Cambridge Structural Database: a quarter of a million crystal structures and rising. Allen, F.H., *Acta Cryst.* (2002), **B58,** 380–388. *ConQuest:* New Software for searching the Cambridge Structural Database and visualizing crystal structures. Bruno, I.J., Cole, J.C., Edgington, P.R., Kessler, M., Macrae, C.F., McCabe, P., Pearson, J., Taylor, R., *Acta Cryst.* (2002), **B58,** 389–397. *IsoStar:* IsoStar: A Library of Information about Nonbonded Interactions. Bruno, I.J., Cole, J.C., Lommerse, J.P.M., Rowland, R.S., Taylor, R., Verdonk, M., *Journal of Computer-Aided Molecular Design* (1997), **11-6,** 525–537.

The Inorganic Crystal Structure Database (ICSD) is produced and owned by Fachinformationszentrum Karlsruhe (FIZ Karlsruhe) and National Institute of Standards and Technology, an agency of the U.S. Commerce Department's Technology Administration (NIST).

Photo Credits

Chapter 14: Opening photo Fritz Goro; **14.1** Chip Clark; **14.2** Chip Clark; **14.3** Chip Clark; **14.4** Chip Clark; **14.23** Thomas Eisner

Chapter 15: Opening photo James Whitlow Delano/Redux; **15.1** Richard Megna/Fundamental Photographs; **p 690** Deutsches Museum, Munich; **15.12** Richard Megna/Fundamental Photographs; **15.13** Richard Megna/Fundamental Photographs

Chapter 16: Opening photo Richard Megna/Fundamental Photographs; **16.13** Richard Megna/Fundamental Photographs; **16.15** Richard Megna/Fundamental Photographs; **16.16** Richard Megna/Fundamental Photographs

Chapter 17: Opening photo Andrew Syred/Photo Researchers; **p 784** Chip Clark; **17.4** Richard Megna/Fundamental Photographs; **17.5** Wesley Vick and Taylor Chung, Baylor College of Medicine, Houston; **17.6a** Martin Siepmann/AGE Fotostock; **17.6b** Chase Studio/Photo Researchers; **17.7** Carl & Ann Purcell/CORBIS; **17.9** Richard Megna/Fundamental Photographs; **17.11** Richard Megna/Fundamental Photographs

Chapter 18: Opening photo Robert Llewellyn/Image State; **18.1** Robert Harding Picture Library Ltd./Photolibrary; **18.2 top** Bettmann/CORBIS; **18.2 bottom** Bettmann/CORBIS; **18.6** Richard Megna/Fundamental Photographs; **18.7** Richard Megna/Fundamental Photographs; **18.11** James Klett, Oak Ridge National Laboratory; **18.12** Olivier Grunewald/Photolibrary; **18.15 left** Dorling Kindersley; **18.15 right** Andrew Lambert Photography/Photo Researchers; **p 829** Library of Congress

Chapter 19: Opening photo Paul Chesley/National Geographic/ Getty Images; **19.2** Richard Megna/Fundamental Photographs; **19.3** Stephen Frisch/Stock Boston; **p 863** Richard Megna/Fundamental Photographs; **19.7** Richard Megna/Fundamental Photographs; **p 880** Hulton Archive/Getty Images; **p 892** Charles O'Rear/CORBIS; **19.22** Charles D. Winters/Photo Researchers; **19.23** Sam Ogden/Photo Researchers

Chapter 20: Opening photo US Dept. of Energy/SPL/Photo Researchers; **20.8a** Michael Collier; **20.9** Fermilab Visual Media Services; **20.10** Dwayne Anthony and the National Insulator Association; **p 928** Bettmann/CORBIS; **p 939** Thomas Seilnacht; **20.18** Robert D. Loss, WAISRC; **20.19** NOVOSTI/SIPA; **20.21a** Plasma Physics Laboratory, Princeton University; **20.21b** Lawrence Livermore National Laboratory; **20.22a** Chris Priest/SPL/Photo Researchers; **20.22b** Simon Fraser/SPL/Photo Researchers; **20.23** International Atomic Energy Agency; **20.27** Space Telescope Science Institute

Index

Page references followed by italicized *f*, *n*, and *t* refer to figures, notes, and tables, respectively.

C

D

E

F

I

N

Q

R

S

U

V

W

Selected Physical Constants

Atomic mass unit	1 amu	$= 1.6605389 \times 10^{-24}$ g
	1 g	$= 6.022142 \times 10^{23}$ amu
Avogadro's number	N	$= 6.022142 \times 10^{23}$/mol
Boltzmann's constant	k	$= 1.380651 \times 10^{-23}$ J/K
Charge on electron	e	$= 1.6021765 \times 10^{-19}$ C
Faraday's constant	F	$= 9.6485338 \times 10^{4}$ C/mol
Gas constant	R	$= 0.0820575$ (L · atm)(mol · K)
		$= 8.31447$ J/(mol · K)
Mass of electron	m_e	$= 5.485799 \times 10^{-4}$ amu
		$= 9.109383 \times 10^{-28}$ g
Mass of neutron	m_n	$= 1.0086649$ amu
		$= 1.6749273 \times 10^{-24}$ g
Mass of proton	m_p	$= 1.0072765$ amu
		$= 1.6726217 \times 10^{-24}$ g
Pi	π	$= 3.1415927$
Planck's constant	h	$= 6.626069 \times 10^{-34}$ J · s
Speed of light (in vacuum)	c	$= 2.99792458 \times 10^{8}$ m/s (exact)

Useful Conversion Factors and Relationships

Length

Si unit: meter (m)

1 km = 0.62137 mi
1 mi = 5280 ft
= 1.6093 km
1 m = 1.0936 yd
1 in. = 2.54 cm (exact)
1 cm = 0.39370 in.
1 Å = 10^{-10} m

Energy (derived)

Si unit: joule (J)

1 J = 1 N · m = 1 (kg · m^2)/s^2
1 J = 0.2390 cal
= 1 V × 1 C
1 cal = 4.184 J (exact)
1 eV = 1.602×10^{-19} J

Mass

SI unit: kilogram (kg)

1 kg = 2.2046 lb
1 lb = 453.59 g
= 16 oz

Pressure (derived)

SI unit: pascal (Pa)

1 Pa = 1 N/m^2
= 1 kg/(m · s^2)
1 atm = 101,325 Pa
= 760 torr
= 14.70 lb/in.^{2}
1 bar = 10^5 Pa

Temperature

Si unit: kelvin (K)

0 K = −273.15°C
= 459.67°F
K = °C + 273.15

$$°C = \frac{5}{9}(°F - 32°)$$

$$°F = \frac{9}{5}°C + 32$$

Volume (derived)

SI unit: cubic meter (m^3)

1 L = 10^{-3} m^3
= 1 dm^3
= 10^3 cm^3
1 gal = 4 qt
= 3.7854 L
1 cm^3 = 1 mL